Metabolomics

Vijay Soni • Travis E. Hartman
Editors

Metabolomics

Recent Advances and Future Applications

Springer

Editors
Vijay Soni
Division of Infectious Diseases
Weill Department of Medicine
Weill Cornell Medicine
New York, NY, USA

Travis E. Hartman
Division of Infectious Diseases
Weill Department of Medicine
Weill Cornell Medicine
New York, NY, USA

ISBN 978-3-031-39093-7 ISBN 978-3-031-39094-4 (eBook)
https://doi.org/10.1007/978-3-031-39094-4

This Springer imprint is published by the registered company Springer Nature Switzerland AG
The registered company address is: Gewerbestrasse 11, 6330 Cham, Switzerland

Paper in this product is recyclable.

Foreword

This book by editors Soni and Hartman, *Metabolomics: Recent Advances and Future Applications*, is an excellent resource for researchers and students interested in this field, as it covers a wide range of topics and provides a comprehensive overview of the current state of the art. Metabolomics is a rapidly growing field of research that has the potential to revolutionize our understanding of biological systems. The development of advanced analytical techniques and computational tools has enabled researchers to explore the metabolome, yet the technologies and their application are in a constant state of flux. This book aims to make these technologies more accessible.

The book is organized into 14 chapters, each covering a different aspect of metabolomics research. The first chapter introduces metabolomics, including definitions, types of metabolomics, sample preparation, and methods of separation. This sets the foundation for the subsequent chapters, which delve deeper into specific topics, including network development and comparison, analysis and interpretation of metabolite associations, biomarker discovery, pharmacometabolomics, microbial metabolomics, and metabolomics in autoimmunity, infections, and physiological diseases. One of the most exciting aspects of metabolomics is its potential to reveal new insights into complex biological systems. For example, in chapter "Network Development and Comparison in Lipidomics and Metabolomics", the authors discuss the importance of network analysis and investigation in lipidomics and metabolomics. By studying the interactions between metabolites in these systems, researchers can gain a better understanding of the underlying mechanisms and identify potential therapeutic targets.

Chapter "Metabolomics Approach to Identify Biomarkers of Epidemic Diseases" focuses on the use of metabolomics for biomarker discovery in epidemic diseases, such as COVID-19, HIV, and tuberculosis. The authors discuss the various applications of metabolomics in disease diagnosis, risk factor characterization, and data integration and management. This chapter highlights the importance of translational research and the potential impact of metabolomics on public health. In chapter "Pharmacometabolomics: General Applications of Metabolomics in Drug Development and Personalized Medicine", the authors explore the use of metabolomics in drug development and personalized medicine. By studying the metabolites present in an individual's body, researchers can predict how they will respond to different drugs, potentially leading to more effective treatments with fewer side effects. The integration of pharmacometabolomics and pharmacogenomics in clinical studies is discussed, emphasizing the importance of a multidisciplinary approach to personalized medicine.

Chapters "Microbial Metabolomics: An Overview of Applications" and "Metabolomics in Autoimmunity, Infections, and Physiological Diseases" focus on microbial metabolomics and metabolomics in autoimmunity, infections, and physiological diseases, respectively. These chapters demonstrate the wide range of applications of metabolomics and its potential to impact many areas of research and medicine. In addition to its applications in biology and medicine, metabolomics also has important implications for nutrition research and natural product discovery, as discussed in chapters "Nutrimetabolomics: Metabolomics in Nutrition Research" and "Metabolomics in Natural Product Discovery and Their Applications". The authors highlight the importance of machine learning approaches and analytical methods for the purification and quantitation of analytes of interest.

Chapter "Metabolomics Approach in Environmental Studies: Current Progress, Analytical Challenges, and Future Recommendations" discusses the use of metabolomics in environmental health research, emphasizing the challenges and opportunities of ecometabolomics and biomonitoring. The authors discuss the technical challenges of environmental metabolomics and the future perspective of this exciting field. The final four chapters of the book explore the technical advances and challenges of metabolomics, including single-cell metabolomics, spatial metabolomics, metabolic sensors, and industrial applications. These chapters provide insights into the future of metabolomics research, including the miniaturization of instruments, automation in sample preparation and data processing, and collaboration and data exchange. Overall, *Metabolomics: Recent Advances and Future Applications* is a comprehensive and informative book that provides an excellent overview of the current state of the art in metabolomics research. It is an important contribution for anyone interested in this exciting field, and it will undoubtedly be a valuable resource for researchers and students alike.

Scripps Research Institute
San Diego, CA, USA

Gary E. Siuzdak

Preface

The study of living organisms has involved an exploration of a broad range of topics. From adaptation and evolution, to the complexity of cellular development and their networks of interacting parts, to metabolism. For millennia, the ancient Indian medical system of Ayurveda has identified *metabolism* as "Agni" (Fire), which was considered one of the five fundamental elements of life. Sir Michael Foster was the first in Western medicine to use the term "metabolism" in his A *Text Book of Physiology*, which was first published in 1883. The term metabolism is broad, covering the various chemical processes of living organisms, the production of energy and essential biomolecules for survival, reproduction, growth, and homeostasis. While metabolites are the chemical entities that participate in this process, regardless of their role. In the late 1940s, Roger Williams first introduced the concept of "metabolic profile." Using paper chromatography, he suggested that the composition of biological fluids (such as urine and saliva) could be associated with diseases like schizophrenia. However, precise measurement of these compounds was not feasible until 1968 when Horning et al. demonstrated the abilities of gas-chromatography-mass spectrometry to identify lipids from urine and tissue samples.

In 1940, nuclear magnetic resonance (NMR) spectroscopy was discovered, and was later used by Seeley et al. for metabolite detection from biological samples. In 1984, J. K. Nicholson et al. first showed the potential of ^{1}H NMR spectroscopy for an inexpensive and non destructive bioanalytical method for chemical structure elucidation and used it to define metabolic patterns for disease diagnosis, such as diabetes. Critical early work in the field of metabolomics also includes the use of liquid chromatography-mass spectrometry by Richard Lerner, Gary Siuzdak, and Benjamin Cravatt in 1994 and 1996, who analyzed cerebral spinal fluid to identify a molecule called "oleamide" in sleep-deprived animals. These developments led to the inception of a novel field called "Metabolomics" which can be referred to as "the quantitative measurement or evaluation of multiparametric and dynamic metabolic processes of living organisms in response to diseases, treatment, or any genetic adjustments."

Metabolomics is a growing field and is widely applicable to all branches of life sciences, biomedical research, agriculture, biotechnology, and environmental science. As innovative research accumulates, and novel technologies are developed, it is now possible to study the global metabolic changes in the patient's biological fluids to delineate disease-related biomarkers and use them to monitor disease progression and treatment success and to develop personalized therapies. This methodology brings new tools to diagnose and treat difficult ailments like cancer, diabetes, Alzheimer's disease, cardiovascular diseases, and rare genetic and metabolic diseases. Further, metabolomics has become an indispensable tool in drug discovery to identify drug-related biochemical pathways and possible modes of action. Metabolic profiles can be employed to define the drug response in patient subgroups in preclinical stages to shortlist the lead drug candidates with high efficacy and low toxicities.

Metabolomics has also revolutionized nutrition by spotlighting the biochemical effects of various dietary components on the human body's metabolism. This information has been used to develop novel methods of personalized dietary interventions and recommendations and their relationship with disease prevention, treatments, and aging. Agriculture is another field that has extensively benefited from metabolomics. It can help researchers to understand plant metabolism, and investigate the biochemical connections between environmental stresses and diseases to increase crop yield and quality. Further, metabolomics has been used in biotechnology to optimize microbial strains and bioproduction. Applications of ecometabolomics have also been applied to monitor the impact of pollutants on biological systems through the identification of biological markers and the development of different methods for bioremediation.

In this book, we want to introduce the extensive applications of metabolomics from a broad range of areas of research and development so that not only can an undergraduate understand the advancement of metabolomics, but an entrepreneur can harness the knowledge to address possible problems to make a perfect tool to address their research question. We intend to present up-to-date research advancements from our diverse research network so that the global research community can learn the mega powers of metabolomics and recognize various metabolomics experts from different parts of the world to extend their collaborations. Availability of metabolomics research and various applications in this book will provide the opportunity for researchers and industries to bridge the gap in diverse areas to develop even better alternatives to tackle global challenges such as drug development, antimicrobial resistance, novel diagnostics, better nutrition, agriculture, etc.

Each chapter is written by experts in the field, providing valuable insights into the current state of metabolomics research and future directions. We started with the introduction of metabolomics in the first chapter and provided an overview of the terminology and technologies currently used in the field. Next, we included a general introduction to network development and comparison in lipidomics and metabolomics. This is followed by chapter "Analysis and Interpretation of Metabolite Associations Using Correlations" discussing various methods of analysis and their applications to interpret metabolites using various mathematical models and network topology. Further, we started covering various applications of metabolomics. Chapter "Metabolomics Approach to Identify Biomarkers of Epidemic Diseases" discusses about metabolic approaches to identify biomarkers of epidemic diseases and their use in disease diagnosis and management. Chapter "Pharmacometabolomics: General Applications of Metabolomics in Drug Development and Personalized Medicine" describes developments in pharmacometabolomics in novel drug discovery and personalized medicine. This is followed by chapter "Microbial Metabolomics: An Overview of Applications", which is concerned with microbial metabolomics – encompassing various biochemical applications of metabolomic profiling to understand bacterial metabolism and their use in a different area of research. Immunometabolomics is another developing avenue and we have covered it in chapter "Metabolomics in Autoimmunity, Infections, and Physiological Diseases" by presenting a metabolic-level understanding of different autoimmune diseases, infections, and other physiological diseases. Then progressively, readers are taken through the application of nutrimetabolomics in chapter "Nutrimetabolomics: Metabolomics in Nutrition Research". It includes different analytical and computational methods to process and identify crucial diet-related biomarkers and their role in precision nutrition. This is followed by chapter "Metabolomics in Natural Product Discovery and Their Applications", which discusses how metabolomics impacts natural product discovery to reveal their medicinal properties. Further, chapter "Metabolomics Approach in Environmental Studies: Current Progress, Analytical Challenges, and Future Recommendations" describes ecometabolomics and exposomics and their use in environmental health monitoring. Chapter "Deciphering Plant-Pathogen Interactions Through Plant Metabolomics: From Technical Advances to Applied Research" provides an overview of plant-pathogen interactions, while chapter "Metabolomics in Fundamental Plant Research" shows how metabolomics can be applied in fundamental plant research. Chapter "Spatial Metabolomics Using Imaging Mass Spectrometry" is a review of modern spatial metabolomics and multi-omics approaches and their applications. Finally, the book concludes with chapter "Future Perspectives of Metabolomics: Gaps, Planning, and Recommendations", describing future perspectives and recommendations on metabolomics.

We intend to compile a concise and application-oriented book as a guideline for metabolomic advancements. We believe that this book will serve university professors, researchers, technocrats, engineers, and advanced-level scientists who are exploring different avenues in metabolomics. Availability of this concise information in one place will aid scientists by expanding their arsenal of techniques and can be helpful to bring more collaborations and to identify the expert at the global level. This book will also be helpful to those industries that are developing novel tools, methodologies, and drugs with the help of metabolomics in different research areas. We believe that metabolomics is not a scientific tool but an advanced research field with immense powers to understand life at the molecular level. With interdisciplinary collaborations and the use of modern techniques, the "Metabolomics" field can change many sectors of science and we want to invite you to join us on this journey.

New York, NY, USA

Vijay Soni
Travis E. Hartman

Acknowledgments

Writing a book is a journey that cannot be accomplished alone. We are deeply grateful for the support, guidance, and inspiration many people have given us throughout this project.

First and foremost, we would like to thank all the contributors and authors who agreed to take on the challenge of writing the chapters. Their consent efforts, patience, time, and enthusiasm made this project possible. Additional thanks to those exceptional authors who agreed to write at the last moment, despite time constrain, and successfully made it on time.

We would like to express our gratitude to the publisher's team at Springer Nature, Tiffany Lu, Merry Stuber, Amrita Unnikrishnan, and Deepak Ravi, for their guidance, patience, expertise, and professionalism. Their commitment to excellence and their tireless efforts to ensure the success of this project has been invaluable.

Special thanks to our panel of reviewers, Yogita Soni, Yogesh Chawla, Sonia Jain, Brendon Lee, Fernanda Teixeira Subtil, and Arka Banerjee, who constantly helped us to access the scientific quality of the manuscripts of every chapter.

We also wanted to thank our family for their unwavering support, encouragement, and patience. Their love and belief in us have been the driving force behind our work, and we could not have completed this project without them. We are also indebted to our friends, colleagues, and mentors who have been a constant source of motivation and inspiration. Their constructive feedback and insightful comments have helped us to refine the structure of the book.

We would like to express our appreciation to the broader metabolomics research community for their ongoing contributions to this dynamic field. Lastly, we would like to thank our readers for their interest in this book. It is an honor to share our ideas with you, and we hope that this book will enrich your understanding and inspire you to explore new horizons in the field of metabolomics.

Division of Infectious Diseases
Weill Department of Medicine
Weill Cornell Medicine
New York, NY, USA

Travis E. Hartman
Vijay Soni

Contents

Contributors

Qassim Alkassir Department of Biochemistry, Microbiology, and Immunology, University of Ottawa, Ottawa, ON, Canada

Neural Regeneration Laboratory, Ottawa Institute of Systems Biology, Brain and Mind Research Institute, University of Ottawa, Ottawa, ON, Canada

Anita Arroyo-Silva Metabolomics and Mass Spectrometry Laboratory, Unidad de Genómica Avanzada, Irapuato, Gto, Mexico

Nicholas Bartelo Department of Physiology and Biophysics, Weill Cornell Medicine, Institute for Computational Biomedicine, Englander Institute for Precision Medicine, New York, NY, USA

Steffany A. L. Bennett Department of Biochemistry, Microbiology, and Immunology, University of Ottawa, Ottawa, ON, Canada

Neural Regeneration Laboratory, Ottawa Institute of Systems Biology, Brain and Mind Research Institute, University of Ottawa, Ottawa, ON, Canada

Department of Chemistry and Biomolecular Sciences, Centre for Catalysis Research and Innovation, University of Ottawa, Ottawa, ON, Canada

Yogesh Chawla Department of Immunology, Mayo Clinic, Rochester, MN, USA

Miroslava Cuperlovic-Culf Department of Biochemistry, Microbiology, and Immunology, University of Ottawa, Ottawa, ON, Canada

Neural Regeneration Laboratory, Ottawa Institute of Systems Biology, Brain and Mind Research Institute, University of Ottawa, Ottawa, ON, Canada

Digital Technologies Research Centre, National Research Council of Canada, Ottawa, ON, Canada

Amber Dahlin Channing Division of Network Medicine, Department of Medicine, Brigham and Women's Hospital and Harvard Medical School, Boston, MA, USA

Láisa Gomes Dias School of Agronomy, Federal University of Goiás (UFG), Goiás, Brazil

Ankita Dutta International AIDS Vaccine Initiative, Gurugram, Haryana, India

Andreia Figueiredo Grapevine Pathogen Systems Lab, Biosystems and Integrative Sciences Institute (BioISI), Plant Biology Department, Science Faculty of Lisbon University, Lisbon, Portugal

Moisés Guerrero-Esperanza Metabolomics and Mass Spectrometry Laboratory, Unidad de Genómica Avanzada, Irapuato, Gto, Mexico

Travis E. Hartman Division of Infectious Diseases, Weill Department of Medicine, Weill Cornell Medicine, New York, NY, USA

Sonia Jain Department of Radiation Oncology, Mayo Clinic, Rochester, MN, USA

Robert S. Jansen Radboud University, Department of Microbiology, Radboud Institute for Biological and Environmental Sciences, Heyendaalseweg, Nijmegen, The Netherlands

Swarna Kanchan INRIA, Bordeaux Sud-Ouest, Talence, France

Boise State University, Boise, ID, USA

Minu Kesheri Boise State University, Boise, ID, USA

Hannah Jane Lees Memorial Sloan Kettering Cancer Center, New York, NY, USA

Marisa Maia Grapevine Pathogen Systems Lab, Biosystems and Integrative Sciences Institute (BioISI), Plant Biology Department, Science Faculty of Lisbon University, Lisbon, Portugal

Pooja Rani Mina Department of Immunology, Mayo Clinic, Scottsdale, AZ, USA

Seema Nath University of Texas Health Science Center San Antonio (UTHSCSA), San Antonio, TX, USA

Thao Nguyen-Tran Department of Biochemistry, Microbiology, and Immunology, University of Ottawa, Ottawa, ON, Canada

Neural Regeneration Laboratory, Ottawa Institute of Systems Biology, Brain and Mind Research Institute, University of Ottawa, Ottawa, ON, Canada

Department of Chemistry and Biomolecular Sciences, Centre for Catalysis Research and Innovation, University of Ottawa, Ottawa, ON, Canada

José Juan Ordaz-Ortiz Metabolomics and Mass Spectrometry Laboratory, Unidad de Genómica Avanzada, Irapuato, Gto, Mexico

Bhupendra Singh Rawat Center for Immunity and Inflammation, Rutgers New Jersey Medical School, Newark, NJ, USA

Suyasha Roy Laboratory of Molecular Immunology, National Heart, Lung and Blood Institute, National Institutes of Health, Bethesda, MD, USA

Edoardo Saccenti Laboratory of Systems and Synthetic Biology, Wageningen University & Research, Wageningen, The Netherlands

Stefania Savoi Department of Agricultural, Forest and Food Sciences, University of Turin, Turin, Italy

Annalise Schweickart The HRH Prince Alwaleed Bin Talal Bin Abdulaziz Alsaud Institute for Computational Biomedicine, Weill Cornell Medicine, New York, NY, USA

Tri-Institutional Computational Biology and Medicine Program, Weill Cornell Medicine, New York, NY, USA

Satendra Singh Department of Computational Biology and Bioinformatics JSSB, Sam Higginbottom Institute of Agriculture, Technology and Sciences (Formerly Allahabad Agriculture Institute), Allahabad, Uttar Pradesh, India

Vijay Soni Division of Infectious Diseases, Weill Department of Medicine, Weill Cornell Medicine, New York, NY, USA

Upasna Srivastava University of California, San Diego, CA, USA

School of Medicine, Yale University, New Haven, CT, USA

Dung Thuy Tran Channing Division of Network Medicine, Department of Medicine, Brigham and Women's Hospital and Harvard Medical School, Boston, MA, USA

Pieter M. M. van der Velden Radboud University, Department of Microbiology, Radboud Institute for Biological and Environmental Sciences, Heyendaalseweg, Nijmegen, The Netherlands

Ramya Venkataraman Laboratory of Innate Immunity, National Institute of Immunology, New Delhi, India

Kanchustambham Vijaya Lakshmi Proteomics Center of Excellence, Northwestern University, Evanston, IL, USA

Introduction of Metabolomics: An Overview

Travis E. Hartman and Hannah Jane Lees

1 Introduction

Any survey of recently published literature would leave the reader with the impression that a sizable majority of the scientific endeavor relies on the use of an omics technology.[1] In support of this, the number of published works containing reference to an omics has continued to rise over the last decade (Fig. 1). There are numerous causes for their popularity, like the drop in cost as a result of recent technological advances and economies of scale, but their widespread adoption is undoubtedly an attestation to their utility across scientific disciplines. The four big omics technologies (genomics, transcriptomics, proteomics, and metabolomics) basically rely on two distinct technologies: nucleotide sequencing and spectroscopy. The switch to next-generation sequencing (NGS) has driven down the costs of deploying genomics methods to a point where they are accessible to more biological and biomedical researchers [1]; unfortunately, the cost of high-end mass spectrometers has hardly budged in the same period. Still, these tools have proven indispensable in many cases and have spawned investigations of new basic science and therapeutic avenues that were once only dreamt of [2, 3].

[1] Note that we refer to omics technologies with reference to the actual methodologies – as opposed to the general concept of the comprehensive classification of a field.

T. E. Hartman (✉)
Division of Infectious Diseases, Weill Department of Medicine, Weill Cornell Medicine, New York, NY, USA
e-mail: teh2002@med.cornell.edu

H. J. Lees
Memorial Sloan Kettering Cancer Center, New York, NY, USA
e-mail: leesh@mskcc.org

V. Soni, T. E. Hartman (eds.), *Metabolomics*,
https://doi.org/10.1007/978-3-031-39094-4_1

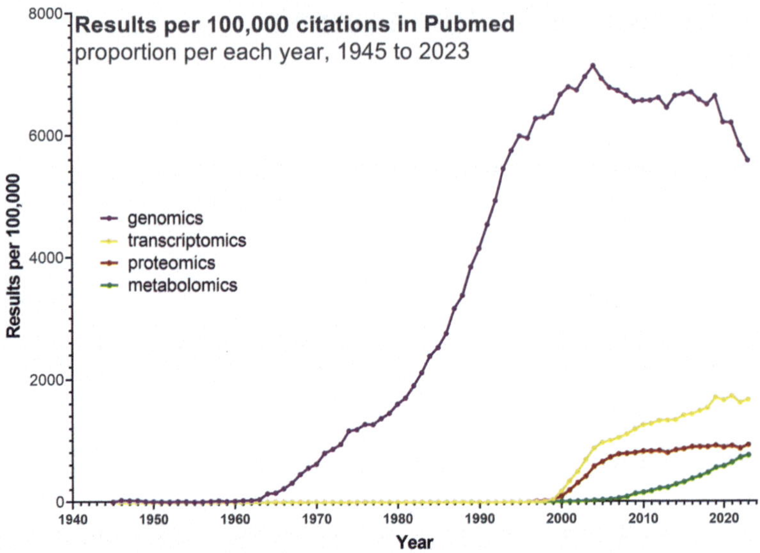

Fig. 1 Trends in publications containing some reference to one of the four big omics technologies

Beginning in the mid-1980s, pivotal advances in sequencing technology propelled studies involving genomics into the mainstream, followed soon by advances in technologies allowing relatively easy access to transcriptome and proteome data [4]. Although genomics continues to eclipse reports of other omics surveys, there has been a discernable increase in the volume of research reporting transcriptome and metabolome data since the turn of the millennium. It is fair to say that it was this early success of genomics that has spawned the revolution in big-data science. Soon after the complete human genome was first published, we came to the realization that genomics alone wasn't going to solve all of our health problems; but the top-down viewpoint that we have become accustomed to was attractive enough to apply to the study of message, protein, and small molecules. Today we refer to these techniques that seek to classify pools of small molecules from experimental samples collectively as *omics*. These omics technologies share the same nominal objective: to provide biological insights via high-throughput analysis of their respective datasets. One might state that the aspirational goal of these approaches is *to provide a complete dataset containing all relevant biological information of a sample in a particular condition or at a particular time*. While this perfect picture remains unrealized, our efforts have certainly they have helped us develop a more thorough understanding of many biological systems. This understanding is at the core of much of modern scientific research [4–6] and has proven useful across a wide array of disciplines.

While there are volumes of curated information about the technology and various methodologies for nucleotide-based omics, we have found fewer resources for the *small molecule-based* omics. This can be partly attributed to the aforementioned boom in genomics, but we posit that there is additionally a gap in basic understanding of the fundamentals for these fields. The sequencing methods that are applied to genomics/transcriptomics and even peptides are stepwise chemical procedures in which each step in the reaction is predetermined and the results are often not open to interpretation. This is not the case for the analytical methodologies used to probe small molecules. The hurdles therein are daunting for newcomers because mastery of these omics requires both a conceptual command of the technology and an exegetical expertise usually borne of experience. We hope that this text will help to foster an understanding of these techniques so that more scientists will feel comfortable employing them for their specific research questions.

2 The "Omics"

While genomics and transcriptomics rely on nucleotide sequencing to provide information on an organism's genetic makeup, proteomics and metabolomics use spectroscopic technologies to monitor the functional activity of the organism. Although they utilize the same analytical instruments (mass spectrometers paired to liquid or gas chromatography (*LC-MS/GC-MS*) or nuclear magnetic resonance spectroscopy (*NMR*)), they are considered complimentary approaches for studying biological systems. *Proteomics* is primarily concerned with the identification, characterization, and quantification of all the proteins present in a given sample. This is accomplished most frequently via the digestion of the protein fraction of a cell lysate followed by quantitative sequencing of the resulting peptides. In contrast, *metabolomics* is concerned with the identification, characterization, and/or quantification of all *small molecules*[2] present in a given sample.[3] We should also mention that the terms *metabolomics* and *metabonomics* are used interchangeably. Since metabolites are the products of metabolic pathways, they reflect the biochemical activity of the organism at a systems level. Although the instrumentation to collect this information is the same (in most cases), there are challenges specific to isolating these samples that we will discuss below.

As is often the case in the annals of scientific inquiry, technological innovations developed by seemingly disparate disciplines have been co-opted to answer biological questions. Such was the case when J.J. Thompson's dogged exploration of charged particles produced by cathode ray tubes (Crookes tubes) led to the discovery of the electron and subsequently the first mass spectrometer in 1912 (https://www.aps.org/publications/apsnews/200104/history.cfm). The machine he developed was the first instrument capable of robust measurement of the *mass-to-charge ratio* (*m/z*) of ions. The m/z is an important measurement in metabolomics and is

[2] This is widely accepted to refer to biomolecules <1500 Da.

[3] We will focus on metabolomics exclusively in this text, but we refer the reader to an excellent text published by Springer for a thorough examination of proteomics [97].

defined as *the mass of the cation divided by its charge*. Thompson described the benefit of assaying charged molecules as such: “An unelectrified atom is so elusive that unless more than a million are present we have no means sufficiently sensitive to detect them, or, to put it another way, unless we had a better test for a man than for an unelectrified molecule, we should be unable to find out that the earth was inhabited” [7]. Thomson’s early work paved way for the development of more sophisticated mass spectrometers. In the 1920s, German physicist Wolfgang Paul and American physicist Arthur J. Dempster independently developed *time-of-flight* (*TOF*) mass spectrometers. The TOF uses electric fields to accelerate ions and each ions time of flight across a known distance is determined by its mass, thus greatly improving the resolution of these instruments. In the 1950s and 1960s, mass spectrometry technology continued to evolve, with the development of quadrupole and ion trap mass spectrometers which allow tandem mass spectrometry experiments to be performed. In the 1980s, *matrix-assisted laser desorption/ionization* (*MALDI*) and *electrospray ionization* (*ESI*) were developed. These soft ionization techniques are useful for ionizing biomolecules without generating fragment ions and have improved our ability to assay important biomolecules. Richard R. Ernst was awarded the Nobel Prize for his contribution to the development of NMR in 1991, and these instruments are now used for metabolomics studies – primarily for their ability to help solve molecular structures. The advent of these technologies has had a great impact on modern biological research and ushered in a new era of small molecule research.

3 Why (and How) We Use Metabolomics to Answer Biological Questions

The Many Applications of Metabolomics

The last 20 years has seen both a precipitous increase in the number of metabolomics studies published and also a clear maturation of the field. Consider (Fig. 2) [8], whereas early metabolomic-centric publications were focused on technological development and the bioinformatic challenges of dealing with big-data – more recent work trends toward the clinical applications of metabolomics and its application to an expanding web of scientific lines of inquiry. A closer look at these keywords (Fig. 3) reveals the breadth of studies that benefit from a metabolomic perspective. This progression of the field is a clear sign of the maturation of metabolomics as a discipline.

Metabolomics is a field in transition. It can act to complement other omics, as we collect metabolome data and add it to databases that are mirrors of big data depositories that were assembled to store DNA sequencing, transcriptome, and proteome data. However, an important distinction must be made between the genetic sequence (which is mostly fixed at birth) or the transcriptome, and the metabolome – which is highly dynamic and under the influence of the physio-chemical milieu at short time

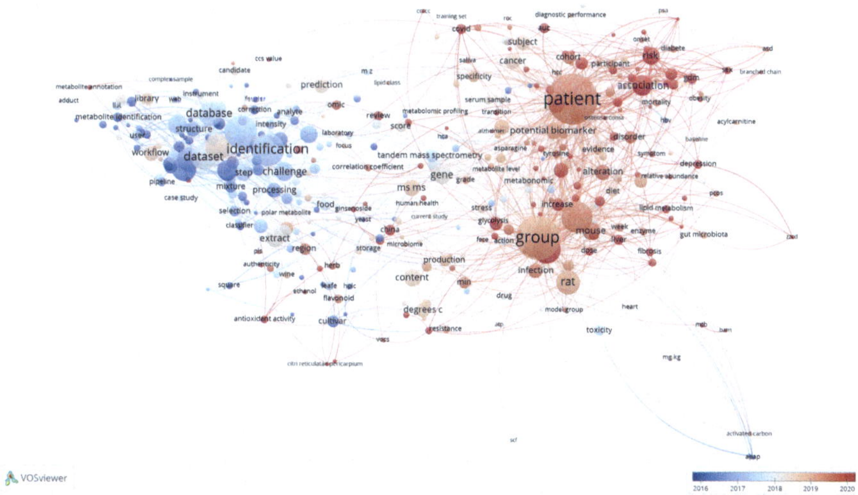

Fig. 2 Keyword trends in published articles containing "metabolomics" in Pubmed [8]

Fig. 3 Keywords associated with "metabolomics" since 2016 in the Web of Science

scales. In this way, it is more responsive to external stimuli and is thus closer to phenotype than the other omics. The metabolome is the product of the other omics and provides insights into the function of DNA/RNA/proteins. But it is also swift to respond to a change in conditions. This characteristic makes it so attractive to apply to biological investigations.

Naturally, the trend in big data science has been followed by more recent efforts to integrate all of these into a multiomic description of phenotype. Tools are being developed via the contributions of talented bioinformaticians to handle information on a scale that was once an impediment to progress. *Multiomic* workflows to integrate data collected for the individual omics methodologies into a comprehensive picture (or as close as possible) are an active area of interest [9]. It is hard to overstate the power that this type of perspective can lend to an investigation of phenotype. Yet, the challenges associated with this integration of three or more (already complex) datasets are formidable. The mechanics of the investigation are complex because they employ have to accommodate diverse analyses methods [10] and apply novel statistical processes [11]. And the interpretation of this dynamic multi-layered data will occupy researchers for decades to come.

Every scientist can appreciate the complexity of the relationships between genetic material, proteins, and metabolites. The discipline of *systems biology* is entirely devoted to understanding each of these parts – but this discipline is sufficiently expansive to deserve its own textbook and will not be otherwise addressed here. There are some areas of research that employ metabolomics methods but rely on the techniques of systems biology that are worth mention, however. An example is the so-called *interactome*, which started as a study of protein–protein interactions [12] but has now developed to encompass others: like the protein-metabolite axis [13], cell-to-cell interactions [14], the signal transduction cascades involved in proliferation [15]. There is a related field called *ecometabolomics* that takes this approach to study the interactions across species, including host and microbes [16], plants and microbes [17], or plants and animals [18]. To assess the interaction between complex systems, tools are being developed that allow us to make sense from apparent chaos.

One other emerging field that will not be covered in detail here is precision medicine. Metabolomics techniques have already begun to play a role in the shifting focus in clinical research toward individualized treatment (reviewed in [3]). From the discovery of specific disease-associated biomarkers (reviewed in [19, 20]) and metabolic risk factors [21–28] to measurements of the bioavailability and pharmacodynamics of drugs [29–31], treatment can be more precisely applied to the unique physiology of the individual. To add to this, work is ongoing to collect metabolite profiles to deconvolute the influence of the external environment (the "exposome") and disease risk [32–34]. The power of metabolomics to simultaneously measure thousands of molecules will allow deep phenotyping of patient samples that can be mined for biomarkers associated with good health, longevity, and physical fitness [35, 36]. In the review above, David Wishart poses the question succinctly: "Metabolomics is changing biomedical research. The question is: can it change health delivery? [3]" We concur; as researchers accumulate and analyze more patient health data, the publication of reviews and textbooks dedicated to the subject to focus attention on the development of our discoveries into clinical applications is required.

As is true for everything we call an omics technology, the vast amount of data generated by even a single experiment can be overwhelming without an efficient strategy for data manipulation and analysis. We have dedicated *Chapter 2: Network Development and Comparison in Lipidomics and Metabolomics* to describe ways to overcome this specific hurdle. Within, the reader will find that – given a well-planned experiment – specific analytical techniques have been developed which can be modified to address almost any question of interest.

We will devote the following chapters in this book to an examination of how metabolomics can benefit a wide range of research questions. The authors of these chapters are leaders in their field of study and have compiled detailed summaries that we hope will open the eyes of the reader to the potential of this field and encourage them to embark on the same journey that we have – once the potential for this type of analysis becomes apparent. This introductory chapter will hopefully acquaint the reader with some key concepts and terminology. We regard this background information to be essential to understanding the work that follows and to familiarize the reader with a foundation that will allow them to avoid certain well-known pitfalls in metabolomic analysis.

4 Methodologies in Metabolomics

4.1 Instrumentation

Below we will describe the analytical methods most commonly used in metabolomics analysis, but the primary instruments used in modern metabolomics studies are the MS and NMR. When we employ these technologies to analyze biological materials in metabolomics, each method has advantages and disadvantages – but the platforms are sufficiently developed to tailor methods to answer specific metabolomic questions. The analytical method and workflow for the user will depend largely on the goal of the study and whether a targeted or untargeted design is appropriate. The choice of instrument will depend on the characteristics of the molecules the researcher expects to analyze. A rough comparison of the most popular instruments for metabolomics is given in Table 1 [37, 38], but experience is often a better guide. It is always worthwhile to search the published literature for work done on similar systems and we will present guidance and best practices in the later chapters of this book.

Table 1 Analytical parameters of metabolomics instrumentation [37, 38]

Instrument	Mass resolution (amu)	Mass accuracy (ppm)	Sensitivity	m/z range	Dynamic range
QqQ	1×10^3	100–1000	fM-aM	10–4000	6×10^6
Qq-linear ion trap	2×10^3	100–500	fM	5–2800	4×10^6
Qq-TOF	2×10^4	<5	fM	No upper limit	1×10^4
FTICR	5×10^4 to 7.5×10^5	<2	fM	50–2000; 200–4000	1×10^3
Orbitrap	~10^5	<5	fM	50–2000; 200–4000	4×10^3
NMR	???	0.2 to <2.5	mM-uM	<2000	Various

4.2 The Mass Spectrometer

Fundamentally all mass spectrometers are based on the same three elements (an *ion source*, a *mass analyzer*, and a *detector*). Figure 4 depicts the workhorse of the metabolomics world – the triple quadrupole MS. The first step in this process is that a charge is imparted with the help of the ion source. As Thompson alluded to, mass spectrometers can only analyze gaseous ions, and the ionization source is where an electrical current is used to generate electrons that then impact the sample to ionize them. The ions generated from target molecules are very reactive, so the mass spectrometer is kept under a hard vacuum (about 760 torr) to prevent unwanted interactions with these short-lived species.

The principle of mass spectrometry is actually very simple when viewed stepwise:

1. *Sample Introduction*: The process begins by introducing the sample into the mass spectrometer. This can be done in various ways depending on the nature of the sample, such as direct injection, gas-phase sampling, or sample vaporization.
2. *Ionization*: When the vaporized sample enters the ionization chamber, it is bombarded with electrons from the ion source. The goal is to convert the sample molecules into ions, which are atoms or molecules with an electric charge. These electrons displace an electron from the sample molecules, resulting in the creation of charged ions called the molecular or parent ions. The parent ion is commonly denoted as [M^+] or [M^-] depending on ion mode. There are different ionization techniques available, including electron ionization (EI), electrospray ionization (ESI), matrix-assisted laser desorption/ionization (MALDI), and others. Each technique has its advantages and is suitable for specific types of samples.
3. *Acceleration and Separation*: The ionized sample ions are then accelerated using electric fields to impart kinetic energy to them. Subsequently, the ions are separated based on their mass-to-charge ratio (m/z) using a mass analyzer (which we will discuss below). Each type of mass analyzer operates on different principles but aims to separate ions based on their mass-to-charge ratio.

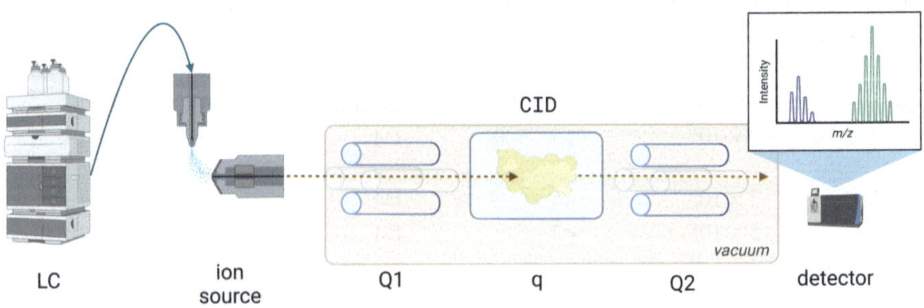

Fig. 4 Triple quadrupole. (Created in Biorender)

4. *Detection*: After the ions are separated based on their mass-to-charge ratio, they are detected by a detector system. The detector measures the abundance or intensity of ions at different m/z values, generating a mass spectrum.
5. *Data Analysis*: The mass spectrum obtained from the detector is then processed and analyzed. The data analysis involves interpreting the mass spectrum to determine the presence of different ions and their relative abundances. This information can provide insights into the molecular composition, structure, and fragmentation patterns of the sample molecules.
6. *Identification and Quantification*: By comparing the mass spectrum with databases or reference spectra, the identified ions can be matched to known compounds. This process allows the identification of the molecules present in the sample. Additionally, the relative abundance of ions can be used to quantify the concentration of specific compounds.

4.2.1 The Ion Source

When the charge applied to the target molecule comes from electrons generated by the ion source itself, spectrometrists refer to the process as a *hard ionization*. Hard ionization is sufficiently energetic (~70 eV) to break the covalent bonds in the target molecule and is usually reserved for nonpolar molecules and compounds that are not thermolabile – thus making it useful for inorganic materials – but not most biomolecules. Because we wish to analyze the molecules of interest in metabolic pathways intact, and these molecules are polar (or moderately polar) and less than 1000 m.wt., a more gentle approach is used [39].

Soft ionization techniques can accomplish ionization of small molecules without fragmentation. There are a number of methods to accomplish both the vaporization (gasification) and the charging of sample material – commonly *electrospray ionization (ESI)*, *chemical Ionization (CI)*, and *matrix-assisted laser desorption ionization (MALDI)* are employed for the purposes of metabolomics. The most popular of these is ESI, in which the sample liquid is pumped through a specialized needle at a high temperature (220–500 °C) and at a high voltage (2–4 kV). The resulting fine spray quickly evaporates into charged droplets that then desolvate as the solvents evaporate, leaving protonated (or deprotonated in the case of negative mode ionization) molecules. CI techniques are even gentler (including *atmospheric pressure chemical ionization (APCI)*) and combine the sample with an ionized reagent gas to form a plasma. Different gases can be used to alter the charge on the molecule (which we denote as *M* – for mass) to form the positive ion $[M + H]^+$ or the negative ion $[M - H]$. While these techniques are suitable for soluble samples, MS can be performed on intact specimens (such as tissue samples) using MALDI, in which a laser is pulsed tangentially to the surface of the sample thus ablating molecules as singly charged ions.

The choice of ionization method depends largely on the information the researcher expects to get from the analysis of the sample (see Fig. 5). Molecular structure determination can be performed using hard ionization techniques in which

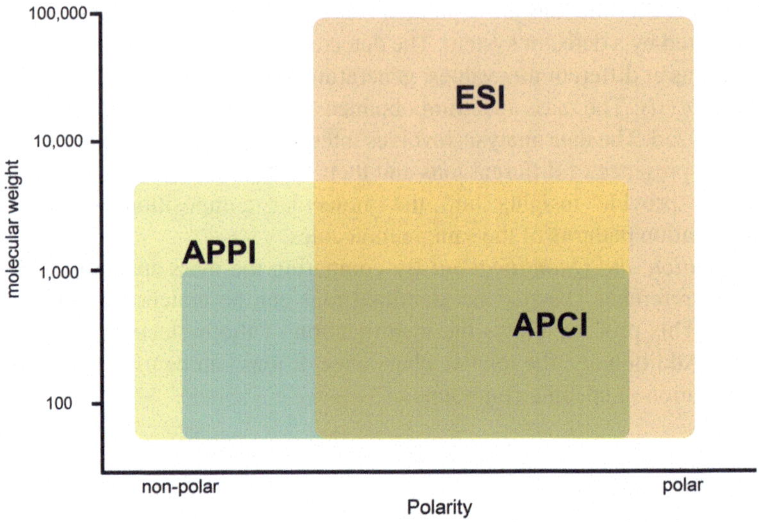

Fig. 5 Ionization techniques used in metabolomics

the fragmentation pattern of the molecule of interest can be interpreted to deduce likely structure. However, soft ionization techniques are better suited to simple identification and quantification of biomolecules from solutions, where fragmentation would create superfluous signals that could obscure or confuse the identity of the molecular species of interest. Because MALDI uses a laser for ionization of matrix-fixed sample, the resulting signal can be used for mapping biomolecules in a tissue slice or even microbial identification.

4.2.2 The Mass Analyzer

The sole job of the mass analyzer in a mass spectrometer is to move the ions coming from the ion source toward the detector, while filtering them based on m/z. This is accomplished by focusing the ions into a beam and applying an electrostatic field across either an arrangement of electrodes or a strong magnetic field. One might surmise that there are many ways to accomplish this (and one would be right), but a thorough description of these is outside of the scope of this text. Instead, we will focus on the analyzers that a practitioner of metabolomics should be familiar with. There are only five of these: the *quadrupole* mass analyzer, the *time of flight* mass analyzer (*TOF*), the *ion trap* mass analyzer, the *ion cyclotron resonance* (*ICR*), and the *Orbitrap*™.[4]

[4] The Orbitrap Mass Analyzer is a version of the ion trap that is licensed exclusively to ThermoFisher Scientific. It is worth mentioning here because of its utility and popularity in metabolomics [98].

The quadrupole MS is the simplest of the instruments used in modern metabolomics studies. The quadrupole itself consists of four parallel cylindrical electrodes inside a vacuum chamber arranged so that ions can be shuttled through the space between them (Fig. 4). The ions flying through the field at their center oscillate based on the current (either DC or AC) or radiofrequency (RF) that are applied to the electrodes. These parameters can be tuned electronically to allow only the ions within a specified m/z range to successfully pass through the quadrupole. Ions that fall outside of this range are shunted from the electric field in the center of the quadrupole and are lost to the vacuum. In this way, a quadrupole works as a *mass filter* – which is a useful property when attempting to identify or quantify specific molecular species.

To take advantage of this filtering ability, a team of brilliant researchers (ja00475a072) arranged a pair of quadrupoles in tandem with an RF – only quadrupole between them in an arrangement now known as the triple quadrupole (abbreviated *QqQ*). The lowercase q is meant to denote that the middle quadrupole functions instead as a *collision cell*, in which ions can be fragmented by the introduction of a neutral gas that transfers translational energy to the ions that collide with it. The fragmentation of molecular ions that occurs within the collision cell is referred to as *collision-induced dissociation* (*CID*). This process makes *tandem mass spectrometry* (or *MS/MS*) possible. In the realm of metabolomics, MS/MS is especially useful when attempting to determine the structure of a compound or to distinguish between molecules with a similar m/z value (216_2021_Article_3425). The QqQ is also often the instrument of choice when attempting quantification of biomolecules in a complex sample (i.e., whole blood or serum samples, urine, cerebrospinal fluid (CSF), etc.) because this mass filtering ability allows the removal of interfering signals from the matrix.

This particular arrangement of quadrupoles enables the QqQ to perform a number of convenient tricks that are useful for the quantification of small molecules or analysis of experimental samples. As mentioned above, a quadrupole can scan a mass range (the *full scan mode*) by allowing all ions in that range to pass through to the detector or act as a mass filter in which only selected ions are allowed to pass. The latter mode is known as *selected ion monitoring* (*SIM*). Now if we apply this to an instrument outfitted as a QqQ, we can perform different scan modes to isolate particular ions. These scan modes are the precursor/product ion scan, the neutral loss scan, and multiple reaction monitoring (MRM). In describing these, it is useful to think of the QqQ as a set of pinholes in two pieces of paper – through which a light shine. If they don't line up, then the light doesn't reach the other side (the detector) (Fig. 6).

The *product ion scan* is perhaps the most intuitive of the scan modes to describe. In the product scan, the first quadrupole Q1 allows only an ion of a specified mass to pass through. This ion (and only this ion) is then fragmented in the collision cell (q2). Then Q3 scans the entire m/z range, revealing the sizes of every molecular ion fragment that was produced by the CID of the specified compound. This is particularly helpful when seeking information about molecular structure, and the specific change in m/z of the fragment ions are known as transitions. Similarly, the

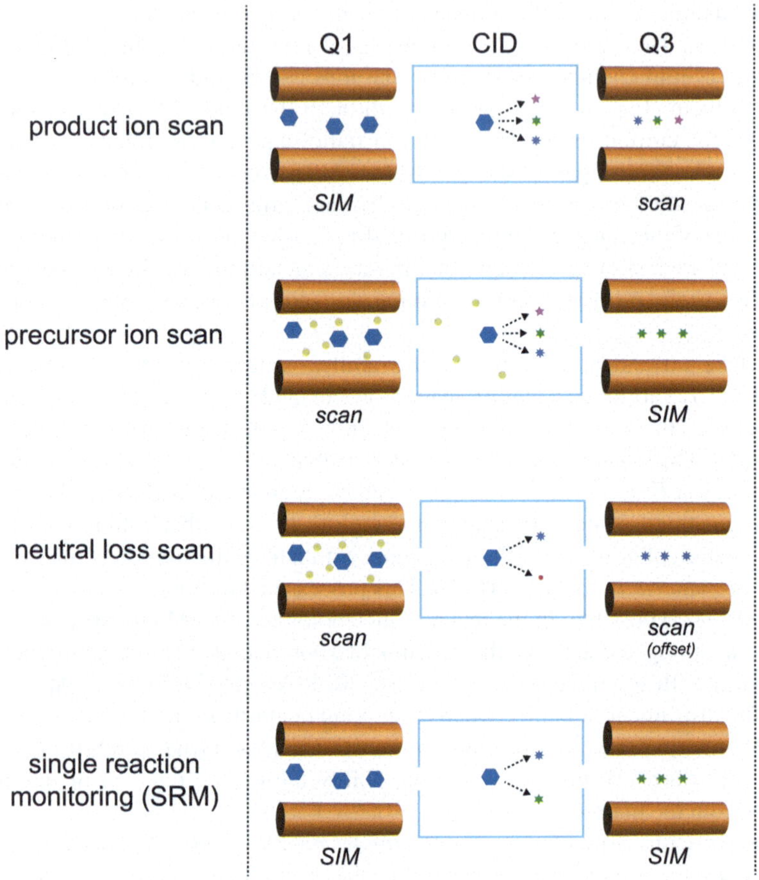

Fig. 6 Overview of scan modes for a triple quadrupole

precursor scan can be viewed as the opposite arrangement. In the precursor scan, a large range of m/z's is allowed to pass Q1, but only a certain product ion is allowed to pass through Q3. This can be used to detect ions with a particular functional group (like a glycosylated peptide).

The *neutral loss scan* method can be particularly useful if one is looking for modifications of a small molecule or peptide, such as a phosphorylation or hydroxylation. To perform this scan, the two quadrupoles (Q1 and Q3) operate in scan mode, with the m/z values of each scan offset by a user-specified amount. For example, if the user were looking for a phosphorylation event, the specific difference between the phosphorylated ion [+HPO3] and the non-phosphorylated ion would be 79.97 m/z. So, to detect the neutral loss of the phosphate, the user would set Q1 to scan all molecules, but when the scan reaches each specific point (e.g., 181.07 m/z), Q3 would be scanning at a specific offset mass (filter out everything that isn't 101.1 m/z) (nihms130718, App-Note-332-Data-Dependent-Neutral-Loss-Mass-Spectrometry-for-the-Identification-of-Protein-Phosphorylation). Remember that the detector only registers a signal when both Q1 and Q3 are permitting ions through the filter. This allows the user to distinguish the molecules in the sample that were modified because only molecules that lost that specific neutral mass – at that time in the scan – are detected.

The other important trick of the QqQ is SIM, also known as *selected reaction monitoring* (SRM). In this scan mode, both Q1 and Q3 are set to allow a specified m/z to pass, but Q3 is set to an offset in which only a distinct fragment ion can be detected. The important advantage of this mode is in increased sensitivity because the quadrupoles filter out all of the signals from interfering ions, allowing noise-free quantification. Another flavor of this mode is called multiple reaction monitoring (MRM), where Q1 and/or Q3 is set to more than 1 m/z to maximize the efficiency of quantification of multiple ions.

4.2.3 The Detector

Regardless of the manipulations we apply to the ions that flow through the vacuum of the MS, our analytical power is only as good as the tool we use to measure them. This measurement takes place at the detector, which must make an electronically readable signal when it registers a collision with an ion. Thankfully we have made great strides in this area over the century since J.J. Thompson's first work using his parabola spectrograph. It is useful to consider though that Thompson first detected his cathode rays (that he would later reveal were actually electrons) using a discharge tube that consisted of a sealed glass container of inert gas [40]. Thompson and his collaborator Francis William Aston soon switched over to photographic film before replacing that with a *Faraday cup (FC)*. These early detectors were able to provide evidence of charged particles but were subject to interference from outside sources. Although Faraday cups are still used, the detectors on modern instruments are usually *electron multipliers (EM)*, photomultiplier conversion dynodes, or array detectors.[5]

The simplest of these, the FC, is essentially a metal cup that intercepts the beam of ions coming through the mass analyzer. When an ion (or packet of ions) impacts the cup, it induces a positive charge as it is neutralized against the metal. That positive charge induces a quantifiable current in the FC equivalent to $+1.602 \times 10^{-19}$ C (or the current carried by a proton), and thus directly measures the number of ions. In the base FC design, there is no amplification – so these detectors are considered to suffer from low sensitivity. However, since there is a direct relationship between the ions that strike the FC dynode and the electrons released, the generated current can be amplified when paired with an EM.

[5]Almost every MS manufacturer freely provides detailed information about the configuration of the detectors fitted to their instruments. There are many variations on the few that are listed here, but an exhaustive description of all the configurations is outside of the scope of this text.

More commonly used is the EM, which consists of a series of aluminum oxide dynodes that are kept at increasing potentials. When an ion strikes the first dynode, it ejects some of the electrons from the material – that then strike the second dynode – and so on. This process cascades and can amplify the ion current by ~10^8. The benefits of this amplification are that sensitivity is greatly increased by the magnified signal intensity gain. In their native state, the EM cannot detect negative ions, so a "conversion dynode" with a high positive voltage is placed in front of the EM. This allows the negative ions to strike the dynode first, thus releasing electrons down the EM chain.

Less common is the photomultiplier conversion dynode detector – a type of scintillation counter. In this configuration, the ions from the mass analyzer strike either a positively charged conversion dynode or a negatively charged conversion dynode – either of which releases secondary electrons. These electrons then strike a phosphor screen which in turn releases photons that pass into an EM. This arrangement is known to have a longer lifespan than simpler detectors and is as sensitive as an EM alone.

4.3 High-Resolution Mass Spectrometry (HRMS)

Despite its robust suite of configurations and practical versatility, the QqQ design does suffer from a drawback in that its m/z resolution is functionally limited at about 1 atomic mass unit (amu). This is because the resolving power of a quadrupole depends on the number of RF cycles the ion is exposed to while it flies through its center (in a perfect scenario). So, if an ion is exposed to ~200 RF cycles during its passage, the resolution would work out to roughly 1000 full width at half maximum (FWHM – more on this later). Importantly, as the mass of the ion increases, its velocity falls, thus increasing the number of RF cycles it is exposed to, which increases the resolution. This resolution is fine for known compounds in relatively clean solutions but is less useful for identification or quantification of isobaric compounds or small molecules in more complex matrices.

Since metabolomics benefits from analysis of both of these, HRMS instruments are more frequently employed. These designs add another dimension to the ion path that helps us discriminate between m/z's to several decimal places. This allows us to measure molecules with *exact mass* (rather than nominal mass), which in turn allows us to assign a chemical formula to our compound/s of interest. The *exact mass* is measured in Daltons (Da) and is determined by summing the individual isotopes of the molecule (it is also known as the monoisotopic mass). For example, to find a signal for glucose in our sample, we know that the molecular formula of glucose is $C_6H_{12}O_6$, so we calculate based on the monoisotopic masses of the elements in the formula (Table 3). So [$6 \times 12.011 + 12 \times 1.008 + 6 \times 16.00$] comes out

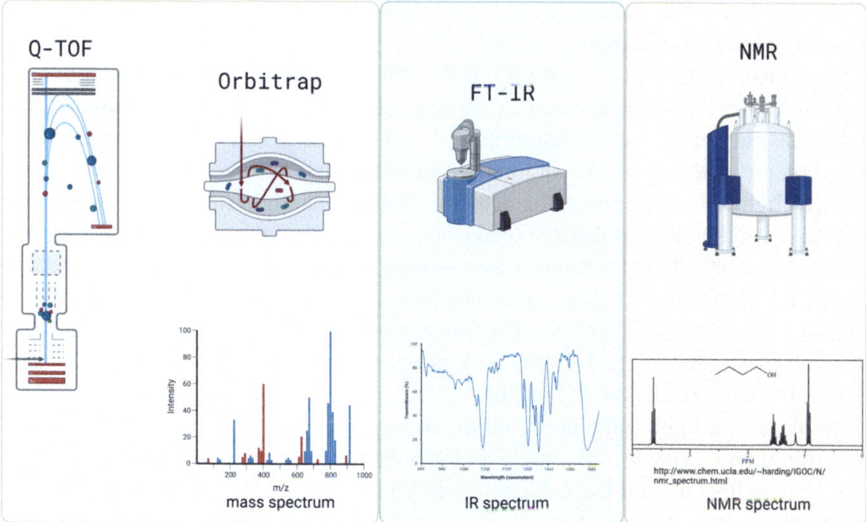

Fig. 7 HR instruments commonly used in metabolomics. (Created in Biorender)

to an exact mass of 180.162 Da for glucose.[6] Vendors offer a number of categories of instruments with different resolving power that fit distinct research needs, but the most commonly used instruments for current metabolomics are the QqQ, the ion trap, the Q-TOF, the Orbitrap, and the FTIR (Fig. 7).

Around the same time, as Wolfgang Paul was working on the quadrupole mass analyzer, he also conceived of the ion trap mass analyzer. The physics behind the two analyzers are the same, but the ion path in an ion trap is three dimensional rather than the linear (corkscrew) pattern in a quadrupole. The ion trap itself can have multiple configurations, but the Paul, Penning, and Kingdon (Orbitrap) traps are the most common. Paul's trap consists of a ring electrode (that looks like a donut) placed between a pair of end-cap electrodes. An oscillating RF field is generated within, and the ions orbit around the center of the trap until their orbit destabilizes. The point at which ions of different m/z will form a stable orbit is determined by the voltage applied to the electrodes. As the voltage increases, smaller ions destabilize more quickly and exit the trap to the detector [41]. The Penning trap uses axial electromagnetic fields in combination with a quadrupole electric field to trap the ions in a hyperboloid shape. Because the fields in a Penning trap are static, the trap can hold an ion for up to several days; this stability offers sufficient sensitivity and precision to measure even sub-atomic particles [42, 43]. The Kingdon design will be discussed below, but there is another version of the ion trap worth mentioning here: the *linear ion trap* (*LIT*). The chief difference between these is that the field applied to the ion trap is a two-dimensional (2D) RF field, so the ion path looks similar to that

[6] In actuality, when looking for a signal for glucose on a QqQ, commonly a signal is found at 73 m/z in negative mode; we will see why in the next section.

of a quadrupole. This configuration is known to have impressive dynamic range and is dependable for quantitative studies.

The quadrupole time-of-flight (Q-TOF) MS and Orbitrap MS are similar in concept but utilize different geometries for separating ions. The Q-TOF is known as a hybrid instrument because it combines Q1 and q2 with a TOF tube to separate ions based on the time it takes for an ion (or packet of ions) to travel through a vacuum-sealed "flight tube" and reach the detector. Within this flight tube, ions are accelerated with a high voltage pulse that sends them through the tube toward an electrostatic mirror that repels them back down the opposite side of the tube toward the detector. In this configuration, smaller molecules pick up more velocity than larger ones – and thus reach the detector first. The amount of time that the ion travels (the flight time) is determined by the formula for kinetic energy and is sensitive to the distance it has to travel through the flight tube.

In place of a flight tube, an Orbitrap has a chamber known as a high-field compact trap with electrode cups sandwiching a central electrode. When ions enter the trap, they deflect toward the detector based on the applied electric field, which then spirals them around the central electrode core (orbiting) until they reach the amplifier and then the detector. Both configurations offer fast sensitive scans and deliver high-resolution data, and both are suited to most metabolomics workflows [44].

4.4 *The Mass Spectrum*

Once a sample has been analyzed, each of the MS instruments listed above will produce a datafile with some proprietary format. Generally, these can be converted by third-party programs to be manageable by different analysis software (*msconvert* and *Rawconvertor* are two examples). The output files are made up of the mass spectrum that was detected at each scan over the course of a run. Each scan of the *mass spectrum* (the m/z's detected over the mass range – usually 100–~3200 Da) can be performed many times per second. An FTIR might scan once per second but with a resolving power of ~1 M while a QTOF might perform up to 500 scans per second with a resolution of 60,000. Note that the term *resolving power* is measured in m/dm where m refers to mass and dm is the peak width necessary for separation measured in full width at half maximum (FWHM). After each scan, the mass spectrum is collected and represented in the datafile for that sample. Since the mass analyzer and detector of the MS are always on during a sample run, a different mass spectrum is collected over user-defined intervals.

The mass spectrum itself is represented as a table or in graphical form where the m/z ratios of the ions are plotted on the x-axis and their intensities are plotted on the y-axis. Each peak in a mass spectrum shows a component of unique m/z in the sample, and heights of the peaks connote the relative abundance of the various components in the sample. If a specific molecule of interest is expected to be represented in the spectrum, the analysis software can query (extract) that specific m/z from the sample run and display the times that it was detected. This is represented as the ion chromatogram.

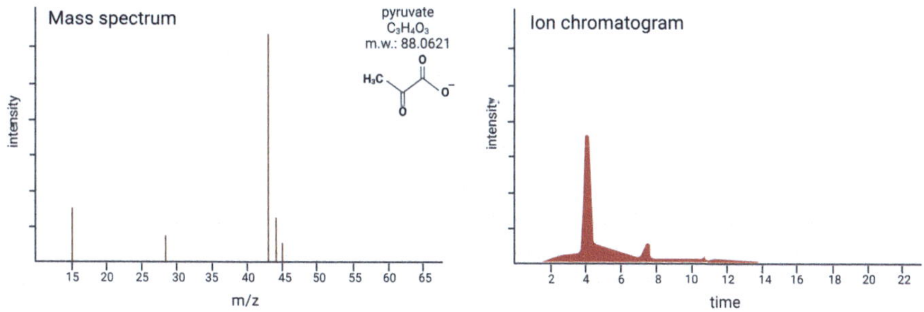

Fig. 8 Stylized mass spectrum and ion chromatogram for alanine. (Created in Biorender)

Figure 8 depicts a stylized mass spectrum for alanine (an amino acid). The parent molecule is composed of three carbons, an amine group, and seven protons. When the molecular ion of alanine is ionized in positive mode, it acquires a proton to yield a mass of 90 (the exact mass is 89.04768 + the mass of the hydrogen adduct is 1.007825 = 90.0560). There are two other ions to note in this spectrum. The ion at ~91 m/z corresponds to alanine where one of the carbons (with an exact mass of 12.00000, aka ^{12}C was substituted with the naturally occurring ^{13}C to yield a fragment that is 1.003355 amu heavier). The natural abundance of ^{13}C in the environment is about 1.07%, so we would expect about that percentage of alanine would contain one of these heavy carbons. The presence of this peak is useful in complicated spectra where it is important to determine if the parent ion is genuine or a fragment of another ion (the fragment won't have a daughter ion that is ~1.07% of its height). The other ion is at 44.0490 m/z. That ion happens to be a fragment ion corresponding to the left half of the molecule (when viewed as depicted in the inset). Fragment ions are created not only intentionally (as in CID) but also unintentionally as a result of the conditions in the ion source

It is important to verify that a peak on the mass spectrum is an *in-source fragment* using a QC step for metabolomics experiments. This is most frequently accomplished using an *authentic standard* consisting of a purified molecule spiked into a sample matrix or solvent and subjected to the same analytical conditions as the sample. If the same fragment appears in the standard as was observed in the complex sample mixture, this gives the user confidence of the identity of the molecule.

4.5 NMR

Nuclear magnetic resonance (NMR) spectroscopy is a key analytical tool used in metabolomic studies, with unique strengths compared to GC- and LC-MS based methods. NMR is non-destructive, allowing samples to be rerun using different pulse programs; it requires limited sample preparation; it is quantitative; cost per sample is low; it is particularly suited for comprehensive untargeted analysis, detecting a wide range of metabolites in a single experiment; it's application in structural elucidation allows for the identification and characterization of novel compounds and positional information can be generated for isotopomers [45]. In addition, NMR can add value for molecules that may not ionize well, require derivatization, or cannot be separated based on mass. Furthermore, NMR experiments are highly reproducible, with very limited batch effects and no carry-over between samples. It is this benefit, combined with the short duration of experimental analyses (a typical 1D ^{1}H NMR analysis can be acquired in a few minutes) and automation of experiments, which make NMR ideal for high-throughput metabolomics studies [46, 47].

The major limitation of NMR is its low sensitivity, which translates to a reduced number of biomarkers in metabolomic analyses (typically 50–200 identified metabolites in metabolomic studies). In addition, sample complexity can hinder interpretation of NMR spectra and quantitation of molecules, with resonances from highly abundant or high molecular weight molecules, such as lipids and lipoproteins, overlapping with and obscuring peaks from other molecules. Furthermore, a greater sample volume is required, instrument costs are high, and the footprint of lab space required is large, compared to LC-MS instruments [48, 49].

The versatility and capabilities of NMR continues to grow, with increases in sensitivity owing to developments in field strength and cryoprobe technology, and the use of hyperpolarized probes (capable of achieving detection limits in the nanomolar range [50]). Neither NMR nor MS used alone has the ability to detect all metabolites in a metabolomics study; however, efforts to integrate both methodologies and harness their collective advantages will offer a more comprehensive coverage of the metabolome and improve the accuracy of metabolite identification [51].

4.6 FTIR

Fourier transform infrared spectroscopy (FTIR) is a vibrational spectroscopy technique that uses infrared radiation to excite a sample. In a modern FTIR, infrared radiation (IR) is generated from a radiation source (usually an inert solid that is heated to 1000–1800 °C) and is then passed through a sample. Some of that radiation is absorbed, and some reflected through to the monochromator (or inferometer depending on the application), which then disperses it to the detector. Either a thermal detector or photon detector measures the heat produced by the IR and generates an electric (IR spectroscopy Hsu) interferogram plot. This IR spectrum can then be mined for information about the constitution of the sample. Generally, a background spectrum is generated containing the solvent that the sample was prepared in, and this background can be subtracted to generate a clean sample spectrum. Data analysis consists of matching the frequency bands of the sample spectrum which is then converted through an (fast Fourier transform) algorithm into a high-resolution molecular fingerprint. FTIR excels at identification of unknown molecules and structure elucidation [52] and is amenable to both solid and liquid samples.

4.7 Other Analytical Techniques

In addition, there are a number of ways to measure aspects of metabolism that we will not be covered in detail in this text. These include enzymatic analyzers that are designed to accurately measure a specific set of biomolecules (often biologically relevant carbohydrates or amino acids) using proprietary technology. Often these methods include a purpose-built instrument, but many are available as multiwell plates prepared with a substrate that functions as a biosensor (fluorescent or chemiluminescent). Also, several fluorescent probes have been developed that can be exogenously applied to cell preparations, or genetically encoded [53] to function as in situ biosensors of aspects of cellular metabolism. These methods are often useful for testing specific hypotheses or for verifying findings made using one of the canonical metabolomics methodologies.

5 Separation Techniques

The mass accuracy and resolution of HRMS, NMR, and FTIR instruments are more than sufficient to accurately detect and quantify most small molecules. However, the metabolome is a series of complex networks of interacting molecules. These networks include biomolecules with chemically diverse structures from organic and amino acids to nucleotides to carbohydrates or lipids. It is this vast chemical diversity that poses the greatest challenge to the analysis of small molecules. We used alanine as an example of what one would see in a mass spectrum above (Fig. 8), but the mass spectrum that is typical of a metabolomics experiment contains hundreds or thousands of biomolecules. Each has the potential to form its own in source fragments. When molecules with similar structural features are fragmented, identical in-source fragments can be produced, complicating positive identification. The solution to this is to separate the sample based on some physical characteristic like size or charge.

Currently, there are no analytical platforms capable of assaying all of the classes of biomolecules simultaneously, but separation methods have been developed that provide fairly extensive distribution of important biomolecules over time. Although variations in instrumentation exist, protocols designed to assay ions with dissimilar molecular characteristics generally rely on biophysical separation techniques like chromatography. There are (of course) an ever-growing number of variations in separation techniques. The methods are sufficiently diverse as to demand dedicated journals, but we will focus on a few popular methods here [54–56].

5.1 Liquid Chromatography

By far the most common separation method for metabolomics is liquid chromatography [57]. It involves the separation of metabolites based on their physicochemical properties, such as polarity and hydrophobicity, using a liquid mobile phase and a stationary phase.

The *mobile phase* is composed of two solvents selected for the solubility of the class of molecules being targeted and their compatibility with the chosen ionization technique. The experimental sample is injected into a stream containing the first mobile phase solvent (in which the sample should be soluble) and it is guided to the LC column were it adsorbs to the material within. The second mobile phase solvent is then introduced at a concentration gradient, which elutes the sample from the column and passes it to the ion source. The *stationary phase* is a solid matrix confined to the inside of an LC column and is usually made of a tightly packed silica or alumina-based adsorbent. It is generally a good practice to refer to the existing literature for help selecting the mobile phase and solid phase components for the specific class of metabolites one wishes to target. There are two points of caution to use when selecting solvents: the first is to seek out MS-grade solvents that have been sufficiently processed to remove impurities that may appear on the mass spectrum. The second is to avoid buffers and salts (Na+, K+, and PO_4), which can reduce the overall signal.

Different LC modes, such as *reverse-phase liquid chromatography* (*RPLC*), *hydrophilic interaction chromatography* (*HILIC*), and *ion-exchange chromatography* (*IEC*), are utilized to achieve specific separation objectives. RPLC is based on the differential partitioning of metabolites between a non-polar stationary phase and a polar mobile phase. It is suitable for separating non-polar and moderately polar metabolites. Unfortunately, highly polar and ionic metabolites are not retained in RPLC and appear in an unseparated bolus prior to the addition of the second solvent. The metabolites that are not compatible with RPLC include nucleotides, some carboxylic acids, and sugar phosphates. Still, when global metabolite coverage is the goal of the researcher, RPLC is often a good place to start because it has been successfully used to separate a wide range of important biomolecules that are soluble in aqueous solvents [57].

HILIC [58] is a version of RP which employs a polar stationary phase and a less polar mobile phase, making it ideal for separating highly polar metabolites [59]. There are still some questions about the exact mechanism of retention in HILIC columns [60], but generally they use an elution solvent that has a high water content. It has become a popular secondary methodology for metabolomics because so many critical metabolites can be resolved using this technique. Critical metabolites like amino acids, carboxylic acids, phosphorylated compounds, nucleotides, and sugars are well separated and many useful workflows have been published for various sample types.

IEC separates metabolites based on their charge by utilizing a charged stationary phase and adjusting the pH and ionic strength of the mobile phase. The separation provided is based on charge and columns with both anionic and cationic resins being available to separate negatively charged or positively charged ions (respectively). The IEC column with a charge opposite to that of the target ions is fitted to the LC system, and when the sample is loaded at a particular pH, appropriately charged ions bind to the resin via electrostatic interactions. Then the column is washed and a salt gradient is introduced as the second mobile phase solvent is added to elute the ions from the resin. This technique is useful for separation of proteins, peptides, amino acids, or nucleotides. IEC has been proven to be useful for excellent coverage of some features that are troublesome for other methods [61], but it is quite sensitive to changes in pH.

5.2 Gas Chromatography

Gas chromatography is a separation technique primarily used for volatile and thermally stable metabolites. It involves the vaporization of metabolites and their separation using a gaseous mobile phase and a stationary phase. In GC, ions are separated based on their volatility and affinity for the stationary phase, that is, either packed or capillary columns (although the latter are more popular). The sample is injected into a heated injection port, where it vaporizes and is pushed by an inert gas through a long, thin capillary column (commonly 30-m-feet long). Metabolites interact with the stationary phase, and their separation is achieved based on differences in their boiling points and chemical interactions. The separated metabolites are then detected by the MS detector, enabling their identification and quantification. Importantly, the amount of time that the individual ions are retained in the capillary (the *retention time* or *RT*) is very stable and predictable, making GC a robust method. In fact, GC-based metabolomics was the cornerstone of the field until LC-based methods supplanted it [62]. The RTs of GC-based workflows are sufficiently standardized that researchers around the world can query existing validated databases to annotate the signals from their samples with a high degree of confidence.

GC-MS is highly sensitive, capable of analyzing a wide range of metabolites, and provides reproducible results. It is particularly useful for metabolite profiling in areas such as lipidomics, volatile metabolites analysis, and stable isotope labeling studies. In practice, metabolite samples are frequently chemically *derivatized* to improve their volatility. Derivitization reagents like N,O-bis(trimethylsilyl)trifluoroacetamide (BSTFA) are added to the sample prior to loading where they add a trimethylsilyl (TMS) functional group to the sample compounds. Using derivitization procedures can increase the coverage of experiments to over 300 unique compounds.

5.3 Capillary Electrophoresis (CE)

Capillary electrophoresis is a separation technique that utilizes an electric field to separate metabolites based on their charge and size [63]. The techniques use hollow glass capillary tubes filled with an electrolyte solution wherein the inner surface of the capillary maintains a negative charge through its silanol groups. Then an electric field is generated by applying a voltage to both ends of the capillary and the metabolite ions migrate based on variations in their charge-to-size ratio. The separation is achieved due to differences in electrophoretic mobility, and various detection methods such as UV-Vis absorbance, fluorescence, or mass spectrometry can be coupled with CE for metabolite detection and quantification. It offers high separation efficiency, short analysis time, and requires minimal sample volume.

CE is particularly suitable for the separation of polar and charged metabolites. Small polar metabolites, amino acids, organic acids, and other charged compounds are readily separated. It offers high resolution and can be used for both qualitative and quantitative analyses. CE-MS coupling further enhances metabolite identification capabilities.

5.4 Ion Mobility Spectrometry (IMS)

IMS is an electrophoretic technique that separates ions based on their size, shape, and charge in the gas phase. In IMS, metabolite ions are introduced into a drift tube filled with a buffer gas, typically nitrogen, and subjected to an electrical field. One advantage of IM is that it can perform separations of ions on a millisecond timescale. It also works to increase the number of detectable features by as much as 20% over an LC-MS experiment [64]. These advantages have made it increasingly popular in metabolomics due to its ability to provide rapid and sensitive separations of metabolites.

6 Sample Preparation

Sample preparation is a critical step in metabolomics, as it can greatly impact the quality and accuracy of results. However, the optimal sample preparation method can vary depending on the type of sample, the metabolites of interest, and the analytical technique used. Proper sample preparation can ensure optimal recovery, stability, and reproducibility. Metabolomics can be performed from a range of specimens including blood, urine, tissues, cultured cells, biofluids, plant exudates, stool, etc. Dedicated protocols exist for each of these, so we will instead focus on general principles of extraction – independent of the first few steps of isolation that differ between samples. It is important to emphasize that our goal for this section is to give an overview of sample preparation to familiarize the reader with the basic concepts of sample preparation. Any researcher is advised to seek out peer-reviewed protocols prior to committing to a specific technique. We will also cover intracellular metabolites independently of extracellular/excreted molecules. Regardless of the sample, the most critical factor in the success of metabolomics studies is how fast the sample is quenched and how well it is kept cold [65]. The quenching step in particular is crucial because many metabolites are unstable or rapidly interconvert [66]. The four steps are: cell isolation, quenching, lysis/homogenization, and aqueous phase collection/storage.

- *Cell isolation*: To extract metabolites into an aqueous solution that can be subjected to one of the separation protocols detailed above, the cells of interest must be separated from the water or media that surrounds them. This step involves removal of mass that the researcher does not wish to measure. Isolating cultured cells from the culture media can be accomplished with a quick, gentle centrifugation step that avoids cell lysis, but fast filtration is preferred. Isolation of tissue might involve resection or laser-microdissection. Stool and urine samples are already sufficiently concentrated and can often be diluted in water or solvent [67].
- *Quenching*: For almost all samples with active metabolism, rapid quenching can be accomplished by plunging the isolated sample into liquid nitrogen or cold methanol. Rapid boiling accomplishes the same goal, but at the expense of the degradation of thermolabile compounds. Either of these is sufficient to stop enzymatic activity, which would otherwise lead to undesirable catalysis that occurs post-isolation.
- *Cell lysis/homogenization*: Metabolite extraction requires the release of metabolites from the sample matrix. Various extraction techniques are employed depending on the sample type and the metabolites of interest. Common extraction methods include liquid-liquid extraction (LLE) and solid-phase extraction (SPE). LLE and SPE can be accomplished by rapidly lysing cells in the extraction solvent using a homogenizer. Because the metabolites of interest reside (mostly) in the cytosol, the cell lysis step releases them into the extraction solvent, which can be stored and measured with less concern for degradation.
- *Sample collection and storage*: Appropriate sample collection and storage methods are essential to maintain metabolite integrity. Samples should be collected under standardized protocols, considering factors such as timing, sample matrix, and storage conditions (temperature, light exposure, freeze-thaw cycles). Storage of samples at temperatures at or lower than-80 °C is preferred. The consequences of improper storage are sample degradation, oxidation, and residual enzymatic activity, the latter of which can lead to artifactual metabolite abundances [68].

Post-extraction, sample clean-up steps are performed to remove unwanted compounds such as proteins, lipids, salts, and other interfering substances that can affect metabolite analysis. Techniques like centrifugation, filtration, solid-phase extraction, and liquid-liquid partitioning are employed for sample cleanup. Methods such as ultrafiltration, immunoaffinity depletion, and size-exclusion chromatography can be employed for protein removal, but using acetonitrile or methanol or a methanol/chloroform precipitation is sufficient [67]. Derivatization also can be performed post-extraction and before storage.

It is important to select sample preparation techniques that are compatible with the analytical methods employed in metabolomics, such as liquid chromatography-mass spectrometry (LC-MS) or gas chromatography-mass spectrometry (GC-MS). Optimization and validation of sample preparation protocols are necessary to ensure consistent and accurate metabolite profiling. It is also a good idea to perform pilot experiments prior to attempting extraction with precious samples.

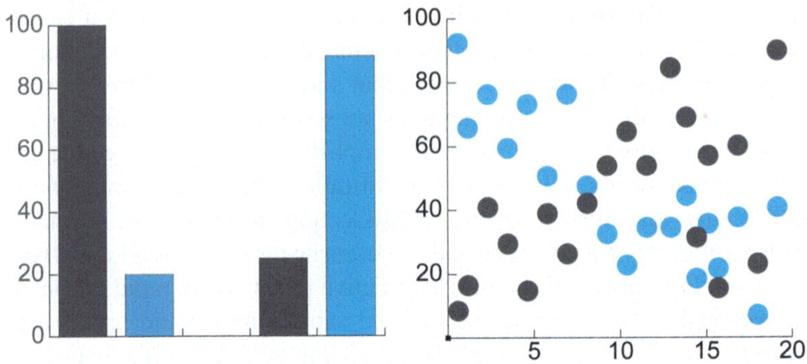

Fig. 9 Targeted vs untargeted data (fictional)

7 Study Design

One of the first considerations one must take when embarking on this path also happens to be the most intuitive. The researcher should consider how they would like to present their data – as a direct comparison of specific metabolites between groups or as a scatterplot (or perhaps a heatmap) that highlights a systemic difference in response. Figure 9 depicts how this data might be represented between a control and experimental group using metabolomics data. The difference between these representations illustrates the two broad categories used to describe distinct metabolomics studies: *targeted metabolomics* and *untargeted* (or "discovery") *metabolomics*.

Targeted methods are focused on specific metabolites and can refer to a single molecule of interest or up to thousands of specific molecules. When referring to targeted methods, researchers are generally referring to *identification* of metabolites rather than their absolute or relative *quantification*, though both can be accomplished with proper consideration of method design. The distinction is that the molecules in targeted metabolomics are known and are usually *validated*[7] prior to performing the analysis. The unequivocal validation of a molecule requires the reporting of an exact mass, so HRMS instruments are preferred. One caveat to this is that good sample preparation along with a validated analytical method is sufficient for quantification of a pre-validated molecule, and this can be accomplished on an LRMS. Structure determination studies are better suited to NMR or an ion trap. With a sufficiently established analytical method, one can acquire reliable targeted information for many hundreds of molecules that have been pre-validated.

Untargeted methods are much more open-ended and are frequently used for hypothesis generation. A typical untargeted workflow involves preparing samples from experimental and control groups and acquiring all of the peak data from

[7] Validation of a biomolecule refers to the unambiguous confirmation of a molecule's identity (including an empirical formula) in a specific method, and it is not a trivial task. Fortunately, guidance exists for small molecules [99, 100], proteins [101], and lipids [102]. Additionally, the FDA offers its own guidance for small molecules [103] but has separate guidelines for validation of analytical methods [104].

whichever assay is chosen. Once the data is processed and analyzed, univariate or multivariate bioinformatical analysis methods are applied for relative quantification of features (putative metabolites) across groups. These features are presumed to be qualitative and require further validation, but there are freely available databases and libraries for >150,000 metabolites (https://www.metabolomicsworkbench.org/data/index.php) to aid in identification. It is then the researchers' job to develop an actionable hypothesis from the data profile. This is the design of choice for a follow-up study after the discovery of an interesting phenotype or phenomenon. As we will discuss later in the chapter, the staggering heterogeneity of biomolecules (in their polarity, solubility, pKa, log P, stability, reactivity, etc.) makes untargeted analysis dependent on discreet analytical techniques. There is a common aphorism uttered in this field that is usually phrased something like: "there is no one method to fit all cases." Most methods are designed for molecules with specific characteristics, but even robust methods that can reliably assay broad categories of molecules will be opaque to other (potentially biologically relevant) factors. Absolutely requisite is the understanding by the researcher that the choice of methodology will influence the results of the study. However, untargeted methods can be powerful discovery tools and metabolite fingerprinting delivers a useful snapshot of the physiological state of the sample at the moment of its preparation. Outside of these two broad categories of study design, there are a number of other precautions that must be addressed when designing a metabolomics study.

8 Challenges in Metabolomics

8.1 Challenges Common to All Omics

The omics all share a similar set of challenges: the rapid pace of technological innovation, the necessity for exacting experimental procedures, the requirement for unique analytical pipelines for big data, and the integration of the findings into a format that is conducive to the formation of biological insight. These concerns are both technical and analytical in nature. Today, diligent work on method development in the individual omics fields has alleviated a large part of the technical hurdles. Protocols for sample handling and preparation, analytical techniques, pre-processing (including library construction in the case of next-generation sequencing (NGS) – or derivitization in the case of proteomics/metabolomics), data storage, and data hosting have been developed and refined. Publicly available free repositories exist for vast majority of published datasets, thanks to the investment and hard work of scientists around the world [69]. The force of scientific perseverance has turned most of the sample preparation techniques into routine protocols that can be accomplished in most laboratories around the world.

The technical challenges that remain are largely related to – or are the direct consequences of – the data dimensionality of omics studies. These are noise and reproducibility. It seems fairly obvious that these are two sides of the same coin, but the intrinsic underlying causal factor for them is amplified in datasets with small sample number and large numbers of measurements. That factor is the spectacular heterogeneity and stochasticity of biological systems. This stochasticity leads to an inconveniently wide distribution of mRNA expression or protein abundance, even in a carefully prepared biological sample [70]. The resulting variation is classified as noise and is compensated for using repeated measurements or biological replicates [71]. We acknowledge that this factor affects genomic studies less than other omics, but even in the case of a next-generation sequencing (NGS) workflow, the quality among reads always varies. The reason for this is also rooted in dimensionality as we will see below.

Reproducibility remains a challenge across all of science [72, 73]. Omics studies are no exception to this concern. The myriad causes for the so-called "replication crisis" don't have a single solution, but the aforementioned biological noise certainly contributes. The publication of noisy datasets is not in any way reflective of a lack of sufficient precautions or scientific expertise, but findings that cannot be replicated lead to suspicion of fraud. Other concerns are the numerous legitimate ways of analyzing a complex dataset, human error in the form of confirmation bias [74], and statistical mismanagement [75]. There is the hope that this concern can be mediated through organizational means [76], and best practices are being developed to address the issue [77].

The analytical challenges for omics are considerably more intractable. Omics studies usually start with the desire to answer a biological question and end with a dataset that can then be analyzed for the generation of a testable hypothesis. The analysis of the dataset is much more complex than that of a classical experiment with a single or a few variables of interest. Classical experimentation comprises dependent, independent, and control variables that are then measured to determine correlation. While the experimental design of omics experiments is conceptually similar, the fundamental structure of the data is different than in classical experimentation [78]. This discrepancy is well known to informaticians; and mathematician Richard Bellman coined a phrase to describe it: The Curse of Dimensionality [79]. It suggests that as the number of dimensions (features or measurements in this case) increases, the volume of Euclidean space (the error) increases exponentially. This has become to be commonly known as the "big-p, little-n" problem (abbreviated $p >> n$), where p represents the number of observed variables and n indicates the sample size. Omics experiments fit this description [80]. In omics experiments, the variables are a large number of metabolites or genes and a small number of samples. The consequence of this sparsity is that observations become harder to cluster making it harder to detect meaningful trends in the data.

Let's take the simple decision tree that we use to determine which statistical analysis is appropriate for our experimental design. Classical experimentation may be interested in a categorical variable where the number of groups would dictate whether we should use (for example) a Fisher test or a t-test to determine our effect size. If we wish to measure multiple variables, we might employ a multinomial logistic regression or apply a Pearson test. These multivariate analyses are designed to find patterns and correlations, but small sample sizes risk overfitting the data. In order to increase power, sample size must increase [81]. This is often infeasible in omics experiments due to cost and complex sample processing. Omics experiments simultaneously test hundreds or thousands of hypotheses. Their analysis relies on a more complicated suite of statistical tools for proper investigation [82] and interpretation [83].

8.2 Challenges Specific to Metabolomics

The goal of metabolomics is to identify and quantify the metabolites present in a sample in order to gain insight into the metabolic pathways and processes occurring in a particular organism or tissue. There are several challenges specific to metabolomics that make this process more difficult compared to other "omics" fields such as genomics or proteomics. Greater chemical diversity and dynamic range, complex profiles, multiple peaks representing individual molecules, standardization and validation, and sample collection bias are all technical hurdles that make metabolomics more difficult to perform properly than the other omics fields.

Genomics and transcriptomics are concerned with four nucleotide bases that are relatively stable (or can be stabilized in solution) and a handful of modifications in epigenomic studies. The chemical diversity of the proteome covers about four orders of magnitude [87] (Fig. 10). An important contrast is that the variety in the properties of small molecules makes them much more difficult to assay in a single experiment or method. A typical metabolomic experiment will provide information about a thousand unique metabolites [88], and the Human Metabolite Database (https://hmdb.ca/) currently references over a quarter of a million molecules.

Whereas one can sequence all of the DNA or RNA or proteins in a cell, complete measurements of all small molecules are problematic for a number of reasons. Cells and tissues are made up of small molecules with substantial chemical diversity. Researchers are interested in biomolecules with widely divergent chemical

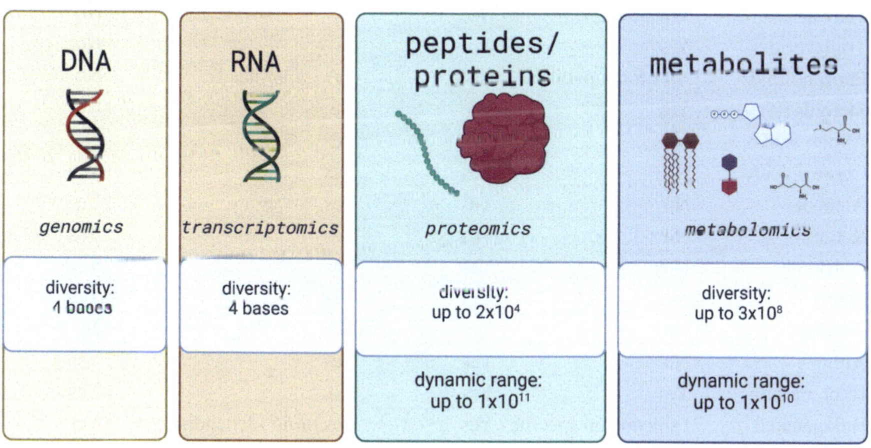

Fig. 10 Omics at a glance [84–86]. (Created with BioRender.com)

properties like lipids, sterols, nucleotides, amino acids, carbohydrates, terpenes, ketones, and organic acids. This diversity makes metabolomics a challenge in analytical chemistry as the solubility and stability of these molecular classes vary greatly. There are several considerations to make regarding the class of biomolecule that the researcher is interested in measuring. Chromatographic columns poorly retain biomolecules that are polar or ionic, such as many amino and organic acids, carbohydrates, and nucleotides [56]. Also, some molecular classes fare poorly in some of the solvents used in LC methods based on their pKa in specific solvents. Fortunately, there is a method to predict the pKa of a target molecule to aid solvent selection [89]. Although they are a special case, volatile molecules may require precautions to be taken at the time the sample is gathered to be properly measured [38]. These attributes can hinder the positive identification of important metabolites (see Table 2), so alternate chromatographic systems were developed to increase their separation. *There are no sample preparation methods that allow for the complete ionization of all biomolecules in a single sample*, so methodologies have been developed to target specific classes.

To add to this analytical challenge, the metabolome is highly complex and dynamic. Metabolites differ in natural abundance with a dynamic range that covers ten orders of magnitude [90]. Physiological concentrations of metabolites in cytosol are present in quantities ranging from femtomolar to millimolar, challenging the detection limits of even sensitive instruments. Moreover, some metabolites have turnover times in the cytoplasm of microseconds, while some are longer lived and enjoy half-lives on the order of seconds [91]. These high turnover rates demand

Table 2 Polarity and volatility of biomolecules

	Nonpolar and nonvolatile	Nonpolar and volatile	Polar and nonvolatile	Polar and volatile
Carbohydrates	No	No	Yes	No
Proteins	No	No	Yes	No
Nucleic acids	No	No	Yes	No
Fatty acids	Yes	No	No	No
Triglycerides	Yes	No	No	No
Phospholipids	No (amphipathic)	No	Yes (polar head)	No
Steroids	Yes	No	No	No
Amino acids	No	No	Yes	No
Organic acids	No	Yes	Yes	Yes
Alcohols	No	Yes	Yes	Yes
Ketones	No	Yes	Yes	Yes
Aldehydes	No	Yes	Yes	Yes
Esters	No	Yes	Yes	Yes
Amides	No	No	Yes	No
Amines	No	Yes	Yes	Yes
Hydrocarbons	Yes	Yes	No	Yes
Halogenated compounds	Depends on specific compound	Yes	Yes or no (depending on the level of halogenation)	Yes

careful sample handling precautions. Analytically they are challenging to measure for another reason: volatile metabolites with high vapor pressure and low molecular weight can diffuse freely in air and avoid capture during sample preparation steps [92].

8.3 *Technical Hurdles to Metabolomics*

In the previous sections, we introduced the instrumentation used in most metabolomics experiments. Some of the fundamental analytical obstacles to metabolomics analysis are due to the instrumentation that we chose to perform our analysis. These include the existence of multiple signals for a single metabolite, the remarkable chemical diversity of metabolites, and their range of physiological concentrations and volatility. If one thinks of the MS/FTIR/NMR as fundamentally a molecule detector with a series of (ingenious) approaches to separate those molecules, it is easy to see how the rate of signal might affect the accuracy and precision of the detection of our signals of interest. Each is a powerful tool for small molecule characterization, but each has strengths and weaknesses that confound interpretation. Fortunately, the pitfalls of each are well characterized and existing methods of sample preparation and separation techniques largely ameliorate the drawbacks. There are several reasons for which an experimental sample might provide a signal too low to be detected or too high, resulting in saturating the detector and masking further signals.

Some terminologies that describe these cases are listed below. First, there is a hard floor on the amount of signal that is necessary to trigger a detector of an instrument resulting in a signal. This is referred to as the *instrument detection limit*. It is determined by analyzing blank samples and determining a background level (usually the signal arising from the matrix itself). Then there is *the limit of quantification* (LOQ); this is the lowest amount of analyte that can be quantified with a degree of certainty. To faithfully measure the LOQ, one must take repeated measurements of the same analyte under the same conditions. If our analyte-of-interest has a signal above the LOQ, we can be confident that we can distinguish between different amounts of that analyte.

Metabolite profiling via MS is complicated by the presence of multiple peaks representing individual molecules, standardization and validation, and sample collection bias. The process of ionizing these already unstable molecules can lead to unwanted fragmentation when the energy transferred by the electrons from the source overcomes the thermal energy of the molecule. This is known as *in-source fragmentation* and is the cause of many confounding extraneous peaks on the mass spectrum. Thanks to work done by the Patti lab, there is a workflow to help detect and credential in-source fragments [93], thus avoiding false positives.

Two more complexities of the standard MS-based metabolomics workflow are the *matrix effect* and *ion suppression*. The matrix effect is the straightforward consequence of components or endogenous impurities in the sample that interfere with the ionization and detection of the compounds of interest. In settings that require rigorous controls and validation data, this effect must be compensated for experimentally [94]. This can be accomplished using reagent blanks or method blanks that contain the extraction solvent used to prepare the samples. Better still is the addition of a matrix spike duplicate (MSD) containing a precise amount of a specific target analyte to determine the level of interference. Note that these blanks and/or MSDs are assayed in addition to the standard quality control samples. If the interference from the sample matrix results in the disruption of detection, precision, and/or accuracy of a target compound, that interference is deemed to be ion suppression. Ion suppression can often be overcome by changing ion modes (from positive to negative or vise-versa) or dilution of the sample.

9 Conclusion

Hopefully the readers of this text will appreciate the utility and potential of metabolomics as a discipline – but to formalize the rationale for its use: the ability of metabolomics to provide a "global" picture of a biological sample is nigh irreplaceable for expanding knowledge of fundamental biological/biochemical questions, discovery of biomarkers, and to investigate connections within metabolic networks. If we encapsulate its definition as the systems-level study of biology at the biochemical level, it is clear how the simultaneous measurement of the small molecules in a sample might be applied to a host of interesting scientific avenues.

Many researchers have found that there are countless benefits of this type of comprehensive measurement. This text, *Metabolomics: Recent Advances and Future Applications*, aims to do keep the reader abreast of the state of the art for the discipline of metabolomics. In this text, we have solicited works by experts in the field that communicates the state of the art for relevant topics but does so in a way that is accessible to both undergraduates and experts. The general application of metabolomics to biomedical and pharmaco-biological research has grown to such an extent that an attempt at comprehensive summation of contemporary research is a fool's errand. Instead, we opted to focus this work on some of the areas which show the greatest focus of recent investigations, while attempting to honor the depth and breadth of fields which benefit from the core techniques described above. These were presented as a broad overview rather than a comprehensive treatment in the hope that the reader would easily grasp the fundamentals to aid in the understanding of the chapters that follow.

In the following chapters, we will focus on the use of metabolomics to develop a comprehensive understanding of metabolic pathways, identifying health/disease biomarkers, predicting drug efficacy and toxicity, and understanding the molecular mechanisms of biological processes and their interactions. Through the integration of metabolomics data with other omics disciplines such as genomics, transcriptomics, and proteomics, we can gain a more holistic view of biological systems and uncover novel insights into their functioning.

Looking ahead, future advancements in metabolomics will focus on enhancing analytical sensitivity, expanding metabolite coverage, and improving data processing and analysis workflows. Metabolomics will be applied to clinically important questions and enable the dream of personalized medicine. Additionally, the integration of metabolomics with systems biology approaches and artificial intelligence algorithms will enable more accurate predictions and modeling of metabolic pathways and networks. There is little doubt that the future of science will rely more heavily on global and systems-level studies to bring us to the goal of predictive modeling for biology (Table 3).

Table 3 Exact mass and isotopic abundance

Element	Isotope	Exact mass	Natural abundance
Hydrogen	H(1)	1.007825	99.99
Hydrogen	H(2)	2.014102	0.015
Carbon	C(12)	12	98.9
Carbon	C(13)	13.003355	1.1
Nitrogen	N(14)	14.003074	99.63
Nitrogen	N(15)	15.000109	0.37
Oxygen	O(16)	15.994915	99.76
Oxygen	O(17)	16.999131	0.038
Oxygen	O(18)	17.999159	0.2
Fluorine	F(19)	18.998403	100
Sodium	Na(23)	22.98977	100
Magnesium	Mg(24)	23.985045	78.9
Magnesium	Mg(25)	24.985839	10
Magnesium	Mg(26)	25.982595	11.1
Phosphorus	P(31)	30.973763	100
Sulfur	S(32)	31.972072	95.02
Sulfur	S(33)	32.971459	0.75
Sulfur	S(34)	33.967868	4.21
Chlorine	Cl(35)	34.968853	75.77
Sulfur	S(36)	35.967079	0.02
Chlorine	Cl(37)	36.965903	24.23
Potassium	K(39)	38.963708	93.2
Calcium	Ca(40)	39.962591	96.95
Potassium	K(40)	39.963999	0.012
Potassium	K(41)	40.961825	6.73
Calcium	Ca(42)	41.958622	0.65
Calcium	Ca(43)	42.95877	0.14
Calcium	Ca(44)	43.955485	2.086
Calcium	Ca(46)	45.953689	0.004
Calcium	Ca(48)	47.952532	0.19
Iron	Fe(54)	53.939612	5.8
Manganese	Mn(55)	54.938046	100
Iron	Fe(56)	55.934939	91.72
Iron	Fe(57)	56.935396	2.2

(continued)

Table 3 (continued)

Element	Isotope	Exact mass	Natural abundance
Iron	Fe(58)	57.933278	0.28
Copper	Cu(63)	62.929599	69.17
Zinc	Zn(64)	63.929145	48.6
Copper	Cu(65)	64.927792	30.83
Zinc	Zn(66)	65.926035	27.9
Zinc	Zn(67)	66.927129	4.1
Zinc	Zn(68)	67.924846	18.8
Zinc	Zn(70)	69.925325	0.6
Selenium	Se(74)	73.922477	0.9
Selenium	Se(76)	75.919207	9
Selenium	Se(77)	76.919908	7.6
Selenium	Se(78)	77.917304	23.5
Selenium	Se(80)	79.916521	49.6
Selenium	Se(82)	81.916709	9.4
Molybdenum	Mo(92)	91.906809	14.84
Molybdenum	Mo(94)	93.905086	9.25
Molybdenum	Mo(95)	94.905838	15.92
Molybdenum	Mo(96)	95.904676	16.68
Molybdenum	Mo(97)	96.906018	9.55
Molybdenum	Mo(98)	97.905405	24.13
Molybdenum	Mo(100)	99.907473	9.63

Exact mass and isotopic abundances for biomolecules and some trace elements. Exact mass measurements for common biomolecules including natural abundance. Table includes some essential trace elements – iron, zinc, fluoride, selenium, copper, chromium, iodine, manganese, and molybdenum [95] – and biomolecules [96].

References

1. Wetterstrand KA. DNA Sequencing Costs: Data from the NHGRI Genome Sequencing Program (GSP). www.genome.gov/sequencingcostsdata (2023).
2. Marshall, J. L. *et al.* The Essentials of Multiomics. *Oncologist* vol. 27 272–284 Preprint at https://doi.org/10.1093/oncolo/oyab048 (2022).
3. Wishart, D. S. Emerging applications of metabolomics in drug discovery and precision medicine. *Nature Reviews Drug Discovery* vol. 15 473–484 Preprint at https://doi.org/10.1038/nrd.2016.32 (2016).
4. Manzoni, C. *et al.* Genome, transcriptome and proteome: The rise of omics data and their integration in biomedical sciences. *Brief Bioinform* **19**, 286–302 (2018).
5. McGuire, A. L. *et al.* The road ahead in genetics and genomics. *Nature Reviews Genetics* vol. 21 581–596 Preprint at https://doi.org/10.1038/s41576-020-0272-6 (2020).
6. Lawrence, J. G. *Why genomics is more than genomes.* http://genomebiology.com/2004/5/12/357 (2004).
7. Alan Chodos. April 1946: First Concept of Time-of-Flight Mass Spectrometer. *APS News* (2001).
8. van Eck, N. J. & Waltman, L. Software survey: VOSviewer, a computer program for bibliometric mapping. *Scientometrics* **84**, 523–538 (2010).

9. Graw, S. *et al.* Multi-omics data integration considerations and study design for biological systems and disease. *Molecular Omics* vol. 17 170–185 Preprint at https://doi.org/10.1039/d0mo00041h (2021).
10. Krassowski, M., Das, V., Sahu, S. K. & Misra, B. B. State of the Field in Multi-Omics Research: From Computational Needs to Data Mining and Sharing. *Frontiers in Genetics* vol. 11 Preprint at https://doi.org/10.3389/fgene.2020.610798 (2020).
11. Vahabi, N. & Michailidis, G. Unsupervised Multi-Omics Data Integration Methods: A Comprehensive Review. *Frontiers in Genetics* vol. 13 Preprint at https://doi.org/10.3389/fgene.2022.854752 (2022).
12. Cusick, M. E., Klitgord, N., Vidal, M. & Hill, D. E. Interactome: Gateway into systems biology. *Hum Mol Genet* **14**, (2005).
13. Doerr, A. Gene factories made of droplets. *Nature Methods* vol. 15 160–161 Preprint at https://doi.org/10.1038/nmeth.4622 (2018).
14. Chen, X., Peng, Z. & Yang, Z. Metabolomics studies of cell-cell interactions using single cell mass spectrometry combined with fluorescence microscopy. *Chem Sci* **13**, 6687–6695 (2022).
15. Gupta, N., Duggal, S., Kumar, A., Saquib, N. M. & Rao, K. V. S. Concurrent interactome and metabolome analysis reveals role of AKT1 in central carbon metabolism. *BMC Res Notes* **11**, (2018).
16. Geier, B. *et al.* Spatial metabolomics of in situ host–microbe interactions at the micrometre scale. *Nat Microbiol* **5**, 498–510 (2020).
17. Gupta, S., Schillaci, M. & Roessner, U. Metabolomics as an emerging tool to study plant-microbe interactions. *Emerging Topics in Life Sciences* vol. 6 175–183 Preprint at https://doi.org/10.1042/ETLS20210262 (2022).
18. Maag, D., Erb, M. & Ga Etan Glauser, &. Metabolomics in plant-herbivore interactions: challenges and applications. https://doi.org/10.7892/boris.72209 (2015).
19. Tolstikov, V., James Moser, A., Sarangarajan, R., Narain, N. R. & Kiebish, M. A. Current status of metabolomic biomarker discovery: Impact of study design and demographic characteristics. *Metabolites* vol. 10 Preprint at https://doi.org/10.3390/metabo10060224 (2020).
20. Rinschen, M. M., Ivanisevic, J., Giera, M. & Siuzdak, G. Identification of bioactive metabolites using activity metabolomics. *Nature Reviews Molecular Cell Biology* vol. 20 353–367 Preprint at https://doi.org/10.1038/s41580-019-0108-4 (2019).
21. Yang, J. *et al.* Early screening and diagnosis strategies of pancreatic cancer: a comprehensive review. *Cancer Communications* vol. 41 1257–1274 Preprint at https://doi.org/10.1002/cac2.12204 (2021).
22. Castiglione, V. *et al.* Biomarkers for the diagnosis and management of heart failure. *Heart Failure Reviews* vol. 27 625–643 Preprint at https://doi.org/10.1007/s10741-021-10105-w (2022).
23. Dubin, R. F. & Rhee, E. P. Proteomics and metabolomics in kidney disease, including insights into etiology, treatment, and prevention. *Clinical Journal of the American Society of Nephrology* **15**, 404–411 (2020).
24. Iida, M., Harada, S. & Takebayashi, T. Application of metabolomics to epidemiological studies of atherosclerosis and cardiovascular disease. *Journal of Atherosclerosis and Thrombosis* vol. 26 747–757 Preprint at https://doi.org/10.5551/jat.RV17036 (2019).
25. Alonso, C., Noureddin, M., Lu, S. C. & Mato, J. M. Biomarkers and subtypes of deranged lipid metabolism in nonalcoholic fatty liver disease. *World Journal of Gastroenterology* vol. 25 3009–3020 Preprint at https://doi.org/10.3748/wjg.v25.i24.3009 (2019).
26. Oskovi Kaplan, Z. A. & Ozgu-Erdinc, A. S. Prediction of Preterm Birth: Maternal Characteristics, Ultrasound Markers, and Biomarkers: An Updated Overview. *Journal of Pregnancy* vol. 2018 Preprint at https://doi.org/10.1155/2018/8367571 (2018).
27. Emamzadeh, F. N. & Surguchov, A. Parkinson's disease: Biomarkers, treatment, and risk factors. *Frontiers in Neuroscience* vol. 12 Preprint at https://doi.org/10.3389/fnins.2018.00612 (2018).

28. Badhwar, A. P. *et al.* A multiomics approach to heterogeneity in Alzheimer's disease: Focused review and roadmap. *Brain* vol. 143 1315–1331 Preprint at https://doi.org/10.1093/brain/awz384 (2020).
29. Fernández-Ochoa, Á. *et al.* Recent Analytical Approaches for the Study of Bioavailability and Metabolism of Bioactive Phenolic Compounds. *Molecules* vol. 27 Preprint at https://doi.org/10.3390/molecules27030777 (2022).
30. Castelli, F. A. *et al.* Metabolomics for personalized medicine: the input of analytical chemistry from biomarker discovery to point-of-care tests. https://doi.org/10.1007/s00216-021-03586-z/Published.
31. López-Yerena, A. *et al.* Metabolomics technologies for the identification and quantification of dietary phenolic compound metabolites: An overview. *Antioxidants* vol. 10 Preprint at https://doi.org/10.3390/antiox10060846 (2021).
32. Beulens, J. W. J. *et al.* Environmental risk factors of type 2 diabetes-an exposome approach. https://doi.org/10.1007/s00125-021-05618-w/Published.
33. Juarez, P. D., Hood, D. B., Song, M. A. & Ramesh, A. Use of an Exposome Approach to Understand the Effects of Exposures From the Natural, Built, and Social Environments on Cardio-Vascular Disease Onset, Progression, and Outcomes. *Frontiers in Public Health* vol. 8 Preprint at https://doi.org/10.3389/fpubh.2020.00379 (2020).
34. Sun, J. *et al.* A review of environmental metabolism disrupting chemicals and effect biomarkers associating disease risks: Where exposomics meets metabolomics. *Environment International* vol. 158 Preprint at https://doi.org/10.1016/j.envint.2021.106941 (2022).
35. Adav, S. S. & Wang, Y. Metabolomics signatures of aging: Recent advances. *Aging and Disease* vol. 12 646–661 Preprint at https://doi.org/10.14336/AD.2020.0909 (2021).
36. Sharma, R. & Ramanathan, A. The Aging Metabolome—Biomarkers to Hub Metabolites. *Proteomics* vol. 20 Preprint at https://doi.org/10.1002/pmic.201800407 (2020).
37. Han, X., Aslanian, A. & Yates, J. R. Mass spectrometry for proteomics. *Current Opinion in Chemical Biology* vol. 12 483–490 Preprint at https://doi.org/10.1016/j.cbpa.2008.07.024 (2008).
38. Tsedilin, A. M. *et al.* How Sensitive and Accurate are Routine NMR and MS Measurements? *Mendeleev Communications* **25**, 454–456 (2015).
39. Wang, Y., Sun, J., Qiao, J., Ouyang, J. & Na, N. A 'soft' and 'hard' Ionization Method for Comprehensive Studies of Molecules. *Anal Chem* **90**, 14095–14099 (2018).
40. Alan Dronsfield. Mass spectrometry – the early days. *Royal Soceity of Chemistry* 1 (2010).
41. Wong, P. S. H. & Cooks, R. G. *Ion Trap Mass Spectrometry*.
42. Dilling, J., Blaum, K., Brodeur, M. & Eliseev, S. Penning-Trap Mass Measurements in Atomic and Nuclear Physics. https://doi.org/10.1146/annurev-nucl-102711 (2018).
43. Eliseev, S. & Novikov, Y. High-precision Penning-trap mass spectrometry for neutrino physics. *European Physical Journal A* vol. 59 Preprint at https://doi.org/10.1140/epja/s10050-023-00946-4 (2023).
44. Schwaiger-Haber, M. *et al.* A Workflow to Perform Targeted Metabolomics at the Untargeted Scale on a Triple Quadrupole Mass Spectrometer. *ACS Measurement Science Au* **1**, 35–45 (2021).
45. Fan, T. W. M. & Lane, A. N. Applications of NMR spectroscopy to systems biochemistry. *Progress in Nuclear Magnetic Resonance Spectroscopy* vols 92–93 18–53 Preprint at https://doi.org/10.1016/j.pnmrs.2016.01.005 (2016).
46. Edison, A. S. *et al.* NMR: Unique Strengths That Enhance Modern Metabolomics Research. *Analytical Chemistry* vol. 93 478–499 Preprint at https://doi.org/10.1021/acs.analchem.0c04414 (2021).
47. Wishart, D. S. *et al.* NMR and Metabolomics—A Roadmap for the Future. *Metabolites* vol. 12 Preprint at https://doi.org/10.3390/metabo12080678 (2022).

48. Nagana Gowda, G. A. & Raftery, D. Can NMR solve some significant challenges in metabolomics? *Journal of Magnetic Resonance* **260**, 144–160 (2015).
49. Emwas, A. H. *et al.* Nmr spectroscopy for metabolomics research. *Metabolites* vol. 9 Preprint at https://doi.org/10.3390/metabo9070123 (2019).
50. Ribay, V., Praud, C., Letertre, M. P. M., Dumez, J. N. & Giraudeau, P. Hyperpolarized NMR metabolomics. *Current Opinion in Chemical Biology* vol. 74 Preprint at https://doi.org/10.1016/j.cbpa.2023.102307 (2023).
51. Bhinderwala, F., Wase, N., Dirusso, C. & Powers, R. Combining Mass Spectrometry and NMR Improves Metabolite Detection and Annotation. *J Proteome Res* **17**, 4017–4022 (2018).
52. Martens, J. *et al.* Molecular identification in metabolomics using infrared ion spectroscopy. *Sci Rep* **7**, (2017).
53. Wang, M., Da, Y. & Tian, Y. Fluorescent proteins and genetically encoded biosensors. *Chemical Society Reviews* vol. 52 1189–1214 Preprint at https://doi.org/10.1039/d2cs00419d (2023).
54. Rodriguez, E. L. *et al.* Affinity chromatography: A review of trends and developments over the past 50 years. *Journal of Chromatography B: Analytical Technologies in the Biomedical and Life Sciences* vol. 1157 Preprint at https://doi.org/10.1016/j.jchromb.2020.122332 (2020).
55. Perez de Souza, L., Alseekh, S., Scossa, F. & Fernie, A. R. Ultra-high-performance liquid chromatography high-resolution mass spectrometry variants for metabolomics research. *Nature Methods* vol. 18 733–746 Preprint at https://doi.org/10.1038/s41592-021-01116-4 (2021).
56. Pezzatti, J. *et al.* Implementation of liquid chromatography–high resolution mass spectrometry methods for untargeted metabolomic analyses of biological samples: A tutorial. *Analytica Chimica Acta* vol. 1105 28–44 Preprint at https://doi.org/10.1016/j.aca.2019.12.062 (2020).
57. Patti, G. J. Separation strategies for untargeted metabolomics. *Journal of Separation Science* vol. 34 3460–3469 Preprint at https://doi.org/10.1002/jssc.201100532 (2011).
58. Alpert, A. J. Hydrophilic-interaction chromatography for the separation of peptides, nucleic acids and other polar compounds. *Journal of Chromatography* vol. 499 (1990).
59. Hosseinkhani, F. *et al.* Systematic Evaluation of HILIC Stationary Phases for Global Metabolomics of Human Plasma. *Metabolites* **12**, (2022).
60. Svec, F. Preparation and HPLC applications of rigid macroporous organic polymer monoliths. *Journal of Separation Science* vol. 27 747–766 Preprint at https://doi.org/10.1002/jssc.200401721 (2004).
61. Walsby-Tickle, J. *et al.* Anion-exchange chromatography mass spectrometry provides extensive coverage of primary metabolic pathways revealing altered metabolism in IDH1 mutant cells. *Commun Biol* **3**, (2020).
62. Fiehn, O. Metabolomics by gas chromatography-mass spectrometry: Combined targeted and untargeted profiling. *Curr Protoc Mol Biol* **2016**, (2016).
63. Neusüß, C. & Jooß, K. *Capillary Electrophoresis-Mass Spectrometry Methods and Protocols Methods in Molecular Biology 2531*. http://www.springer.com/series/7651.
64. Paglia, G., Smith, A. J. & Astarita, G. Ion mobility mass spectrometry in the omics era: Challenges and opportunities for metabolomics and lipidomics. *Mass Spectrometry Reviews* vol. 41 722–765 Preprint at https://doi.org/10.1002/mas.21686 (2022).
65. Lu, W. *et al.* Metabolite measurement: Pitfalls to avoid and practices to follow. *Annual Review of Biochemistry* vol. 86 277–304 Preprint at https://doi.org/10.1146/annurev-biochem-061516-044952 (2017).
66. Gil, A. *et al.* Stability of energy metabolites-An often overlooked issue in metabolomics studies: A review. *Electrophoresis* vol. 36 2156–2169 Preprint at https://doi.org/10.1002/elps.201500031 (2015).
67. Martias, C. *et al.* Optimization of sample preparation for metabolomics exploration of urine, feces, blood and saliva in humans using combined nmr and uhplc-hrms platforms. *Molecules* **26**, (2021).
68. Vuckovic, D. Current trends and challenges in sample preparation for global metabolomics using liquid chromatography-mass spectrometry. *Analytical and Bioanalytical Chemistry* vol. 403 1523–1548 Preprint at https://doi.org/10.1007/s00216-012-6039-y (2012).

69. Conesa, A. & Beck, S. Making multi-omics data accessible to researchers. *Scientific Data* vol. 6 Preprint at https://doi.org/10.1038/s41597-019-0258-4 (2019).
70. Elowitz, M. B., Levine, A. J., Siggia, E. D. & Swain, P. S. Stochastic Gene Expression in a Single Cell. *Science (1979)* **297**, 1183–1186 (2002).
71. Eling, N., Morgan, M. D. & Marioni, J. C. Challenges in measuring and understanding biological noise. *Nature Reviews Genetics* vol. 20 536–548 Preprint at https://doi.org/10.1038/s41576-019-0130-6 (2019).
72. National Academies of Sciences, E., National Academies of Sciences, E., National Academies of Sciences, E., National Academies of Sciences, E. & Committee on Science, E. *Reproducibility and replicability in science.*
73. *Harvard Data Science Review • Issue 2.4, Fall 2020 Reproducibility and Replicability in Science: Report Highlightsnas-report-highlights License: Creative Commons Attribution 4.0 International License (CC-BY 4.0).* (2020).
74. Munafò, M. R. *et al.* A manifesto for reproducible science. *Nature Human Behaviour* vol. 1 Preprint at https://doi.org/10.1038/s41562-016-0021 (2017).
75. Fanelli, D. Is science really facing a reproducibility crisis, and do we need it to? doi:10.1073/pnas.1708272114/-/DC Supplemental.
76. Kolker, E. *et al.* Toward More Transparent and Reproducible Omics Studies Through a Common Metadata Checklist and Data Publications. *OMICS A Journal of Integrative Biology* vol. 18 10–14 Preprint at https://doi.org/10.1089/omi.2013.0149 (2014).
77. Perng, W. & Aslibekyan, S. Find the needle in the haystack, then find it again: Replication and validation in the 'omics era. *Metabolites* vol. 10 1–13 Preprint at https://doi.org/10.3390/metabo10070286 (2020).
78. Liu, C. *et al.* High-dimensional omics data analysis using a variable screening protocol with prior knowledge integration (SKI). *BMC Syst Biol* **10**, (2016).
79. Altman, N. & Krzywinski, M. The curse(s) of dimensionality this-month. *Nature Methods* vol. 15 399–400 Preprint at https://doi.org/10.1038/s41592-018-0019-x (2018).
80. Chattopadhyay, A. & Lu, T.-P. Gene-gene interaction: the curse of dimensionality. *Ann Transl Med* **7**, 813–813 (2019).
81. Riley, R. D. *et al.* Minimum sample size for developing a multivariable prediction model: PART II – binary and time-to-event outcomes. *Stat Med* **38**, 1276–1296 (2019).
82. Krassowski, M., Das, V., Sahu, S. K. & Misra, B. B. State of the Field in Multi-Omics Research: From Computational Needs to Data Mining and Sharing. *Frontiers in Genetics* vol. 11 Preprint at https://doi.org/10.3389/fgene.2020.610798 (2020).
83. Yamada, R., Okada, D., Wang, J., Basak, T. & Koyama, S. Interpretation of omics data analyses. *Journal of Human Genetics* vol. 66 93–102 Preprint at https://doi.org/10.1038/s10038-020-0763-5 (2021).
84. Ponomarenko, E. A. *et al.* The Size of the Human Proteome: The Width and Depth. *International Journal of Analytical Chemistry* vol. 2016 Preprint at https://doi.org/10.1155/2016/7436849 (2016).
85. Bennett, B. D. *et al.* Absolute metabolite concentrations and implied enzyme active site occupancy in Escherichia coli. *Nat Chem Biol* **5**, 593–599 (2009).
86. Zenobi, R. Single-cell metabolomics: Analytical and biological perspectives. *Science* vol. 342 Preprint at https://doi.org/10.1126/science.1243259 (2013).
87. Ong, S. E. & Mann, M. Mass Spectrometry–Based Proteomics Turns Quantitative. *Nat Chem Biol* **1**, 252–262 (2005).
88. Mahieu, N. G. & Patti, G. J. Systems-Level Annotation of a Metabolomics Data Set Reduces 25 000 Features to Fewer than 1000 Unique Metabolites. *Anal Chem* **89**, 10397–10406 (2017).
89. Busch, M., Ahlberg, E., Ahlberg, E. & Laasonen, K. How to Predict the pKa of Any Compound in Any Solvent. *ACS Omega* (2022) https://doi.org/10.1021/acsomega.2c01393.
90. Lisec, J., Hoffmann, F., Schmitt, C. & Jaeger, C. Extending the Dynamic Range in Metabolomics Experiments by Automatic Correction of Peaks Exceeding the Detection Limit. *Anal Chem* **88**, 7487–7492 (2016).
91. Yuan, J., Fowler, W. U., Kimball, E., Lu, W. & Rabinowitz, J. D. Kinetic flux profiling of nitrogen assimilation in Escherichia coli. *Nat Chem Biol* **2**, 529–530 (2006).

92. Rowan, D. D. Volatile metabolites. *Metabolites* vol. 1 41–63 Preprint at https://doi.org/10.3390/metabo1010041 (2011).
93. Mahieu, N. G., Huang, X., Chen, Y. J. & Patti, G. J. Credentialing features: A platform to benchmark and optimize untargeted metabolomic methods. *Anal Chem* **86**, 9583–9589 (2014).
94. Matuszewski, B. K., Constanzer, M. L. & Chavez-Eng, C. M. Strategies for the assessment of matrix effect in quantitative bioanalytical methods based on HPLC-MS/MS. *Anal Chem* **75**, 3019–3030 (2003).
95. Rosman' And, K. J. R. & Taylor, P. D. P. *INTERNATIONAL UNION OF PURE AND APPLIED CHEMISTRY INORGANIC CHEMISTRY DIVISION COMMISSION ON ATOMIC WEIGHTS AND ISOTOPIC ABUNDANCES* SUBCOMMITTEE FOR ISOTOPIC ABUNDANCE MEASUREMENTS** ISOTOPIC COMPOSITIONS OF THE ELEMENTS 1997 (Technical Report) Prepared for publication by Isotopic compositions of the elements 1997 (Technical Report). Pure & Appl. Chern* vol. 70 (1998).
96. Goldfarb, D., Lafferty, M. J., Herring, L. E., Wang, W. & Major, M. B. *Approximating isotope distributions of biomolecule fragments.*
97. Comai, L., Katz, J. E. & Mallick, P. *Proteomics Methods and Protocols Methods in Molecular Biology 1550.* http://www.springer.com/series/7651.
98. Zubarev, R. A. & Makarov, A. Orbitrap mass spectrometry. *Anal Chem* **85**, 5288–5296 (2013).
99. Sreekumar, J., Hogan, T. J., Taylor, S., Turner, P. & Knott, C. A quadrupole mass spectrometer for resolution of low mass isotopes. *J Am Soc Mass Spectrom* **21**, 1364–1370 (2010).
100. Schymanski, E. L. *et al.* Identifying small molecules via high resolution mass spectrometry: Communicating confidence. *Environmental Science and Technology* vol. 48 2097–2098 Preprint at https://doi.org/10.1021/es5002105 (2014).
101. Nakayasu, E. S. *et al.* Tutorial: best practices and considerations for mass-spectrometry-based protein biomarker discovery and validation. *Nature Protocols* vol. 16 3737–3760 Preprint at https://doi.org/10.1038/s41596-021-00566-6 (2021).
102. Köfeler, H. C. *et al.* Recommendations for good practice in ms-based lipidomics. *Journal of Lipid Research* vol. 62 Preprint at https://doi.org/10.1016/j.jlr.2021.100138 (2021).
103. Hfv-, C. *Guidance for Industry #118 – Mass Spectrometry for Confirmation of the Identity of Animal Drug Residues – Final Guidance, May 1, 2003.* http://www.fda.gov/AnimalVeterinary/GuidanceComplianceEnforcement/GuidanceforIndustry/default.htm (2003).
104. FDA & CDER. *Bioanalytical Method Validation Guidance for Industry Biopharmaceutics Bioanalytical Method Validation Guidance for Industry Biopharmaceutics Contains Nonbinding Recommendations.* http://www.fda.gov/Drugs/GuidanceComplianceRegulatoryInformation/Guidances/default.htm and/or http://www.fda.gov/AnimalVeterinary/GuidanceComplianceEnforcement/GuidanceforIndustry/default.htm (2018).

Network Development and Comparison in Lipidomics and Metabolomics

Thao Nguyen-Tran, Qassim Alkassir, Steffany A. L. Bennett, and Miroslava Cuperlovic-Culf

Abbreviations

AI	Artificial intelligence
GNN	Graph Neural Network
KNC	Known node correspondence methods
LASSO	Least absolute shrinkage and selection operator
MCMC	Markov-Chain-Monte-Carlo
ML	Machine learning
NLP	Natural Language Processing
PLS	Partial least squares
RF	Random forest classification method
RNA-seq	RNA sequencing

T. Nguyen-Tran · S. A. L. Bennett
Department of Biochemistry, Microbiology, and Immunology, University of Ottawa, Ottawa, ON, Canada

Neural Regeneration Laboratory, Ottawa Institute of Systems Biology, Brain and Mind Research Institute, University of Ottawa, Ottawa, ON, Canada

Department of Chemistry and Biomolecular Sciences, Centre for Catalysis Research and Innovation, University of Ottawa, Ottawa, ON, Canada

Q. Alkassir
Department of Biochemistry, Microbiology, and Immunology, University of Ottawa, Ottawa, ON, Canada

Neural Regeneration Laboratory, Ottawa Institute of Systems Biology, Brain and Mind Research Institute, University of Ottawa, Ottawa, ON, Canada

M. Cuperlovic-Culf (✉)
Department of Biochemistry, Microbiology, and Immunology, University of Ottawa, Ottawa, ON, Canada

Neural Regeneration Laboratory, Ottawa Institute of Systems Biology, Brain and Mind Research Institute, University of Ottawa, Ottawa, ON, Canada

Digital Technologies Research Centre, National Research Council of Canada, Ottawa, ON, Canada
e-mail: Miroslava.Cuperlovic-Culf@nrc-cnrc.gc.ca

V. Soni, T. E. Hartman (eds.), *Metabolomics*,
https://doi.org/10.1007/978-3-031-39094-4_2

scRNAseq	Single-cell RNA sequencing
UNC	Unknown node correspondence methods
WGCNA	Weighted network approach

1 Introduction

Biological networks can be described as a set of biological molecules represented as nodes (also called vertices) connected, via a measure of the strength of a biomolecular interaction, by edges. Metabolic and lipidomic networks connect metabolites and lipids as nodes through edges representing the chemical or metabolic reactions that generate product from substrate (or reactant). The association (edges) between molecules (nodes) can stem from different types of relationships or interactions that provide information about the chemical or metabolic reactions that link two nodes, their correlation or co-behavior in a specific condition, or their chemical properties that define node relationships. Metabolic reaction networks, connecting metabolites as nodes through chemical reactions as edges, describe systems that are responsible for maintaining homeostasis and regulating cellular functions. The construction of complete chemical reaction networks of metabolism are informed by the following:

(i) The precursor and product of an enzymatic reaction
(ii) Reaction stoichiometry and enzyme kinetics
(iii) Reaction directionality
(iv) Subcellular localization of reaction

Pathway is a set of context-dependent interactions with clear beginning and end and often delineated directionality.

Network is any structure of nodes connected with edges. Metabolic network aims to provide context-free representation of the complete process often by combining pathways.

Inclusion of all these properties would allow both network analysis and simulation. Alternatively, network development that only includes undirected, correlation information provide networks of relationships without causality or modeling utility.

Another layer of metabolic network complexity is that the subcellular membrane-bound compartments allow for the separation of different environments within the cell while at the same time bringing enzymes and their corresponding substrates in close proximity. Consequently, compartmentalization provides the optimal condition for enzymatic reaction. In constructing metabolic networks, information associated to subcellular compartmentalization of enzymatic reaction can be incorporated when either data or knowledge is available. Compartments provide optimal conditions for function of enzymes and additionally allow equivalent chemicals to be utilized for different purposes. Enzymes from the same family can reside across different compartments with members possibly functioning under slightly different conditions.

Networks or graphs, consist of nodes, i.e. vertices that correspond to objects, for example metabolites or lipids, and edges that show connections between objects. Edges can have *weights* indicating strengths of connections. In *undirected* graphs, edges have no direction, and only show relationship not causality. In *directed* graphs, edges have direction indicating one-way relationships showing that edge can only transverses in a single direction. In a *bipartite* graph, vertices are separated into two sets, where nodes from one set can be unidirectionally connected to any node in the other set, but there are no edges within nodes of the same set. Graphs can be also *multi-edge* containing multiple edges between same two nodes, for example, edge for data and edge for knowledge. *Hypergraph* consists of nodes and hyperedges, where an edge can join any number of vertices. Graph is called *connected* if there is a connection between any two points and *complete* if every pair of vertices is connected by a unique edge.

Through the web of metabolic reactions, biological systems are in constant flux with metabolites in a dynamic interaction with other biological molecules, experiencing continual chemical change. Metabolites concentrations, their destiny in a system, are thus determined by other members of the network, and greater understanding of either individual metabolites' behavior or the biological systems can only come from the analysis of the network of associates.

2 Network Development Methods

High-throughput bioanalytical methods are providing increasingly detailed molecular coverage in a variety of sample types delivering range of datasets that can be explored through network investigation. At the same time, our knowledge about metabolite and lipid functions and processes is increasing, and there is a growing appreciation of the importance of their relationships within pathways and networks. Metabolic networks can be derived from data – data-driven networks, from biological information in knowledge-driven networks or in a hybrid approach combining knowledge and data. Each of these approaches comes with its own set of advantages and disadvantages and the road taken has to be optimized based on the application of interest.

Knowledge-based network development includes combining enzymatic reaction information from databases or known pathways as well as literature derivation of, for example, possible enzyme-metabolite relationships. Derivation of this information can be done directly from existing databases (some examples shown in Table 1) or from literature search, manually or with the help of Natural Language Processing methods (NLP). Clearly, in this approach, the network depends on the level of prior biological knowledge provided in selected information resource. These types of networks provide a map for observing interactions from the data in a context of predefined possibilities. Knowledge-based networks can be small, for example, only observing individual pathways and showing relationships between only small

Table 1 Examples of resources for knowledge-based network and pathway information

Pathway and network information databases	Brief information	Number of metabolic pathways	Reference and site
Kyoto Encyclopedia of Genes and Genomes (KEGG)	Major database of metabolic pathways	~6500 (Nov 10, 2022)	[18] https://www.kegg.jp/
LIPID MAPS Reaction Explorer	Lipid reactions database is linked to the LIPID MAPS® Structure Database (LMSD)	~47,800 unique lipid structures	https://www.lipidmaps.org/
SwissLipids	Database and searchable site for lipid metabolic reactions	779,688 lipid species	[1] https://www.swisslipids.org
Small Molecule Pathway Database (SMDB)	Database containing detailed information about small molecule metabolites found in the human body	~29,011	http://www.smpdb.ca/
Rhea	Expert-curated knowledgebase of chemical and transport reactions of biological interest	15,453 reactions with 12,972 unique compounds	Rhea, the reaction knowledgebase in 2022 [4] https://www.rhea-db.org/
Reactome	Relational database of signaling and metabolic molecules and their relations organized into biological pathways and processes	Overall metabolic map	https://reactome.org/ [13]
MetaCyc (bioCyc – add separately)	Curated database of experimentally elucidated metabolic pathways from all domains of life	~20,000 across different species	https://metacyc.org/
Biocarta	Database of maps of biochemical pathways including metabolic and signaling transduction pathways	Metabolic maps	https://maayanlab.cloud/Harmonizome/dataset/Biocarta+Pathways
BioPAN	Web-based application for the visualization of lipidomics data on a mammalian lipidome metabolic pathways	Data drive visual representation of networks	https://lipidmaps.org/biopan/ [14]
LINEX	Lipid Network Explorer is a Webapp used to analyze lipid metabolic networks and provides data-specific network from user data	Data drive visual representation of networks	https://exbio.wzw.tum.de/linex/ [21]

subset of molecules based on some property of interest. Alternatively, these networks can represent a large-scale efforts in building genome-scale, i.e. complete, metabolic processes networks for subsequent system modeling [11]. Some examples of metabolic knowledge resources as well as knowledge-based networks are listed in Table 1. An advantage of knowledge-based networks is that they do not rely on the quality, quantity, or accuracy of the data in user's possession and thus will not be biased by small, sparse, or erroneous datasets. However, as knowledge-based networks depend on the current biological knowledge, it is natural to assume that they still have gaps or missing links in information that have not been discovered yet or is not readily available in resources utilized for creation of a network. Also, with different level of knowledge available for distinct biological systems number of known interactions varies between species as well as metabolic processes. As an example, Recon3D represents the most comprehensive human metabolic network model to date. The version of this genome-scale model made available on MetabolicAtlas has 13,070 reactions and 8369 metabolites (current version available in metabolicatlas.org [32]. When building metabolic networks for other species, a number of methods are made available including PathoLogic [20], which is used to build pathways from predicted enzymes that can be further validated using Semi-Automated Validation infrastructure (SAVI) software applying range of curation decisions [34]. With lipidomics methods becoming increasingly powerful and providing concentrations for hundreds of individual lipid molecules, there is an increasing effort to also deliver corresponding knowledge-based networks for lipidome. BioPAN [14] and LINEX [21] provide users with the pathway mapping for lipidomic dataset, where in both cases knowledge-based lipid networks are used to provide insights about functional lipid associations.

An alternative knowledge-based approach relies on the knowledge of properties of the molecular set rather than knowledge of biological interactions. In this case, chemical ontologies or molecular characteristics can be used to educate building of relationships among compounds. In the example of ChEBI approach [15], users' entry is represented in a network based on chemical properties information. This is a very useful approach for exploring known properties for molecules of interest; however, in this case network is developed for representation of properties, not for further utilization and can be a useful way to analyze chemical properties of a selected set of compounds. However this approach does not provide avenue for analysis of network interactions, only organization of known information. An alternative way to explore statistical enrichment of molecular groups based on chemical ontologies as well as structural similarities is provided by ChemRICH [6]. In this method, the goal is to step away from reliance on often limiting pathway information in obtaining representation of the set and instead use structural similarities and chemical ontologies to map molecules (metabolites or lipids). This approach follows the notion that chemically similar compounds remain in biochemical proximity [7], thus possibly providing a way for assigning unknown molecules based on their properties and network clusters. Chemical ontology or properties networks can

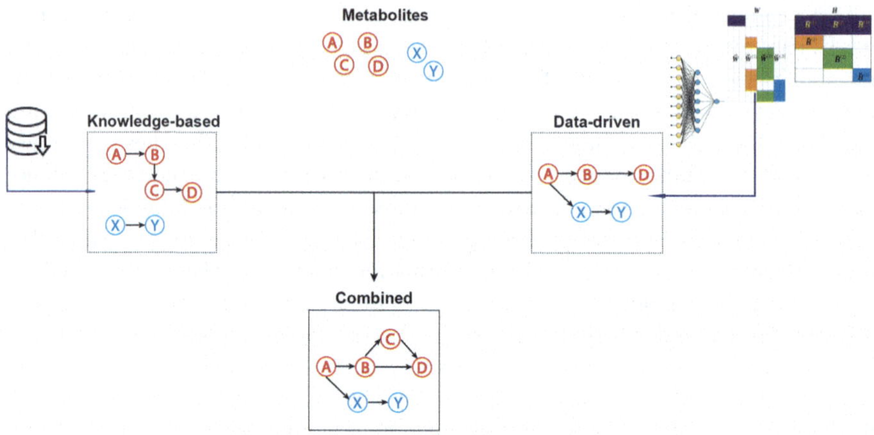

Fig. 1 Most methods for network development are combining knowledge and data to give the most accurate network for features of interest. Based on information in databases and literature in knowledge-based approach, the user creates connections between features. In data-driven approach, the user relies on data and mathematical procedure to establish relationships. Combining the results of two approaches is done in order to reduce errors of the two methods and establish closest network to reality

be utilized to determine related compounds through their shared class belongings, where graph distance between molecular node and a class node can be used to quantify relatedness between pairs of compounds (Fig. 1).

As a rule, *data-driven methods* for network derivation depend on the availability of datasets of sufficiently high quality and quantity and rely on a variety of mathematical tools to build network directly from the data. Network edge determination, in this case, searches for dependences or similarities between node behaviors in samples or similarities in node properties based on a measure of choice. Applications for these highly versatile approaches range from spectral assignments [35] to derivation of metabolic or signaling processes functions or dependencies between features or samples (reviewed recently in [3]. A number of methods have originally been developed for general graph theory and a number of them are applied to other omics datasets, but they are in general also appropriate for metabolomics or lipidomics data as well. Data-driven methods can be further combined with knowledge-based networks in *hybrid approaches.* In hybrid methods the attempt is to take advantage of the available knowledge to either initiate network development from the data by using known interactions as a base for growing more extensive networks from data or to threshold fully data-derived networks at the end of the process. Several different approaches have been developed and tested for data-driven applications, without or with combination to knowledge-based networks. These methodologies can be broadly divided into correlation and classification-based network development methods.

3 Correlation-Based Methods for Molecular Network Derivation

Metabolites and lipids that are linked through enzymatic or signaling pathways often show co-dependencies in the values that are represented through correlated changes in their concentrations across samples. Calculation of these pairwise correlations based on the data from metabolomics or lipidomics measurements can provide data-derived adjacency matrix, where it is hypothesized that two metabolites are linked if their correlation value is statistically significant and larger than a user-defined threshold. Correlation between molecules can then be viewed as the network edge value and a way to construct molecular network. General steps in constructing network from correlation analysis of the data are shown in Fig. 2.

The standard approach for correlation derivation is Pearson method, where correlation is calculated as $r = \frac{\sum_{i=1}^{N}\left(X_i - \overline{X}\right)\left(Y_i - \overline{Y}\right)}{\sqrt{\sum_{i=1}^{N}\left(X_i - \overline{X}\right)^2}\sqrt{\sum_{i=1}^{N}\left(Y_i - \overline{Y}\right)^2}}$ for features $\overline{X}$ and $\overline{Y}$ measured across N samples and having mean values of $\overline{X}$ and $\overline{Y}$, respectively. Pearson's correlations are easy to interpret and calculate; however, this method does not accurately determine nonlinear dependencies. Alternative methods include Spearman – rank-based method, distance correlation [38] – calculating distance covariance, or mutual information [37] – machine learning-based correlation analysis method. Correlation network design with any of these approaches does not guarantee the capture of biologically relevant mechanisms nor does it ensure selection of only direct relationships. A number of additional approaches have been developed in order to help narrow correlations down to only significant ones. The

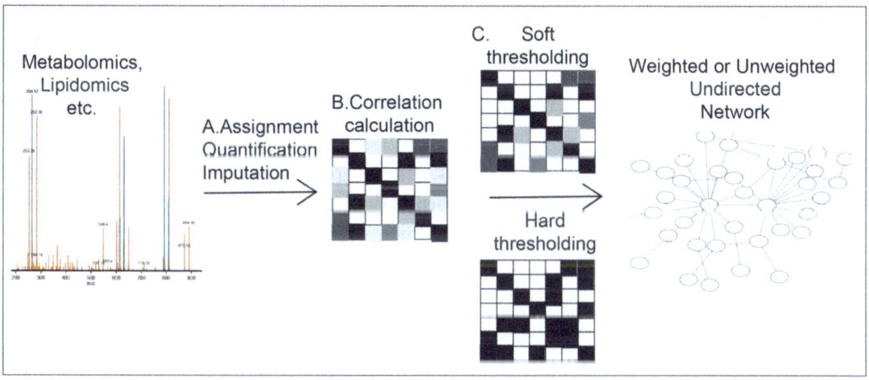

Fig. 2 Steps involved in the construction of molecular network from data using correlation-based methods including (**a**) data quantification and preprocessing; (**b**) determination of pairwise correlations; (**c**) selection of correlations that are statistically significantly different from zero through comparison of p-value for significance of the difference of correlation from zero with a significance level observed after appropriate multiple hypothesis testing. Finally, remaining edges can be represented in a network plot with either binary or weighted edges

simplest approach utilizes threshold parameters for selection of relevant edges through *p*-value of correlation level significance as well as direct correlation level thresholding. In addition, regardless of the method of choice, correlation values are sample size dependent and thus networks have to be constrained by appropriate thresholds for statistical significance (*p*-value) and/or correlation level for reduction of the effect of sample size as shown by [40]. Once the significant correlation values are selected in a hard-threshold approach, they can be combined in an Adjacency matrix $A = [a_{ij}]$ with entries that are either 1 or 0 decided up on using threshold values as

$$a_{ij} = \begin{cases} 1 \text{ if } r_{ij} \geq r_{\text{threshold}} \text{ and } p_{ij} \leq p_{\text{threshold}} \\ 0 \text{ otherwise} \end{cases}$$

Alternatively, actual correlation values can be kept as edge weights showing pairs' "closeness" level in the soft threshold application.

The statistical significance can be determined using T value calculation forcorrelation as $T = r_{ij}\sqrt{\frac{N-2}{1-r_{ij}^2}}$, where r_{ij} is the correlation value between nodes i and j and N is a number of samples used for correlation calculation. *P*-value can be determined from T using Student's t cumulative distribution function if normal distribution can be assumed for correlations. Fisher's z-transformation of correlation levels establishes normal distribution for correlation values. Fisher's z-transformed correlation is obtained as $z_{ij} = \frac{1}{2}\ln\frac{1+r_{ij}}{1-r_{ij}}$ and corresponding *p*-values are calculated as $p_{ij} = 2\left(1-\theta(z_{ij}\sqrt{(N-3)-(M-2)}\right)$ for a sample set with N samples and M features; θ corresponds to cumulative distribution function of standard normal distribution. It is important to notice that *p*-value in both approaches depend on the sample size or both size of sample and feature space. Correlation values are generally sample size dependent, where in small sample sizes, it is more likely to get spuriously large correlation values due to random sample variations, while with sample size increase individual sample variations contribute less and correlations become a better reflection of the population levels. In smaller populations p-value for correlations is larger, thereby making higher correlation values statistically insignificant and reducing the error caused by artificially large correlations of small sets. Although thresholding to the significance level helps reduce number of low significance edges, it does not guaranty that only direct relationships are kept in the network.

Selection of an optimal *p*-value correlation coefficient threshold depends on the distribution of the number of edges at different *p*-values [40] similarly to *p*-value threshold selection performed based on the *p*-value distribution in Benjamini-Hochberg FDR multiple hypotheses testing corrected set [9]. Therefore, multiple testing correction should be applied to significance thresholding of correlations, where Bonferroni correction, as the most conservative approach, has been utilized to determine nominal significance level of $a = 0.01$ for a given sample size and can then be used to determine corrected p-value correlation threshold. In this way, p-value thresholding follows statistical rules and is determined appropriately for the sample size. Setting of correlation value threshold is less clearly defined. Shen et al. [36] have proposed a theoretically derived threshold for distance correlation that depends exclusively on sample size and is determined as $\tau_N = 2F_{\beta}^{-1}\left(1-\frac{0.02}{N}\right)-1$, where F_{β}^{-1} is the inverse cumulative distribution function with symmetric Beta distribution with shape parameter equal to $\frac{1}{2}\left(\frac{1}{2}N(N-3)-1\right)$. This approach provides an interesting, theory-based method for general threshold dependence on sample size in a random dataset giving theoretical lower bound for the threshold; however, it does not include any specific properties of data. Toubiana and Maruenda [40] proposed an iterative approach where topologies of the correlation networks constructed at different levels of threshold are compared and the point of significant change is selected as an analysis threshold. In principle, this approach can be applied to any correlation analysis methodology. In the vicinity of the optimal threshold level for correlation value, the number of network edges is expected to remain stable with gradual increase in p-value stringency, going from 0.05 to 0.01 as a cut-off point for statistical significance. Following this assumption, the analysis of the significance of changes in the edge number, using statistical methods such as modified Cox method can be implemented to determine optimal thresholds for both correlation and p-value [40].

Edges between nodes in correlation matrix should ideally correspond to metabolic fluxes, that is, reactions in the metabolic network. However, a number of factors influence the correlation results including short-term changes in enzymatic activity due to inhibitors or activations, random fluctuations due to noise in the data or reactions, metabolic processes compartmentalization in cells and organs, involvement of metabolites in multiple pathways, or incomplete experimental coverage. Additionally, through the network, interaction between metabolites could result in indirect correlations leading to highly dense networks. Pairwise correlations that are caused by the presence of mediators can be high and thus remain after correlation thresholding. Issue of indirect correlations is addressed in the Gaussian graphical model approach used to determine partial correlations, that is, remove indirect dependencies. Briefly, Krumsiek et al. [23] showed that when dataset includes many more samples then features correlation matrix has a full rank andit is possible to calculate an inverse of correlation matrix. Following our nomenclature above, partial correlation coefficients can be calculated as $\rho_{ij} = -\vartheta_{ij} / \sqrt{\vartheta_{ii}\vartheta_{jj}}$, where $(\vartheta_{ij}) = R^{-1}$ and R is the correlation adjacency matrix. Partial correlation values correspond to pairwise correlation of metabolites i, j after correction for the correlation through all the other metabolites. Partial correlation in this way accounts for the presence of confounders and covariates, that is, correlation between features through the network. This is a very powerful approach when significant number of samples is made available (Fig. 3 shows graphical explanation of the approach).

In a case when the number of features is larger than the number of samples or if any of the features are a linear combination of other features in the set, the resulting

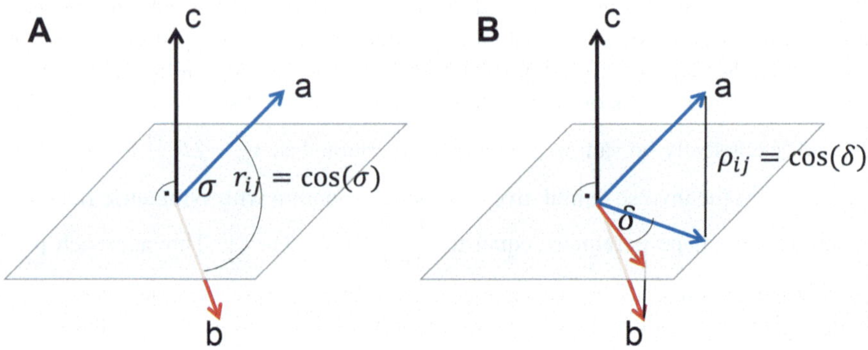

Fig. 3 Geometrical representation of the partial correlation through removal of orthogonal interactions. (**a**) Correlation between two features, *a* and *b* with values across measurements represented vectors in the figure. *a* and *b* are not orthogonal to vector representing feature c and are thus correlated with feature *c*. (**b**) Projections of vectors *a* and *b* onto a plane orthogonal to vector of feature *c* provide value for correlation between and *b* with contributions of *c* removed represented as a cosine of angle between projects

covariance matrix is ill-conditioned, that is, singular and matrix inversion and thus partial correlation calculation through inversion is not possible. GeneNet [30, 33] is initially built for analysis of genomics data; however, it has been now successfully applied to metabolomics as well [12]. In order to allow determination of partial correlations in smaller sample sizes, GeneNet utilizes novel algorithm for shrinking correlation (covariance) matrix making it nonsingular and allowing inversion and derivation of partial correlation matrix possible for all sample sizes. The methodology used for correction of covariance matrix in this approach is analytical shrinkage estimation of covariance and partial correlation matrices on model selection using local FDR multiple testing [33]. In GeneNet, authors decided to shrink correlation matrix toward identity matrix while leaving empirical variances unchanged. The goal of GeneNet is to provide a graph, where edges show direct dependencies between nodes. Alternative methods for covariance matrix shrinkage have been proposed in order to provide improvement in network reconstruction performance [10].

Benedetti et al. [8] proposed an algorithm that optimizes correlation level cutoff selection through maximization of the overlap between the inferred network and available biological, prior, knowledge. With this approach, the focus is on finding network threshold that has the highest overlap with the known biological network, rather than utilizing predefined *p*-value threshold. Several methods have been compared in the analysis including Pearson correlation network, inversion of covariance matrix, and GeneNet, where the approach of Benedetti et al. showed the best performance overall. Interestingly, even in the optimization using a very limited knowledge-based network, authors were able to obtain significantly better network overall.

An alternative approach replaces need for user-defined cut-off point with the user-specified power value for correlations in weighted network approach (WGCNA)

[25]. WGCNA produces a fully connected network with edge weights and can be utilized to determine clusters of co-regulated molecules. In the original version, WGCNA is based on Pearson's correlation matrix; however, recently other correlation approaches have been tested with this method [42]. In WGCNA, the initial step is the calculation of correlation matrix that is transformed by rising all values to a "soft threshold power," that is, value that is used to power the correlation of feature thereby emphasizing strong correlations. Soft power and threshold for the transformed correlation matrix are optimized in this approach for maximal scale-free properties of the selected network. The scale-free network property is optimized by selecting threshold for soft power transformed correlation values that leads to a best linear fit for $\log_{10}(H(d))$ vs $\log_{10}(d)$, where d corresponds to a degree and $H(d)$ is distribution of a degree d across the network. Soft power and threshold are determined for each sample set separately by maximizing R^2 value for the scale-free plot. Specific criteria in selecting correlation exponent in soft thresholding with weights are as follows (following recommendations of [41]: (a) power leads to a network satisfying scale-free topology at least approximately; (b) the mean connectivity should be high so that the network contains enough information (e.g., for module detection); (c) the slope of the regression line between $\log(p(k))$ and $log(k)$ should be negative (typically smaller than −2). The main result of a WGCNA method is network that is used for the determination of clusters, that is, node modules. These modules often represent specific processes, and highly connected modules have been shown to have, for example, major regulatory role. Although this is an interesting approach for selection of major nodes through correlation analysis as well as clustering, WGCNA does not focus on the determine of single-step enzymatic reactions although they are of major interest in metabolomics and lipidomics analysis. Additionally, WGCNA method's assumption of a scale-free network topology is not always appropriate, particularly in metabolic networks.

It should be underlined that all correlation approaches thus far can only be viewed as exploratory methods developed to identify functionally related groups of metabolites or lipids and are not guaranteed to provide only direct mechanistic interactions. Even in the case of partial correlation analysis, it cannot ensure presentation of only direct interaction particularly in a case of partial metabolite or lipid coverage with a range of latent variables. Results of these networks need to be further validated and interpreted using biological knowledge and focused experimental analysis; however, they provide very valuable information to guide future experiments. In spite of their approximate nature, they provide valuable information about network changes across experimental conditions or phenotypes.

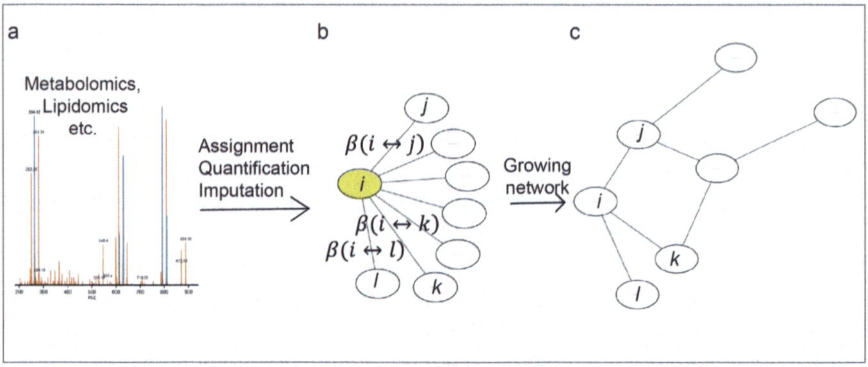

Fig. 4 Steps involved in the construction of molecular network from data using general classification-based methods including (**a**) data quantification and preprocessing; (**b**) determination of regression relationship between features; (**c**) selection of significant edges based on regression analysis and continuing to grow network for all features

4 Classification-Based Methods

Network derivation using classification approaches is once again developed extensively for genomics with metabolomics and lipidomics only starting to benefit from these methods. In this type of approaches, regression of each feature $i \in \{1, \ldots, N\}$ is estimated against all the remaining N-1 features. Edge between pair of features is calculated as $r_{ij} = sign\left(\beta_i^{(j)}\right)\sqrt{\beta_i^{(j)}\beta_j^{(i)}}$, where $\beta_i^{(j)}$ is the regression coefficient of predictor variable x_i for the response x_j (Fig. 4). This approach can be used for the determination of regression-based edges using a variety of methods, outlined in great detail in [33].

Random forest (RF), a classification method, has been proposed by several authors as a base for data-driven network derivation. Two interesting examples are GENIE3 [17], directly applying RF to the dataset and iRafNet [31], combining different data types under a unified RF framework. Both approaches have been developed for genomics but are directly applicable to metabolomics or lipidomics data as well. The GENIE3 model considers characteristics (e.g., concentration or expression level) of each feature as a function of values for all other features sampled randomly from the complete dataset. iRafNet, a subset of potential network partners, is selected based on the information in other provided datasets. When additional data or information is available, iRafNet generally performs better, as it includes prior knowledge, but in the case of fully unique dataset-driven dataset, two methods are equivalent. In this approach, determination of network is viewed as a collection of M subproblems trying to find regulators for M features, where determination of regulators is viewed as a classification equivalent to feature selection problem in classification.

In both approaches, the measure of feature x_i is modeled as a function of the values for other features using RF, that is, tree ensemble. Features that are strong predictors of x_i are considered as regulators of this feature. Specifically, the importance score for feature x_k as a predictor of x_i, S_{ki} is equal to the total decrease in node impurity following the splitting of samples based on the measurements of feature x_k.

Partial least squares (PLS) regression is also presented as a powerful method for exploring relationships between biological molecules, with application for lipid network derivation presented by Kujala et al. [24]. Connectivity score in this approach is based on the fitting of n PLS models one for each lipid, where each lipid measurements are predicted with n-1 remaining lipids. The latent factors $t_j^{(l)}$ for lipid j are a linear combination of values for all other lipids with PLS determined regression coefficient $c_j^{(l)}$ such as $x_j = \sum_{\vartheta}^{l=1} \beta_l t_j^{(l)} + \varepsilon$; ϑ is number of orthogonal latent factors used for the fitting, that is, number of PLS components used in the model. The connectivity score that can be viewed broadly as the edge between pair of lipids is calculated from PLS parameters as $s_{jk} = \frac{\sum_{l=1}^{\vartheta} \beta_{jl} c_{jk}^{(l)} + \sum_{l=1}^{\vartheta} \beta_{kl} c_{kj}^{(l)}}{2}$ providing weights for the network.

A number of other supervised machine learning methods have been explored as possibly powerful ways to derive feature edges. Applications of LASSO regularization method and Bayesian network inference have been recently reviewed [16]. With LASSO, regularization is explored as a way to reduce model complexity. L1 regularization of LASSO is used as a way to push edge coefficients toward zero, in a way providing a variable selection thereby reducing model complexity. Bayesian networks are directed acyclic graphs providing both dependence and causality between features. In this approach, Markov-Chain-Monte-Carlo (MCMC) procedure is used to estimate precision matrix by searching for the best fit with the data of large space of possible graph configurations in total $2^{\frac{M(M-1)}{2}}$, where M is number of nodes, features. Recently, Graph Neural Network (GNN) approach was used by Alghamdi et al. [2] to model cell-wise metabolic flux from single-cell RNA-seq data. The scFEA method assumes that the modeling of the flux variations of metabolic modules can be performed using nonlinear function of the changes in enzyme levels obtained using transcriptomics and that in all single cells total intermediate substrates flux imbalance is minimized. Using scRNAseq data and GNN, this approach can model flux through metabolic network from transcriptomics data while at the same time providing graph of metabolic modules. Application of neural network analysis directly to metabolomics and lipidomics data for network derivation or analysis is thus far only done by a handful of authors, with an example of deep learning use presented by [5]. Further applications of modern classification, machine learning, and neural network methodologies in metabolomics and lipidomic network derivation and analysis are desired.

5 Analysis of a Network

Once network is built, it motivates exploration using a variety of methods from graph analysis and data mining. Generally, learning based on networks can be broadly divided into node classification, link, that is, edges prediction, network classification, and embedding [29].

Node classification can find a role in the prediction of function of biomolecules using semi-supervised learning by grouping nodes within the entire network. In the context of metabolic or lipidomic networks, node classification can be used to obtain functional similarities between metabolites in one network, for example, biological system, using information known from other biosystems. Similarly, link, that is, edge prediction can be performed as an ML task where known edges are used to train the model that is then used to predict additional, missing links from network data. Graph classification or regression is utilized to predict properties of graphs. When graph is a representation of a molecule, this approach can be used to predict molecular properties. In the context of lipidomic or metabolomics network, this approach can be used to determine similar metabolic outcomes. Graph embedding is most often a preprocessing step that is used to devise representation of nodes or graphs as fixed size vectors making subsequent machine learning analysis easier. Graph Neural Networks (GNNs) are a class of deep learning AI methods designed to analyze network, graph data, unlike regular deep learning approaches appropriate for analysis of vector data. Examples of some GNNs used in biological network data analysis are recently reviewed in [29].

6 Network Comparison Methods

Increasing sizes of datasets and abundance of network development methods and models introduces the next challenge of trying to derive biological information from networks. Analysis can be either aimed at specific characteristics of nodes and edges or overall network structure or, more often, investigation of similarities and differences between networks in different conditions, that is, health and disease or treatment and placebo or changes in the network during time course analysis. Comparison between networks in the context of metabolome or lipidome can be broadly divided into three different goals:

- Comparison of the overall network equality through analysis of the distance between complete networks. In this approach, the goal is to provide numeric estimate of the change for the whole network.
- Determination of the major changes between nodes through either/or analysis of differences in the number of edges or edge weights, where in the context of metabolic network, this would be an indication of changes in metabolite or lipid behavior.

- Determination of the major changes in edges through analysis of the changes in the edge weights. In a metabolic network, this would be an indication of changes in reactions, that is, enzymatic functions.

7 Overall Network Comparison

With network analysis becoming a staple in variety of areas, there is an abundance of methods for network comparison, and the main issue is the selection of the most appropriate approach for the dataset and analysis goals. In an effort to help select optimal methodology, Tantardini et al. [39] have recently presented an appraisal of several popular network comparison methods for mostly undirected, unweighted graphs as well as few methodologies for comparison of directed or weighted networks. Although most methods have been developed to deal with significantly larger networks than what is generally seen in metabolomics and lipidomics, they can easily be applied in these areas as well. Network analysis methods deal with either networks with the same node sets (known node correspondence methods, KNC) or networks with possibly different node sets (unknown node correspondence methods, UNC). UNC methods can be of interest in, for example, comparison of metabolic networks between different species, while KNC methods provide direct comparison of networks derived from data exploring related sample types (e.g., disease vs. control).

The initial task in network comparison is the determination of optimal distance metric for graph analysis. An obvious approach is to directly compute differences between adjacency matrices between networks using any of the distance calculation methods (Euclidean, Jaccard, weighted Jaccard, etc.). Alternatively, the method based on the direct node distance comparison is DeltaCon [22]. DeltaCon compares similarity between all node pairs in two graphs using Matusita distance:

$d = \left(\sum_{N}^{i,j=1} \left(\sqrt{s_{ij}^{A}} - \sqrt{s_{ij}^{B}} \right)^2 \right)^{1/2}$, where $S^A = \left[s_{ij}^{A} \right]$ and $S^B = \left[s_{ij}^{B} \right]$ are similarity matrices for network A and B defined as $S = [I + \epsilon^2 D - \epsilon A]^{-1}$ and A is network adjacency matrix, $D =$ diag (k_i) is degree matrix of node degree (k_i), and ϵ is a small positive constant. Computational cost of DeltaCon is (N^2) thus for networks in metabolomics and lipidome that rarely have more than few hundred nodes this is acceptable. This approach provides more significant change for larger weight changes or for removing edges, while random changes favorably lead to a smaller impact on distance measure.

An alternative method to direct distance analysis is the network alignment or graph isomorphism analysis used to directly compare networks in order to determine conserved and missing nodes and edges across two, pairwise, or multiple network comparisons. Alignment can be performed locally or globally, where local alignment tries to align small regions accurately risking failing in finding large, conserved connections between subgraphs. Global alignment searches for one-to-one mapping of nodes in different networks aiming to overcome shortcomings of local alignment methods. In general, all alignment methods define an objective function, measure, or a score of alignment quality and utilize a search algorithm that tries to find an optimal solution.

An interesting method developed for metabolic network alignment is H-GRAAL [28] specifically designed for the comparison of metabolic networks between different species. H-GRAAL and a number of related methods are based on the original GRAAL algorithm, which detects statistically significantly similar topological regions in large networks in order to highlight conserved or missing nodes and edges between two or multiple networks. GRAAL approach introduces concept of graphlets that include more detailed description of nodes by incorporating consideration of its degree based on its local neighborhood of connections. Graphlet similarity search is performed over a pair of aligned nodes independently, locally, of all other nodes. Large-scale networks make prioritizing of curation challenging and with uncertainty in the parts of the network that need further consideration and make comparison of networks as well as simulation of systems difficult. Medlock and Papin [27] have recently introduced a ML-based approach for automated metabolic model ensemble-driven uncertainty elimination using statistical learning (AMMEDEUS) as a way to guide curation of genome-scale metabolic models as well as databases.

Clustering of network provides modules of nodes, in our case modules of lipids or metabolites, where molecules within a same module are connected by a short edge paths and strong connections. Node modules can be determined for any network regardless of the method for network derivation. Kujala et al. have shown module cluster comparison method based on PLS-derived association scores [24]. The differences between modular structures in two networks is calculated as $N = 1 - \frac{1}{|L_0|} \sum_{j \in L_0} \frac{|F_{1r(j)} \cap F_{2r(j)}|}{|F_{1r(j)} \cup F_{2r(j)}|}$, where $F_{kr(j)}$ is the module, L_0 in network k that contains lipid j. $|L_0|$ shows the number of lipids that belong to modules in both networks. If $N = 0$, modular structures of the two networks are identical; otherwise, p-value for the statistical significance of the modules difference can be calculated as $p(N) = \frac{1}{P} \sum_{\pi} I(N(\pi) \geq N)$, with sum taken over all P permutations. This approach provides information about the changes in network modules. Differential connectivity for a single node can be obtained using mean absolute distance statistic as $d(j) = \frac{1}{p-1} \sum_{j' \in L, j' \neq j} |\hat{s}^1_{jj'} - \hat{s}^2_{jj'}|$. With this approach, it is possible to obtain differences bws of nodes as well as for each individual node.

Analysis of maximally dysregulated subnetworks, proposed by Mamano and Hayes [26], uses simulated annealing-supported local search for biological network alignment. In this approach, simulated annealing was shown to provide the optimal solution with better node pairing between networks and good topological and functional similarity scores. Simulated annealing, as a metaheuristic algorithm, is not developed for any specific problem and can be applied to any optimization problem as long as there is a defined objective function and neighbor relationship and there are solutions for different states. Although the example presented by [26] is in protein-protein interaction network, similar approach can be utilized in metabolomics and lipidomics network comparisons, where score function can combine various topological and biological similarity measures and simulated annealing approach can provide global optimization solution. Once again, methods developed for network comparison in different applications present themselves to metabolomics and lipidomics applications, but it is up to the analyst to select the most appropriate comparison approaches for the network type, size, and analytical question.

8 Network Visualization

Visualization of networks and network components is an extremely important, intuitive way for the interpretation of results, but with variety of network sizes, data types, and applications, there is no single solution. Visualization methods range from the simplest ones showing adjacency matrices to more complex methods that are visualizing force directed layouts in 2D or 3D. Some examples of freely available network visualization software application are listed in Table 2 in addition to many libraries available in different programming languages dedicated to network visualization.

We recommend to the reader to freely explore many possible ways for the visual presentation of networks as the complexity and size of metabolic network necessities optimization of visualization for each application.

9 Conclusions

A number of methods for knowledge-based, chemical ontology, or data analyses network derivation combined with methods for network analysis, comparison, and visualization provide abundance of possibilities, all with their strengths and weaknesses. Knowledge-based networks are clearly limited by gaps in current

Table 2 Examples of free software tools for network visualization

Software application	Brief outline	Reference and Site
Cytoscape	Network visualization and analysis tool with a number of applications developed for bioinformatics	https://cytoscape.org/
OmicsNet	WebGL-based method	https://omicsnet.ca [43]
Gephi	General open graph visualization platform.	https://gephi.org/
Arena3D	Interactive 3D visualization of multimodal networks particularly appropriate for polyomics datasets	http://bib.fleming.gr:3838/Arena3D [19]

information and can create incomplete networks. Ontologies can possibly represent different concepts at different levels of representation. Data-driven networks in general only show co-behaviors and cannot ensure representation of metabolic relationships or direct interdependencies. Major developments are under way and further improvements are absolutely required before modeling of complete metabolic network becomes possible.

Only by combining knowledge, large and diverse datasets and appropriate statistical, machine learning, and modeling tools, we will be able to ultimately obtain truly a complete in silico representation of biological systems' weird and wonderful metabolic network.

References

1. Aimo, L., *et al.* The SwissLipids knowledgebase for lipid biology. *Bioinformatics* 2015;31(17):2860–2866.
2. Alghamdi, N., *et al.* A graph neural network model to estimate cell-wise metabolic flux using single-cell RNA-seq data. *Genome Res* 2021;31(10):1867–1884.
3. Amara, A., *et al.* Networks and Graphs Discovery in Metabolomics Data Analysis and Interpretation. *Frontiers in Molecular Biosciences* 2022;9.
4. Bansal, P., *et al.* Rhea, the reaction knowledgebase in 2022. *Nucleic Acids Res* 2022;50(D1):D693–d700.
5. Baranwal, M., *et al.* A deep learning architecture for metabolic pathway prediction. *Bioinformatics* 2020;36(8):2547–2553.
6. Barupal, D.K. and Fiehn, O. Chemical Similarity Enrichment Analysis (ChemRICH) as alternative to biochemical pathway mapping for metabolomic datasets. *Scientific Reports* 2017;7(1):14567.
7. Barupal, D.K., *et al.* MetaMapp: mapping and visualizing metabolomic data by integrating information from biochemical pathways and chemical and mass spectral similarity. *BMC Bioinformatics* 2012;13(1):99.
8. Benedetti, E., *et al.* A strategy to incorporate prior knowledge into correlation network cutoff selection. *Nature Communications* 2020;11(1):5153.
9. Benjamini, Y. and Hochberg, Y. Controlling the False Discovery Rate: A Practical and Powerful Approach to Multiple Testing. *Journal of the Royal Statistical Society: Series B (Methodological)* 1995;57(1):289–300.
10. Bernal, V., *et al.* Exact hypothesis testing for shrinkage-based Gaussian graphical models. *Bioinformatics* 2019;35(23):5011–5017.
11. Brunk, E., *et al.* Recon3D enables a three-dimensional view of gene variation in human metabolism. *Nature Biotechnology* 2018;36(3):272–281.
12. Buyukozkan, M., *et al.* Integrative metabolomic and proteomic signatures define clinical outcomes in severe COVID-19. *iScience* 2022;25(7):104612.
13. Fabregat, A., *et al.* The Reactome Pathway Knowledgebase. *Nucleic Acids Res* 2018;46(D1):D649–d655.
14. Gaud, C., *et al.* BioPAN: a web-based tool to explore mammalian lipidome metabolic pathways on LIPID MAPS. *F1000Res* 2021;10:4.
15. Hastings, J., *et al.* The ChEBI reference database and ontology for biologically relevant chemistry: enhancements for 2013. *Nucleic Acids Research* 2012;41(D1):D456–D463.
16. Hawe, J.S., Theis, F.J. and Heinig, M. Inferring Interaction Networks From Multi-Omics Data. *Front Genet* 2019;10:535.
17. Huynh-Thu, V.A. and Geurts, P. dynGENIE3: dynamical GENIE3 for the inference of gene networks from time series expression data. *Sci Rep* 2018;8(1):3384.

18. Kanehisa, M. and Goto, S. KEGG: kyoto encyclopedia of genes and genomes. *Nucleic Acids Res* 2000;28(1):27–30.
19. Karatzas, E., *et al.* Arena3Dweb: interactive 3D visualization of multilayered networks. *Nucleic Acids Research* 2021;49(W1):W36–W45.
20. Karp, P.D., *et al.* Pathway Tools version 23.0 update: software for pathway/genome informatics and systems biology. *Briefings in Bioinformatics* 2019;22(1):109–126.
21. Köhler, N., *et al.* Investigating Global Lipidome Alterations with the Lipid Network Explorer. *Metabolites* 2021;11(8).
22. Koutra, D., *et al.* DeltaCon: Principled Massive-Graph Similarity Function with Attribution. *ACM Trans. Knowl. Discov. Data* 2016;10:28:21–28:43.
23. Krumsiek, J., *et al.* Gaussian graphical modeling reconstructs pathway reactions from high-throughput metabolomics data. *BMC Systems Biology* 2011;5(1):21.
24. Kujala, M., *et al.* Differential network analysis with multiply imputed lipidomic data. *PLoS One* 2015;10(3):e0121449.
25. Langfelder, P. and Horvath, S. WGCNA: an R package for weighted correlation network analysis. *BMC bioinformatics* 2008;9(1):559.
26. Mamano, N. and Hayes, W.B. SANA: simulated annealing far outperforms many other search algorithms for biological network alignment. *Bioinformatics* 2017;33(14):2156–2164.
27. Medlock, G.L. and Papin, J.A. Guiding the Refinement of Biochemical Knowledgebases with Ensembles of Metabolic Networks and Machine Learning. *Cell Syst* 2020;10(1):109–119 e103.
28. Milenkovic, T., *et al.* Optimal network alignment with graphlet degree vectors. *Cancer Inform* 2010;9:121–137.
29. Muzio, G., O'Bray, L. and Borgwardt, K. Biological network analysis with deep learning. *Brief Bioinform* 2021;22(2):1515–1530.
30. Opgen-Rhein, R. and Strimmer, K. From correlation to causation networks: a simple approximate learning algorithm and its application to high-dimensional plant gene expression data. *BMC Systems Biology* 2007;1(1):37.
31. Petralia, F., *et al.* Integrative random forest for gene regulatory network inference. *Bioinformatics* 2015;31(12):i197–205.
32. Robinson, J.L., *et al.* An atlas of human metabolism. *Science Signaling* 2020;13(624):eaaz1482.
33. Schafer, J. and Strimmer, K. A Shrinkage Approach to Large-Scale Covariance Matrix Estimation and Implications for Functional Genomics. *Statistical Applications in Genetics and Molecular Biology* 2005;4(1).
34. Schläpfer, P., *et al.* Genome-Wide Prediction of Metabolic Enzymes, Pathways, and Gene Clusters in Plants. *Plant Physiology* 2017;173(4):2041–2059.
35. Schmid, R., *et al.* Ion identity molecular networking for mass spectrometry-based metabolomics in the GNPS environment. *Nature Communications* 2021;12(1):3832.
36. Shen, C., Priebe, C.E. and Vogelstein, J.T. From Distance Correlation to Multiscale Graph Correlation. *Journal of the American Statistical Association* 2020;115(529):280–291.
37. Song, L., Langfelder, P. and Horvath, S. Comparison of co-expression measures: mutual information, correlation, and model based indices. *BMC Bioinformatics* 2012;13(1):328.
38. Székely, G.J. and Rizzo, M.L. Brownian distance covariance. *The annals of applied statistics* 2009;3(4):1236–1265.
39. Tantardini, M., *et al.* Comparing methods for comparing networks. *Sci Rep* 2019;9(1):17557.
40. Toubiana, D. and Maruenda, H. Guidelines for correlation coefficient threshold settings in metabolite correlation networks exemplified on a potato association panel. *BMC Bioinformatics* 2021;22(1):116.
41. Zhang, B. and Horvath, S. A general framework for weighted gene co-expression network analysis. *Stat Appl Genet Mol Biol* 2005;4:Article17.
42. Zhang, T. and Wong, G. Gene expression data analysis using Hellinger correlation in weighted gene co-expression networks (WGCNA). *Computational and Structural Biotechnology Journal* 2022;20:3851–3863.
43. Zhou, G., *et al.* OmicsNet 2.0: a web-based platform for multi-omics integration and network visual analytics. *Nucleic Acids Res* 2022;50(W1):W527–533.

Analysis and Interpretation of Metabolite Associations Using Correlations

Edoardo Saccenti

Abbreviations

COVSA	Covariance Simultaneous Component Analysis
PCA	Principal Component Analysis

1 Introduction

Metabolomics is the study of small molecules, called metabolites, produced by living organisms. These molecules are important indicators of the state of a biological system, providing insights into the metabolic pathways involved in various processes. The complexity of metabolomics data, which can involve hundreds or thousands of metabolites, requires statistical methods to identify relationships between them.

Correlation analysis is an essential tool in metabolomics research for identifying metabolite associations and gaining insights into underlying biological processes. By measuring the strength and direction of the relationship between two metabolites, correlation analysis can provide insights into metabolic pathways and biological systems.

As the field of metabolomics continues to expand, new techniques and approaches are being developed to better understand the complex metabolic pathways involved in various biological systems.

This chapter provides a detailed overview of how correlation analysis can be used to identify metabolite associations in complex metabolomics data. The advantages and limitations of each method are discussed, providing guidance on when to use each approach.

It discusses the correlation between metabolites in biochemical networks and explores possible reasons behind these correlations, including global perturbations,

E. Saccenti (✉)
Laboratory of Systems and Synthetic Biology, Wageningen University & Research, Wageningen, The Netherlands
e-mail: edoardo.saccenti@wur.nl

V. Soni, T. E. Hartman (eds.), *Metabolomics*,
https://doi.org/10.1007/978-3-031-39094-4_3

local perturbations, intrinsic variability, experimental noise, and data pre-processing. Molecular mechanisms that lead to correlations in replicated experiments of metabolic networks are also discussed, providing an overview of how equilibrium and mass conservation, asymmetric control, enzyme variability, and the systemic nature of metabolic control can lead to patterns of correlation.

The chapter covers diverse types of correlation analysis, such as Pearson's, Spearman's, and Kendall's correlation, and how to assess the significance of correlation analysis results. It also discusses the importance of correcting for multiple testing and taking biological factors into account when interpreting correlation analysis results. The chapter presents mutual Information and its relationship with correlation coefficients.

The chapter explores how correlation analysis can be integrated with other statistical methods, such as network analysis. Analysis of a set of correlation matrices using component methods is presented. The chapter includes several case studies that demonstrate the practical applications of correlation analysis in metabolomics research. These case studies illustrate how correlation analysis can be used to identify metabolite associations in different biological systems.

By covering the principles, methods, and applications of correlation analysis, this chapter serves serves as a resource for researchers seeking to identify metabolite associations and gain insights into the complex metabolic pathways and biological systems involved in various biological processes.

2 Correlation Among Metabolites in Metabolomic Data

In metabolomics, correlation refers to the relationship between the levels of different metabolites in a biological sample. Quantifying and analyzing correlations among metabolites is important because it can help to identify groups of metabolites that are co-regulated or interact with each other in a biological pathway or process. This information can be used to gain insights into the underlying biological mechanisms that are responsible for a particular metabolic phenotype or disease state. Additionally, correlation analysis can be used to identify potential biomarkers or therapeutic targets and to prioritize features for further investigation in subsequent studies. Therefore, understanding the correlations between metabolites can provide valuable insights into the underlying biology and aid in the interpretation of metabolomics data.

A typical experimental metabolomic study involves collecting multiple samples from replicates of a biological system and measuring a large number of metabolites. Metabolites are organic molecules that are not directly encoded in the DNA sequence of an organism. Instead, they are produced by other metabolites through chemical reactions, which are mostly mediated by a vast array of different enzymes [20]. Metabolite concentrations are regulated by a network of biochemical reactions that result in a level of interdependence that is not observed in proteins and transcripts [20].

In metabolomic data, it is often observed that some metabolites show high correlation when correlations are calculated on replicate samples (i.e. biological replicates of a system under identical conditions), while the majority show little or no correlation. One of the first observations of this phenomenon was reported in Roessner et al. [69] and later observed in subsequent studies [17, 29, 56, 104]. These early metabolomics studies were conducted on plants or unicellular organisms. However, such correlation patterns are easily observed when metabolite concentrations are measured in biofluids such as blood, urine, and saliva. When analyzing correlations in metabolomics data, it is important to not only examine the strength and direction of the correlations but also the distribution of these correlations. By examining the distribution of correlations, one can identify outlier correlations that may not reflect true biological relationships or correlations that are highly dependent on a small subset of samples.

Figure 1 shows the distribution of pairwise correlation coefficients between metabolite concentrations measured in biofluids (urine and blood) in animal and human metabolomic studies. The two distributions are skewed towards low values, with a few pairs of metabolites strongly positively correlated and relatively few negative correlations. These patterns are typical of the correlations observed in metabolomics data.

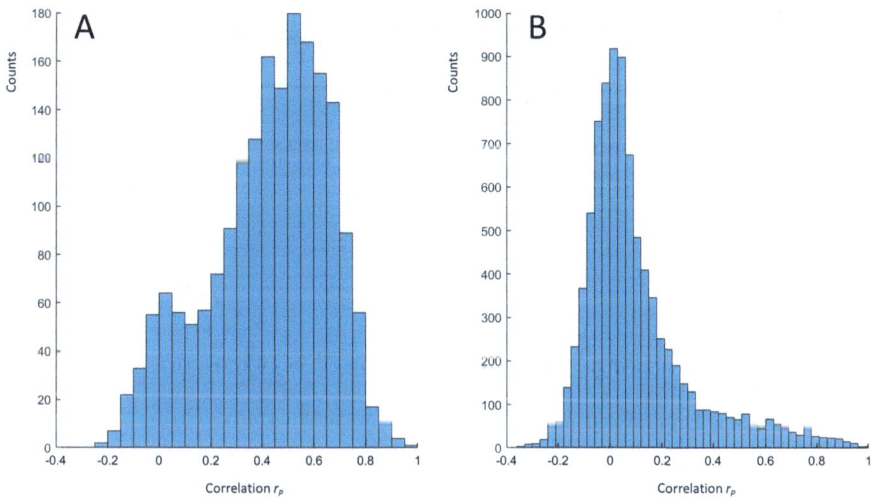

Fig. 1 Distribution of metabolite correlations (Pearson's coefficient r_P) observed in two metabolomic studies. (**a**) Correlation between the 1891 metabolites pairs obtained from 62 metabolites measured using NMR on 79 urine samples collected from pigs. (Data from Lusczek et al. [54] obtained from https://www.ebi.ac.uk/metabolights/with accession number MTBLS123). (**b**) Correlation between the 8778 metabolites pairs obtained from 133 metabolites measured using UPLC-MS on 2139 blood samples collected from humans. (Data from Magnusson et al. [55] obtained from https://www.ebi.ac.uk/metabolights/with accession number MTBLS93)

3 Why Are Metabolites Correlated?

A legitimate question to ask is from where and why these correlation patterns arise. A good starting point would be to consider Fig. 2, where the correlation among three metabolites is shown [20]: Glutamine is produced from glutamate through a biochemical reaction enzymatically mediated by glutamine synthetase, and thus, they are direct neighbours in the metabolic map (see Fig. 3, top). However, their concentrations are not correlated, as shown in Fig. 2a. On the contrary, the concentrations of methionine and valine, which belong to two different metabolic pathways (see Fig. 3, bottom), are highly correlated (see Fig. 2b).

Since metabolites are organized in interdependent biochemical networks, the naive explanation is that metabolites that are directly connected in the reaction chain are highly correlated, while metabolites far away in the network are less correlated. However, as this example shows, this is generally not true [20].

After establishing that concentrations between metabolites do not arise due to topological characteristics of the metabolic network, the question remains about what mechanisms generate the observed correlation patterns. The Pearson's correlation (Eq. 1) used to quantify metabolite associations in Fig. 1 is a standardized measure of co-variation. The existence of a correlation presupposes the existence of a source of variability for some of the metabolites [90]. Steuer [90] defined three scenarios in which variation (or perturbations) can be induced in a biological system, resulting in correlations among metabolite concentrations:

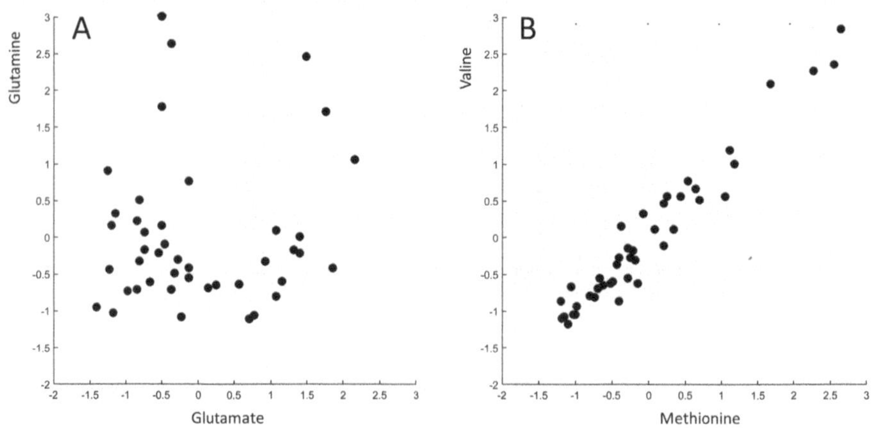

Fig. 2 Scatter plot of standardized concentrations of (**a**) Glutamate and glutamine and (**b**) Methionine and valine. Glutamate and glutamine are direct neighbours in the metabolic map, being glutamine directly produced from glutamate: Nonetheless their concentrations are uncorrelated ($r_S = 0.02$). On the contrary, methionine and valine are far apart in the metabolic network (see Fig. 3): nonetheless their concentrations are highly correlated ($r_S = 0.95$). (Data extracted from Fig. 1 in Camacho et al. [20]. Data originally published in Weckwerth et al. [104])

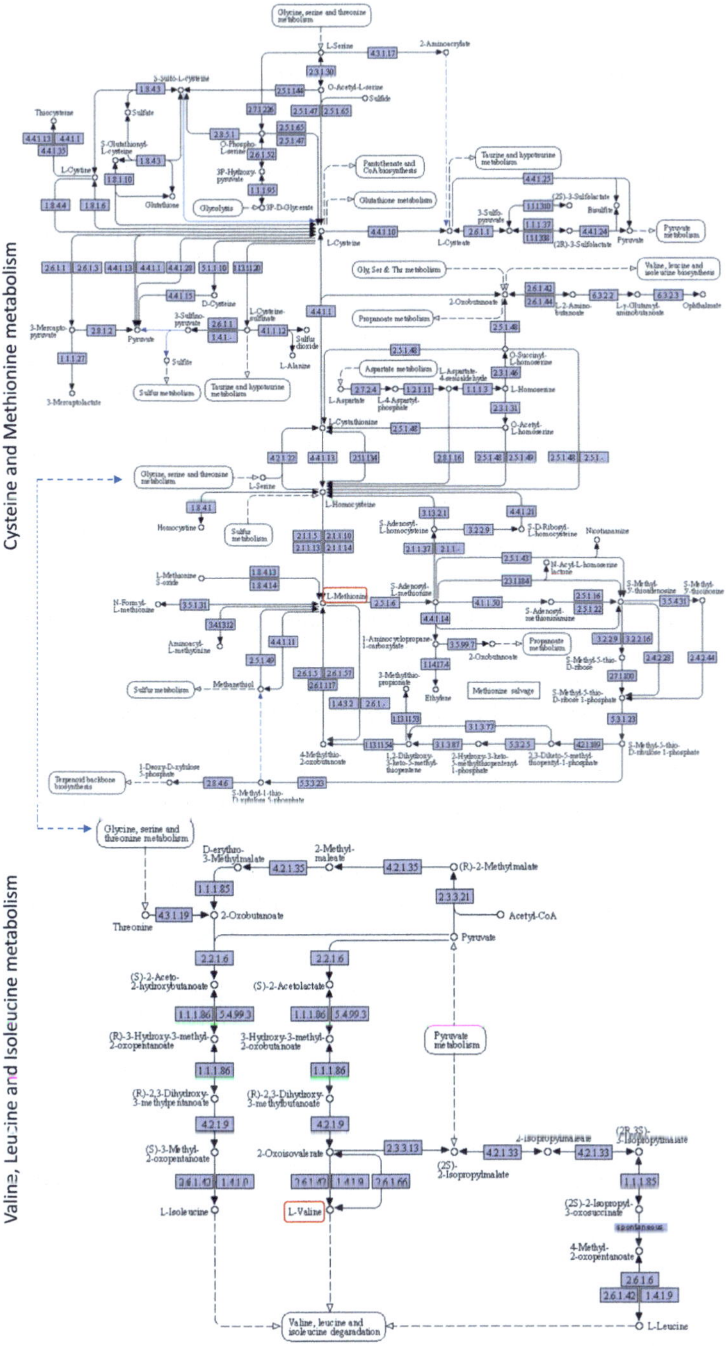

Fig. 3 Metabolic map of the cysteine and methionine metabolism (top) and of valine, leucine, and isoleucine metabolism (bottom). Methionine and valine (highlighted in red) are far apart in the metabolic network; nonetheless, their concentrations are highly correlated, as shown in Fig. 2b. (Figure assembled with metabolic maps map00270 and map00290 from KEGG (https://www.genome.jp/kegg/) [47, 48]. Reproduced with permission a.)

1. Global perturbations: If the metabolic network is perturbed globally such that many metabolite concentrations are affected, this results in correlation among metabolites. This can happen when external factors like temperature, light, or stress are manipulated or when a treatment is imposed on the system, influencing many metabolites simultaneously.
2. Local perturbations: This happens when a specific point of the network (or a specific reaction) is perturbed, for instance, by inactivating an enzyme.
3. Intrinsic variability: Correlation can also arise in the absence of external perturbations due to variability and fluctuations inherent in all biological systems. For example, small differences in enzyme concentrations that depend on (slightly) different gene expression levels result in the correlation patterns observed in biological replicates, that is, when copies of systems are measured under identical (experimental) conditions. Such intrinsic variability is what causes the correlation patterns observed in Fig. 2b, bottom.

To these three biological scenarios, we can add two further possible technical, non-biological, reasons for the observed correlation among metabolites:

1. Experimental noise: Metabolites are usually measured and quantified using comprehensive *omic* experimental platforms like nuclear magnetic resonance (NMR) [27, 101] or mass spectrometry (MS) [23, 33], both of which require sample preparation before measurements can be performed. Any error in the sample preparation may affect all metabolite in the same direction, introducing spurious correlation [59]. Internal standards are often added with the aim of achieving exact quantification. Any error in the quantity of internal standard added will result in a spurious correlation between metabolite concentrations. Metabolomics (and proteomics) data acquired with MS are often processed using deconvolution tools to enable mathematical separation and quantification of co-eluting peaks. Any positive error on one of the deconvoluted peaks will result in a negative error on the other(s) since the area under the peaks is constant. These errors will introduce negative correlations among concentrations [76].
2. Metabolomics data are often pre-processed before statistical analysis: This may entail scaling, normalization, or transformation [98]. Normalization, in particular, is a very critical step since it can induce spurious correlations [73]. Some examples are discussed in Sect. 5.4.

As discussed in Sect. 5.1, the calculation of correlation between two or more metabolites is appropriate and statistically meaningful only if all samples come from the same distribution and are identically distributed (see Fig. 6). This implies that correlations are meaningful only when taken over homogeneous biological replicates. This assumption resonates with the observation that high correlations are often observed in metabolomic data measured on replicate experiments pertaining to replicas of the same system under identical conditions (see Figs. 1 and 3). Therefore, it is of interest to understand where these correlation patterns arise from.

Figure 4a shows a metabolic network consisting of five metabolites involved in five reactions (Fig. 4b). The reaction network includes a conserved moiety (i.e. the total concentration $[A_1] + [A_2]$ of metabolites A_1 and A_2 is constant) and a rapid reversible equilibrium reaction ($S_1 \leftrightarrow S_2$). The correlation among the concentrations at steady state of these metabolites is shown in Fig. 4c. The negative correlation between the concentration of A_1 and A_2 is expected since the sum must be constant. If one increases, the other must decrease and vice versa. S_1 and S_2 are neighbours in the reaction network and are also correlated. However, S_1 and S_3 are not neighbours, and still, they are highly correlated. These correlations arise as a consequence of the data generation procedure: Small differences (random perturbations) among the parameters of the dynamic system (i.e. in the parameters of the system of differential equations) have been introduced between different runs, resulting in slightly different time profiles and steady-state concentrations. This mimics the differences observed between similar, albeit not identical, biological systems, like two bacteria or plants of the same species or two humans. For more details, see the Methods section of Steuer [90]. Strategies for the simulation of metabolomics data can be found in Jahagirdar et al. [42].

This example shows how correlation patterns are a systemic characteristic of the metabolic network. All reactions contribute to the observed patterns of correlations, and any two metabolites can be correlated or not depending on the entire complex of biochemical reactions, regulatory interactions, and fluctuations characterizing the system [90].

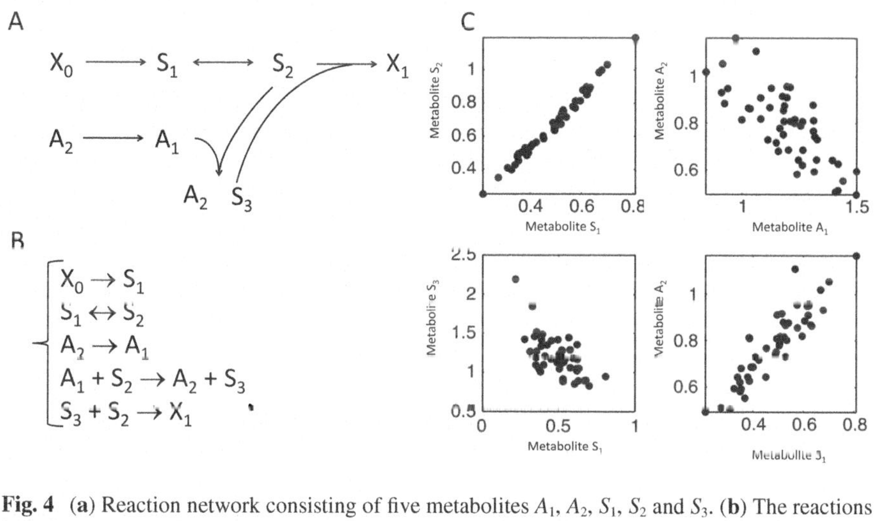

Fig. 4 (**a**) Reaction network consisting of five metabolites A_1, A_2, S_1, S_2 and S_3. (**b**) The reactions include a reversible rapid equilibrium reaction ($S_1 \leftrightarrow S_2$) and conserved moiety $[A_1] + [A_2] = \text{const.}$ All reaction rates follow mass-action kinetics. For more details see caption of Figure 3 in Steuer [90]. (**c**) Correlation patter among metabolite concentrations measured at steady state. Replicate samples have been obtained introducing small perturbations in the model parameters. (Figure adapted from Steuer [90]. Reproduced with permission)

4 Molecular Mechanisms Responsible for Observed Correlations in Replicate Experiments

Camacho et al. [20] used a fairly complex model of yeast glycolysis containing 17 reactions [67, 95] (in particular, see Figure 1 in Pritchard and Kell [67]) to investigate the underlying characteristics of the metabolic network that contribute to the observed correlation patterns in replicated experiments. The authors introduced variability among replicates by slightly altering the initial enzyme concentrations, ranging between 90% and 110% of a base model, and employed co-response profile analysis to identify the enzymes that influenced the system and the extent of their impact [38, 46]. They found and discussed several mechanisms:

1. *Equilibrium.* If two metabolites are near to or in chemical equilibrium (like S_1 and S_2 in the reaction network of Fig. 4a, b), their concentrations at steady state will exhibit positive correlation. This is because at equilibrium, the ratio $[S_1]/[S_2]$ of the two concentrations at steady state is (approximately) equal to the equilibrium constant, $K_e = [S_1]/[S_2]$. Hence, the concentration of S_1 and S_2 is (almost) perfectly correlated. Any change in the concentration of S_1 results in a change in the concentration of S_2 in the same direction.
2. *Mass conservation.* If two metabolites are involved in a mass conservation relation, they exhibit correlated concentration. This happens, for instance, in moiety-conserved species, i.e. chemical subunits that are conserved throughout the chemical cycle. The constraint of mass conservation imposes that the correlation must be negative. Metabolites A_1 and A_2 in the reaction network in Fig. 4a are in a conserved moiety; thus, the sum $[A_1] + [A_2]$ is constant, and the concentration of A_1 and A_2 are necessarily anticorrelated. Any change in the concentration of A_1 results in a change in the concentration of A_2 in the opposite direction.

   Correlations arising from equilibrium and mass conservation are necessarily large, but metabolomics data analysis shows that they are also quite uncommon. Referring to the examples in Fig. 1, the fraction of large correlations ($|r_P| > 0.8$) is relatively small: 1.8% (Example A) and 1.2% (Example B).
3. *Asymmetric control.* If the concentrations of two metabolites are controlled by the same enzyme(s), they will exhibit correlated behaviour, no matter how far or close they are in the metabolic network. The correlation arises from the amplification of the variation of that enzyme as a consequence of the summation theorem for concentration control [38].
4. *High variability of one enzyme.* If the concentration of one enzyme differs greatly between replicates, that is, if it is poorly controlled, its high variance will lead to a negative correlation between its substrate and the product metabolites [90].

   Correlations arising because of these two latter mechanisms are usually moderate in magnitude: $0.6 < |r_P| < 0.8$. Despite being more common than strong correlations, they are far from being abundant: With reference to the data sets from Fig. 1, the fraction of moderate correlations is 23.4% (Example A) and 3.5% (Example B). In fact, the vast majority of correlations is small, with $|r_P| <$

0.6: 95.3% (Example A) and 74.8% (Example B). This a results of the Systemic nature of metabolic control

5. *Systemic nature of metabolic control.* Two metabolites can be produced by reac tions catalyzed by the same enzyme, and despite this metabolic constraint, their concentration can be uncorrelated. This is because at the overall systemic level, they can be influenced by other enzymes that affect them with the same strength but in different directions. This explains why metabolites that are close in the metabolic network may show little or no correlation despite being directly related.

5 Measures of Association

Correlation and associations among metabolite concentrations can be quantified using different measures (or indexes) of correlations. This section reviews the most commonly used correlation coefficients (Pearson's, Spearman's, and Kendall's) together with mutual information. Characteristics and assumptions of each index are discussed together with pitfalls. An overview is given in Table 1, together with other approaches used to analyze correlation matrices.

5.1 Pearson's Correlation

The Pearson's correlation coefficient r_P (also called moment-product correlation) is probably the oldest and most used measure of association between two variables. It dates back to the work of Bravais [15] and Galton [31] and was defined, in modern terms, by Pearson [63, 65]. Given two variables x and y (representing metabolite concentrations, or any other biological feature), measured on n observations, the Pearson's correlation coefficient r_P is defined as:

$$r_P = \frac{\text{cov}(x,y)}{\text{std}(x)\,\text{std}(y)} \tag{1}$$

where cov(x,y) is the covariance between x and y and std(x) and std(y) are the standard deviation of x and y, respectively. Variance and covariance are estimated from the sample observations. The Pearson's coefficient r_P can be interpreted in various ways, with different, albeit equivalent, algebraic, geometric, and trigonometric reformulations [53].

The Pearson's correlation is a measure of linear association between x and y (Fig. 5a) and, as such, should not be used when the relationship is not linear: This will result, in general, in correlation coefficients which are biased downwards, as shown in Fig. 5b. Non-linear, monotonic relationships are, in general, better

Table 1 Overview of methods for analyzing correlations discussed in this chapter

Method	Description	Equation in text	References
Pearson's correlation	Quantifies the degree of linear association between two variables. Not suitable for not linear relationship.	Eq. (1)	Pearson [63, 65]
Spearman's correlation	It is a non-parametric measure of association and is often used when the relationship between two variables is not linear, but rather monotonic. It is calculated by converting the values of each variable to their ranks and then calculating the Pearson's correlation coefficient between the two sets of ranks. This allows for the detection of non-linear relationships between the variables.	Eqs. (2), (3), and (4)	Spearman [87]
Kendall's tau corelation	Like Spearman's correlation, it is often used when the relationship between variables is not linear, but rather monotonic. It is calculated by comparing the number of concordant and discordant pairs of observations in the two variables being compared.	Eqs. (5), (6), and (7)	Kendall [49]
Mutual information	It is a measure of the dependence between two variables. It quantifies the amount of information that is shared between two variables and measures how much knowing the value of one variable reduces the uncertainty about the other variable.	Eqs. (8), (9), (11), and (12)	Kullback and Leibler [52]
Winsorization	A data pre-processing technique that involves limiting extreme values in a dataset by replacing them with less extreme values. It is often used to mitigate the effect of outliers on statistical analysis.		Hastings Jr et al. [36]
COVSCA	Covariance Simultaneous Component Analysis. A component model used to analyze sets of covariance/correlation analysis. It transform each matrix into a point in low dimensional space.	Eq. (16)	Smilde et al. [85]

described by Spearman's and Kendall's correlation (see Sects. 5.2 and 5.3) or mutual information. For a discussion on the use of mutual information (see Sects. 6 and 5.5).

The calculation and use of the Pearson's correlation coefficient rest on several assumptions [60]:

1. The covariation between x and y is linear.
2. The observations of x and y are normally distributed.
3. The observations of x and y are identically distributed.
4. The observations of x and y are independently distributed.

Apart these fundamental "statistical" assumptions, there are other practical assumptions that should be considered:

1. The value of x is not used to calculate y (and vice versa).
2. There are no outliers in the observations.
3. The value of x (or y) is not experimentally controlled.

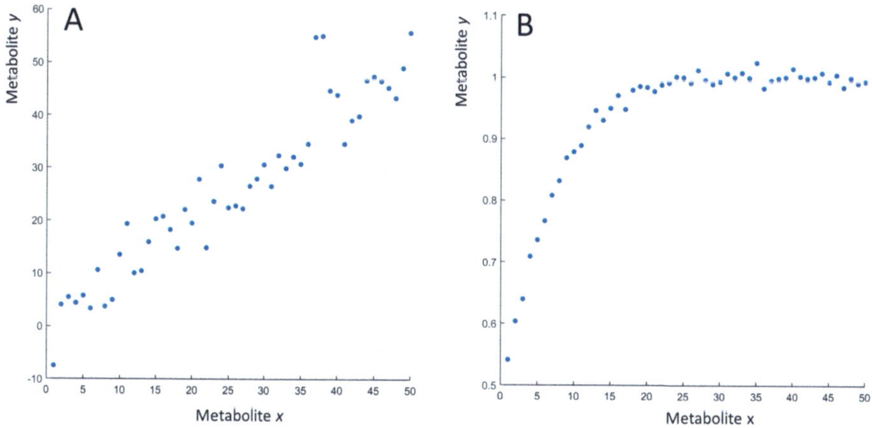

Fig. 5 (**a**) Concentrations [a.u.] of two metabolites, x and y, are linearly associated: Pearson's correlation $r_P = 0.74$. (**b**) Two metabolites that are monotonically associated: $r_P = 0.75$ and Spearman's correlation $r_S = 0.81$

From the definition of r_P, it follows that the Pearson's correlation coefficient is a standardized form of covariance: The covariance between x and y can vary, in principle, between 0 and $\pm\infty$ (where the sign depends on the sign of the relationship). When the correlation is taken, this variability is mapped to the interval -1, 1. Since the standard covariance is a measure of linear relationship, also r_P is a measure of linear relationship. In fact, r_P relates to the correlation parameter of the bivariate normal distribution, which is, in most cases, assumed to be linear [99]. The Pearson's coefficient r_P is also related to regression, being the standardized slope of the regression of x on y (or vice versa).

The assumption of normal distribution of the observation is often stated, and it is here repeated, but has been often debated in literature. Kowalski [51] reviewed existing literature which provided contrasting results: However, based on simulations, it was shown that the distribution of r_P can rather sensitive to deviation from normality and that Pearson's correlation analyses should be limited to situations in the distribution of x and y is almost normal. The key point here is that the non-normality of the observations distorts the distribution of r_P: This has direct implication for hypothesis testing. Most software tools used to calculate r_P also output a P-value, which results from the testing of the null hypothesis

$$H_0 : \rho_P = 0$$

against the alternative

$$H_1 : \rho_P \neq 0,$$

where the observed r_P is compared with its expected reference distribution under H_0 being true. Since the reference distribution of r_P is defined under the condition of normality of the observations, if the correlation is calculated for non-normally distributed data, the *P*-value cannot be interpreted at its face value: In other words, while r_P can be used, in principle, as a measure of linear association between variables that may not be normally distributed, the correct interpretation of its statistical significance depends on data being normally distributed. In general, deviation from normality increases the number of false positives [8].

Hypothesis testing for r_P under non-normality of data is discussed in Edgell and Noon [26] and Bishara and Hittner [8]; confidence intervals are discussed in Bishara and Hittner [9]. The effect of non-normality is also explored and discussed in Calkins [19], Havlicek and Peterson [37], Wilcox [108], and Ventura-León et al. [100]. When data are not normal, transformations can be applied like log-scaling or square-rooting (see Sect. 5.4).

Calculation of r_P assumes that the observations of x (and of y, respectively) come from the same population, that is, all follow the same distribution. If observations from different distributions are mixed, this can severely affect the correlation coefficient. Figure 6 shows an example where two sets of observations two uncorrelated variables, x and y, coming from two different distributions: If the observations are taken all together to calculate the correlation coefficient, the result is that the two variables are correlated when actually there are not. The contrary can also happen: Correlated variables may result to be uncorrelated (see Figure 4 in Saccenti [74]). Havlicek and Peterson [37] also briefly discussed this topic.

The assumption of independence of observations is violated, for instance, when repeated measures are considered, such as in the case of metabolite measured on multiple occasions on the same subject(s) [5] or when time series are considered

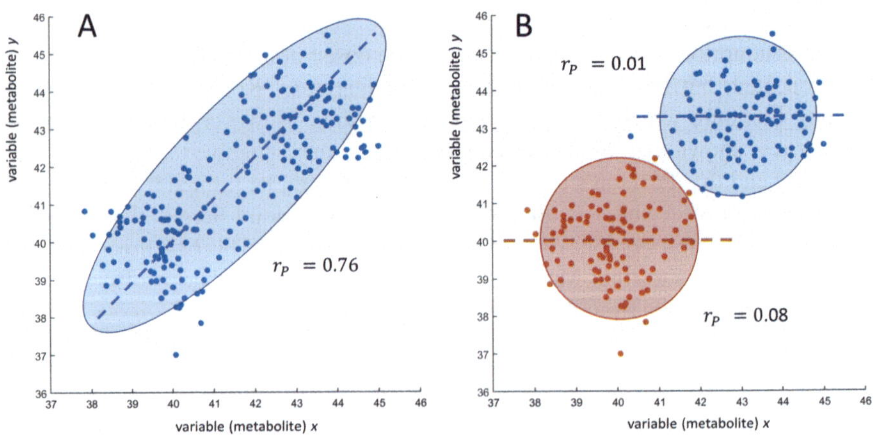

Fig. 6 (**a**) Correlation plot of 200 observations of two variables x and y: the Pearson's sample correlation between x and y is $r_P = 0.76$. (**b**) The same scatter plot as in Panel (**a**), but with data points colour-coded to highlight the actual data structure: When taken separately, the $n = m = 100$ observations of x and y (condition A: blue; condition B: red)are uncorrelated: $r_P = 0.01$ (data set A) and $r_P = = 0.08$ (data set B). (Figure reproduced and adapted from Saccenti [74] under the CC BY 4.0 license)

[113]. Caution should be taken when dealing with these kinds of measurements, since special statistical tools are needed.

A common mistake that violates this assumption is when technical replicates are used to increase the sample size, under the wrong presumption of increasing the sample size. A simulated example is shown in Fig. 7a.

We have already mentioned how experimental error can introduce correlation across measurements. The effect of measurement error on correlation is very complicated: a discussion can be found in [76]. Figure 7b shows an example of independent variables which become dependent because of additive correlated experimental noise.

The r_P is extremely sensitive to outliers: Just one outlier is sufficient to steer the correlation from 0.014 to 0.82, as shown in Fig. 8a. For this reason, it is imperative not to trust blindly numerical values but also to explore visually the relationship between (measured) variables to protect against the effect of outliers. For extended discussion about the impact of outliers on statistical analysis see [1, 40, 100].

When analyzing (metabol)omic data and visual exploration of all possible correlation plots is not feasible, Principal Component Analysis (PCA) [39, 43, 64] can be used to detect the presence of outlying observations or unwanted data structure and variation (see, for instance, Figure 9 in Saccenti [74]). Jolliffe and Cadima [45], Jolliffe [44], Bro and Smilde [16], and Sainani [79] provide a good introduction to this topic.

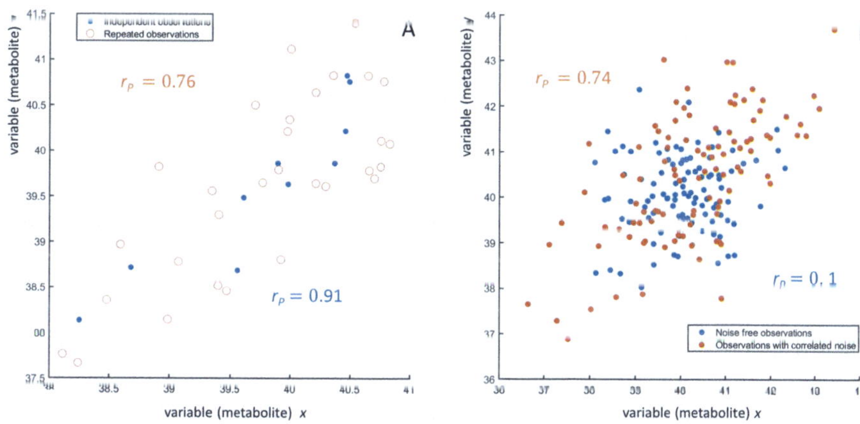

Fig. 7 (**a**) Effect of the use of replicated observations on the Pearson's correlation coefficient: $n = 10$ observations of two variables, x and y (blue dots •), and of three replicates for each observation (red open circles °). The sample correlation between the $n = 10$ observations is $r_P = 0.91$. If the replicates are considered using a total of 30 observations, the sample correlation reduces to $r_P = 0.76$. (**b**) Effect of correlated measurement noise: Uncorrelated variables becomes correlated in presence of additive correlated noise. (Figure reproduced and adapted from Saccenti [74] under the CC BY 4.0 license)

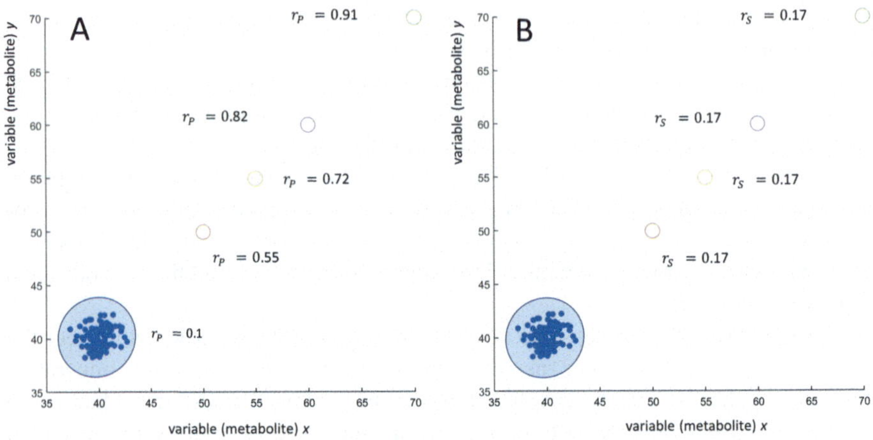

Fig. 8 Effect of outlier(s) on the correlation coefficient: $n = 100$ observations of two uncorrelated metabolites, x and y (blue dots •, encircled), to which 1 outlier is added. (**a**) Pearson's correlation (r_P): The larger the outliers (i.e. the large its distance from the bulk of observation), the larger the correlation coefficient. (**b**) Spearman's correlation r_S: The outlier has less impact on the correlation. Since Spearman r_S is a rank measure, it does not depend on the magnitude of the outlier. However, it must be reminded that r_S depends on the number of outliers: See, for instance, Figure 6A in Saccenti (2023). (Figure reproduced and adapted from Saccenti [74] under the CC BY 4.0 license)

5.1.1 Testing the Significance of Correlations and Multiple Testing

In the *omic* setting, testing the significance of the correlation coefficients will usually result in a large number of tests to be performed: This naturally leads to the problem of multiple testing and Type 1 errors, hence the necessity of correcting for false positives. Unfortunately, no method for correction is perfect and there is always the risk of increase the false negative rate, which usually increases with the number of tests performed: If p metabolites are measured, there are $0.5p(p-1)$ tests to be performed, which can be exceptionally large. The Bonferroni correction (which is a familywise error rate (FWER) controlling procedures) [12] is often suggested as a conservative approach, but precedence should be given to the Benjamini-Hochberg [7] or to Storey's [92, 93] methods (which are False Discovery Rate controlling procedures designed to control the expected proportion of wrongly rejected null hypotheses). The reason is twofold. The Bonferroni correction has basically no power when the number of test is larger and results in a dramatic inflation of false negatives [25]: With more than 70 tests performed (a threshold attained when more than thirteen metabolites are measured), the probability of a false negative with Bonferroni is twice as with Benjamini-Hochberg [105]. The Bonferroni correction is recommended for testing a single universal null hypothesis that *all* correlations are not significant [3], a situation that is seldom of interest in the case for *omic* studies. There are no consolidated strategies to alleviate this problem: As usual, there is a trade-off between the number of missed new discoveries and the necessity of protecting against false discoveries, and such trade-off depends on circumstances and consequences of committing an error. A recommendation is to provide, if possible, the analysis of both corrected and uncorrected results.

An optimal solution would be to validate observed significant correlations in an independent data set. If this is not available, data splitting strategies could be employed [21, 72]. Basically, a portion of the data is used to suggest a hypothesis (i.e. to calculated metabolite correlations), and a second independent portion is used to test it, i.e. to assess it significance. This approach can be rephrased in an inferential setting [72] after correction for multiple testing. Reducing the sample size can potentially reduce the power, but this approach is effective in giving valid inference after the selection of a hypothesis, estimating nuisance parameters, and avoiding overfitting [72]. An application is given in Di Cesare et al. [24], where data splitting was used to estimate the reproducibility and robustness of metabolite correlations with age.

5.1.2 Robust Correlation Coefficients

The effect of outliers can be counteracted by using robust methods for the calculation of the correlation coefficient. The Spearman's correlation r_S, being a rank measure (see Sect. 4.3), is less sensitive to outliers than the Pearson's coefficient r_P, as shown in Fig. 7b: However, it can be distorted if there are several outliers: see, for instance, Fig. 6a in Saccenti (2023). The same is true for Kendall's tau (see Sect. 4.4). Moreover, it is known that these two measures do not deal with outliers considering the overall structure of the data [106, 109]. For example, see Fig. 1 and the discussion, thereof, in Wilcox [110].

Alternative robust approaches that can be considered are Winsorized correlations, skipped correlations, and bootstrapping.

Winsorization [36] consists in replacing the k smallest observations with the $(k + 1)$st smallest observation, and the k largest observations with the $(k + 1)$st largest observation: This procedure winsorizes both left and right tails of the data distribution: The Pearson's correlation coefficient is then calculated on the Winsorized variables [107]. For an application to metabolomics data, see Di Cesare et al. [24]. The problem of testing for the null hypothesis for Winsorized correlations is addressed in Wilcox [107].

The term skipped correlation refers to the procedure of using some outlier detection method that takes into account the overall structure of the data, remove any outliers that are found, and then compute Pearson's correlation using the remaining data [106]. Wilcox [106] also discusses projection in combination with bootstrapping [66] to obtain skipped correlations.

Bootstrapping pertains to resampling with replacement of the data and the calculation of a measure of interest; in this case, correlation coefficient, from this sample. When the procedure is repeated many times, the properties of the obtained distribution of the measure of interest can be explored to investigate the effect of outliers considering the overall data structure. Common approaches are the classical percentile bootstrap method [96] and the method proposed by Zou [115]. These approaches are reviewed and compared in Wilcox [110].

5.2 Spearman's Correlation

The Spearman's rank correlation r_S [87] is a measure of monotonic association between two variables (metabolite concentration), x and y, based on the ranks of each observation. It is defined as

$$r_S = \frac{\text{cov}(R(x),R(y))}{\text{std}(R(x))\text{std}(R(y))}, \tag{2}$$

where $R(*)$ indicated the rank(s) of the observations of x and y and amounts to the Pearson's correlation among the ranks of the observations. If the ranks are all integer numbers, r_S can be expressed as

$$r_S = 1 - \frac{6\sum_i d_i^2}{n(n^2-1)}, \tag{3}$$

where d_i is the difference between the two ranks of each observation and n is the number of observations.

Since the correlation is taken among the ranks of the observations, rather than on the value of the observations themselves, the Spearman's r_S can be used to measure non-linear associations as long they are monotonic, i.e. strictly increasing or decreasing. An example is shown in Fig. 5b.

The Spearman's correlation is less sensitive to outliers, as shown in Fig. 6b: In this case, $r_S = 0.06$ as it should be since x and y are uncorrelated, while the Pearson's correlation is $r_S = 0.79$. This happens because each observed value is transformed into its rank, and this reduces its overall impact. However, if the number of outliers is substantial, also this measure is distorted (see, for instance, Figure 6 in Saccenti [74]).

Spearman's and Pearson's coefficients are mathematically related. For n normally distributed observations and in absence of ties, the relationship is [50]:

$$r_S = \frac{6}{(n+1)}\left[\arcsin(r_P) + (n-2)\arcsin\left(\frac{r_P}{2}\right)\right] \tag{4}$$

If the two variables, x and y, are positively and linearly correlated, the Spearman's correlation is biased downwards (in absolute value). The difference $|r_S - r_P|$ between the two measures is maximal for $r_P = \pm 0.577$ (where the sign depends on the sign of the correlation). The difference depends on the number n of observations, however for large n ($n > 50$) the difference between the two is negligible [114].

5.3 Kendall's τ Correlation Coefficient

The Kendall's τ (tau) correlation coefficient [49] between n observations of two variables, x and y, is also a rank statistic. If the ranks of the observations (x_i, x_j) and (y_i, y_j) agree, i.e. if the sorting orders $x_i > x_j$ and $y_i > y_j$ or $(x_i < x_j$ and $y_i < y_j)$ hold, the pair (x_i, x_j) and (y_i, y_j) is said to be concordant; otherwise, it is said to be discordant. Given c the number of concordant pairs and d the number of discordant pairs, τ is given by:

$$\tau = \frac{c-d}{c+d}. \tag{5}$$

An alternative formulation is

$$\tau = \frac{2}{n(n-1)} \sum_{i<j}^{2} sign\left(x_i - x_j\right) sign\left(y_i - y_j\right). \tag{6}$$

Kendall's and Pearson's coefficients are mathematically related. For n normally distributed observations and in the absence of ties, the relationship is:

$$\tau = \frac{2}{\pi} \arcsin\left(r_P\right). \tag{7}$$

The use of the Kendall's correlation coefficient correlation is not commonly observed in the metabolomic literature. However, it has been shown that this measure possess an adequate control of type I errors, was nearly as powerful as Pearson's r, provided much tighter confidence intervals and had a clear interpretation than the Spearman's index [4] and appears also to be more robust to outliers than Spearman's.

5.4 Correlations and Data Pre-processing: Transformation and Normalization

Pearson's correlations are scale independent: the correlation between x and y is the same of that between $x + a$ and $y + a$. If variables are transformed with an increasing monotonic transformation which maintains ordering (like logarithmic, square rooting, squaring, etc...) the Pearson's correlation is affected. With reference to the examples shown in Fig. 5a, the correlation r_P between the original variables is $r_P = 0.74$, for log-transformed is $r_P = 0.93$, for square-rooted is $r_P = 0.83$, for squared is $r_P = 0.58$. Spearman's and Kendall's indexes are not affected: $r_P = 0.63$ and $\tau = 0.48$ in all cases.

As mentioned in Sect. 4.1, data transformation can be applied to correct for non-normality of the data when Pearson's correlation is used. Log-scaling can be used and are almost always applicable since concentration are positive by definition.

However, Log-scaling is not applicable in presence of zeros. Square-rooting can be used as an alternative. Cubic-rooting can be used in presence of negative values when necessary. Square-rooting and Log-transformation are two examples of Box-Cox transformations [14, 80].

Metabolomics data are usually normalized to correct for variability between samples arising from instrumental drift and/or sample preparation or unwanted biological variation, like different dilution factors in urine samples [73]. However, normalization affects correlation patterns [73]: Fig. 9 shows correlation profiles of some metabolite pairs in the raw data and after normalization to creatinine concentration. Panels c and d show how the overall patterns of correlations (small, medium, large) and their significance change when different normalization methods are applied. Care should be taken when calculating and interpreting correlations on normalized data. An in-depth discussion can be found in Saccenti [73].

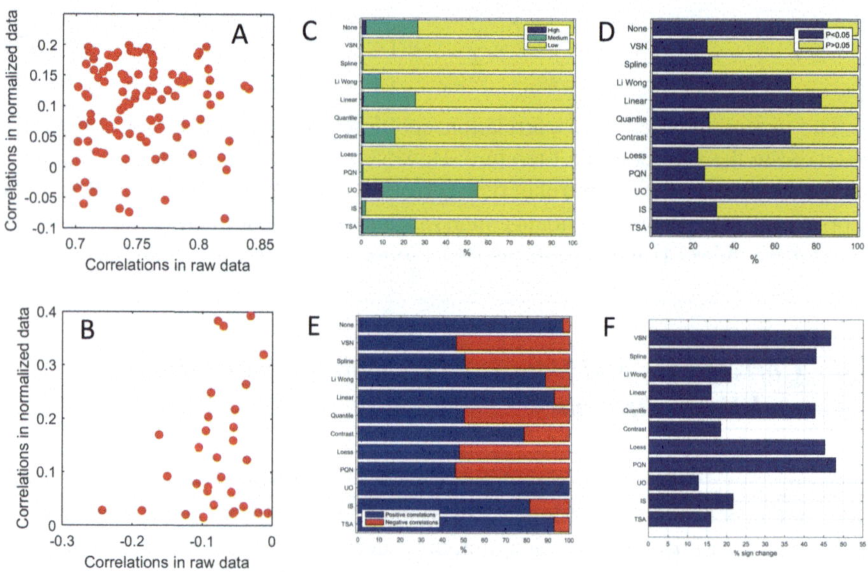

Fig. 9 Scatter plots of the correlation values (raw data vs creatinine normalized). Creatinine normalization affects (**a**) the magnitude of correlations and (**b**) their sign. Percentage (%) of (**c**) high, medium, and low correlations; (**d**) statistically significant correlations; (**e**) positive and negative correlations; and (**f**) correlations that change sign (with respect to raw data) after different normalization methods have been applied to the data. (Data from Lusczek et al. [54]). For details and discussion on the normalization methods applied, see Saccenti [73]. (Figure adapted and reproduced with permission from Saccenti [73])

5.5 *Mutual Information*

The mutual information MI(x,y) of two (random) variables, x and y, describes the mutual dependence between x and y and can be defined in term of the entropy of the two variables as

$$\mathrm{MI}(x,y) = H(x) + H(y) - H(x,y), \tag{8}$$

where $H(x)$ and $H(y)$ are the entropy of x and y, respectively, and $H(x, y)$ is the entropy of x and y.

The entropy H is a measure of the uncertainty about the values that random variable x, distributed with probability distribution $p(x)$, can assume. It is given, for discrete variables, by

$$H(x) = -\Sigma\, p(x)\log(x). \tag{9}$$

The higher the entropy, the higher the uncertainty on that variable: If a metabolite shows little variability, its entropy is also low(er), while if it has large variability, it has also large(er) entropy. Thus, the entropy describes the content of information of a random variable: The higher the entropy, the higher the information content. If (the concentration of a) metabolite that does not vary, thus, is $x = c$ with probability 1, $p(x = c) = 1$, its entropy is $H(x) = 0$, which nullifies the information associated to it [41]. Recall that in statistics and data analysis, the absence of variation translates to the absence of information.

As a numerical illustrative example [41], we can consider a metabolite whose concentration x can assume only the values $x_1 = 0.4$, $x_2 = 0.9$, and $x_3 = 1.3$ with probability $p(x = x_1) = 0.2$, $p(x = x_2) = 0.7$, and $p(x = x_3) = 0.1$: Its entropy $H(x)$ is:

$$H(x) = -\sum\nolimits_{x_1,x_2,x_3} p(x)\log(x) = -0.4\times -0.9\times 0.7 - 1.3\times 0.1 = 0.8018 \tag{10}$$

In real-life situations, the theoretical distribution of x is not known and must be estimated from the data, for instance by computing the relative frequency of the occurrence of each value (i.e. estimation from the empirical probability distribution) [57]. Other commonly used approaches are the Miller-Madow Asymptotic Bias Corrected Empirical Estimator [62], shrinkage estimation [81, 82], or the Schürmann-Grassberger estimation [83].

The mutual information between x and y is commonly expressed by taking the distance between the joint distribution $p(x, y)$ and product distribution $p(x)p(y)$ using the Kullback-Leibler divergence [52]:

$$\mathrm{MI}(x,y) = \sum_x \sum_y p(x,y)\log\left(\frac{p(x,y)}{p(x)p(y)}\right), \tag{11}$$

from which Eq. (9) can be derived (see Jahagirdar and Saccenti [41] for calculations).

The mutual information is well suited to capture non-linear relationships as those often observed among metabolite concentrations, and it is expected to be a better metric of dependence than the Pearson's correlations correlation coefficient, which can underestimate the dependence between variables when the dependence translates into non-linear relationships [86]. Figure 10 shows four simulated metabolite concentration patterns of two metabolites A and B that have the same mutual information 1.32 nats: 1 nat is the information content of the uniform distribution on the interval [0,*e*], where *e* is the basis of the natural logarithm. In all four cases, there is obvious dependence between the two metabolites, but the Pearson's correlation can only capture the linear dependence (Panel a and b), while mutual information can capture even the highly non-linear relationship (Panel c and d). Also the Spearman's and Kendall's correlations fail to capture the latter two situations since the relationship is not monotonic.

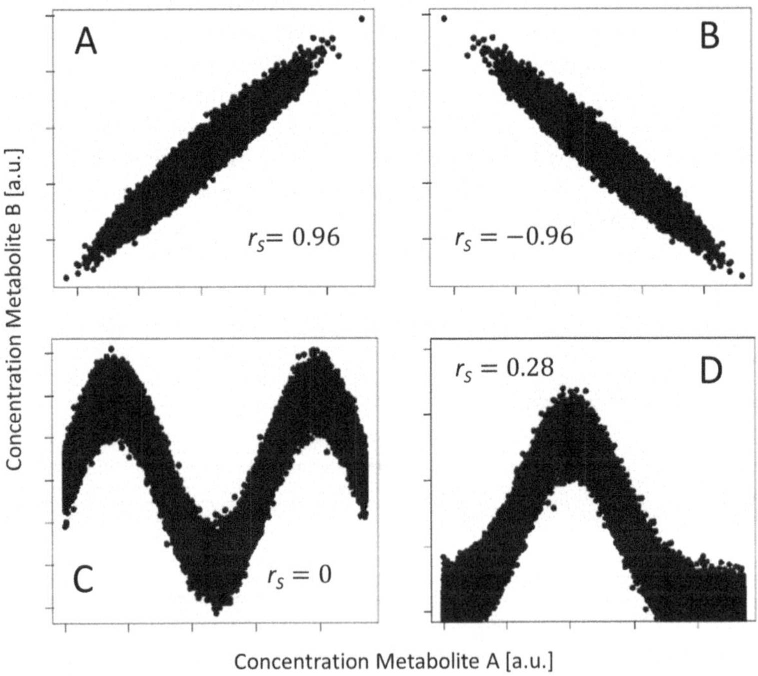

Fig. 10 Mutual information can capture highly non-linear and non-monotonic relationships whereas the Pearson's correlation coefficient cannot. Simulated concentrations of two metabolites A and B showing: (**a**) Positive linear relationship, $r_S = 0.96$ (Pearson's correlation); (**b**) negative linear relationship, $r_S = -0.96$; (**c**) Sine-wave relationship, $r_S = 0$; (**d**) bell-shaped relationship, $r_S = 0.28$. In all four cases, the mutual information MI is 1.32 nats (or 1.90 bits). (Figure adapted from Table 1 from Smith [86] and reproduced from Jahagirdar and Saccenti [41] under the CC BY 4.0 license)

Under certain conditions, an exact relationship exists linking mutual information and the Pearson's correlation coefficient r_S: If two variables are linearly correlated and normally distributed, it holds that

$$\mathrm{MI}(x,y) = -\frac{1}{2}\log\left(1 - r_S^2\right), \tag{12}$$

from which follows that the mutual information is almost always smaller than the correlation coefficient. Precisely,

$$\mathrm{MI}(x,y) \rightarrow \begin{cases} < r_S & \text{if } |r_S| < 0.916 \\ = r_S & \text{if } |r_S| = 0.916 \\ > r_S & \text{if } |r_S| > 0.916. \end{cases} \tag{13}$$

Figure 11a presents a plot of the theoretical relationship in Eq. (12), showing the intersection at approximately 0.916. This relationship is maintained with good approximation for variables generated under a linear correlative model (Fig. 11b). In case of experimental data (Fig. 11c), the approximation is less good because the correlation among measured metabolite abundances is, in most cases, not linear and Eq. (12) does not hold exactly anymore.

Equations (4), (7), and (12) link mutual information and the Spearman's and Kendall's correlation coefficients. Formulas have been derived that relate the Kendall's τ (tau) correlation and other families of bivariate distributions [11], including t- [35], Cauchy- [34], or elliptical distributed variables.

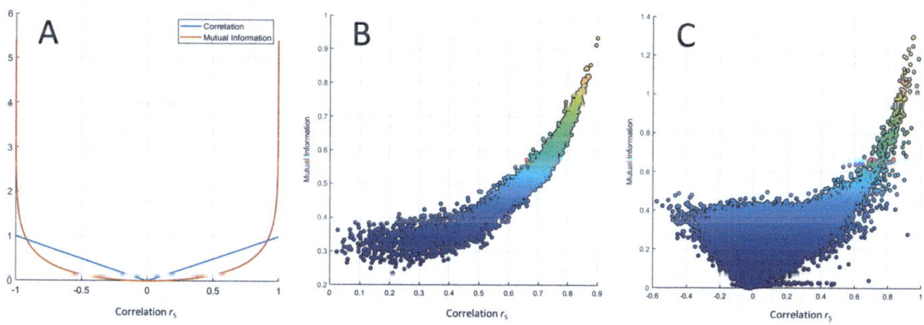

Fig. 11 Relationship between mutual information and the Pearson's correlation coefficient for (**a**) bivariate linearly correlated variables x and y, as given by Eq. 1. (**b**) Variable with an average correlation of 0.6 (for details on the simulation strategy see Jahagirdar and Saccenti [41], Sect. 3.4.3). (**c**) Metabolite concentrations from experimental data set [91] containing 669 observations of 371 metabolites, measured using LC-MS on subjects using oestrogen-only or oestrogen plus progestin. (Figure adapted and reproduced from Jahagirdar and Saccenti [41] under the CC BY 4.0 license)

6 Which Measure of Association to Use?

The central tenet of statistics and data analysis has always been that the Pearson's correlation index is a measure of liner to linear association, while Spearman's and Kendall's indices are measures of monotonic associations and should be used as such, with Pearson's correlation not to be used in presence of non-linear associations. There is indeed ample literature agreeing on this point [51, 58, 88], and the converse has also been highly supported [13, 28, 32, 51].

In the null hypothesis test setting, the power (i.e. the probability of rejecting the null hypothesis of the population correlation being zero) of the different correlation indexes depends heavily on the underlying data distribution [68]. For normally distributed data the Pearson's index has been found to have the largest power. For non-symmetrical and low-peaked distribution, the Pearson's correlation has larger power than the Spearman's correlation. For non-symmetrical and high-peaked distributions, the Spearman's index performed better that the Pearson's index. The same authors [68] explored different interesting association indexes, like the Gini's correlation coefficients [10, 112], which are not discussed here.

Van den Heuvel and Zhan [99] have presented empirical and theoretical evidence that for certain families of bivariate distribution functions with non-linear monotonic associations, the Pearson's correlation is a better measure than Spearman's and Kendall's indices; conversely they also presented families of bivariate distribution functions with linear associations for which Spearman's and Kendall's are to be preferred to the Pearson's index. How this translates and applies to metabolomic data has to be ascertained, since it is not yet clear if these distribution families may represent data pattern observed in metabolomic data.

From theoretical considerations, mutual information should be able to provide more information about metabolite associations than standard correlation measures like Pearson's and Spearman's correlation. However, the advantage of its use for the analysis of metabolomic data is not clear. Jahagirdar and Saccenti [41] found that using mutual information does not provide better results in comparison with Pearson's correlation when used for differential association analysis in a differential network analysis setting.

Which index to use will depend ultimately on the type of data at hand and its distributional properties and presence/absence of outliers. The recommendation is that any index used should not be taken at its face value but considered and interpreted in context. We have shown how correlation indexed are sensitive to outliers, in particular the Pearson's correlation. We have also shown how and why metabolomic data do usually contain a limited fraction of large correlations (i.e. larger than 0.8), so large observed correlation should always be carefully double-checked to avoid the risk that they arise from outliers or other distortion in the data.

When analyzing *omic* data and a large number of correlation pair are calculated, a simple strategy is to calculate both Pearson's and Spearman's or Kendall's indexes and compare them for the same metabolite pair: If the two measures diverge greatly, there is a chance that outliers are present OR that non-linearity exists, and as such, correlation plots should be explored visually to detect problems in the data. Outlier detection is a complicate matter as it is outlier removal. The physics approach (that we recommend) is that a sample/observation is an outlier is such only when there is an explanation for its outlying behaviour (for instance: Is the outlier due to a machine breakdown or an error in the sample preparation?). In the absence of an apparent reason, removing the sample/observation is, in principle, not justified. For some strategy for outlier detection applicable to metabolomic data, see, for instance, Walach et al. [103], Rousseeuw and Bossche [71], and Sun et al. [94].

7 Comparing Correlations

Correlation patterns reflect aspects of regulation, and as such, comparing correlation between two (or more) multiple states can provide insights on modifications and/or disruptions of the underlying molecular mechanisms. Correlation patterns can be different between different conditions, with two or more metabolites being correlated in one condition and uncorrelated in the other, as in the example of Fig. 12a, b; in other situations, the sign of the correlation can be opposed in two different conditions, as shown in Fig. 12c.

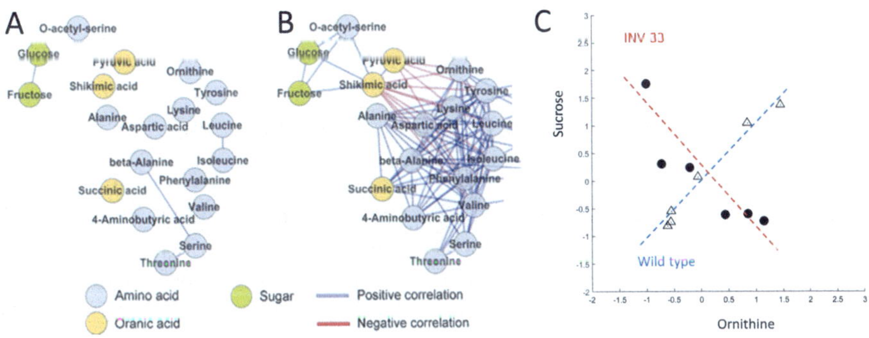

Fig. 12 Correlation between amino acids and metabolites participating in the tricarboxylic acid (TCA) cycle measured in a plant (*Arabidopsis thaliana*) under different environmental conditions. (**a**) Heat and (**b**) dark (no light). Blue and red lines depict positive and negative correlations, respectively. Under dark conditions, metabolite and amino acid concentrations are highly correlated, which is not the case under light conditions. (Adapted and reproduced with permission from Figure 8 in Caldana et al. [18]). (**c**) Correlation between sucrose and ornithine in in a wild type and a transgenic potato tuber (INV-33). (Adapted and reproduced with permission from Figure 8 in Camacho et al. [20]. Data from Roessner et al. [69]). In both cases, the remodulation of correlation patterns indicate different regulation modes

7.1 Comparing Two Correlations

The statistical significance of the difference between two correlation coefficients measured over two different conditions like those showed in Fig. 12 can be assessed using hypothesis testing. Given two sample correlation (Pearson's) coefficients r_1 and r_2 measured on two different conditions, the testing procedure is

$$H_0 : \rho_1 = \rho_2$$

against the alternative

$$H_1 : \rho_1 \neq \rho_2.$$

The testing procedure is usually carried on after z-transformation (Fisher's transformation [30]) of r_1 and r_2 since after z-scoring the z statistics can compared in a 1-tailed or 2-tailed fashion using a standard normal distribution [89]. Such transformation is defined as

$$z = \frac{1}{2}\log\frac{1+r}{1-r}. \tag{14}$$

Care should be taken, when performing statistical testing, that all underlying hypotheses are met and that the appropriate test is chosen. Most statistical software packages offer testing procedures for different correlation types.

7.2 Averaging Correlations

Sometimes it is of interest to average two or more corelations. Correlation coefficients cannot be directly averaged because they are not additive. The standard approach to average correlations is first to z-transform them (Eq. 14), average them, and then back transform the average value using the inverse z-transformation [84]:

$$r = \frac{e^{2z} - 1}{e^{2z} + 1} \tag{15}$$

For instance, the average correlation between $r_1 = 0.1$ and $r_2 = 0.7$ is not $r = 0.4$. Z-transforming gets $z_1 = 0.1003$ and $z_2 = 0.8673$, whose average is $z = 0.4838$ which back transformed (Eq. 15) gives $r = 0.4493$, which is quite different from 0.4. Other approaches for averaging have been also suggested [2].

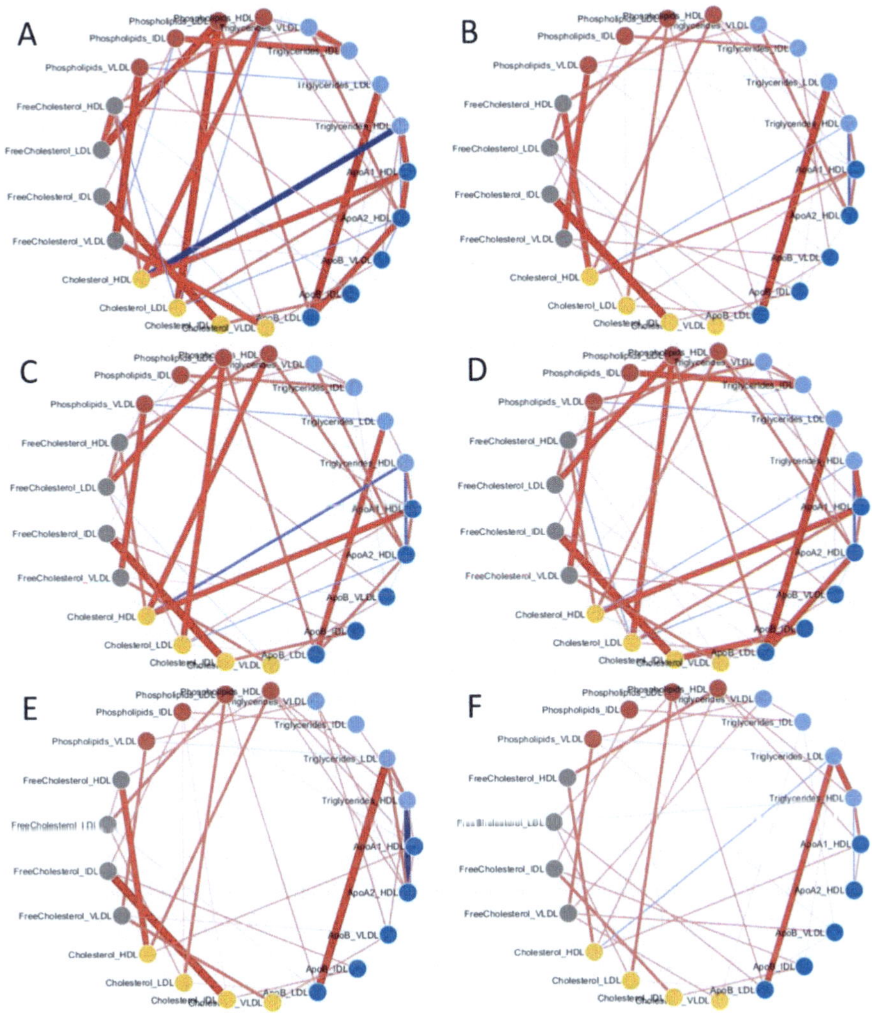

Fig. 13 Correlation networks of lipoprotein and lipid main fractions in healthy subjects. Each node represents a lipid/lipoprotein specie. Edges connecting the nodes represent the existence of a correlation between two nodes. Partial correlation estimated with a Gaussian Graphical Model [61] in combination with the PCLRC probabilistic algorithm [78]. (**a**) Women, (**b**) men, (**c**) young men, (**d**) old men, (**e**) young women, and (**f**) old women. Positive associations are coloured in red; negative correlations are coloured in blue. The edge weights are proportional to the correlation magnitude. Nodes are colour-coded according to the five lipid groups considered: Apo (dark blue), cholesterol (yellow), free cholesterol (grey), phospholipids (red), and triglycerides (light blue). (Figure reproduced from Balder et al. [6] under the CC BY 4.0 license)

7.3 Analysis and Comparison of Correlation Matrices

When analyzing multivariate data, of which metabolomics data are just an example, all the possible pairwise correlations are collected in a correlation matrix (see Fig. 14 for two examples). For a data set containing n observations of p metabolites (variables), the correlation matrix $\mathbf{C}$ has dimension $p \times p$, it is symmetric and contain $p(p-1)/2$ unique correlation pairs. In some situations, there are more than one correlation matrix, with different matrices pertaining to different conditions: The interest is then in analyzing this set of matrices in a comprehensive way and visualizing the relationship existing among them.

A convenient approach to analyze a set of K correlation (covariance matrices) is to use COVSCA (Covariance Simultaneous Component Analysis) [85]. Loosely speaking, COVSCA is a dimensionality reduction technique which reduces the set of K (high- dimensional) correlation matrices to a set of points on a low-dimensional component space, each point representing one of the correlation matrices. When the components are plotted against each other, patterns of (dis)similarity between the points (i.e. correlation matrices) can be highlighted like in a conventional Principal Component Analysis: Points close in the COVSCA space indicate matrices containing similar correlation patterns. The variables (metabolites) contributing to those patterns can be obtained through the analysis of the associated loadings.

More specifically, the K matrices are modeled as a combination of L low dimensional prototypes ($L \ll K$) matrices. The k-th correlation matrix $\mathbf{C}_k$ is modelled as

$$\mathbf{C}_k = \sum_{l=1}^{L} c_{lk} \mathbf{Z}_l \mathbf{Z}_l^T \tag{16}$$

where $c_{kl} \geq 0 (l = 1, 2, \ldots, L)$ are weight coefficients and $\mathbf{Z}_l$ are prototypical symmetric matrices consisting of loading $\mathbf{Z}$ of size $J \times R_l$ that hold simultaneously for all S_k. A brief introduction to the fitting procedure of a COVSCA model and its analysis can be found in Saccenti and Camacho [75].

Figure 15 shows correlation networks obtained from lipid and lipoprotein profiles measured on six groups of healthy subjects (women, men, young women, young men, old women, old men). The overall structure of the six networks is similar across and differences mostly concern variation of the strength of the correlation: For instance, the correlation between cholesterol and triglycerides (HDL) is stronger in the association network for men (Fig. 15b) than in that for women (Fig. 15a). These six networks can be analyzed comprehensively using COVSCA. The score plot is shown in Fig. 15a and can be interpreted in a PCA-like fashion. Every point on the plot correspond to one of the networks in Fig. 15: Points (networks) close in the space have similar characteristics; thus, networks have similar characteristics. The loadings shown in Fig. 15b, c quantify the importance of each lipoprotein to explain the different pattern of correlation observed. Other examples of application of COVSCA to the analysis of sets of correlation matrices in

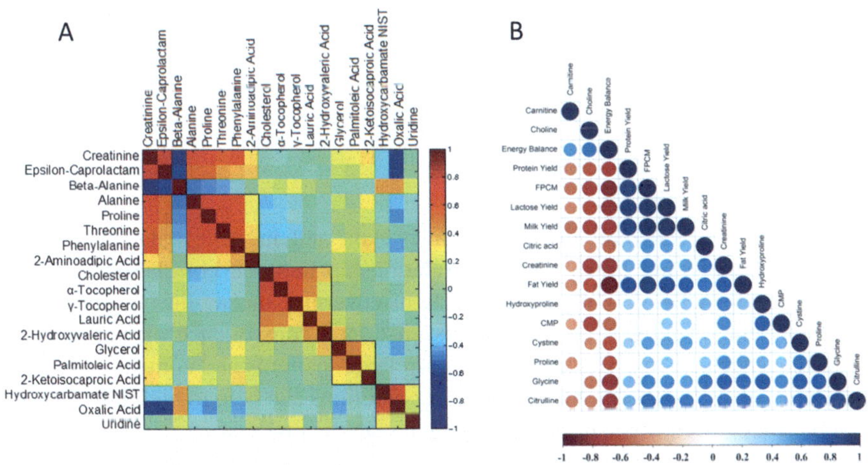

Fig. 14 Visualization of correlation matrices from metabolites correlations. When p metabolites measured the pairwise correlation are arranged in a $p \times p$, symmetric matrix and containing p^2 correlations values of which $p(p-1)/2$ are unique. (**a**) Classical visualization of the full correlation matrix. (Adapted from Trupp et al. [97] under the CC BY 4.0 license). (**b**) Visualization of the unique part of the correlation matrix calculated on metabolomic and in farm data. (Adapted from Xu et al. [111]. Reproduced from Balder et al. [6] under the CC BY 4.0 license)

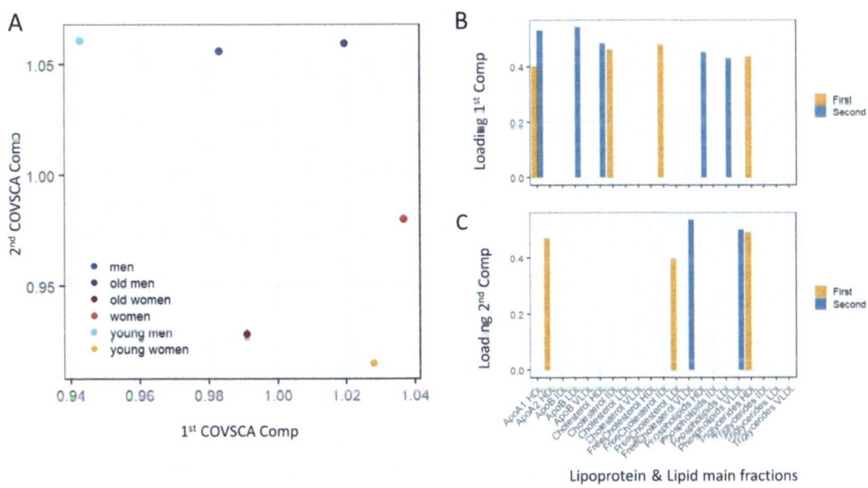

Fig. 15 Covariance Simultaneous Component Analysis (COVSCA) of the lipoprotein and lipid fractions correlation networks shown in Fig. 13. (**a**) COVSCA score plot: Each dot is a low-dimensional representation of lipid association network. (**b, c**) Loadings of the COVSCA model quantifying the importance of each lipoprotein and lipid fraction to describe the different correlation structure observed in the six correlation networks specific to each sex and age group. (Reproduced from Balder et al. [6] under the CC BY 4.0 license)

metabolomics can be found in Saccenti and Camacho [75], Smilde et al. [85], Dekker et al. [22], Vignoli et al. [102], and Saccenti [73].

8 Conclusions

Correlations are abundant in metabolomics data and in omic data, in general. Although there is not always a direct relationship between correlations and the structure of the network of metabolic reactions governing metabolite concentrations and their relationships, such correlations arise as a consequence of well-defined molecular mechanisms. The analysis of correlation patterns and comparison of the variability of such patterns across different conditions can help to highlight modifications and/or alterations of the underlying metabolic processes.

When estimating and analyzing correlations, it is necessary to carefully consider the type of correlation analysis that is most appropriate for the specific research question and dataset. As discussed in the article, the Pearson's correlation coefficient is most commonly used to measure linear relationships, while Spearman's and Kendall's correlations are better suited for non-linear, monotonic relationships. Therefore, researchers should carefully consider the nature of the relationship they expect to observe in their data before choosing a correlation analysis method. Another critical point is also to understand the potential for confounding factors, such as outliers and batch effects, and, when possible, correct for them.

Correlations are also exploited by many multivariate tools [77] and form the basis for the construction of associations networks [70]. For this reason, the analysis of correlations can provide intriguing biological hypotheses. However, there is a need to shift from using correlations as hypothesis-generating tools to using them as tools that can unlock the full potential of metabolomic data. This can only be achieved by a full understanding of the characteristics, advantages, and limitations of the different correlation indexes used to quantify metabolite associations and of the tools used to analyze and manipulate them within a systems biology context of which metabolomics is one of the most powerful and versatile tools [70].

Finally, it is important to recognize that correlation analysis alone cannot establish causality. While correlation can provide insight into potential relationships between metabolites, additional experimental evidence is necessary to establish a causal relationship. Therefore, researchers should carefully interpret their correlation results in the context of existing biological knowledge and other lines of evidence.

Acknowledgments Robert van Vorstenbosch is acknowledged for the fruitful discussion and comments on the content and presentation of this chapter. The writing of this chapter was partially supported by the financial contribution of the the Netherlands Organization for Health Research and Development (ZonMW) under the frame of ERA PerMed (Project 2018-151, PerMIT).

References

1. Aggarwal, C. C. 2017. An introduction to outlier analysis. *Outlier analysis*. Springer.
2. Alexander, R. A. 1990. A note on averaging correlations. *Bulletin of the Psychonomic Society,* 28, 335–336.
3. Armstrong, R. A. 2014. When to use the B onferroni correction. *Ophthalmic and Physiological Optics,* 34, 502–508.
4. Arndt, S., Turvey, C. & Andreasen, N. C. 1999. Correlating and predicting psychiatric symptom ratings: Spearmans r versus Kendalls tau correlation. *Journal of Psychiatric Research,* 33, 97–104.
5. Bakdash, J. Z. & Marusich, L. R. 2017. Repeated measures correlation. *Frontiers in Psychology,* 8, 456.
6. Balder, Y., Vignoli, A., Tenori, L., Luchinat, C. & Saccenti, E. 2021. Exploration of blood lipoprotein and lipid fraction profiles in healthy subjects through integrated univariate, multivariate, and network analysis reveals association of lipase activity and cholesterol esterification with sex and age. *Metabolites,* 11, 326.
7. Benjamini, Y. & Hochberg, Y. 1995. Controlling the False Discovery Rate: A Practical and Powerful Approach to Multiple Testing. *Journal of the Royal Statistical Society. Series B (Methodological),* 57, 289–300.
8. Bishara, A. J. & Hittner, J. B. 2012. Testing the significance of a correlation with nonnormal data: comparison of Pearson, Spearman, transformation, and resampling approaches. *Psychological Methods,* 17, 399.
9. Bishara, A. J. & Hittner, J. B. 2017. Confidence intervals for correlations when data are not normal. *Behavior Research Methods,* 49, 294–309.
10. Blitz, R. & Brittain, J. 1964. An extension of the Lorenz diagram to the correlation of two variables. *Metron,* 23, 137–143.
11. Bolbolian Ghalibaf, M. 2020. Relationship between kendall's tau correlation and mutual information. *Revista Colombiana de Estadística,* 43, 3–20.
12. Bonferroni, C. E. 1935. Il calcolo delle assicurazioni su gruppi di teste. *Studi in Onore del Professore Salvatore Ortu Carboni.* Rome: Bardi.
13. Borkowf, C. B. 2002. Computing the nonnull asymptotic variance and the asymptotic relative efficiency of Spearman's rank correlation. *Computational Statistics & Data Analysis,* 39, 271–286.
14. Box, G. E. & Cox, D. R. 1964. An analysis of transformations. *Journal of the Royal Statistical Society: Series B. (Methodological),* 26, 211–243.
15. Bravais, A. 1844. *Analyse mathématique sur les probabilités des erreurs de situation d'un point*, Impr. Royale.
16. Bro, R. & Smilde, A. K. 2014. Principal component analysis. *Analytical Methods,* 6, 2812–2831.
17. Broeckling, C. D., Huhman, D. V., Farag, M. A., Smith, J. T., May, G. D., Mendes, P., Dixon, R. A. & Sumner, L. W. 2005. Metabolic profiling of Medicago truncatula cell cultures reveals the effects of biotic and abiotic elicitors on metabolism. *Journal of Experimental Botany,* 56, 323–336.
18. Caldana, C., Degenkolbe, T., Cuadros-Inostroza, A., Klie, S., Sulpice, R., Leisse, A., Steinhauser, D., Fernie, A. R., Willmitzer, L. & Hannah, M. A. 2011. High-density kinetic analysis of the metabolomic and transcriptomic response of Arabidopsis to eight environmental conditions. *The Plant Journal,* 67, 869–884.
19. Calkins, D. S. 1974. Some effects of non-normal distribution shape on the magnitude of the Pearson product moment correlation coefficient. *Revista Interamericana de Psicología.*
20. Camacho, D., De La Fuente, A. & Mendes, P. 2005. The origin of correlations in metabolomics data. *Metabolomics,* 1, 53–63.
21. Cox, D. R. 1975. A note on data-splitting for the evaluation of significance levels. *Biometrika,* 62, 441–444.

22. Dekker, P. M., Azad, M. B., Boeren, S., Mandhane, P. J., Moraes, T. J., Simons, E., Subbarao, P., Turvey, S. E., Saccenti, E. & Hettinga, K. A. 2022. The human milk proteome and allergy of mother and child: Exploring associations with protein abundances and protein network connectivity. *Frontiers in Immunology,* 13.
23. Dettmer, K., Aronov, P. A. & Hammock, B. D. 2007. Mass spectrometry-based metabolomics. *Mass Spectrometry Reviews,* 26, 51–78.
24. Di Cesare, F., Luchinat, C., Tenori, L. & Saccenti, E. 2022. Age-and sex-dependent changes of free circulating blood metabolite and lipid abundances, correlations, and ratios. *The Journals of Gerontology: Series A,* 77, 918–926.
25. Dunnett, C. W. 1955. A multiple comparison procedure for comparing several treatments with a control. *Journal of the American Statistical Association,* 50, 1096–1121.
26. Edgell, S. E. & Noon, S. M. 1984. Effect of violation of normality on the t test of the correlation coefficient. *Psychological Bulletin,* 95, 576.
27. Emwas, A.-H., Szczepski, K., Poulson, B. G., Mckay, R., Tenori, L., Saccenti, E., Lachowicz, J. & Jaremko, M. 2022. Nuclear magnetic resonance in metabolomics. *Metabolomics Perspectives.* Elsevier.
28. Farlie, D. J. 1960. The performance of some correlation coefficients for a general bivariate distribution. *Biometrika,* 47, 307–323.
29. Fiehn, O. 2003. Metabolic networks of Cucurbita maxima phloem. *Phytochemistry,* 62, 875–886.
30. Fisher, R. A. 1915. Frequency distribution of the values of the correlation coefficient in samples from an indefinitely large population. *Biometrika,* 507–521.
31. Galton, F. 1889. I. Co-relations and their measurement, chiefly from anthropometric data. *Proceedings of the Royal Society of London,* 45, 135–145.
32. Genest, C. & Verret, F. 2005. Locally most powerful rank tests of independence for copula models. *Nonparametric Statistics,* 17, 521–539.
33. Gowda, G. & Djukovic, D. 2014. Overview of mass spectrometry-based metabolomics: opportunities and challenges. *Mass Spectrometry in Metabolomics,* 3–12.
34. Guerrero-Cusumano, J.-L. 1996a. An asymptotic test of independence for multivariate t and Cauchy random variables with applications. *Information sciences,* 92, 33–45.
35. Guerrero-Cusumano, J.-L. 1996b. A measure of total variability for the multivariate t distribution with applications to finance. *Information Sciences,* 92, 47–63.
36. Hastings Jr, C., Mosteller, F., Tukey, J. W. & Winsor, C. P. 1947. Low moments for small samples: a comparative study of order statistics. *The Annals of Mathematical Statistics,* 18, 413–426.
37. Havlicek, L. L. & Peterson, N. L. 1976. Robustness of the Pearson correlation against violations of assumptions. *Perceptual and Motor Skills,* 43, 1319–1334.
38. Heinrich, R. & Rapoport, T. A. 1974. A linear steady-state treatment of enzymatic chains: general properties, control and effector strength. *European Journal of Biochemistry,* 42, 89–95.
39. Hotelling, H. 1933. Analysis of a complex of statistical variables into principal components. *Journal of Educational Psychology,* 24, 417.
40. Iglewicz, B. & Hoaglin, D. C. 1993. *How to detect and handle outliers,* Asq Press.
41. Jahagirdar, S. & Saccenti, E. 2020. On the use of correlation and mi as a measure of metabolite—metabolite association for network differential connectivity analysis. *Metabolites,* 10, 171.
42. Jahagirdar, S., Suarez-Diez, M. & Saccenti, E. 2019. Simulation and reconstruction of metabolite-metabolite association networks using a metabolic dynamic model and correlation based-algorithms. *Journal of Proteome Research.*
43. Jolliffe, I. 2002. *Principal component analysis,* New York, Springer.
44. Jolliffe, I. T. 2012. Principal component analysis: a beginner's guide—I. Introduction and application. *Weather,* 45, 375–382.

45. Jolliffe, I. T. & Cadima, J. 2016. Principal component analysis: a review and recent developments. *Philosophical Transactions Series A, Mathematical, Physical, and Engineering Sciences,* 374, 20150202–20150202.
46. Kacser, H., Burns, J. A., Kacser, H. & Fell, D. 1995. The control of flux. *Biochemical Society Transactions,* 23, 341–366.
47. Kanehisa, M. & Goto, S. 2000. KEGG: kyoto encyclopedia of genes and genomes. *Nucleic Acids Research,* 28, 27–30.
48. Kanehisa, M., Goto, S., Sato, Y., Furumichi, M. & Tanabe, M. 2012. KEGG for integration and interpretation of large-scale molecular data sets. *Nucleic Acids Research,* 40, D109–D114.
49. Kendall, M. G. 1938. A new measure of rank correlation. *Biometrika,* 30, 81–93.
50. Kendall, M. G. 1948. *Rank correlation methods,* London, Gr.
51. Kowalski, C. J. 1972. On the effects of non-normality on the distribution of the sample product-moment correlation coefficient. *Journal of the Royal Statistical Society: Series C (Applied Statistics),* 21, 1–12.
52. Kullback, S. & Leibler, R. A. 1951. On information and sufficiency. *The Annals of Mathematical Statistics,* 22, 79–86.
53. Lee Rodgers, J. & Nicewander, W. A. 1988. Thirteen ways to look at the correlation coefficient. *The American Statistician,* 42, 59–66.
54. Lusczek, E., Lexcen, D., Witowski, N., Mulier, K. & Beilman, G. 2013. Urinary metabolic network analysis in trauma, hemorrhagic shock and resuscitation. Metabolomics, 9, 223–235.
55. Magnusson, P. K., Almqvist, C., Rahman, I., Ganna, A., Viktorin, A., Walum, H., Halldner, L., Lundström, S., Ullén, F. & Långström, N. 2013. The Swedish Twin Registry: establishment of a biobank and other recent developments. *Twin Research and Human Genetics,* 16, 317–329.
56. Martins, A., Camacho, D., Shuman, J., Sha, W., Mendes, P. & Shulaev, V. 2004. A systems biology study of two distinct growth phases of Saccharomyces cerevisiae cultures. *Current Genomics,* 5, 649–663.
57. Meyer, P. E. 2008. Information-theoretic variable selection and network inference from microarray data. *Universite Libre de Bruxelles.[Google Scholar].*
58. Monroe, W. S. & Stuit, D. B. 1933. The interpretation of the coefficient of correlation. *The Journal of Experimental Education,* 1, 186–203.
59. Moseley, H. N. 2013. Error analysis and propagation in metabolomics data analysis. *Computational and Structural Biotechnology Journal,* 4, e201301006.
60. Motulsky, H. 2014. *Intuitive biostatistics: a nonmathematical guide to statistical thinking,* Oxford University Press, USA.
61. Opgen-Rhein, R. & Strimmer, K. 2007. From correlation to causation networks: a simple approximate learning algorithm and its application to high-dimensional plant gene expression data. *BMC Systems Biology,* 1, 1–10.
62. Paninski, L. 2003. Estimation of entropy and mutual information. *Neural Computation,* 15, 1191–1253.
63. Pearson, K. 1895. VII. Note on regression and inheritance in the case of two parents. *Proceedings of the Royal Society of London,* 58, 240–242.
64. Pearson, K. 1901. On lines and planes of closest fit to systems of points in space. *The London, Edinburgh, and Dublin Philosophical Magazine and Journal of Science,* 2, 559–572.
65. Pearson, K. 1920. Notes on the history of correlation. *Biometrika,* 13, 25–45.
66. Pernet, C. R., Wilcox, R. & Rousselet, G. A. 2013. Robust correlation analyses: false positive and power validation using a new open source matlab toolbox. *Frontiers in Psychology,* 3, 606.
67. Pritchard, L. & Kell, D. B. 2002. Schemes of flux control in a model of Saccharomyces cerevisiae glycolysis. *European Journal of Biochemistry,* 269, 3894–3904.
68. Riaz, M., Munir, S. & Asghar, Z. 2014. On the performance evaluation of different measures of association. *Revista Colombiana de Estadística,* 37, 1–24.
69. Roessner, U., Luedemann, A., Brust, D., Fiehn, O., Linke, T., Willmitzer, L. & Fernie, A. R. 2001. Metabolic profiling allows comprehensive phenotyping of genetically or environmentally modified plant systems. *The Plant Cell,* 13, 11–29.

70. Rosato, A., Tenori, L., Cascante, M., De Atauri Carulla, P. R., Martins Dos Santos, V. a. P. & Saccenti, E. 2018. From correlation to causation: analysis of metabolomics data using systems biology approaches. *Metabolomics,* 14, 37.
71. Rousseeuw, P. J. & Bossche, W. V. D. 2018. Detecting deviating data cells. *Technometrics,* 60, 135–145.
72. Rubin, D., Dudoit, S. & Van Der Laan, M. 2006. A method to increase the power of multiple testing procedures through sample splitting. *Statistical Applications in Genetics and Molecular Biology,* 5.
73. Saccenti, E. 2016. Correlation Patterns in Experimental Data Are Affected by Normalization Procedures: Consequences for Data Analysis and Network Inference. *Journal of Proteome Research.*
74. Saccenti, E. 2023. What can go wrong when observations are not independently and identically distributed: A cautionary note on calculating correlations on combined data sets from different experiments or conditions. *Frontiers in Systems Biology,* 3.
75. Saccenti, E. & Camacho, J. 2020. Multivariate exploratory data analysis using component models. *Comprehensive foodomics.* Elsevier.
76. Saccenti, E., Hendriks, M. H. & Smilde, A. K. 2020. Corruption of the Pearson correlation coefficient by measurement error and its estimation, bias, and correction under different error models. *Scientific Reports,* 10, 1–19.
77. Saccenti, E., Hoefsloot, H. C., Smilde, A. K., Westerhuis, J. A. & Hendriks, M. M. 2014a. Reflections on univariate and multivariate analysis of metabolomics data. *Metabolomics,* 10, 361–374.
78. Saccenti, E., Suarez-Diez, M., Luchinat, C., Santucci, C. & Tenori, L. 2014b. Probabilistic Networks of Blood Metabolites in Healthy Subjects As Indicators of Latent Cardiovascular Risk. *Journal of Proteome Research,* 14, 1101–1111.
79. Sainani, K. L. 2014. Introduction to principal components analysis. *PM&R,* 6, 275–278.
80. Sakia, R. M. 1992. The Box-Cox transformation technique: a review. *Journal of the Royal Statistical Society: Series D (The Statistician),* 41, 169–178.
81. Schäfer, J. & Strimmer, K. 2005a. An empirical Bayes approach to inferring large-scale gene association networks. *Bioinformatics,* 21, 754–764.
82. Schäfer, J. & Strimmer, K. 2005b. A shrinkage approach to large-scale covariance matrix estimation and implications for functional genomics. *Statistical Applications in Genetics and Molecular Biology,* 4.
83. Schürmann, T. & Grassberger, P. 1996. Entropy estimation of symbol sequences. *Chaos: An Interdisciplinary Journal of Nonlinear Science,* 6, 414–427.
84. Silver, N. C. & Dunlap, W. P. 1987. Averaging correlation coefficients: should Fisher's z transformation be used? *Journal of Applied Psychology,* 72, 146.
85. Smilde, A. K., Timmerman, M. E., Saccenti, E., Jansen, J. J. & Hoefsloot, H. C. J. 2015. Covariances Simultaneous Component Analysis: a new method within a framework for modeling covariances. *Journal of Chemometrics,* 29, 277–288.
86. Smith, R. 2015. A mutual information approach to calculating nonlinearity. *Stat,* 4, 291–303.
87. Spearman, C. 1904. The proof and measurement of association between two things (2010). *International Journal of Epidemiology,* 39, 1137–1150.
88. Speed, T. 2011. A correlation for the 21st century. *Science,* 334, 1502–1503.
89. Steiger, J. H. 1980. Tests for comparing elements of a correlation matrix. *Psychological Bulletin,* 87, 245.
90. Steuer, R. 2006. Review: On the analysis and interpretation of correlations in metabolomic data. *Briefings in Bioinformatics,* 7, 151–158.
91. Stevens, V. L., Wang, Y., Carter, B. D., Gaudet, M. M. & Gapstur, S. M. 2018. Serum metabolomic profiles associated with postmenopausal hormone use. *Metabolomics,* 14, 1–14.
92. Storey, J. D. 2002. A direct approach to false discovery rates. *Journal of the Royal Statistical Society. Series B, Statistical Methodology*, 479–498.

93. Storey, J. D. & Tibshirani, R. 2003. Statistical significance for genomewide studies. *Proceedings of the National Academy of Sciences,* 100, 9440–9445.
94. Sun, H., Cui, Y., Wang, H., Liu, H. & Wang, T. 2020. Comparison of methods for the detection of outliers and associated biomarkers in mislabeled omics data. *BMC Bioinformatics,* 21, 1–23.
95. Teusink, B., Passarge, J., Reijenga, C. A., Esgalhado, E., Van Der Weijden, C. C., Schepper, M., Walsh, M. C., Bakker, B. M., Van Dam, K. & Westerhoff, H. V. 2000. Can yeast glycolysis be understood in terms of in vitro kinetics of the constituent enzymes? Testing biochemistry. *European Journal of Biochemistry,* 267, 5313–5329.
96. Tibshirani, R. J. & Efron, B. 1993. An introduction to the bootstrap. *Monographs on Statistics and Applied Probability,* 57, 1–436.
97. Trupp, M., Zhu, H., Wikoff, W. R., Baillie, R. A., Zeng, Z.-B., Karp, P. D., Fiehn, O., Krauss, R. M. & Kaddurah-Daouk, R. 2012. Metabolomics reveals amino acids contribute to variation in response to simvastatin treatment. *PloS One,* 7, e38386.
98. Van Den Berg, R. A., Hoefsloot, H. C. J., Westerhuis, J. A., Smilde, A. K. & Van Der Werf, M. J. 2006. Centering, scaling, and transformations: improving the biological information content of metabolomics data. *BMC Genomics,* 7, 142.
99. Van Den Heuvel, E. & Zhan, Z. 2022. Myths about linear and monotonic associations: Pearson's r, Spearman's ρ, and Kendall's τ. *The American Statistician,* 76, 44–52.
100. Ventura-León, J., Pena-Calero, B. N. & Burga-León, A. 2022. The effect of normality and outliers on bivariate correlation coefficients in psychology: A Monte Carlo simulation. *The Journal of General Psychology*, 1–18.
101. Vignoli, A., Ghini, V., Meoni, G., Licari, C., Takis, P. G., Tenori, L., Turano, P. & Luchinat, C. 2019. High-throughput metabolomics by 1D NMR. *Angewandte Chemie International Edition,* 58, 968–994.
102. Vignoli, A., Tenori, L., Giusti, B., Valente, S., Carrabba, N., Balzi, D., Barchielli, A., Marchionni, N., Gensini, G. F., Marcucci, R., Gori, A. M., Luchinat, C. & Saccenti, E. 2020. Differential Network Analysis Reveals Metabolic Determinants Associated with Mortality in Acute Myocardial Infarction Patients and Suggests Potential Mechanisms Underlying Different Clinical Scores Used To Predict Death. *Journal of Proteome Research.*
103. Walach, J., Filzmoser, P., Kouřil, Š., Friedecký, D. & Adam, T. 2020. Cellwise outlier detection and biomarker identification in metabolomics based on pairwise log ratios. *Journal of Chemometrics,* 34, e3182.
104. Weckwerth, W., Loureiro, M. E., Wenzel, K. & Fiehn, O. 2004. Differential metabolic networks unravel the effects of silent plant phenotypes. *Proceedings of the National Academy of Sciences,* 101, 7809–7814.
105. White, T., Van Der Ende, J. & Nichols, T. E. 2019. Beyond Bonferroni revisited: concerns over inflated false positive research findings in the fields of conservation genetics, biology, and medicine. *Conservation Genetics,* 20, 927–937.
106. Wilcox, R. 2015. Inferences about the skipped correlation coefficient: Dealing with heteroscedasticity and non-normality. *Journal of Modern Applied Statistical Methods,* 14, 4
107. Wilcox, R. R. 1993. Some results on a Winsorized correlation coefficient. *British Journal of Mathematical and Statistical Psychology,* 46, 339–349.
108. Wilcox, R. R. 2009. Comparing Pearson correlations: Dealing with heteroscedasticity and nonnormality. *Communications in Statistics-Simulation and Computation,* 38, 2220–2234.
109. Wilcox, R. R. 2011. *Introduction to robust estimation and hypothesis testing*, Academic Press.
110. Wilcox, R. R. 2016. Comparing dependent robust correlations. *British Journal of Mathematical and Statistical Psychology,* 69, 215–224.
111. Xu, W., Vervoort, J., Saccenti, E., Van Hoeij, R., Kemp, B. & Van Knegsel, A. 2018. Milk metabolomics data reveal the energy balance of individual dairy cows in early lactation. *Scientific Reports,* 8, 1–11.
112. Yitzhaki, S. 2003. Gini's mean difference: A superior measure of variability for non-normal distributions. *Metron,* 61, 285–316.

113. Yule, G. U. 1926. Why do we sometimes get nonsense-correlations between Time-Series?--a study in sampling and the nature of time-series. *Journal of the Royal Statistical Society,* 89, 1–63.
114. Zimmerman, D. W., Zumbo, B. D. & Williams, R. H. 2003. Bias in estimation and hypothesis testing of correlation. *Psicológica,* 24.
115. Zou, G. Y. 2007. Toward using confidence intervals to compare correlations. *Psychological Methods,* 12, 399.

Metabolomics Approach to Identify Biomarkers of Epidemic Diseases

Pooja Rani Mina

Abbreviations

ACE2	Angiotensin-converting enzyme 2
AIDS	AIDS
ATD-GC–MS	Automated Thermal Desorption Gas Chromatography Mass Spectrometry
AUC	Area under curve
BCG	Bacille Calmette-Guerin
CD	Cluster of differentiation
CD	Crohn's Disease
CE-MS	Capillary electrophoresis-mass spectrometry
COVID19	Corona virus disease
DHEAS	Dehydroepiandrosterone sulfate
FIA	Flow injection analysis
FMO3	Flavin-containing monooxygenase 3
FTICR-MS	Fourier-transform ion cyclotron resonance mass spectrometry
FTIR	Fourier transform infrared
GABA	Gamma-aminobutyric acid
GC-MS	Gas chromatography mass spectrometry
HIF1α	hypoxia inducible factor 1 subunit alpha
HIV	Human immunodeficiency virus
IL	Interleukin
iRBCs	Infected Red Blood Cell
LC	Liquid chromatography
LS-SVM	Least-squares support-vector machines

P. R. Mina (✉)
Department of Immunology, Mayo Clinic, Scottsdale, AZ, USA
e-mail: mina.poojarani@mayo.edu

V. Soni, T. E. Hartman (eds.), *Metabolomics*,
https://doi.org/10.1007/978-3-031-39094-4_4

LTB4	Leukotriene B4
m/z	Mass/charge
mRNA	Messenger ribonucleic acid
MS	Mass spectrometry
NAD	Nicotinamide adenine dinucleotide
NF-κB	Nuclear factor kappa-light-chain-enhancer of activated B cells
NMR	Nuclear magnetic resonance
OPLS-DA	Orthogonal Projections to Latent Structures Discriminant Analysis
PC	Phosphatidyl choline
PCR	Polymerase chain reaction
PG	1-palmitoyl-2-oleoyl-sn-glycero-3-phosphoglycerol
PUFA	Polyunsaturated fatty acids
Q-TRAP	Quadropule Ion Trap
ROC	Receiver operating characteristic
SARS	Severe Acute Respiratory Syndrome
SCD1, ROS VEGF, ATP, AMPK, MAPK, EGFR, UCP2 and PLA2	
SM	Sphingomyelin
SPME GC–MS	Solid-phase microextraction followed by gas chromatography-mass spectrometry method
TB	Tuberculosis
TBSA	Tuberculostearic acid
TCA	Tricarboxylic acid
TG	Triglyeceride
TLR	Toll like receptor
TMAO	Trimethylamine N-oxide
UC	Ulcerative colitis
UHPLC-HRMS	Ultra-High-Performance Liquid Chromatography-High-Resolution Accurate Masses Spectrometry
VEGF	Vascular endothelial growth factor
WHO	World Health Organization

1 Introduction

Diseases characterised by a gradual or sudden increase in cases within certain global areas for a short time period are referred to epidemics. Epidemic diseases could be infectious or not, but most of them concern with infectious disease spread. Epidemic diseases are restricted to locations of certain countries or can spread to other areas or different countries and continents too, if uncontrolled. For an instance, the COVID-19 epidemic, which is highly spreadable (2019 to October 2022 6.2 billion cases recorded), was declared a pandemic, which have been recorded higher than any previously reported infectious disease [1]. Each disease encounter in a different illness phases initial mild infection to acute and chronic by the time. If a disease is can be identifiedy at an initial stage, its severity can be managed by initial medication. Hence some biological molecule that could predict the disease onset condition of a subject always remain to interest for researchers. Disease phases can be distinguished well by either certain disease specific gene or protein or metabolite changes. The omics approach provided large advanced technologies for analysing high-throughput biological molecules in a cost effective manner [2]. System biology approaches (i.e. proteomics, metabolomics and transcriptomics) are useful in finding the biomarkers in the prediction of predisease, acute and chronic diseases changes at the mRNA, protein and metabolite level, including susceptible and tolerant individual identification [3]. Multiomics researchers are moving now from genomics to phenomics studies. Proteome and metabolome of cells or tissues are two main components of phenome and are equally important for scientific discovery of disease-based biomarkers. Metabolome study includes characterisation of metabolites of all the low-molecular weight molecules (typically, 3000 m/z) present in cells in a particular physiological or developmental state. Metabolites are the final products in a biological system and intermediate molecules in pathways of lipids, nucleotides, sugars, and amino acids that govern and regulate several essential biochemical pathways, either directly or indirectly [4]. Multiple factors like genetics, the external and internal environment, drugs, and diet are associated with metabolome variation. Metabolites play important roles in signalling molecules, immune modulators, endogenous toxins, and environmental sensors. These metabolomics variations could create inaccuracy in width and depth of metabolomics. Along with studying metabolome, defining their accuracy is also a challenge, which requires a highly sensitive and reproducible methodology to maximize coverage of metabolomics variations. It is remarkable to uncover changes in metabolites at different disease stages in the blood, stool and tissue with their identification and validate for disease progression biomarkers (Fig. 1), which render crucial prognostic and diagnostic tool investigating toward impact of disease [5], along with data accuracy to imply at clinical level. Epidemiological studies have also been served with applied metabolomics, including different kind of study design. The majority of epidemic diseases have been case control or cross-sectional studies that allow comparison of metabolomics phenotype of an individual to gain exposure to potential metabolite change. Prospective studies require a large follow up time upon large sample size, however providing biomarker identification opportunity sensitive toward disease, predecease phase elaborating disease etiology [6]. Presently, studies that comprise efficient nested-designs case-cohort are increasing, providing epidemiological studies using metabolomics data. Advanced development of metabolomics technologies for a large set data study, analytical platform and data refining tools has made metabolomics studies possible [7]. There is no distrust considering application of metabolomics to growing epidemiological studies to better characterize exposures, markers to detect early disease, improve diagnosis of disease, tracking response towards treatment or disease progression and make disease etiology understandable [8, 9]. There are approaches for metabolomics, including nuclear magnetic resonance (NMR) spectroscopy, mass spectrometry (MS) and Fourier-transform infrared (FTIR) spectroscopy, to achieve metabolic finger prints, foot prints and target

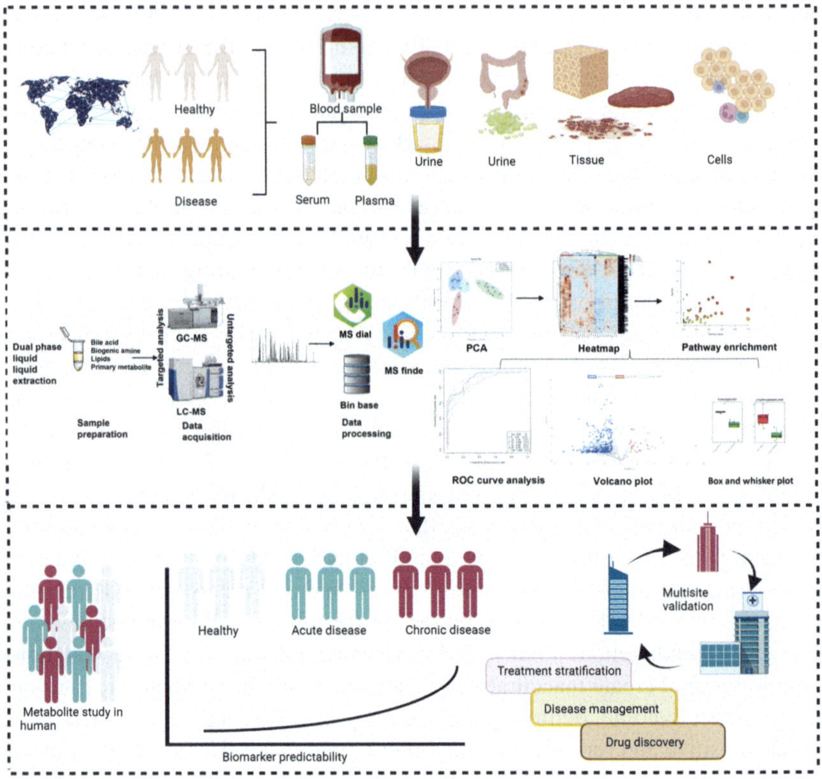

Fig. 1 Depicting the method for analysis of biological samples, blood (plasma, serum), urine, stool, cells and tissue from organs, for metabolomics analysis. Extracted sample is then subjected to different types of suitable platforms based on the chemical class of metabolites. Raw MS data is statistically analysed, followed by target validation and biomarker discovery (Biorender.com licenced)

analys. Metabolomics has proven to be valuable for new drug discovery targets, and diagnostic biomarker discovery like in the differences in host response observed in vitro and in vivo in energy producing pathways (glycolytic, kreb's cycle) and inflammatory pathways [10]. Bacteria-infected (*Streptococcus pneumoniae* and *Staphylococcus aureus* pneumonia) mouse model showed distinct metabolic pattern between infected and non-infected mice while using different biological samples [11]. Hence, infection progression is evident from publication, including fungal [12], parasitic [13, 14], protist [15] and viral infections [16]. For non-communicable diseases (NCD), high degree of interconnectivity with cardiovascular and type 2 diabetes has been shown. A meta-analysis study which compared CD64 and IL-6 in adult sepsis patient predict as a biomarker for bacterial infection along with highest diagnostic value [17]. Multifactorial diseases could not be addressed by diet promotion or physical activities, because lifestyle intervention requires wide-ranging evaluation of an individual's diagnostic biomarkers. This chapter will summarise yielding of metabolites in a number of epidemics categorised into communicable and non-communicable diseases. We elaborate how metabolic modelling, together with the integration of metabolomics and other multiomics datasets, can be used as a tool to understand cellular metabolism.

2 Top Spread Epidemics

For international agencies and national government organisation, public health is a major concern which has been affecting their financial growth. According to the different timelines of the last few decades, several new pathogens have adapted to current therapies and come with more severity. Cholera, plague, and yellow fever resurgence accompanied with immunity change highly affected by biological and environmental changes in the current lifestyle [18]. These changes have enabled epidemics to spread and faster than before, due to decreased immune response to disease phenotype. The recent wreaking havoc of COVID-19 is a supportive example, which caused 3.3 million deaths and affected everyday lives of people globally. To control the increase in cases of global diseases, increased efforts are efficiently made for controlling infectious diseases and designing new mitigation strategies. For viral diseases like SARS in 2003, H1N1 influenza in 2005, Ebola epidemic in 2019 along with corona virus disease in 2020, accelerate disease outbreak was accelerated by increased social connections [19]. This increased disease burden brought global crisis to national economy, disaster to people lives and colossal turmoil to international community [20].

3 Communicable

Communicable diseases are contagious diseases transmitted from one person to another in a short time span by contact with body fluid, blood, airborne virus particles or insect bites [21]. Reporting of disease is important for evaluation of disease prevention and control for assuring appropriate medical therapy in the locality of the outbreak. Tuberculosis, polio, small pox, measles, hepatitis and diphtheria are some of the human diseases having adverse effects in the form of substantial mortality and morbidity [22]. The media by which diseases spread are the fecal-oral route, intercourse, skin contact, contaminated fomites, insect bites (ref), etc.. For example, hepatitis is spread through the oral fecal route, where a person is exposed to contaminated water or blood product or pre-infected food consumption. Some of the communicable diseases require serious attention of health departments. The overarching goal of metabolomics is to assess metabolic changes quantitatively and qualitatively for their diagnostic, therapeutic and prognostic potentials.

3.1 COVID-19

COVID-19 was reported first to WHO on 31 Dec 2019; it caused a major global crisis with economic burden to several countries. From 2019 to July 2022, 567 million cases and 6.3M deaths reported globally, which highest data has been recorded [23]. This disease ranges in severity, from asymptomatic to critical illness and mortality. Pre-existing conditions like diabetes, asthma, obesity and heart disorders increase disease complications. All these disorders make an individual disease-prone because of pre-altered metabolic pathways [24]. Quantitative PCR (polymerase chain reaction) was the only successful technique for qualitative detection of COVID. Now, metabolic profiling has become a diagnostic and prognostic tool critical for future epidemics [25]. COVID-19-positive patient data showed elevated levels of kynurenine and tryptophan [26]. Ratio of arginine/ornithine got elevated, while glutamine/glutamate ratios has been seen to be reduced and increased glutamic acid levels in COVID-19 patients, indicative of severity scaling, triglycerides (TG 20:1_32:3) and TG (22:4_32:2) were consistent with initial 2 waves [27]. In COVID-19 patients, bile acid level was also depleted compared to healthy control. Studies analysed markers associated with different waves of covid and showed elevation of glutamine in wave 1. Elevated methionine sulfoxide (Met-SO) in wave 2 was not common in wave 1. Level of cortisol, the stress chemical, was significantly lower in wave 2 compared to wave1. A set of metabolites including glycolithocholic acid (GLCA), glutamic acid (Glu), aspartic acid (Asp) TG (22:1_32:5), TG (18:0_36:3), and methionine sulfoxide (Met-SO), were able to distinguish healthy vs infected wave 1 and 2 patients. (TG (18:0_36:3 and aspartic acid could be used as healthy and disease state markers as they recover with return to a health state. Metabolites changed differentially across time and phase of infection but are still able to distinguish covid disease group. SARS-CoV-2 patients showed acylcarnitines accumulation, which came to be decreased in recovery phase [28]. Lineloic acid could reduce the interaction of spike protein with angiotensin-converting enzyme receptor (ACE2), which hydrolyses angiotensin II (a vasoconstrictor peptide) into angiotensin [29]. A set of metabolites ratio is introduced for rapid routine practice; glutamine/glutamate ratio, kynurenine/ tryptophan ratio (valine + leucine + isoleucine)/(phenylalanine + tyrosine), also known as BCAA/aromatic amino acids ratio [30]. Some of the markers could be used to identify disease severity and recovered infection state, but they need to be precise based on the time span of the infection. Some metabolites could be utilised as distinct biomarker.

3.2 Tuberculosis

In 2021 WHO reports 9.9M new cases of *Mycobacterium tuberculosis (M.tb)* and 1.5 million deaths [31]. The death rate of *M.tb* is higher in Asian countries. A rapid true diagnostic test plays an important role in identifying the infection at acute level, as it's hard to control chronic stage infection. Currenty, few tests are widely used, such as purified protein derivative (PPD), *M.tb* nucleic acid test and Xpert/MTB system. These tests are old and can distinguish positive and negative phenotypes. More studies on metabolites are required due to increased resistance reports that could also affect host metabolic pathways that are quite different to sensitive strain of TB. Metabolites could distinguish different species and infection type specific to mycobacterium and other genus. Inositol and myo-inositol characterise the Mycobacterium species *M. tuberculosis, M. avium, M. bovis, M. kansasiiand P. aeruginosa.* Indole-acetic acid, cadaverine, purine and putrescine were detected exclusively in *P. aeruginosa* [32]. Succinic acid was uniquely identified in *M. kansasii.* Sputum is the commonly used matrix for diagnosing pulmonary TB; viscosity and uneven consistency of sputum bring about the need for additional, time-consuming sample pre-processing steps before metabolomics analyses. Plasma/serum and urine analysis require less effort in sample preparation. Human subject metabolites research showed that The OPLS-DA models showed two groups were able to clearly separate between comparison groups. Results showed a set of upregulated and downregulated metabolites. Phosphatidylethanolamine (PE), sphingomyelin (SM), phosphatidylcholine (PC) and ceramide (Cer) upregulated at higher extent compared to other group metabolites [33]. Metabolites that are of interest for development as biomarker are those which are elevated and have more chances of true positive result rather than result by error. Amino acid, ceramides and fatty acids downregulated in patient [34]. ROC curve analysis of metabolites specific to TB at 95% CI value showed clear sensitivity and specificity in diagnostic test. Based on clinical practicability and feasibility metabolites specifically increased in osteoarticular tuberculosis are PC(o-16:1 (9Z)/18:0), PC(20:4(8Z,11Z,14Z,17Z)/18:0), PC(18:0/20:3 (5Z,8Z,11Z)), PC(18:0/22:5(4Z,7Z,10Z,13Z,16Z)), SM(d18:1/ 20:0), SM(d18:1/24:1(15Z)), SM(d18:0/16:1(9Z)), SM(d18:0/ 18:1(11Z)), and SM(d18:1/18:1(11Z)) which have an AUC of 0.7, indicating better diagnostic performance at clinical level. Combining AUC and ROC curve data, PC (o-16:1(9Z)/18:0), PC (20:4). (8Z,11Z,14Z,17Z)/18:0), PC(18:0/22:5(4Z,7Z,10Z,13Z,16Z)), SM (d18:1/20:0), and SM(d18:1/18:1(11Z)) may be potentially relevant metabolic biomarkers for the diagnosis of osteoarticular tuberculosis [35]. Two diagnostic models were established (Models A and B), and AUC values of these two models were 0.8820 and 0.7940, respectively. Metabolites specific to *M.tb* infection were involved in necroptosis, retrograde endocannabinoid signaling, sphingolipid metabolism, choline metabolism, sphingolipid signaling and glycerophospholipid metabolism and are maybe the main pathways. In most cases, these metabolites are related to lipid accumulation and obesity in the body. About ¼ th of the population has latent TB, and less than 10% develop active in their life [36]. A rapid test that could detect latent to active TB stage could save many lives with a pre-treatment stage. A set of metabolite including valine, hypoxantine, inosine, creatine, citrate fucose and fructose was significantly elevated, but less abundant in patients not parallel with TB patient. In active TB fatty acids, amino acids and lipids were identified as biomarkers, (lysophosphatidylcholine (18:0), behenic acid, threoninyl-γ-glutamate and presqualene diphosphate) allowed for discrimination between active TB and control samples with an AUC value of 0.991 [37]. Increased lactate in patient sera correspond to high anaerobic glycolysis, which could be an index of tissue hypoxia and an extent of necrosis as the infection progresses [38]. A logistic regression model between patients with active *M.tb* [PG (16:0_18:1), Lyso-PI (18:0) and *m/z* 1321.9177] provided excellent classification accuracy of active TB cases [39]. The ROC curve demonstrated an area under the curve (AUC) of 0.97 (95% CI 0.93–1). The AUC was reduced to 0.94 when using only phosphatidylglycerol (PG 16:0_18:1) and Lyso-PI (18:0) and 0.82 when PG16:0_18:1 alone. Host generate hydrogen peroxide to eliminate infection, in clinical practice elevated hydrogen peroxide level could be used for diagnosis. Metabolites related to increased oxidative stress and glutaminolysis in lung lesions, like Aspartate, glutathione, betaine and trim ethylamine N-oxide, are in the easy detectable range [40]. Elevated levels of isopropyl acetate and o-xylene, accompanied by reduced levels of 3-pentanol, dimethyl styrene and cymol in urine sample discriminate the TB-positive patient from healthy controls and from patients with other lung associated disorders [41].

3.3 *Malaria*

Malaria is a serious global health problem affecting 40% of the global population every year. Malaria has evolved as a colossal infectious killer posing health challenges for many developing countries, affecting socioeconomic growth. Malaria affects about 229 million people globally, with around 409,000 deaths annually [42], most of which comprise children below 5 years of age. *P. falciparum* is the deadliest among the *Plasmodium* species that are known to cause malaria in humans; hence, combined global efforts are needed for its prevention and control [43]. The increasing burden of malaria and longtime ineffectiveness of current drugs pose challenges in malaria elimination programmes [44]. In a human body, *Plasmodium* can cause asymptomatic to severe disease spectrum and, hence, have an interconnection between hosts. Parasite share nutrient and signalling molecules, from host, cause perturbations in the amino acids, lipids, fatty acids, sugars and heme metabolites [45]. Arginine, tryptophan and glutamine have attention as these metabolites successively decrease in infected patients, coming from reduced nitric oxide, endothelial disruption, and vasodilation [46]. Diminished levels of arginine and its biosynthetic pathway metabolites (e.g. ornithine and citrulline) also affect the blood of malaria-infected host. *P. falciparum*-infected hosts have elevated lactate in bloodstream, causing a condition called metabolic acidosis [47], and glycolysis metabolites perturbed in the bloodstream. Glutamine levels are depleted in the human host in both falciparum and vivax malaria, which cause severe malarial anaemia in children. Glutamine is a ‘double-edged sword’ in malaria pathogenesis, because of its opposing disease manifestation effect. In cerebral malaria production of neurotoxic metabolites (e.g., quinolinic and kynurenic acid) from conversion of tryptophan to kynurenine are elevated and a decrease in indolepropionate which is derived from tryptophan [48]. Metabolite corresponds to beta oxidation of fatty acids in mitochondria as a means of energy production. Host derived metabolites are of quite importance while a bunch of metabolites also derived from *Plasmodium* species. Pipecolic acid is detectable in in vitro *P. falciparum* cultures, murine malaria models, and humans with *P. falciparum*, but not in healthy uninfected control [49]. Metabolites of alpha-linoleic acid pathway, which are common to plants, are found in *P. falciparum* culture and plasma of infected patient as well [50]. During *Plasmodium* infection, microbes from gut also translocate from gut barrier and break into the bloodstream. Elevated diaminopimelic acid was observed concomitant with depletion in L-citrulline [51]. Rapid diagnostic tests that include both *Plasmodium* infection markers (elevated pipecolic acid and pinene) and disease severity markers (e.g. depleted arginine, glutamine, and citrulline) could have diagnostic and prognostic benefit.

3.4 Dengue

Dengue is a mosquito-spread disease caused by a virus from family Flaviviridae [52]. Dengue has different antigenicity 4 serotype in humans which infect 390 million people worldwide every year [53], along with a billion people at risk. It cause sudden fall in platelets and serious bleeding in the jaw and other tissue and blood pressure shock, causing death. Its affects the liver, reported by change in liver enzymes like aspartate aminotransferase (AST) and Alanine transaminase (ALT), which can be higher during the critical phase [54]. Kynurenine metabolite from liver elevated in early febrile phase and was higher in hemmoregic dengue fever compared to fever caused by dengue [55]. Kynurenine could be a distinguishing marker, suggesting its potential as a predictor of severe dengue. In the early stage of dengue fever phenylalanine levels increased, for which Tetrahydrobiopterin (BH4) works as a cofactor, which further changed to tyrosine [56]. This metabolic change aggravated into elevated oxidative stress in dengue fever patient. Acylcarnitines are an intermediate of fatty acid β-oxidation (FAO) regulation suppressed during infection or inflammation [57]. Polyunsaturated fatty acids are released from cell membrane of infected dengue patients, representing elevated arachidonic acid, linoleic acid, docosahexaenoic acid and α-linoleic acid [58]. Another human study on urine sample highlighted elevated acetoacetic acid, valerylglycine, fructose, 4-hydroxyphenylpyruvic acid, Ssulfocysteine and betaine compared to healthy individuals [59]. Majority of dysregulated metabolites are involved in the tricarboxylic acid intermediates cycle and β-oxidation, amino acid metabolism, fatty acids related to dengue infection [60]. In same study creatinine, myo-inositol, creatine, succinic acid, citrate and 3-hydroxy-3- methylglutarate observed to be depleted (with significance of $p < 0.001$). A serum metabolite study on humanized mice having dengue infection showed deregulated acylcarnitines, phosphatidylcholines, lysophosphatidylcholines, phosphatidylethanolamines, acylglycines, sphingomyelins, and bile acids, which all showed a parallel increse like dengue patients at the early or acute stage of DENV infection and gradually returned to the control levels at the late stage [61].

3.5 *Typhoid*

Salmonella infection results from consumption of contaminated food and consequent *E. coli* infection. Gastroenteritis, septicaemia and enteric fever are the major symptoms caused by Salmonella in humans as well as in animals. Humans and other chordate cattle, such as pig, showed delayed clearance due to increased resistance to different classes of antibiotics (ampicillin, chloramphenicol, sulfamethoxazole and tetracycline). Highlighting metabolic pathways differences in antibiotic sensitive and resistance strains, along with scientific evidence of metabolite expression from various hosts. An untargeted metabolomics approach identified distinctive biomarker from *S. Typhimurium*, which was variable among host species. *A* significant change in methionine, pantothenate, nicotinamide, pyroglutamic acid, nicotinate, phenylalanine, proline, pyruvate, serine, threonine, tryptophan, tyrosine, uracil and valine and decreases in alanine, aspartate, citrate, cysteine, glutamate, glycerate observed in patient [62]. Majority of affected pathways include glycine, serine, and threonine metabolism; alanine, aspartate and glutamate metabolism; amino acyltRNA biosynthesis; pantothenate and CoA biosynthesis. A study found that pathways which enriched in drug resistant strain confer to amino acid biosynthesis, phenyl propanoid and nucleotide metabolism and concurred antibiotic resistance in *Salmonella* [63]. Increased expression of aspartate is critical for β-lactamases [64]. Increased expression of serine and the decreased expression of glutamate are also critical in resistance development. Exogenous glucose and alanine increase susceptibility to antibiotic treatment by increasing TCA flux and thereby increasing drug uptake by the cell [65]. Decreased concentration of glutamate and pyruvate could be a lead marker for developing diagnostic biomarker for sensitive and resistance strain differentiation, which are involved into tricarboxylic acid cycle [66]. Coinfection of *Salmonella Typhi* and *Salmonella Paratyphi* causes chronic carriage in the gallbladder formed by biofilm formation [67]. A study conducted to discriminate Salmonella carriage samples from non-carriage control samples. OPLS-DA model obtained when comparing S. Typhi and *S. Paratyphi* A carriage samples resulted in a glutaric acid and caproic acid ROC curve with an AUC value of 0.833 [68]. Two different cohort studies of Nepal and Bangladesh showed 24 metabolites which were able to potentially to identify typhoid fever patients.Phospholipid biosynthesis precursor glycerol-3-phosphate, liposome component stearic acid and linoleic acid, creatinine and pyruvic acid [69] . Furthermore, leucine and phenylalanine were consistently up- or downregulated between all collections (Table 1).

Table 1 List of reported metabolic biomarker for communicable diseases

Disease	Pathogen	Metabolite/pathway	Fluid origin	Analytical platform	Ref.
COVID-19	SARS-CoV-2	Lactic acid, trigonelline, xanthine, β-alanine, arachidonic acid, aspartate, C18:1, C5:1, choline, ornithine, phenylalanine, serine serotonin, Succinic acid, hippuric acid, deoxycholic acid, DHEAS, glutamate,	Serum	Triple Quad™ 5500 + System–QTRAP	[70]
		L-kynurenine and L-tryptophan, nucleotide metabolism, energy metabolism and nitrogen metabolism	Plasma	CE-MS	[71]
		Phenylalanine and omega-6 fatty acids, choline, increased triglyceride	Blood	NMR	[72]
Tuberculosis	*M.tb*	1-methyl-naphthalene 1,4-dimethyl-cyclohexane	Culture	ATD-GC–MS	[73]
	M.tb and *M. avium* complex *M. tuberculosis, M. avium, M. bovis and M. kansasii*	Methyl nicotinate	Culture	SPME GC–MS	[74]
		15-tetracosenoic acid, 9-hexadecenoic acid; 10-heptadecenoic acid; heptadecanoic acid; TBSA;11-eicosanoic acid; 13-docosenoic acid;	Culture	GC–MS	[32]
	6 × TB- spiked with *M. tuberculosis* 6 × TB- (not spiked)	Eicosanoic acid; D-glucopyranoside, d-glucose; heptadecanoic acid; TBSA; myo-inositol; d-mannose, d-glucosamine; uridine, cadaverine, l-threonine, docosanoic acid; 9-octadecenoic acid; Nonadecanoic acid;	Sputum	Cx GC-TOF-MS	[75]
Malaria	*P. falciparum*	O-arachidonoyl glycidol, 3-methylindole and succinylacetone, S-methyl-L-thiocitrullin	Culture	LC-MS	[76]
		Pipecolic acid, taurine, N-acetylspermidine, N-acetylputrescine and 1 3-diacetylpropane	Urine	HPLC-HRMS	[77]
		formate, urea, trimethylamine, threonine, choline, myo-inositol and acetate, pyruvate, 3-methylhistidine and dimethylglycine	Urine	NMR	
		Pipecolic acid, GABA, alpha ketoglutarate and NAD	iRBC	NMR	[78]
Typhoid	S. *typhi*	Glycerol-3-phosphat,e stearic acid, and linoleic acid, pyruvic acid and creatinine	Plasma & urine	GCxGC-TOFMS)	[68]

Table 1 (continued)

Disease	Pathogen	Metabolite/pathway	Fluid origin	Analytical platform	Ref.
AIDS	HIV	PS (O-18:0/0:0), sphingosine, PE (21:0/0:0) and 1-Linoleoyl Glycerol. sphingosine and 1-Linoleoyl Glycerol	Plasma	LC-MS/MS	[79]
		Acylcarnitine, sphingomyelin	Plasma	Tandem mass spectrometry	[80]
Pneumonia	*Streptococcus pneumoniae*	Phosphatidic acid, N-acetyl-neuraminic acid, lysophosphatidylcholines (LPC (12:0/0:0), dimethyldisulfide, choline, pyrimidine and oleic acid	Bronchoalveolar lavage fluid	LC-MS/MS	[81]
	Pseudomonas aeruginosa	Histidine, nicotine, uric acid, hypoxanthine and glutamic acid	Urine	LC-MS	[82]
		Uric acid, hypoxanthine and glutamic acid	Plasma	UPLC-TOFMS	[83]
		L-tryptophan and adenosine-59-diphosphate (ADP), uric acid and L-histidine	Urine		
		Methylcysteine, L-Phenylalanine, 2-Hydroxy-3-methylbutyric acid, L-Alanine, 2-Hydroxybutyric acid, Palmitoleic acid, Decanoylcarnitine, Aminoadipic acid, L-Tryptophan, Rhamnose and Oxoglutaric acid	Serum	(GC-TOF/MS	[84]

4 Non-communicable Diseases

Unlike communicable disease, non-communicable diseases are not mediated by any external agent/pathogen or transmitted from one person to another. Non-communicable diseases are chronic disease that reside in human body for a longer duration and take time to eliminate. Non-communicable diseases are the consequences of genetic, environmental, behavioural and physiological disturbance. NCDs such as diabetes, heart disease, cancer, stroke and lung disease contribute to 74% mortality worldwide. Seventy percent of deaths due to NCDs, responsible for premature death at the age of 70s occurs in low-/middle-income countries. The field of non-communicable diseases is immense. The major causes of NCD-attributable mortality are cardiovascular diseases (30% of total global mortality), cancers (13%), chronic respiratory disease (7%) and diabetes (2%). Metabolic factors contribute to risks associated with metabolite change and can be responsible for obesity, high blood pressure, hyperglycaemia and hyperlipidaemia. Some non-communicable diseases are aging disorders, such as Parkinson's, Alzheimer's, arthriti and, other neurological disorders.

4.1 *Chronic Respiratory Diseases*

Chronic respiratory diseases (CRDs) are the greatest risk to the population with the global estimated cases, 262 Million people suffer with CRD and asthama is more common in children (WHO 2021). This lies in both communicable and noncommunicable category and is leading causes of mortality, morbidity for causing economic and societal burden to society. CRDs include infection in different respiratory system of airways, pulmonary vasculature and parenchyma. Based on the mode of transmission and disease pathology CRD categorised broadly in communicable (e.g. tuberculosis and pneumonia) and non-communicable (asthma, interstitial lung disease, cystic fibrosis and lung cancer) disease. Total 400 million individuals across the globe have mild to moderate conditions of Asthma and COPD (chronic obstructive pulmonary disease) [85]. Unlike other diseases, metabolites had significant difference among COPD and pulmonary langerhans cells histiocytosis (PLCH), a type of lung disease. Isobutyrate and 2-propanol are able to characterise COPD with respect to PLCH, high/low in COPD and low/high in PLCH [86].

In pediatric study 1-methylnicotinamide and trimethylamine N-oxide (TMAO) were significantly lower in urine samples of asthamatic children compared healthy children [87]. These detected metabolites changed upto higher extent in severe asthama cases. Dimethylamine has an important role in asthama development, along with guanidoacetic acid, allantoin and 1-methylnicotinamide correlated with asthama [88]. The major dysregulated pathways in asthama children urine samples are sphingolipid metabolism, protein biosynthesis and citric acid cycle [89], transient wheezing phenotype showed different urinary metabolites compared to children with early-onset asthma. Aspartic acid, stearic acid, heptadecanoic acid, threitol, acetylgalactosamine, xanthosine, hypoxanthine and uric acid) have a good ability to distinguish asthma in urine samples [90]. In bronchial epithelial cells (BECs) of asthama patients purine metabolism, amino acid biosynthesis, and glycolysis were dysregulated [91]. Clinical differential diagnosis of asthma and COPD remains a major challenge, especially for individuals who smoke. In COPD, ethanol and methanol are high while acetone/acetoin is low compared to asthma [92]. N-acetyl-glycoprotein (NAG), lipoprotein are important metabolites to differentiate control and COPD [93]. Linear and polynomial LS-SVM classifiers can achieve the total accuracy rates of 80.77% and 84.62% and the AUC values of 0.87 and 0.90 for COPD diagnosis using these metabolites. In another study on plasma, it was found that glutamylphenylalanine and taurine decreased significantly in Chinese COPD patients but were highly expressed in the healthy group [84]. These metabolites could be used as key involver in disease onset and progress. Acylcarnitine, an intermediate of fatty acid β-oxidation (FAO), displayed an increased tendency in COPD which activate inflammatory signal pathways via inducing IL-8 production and nuclear factor kappa-light-chain-enhancer of activated B cells (NF-κB) activation [94]. Ceramide, an intermediate product of sphingomyelin metabolism, abnormally accumulates in lung tissue and may damage endothelial-defense, induce alveolar epithelial cell apoptosis, promote inflammatory response and cause macrophage dysfunction [95]. Leukotriene A4 (LTA4), prostaglandin E2 (PGE2) and 5-hydroxy-eicosatetraenoic acid are initial markers of inflammation that increased in COPD, while low concentration of eicosapentaenoic acid and docosahexaenoic acid prevent inflammation. Inflammation marker (LTB4, TNF- α, IL-8) significantly reduced when patient took long-term omega-3 PUFAs supplementation [96]. Three biomarkers, 411.3208 (4a-formyl-5a-cholesta-8,24-dien-3b-ol), 459.3493 and 568.5661 (Cer (d18:0/18:0)), were verified to show good diagnostic of *Mycoplasma pneumonia* against healthy control and infectious disease cases in children plasma sample analysed using liquid chromatography-quadrupole time-of-flight mass spectrometry. A group of compounds including glycolic acid, glyceric acid and xanthine were elevated in the cystic fibrosis group. A large group of acylcarnitines and aldehydes were found to be decreased in cystic fibrosis [97]

4.2 Cardiac Disorders

Cardiovascular diseases and respiratory diseases are the major causes of mortality throughout the world. Cardiac disorders cover 32% of deaths, with an estimated 17.9 million deaths each year (WHO 2021). Sudden heart attack is the major cause of death among men and women. Cardiovascular disorders are associated with blood vessel structural changes which are responsible for disruptive blood supply to the body organ and causing heart stroke [98]. Along with blood carrying vessels, muscles of heart also get out of rhythm (cardiac myopathy). A routine diagnosis is important to avoid the development risk associated with heart failure. Elevated deposition of unwanted lipids and other fat particles are responsible for blocking of blood vessels as well as in muscles [99]. Metabolite investigations associated with prevention and development of treatment strategies could support lower mortality rate. Surgical examination of vessel and heart showed low density lipoproteins, which are part of lipid oxidation form plaque [100, 101], which is responsible for irregular or discontinuous supply of blood to heart or clogging Whereas measurement of these parameters is not possible to detect by surgery, biological fluids urine/blood are always options to detect changes in healthy and diseased person. Untargeted metabolomics study of 65 coronary heart disease (CHD) patients detected 3069 molecular features. Oxyneurine, acetylcarnitine, PC(17:0/0:0) and isoundecylic acid were four most dysregulated metabolites with AUC range 0.610–0.779 in patient vs control, with highest sensitivity and specificity. Oxyneurine is more sensitive and specific for CVD detection with 69.7 sensitivity and 78.8 specificity [102]. In a targeted study of 101 CVD patient, oxyneurine along with acetylcarnitine,and PC (17:0/0:0)] significantly differentiated from control. In an early diagnosis model, a clinical model of these metabolites showed AUC value of combined diagnostic model only 0.579; the sensitivity was 75.5% and the specificity was 46.5%. Combining oxyneurine, TG and weight considerably increased the accuracy of the clinical model with AUC 0.731 sensitivity 83% and specificity 64%.

4.3 *Inflammatory Bowel Disease*

Crohn's disease and ulcerative colitis are in the spectrum of inflammatory bowel disease (IBD), which has inflammation and ulcers in gastrointestinal tract, including symptoms like diarrhea, bloody stool, abdominal pain and increased influx of immune cells (neutrophils, macrophage) which produce inflammatory cytokines and proteolytic enzymes [103]. IBD is more common in Western countries, including 1.3% of the adult population of the United States. It is one of the five pathologies with the greatest social burden, with a mean annual cost (1998–2000) of 1.7 billion USD in healthcare services. The intestine metabolome is a product of host and gut microbiota derived enzymatic conversion of food nutrients and is better represented in a stool sample. Any change in metabolome directly confers to change in mucosal homeostasis and microbial dysbiosis [104]. Previous studies showed perturbations of the gut microbiota by antibiotic administration to mice affected the host's systemic metabolic phenotype [105]. The detected similar co metabolite can also be utilized for monitoring human intestine inflammation stage in human stool, as supported by mice ileum tissue [106]. Stool samples contain large and small molecules produced by intestinal bacteria from consumed food that is subsequently absorbed in the gastrointestinal tract (GIT) [107]. Hence, stool metabolites are a co-presentation of gut microbiota as well as intestinal tract joint activity in both healthy and diseased state. Using tissue biopsies to test metabolites is a time consuming and uncomfortable procedure for patients, but biological fluid collection is quite feasible. Fecal and serologic biomarkers can be used in the diagnosis and management of inflammatory bowel disease. Fecal calprotectin and lactoferrin are current biomarker for IBD identification [108]. These markers can detect IBD but not its subtype to CD or UC, in current report there are other liver complications with UC called primary sclerosing cholangitis, which also needs to be detect for better treatment [109]. In a study faecal volatile metabolite analysis, heptanal, 1-octen-3-ol, 2-piperidinone and 6-methyl-2-heptanone were able to separate based on disease location, small bowel CD from healthy controls and those with colonic CD from UC ($P < 0.001$) [110]. Citrate levels in serum can differentiate between CD and UC, since CD patients showed higher levels compared to UC [111]. Daniluk and colleagues reported for the first time that there were increased levels of the sphingolipid lactosylceramide in CD patients compared to the UC cohort, a promising biomarker, which might help in the specific diagnosis of CD or UC [112]. Hence, ergothioneine seems to be a potential biomarker to distinguish between CD and UC, since its transporter is only expressed in the small intestine [113]. Dopaquinone was increased in fecal samples of CD patients in comparison to healthy controls [114]. Urinary hippurate is a product of the microbial metabolism of certain dietary compounds to benzoic acid, with subsequent renal and hepatic conjugation of benzoic acid with glycine [115]. Hippurate levels have been found to be significantly lower in CD and UC patients compared to controls [115]. Tryptophan level is significantly lower in IBD while sialic acid is elevated in patient serum and has a stronger association with C-reactive protein in CD than in UC [116]. Significant decrease in Kreb's cycle metabolites including citrate, aconitate, α-ketoglutarate, succinate, fumarate and malate are observed in IBD compared to controls [117]. A significant increase in succinate is found in CD patients compared with control subjects. Succinate level is increased in stool samples of paediatric CD patients infected with *Clostridioides difficile* compared with paediatric CD patients without infection [118]. Docosahexaenoic acid, linolenic acid and arachidonic acid and MCFAs such as pelargonic acid and caprylic acid in serum of CD patients. In a study authors discriminate neutrophil to albumin ratio (NAR) discriminate UC (n = 146) from controls, in ROC analysis, showing that NAR had larger AUC (AUC = 0.8670) compared to neutrophil (AUC = 0.7750) or albumin (AUC = 0.7569) alone [119]. Taken together, these data suggest that NAR could be a practical, rapid and easily accessible biomarker for UC diagnosis. In stool, neoptarin is abundant in active CD patient compared to active CD patient, while in urine able to differentiate active IBD from IBD. Along with studying metabolomics changes, researchers are interested in analysing the link between the microbiota and the metabolome through integrated approaches. Studies reported that abundance of decline in certain bacterial population cause significant alteration in host-microbial co-metabolite. The category of co-metabolites including amino acids, biogenic amines and lipids increased significantly in IBD and B group vitamin B3 and B12 significantly decreased [120]. Mingxiao Li et al. in their study on serum metabolites of UC patient and in experimental colitis mouse model showed receiver operating characteristic (ROC) analysis of palmitoyl glucuronide, isobutyrylglycine, PC (20:3 (5Z, 8Z, 11Z)/15:0) and L-arginine under AUC >0.80 which was specific to DSS remission between control and model groups [121].

4.4 *Liver Disorder*

The liver is one of the vital organs in the human body which plays several critical roles in maintaining human health and well-being. Its main functions are metabolism of nutrients and excretion of toxic substances [122]. Diseases of the liver are complex and require extensive and often invasive investigations, like liver biopsy, which remains, currently, the most reliable test, if not the "gold standard'. Uncovering non-invasive diagnostic tools and scientific research into these has yielded multiple scoring systems, formulae and imaging modalities to diagnose liver disease and to monitor the disease activity. Due to vital role of liver in various complex metabolic and synthetic functions, any pathological insult to hepatic parenchyma leads to altered concentration of toxic metabolites in the systemic circulation [123]. Cirrhosis patients develop high risk of acute decompensation and require hospital admission. Early diagnosis of high risk patients could save their lives if diagnosed on time [124]. Serum samples study of cirrhosis patient showed metabolites change which were related to amino acid metabolism, linolcic acid metabolism, glyoxylate and dicarboxylate metabolism, fatty acid metabolism, α-linolenic acid metabolism, and arachidonic acid metabolism [125]. Panel of metabolite associated with hepatitis B virus (HBV) is associated with cirrhosis development are glutamine, acetate, formate and pyruvate in the serum [126]. A predictive model development used to differentiate fibrosis and cirrhosis at early and advanced stage has high sensitivity to taurocholate, tyrosine, valine and linoelaidic acid. These metabolites were validated in another cohort showing AUC >0.8, which confirmed the strength of their predictive models. In Hepatitis C virus (HCV) and human immunodeficiency virus (HIV) coinfected patient plasma Glycolic acid, LPC (16:0), and taurocholic were he most discriminated metabolites for decompensation stage. The AUC value showed higher sensitivity of set of metabolite compared to individual metabolite [127]. Other studies showed increased taurochenodeoxycholate, phosphatidylcholine, taurocholate, glycocholate, PC (16:0/16:0) and tyrosine in patients with advanced cirrhosis [128]. Taurocholate increased in cirrhosis due to increased bile acid synthesis because of cholestasis. In urine samples where metabolites represents host and microbiome metabolites a study found decreased level of 1-methyluric acid, cinnamic acid, N6-methyladnosine, deconoylcarnitine and phenacetylglutamine derived from decreased microbiota associated sugar fermentation [129]. Study employed GC-MS and targeted LC-MS of serum samples for hepatocellular carcinoma (HCC) and cirrhosis. Methionine, ornithine, proline, octanoylcarnitine and pimelylcarnitine demonstrated higher AUC (AUC: 0.75), which was validated showing reliability of the established model [130]. Sensitivity of a large study of HCC increased when combined with alpha-fetoprotein (AFP) with AUC of 0.946 and 0.829 [131] (Table 2).

Table 2 List of reported metabolic biomarker for noncommunicable diseases

Disease	Metabolite/pathway	Fluid origin	Analytical platform	Ref.
Type 2 diabetes	glycine, lysophosphatidylcholine (LPC) (18:2) and acetylcarnitine	Serum	LC-MS	[132]
Pre-diabetic	valine, palmitic acid, 2-aminoadipic acid andproline, tyrosine, lysine and glutamate, isoleucine, alanine, proline, glutamate, lysine, leucine, isoleucine glycine, serine, and citrulline	Serum and plasma	LC- and FIA-ESI-MS/MS	[133]
	alanine, glutamate and palmitic acid (C16:0)	Blood	UHPLC-MS-MS	[134]
Autosomal dominant polycystic kidney disease (ADPKD)	creatinine, lactate, pyruvate, and succinate	Urine	Analytical enzyme assay	[135]
Chronic Renal Failure	7,8-Heptahydroxyflavone, threoninyl-aspartate, paraquat dichloride, Azelaic acid, (10E, 12Z)-9-HODE, N-acetylglutamine, 4-acetamidobutanoic acid, Isoleucyl-alanine, beta-solamarine, kynurenic acid, alcophosphamide,1-(Beta-D-ribofuranosyl)-1, 4-dihdronicotinamide, Thelephoric acid, 5′-Methylthioadenosine, 3-Methylglutarylcarnitine, formiminoglutamic acid, solacauline, PC (20: 5 (5Z, 8Z, 11Z, 17Z)/(20: 5 (5Z, 8Z, 11Z, 17Z)), serylalanine, 5,6-dihydrouridine, l-beta-aspartyl-l-threonine, kynurenic acid, tiglic acid, N-acetylserine, glutamyltheronine, 1-methylhypoxanthine, beta-carboline, arabinofuranobiose, valdecoxib, glycerol tripropanoate, 4-Guanidinobutanoic acid, 3-methoxy-4-hydroxyphenylethyleneglycol sulfate presqualene diphosphate, lycoperoside D,2-O-(6-Phospho-alpha-mannosyl)-D-glycerate, 2, 8-Di-O-methylellagic acid perlolyrine, 3, 3′, 4′, 5, 6, N-acetyl-L-alanine, mycophenolic acid, formylanthranilic acid, trehalose prostaglandin F3a, thymine and creatinine	Serum	Q-Exactive HFX Orbitrap LC-MS/MS	[136]
Nonalcoholic fatty liver disease	Ratio of 13-hydroxyoctadecadienoic acid (13-HODE) to linoleic acid,	Serum	LC/ESI/ MS/MS)	[137]
Early nonalcoholic fatty liver disease	L-lysine,leukoterine C5, oleic acid, indole, succinic acid, lysoPC(20:3(5Z,8Z,11Z)	Serum	UPLC-Orbitrap mass spectrometry	[138]

(continued)

Table 2 (continued)

Disease	Metabolite/pathway	Fluid origin	Analytical platform	Ref.
Alzheimer disease	Gamma-aminobutyric acid, 17-ββ estradiol, homocysteine to methionine, choline, lecithin, L-carnitine and betaine	Urine	NMR spectroscopy and UHPLC-MS	[139]
	Lithocholic acid	Plasma	UHPLC-QTRAP	[140]
	3-hydroxykynurenine, homogentisate and allantoin	Urine	NMR	[141]
Parkinson	Alanine, valine pyruvate, serine, betaine, β-hydroxybutyrate, dimethyl sulphone, glycine, lactate and threonine	Serum	HNMR	[142]
Cardiovascular	Acylcarnitines, TCA cycle intermediates, Branched chain amino acid, cardiolipins, succinate, short-chain dicarboxylacylcarnitine and rimethylamine N-oxide	Heart, plasma	LC-MS/MS; GC-MS	[143]
Respiratory disease	Urate, tyrosine Amino Acid,3-(hydrocinnamate) Amino Acid,pseudouridine, phenylpropionate, 3-(4-hydroxyphenyl) lactate, 1 glutamine amino acid	Serum	Meta-analysis by multivariate regression	[144]

5 Analytical Methods for Metabolomics

In metabolomics, low molecular weight compounds serve as critical biological regulators including cell signalling and networking with other metabolites [145]. These metabolites could be utilised as a disease distinguishing biomarker and target for drug development. Spectroscopic techniques have been successful in identify and quantifying the metabolite in biological samples, though they have certain limitations. NMR is a pioneer method for metabolites study, in which principle of energy absorption and re-emission employed by atomic nuclei due to variations in an external magnetic field are applied [146]. The advantage of NMR is that it is a non-destructive and highly reproducible technique, and does not require extensive sample preparation. On the other hand, NMR has a lower sensitivity, it requires high concentration of compound to detect and compounds less than micro or pico molar levels are out of its detection limit; hence, a range of desired or significant metabolites might be masked by larger peaks. Mass spectroscopy (MS) has advantage over NMR; MS is able to detect metabolite which are even present at nano or pico level, but require more précised sample preparation compared to NMR [147]. LC MS is the most the widely used platform for polar, non-polar, and moderately polar compounds with more specificity to chemical class like fatty acids, alcohols, phenols, vitamins, organic acids, polyamines, nucleotides, polyphenols, terpenes, flavonoids, lipids and other compounds [148]. MS platforms commonly used for metabolomics study include low-resolution techniques such as 5 triple quadrupole

(QQQ) and quadrupole-ion trap (QIT) and high-resolution techniques such as 6 quadrupole-time of flight (Q-TOF), quadrupole-Orbitrap (Q-Orbitrap) and Fourier transform ion 7 cyclotron resonance mass spectrometry (FTICR-MS) [149]. Some of the biological samples are out of LC-MS detection because of their volatile property; so for such compounds, MS is paired with gas chromatography, but before detection, compounds needs to be derivatised [150]. GC-MS can detect amino acids, organic acids, fatty acids, sugars, polyols, amines, sugar phosphates and other substances. Based on type of detection platforms choose like capillary electrophoresis (CE), and supercritical fluid chromatography (SFC) are other platforms [151]. LC separation works on reverse phase liquid chromatography employed for separation of hydrophobic/hydrophilic metabolites [152]. Along with different platform selection based on their chemical classes, ionization mode can be chosen from electrospray ionization, atmosphere pressure chemical ionization (APCI) and atmospheric pressure photoionization (APPI) [153]. Large-scale targeted metabolomics quantification and untargeted metabolomics profiling is the primary strategies for metabolomics study. The untargeted metabolomics profiling strategy provides high-resolution full-scan MS data covering all metabolites in a sample. To manage large spectra and their reliable annotation, data-dependent acquisition (DDA) and MSE36 strategies have been targeted for metabolite quantification [154]. DDA permits reliable identification of unknown metabolites, but quantitative analysis is not accurate. Consecutive isolation window based data independent acquisition strategy (SWATH) has also been applied in untargeted metabolomics study. Untargeted metabolomics profiling has the best metabolite coverage, but its reliability is suboptimal due to poor reproducibility and quantitative performance. Data processing such as metabolite identification and data mining is complicated and requires significantly more effort than targeted quantification methods.

Targeted metabolomics is considers as gold standard because of its high sensitivity, wide dynamic range, reliable quantification accuracy and stability, but it is only able to measure limited number of known pre-selected analytes. Coverage by global profiling assay is still outpaces to characterise metabolite by annotating signal with their nearest metabolite identity.

In metabolomics-based study, there is a high demand of reliable quantitative assays of large number of metabolites. Multiple reaction monitoring (MRM) and parallel reaction monitoring (PRM) are frequently used for large-scale targeted quantification [155]. MRM strategy is capable of monitoring both specific precursor ions and characteristic product ions for each metabolite. MRM is most popular on triple quadrapole LC-MS system for simultaneous quantitation. MRM metabolite coverage is low which limits its use in complex biological system. Continuous efforts are attempting to improve large scale metabolomics analyses with enhance ease and stability. PRM-based targeted metabolomics has also been developed for Q-Orbitrap instrument. PRM scan entire group fragment ions at high resolution in targeted metabolomics and proven to be a powerful means of targeted quantification.

6 MS Strategies for Large-Scale Metabolite Analysis

Interpretation of global profiling data is fruitless without considering effort toward conversion into biochemical interpretation. This represents a substantial barrier to the success of LC-MS-based metabolomics in clinical and translational research. LC-MS instrument based detection needs marked improvement is in progress for making powerful biological study. Multiple reactions monitoring (MRM) mode has been successful for high-quality metabolite quantitation, but is limited by its relatively low metabolite coverage and throughput capacity. A strategy is needed that enables extended coverage of metabolite, while maintaining reliable quantification performance. The Partner relationship management (PRM) strategy allows a greater range of choices for large-scale untargeted metabolomics quantification. In PRM-based metabolites detection m/z (metabolite precursor) is first selected quadrapole and then fragmented by HCD/CID, altering all fragments of the precursor detected by HCD/CID in a single scan. However, in order to maximize the analytical power of large-scale targeted metabolomics strategies, some issues must be addressed.

7 Data Integration and Management Across Studies

Biology inside cells and organisms is heterogeneously structured, and can be studied at the gene, transcript, protein and metabolite level. Different omics based approaches are available for these studies named genomics, transcriptomics, proteomics and metabolomics. In biological systems, information is transferred from nucleic acids to proteins and then metabolites in order to shape function and phenotype. These macromolecules and small metabolites could provide data about the phenotypes of healthy and diseased individuals. Integrating two or more omics approach provide understanding of relationship among molecules which provide redundancy free information from data sources with overlapping contents. Data mining from knowledge databases provide integration of existing data files. In a study of 16 TB and 32 Mycobacterium tuberculosis-exposed plasma samples, metabolite integrated with whole blood transcriptomics 3 metabolites pyridoxate, N-acetylneuraminate, quinoline were correctly able to identify TB and its duration during treatment ranging AUC value 0.66–0.87 [156] with additional 4 metabolite set gamma-glutamylglycine, gammaglutamylalanine, glutamine and pyridoxate. Transcriptional data correlated pyridoxate with p53-regulated metabolic genes in mitochondria and N-acetylneuraminate with immunoregulation of lymphoid and non-lymphoid cells. Based on decision tree analysis mRNA transcript increased upto the extent of metabolite. Transcript metabolite integration data identified high-yield tests for both diagnosis and response to anti-tubercular treatment. COVID-19 patient immunometabolites correlated with circulating cytokines, showed elevated succinate inter acted with inflammatory cytokine production [157] with auxiliary involvement of tryptophan, arginine, and purine metabolism in proinflammatory responses. Elevated succinate has been crucial for SARS-CoV-2 replication [158] and cytokine production in monocyte. IL-1β and IL-18 production in stimulated macrophages manifest by choline uptake. Dietary supplement of α-ketoglutarate was able to suppress chronic inflammation inducing IL-10 production. Serum kynurenine and tryptophan correlate with IL-6 in COVID-19 patients. Downstreaming of arginine;, ornithine and citrulline are essential for T cell activation and regulate innate and adaptive immunity [159]. Arginase metabolism upon infection restricts inflammation by negative-feedback loop. Integrated comparative genomics and machine learning of saliva, serum, and sebum metabolite identified 11 regions in the SARS-CoV-2 genome, which can predict high fatality unlike other diseases metabolome IBD is a representation of microbiome, which produces metabolites using host enzymes. Microbial dysbiosis is reported to affect the host genetics and linked with a variety of diseases. Studies are being conducted to find the actual cause of IBD, linking microbiome and inflammation as well as multiple sequencing assays for data integration [160]. Reduced antimicrobial peptides E-cadherin and claudins have been observed in IBD patients, supporting excess microbial translocation to epithelium junction through impaired epithelial barrier [161]. CpG oligodeoxynucleotides to stimulate TLR9 signalling and induce the production of IFN-γ in the lamina propria. Proteomics data of IBD patient showed reduced tight junction protein expression like E-cadherin and claudins. Loss of tight junction protein integrity allows microbial antigen translocation to host bowel aberruptive mucosal immune response [162]. Membrane-bound molecules having pivotal role in cell signalling and sphingomyelin are processed by gut microbiota [163]. These results have been proven in mice model administrating PC or SM effect on microbiome and metabolome [164]. Another study showed strong correlation between indolepyruvate (aryl hydro-carbon receptor agonist) and Lactobacillus, Klebsiella responsible to increased gut inflammation [165]. Phosphatidylcholine administration ameliorates this metabolic imbalance greater than sphingomyelin. Dysregulated bile acid in ileum of CD patient have a link to host intestine permeability observed which consequences of bile acid processing microbial gene and alters host Angiopoietin-like 4 transcripts [166]. *Bifidobacterium* sp. produced anti-inflammatory metabolite indoleacetic acid, lower levels of which strongly correlated to host immune tone. Increased levels of benzoic acids, and phenols have been associated with enriched in *Bacteroides sp* dominant in ulcerative colitis patient [167]. Stool metaproteomics has shown high fidelity clinical marker for gut diseases. Metatranscriptomics and metabolomics data revealed decreased bile salt hydrolases, highlighting the role of bile acids in microbial metabolic output and host physiology. The role of gut microbiota derived metabolites not only has importance to gut vicinity, but to other diseases originating at different tissue sites. Integration provides us the microbial strain, and metabolite to manifest the disease condition. A large set of metabolites are common to inflammatory conditions and overlap to each other, but do not provide contrasting result toward disease specificity. In dengue haemorrhagic fever (DHF), elevated IFN-γ is enriched with kynurenin, tryptophan and serotonin which could be used in combination for accurate early disease progression metrics [168]. Dengue virus use mosquito feed blood for de novo phospholipid precursor synthesis which is required for viral genome replication to modify the endomembrane to facilitate replication complex, a process unique to dengue [169]. Multiomics integration identified eight target gene (ACTG1, CALR, ERC1, HSPA5, and SYNE2) involved in protein-protein interaction that could be used as drug repositioning to treat DHF. Valparoic acid resveratrol, vorinostat, sirolimus andY-27632 drug have been reported effective previously against flavivirus-induced diseases [170]. Combined transcriptomics and metabolomics analyses in malaria provide profound approaches to link biological responses to oxidative stress and corroborate the transcriptomic response at the time of malaria diagnosis. In malaria, myeloid cells metabolites have substantial effect and reprogramming of metabolites could have clinical tolerance to *P. vivax* relapse [171]. The metabolite set elicited during infection modulate genes related to platelet activation, interferon and innate immunity and for chemokines and T cell signalling associated with linoleate and glycerophospholipid metabolism in human [172]. Plasma metabolites associated with platelet activation genes in blood were enriched in platelets in blood. Studies proved *P. vivax* infection associated with host metabolic response governed immune response later on. Metabolomics and transcriptomics revealed concerted events during infection which are associated with platelet activation genes were indeed enriched in the platelet metabolome. Metabolite interactions and biomarkers relationship between diseases provide disease understanding and shed light on potential mechanism behind diseases. Pathways and network connections of candidate transcripts provide tool to researcher increasing knowledge on the molecular interaction. Although metabolomics provide metabolite alteration which indeed a result of change in proteome and transcriptome along with affected environmental factors, epigenetics, genetics and influence cellular biochemistry. In Chronic lung disease lipids are leading factors in TH1 immune response [173]. Impaired lipolysis/lipogenesis and hypoxia response is associated with SCD1, ROS VEGF, ATP, AMPK, MAPK, EGFR, UCP2 and PLA2 which are key crosstalk molecules [174]. Upregulation of sphingolipids has a negative impact on VEGF signalling where ceramide inhibit VEGF and upregulate sphingosine-1-phosphate (S1P) signalling [175]. Decreased HIF1α and VEGF lead to pulmonary vasculature impairment during hypoxia [176]. As a result of increased hypoxia, ceramide inhibits FABP4, PPAR, SCD1, ACOT4 and UCP2 genes, which are key regulator of lipid trafficking and energy metabolism. Dysregulation of UCP2 expression is related to the polyunsaturated fatty acids linoleic acid and arachidonic acid [177]. Aggregation of the features that correlate with genotypic data and can be matched to one or several metabolites could function as a novel type of quantitative biomarker for cardiovascular disease as well. In ischemia and idiopathic heart failure patient, two metabolites functionally associated with gene expression change are nitric oxide (NO) and the synthesis of N-acetylneuraminic acid (Neu5Ac) [178]. NO and Neu5Ac metabolic function associated with genes involved in heart failure evidenced across studies and also serve as biomarker and potential target for disease treatment. Neu5Ac de novo synthesis data showed decreased expression of five genes in ischemic strock. In plasma samples increased Neu5Ac is attributed to glycosylation of circulating lipoproteins, which is correlated with increased triglyceride levels [179]. A mouse study demonstrated that protein glycosylation inhibition through a sialyltransferase (ST3Gal4), caused development of dilated cardiomyopathy (DCM) [180]. These results were further validated in a genetic knockout in zebrafish. Increased trimethylamine N-oxide (TMAO) is linked to micro biome metabolism, 2 flavin monoxygenase class of enzyme oxidise trimethylamine to TMAO with FMO3 expression change [181]. Mice expressing high variants of these genes have increased susceptibility to atherosclerosis. Genetic biomarkers are heritable components responsible for disease manifestation in early childhood; molecular detection may serve lifestyle control for delaying disease response.

8 Conclusion

Metabolomics is a rapidly growing field of study that aims to quantify the small molecule metabolites present in biological samples. The integration of metabolomics data with other omics data, such as transcriptomics, proteomics and genomics, has become increasingly important in order to gain a more comprehensive understanding of biological systems. There are several emerging technologies that are playing a key role in the integration of metabolomics data, including machine learning algorithms such as random forest, support vector machines and neural networks, which are being used to integrate metabolomics data with other omics data and identify new biomarkers for disease diagnosis and prognosis. Integration of metabolomics data with electronic health record (EHRs) can provide a more complete picture of patient health and enable personalized medicine. MetaboAnalyst and OmicsDI are data integration platforms being developed to enable the integration of multiple omics data sets, including metabolomics data, in a single platform. RNA-seq and metagenomics are being used to generate large amounts of transcriptomics and metagenomics data, respectively, which can be integrated with metabolomics data to gain a more comprehensive understanding of biological systems. Mass spectrometry imaging (MSI) is a rapidly growing field that enables the spatially resolved analysis of metabolites in tissues. The integration of MSI with other omics data is providing new insights into the relationship between metabolite distribution and disease development. The Importance of integrating multi-omics data is evident in real life disease research as it aids the discovery of fundamental causes of serious health conditions. Genome, proteome to phenotype governed by metabolome build string attached genotype to phenotype and offered improved prediction of disease for future treatment strategies. Hence, the emerging technologies are advancing the field of metabolomics and improving our understanding of biological systems. They are expected to have a significant impact on the development of new diagnostics and therapies for a wide range of diseases.

References

1. Allan M, Lièvre M, Laurenson-Schaefer H, de Barros S, Jinnai Y, Andrews S, et al. The World Health Organization COVID-19 surveillance database. International Journal for Equity in Health. 2022;21(Suppl 3):167.
2. Ocak S, Sos ML, Thomas RK, Massion P. High-throughput molecular analysis in lung cancer: insights into biology and potential clinical applications. European Respiratory Journal. 2009;34(2):489–506.
3. Kussmann M, Raymond F, Affolter M. OMICS-driven biomarker discovery in nutrition and health. Journal of biotechnology. 2006;124(4):758–87.
4. Scott JM, Weir DG. Folic acid, homocysteine and one-carbon metabolism: a review of the essential biochemistry. Journal of cardiovascular risk. 1998;5(4):223–7.
5. Johnson CH, Ivanisevic J, Siuzdak G. Metabolomics: beyond biomarkers and towards mechanisms. Nature reviews Molecular cell biology. 2016;17(7):451–9.
6. Zanetti KA, Mette E, Maruvada P, Milner J, Moore SC, Nicastro HL, et al. The future of metabolomic profiling in population-based research: opportunities and challenges. J Anal Bioanal Tech. 2014;5(203):2.
7. Bictash M, Ebbels TM, Chan Q, Loo RL, Yap IK, Brown IJ, et al. Opening up the" Black Box": metabolic phenotyping and metabolome wide association studies in epidemiology. Journal of clinical epidemiology. 2010;63(9):970–9.
8. Liesenfeld DB, Habermann N, Owen RW, Scalbert A, Ulrich CM. Review of Mass Spectrometry–Based Metabolomics in Cancer ResearchMetabolomics in Cancer Research. Cancer epidemiology, biomarkers & prevention. 2013;22(12):2182–201.
9. Su LJ, Fiehn O, Maruvada P, Moore SC, O'Keefe SJ, Wishart DS, et al. The use of metabolomics in population-based research. Advances in Nutrition. 2014;5(6):785–8.
10. Waltz P, Carchman E, Gomez H, Zuckerbraun B. Sepsis results in an altered renal metabolic and osmolyte profile. Journal of Surgical Research. 2016;202(1):8–12.
11. Slupsky CM, Cheypesh A, Chao DV, Fu H, Rankin KN, Marrie TJ, et al. Streptococcus pneumoniae and Staphylococcus aureus pneumonia induce distinct metabolic responses. Journal of proteome research. 2009;8(6):3029–36.
12. Dessì A, Liori B, Caboni P, Corsello G, Giuffrè M, Noto A, et al. Monitoring neonatal fungal infection with metabolomics. The Journal of Maternal-Fetal & Neonatal Medicine. 2014;27(sup2):34–8.
13. Ghosh S, Sengupta A, Sharma S, Sonawat HM. Early prediction of cerebral malaria by 1H NMR based metabolomics. Malaria Journal. 2016;15(1):1–10.
14. Ghosh S, Sengupta A, Sharma S, Sonawat HM. Metabolic fingerprints of serum, brain, and liver are distinct for mice with cerebral and noncerebral malaria: a 1H NMR spectroscopy-based metabonomic study. Journal of proteome research. 2012;11(10):4992–5004.
15. Balog CI, Meissner A, Göraler S, Bladergroen MR, Vennervald BJ, Mayboroda OA, et al. Metabonomic investigation of human Schistosoma mansoni infection. Molecular Biosystems. 2011;7(5):1473–80.
16. Birungi G, Chen SM, Loy BP, Ng ML, Li SFY. Metabolomics approach for investigation of effects of dengue virus infection using the EA. hy926 cell line. Journal of proteome research. 2010;9(12):6523–34.
17. Patnaik R, Azim A, Agarwal V. Neutrophil CD64 a diagnostic and prognostic marker of sepsis in adult critically ill patients: A brief review. Indian Journal of Critical Care Medicine: Peer-reviewed, Official Publication of Indian Society of Critical Care Medicine. 2020;24(12):1242.
18. Høiby N. Pandemics: past, present, future: that is like choosing between cholera and plague. Apmis. 2021;129(7):352–71.
19. Ortiz-Prado E, Simbaña-Rivera K, Gomez-Barreno L, Rubio-Neira M, Guaman LP, Kyriakidis NC, et al. Clinical, molecular, and epidemiological characterization of the SARS-CoV-2 virus and the Coronavirus Disease 2019 (COVID-19), a comprehensive literature review. Diagnostic microbiology and infectious disease. 2020;98(1):115094.

20. Simou E, Koutsogeorgou E. Effects of the economic crisis on health and healthcare in Greece in the literature from 2009 to 2013: a systematic review. Health policy. 2014;115(2–3):111–9.
21. Connolly MA, Gayer M, Ryan MJ, Salama P, Spiegel P, Heymann DL. Communicable diseases in complex emergencies: impact and challenges. The Lancet. 2004;364(9449):1974–83.
22. Hajj Hussein I, Chams N, Chams S, El Sayegh S, Badran R, Raad M, et al. Vaccines through centuries: major cornerstones of global health. Frontiers in public health. 2015;3:269.
23. Guidotti E, Ardia D. COVID-19 data hub. Journal of Open Source Software. 2020;5(51):2376.
24. Lewek J, Sosnowska B, Adach W, Gryglewska K, Bielecka-Dabrowa A, Maciejewski M, et al. Post-COVID-19 cardiovascular complications and their association with clinical characteristics, symptoms and comorbidities–the LATE-COVID study. European Heart Journal. 2022;43(Supplement_2):ehac544. 2335.
25. Priori R, Scrivo R, Brandt J, Valerio M, Casadei L, Valesini G, et al. Metabolomics in rheumatic diseases: the potential of an emerging methodology for improved patient diagnosis, prognosis, and treatment efficacy. Autoimmunity reviews. 2013;12(10):1022–30.
26. Lionetto L, Ulivieri M, Capi M, De Bernardini D, Fazio F, Petrucca A, et al. Increased kynurenine-to-tryptophan ratio in the serum of patients infected with SARS-CoV2: An observational cohort study. Biochimica et Biophysica Acta (BBA)-Molecular Basis of Disease. 2021;1867(3):166042.
27. Lewis H-M, Liu Y, Frampas CF, Longman K, Spick M, Stewart A, et al. Metabolomics Markers of COVID-19 Are Dependent on Collection Wave. Metabolites. 2022;12(8):713.
28. Yang Z, Wu D, Lu S, Qiu Y, Hua Z, Tan F, et al. Plasma metabolome and cytokine profile reveal glycylproline modulating antibody fading in convalescent COVID-19 patients. Proceedings of the National Academy of Sciences. 2022;119(34):e2117089119.
29. Guang C, Phillips RD, Jiang B, Milani F. Three key proteases–angiotensin-I-converting enzyme (ACE), ACE2 and renin–within and beyond the renin-angiotensin system. Archives of cardiovascular diseases. 2012;105(6–7):373–85.
30. Mussap M, Fanos V. Could metabolomics drive the fate of COVID-19 pandemic? A narrative review on lights and shadows. Clinical Chemistry and Laboratory Medicine (CCLM). 2021;59(12):1891–905.
31. Organization WH. Global tuberculosis report 2021: supplementary material. 2022.
32. Olivier I, Loots DT. A comparison of two extraction methods for differentiating and characterising various Mycobacterium species and Pseudomonas aeruginosa using GC-MS metabolomics. 2012.
33. Chen X, Ye J, Lei H, Wang C. Novel Potential Diagnostic Serum Biomarkers of Metabolomics in Osteoarticular Tuberculosis Patients: A Preliminary Study. Frontiers in cellular and infection microbiology. 2022:90.
34. Vrieling F, Ronacher K, Kleynhans L, Van Den Akker E, Walzl G, Ottenhoff TH, et al. Patients with concurrent tuberculosis and diabetes have a pro-atherogenic plasma lipid profile. EBioMedicine. 2018;32:192–200.
35. Shembekar A, Babhulkar S. Chemotherapy for osteoarticular tuberculosis. Clinical Orthopaedics and Related Research®. 2002;398:20–6.
36. Badawi A, Liu CJ. Obesity and Prevalence of Latent Tuberculosis: A Population-Based Survey. Infectious Diseases: Research and Treatment. 2021;14:1178633721994607.
37. Feng S, Du Y-Q, Zhang L, Zhang L, Feng R-R, Liu S-Y. Analysis of serum metabolic profile by ultra-performance liquid chromatography-mass spectrometry for biomarkers discovery: application in a pilot study to discriminate patients with tuberculosis. Chinese medical journal. 2015;128(02):159–68.
38. James JH, Luchette FA, McCarter FD, Fischer JE. Lactate is an unreliable indicator of tissue hypoxia in injury or sepsis. The Lancet. 1999;354(9177):505–8.
39. Collins JM, Walker DI, Jones DP, Tukvadze N, Liu KH, Tran VT, et al. High-resolution plasma metabolomics analysis to detect Mycobacterium tuberculosis-associated metabolites that distinguish active pulmonary tuberculosis in humans. PloS one. 2018;13(10):e0205398.

40. Bamji-Stocke S, van Berkel V, Miller DM, Frieboes HB. A review of metabolism-associated biomarkers in lung cancer diagnosis and treatment. Metabolomics. 2018;14(6):1–16.
41. Preez Id, Luies L, Loots DT. Metabolomics biomarkers for tuberculosis diagnostics: current status and future objectives. Biomarkers in medicine. 2017;11(2):179–94.
42. Organization WH. WHO Report on the Global Tobacco Epidemic, 2021: Addressing new and emerging products: World Health Organization; 2021.
43. Mueller I, Galinski MR, Baird JK, Carlton JM, Kochar DK, Alonso PL, et al. Key gaps in the knowledge of Plasmodium vivax, a neglected human malaria parasite. The Lancet Infectious Diseases. 2009;9(9):555–66.
44. Breman JG, Alilio MS, Mills A. Conquering the intolerable burden of malaria: what's new, what's needed: a summary. The American journal of tropical medicine and hygiene. 2004;71(2 Supp):1–15.
45. McConville MJ, De Souza D, Saunders E, Likic VA, Naderer T. Living in a phagolysosome; metabolism of Leishmania amastigotes. Trends in parasitology. 2007;23(8):368–75.
46. Colvin HN, Joice Cordy R. Insights into malaria pathogenesis gained from host metabolomics. PLoS Pathogens. 2020;16(11):e1008930
47. Leopold SJ, Ghose A, Allman EL, Kingston HW, Hossain A, Dutta AK, et al. Identifying the components of acidosis in patients with severe Plasmodium falciparum malaria using metabolomics. The Journal of Infectious Diseases. 2019;219(11):1766–76.
48. Medana IM, Day NP, Salahifar Sabet H, Stocker R, Smythe G, Bwanaisa L, et al. Metabolites of the kynurenine pathway of tryptophan metabolism in the cerebrospinal fluid of Malawian children with malaria. The Journal of Infectious Diseases. 2003;188(6):844–9.
49. Keswani T, Obeidallah A, Nieves E, Sidoli S, Fazzari M, Taylor T, et al. Pipecolic acid, a putative mediator of the encephalopathy of cerebral malaria and the experimental model of cerebral malaria. The Journal of Infectious Diseases. 2022;225(4):705–14.
50. Lakshmanan V, Rhee KY, Wang W, Yu Y, Khafizov K, Fiser A, et al. Metabolomic analysis of patient plasma yields evidence of plant-like α-linolenic acid metabolism in Plasmodium falciparum. The Journal of Infectious Diseases. 2012;206(2):238–48.
51. Bawa R. Advances in Clinical Immunology, Medical Microbiology, COVID-19, and Big Data. 2021.
52. Rodenhuis-Zybert IA, Wilschut J, Smit JM. Dengue virus life cycle: viral and host factors modulating infectivity. Cellular and molecular life sciences. 2010;67:2773–86.
53. Kalita JM, Aggarwal A, Yedale K, Gadepalli R, Nag VL. A 5-year study of dengue seropositivity among suspected cases attending a teaching hospital of North-Western region of India. Journal of Medical Virology. 2021;93(6):3338–43.
54. Samanta J, Sharma V. Dengue and its effects on liver. World Journal of Clinical Cases: WJCC. 2015;3(2):125.
55. Geisler S, Lytton SD, Toan NL, Nghia TH, Nam NM, Hung HV, et al. Neopterin levels and Kyn/Trp ratios were significantly increased in dengue virus patients and subsequently decreased after recovery. International Journal of Infectious Diseases. 2020;91:162–8.
56. Cui L, Lee YH, Kumar Y, Xu F, Lu K, Ooi EE, et al. Serum metabolome and lipidome changes in adult patients with primary dengue infection. PLoS neglected tropical diseases. 2013;7(8):e2373.
57. McCann MR, George De la Rosa MV, Rosania GR, Stringer KA. L-carnitine and acylcarnitines: mitochondrial biomarkers for precision medicine. Metabolites. 2021;11(1):51.
58. Villamor E, Villar LA, Lozano-Parra A, Herrera VM, Herrán OF. Serum fatty acids and progression from dengue fever to dengue haemorrhagic fever/dengue shock syndrome. British Journal of Nutrition. 2018;120(7):787–96.
59. Shahfiza N, Osman H, Hock T, Abdel-Hamid A-H. Metabolomics approach for multibiomarkers determination to investigate dengue virus infection in human patients. Acta Biochimica Polonica. 2017;64(2):215–9.
60. Byers NM, Fleshman AC, Perera R, Molins CR. Metabolomic insights into human arboviral infections: Dengue, chikungunya, and zika viruses. Viruses. 2019;11(3):225.

61. Cui L, Hou J, Fang J, Lee YH, Costa VV, Wong LH, et al. Serum metabolomics investigation of humanized mouse model of dengue virus infection. Journal of virology. 2017;91(14):e00386–17.
62. Overton JM, Linke L, Magnuson R, Broeckling CD, Rao S. Metabolomic Profiles of Multidrug-Resistant Salmonella Typhimurium from Humans, Bovine, and Porcine Hosts. Animals. 2022;12(12):1518.
63. Lin Y, Li W, Sun L, Lin Z, Jiang Y, Ling Y, et al. Comparative metabolomics shows the metabolic profiles fluctuate in multi-drug resistant Escherichia coli strains. Journal of proteomics. 2019;207:103468.
64. Franklin TJ, Snow G. Biochemical mechanisms of resistance to antimicrobial drugs. Biochemistry and molecular biology of antimicrobial drug action. 2005:149–74.
65. Su Y-b, Peng B, Li H, Cheng Z-x, Zhang T-t, Zhu J-x, et al. Pyruvate cycle increases aminoglycoside efficacy and provides respiratory energy in bacteria. Proceedings of the National Academy of Sciences. 2018;115(7):E1578–E87.
66. Grube CD, Roy H. A quantitative spectrophotometric assay to monitor the tRNA-dependent pathway for lipid aminoacylation in vitro. SLAS Discovery. 2016;21(7):722–8.
67. Crawford RW, Rosales-Reyes R, Ramirez-Aguilar MdlL, Chapa-Azuela O, Alpuche-Aranda C, Gunn JS. Gallstones play a significant role in Salmonella spp. gallbladder colonization and carriage. Proceedings of the National Academy of Sciences. 2010;107(9):4353–8.
68. Nässtrôm E, Jonsson P, Johansson A, Dongol S, Karkey A, Basnyat B, et al. Diagnostic metabolite biomarkers of chronic typhoid carriage. PLoS neglected tropical diseases. 2018;12(1):e0006215.
69. Nässtrôm E, Parry CM, Vu Thieu NT, Maude RR, de Jong HK, Fukushima M, et al. Reproducible diagnostic metabolites in plasma from typhoid fever patients in Asia and Africa. Elife. 2017;6:e15651.
70. Caterino M, Costanzo M, Fedele R, Cevenini A, Gelzo M, Di Minno A, et al. The serum metabolome of moderate and severe COVID-19 patients reflects possible liver alterations involving carbon and nitrogen metabolism. International Journal of Molecular Sciences. 2021;22(17):9548.
71. Ceballos FC, Virseda-Berdices A, Resino S, Ryan P, Martínez-González O, Peréz-García F, et al. Metabolic Profiling at COVID-19 Onset Shows Disease Severity and Sex-Specific Dysregulation. Frontiers in Immunology. 2022:3155.
72. Dierckx T, Van Elslande J, Salmela H, Decru B, Wauters E, Gunst J, et al. The metabolic fingerprint of COVID-19 severity. MedRxiv. 2020:2020.11. 09.20228221.
73. Phillips M, Cataneo RN, Condos R, Erickson GAR, Greenberg J, La Bombardi V, et al. Volatile biomarkers of pulmonary tuberculosis in the breath. Tuberculosis. 2007;87(1):44–52.
74. Syhre M, Chambers ST. The scent of Mycobacterium tuberculosis. Tuberculosis. 2008;88(4):317–23.
75. Schoeman JC, du Preez I. A comparison of four sputum pre-extraction preparation methods for identifying and characterising Mycobacterium tuberculosis using GCxGC-TOFMS metabolomics. Journal of microbiological methods. 2012;91(2):301–11.
76. Park YH, Shi YP, Liang B, Medriano CAD, Jeon YH, Torres E, et al. High-resolution metabolomics to discover potential parasite-specific biomarkers in a Plasmodium falciparum erythrocytic stage culture system. Malaria Journal. 2015;14:1–9.
77. Abdelrazig S, Ortori CA, Davey G, Deressa W, Mulleta D, Barrett DA, et al. A metabolomic analytical approach permits identification of urinary biomarkers for Plasmodium falciparum infection: a case–control study. Malaria Journal. 2017;16(1):1–8.
78. Beri D, Ramdani G, Balan B, Gadara D, Poojary M, Momeux L, et al. Insights into physiological roles of unique metabolites released from Plasmodium-infected RBCs and their potential as clinical biomarkers for malaria. Scientific reports. 2019;9(1):2875.
79. Lao D-H, Liu R, Liang J. Study on plasma metabolomics for HIV/AIDS patients treated by HAART based on LC/MS-MS. Frontiers in Pharmacology. 2022:3195.

80. Scarpelini B, Zanoni M, Sucupira MCA, Truong H-HM, Janini LMR, Segurado IDC, et al. Plasma metabolomics biosignature according to HIV stage of infection, pace of disease progression, viremia level and immunological response to treatment. PloS one. 2016;11(12):e0161920.
81. Xiao Q, Tan S, Liu C, Liu B, Li Y, Guo Y, et al. Characterization of the Microbiome and Host's Metabolites of the Lower Respiratory Tract During Acute Community-Acquired Pneumonia Identifies Potential Novel Markers. Infection and Drug Resistance. 2023:581–94.
82. Jongers Bs, Hotterbeekx A, Bielen K, Vervliet P, Boddaert J, Lammens C, et al. Identification of Potential Urinary Metabolite Biomarkers of Pseudomonas aeruginosa Ventilator-Associated Pneumonia. Biomarker Insights. 2022;17:11772719221099131.
83. Laiakis EC, Morris GA, Fornace Jr AJ, Howie SR. Metabolomic analysis in severe childhood pneumonia in The Gambia, West Africa: findings from a pilot study. PloS one. 2010;5(9):e12655.
84. Zhou J, Li Q, Liu C, Pang R, Yin Y. Plasma metabolomics and lipidomics reveal perturbed metabolites in different disease stages of chronic obstructive pulmonary disease. International Journal of Chronic Obstructive Pulmonary Disease. 2020:553–65.
85. Almalki WH. Introduction to Lung Disease. Microbiome in Inflammatory Lung Diseases: Springer; 2022. p. 1–12.
86. Maniscalco M, Paris D, Cuomo P, Fuschillo S, Ambrosino P, Tramice A, et al. Metabolomics of COPD pulmonary rehabilitation outcomes via exhaled breath condensate. Cells. 2022;11(3):344.
87. Augustine T, Al-Aghbar MA, Al-Kowari M, Espino-Guarch M, Van Panhuys N. Asthma and the Missing Heritability Problem: Necessity for Multiomics Approaches in Determining Accurate Risk Profiles. Frontiers in Immunology. 2022;13:822324.
88. Papamichael MM, Katsardis C, Sarandi E, Georgaki S, Frima E-S, Varvarigou A, et al. Application of metabolomics in pediatric asthma: prediction, diagnosis and personalized treatment. Metabolites. 2021;11(4):251.
89. Li S, Liu J, Zhou J, Wang Y, Jin F, Chen X, et al. Urinary metabolomic profiling reveals biological pathways and predictive signatures associated with childhood asthma. Journal of asthma and allergy. 2020:713–24.
90. Tao JL, Chen YZ, Dai QG, Tian M, Wang SC, Shan JJ, et al. Urine metabolic profiles in paediatric asthma. Respirology. 2019;24(6):572–81.
91. Ravi A, Goorsenberg AW, Dijkhuis A, Dierdorp BS, Dekker T, van Weeghel M, et al. Metabolic differences between bronchial epithelium from healthy individuals and patients with asthma and the effect of bronchial thermoplasty. Journal of Allergy and Clinical Immunology. 2021;148(5):1236–48.
92. Paris D, Palomba L, Mirra V, Borrelli M, Corcione A, Santamaria F, et al. NMR profiling of exhaled breath condensate defines different metabolic phenotypes of non-cystic fibrosis bronchiectasis. International Journal of Molecular Sciences. 2020;21(22):8600.
93. Zheng H, Hu Y, Dong L, Shu Q, Zhu M, Li Y, et al. Predictive diagnosis of chronic obstructive pulmonary disease using serum metabolic biomarkers and least-squares support vector machine. Journal of clinical laboratory analysis. 2021;35(2):e23641.
94. Jiang Z, Knudsen NH, Wang G, Qiu W, Naing ZZC, Bai Y, et al. Genetic control of fatty acid β-oxidation in chronic obstructive pulmonary disease. American journal of respiratory cell and molecular biology. 2017;56(6):738–48.
95. Ran N, Pang Z, Gu Y, Pan H, Zuo X, Guan X, et al. An updated overview of metabolomic profile changes in chronic obstructive pulmonary disease. Metabolites. 2019;9(6):111.
96. Kotlyarov S, Kotlyarova A. Anti-Inflammatory function of fatty acids and involvement of their metabolites in the resolution of inflammation in chronic obstructive pulmonary disease. International Journal of Molecular Sciences. 2021;22(23):12803.
97. Weber R, Perkins N, Bruderer T, Micic S, Moeller A. Identification of Exhaled Metabolites in Children with Cystic Fibrosis. Metabolites 2022, 12, 980. s Note: MDPI stays neutral with regard to jurisdictional claims in published …; 2022.

98. Lusis A. Atherosclerosis Nature, 2000, 407 (6801), 233–241. CrossRef Link PubMed Link [PMC Link].
99. Rossi MA, Celes MR, Prado CM, Saggioro FP. Myocardial structural changes in long-term human severe sepsis/septic shock may be responsible for cardiac dysfunction. Shock. 2007;27(1):10–8.
100. Ambrose JA, Winters SL, Arora RR, Haft JI, Goldstein J, Rentrop KP, et al. Coronary angiographic morphology in myocardial infarction: a link between the pathogenesis of unstable angina and myocardial infarction. Journal of the American College of Cardiology. 1985;6(6):1233–8.
101. Pongrac Barlovic D, Harjutsalo V, Sandholm N, Forsblom C, Groop P-H, Group FS. Sphingomyelin and progression of renal and coronary heart disease in individuals with type 1 diabetes. Diabetologia. 2020;63:1847–56.
102. Craig SA. Betaine in human nutrition. The American journal of clinical nutrition. 2004;80(3):539–49.
103. Hendrickson BA, Gokhale R, Cho JH. Clinical aspects and pathophysiology of inflammatory bowel disease. Clinical microbiology reviews. 2002;15(1):79–94.
104. Rooks MG, Garrett WS. Gut microbiota, metabolites and host immunity. Nature reviews immunology. 2016;16(6):341–52.
105. Yap IK, Li JV, Saric J, Martin F-P, Davies H, Wang Y, et al. Metabonomic and microbiological analysis of the dynamic effect of vancomycin-induced gut microbiota modification in the mouse. Journal of proteome research. 2008;7(9):3718–28.
106. Baur P, Martin F-P, Gruber L, Bosco N, Brahmbhatt V, Collino S, et al. Metabolic phenotyping of the Crohn's disease-like IBD etiopathology in the TNFΔARE/WT mouse model. Journal of proteome research. 2011;10(12):5523–35.
107. Morowitz MJ, Carlisle EM, Alverdy JC. Contributions of intestinal bacteria to nutrition and metabolism in the critically ill. Surgical Clinics. 2011;91(4):771–85.
108. Lamb C, Mansfield J. Measurement of faecal calprotectin and lactoferrin in inflammatory bowel disease. Frontline Gastroenterology. 2011;2(1):13–8.
109. Singh S, Talwalkar JA. Primary sclerosing cholangitis: diagnosis, prognosis, and management. Clinical Gastroenterology and Hepatology. 2013;11(8):898–907.
110. Khan T, Loftus TJ, Filiberto AC, Ozrazgat-Baslanti T, Ruppert MM, Bandhyopadyay S, et al. Metabolomic profiling for diagnosis and prognostication in surgery: a scoping review. Annals of surgery. 2021;273(2):258.
111. Manzella CR, Jayawardena D, Pagani W, Li Y, Alrefai WA, Bauer J, et al. Serum serotonin differentiates between disease activity states in Crohn's patients. Inflammatory Bowel Diseases. 2020;26(10):1607–18.
112. Filimoniuk A, Blachnio-Zabielska A, Imierska M, Lebensztejn DM, Daniluk U. Sphingolipid analysis indicate lactosylceramide as a potential biomarker of inflammatory bowel disease in children. Biomolecules. 2020;10(7):1083.
113. Bauset C, Gisbert-Ferrándiz L, Cosín-Roger J. Metabolomics as a promising resource identifying potential biomarkers for inflammatory bowel disease. Journal of Clinical Medicine. 2021;10(4):622.
114. Jansson J, Willing B, Lucio M, Fekete A, Dicksved J, Halfvarson J, et al. Metabolomics reveals metabolic biomarkers of Crohn's disease. PloS one. 2009;4(7):e6386.
115. Williams HR, Cox JI, Walker DG, North BV, Patel VM, Marshall SE, et al. Characterization of inflammatory bowel disease with urinary metabolic profiling. Official journal of the American College of Gastroenterology| ACG. 2009;104(6):1435–44.
116. Nikolaus S, Schulte B, Al-Massad N, Thieme F, Schulte DM, Bethge J, et al. Increased tryptophan metabolism is associated with activity of inflammatory bowel diseases. Gastroenterology. 2017;153(6):1504–16. e2.
117. Korwar AM, Hossain A, Lee T-J, Shay AE, Basrur V, Conlon K, et al. Selenium-dependent metabolic reprogramming during inflammation and resolution. Journal of Biological Chemistry. 2021;296.

118. Kolho K-L, Pessia A, Jaakkola T, de Vos WM, Velagapudi V. Faecal and serum metabolomics in paediatric inflammatory bowel disease. Journal of Crohn's and Colitis. 2017;11(3):321–34.
119. Zhou Z, Zhang Y, Pan Y, Yang X, Li L, Gao C, et al. A Novel Neutrophil-Based Biomarker to Monitor Disease Activity and Predict Response to Infliximab Therapy in Patients With Ulcerative Colitis. Frontiers in Medicine. 2022;9.
120. Van Treuren W. Advancing Microbial Metrology to Reveal Microbe-Microbe, Microbe-Host, and Microbe-Environment Interactions: Stanford University; 2020.
121. Liu K, Jia B, Zhou L, Xing L, Wu L, Li Y, et al. Ultraperformance liquid chromatography coupled with quadrupole time-of-flight mass spectrometry-based metabolomics and lipidomics identify biomarkers for efficacy evaluation of mesalazine in a dextran sulfate sodium-induced ulcerative colitis mouse model. Journal of proteome research. 2020;20(2):1371–81.
122. Stewart CM. Excretion and Heartwood Formation in Living Trees: The death of many cells in secondary tissues seems to result from the accumulation of waste metabolites. Science. 1966;153(3740):1068–74.
123. Williams GM, Iatropoulos MJ. Alteration of liver cell function and proliferation: differentiation between adaptation and toxicity. Toxicologic pathology. 2002;30(1):41–53.
124. Yoo HJ, Jung KJ, Kim M, Kim M, Kang M, Jee SH, et al. Liver cirrhosis patients who had normal liver function before liver cirrhosis development have the altered metabolic profiles before the disease occurrence compared to healthy controls. Frontiers in Physiology. 2019;10:1421.
125. Kano N, Want EJ, McPhail MJ. Metabolomics in Advanced Liver Disease. Current Treatment Options in Gastroenterology. 2021;19:380–97.
126. Xie G, Wang X, Wei R, Wang J, Zhao A, Chen T, et al. Serum metabolite profiles are associated with the presence of advanced liver fibrosis in Chinese patients with chronic hepatitis B viral infection. BMC medicine. 2020;18:1–15.
127. Salgüero S, Rojo D, Berenguer J, González-García J, Fernández-Rodríguez A, Brochado-Kith O, et al. Plasma metabolomic fingerprint of advanced cirrhosis stages among HIV/HCV-coinfected and HCV-monoinfected patients. Liver International. 2020;40(9):2215–27.
128. Cano A, Mariño Z, Millet O, Martínez-Arranz I, Navasa M, Falcón-Pérez JM, et al. A metabolomics signature linked to liver fibrosis in the serum of transplanted hepatitis C patients. Scientific reports. 2017;7(1):10497.
129. Shao L, Ling Z, Chen D, Liu Y, Yang F, Li L. Disorganized gut microbiome contributed to liver cirrhosis progression: a meta-omics-based study. Frontiers in microbiology. 2018;9:3166.
130. Kim DJ, Cho EJ, Yu K-S, Jang I-J, Yoon J-H, Park T, et al. Comprehensive metabolomic search for biomarkers to differentiate early stage hepatocellular carcinoma from cirrhosis. Cancers. 2019;11(10):1497.
131. Luo P, Yin P, Hua R, Tan Y, Li Z, Qiu G, et al. A Large-scale, multicenter serum metabolite biomarker identification study for the early detection of hepatocellular carcinoma. Hepatology. 2018;67(2):662–75.
132. Wang-Sattler R, Yu Z, Herder C, Messias AC, Floegel A, He Y, et al. Novel biomarkers for pre-diabetes identified by metabolomics. Molecular systems biology. 2012;8(1):615.
133. Tulipani S, Palau-Rodriguez M, Alonso AM, Cardona F, Marco-Ramell A, Zonja B, et al. Biomarkers of morbid obesity and prediabetes by metabolomic profiling of human discordant phenotypes. Clinica Chimica Acta. 2016;463:53–61.
134. Cobb J, Gall W, Adam K-P, Nakhle P, Button E, Hathorn J, et al. A novel fasting blood test for insulin resistance and prediabetes. Journal of Diabetes Science and Technology. 2013;7(1):100–10.
135. Hallows KR, Althouse AD, Li H, Saitta B, Abebe KZ, Bae KT, et al. Association of baseline urinary metabolic biomarkers with ADPKD severity in TAME-PKD clinical trial participants. Kidney360. 2021;2(5):795.
136. Su X, Ren R, Yang L, Su C, Wang Y, Lu J, et al. Serum Biomarkers for Chronic Renal Failure Screening and Mechanistic Understanding: A Global LC-MS-Based Metabolomics Research. Evidence-Based Complementary and Alternative Medicine. 2022;2022.

137. Feldstein AE, Lopez R, Tamimi TA-R, Yerian L, Chung Y-M, Berk M, et al. Mass spectrometric profiling of oxidized lipid products in human nonalcoholic fatty liver disease and nonalcoholic steatohepatitis [S]. Journal of lipid research. 2010;51(10):3046–54.
138. Hu C, Wang T, Zhuang X, Sun Q, Wang X, Lin H, et al. Metabolic analysis of early nonalcoholic fatty liver disease in humans using liquid chromatography-mass spectrometry. Journal of translational medicine. 2021;19:1–10.
139. Kurbatova N, Garg M, Whiley L, Chekmeneva E, Jiménez B, Gómez-Romero M, et al. Urinary metabolic phenotyping for Alzheimer's disease. Scientific reports. 2020;10(1):21745.
140. Marksteiner J, Blasko I, Kemmler G, Koal T, Humpel C. Bile acid quantification of 20 plasma metabolites identifies lithocholic acid as a putative biomarker in Alzheimer's disease. Metabolomics. 2018;14:1–10.
141. Fukuhara K, Ohno A, Ota Y, Senoo Y, Maekawa K, Okuda H, et al. NMR-based metabolomics of urine in a mouse model of Alzheimer's disease: identification of oxidative stress biomarkers. Journal of clinical biochemistry and nutrition. 2013;52(2):133–8.
142. Mallet D, Dufourd T, Decourt M, Carcenac C, Bossù P, Verlin L, et al. A metabolic biomarker predicts Parkinson's disease at the early stages in patients and animal models. The Journal of clinical investigation. 2022;132(4).
143. McGarrah RW, Crown SB, Zhang G-F, Shah SH, Newgard CB. Cardiovascular metabolomics. Circulation research. 2018;122(9):1238–58.
144. Yu B, Flexeder C, McGarrah III RW, Wyss A, Morrison AC, North KE, et al. Metabolomics identifies novel blood biomarkers of pulmonary function and COPD in the general population. Metabolites. 2019;9(4):61.
145. Vinayavekhin N, Homan EA, Saghatelian A. Exploring disease through metabolomics. ACS chemical biology. 2010;5(1):91–103.
146. Wishart DS. Quantitative metabolomics using NMR. TrAC trends in analytical chemistry. 2008;27(3):228–37.
147. Tugizimana F, Piater L, Dubery I. Plant metabolomics: A new frontier in phytochemical analysis. South African Journal of Science. 2013;109(5–6):01–11.
148. Pitt JJ. Principles and applications of liquid chromatography-mass spectrometry in clinical biochemistry. The Clinical Biochemist Reviews. 2009;30(1):19.
149. Gowda GN, Djukovic D. Overview of mass spectrometry-based metabolomics: opportunities and challenges. Mass Spectrometry in Metabolomics: Methods and Protocols. 2014:3–12.
150. Rieley G. Derivatization of organic compounds prior to gas chromatographic–combustion–isotope ratio mass spectrometric analysis: identification of isotope fractionation processes. Analyst. 1994;119(5):915–9.
151. Zhao Y, Woo G, Thomas S, Semin D, Sandra P. Rapid method development for chiral separation in drug discovery using sample pooling and supercritical fluid chromatography–mass spectrometry. Journal of Chromatography A. 2003;1003(1–2):157–66.
152. Alexander A, Zhang L, Hooker T, Tomasella F. Comparison of supercritical fluid chromatography and reverse phase liquid chromatography for the impurity profiling of the antiretroviral drugs lamivudine/BMS-986001/efavirenz in a combination tablet. Journal of pharmaceutical and biomedical analysis. 2013;78:243–51.
153. Li D-X, Gan L, Bronja A, Schmitz OJ. Gas chromatography coupled to atmospheric pressure ionization mass spectrometry (GC-API-MS). Analytica chimica acta. 2015;891:43–61.
154. Ten-Doménech I, Martínez-Sena T, Moreno-Torres M, Sanjuan-Herráez JD, Castell JV, Parra-Llorca A, et al. Comparing targeted vs. untargeted ms2 data-dependent acquisition for peak annotation in LC–MS metabolomics. Metabolites. 2020;10(4):126.
155. Ronsein GE, Pamir N, von Haller PD, Kim DS, Oda MN, Jarvik GP, et al. Parallel reaction monitoring (PRM) and selected reaction monitoring (SRM) exhibit comparable linearity, dynamic range and precision for targeted quantitative HDL proteomics. Journal of proteomics. 2015;113:388–99.
156. Dutta NK, Tornheim JA, Fukutani KF, Paradkar M, Tiburcio RT, Kinikar A, et al. Integration of metabolomics and transcriptomics reveals novel biomarkers in the blood for tuberculosis diagnosis in children. Scientific reports. 2020;10(1):19527.

157. Xiao N, Nie M, Pang H, Wang B, Hu J, Meng X, et al. Integrated cytokine and metabolite analysis reveals immunometabolic reprogramming in COVID-19 patients with therapeutic implications. Nature Communications. 2021;12(1):1618.
158. Codo AC, Davanzo GG, de Brito Monteiro L, de Souza GF, Muraro SP, Virgilio-da-Silva JV, et al. Elevated glucose levels favor SARS-CoV-2 infection and monocyte response through a HIF-1α/glycolysis-dependent axis. Cell metabolism. 2020;32(3):437–46. e5.
159. Ley K. M1 means kill; M2 means heal. The Journal of Immunology. 2017;199(7):2191–3.
160. Sudhakar P, Verstockt B, Cremer J, Verstockt S, Sabino J, Ferrante M, et al. Understanding the molecular drivers of disease heterogeneity in Crohn's disease using multi-omic data integration and network analysis. Inflammatory Bowel Diseases. 2021;27(6):870–86.
161. Wlodarska M, Finlay B. Host immune response to antibiotic perturbation of the microbiota. Mucosal immunology. 2010;3(2):100–3.
162. Pickard JM, Zeng MY, Caruso R, Núñez G. Gut microbiota: Role in pathogen colonization, immune responses, and inflammatory disease. Immunological reviews. 2017;279(1):70–89.
163. Maceyka M, Spiegel S. Sphingolipid metabolites in inflammatory disease. Nature. 2014;510(7503):58–67.
164. Li Q, Chen G, Zhu D, Zhang W, Qi S, Xue X, et al. Effects of dietary phosphatidylcholine and sphingomyelin on DSS-induced colitis by regulating metabolism and gut microbiota in mice. The Journal of Nutritional Biochemistry. 2022;105:109004.
165. Ye X, Li H, Anjum K, Zhong X, Miao S, Zheng G, et al. Dual role of indoles derived from intestinal microbiota on human health. Frontiers in Immunology. 2022;13.
166. Hernández-Rocha C, Borowski K, Turpin W, Filice M, Nayeri S, Raygoza Garay JA, et al. Integrative analysis of colonic biopsies from inflammatory bowel disease patients identifies an interaction between microbial bile acid-inducible gene abundance and human angiopoietin-like 4 gene expression. Journal of Crohn's and Colitis. 2021;15(12):2078–87.
167. Bosscher D, Breynaert A, Pieters L, Hermans N. Food-based strategies to modulate the composition of the microbiota and their associated health effects. Journal of physiology and pharmacology/Polish Physiological Society-Kraków, 1991, currens. 2009;60(S: 6):5–11.
168. Cui L, Lee YH, Thein TL, Fang J, Pang J, Ooi EE, et al. Serum metabolomics reveals serotonin as a predictor of severe dengue in the early phase of dengue fever. PLoS neglected tropical diseases. 2016;10(4):e0004607.
169. Vial T, Tan W-L, Deharo E, Missé D, Marti G, Pompon J. Mosquito metabolomics reveal that dengue virus replication requires phospholipid reconfiguration via the remodeling cycle. Proceedings of the National Academy of Sciences. 2020;117(44):27627–36.
170. Amemiya T, Gromiha MM, Horimoto K, Fukui K. Drug repositioning for dengue haemorrhagic fever by integrating multiple omics analyses. Scientific reports. 2019;9(1):1–13.
171. Dos Santos RO, Gonçalves-Lopes RM, Lima NF, Scopel KK, Ferreira MU, Lalwani P. Kynurenine elevation correlates with T regulatory cells increase in acute Plasmodium vivax infection: a pilot study. Parasite Immunology. 2020;42(3):e12689.
172. Gardinassi LG, Arévalo-Herrera M, Herrera S, Cordy RJ, Tran V, Smith MR, et al. Integrative metabolomics and transcriptomics signatures of clinical tolerance to Plasmodium vivax reveal activation of innate cell immunity and T cell signaling. Redox biology. 2018;17:158–70.
173. Ito K, Caramori G, Adcock IM. Therapeutic potential of phosphatidylinositol 3-kinase inhibitors in inflammatory respiratory disease. Journal of Pharmacology and Experimental Therapeutics. 2007;321(1):1–8.
174. Jamalkandi SA, Mirzaie M, Jafari M, Mehrani H, Shariati P, Khodabandeh M. Signaling network of lipids as a comprehensive scaffold for omics data integration in sputum of COPD patients. Biochimica et Biophysica Acta (BBA)-Molecular and Cell Biology of Lipids. 2015;1851(10):1383–93.
175. Diab KJ, Adamowicz JJ, Kamocki K, Rush NI, Garrison J, Gu Y, et al. Stimulation of sphingosine 1-phosphate signaling as an alveolar cell survival strategy in emphysema. American journal of respiratory and critical care medicine. 2010;181(4):344–52.

176. Shimoda LA, Laurie SS. HIF and pulmonary vascular responses to hypoxia. Journal of applied physiology. 2014;116(7):867–74.
177. Armstrong MB, Towle HC. Polyunsaturated fatty acids stimulate hepatic UCP-2 expression via a PPARα-mediated pathway. American Journal of Physiology-Endocrinology and Metabolism. 2001;281(6):E1197–E204.
178. Dougherty BV, Rawls KD, Kolling GL, Vinnakota KC, Wallqvist A, Papin JA. Identifying functional metabolic shifts in heart failure with the integration of omics data and a heart-specific, genome-scale model. Cell Reports. 2021;34(10):108836.
179. Lee SE, Schulze K, Stewart CP, Cole RN, Wu LS, Eroglu A, et al. Plasma proteome correlates of lipid and lipoprotein: biomarkers of metabolic diversity and inflammation in children of rural Nepal [S]. Journal of lipid research. 2019;60(1):149–60.
180. Deng W, Ednie AR, Qi J, Bennett ES. Aberrant sialylation causes dilated cardiomyopathy and stress-induced heart failure. Basic research in cardiology. 2016;111:1–14.
181. Schugar RC, Shih DM, Warrier M, Helsley RN, Burrows A, Ferguson D, et al. The TMAO-producing enzyme flavin-containing monooxygenase 3 regulates obesity and the beiging of white adipose tissue. Cell Reports. 2017;19(12):2451–61.

Pharmacometabolomics: General Applications of Metabolomics in Drug Development and Personalized Medicine

Dung Thuy Tran and Amber Dahlin

1 Introduction to Pharmacometabolomics

Metabolomics is the study of small molecules (metabolites) present in cells, organs, and body fluids. In contrast to genomic, proteomic, transcriptomic, and other "omics" data types, metabolomics is representative of the actual physiological status of patients in real time, as it profiles biochemicals that are most closely related to the metabolic status of cells or tissues [32].

The responses of patients to xenobiotics (drugs) are influenced by a variety of factors (genetic and environmental) that can be interrogated using various "omics"-based approaches. *Pharmacogenomics* is the study of how genetic information influences patient responses to drug treatment, or the extent of contribution of genetic variation to a drug response phenotype. "Drug response" broadly refers to the net effect of drug treatment or exposure on the disease, symptom, target organ or tissue, and it is influenced by numerous host and environmental factors. Through the development of methodology to conduct high-throughput genetic studies in patients, and the availability of large-scale population genomic data through the Human Genome Project and related resources, scientists have identified and characterized numerous genetic variants that contribute to patients' drug responses. Although a number of clinically relevant pharmacogenomic relationships underlying drug responses have been identified to date, leading to 500 drug label warnings for FDA approved drugs or therapies (Table of Pharmacogenomic Biomarkers in Drug Labeling; https://www.fda.gov/drugs/science-and-research-drugs/table-pharmacogenomic-biomarkers-drug-labeling), the ability of clinicians to predict patient responses to drug therapies has been limited. Host genetics accounts for

D. T. Tran (✉) · A. Dahlin
Channing Division of Network Medicine, Department of Medicine,
Brigham and Women's Hospital and Harvard Medical School, Boston, MA, USA
e-mail: dung.t.tran@channing.harvard.edu; amber.dahlin@channing.harvard.edu

V. Soni, T. E. Hartman (eds.), *Metabolomics*,
https://doi.org/10.1007/978-3-031-39094-4_5

20–40% of the variation of patients' responses to drugs [116], with the remaining 60–80% resulting from epigenetic and environmental factors including diet, lifestyle, and microbiome [9]. Therefore, well-defined strategies that account for the influence of non-genetic factors is necessary, and pharmacometabolomics represents a promising approach to better understand and predict drug response.

Pharmacometabolomics involves the collection of quantitative data on a large number of metabolites to discover biochemicals and biochemical pathways related to drug response outcomes [57]. Pharmacometabolomics was firstly defined as "the prediction of the outcome of a drug or xenobiotic intervention in an individual based on a mathematical model of pre-intervention metabolite signatures" [17]. In this chapter, we will discuss the application of pharmacometabolomics in drug discovery and development, describe methodological details and challenges, present compelling evidence from pharmacometabolomics studies published to date, and highlight examples of these applications in treatment of different diseases.

2 Methods and Resources for Pharmacometabolomics Investigations

From the perspective of metabolite data generation, a targeted or untargeted approach can be considered. The targeted approach is applicable in profiling a set of specific, well-defined metabolites [83]. This approach has been used in drug development when the target of a drug or disease process is relatively understood [9]. In contrast, the untargeted approach profiles the global pool of metabolites, many of which are novel or un-named, in a biological fluid, tissue, or cell [83]. Targeted metabolomics is useful for evaluating metabolites associated with specific conditions, such as comparisons of healthy vs. pathological groups [9]. Targeted metabolomics could be considered hypothesis-driven, whereas untargeted metabolomics is discovery-based. In both cases, the resulting dataset will include a list of identified metabolites and concentrations or fold changes across samples, which can then be investigated using statistical models to discern relationships between exposure (metabolite) and outcome (e.g., disease outcome or drug response).

2.1 Analytical Techniques

The development of analytical methods routinely employed in metabolomics studies, including nuclear magnetic resonance (NMR) and mass spectrometry (MS) coupled with different separation techniques, has rapidly accelerated pharmacometabolomics discoveries (Fig. 1) (Zeki et al. [135]). Liquid and gas chromatography, and electrophoresis (LC, GC, and CE, respectively), are the leading technologies for metabolite separation, while NMR and MS are state-of-the-art methods for

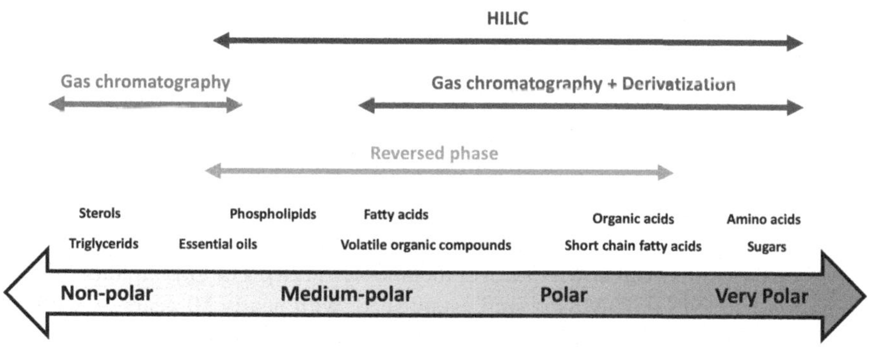

Fig. 1 Analysis methods of various molecular metabolites. (Adapted from Zeki et al. [135])

Table 1 The advantages and disadvantages of different analysis techniques in metabolomics studies [30, 59]

Analysis technique	Method	Application	Advantages	Disadvantages
GC-MS	Causes high-resolution separation of compounds in the gas phase using gas chromatography	Analysis of volatile metabolites	High resolution, able to resolve complex samples, can analyze different classes simultaneously	Cannot analyze thermolabile compounds, non-volatile metabolites must be derivatized, hard to identify unknown compounds after derivatizations
LC-MS	Combines high pressure liquid chromatography (HPLC) and MS to separate then quantify molecules	Metabolite profiling studies	High sensitivity, good chromatographic resolution	Not appropriate for polar and ionic species including metabolites that are components of biochemical pathways
CE-MS	Combines the liquid separation process of capillary electrophoresis with MS	Analysis of metabolites in small volumes	High resolution, small volumes	High complexity, still in development phase
NMR	Measures molecular structures by identifying signals from protons resonating within a magnetic field	Untargeted metabolomics methods	Can detect high number of metabolites quickly, allows absolute quantification with less steps	Costly, low sensitivity, requires large sample size, high complexity

metabolite detection [50]. NMR is generally a stand-alone technology, while MS is used with LC or GC separations to reduce sample complexity [78]. Each analytical technique has respective advantages and disadvantages (Table 1) [30, 59], so it is often recommended to analyze samples using different analytical methods in order to obtain comprehensive and reliable data.

2.1.1 Mass Spectrometry (MS)-Based Approaches

MS-based methods have high sensitivity, facilitating a wide range of metabolomics studies, by monitoring mass over charge number (m/z) values [128]. It is a powerful technique to identify unknown or novel metabolites [128]. The combination of GC separation with MS (GC-MS) is well established, with nearly 50 years in application, and is capable of analyzing amino acids, steroids, hormones, fatty acids, and intermediates of primary metabolism [33]. It confers several advantages including low cost, ease of operation, and high separation efficiency, making it the leading technology in metabolomics. Generally, the sample transits a gas chromatography unit resulting in a high-resolution separation of volatile compounds in the gas phase. As a result, the method is mainly utilized for the analysis of volatile metabolites [136]. Non-volatile samples need to be derivatized prior to analyzing to reduce polarity and increase stability, which can be tedious for sample processing [136].

LC-MS combines liquid chromatography (LC) and MS into a complete process from separation to quantification of molecules. LC-MS changes the mobile phase from gas to liquid to overcome the limitations of GC-MS. This change means that LC-MS does not need metabolites to be volatile before analyzing; hence, sample derivatization is not required. It has several advantages including column separation, the ability to detect many chemicals, need for smaller sample volumes, and is less expensive compared to NMR [62]. It is also suitable for relatively polar compounds from low to high molecular weights [31]. For flexible and efficient separation, high performance liquid chromatography (HPLC) allows separation of compounds with different polarity ranges using isocratic or gradient elution methods (Yukta [132]). HPLC first separates molecules based on different physical and chemical properties such as molecular size, polarity, and affinity, using the reversed phase mode. Next, in the stationary phase, the movement of molecules based on their molecular sizes is limited, facilitating separation. Column chromatography then can be used to purify individual chemical compounds from the mixture [21]. Currently, HPLC is widely used in different fields including medicine and biochemistry. This technique can analyze over 70% of organic compounds with an average analysis time of 15–30 minutes [121]. HPLC can also couple with fluorescent detection (FLD) with high sensitivity, high selectivity, and repeatability. For example, HPLC-FLD performed through direct or indirect methods can be applied for quantification of steroids, as shown in Fig. 2 [44]. Some non-native steroids such as trenbolon can be quantified without the derivatization due to the fluorescence properties while substances which do not exhibit fluorescence need to undergo derivatization before analysis.

In 2004, ultra-high-performance LC (UHPLC) was introduced [89]. Similar to HPLC, UHPLC is a liquid chromatography technique, but UHPLC operates at higher pressure (15,000 psi) and is suitable for smaller particle sizes [89], while

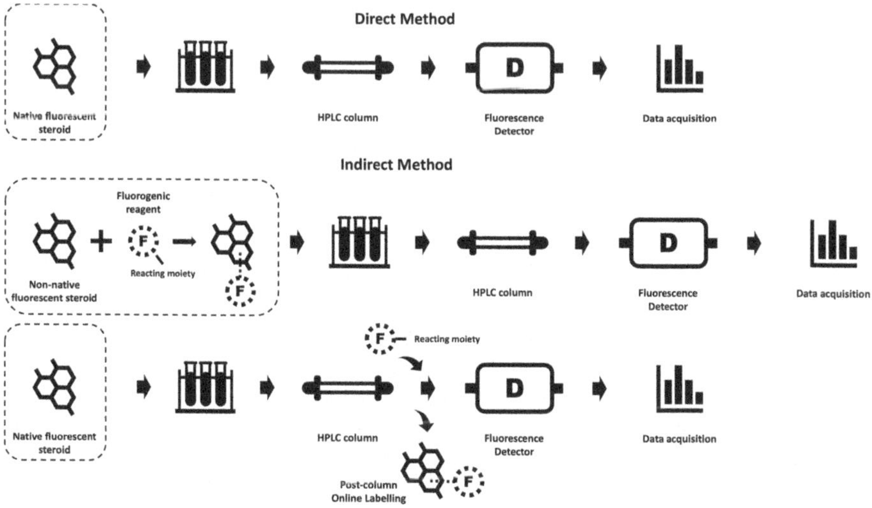

Fig. 2 High performance liquid chromatography with fluorescence detection for quantification of steroids through direct or indirect method. (Adapted from Hameedat et al. [44])

HPLC performs at lower pressure (< 6000 psi) [122]. Therefore, compared to HPLC, UHPLC is faster, which decreases solvent consumption and shortens run times. As a result, UHPLC can potentially increase the number of metabolites detected by MS. Wilson et al. [123] reported that UHPLC provided more than double peak capacity, tenfold increase in speed, and up to fivefold increase in sensitivity. Additionally, another study also indicated UHPLC-MS offered a 20% increase in detected components in human serum, compared to HPLC [82]. Currently, it has become more popular in metabolomic profiling and is considered a novel LC technique for various human biofluids [24]. Hydrophilic interaction liquid chromatography (HILIC) is another new and popular LC-based separation technique that couples with MS [64]. HILIC combines a silica- or polymer-based stationary phase, a mobile phase in reversed phase separation mode, and ion exchange chromatography (IEX) [65]. The HILIC separation technique confers several advantages, with the capability of analyzing polar metabolites due to improved retention capability. HILIC has been used to separate proteins including histones, membrane proteins, and neoglycoproteins [37]. Additionally, HILIC can also separate different lipid classes based on their polar head and charge. For instance, positional isomers of polar lysophospholipids, including lysophosphoglycerols, lysophosphocholines, and lysophosphoethanolamines, were analyzed using HILIC-MS [72].

In addition to GC and LC separation techniques, the capillary electrophoresis (CE-MS) method was introduced in 2003 by Soga et al. [101]. CE is a separation technique based on the movement of ions through the capillary at different rates due to different electrophoretic charges [39]. Therefore, compared to other methods, the CE-MS method has greater resolution for quantifying polar metabolites including amino acids and carbohydrates [101]. Unlike GC-MS, CE-MS does not need derivatization. CE-MS was employed to analyze large-scale studies with high-resolution power and high reproducibility, and the advantage of requiring lower sample and solvent volumes [11]. For example, CE-MS analyzed polar metabolites in plasma samples from 11,002 individuals, providing absolute quantification of 94 polar and charged metabolites [45]. Additionally, CE-MS can quantify metabolites from different biofluids [92]. Saliva samples are a preferred fluid for metabolomic profiling due to simplicity, non-invasiveness, and lower cost [92]. In a metabolomics study of oral cancer patients, CE-MS was utilized to analyze saliva samples to determine optimal sample collection times. Concentrations for 58 metabolites were significantly different between control and oral cancer patients at the 12-hour fasting time point [49]. Despite its many advantages, CE-MS has not been as widely used compared to LC-MS. Several disadvantages of this technique include limited loading capacity (which prohibits the loading of large sample volumes), limits of detection [127], and unsuitability for analyzing high molecular weight proteins (>20 kDA) [10].

2.1.2 Nuclear Magnetic Resonance (NMR)

NMR is a powerful spectroscopic method used to determine the molecular structures of metabolites by measuring signals from protons resonating within a magnetic field [85]. NMR is able to detect a large number of metabolites (<1 kDa) in different biological samples including urine, plasma, saliva, and tissues [36]. NMR can detect many metabolites over a short time period; for example, a single proton spectrum quantified 100 metabolites in human urine samples [20]. NMR is applicable for both targeted and untargeted metabolomics approaches [30]. The technique is also suitable for detection of amino acids, carbohydrates, alcohols, and organic acids [118]. A major advantage of NMR is that it does not require additional steps of sample preparation including separation and derivatization [29]. However, compared with MS, NMR has lower sensitivity and is limited to identifying 50–200 metabolites with concentrations <1 μM, while MS is able to identify >1000 metabolites with concentrations >10 nM [30]. NMR also requires highly skilled and trained operators, and machine costs are higher than MS [29].

The two major types of NMR include ^{1}H NMR and ^{13}C NMR; these techniques differ in their detection of the type and number of hydrogen (proton) or carbon nuclei, respectively, that are present within a molecule. Both also differ in their methodologies used to obtain NMR spectra. The continuous wave method (slower) and Fourier transform (faster) are utilized by ^{1}H NMR and ^{13}C NMR, respectively. The two techniques also differ in the type of spectral data generated [90]. Metabolomics studies generally utilize ^{1}H NMR due to its superior sensitivity. However, ^{13}C NMR is also performed in tandem with ^{1}H NMR due to advantages it confers over ^{1}H NMR in the breadth and type of molecules containing carbon backbones that can be detected [18].

2.2 Repositories and Databases for Pharmacometabolomics Studies

Publicly available, online databases of metabolites and metabolite reactions condense complex information for thousands of metabolites from multiple species into a single resource. These databases are useful for annotating and characterizing metabolites of interest and performing straightforward bioinformatic or statistical analyses. A list of current resources is presented in Table 2.

Currently, HMDB is the largest and most comprehensive database, containing 217,920 annotated metabolites [125]. A new version, HMDB 5.0, includes a new Chemical Functional Ontology (ChemFOnt), which describes the biological and industrial function of metabolites [125]. Each metabolite entry provides information including chemical structures, identifiers, descriptions, and other chemical and biological information, which are stored in a "MetaboCard" web page. Each MetaboCard page also links to many other databases (including KEGG, BioCyc, PubChem, ChEBI, PubMed, Genbank, and dbSNP) [126]. The Small Molecular Pathway Database (SMPDB) for human metabolites consists of over 600 hand-drawn pathways, 280 of which are unique to SMPDB, describing metabolic and physiological action pathways [35, 51]. SMDB 2.0 also includes information about reactions occurring in cellular compartments, organs, and tissues, and the organs targeted by drugs or toxic metabolites. SMDB 2.0 links to other databases including HMDB, DrugBank, UniProt (protein search), TextQuery (text searching utilities),

Table 2 Metabolomic databases

Database	Description	Link
BioCyc	20,024 pathway/genome database for eukaryotes providing reference to genome and metabolic pathway	https://biocyc.org/
Metabolite database	Structures and annotations of 167,000 metabolites	https://www.metabolomicsworkbench.org/
Human Metabolome Database (HMDB)	Comprehensive collection of human metabolites with a total of 220,945 metabolites	https://hmdb.ca/
Human fecal metabolome database (HFMDB)	Freely database with ~6000 human fecal metabolites	https://fecalmetabolome.ca/
Metabolite and tandem MS database (METLIN)	Over a million molecules ranging such as lipids, steroids, small peptides. From MS/MS analysis	https://metlin.scripps.edu/l
Massbank	Public database of mass spectra of small chemical compounds	https://massbank.eu/MassBank/
LipidMaps	Structures and annotations of biological relevant lipids	https://www.lipidmaps.org/
Chemical Entities of Biological Interest (ChEBI)	Freely resources of small chemical compounds	https://www.ebi.ac.uk/chebi/

ChemQuery (chemical searching), Sequence Search, SNP-Browse (pathway browsing), and others [51]. Another helpful online metabolomics resource is the Metabolomics Workbench Metabolite Database (www.metabolomicsworkbench.org), a public repository containing 60,000 chemical structures and annotations of biologically relevant metabolites [103]. Currently, it includes over 130,000 entries collected from over 2000 public sources. Other similar metabolic data repositories and databases include MetaboLights, KEGG, LIPID MAPS, CheBI, HMDB, ChemSpider, and MetaCy. As new pharmacometabolomics information is generated, published online, and curated within these repositories, these useful resources will continue to develop.

In addition to these databases, web-based software tools are available for analyzing metabolites and performing statistical procedures to evaluate outcomes and infer biological relevance. MetaboAnalyst (https://www.metaboanalyst.ca/) is the most widely used and comprehensive platform for high-throughput metabolite analysis informed by biological knowledge, and it is relevant for analyzing data from both targeted and untargeted assays. The most current version (V5.0) facilitates multiple analytical approaches from processing of raw MS spectra to performing statistical tests including meta-analyses and integration of multiple "omics" data types [84]. Multiple types of statistical analyses can be conducted using a list of metabolite identifiers and their concentrations or fold change, from simple (e.g., t-test, correlation analysis) to complex, including machine learning approaches such as supervised classification. MetaboAnalyst also facilitates biomarker identification through receiver operating characteristic (ROC) curve analysis, in addition to metabolite set enrichment and pathway analysis. Molecules of interest can also be visually described in the context of biological and biochemical networks using the network analysis function. A limitation of MetaboAnalyst is that it currently accepts only one kind of metabolite identifier (e.g., KEGG or HMDB identifier) at a time as input data, which can reduce the number of metabolites included in the analysis if the dataset includes identifiers from different sources. Chemical Similarity Enrichment Analysis for Metabolomics (ChemRICH) (https://chemrich.fiehnlab.ucdavis.edu/) is a pathway enrichment approach to metabolite set analysis that can overcome this limitation as it facilitates multiple input identifiers, yielding sets of metabolites that do not overlap [7]. ChemRICH relies on chemical ontologies and structural similarities rather than pathway or biochemical annotations to group metabolites. Because P values are not derived using a background database (as for pathway mapping), the statistics generated through enrichment analysis are superior to those derived from annotation-driven pathway enrichment. The output of ChemRICH takes the form of multiple enrichment plots of metabolite clusters ordered by ontology and chemical similarity; interactive tables also include metabolite annotations to clusters, representations of metabolites within clusters, and the direction of change in metabolite effect (increase or decrease), with raw and adjusted P values [7].

3 Pharmacometabolomics as an Innovative Solution for Enhancing the Discovery and Development of Novel Drugs

The processes of drug discovery and development aim to identify new medications with novel modes of action or clinical indications [77]. Drug development comprises five stages: discovery and development, preclinical research, clinical research, FDA review, and FDA post-market safety monitoring (Fig. 3) [3]. The discovery step begins with the identification and validation of a biological target of interest (e.g., a receptor or gene) [93]. Target validation involves the application of various techniques including functional analysis (using in vitro assays or animal models) and expression profiling (of mRNA or protein) to demonstrate specificity and establish that a therapeutic benefit exists within an adequate range of safety [48]. After successfully validating the target, the next process is hit identification and characterization. A hit is a compound that interacts with the validated target and has desired activities. A hit can be identified by high-throughput screening using a large number of compounds or other screening approaches [48]. After identifying promising compounds, information related to how they are absorbed, distributed, metabolized, and excreted is determined, as is any presence of off-target effects or potential interactions of the compounds with other drugs. Prior to human testing, candidate drugs are investigated in vitro as well as in vivo using appropriate cellular and animal models. Both provide detailed information regarding activity, dosing, and toxicity, which can help to determine whether the drug can be prioritized for clinical research. After preclinical and clinical trials, if the candidate drug is determined to be safe and effective, a company can file an application to market the drug and the FDA review team examines all submitted information to make the decision to approve or decline. For the last stage, drugs are reviewed post-market to determine additional information including dosage or other clinical indications [93].

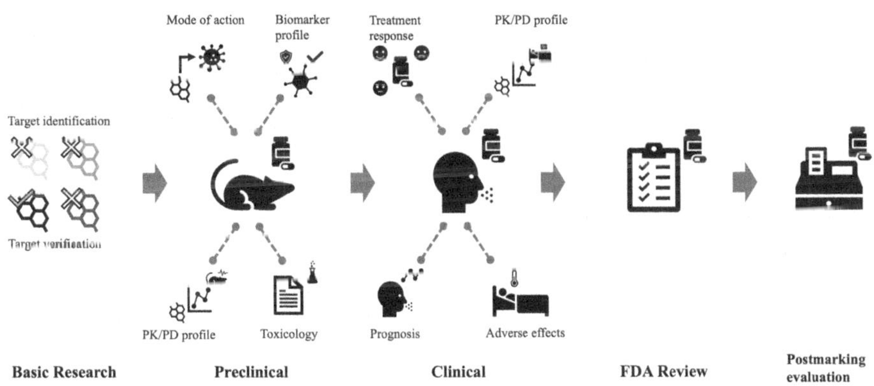

Fig. 3 The application of metabolomics in five stages of drug discovery and development phases. (Adapted from Alarcon-Barrera et al. [3])

Time and expense remain the greatest challenges for drug development. Approximately 12 years, on average, are required to develop a new medication from the time it is discovered to when it is available to patients [77]. The average cost for research and development for an approved medication is over $2 billion USD [77]. Not every drug can be marketed successfully, and the average rate of approval of new drugs by the FDA was ~40% over the past decade [1]. After entering clinical studies, generally, nine out of ten drug candidates will fail during the phase I-III clinical trials and drug approval stages [104]. Possible reasons for failure include a lack of clinical efficacy, unmanageable toxicity, issues with formulation, insufficient commercial need, and lack of efficient strategic planning [25, 46]. To account for these limitations, pharmacometabolomics approaches have become increasingly informative and valuable in minimizing development failure and saving resources from the initial drug discovery to preclinical and clinical stages [110]. We discuss the application of pharmacometabolomics in the drug discovery and development processes below.

3.1 Pharmacometabolomics as a Resourceful Tool in Drug Discovery and Development

3.1.1 Target Identification

Most drug discovery programs initially pursue identification of the molecular target(s) (mRNA, gene, protein, or receptor) directly related to disease pathogenesis, followed by screening for possible interactions [77]. Ideal drug targets have the property of "druggability,"that is, the ability to efficiently bind to small molecules, leading to therapeutic benefit [48]. This property is investigated through screening of chemical libraries and relevant cell lines or in vitro models. In traditional screening, a macromolecular target is selected, and an assay is performed to monitor the ability of small molecules to perturb the target [98]. One of the problems of this approach is that it can lead to misidentification of drug targets; in addition, traditional chemical screening can be time-consuming due to the large number of compounds that are interrogated. Pharmacometabolomics has the potential to overcome some of these limitations. Tiziani et al. [111] demonstrated an NMR-based metabolomics approach to develop a method for high-throughput screening of drug libraries in cancer cell lines. The authors monitored the global, intracellular, and extracellular metabolomic profiles following treatment of the cell lines with 56 kinase inhibitors [111]. PCA models of the metabolomic data proved clear group separation based on drug treatment or cell line, demonstrating the feasibility of a metabolomics-based approach for large-scale screening of drug-treated samples [111]. Recently, Holbrook-Smith et al. [47] also demonstrated the effectiveness of high-throughput metabolomics approaches for rapidly and accurately predicting drug-target relationships to speed the rate of discovery. This method required less than one minute per sample using flow injection time of flight mass spectrometry;

therefore, thousands of samples could be profiled in a single assay, facilitating large amounts of data for candidate drug identification [47]. The authors compared the metabolomic profiles of *Saccharomyces cerevisiae* samples that were subjected to either chemical perturbation or inducible overexpression of six intracellular and membrane bound proteins. Metabolomic profiles were collected from two groups of yeast that were either treated with 1280 drugs or were subjected to gene overexpression by a β-estradiol inducible overexpression system (Fig. 4). The authors implemented a synthetic promoter system, Z_4EV, that replaced the native gene promoter. In the absence of an inducer, Z_4EV was inactive. Following treatment of the samples with inducers, mRNA transcripts and intracellular metabolites were profiled and compared with profiles from the pre-treatment state [47]. If the post-treatment metabolite and gene overexpression profiles were highly correlated, a drug was considered a candidate for target prioritization. Using this approach, the authors identified 86 druggable genes for validation of the method, and validated five novel antagonists for the G protein coupled receptor, *GPR1* [47]. These results revealed the potential of high-throughput metabolomics in the identification of drug-target relationships [47].

The use of metabolomics data to map changes in biochemical characteristics onto canonical metabolic pathways and networks is also valuable for drug-target identification. With this approach, it is possible to explore the appropriateness of a particular drug or target before further investment in development. A detailed map of the relationships between known chemicals and metabolic networks for several species was provided by Adams et al. [2]. The authors used similarity ensemble approach (SEA) to link metabolic reactions and drug classes by their chemical

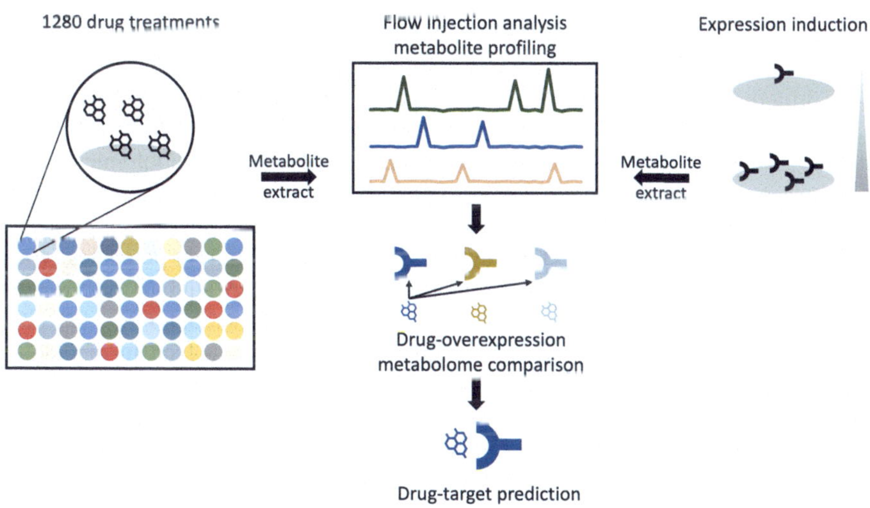

Fig. 4 High-throughput metabolomics for prediction of drug-target relationships in *Saccharomyces cerevisiae* by comparison of metabolomics profile between induced overexpression genes and drug treatments. (Adapted from Holbrook-Smith et al. [47])

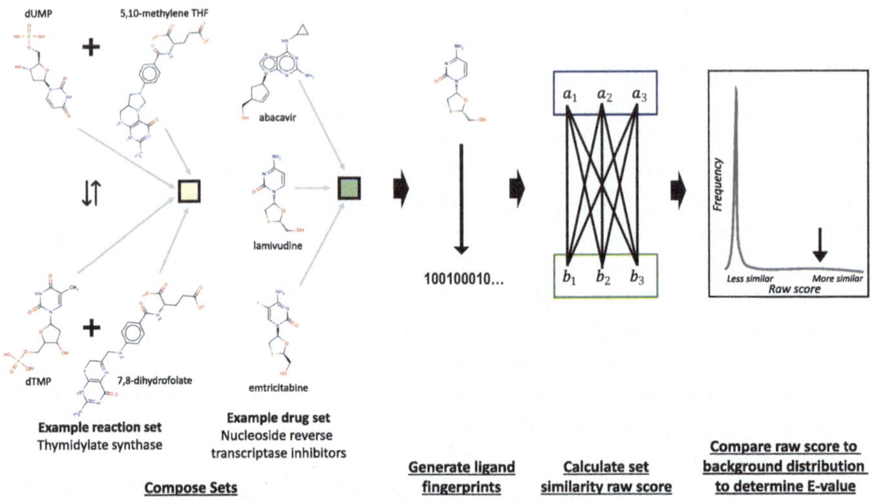

Fig. 5 Workflow of similarity ensemble approach (SEA) to link metabolic reactions and drug classes by chemical similarity among ligands. (Adapted from Adams et al. [2])

similarity, which were measured by comparing patterns of bonding between sets of molecules (Fig. 5) [2]. The map was then used to predict interactions between drug classes and metabolic targets.

3.1.2 Determining Mode of Action

Understanding the mode of action (MOA), which describes the functional cellular-level changes induced by exposure to a substance, is an essential component of the early drug discovery process [112]. Traditionally, the study of MOA has been based on biochemical and functional genomic approaches, but these are limited by reliability, scalability, and time/labor consumption [133]. Metabolomics-based approaches show great potential for assessing MOA. For example, Yao et al. [130] introduced a novel metabolomic workflow to investigate dose-response relationships for compounds that targeted multiple proteins with different potencies [100, 130]. A second example of applying metabolomics in MOA studies is shown in Fig. 6. An interaction network analysis of lipidomic and metabolomic data, performed by Xu et al. [129], identified regulatory enzymes involved in metabolic pathways affected by triphenyl phosphate (TPhP) (Fig. 6). Global metabolomics and lipidomics were applied to screen dysregulated metabolites and lipids; an interaction network analysis for the metabolites and lipids was conducted to target the most affected pathways and enzymes. The enzymes connected to two dysregulated metabolites or enzymes related to significantly dysregulated metabolic pathways were selected for docking. Next, molecular docking analysis was applied to model the interaction between small molecules at the atomic level by scoring binding

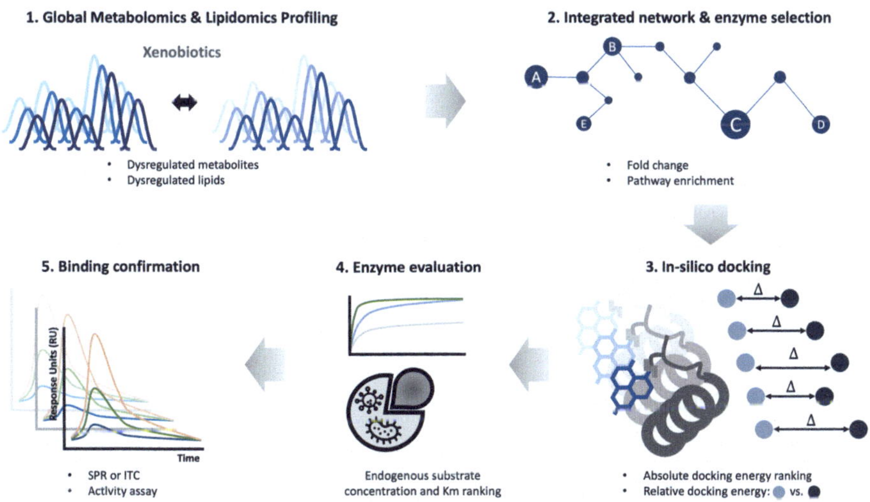

Fig. 6 "Bottom up" oriented target finding approach based on metabolomics and in silico docking for novel molecular target discovery. (Adapted with permission from Xu et al. [129]. Copyright 2022 American Chemical Society)

affinities and comparing binding energies. The next step involved the evaluation of selected enzymes based on their endogenous substrate concentration, followed by validation of interactions between molecules using biophysical and in vitro activity assays. This analysis revealed that TPhP showed relatively strong binding affinity [129].

Untargeted metabolomics was applied to investigate the MOA and short-term responses of *Escherichia coli* (*E.coli*) to nine antibiotics [134]. Changes in the levels of ~750 intracellular metabolites in *E. coli* cultures were observed after antibiotic exposure from 1 to 60 minutes. Post-treatment metabolic changes were both dose- and drug-dependent; in particular, levels of arginine were increased. Short-term responses of *E.coli* showed dynamic metabolic changes that were time dependent and reliant upon a sequence of events related to cell death through antibiotic-induced stress [134]. Later, the same group combined untargeted MS-based metabolomic profiling with ad hoc data mining to predict the MOA of small molecules [133]. The authors first created a reference database of metabolic responses following treatment of *Mycobacterium smegmatis* with 62 compounds, and then classified the responses [133]. The approach was validated against 212 anti-tuberculosis compounds with unknown MOAs. Of the 212 compounds, 77% could be classified and the MOA of seven compounds could be experimentally validated [133]. In another study, [6] used untargeted metabolomics to investigate the MOA of pretomanid, an antibiotic used to treat multi-drug-resistant tuberculosis, in comparison to eight different anti-tubercular drugs with known MOAs. Principal component analysis (PCA) indicated that pretomanid treatment generated a distinct metabolomic profile at all time points in *Mycobacterium smegmatis* compared with other known antibiotics [6]. A high level of phosphate sugars related to the accumulation of a toxic metabolite, methylglyoxal, was also observed following pretomanid treatment [6].

Metabolomics data may be combined with other "omics" datasets to investigate MOA. In a recent study from Campos and Zampieri [12], the metabolic response of *E.coli* to exposure to 1279 drugs was monitored by high-throughput metabolomic analysis. Compounds that induced specific metabolic changes were mapped to previously generated genomic network to identify genes associated with drug response [12]. Together, these studies collectively demonstrate how metabolomic analysis can elucidate MOA for various candidate drugs.

3.1.3 Drug Repurposing

Drug repurposing is a strategy for investigating new uses for approved drugs [91]. Drug repurposing confers numerous advantages including a lower risk of failure, and reduced time and cost for development [91]. Drug repurposing is a relatively new focus for pharmaco-"omics" investigations. Khosravi et al. integrated genetic, biomedical, transcriptomic, and metabolomic data with information from drug databases in a drug repurposing strategy to identify new targets and potential drug candidates for melanoma. Thirty-five drugs approved for other indications were identified that had promise for treatment of melanoma [63]. Recently, two compounds, sertraline and thimerosal, were repurposed for serine/glycine synthesis-addicted breast cancer treatment using isotope tracer metabolomics, enzymatic assays, and drug-target interaction studies; the authors found that sertraline, an antidepressant, can inhibit the serine/glycine synthesis enzyme, serine hydroxymethyltransferase (SHMT) [38]. Together, these studies demonstrate the potential of integration of metabolomics with other kinds of data in pursuit of advancing drug repurposing strategies.

3.1.4 Drug Safety Evaluation

Drug safety evaluation is a critical component of drug development and a top priority for regulatory agencies. Safety evaluation begins in preclinical studies and continues in the post-approval stages. Safety evaluation includes performing biochemical and histological analyses, but these approaches have disadvantages including toxicities originating from off-target effects, lack of information regarding the mechanisms of drug toxicity, and the fact that not all adverse effects are directly related to drug exposure or dosage. As an alternative, metabolomics has been successfully applied to investigate drug toxicity. Animal toxicology screens are among the earliest applications of metabolomics in pharmaceutical studies. The best known example is provided by the Consortium for Metabonomic Toxicity (COMET), a consortium of five major pharmaceutical companies and Imperial College London, UK, which predicted kidney and liver toxicity of drug-like compounds in rodents

[71]. Below, we highlight various applications of metabolomics in profiling drug metabolism and drug toxicity.

Based on the pharmacological and toxicological properties of a drug, its metabolite(s) are classified into three types: active, inactive, and reactive. Active metabolites arise through biotransformation of a parent drug into a modified form(s), which produces the desired therapeutic effects. In some cases, drugs are metabolized into an inactive form that either has little or no direct effects, or requires biotransformation to an active form to exert its effects. Finally, reactive metabolites are biotransformation products that play a role in toxicity [88]. Reactive metabolites bind to cellular macromolecules including protein and DNA, thereby altering cellular function and leading to toxic effects. The earlier such reactivity is detected, the greater the chances that toxic effects can be mitigated or avoided [5]. However, as reactive metabolites are unstable, they are often difficult to directly detect [70].

Three techniques are used to evaluate reactive metabolite formation, including (i) evaluation of covalent binding of metabolites to proteins; (ii) trapping and characterizing reactive metabolites; and (iii) time- and co-factor-dependent cytochrome P450 (CYP) inhibition [74]. Of these, trapping reagents are more commonly used to identify reactive metabolites. Nucleophilic chemical trapping reagents, such as glutathione tri-peptide (GSH), potassium cyanide (KCN), or semicarbazide, are able to form stable adducts with many reactive species. GSH is a tripeptide in mammalian systems and its nucleophilic cysteinyl thiol group can trap electrophilic species to form GSH conjugates [81]. LC-MS has been used to screen trapped reactive metabolites [70]. In this investigation, GSH was used to trap reactive metabolites in human liver microsomes, and the trapped metabolites were analyzed with UPLC and time of flight mass spectrometry (TOFMS) [70]. This method identified a large number of stable and reactive metabolites formed through bioactivation of xenobiotics [70].

3.2 *Pharmacogenomic and Pharmacometabolomic Approaches in Studies of Drug Response Outcomes*

The recent inclusion of pharmacometabolomics with pharmacogenomics research increases the potential to better predict treatment outcomes for patients. Below, we discuss how pharmacogenomics and pharmacometabolomics can provide new insights into drug responses, using the treatment of asthma as an example.

Asthma is chronic inflammatory disorder of the airways and is a common complex disease worldwide, affecting >300 million people [8]. The goal of asthma treatment is to achieve and maintain symptomatic control using commonly prescribed medications including inhaled corticosteroids, long-acting agonists, leukotriene modifiers, methylxanthines, and omalizumab. However, treatment response demonstrates significant inter-individual variation among patients, which can be attributed to genetics and other host factors. Patient genetics may actually be responsible for 60–80% of the observed variation in response to asthma medications [26]. Poor responsiveness to asthma medications increases the severity of asthma symptoms and can lead to increased risk of exacerbations and risk of death. For example, reduced responsiveness to inhaled short-acting β_2 agonist was associated with a 45% increase in risk of asthma exacerbations after 3 months of treatment [34]. In 2019, the updated Global Strategy for Asthma Management and Prevention (GINA) guideline recommended replacing short-acting β_2 agonist with other treatments such as inhaled corticosteroids (ICS) for safety [4]. ICS is a preferred medication for asthma patients, but significant variability in response to ICS has been reported, with more than 30% of patients demonstrating poor or non-response [13]. One study showed that only 15% of patients taking ICS for asthma symptom control had adequate response [106]. The PRICE study also reported 40% of patients were ICS non-responders after 6-week ICS treatment [76]. Therefore, a fixed treatment regimen could not reliably treat all individuals with the same diagnosis. Pharmacogenomics approaches have been employed to obtain a deeper understanding of variation within drug response, with the ultimate goal of individualizing asthma therapy. In recent years, an increasing number of studies have incorporated pharmacometabolomics to address this issue.

The relationship between genetic variability in patients and differential responsiveness to ICS has been widely investigated, with over 150 studies related to this outcome reported to date. The first large-scale GWAS to identify determinants of ICS response in asthma patients was performed in 2011. The study combined genotype and phenotype data from multiple cohorts including Childhood Asthma Management Program (CAMP), Childhood Asthma Research and Education (CARE), and Asthma Clinical Research Network (ACRN) to investigate the potential association of genetic variation with measures of lung function in asthma patients taking budesonide, an ICS [109]. GWAS identified a SNP (rs37972) located in *glucocorticoid induced transcript 1 (GLCCI1)* that was significantly associated with changes in forced expiratory volume in one second (FEV_1) [109]. This study was one of the first to directly implicate a role of genetic variation within the glucocorticoid pathway on ICS-related changes in lung function. Another study by same group reported results from a GWAS to evaluate ICS response in 418 children and adults with asthma participating in the Single Nucleotide Polymorphism Health Association Asthma Resource Project (SHARP) [108]. The authors identified 47 SNPs significantly associated with ICS response [108]. Two replicating SNPs (rs6456042 and rs3127412) were annotated to a haplotype block with three SNPs present in functional regions of T gene [108]. Of the three SNPs, patients with rs1134481-T demonstrated an improved response to ICS (>10% increase in FEV_1 at the end of 8th week) [108]. In 2014, these authors performed a GWAS evaluating asthma symptoms (that were tracked using diary cards) in 124 children on ICS from the CAMP trial, and identified three significantly associated SNPs (rs1558726, rs2388639, and rs10044254) that also independently replicated [87]. Notably, rs10044254 was annotated to *F-box and leucine-rich repeat protein 7* (*FBXL7*) gene, which also showed decreased expression in immortalized B lymphocytes derived from CAMP subjects [87]. In 2020, a genome-wide interaction study (GWIS) reported the interaction of genetic variation with age on the outcome of frequency of exacerbations occurring on ICS, in over 1000 adult and pediatric patients with asthma [22]. Two SNPs annotated to the *THSD4* and *HIVEP2* genes were significantly associated with poorer ICS response with increasing age [22]. Recently, four novel SNPs on chromosome 6 were significantly associated with differential ICS response in older adult asthma patients from the Genetic Epidemiology on Adult Heath and Aging (GERA) cohort of 5710 participants [120].

These asthma pharmacogenomic studies established multiple genetic loci associated with different measures of ICS response (and age). However, the discovery of metabolites associated with ICS response in asthma is currently in its infancy. In a recent study, Kachroo et al. [53] investigated an association of plasma metabolomic profiles with asthma exacerbations in patients taking ICS, and also determined whether these profiles changed with age from adolescence to adulthood, using data from 170 asthma patients on ICS from the Mass General Brigham Biobank [53]. The authors evaluated metabolites associated with exacerbations as well as the potential effect modification of age, and identified eight metabolites with an effect modification by sex and 38 metabolites demonstrating a significant interaction with age at a nominal p-value threshold of 0.05 [53]. In particular, hexadecanedioate and tetradecanedioate were the most highly significant metabolites associated with exacerbations. Amino acid metabolism pathways including valine and arginine were also implicated [53]. A subsequent study by the same group also indicated that ICS use was linked to adrenal suppression, an adverse effect of long-term exogenous corticosteroid use, providing rationale for recommendation of treatment monitoring. Indeed, from blood samples of over 14,000 individuals from multiple asthma cohorts, 17 steroid metabolites including as DHEA-S and cortisol were significantly reduced in asthma patients and patients taking low-dose ICS [54]. Together with pharmacogenomic markers, these pharmacometabolomics studies also provide evidence of metabolites contributing to variation in ICS response, expanding knowledge of affected molecular pathways driven by specific genes and metabolites, and increasing the number of potential therapeutic targets for asthma treatment.

3.3 *The Importance of Pharmacometabolomics for Predicting Patients' Responses to Drugs*

Non-response or poor response to drugs is a serious issue in clinical care as well as clinical trials. Approximately 40–50% of clinical development programs are discontinued due to treatment failures [104]. In clinical care, several routinely used drugs including anti-inflammatories, anti-depressants, and anti-virals demonstrate limited efficacy within patients. A randomized trial of 326 patients taking aspirin showed that 5% developed resistance to aspirin [43]. Chen et al. [15] investigated the effect of aspirin resistance on myonecrosis among 151 patients taking clopidogrel; 29 developed aspirin resistance and increased risk of myonecrosis. Further, among patients taking peginterferons for hepatitis C treatment in two clinical trials,

18–20% of patients failed to respond and needed to retreat within 48 weeks [131]. Responses to drug treatment of major depressive disorder are often inconsistent, with over 30% of patients failing to respond to the first antidepressant prescribed and fewer than 50% of treated patients resolving symptoms [68]. In summary, as patients can respond differently to the same medication and treatment regimen, it becomes increasingly important to be able to predict a patient's response and tailor their treatment, saving time and healthcare costs.

Over the past decade, 3196 pharmacometabolomics studies have been published investigating the impact of many drug classes on metabolite profiles in patients [41]. Most of these studies have focused on identifying biomarkers, metabolic patterns associated with drug metabolism, characterizing responders vs. non-responders, and adverse drug effects. Samples can be collected before, during, and/or after drug treatment, and statistical models can then be built to evaluate treatment outcomes. The baseline metabolomics profile is useful for discerning the underlying metabolite profiles of subjects in the absence of treatment, while comparison of baseline metabolomics profiles with profiles collected during and after drug treatment can be used to predict whether a patient may have favorable or unfavorable drug responses [41].

A number of recent studies have identified biomarkers or metabolic patterns associated with responders and non-responders [9]. A paper published in 2010 firstly demonstrated that pharmacometabolomics could be used to predict clinical response [55]. This lipidomics study profiled plasma samples from 944 patients with differential responses to simvastatin. Simvastatin is used to treat cardiovascular disease, which one of the leading causes of death in the USA. In this study, participants took simvastatin for 6 weeks, after which their plasma lipid profiles were interrogated. Approximately 40 metabolites were altered by simvastatin in the adequate responders, but not in the poor responders [55]. Later, Kapoor et al. [60] used NMR spectroscopy to evaluate urine samples from patients undergoing treatment with anti-tumor necrosis factor α. The response rate of patients was variable and showed strong separation in baseline urine metabolites between patients with or without adequate response at 12 months of treatment. The authors identified key metabolites responsible for the differences between the two groups, including histamine, glutamine, xanthurenic acid, and ethanolamine; of these, histamine was the strongest discriminator for the baseline metabolites. Of these key metabolites, the concentrations of glutamine and xanthurenic acid were increased in the urine samples from responders. Nam et al. [80] used UPLC-MS techniques to determine pre-dose biomarkers in sera that could distinguish differential drug responses to oral acamprosate [80]. Samples from 120 participants (71 responders vs. 49 non-responders) were collected at baseline and 12 weeks following acamprosate treatment. This study revealed that pre-dose glutamate levels were significantly higher in responders than non-responders, which was also consistent with findings from a replication cohort of 20 responders and 10 non-responders. Glutamate levels decreased after 12 weeks of treatment relative to baseline, while there was no change in glutamate levels in the non-responders. In responders, ammonia levels also increased at baseline and reduced after treatment, suggesting that it has an important role in the pharmacological effect of acamprosate. Another study identified urinary biomarkers related to variation in metformin responses within patients with newly diagnosed type 2 diabetes mellitus [86]. Twenty-two patients used metformin for 6 months; urine samples were collected at baseline and at 3 and 6 months after treatment. Multivariate analyses including principal component analysis (PCA) and partial least squares-discriminant analysis (PLS-DA) models showed differences between responders and non-responders after 3 months of treatment [86]. The concentrations of three metabolites including citric acid, myoinositol, and hippuric acid were significantly different between two groups at baseline. For instance, the amount of citric acid was 54.6% lower in non-responders, while myoinositol was 18% higher in non-responders compared to responders. These three metabolites might be potential to be diagnostic biomarkers that could predict metformin response in patients with type 2 diabetes mellitus [86].

As a final example, although there was no difference between responders and non-responders at baseline, healthy patients showed variation in their metabolomic profiles soon after dosing with acetaminophen [124]. Seventy-one participants were treated with acetaminophen for a week, and then urine and serum samples were collected for metabolomic analysis. Using PCA, two clusters corresponding to metabolic profiles of responders and non-responders were separated after treatment, suggesting that treatment promoted changes in individual metabolism in response to acetaminophen [124].

3.4 The Role of Pharmacometabolomics in Adverse Drug Reactions

Adverse drug reactions (ADRs) also represent a significant barrier for clinical care, with more than two million cases (including 329,838 serious cases and 187,949 mortalities) reported in 2021, based on the US Food and Drug Administration (FDA) Adverse Event Reporting System (FAERS) dataset (available at the link: https://www.fda.gov/drugs/questions-and-answers-fdas-adverse-event-reporting-system-faers/fda-adverse-event-reporting-system-faers-public-dashboard). ADRs, also known as unintended and undesirable side effects related to drug usage [27], are ranked as the fourth significant cause of death and serious illness in patients undergoing medical care in the USA [79]. The FAERS dataset also indicates an increasing trend of ADR cases in the last decade. More specifically, the annual numbers of ADR cases have increased from 930,211 in 2012 to over two million cases in 2021(FAERS database US FDA, 2022) (Fig. 7). While some ADRs are the result of prescription or treatment errors, others may occur despite correct medication use [61], potentially due to variable responses to drugs among patients [114].

The first pharmacometabolomics study investigating adverse drug reactions was conducted in 2006 [17]. The pre- and post-treatment urinary metabolite profiles of rats were used to differentiate treatment responses to acetaminophen [17]. The study

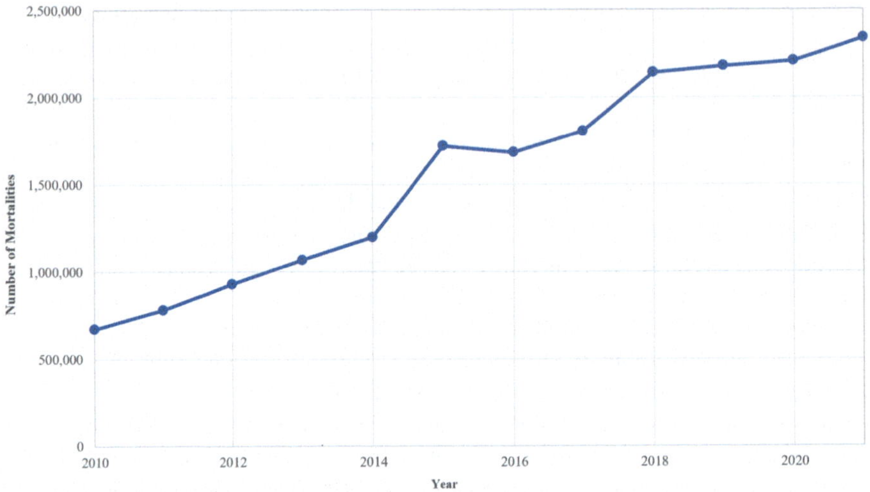

Fig. 7 The annual number of adverse drug reactions cases from 2012 to 2021 (FAERS database US FDA, 2022)

identified a significant change in concentrations of four acetaminophen-related metabolites after treating rats with a single dose [17]. Baseline urinary metabolites related to acetaminophen-induced liver injury were also identified. Higher concentrations of taurine were associated with a lower degree of hepatotoxicity, while increased concentrations of trimethylamine-N-oxide and betaine were associated with greater severity of liver damage [17]. This study demonstrated how metabolomic analysis before drug exposure could predict drug-related side effects [17]. In 2009, the first pharmacometabolomics study in humans was conducted to identify risk factors related to acetaminophen-induced liver injury. NMR spectroscopy identified p-cresol as a pre-treatment urinary biomarker of acetaminophen metabolism [16]. P-cresol is derived from tyrosine and phenylalanine via gut microbial metabolism, and high baseline levels of p-cresol were associated with a lower acetaminophen sulfate:acetaminophen glucuronide ratio after treatment [16]. After this first demonstration that endogenous metabolites could predict susceptibility to adverse effects of a drug, several additional studies have reported other drug-related adverse effects, including cardiovascular events, hyperglycemia, and dyslipidemia, caused by atenolol and hydrochlorothiazide, and kidney disease caused by metformin [96]. Pregnant women with epilepsy require continuous treatment with anti-epileptic drugs (AED), including lamotrigine and levetiracetam, to avoid fetal risk and maternal seizure, but treatment also confers an increased risk of abnormal fetal neurodevelopment. Comparison of metabolomic profiles of plasma samples from treated vs. untreated women identified changes in metabolic pathways including one carbon metabolism, neurotransmitter biosynthesis, and steroid metabolism [119]. Folate metabolites were also decreased following lamotrigine treatment [119]. Another study found key metabolites that could serve as biomarkers for adverse effects caused by anti-hypertensive drugs [94]. An untargeted metabolomics study identified 489 metabolites in patients taking hydrochlorothiazide (HCTZ) and atenolol (ATEN), respectively, of which 29 baseline metabolites were significantly associated with effects of ATEN and 170 metabolites were associated with HCTZ treatment [94]. In particular, palmitoleic acid was positively associated with home diastolic blood pressure (HDBP) in patients treated with HCTZ, while baseline 5-methoxytryptamine was negatively associated with HDBP response in patients treated with ATEN. Using these data, a predictive model of metabolite profiles for HDMP treatment was generated [94]. Through pathway analysis, metabolic pathways for gluconeogenesis, plasmalogen synthesis, and tryptophan metabolism were also enriched in patients treated with HCTZ [94].

3.5 Pharmacometabolomics and the Microbiome

The human microbiome plays a significant role in the biotransformation and metabolism of xenobiotics, and a variety of microbial species with an established impact on drug disposition and activity have been identified. The composition of bacteria within the microbiome also varies tremendously among individuals, contributing to significant differences in drug metabolism. The first human pharmacometabolomics investigation to demonstrate the impact of the microbiome on drug response revealed that, in individuals taking acetaminophen, baseline levels of p-cresol sulfate in urine were higher and the ratio of acetaminophen sulfate to acetaminophen glucuronide was lower after treatment [16]. P-cresol is derived from tyrosine and phenylalanine metabolism by species within the *Firmicutes*, *Bacteroidetes*, *Actinobacteria*, and *Fusobacterium* phyla, indicating a relationship between gut microbiome activity and acetaminophen metabolism [97]. The gut microbiome was subsequently proven to be extensively involved in the metabolism of a number of xenobiotics. In a study by Zimmermann et al. [138], 75% of 271 drugs that were screened for interaction with 76 human gut microbes demonstrated some degree of metabolism by these bacteria.

The exposure of a drug to the microbiome often results in inactivation of the drug through metabolism, resulting in reduced activity and efficacy of the drug (or its active metabolites). The gut microbiome significantly impacts the efficacy of digoxin, a cardiac glycoside used to treat atrial fibrillation and congestive heart failure. As digoxin has a very narrow therapeutic window (with a target concentration range of 0.5–2 mcg/L), it requires careful monitoring to ensure sufficient concentrations for adequate response while avoiding cardiotoxicity. Strains of the Actinobacterium *Eggerthella lenta* produce an enzyme associated with reductive metabolism of digoxin leading to its inactivation, thereby contributing to the differential responsiveness among patients [66]. The immunosuppressive drug tacrolimus is metabolized extensively by gut bacteria including *Faecalibacterium prausnitzii*; a positive correlation was found among kidney transplant patients requiring higher doses of tacrolimus and the presence of larger populations of this microbe in the gut [69]. The efficacy of levodopa, which is used to treat symptoms of Parkinson's disease, is highly variable among patients, and genetic and other host-related factors do not entirely explain this variability. Levodopa is a prodrug that must first cross the blood-brain barrier and be converted to dopamine, its active metabolite. The gut bacterium *Enterococcus faecalis* reduces the activity of levodopa through a conserved tyrosine decarboxylase with the ability to metabolize levodopa to dopamine prior to the drug crossing the blood-brain barrier, reducing the effectiveness of this medication [75]. Additional bacterial species also contribute to the inactivation of levodopa. Recently, *Clostridium sporogenes* was shown to be capable of deaminating levodopa in the gut [115]. Collectively, these studies provide evidence of the importance of the gut microbiome as a major driver of inter-individual variation in response to numerous drugs.

4 Applications of Pharmacometabolomics in Complex Diseases

4.1 Breast and Lung Cancer

The most recent data from the Global Cancer Observatory (GLOBOCAN (https://gco.iarc.fr)) reports ~19 million new cancer cases and nearly ten million deaths related to cancer, in 2020. Of 36 reported cancer types in GLOBOCAN, the top 10 with highest mortality rates include cancers of the lung, liver, stomach, breast, colon, esophagus, pancreas, prostate, rectum, and cervix (Fig. 8). Breast cancer is the leading cause of global cancer cases, with an estimate of 2.3 million new cases representing 11.7% of all cancer incidences (Fig. 8a). It is ranked as the fifth leading cause of cancer mortality. Breast cancer treatment includes surgery, radiation therapy, chemotherapy, hormone therapy, and targeted therapy. In breast cancer treatment, drug resistance is a serious concern leading to disease progression [107]. Over the past decade, four randomized controlled trials related to metabolomic profiling of drug response outcomes in breast cancer were published. Paclitaxel is routinely used for breast cancer treatment, but approximately 25% of patients experience treatment delays or discontinuation due to the development and progression of paclitaxel-induced peripheral neuropathy [102]. Paclitaxel-induced peripheral neuropathy can reduce patients' quality of life, and up to 80% of affected patients still experience symptoms of neuropathy after discontinuation of treatment. Sun et al. [105] analyzed blood samples from 60 breast cancer patients (stage I to stage III) to characterize the baseline and post-first-treatment metabolomics profiles associated with the occurrence of paclitaxel-induced peripheral neuropathy. Three metabolites – histidine, phenylalanine, and threonine – were potentially predictive for the severity of paclitaxel-induced peripheral neuropathy [105]. Another study related to the response to gemcitabine-carboplatin therapy was conducted by Jiang et al. [52]. In this work, serum samples of 29 metastatic breast cancer patients were collected

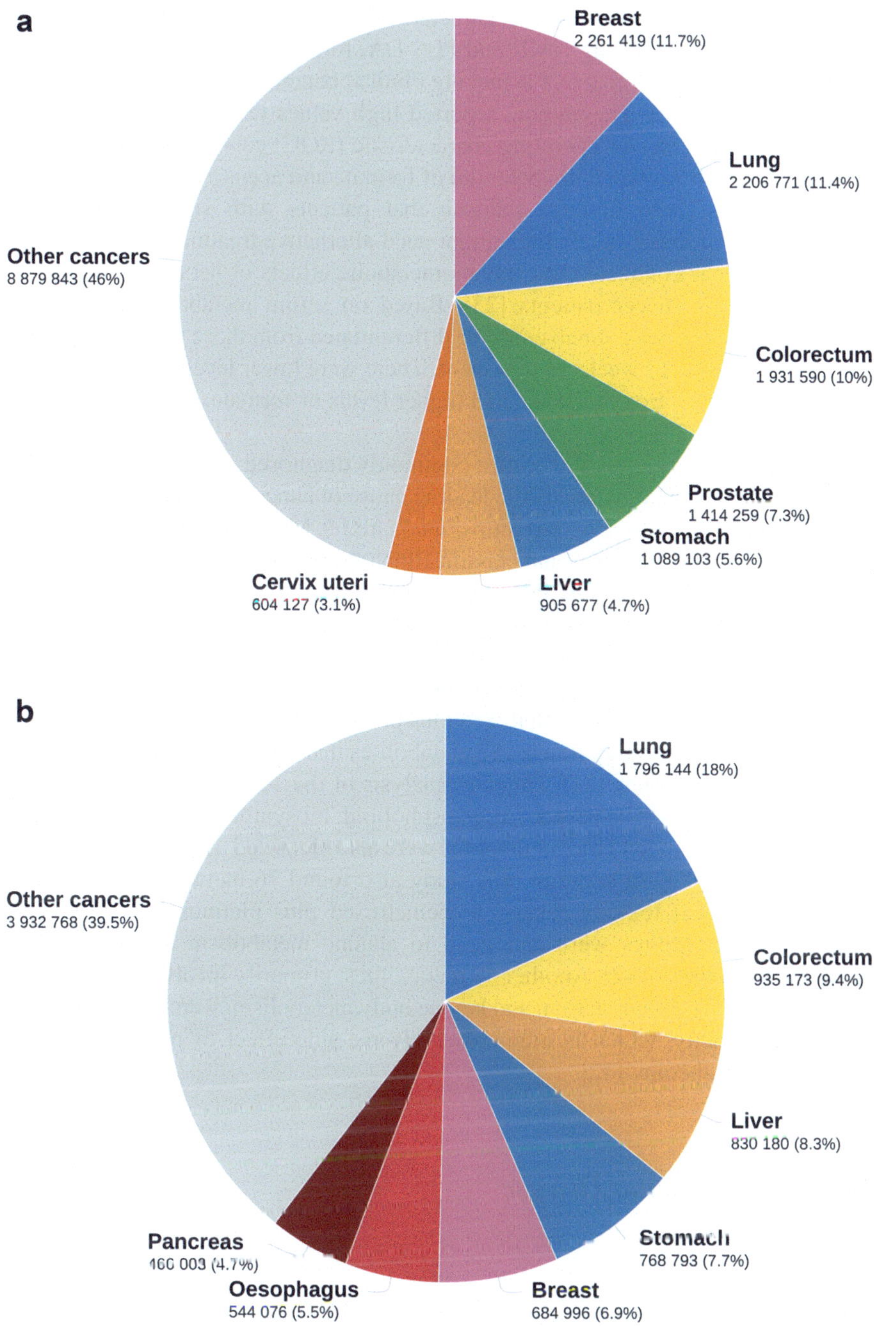

Fig. 8 Top 10 with high mortality rates. (**a**) Proportions of new cases in 2020. (**b**) Proportions of mortality in 2020. (Data and graphs were accessed from https://gco.iarc.fr)

and metabolomic profiles were characterized before and after gemcitabine-carboplatin treatment. Using NMR and PLS-DA, formate and acetate significantly differentiated groups of patients achieving clinical benefit and those who had progressive disease ([52]. The authors reported high values (>0.8) of area under the curve (AUC) and receiver operating characteristic (ROC) curves for the classification model, which validated the potential of formate and acetate in predicting treatment response. These findings indicate that patients with significantly lower concentrations of these metabolites might need alternative treatment [52]. Another clinical study was conducted to evaluate metabolic effects of neoadjuvant chemotherapy in breast cancer patients [23]. Based on serum metabolomics profiles, patients receiving bevacizumab could be differentiated from those treated only with chemotherapy after 12 weeks of treatment. There were lower levels of leucine, acetoacetate, and tri-hydroxybutyrate, and higher levels of formate, in patients receiving bevacizumab [23].

Lung cancer is also one of the most commonly diagnosed cancer types, with 2.3 million new cases in 2020, and is the leading cause of cancer mortality with an estimated 1.8 million deaths (Fig. 8b) (https://gco.iarc.fr). Metabolomics studies to date have focused on the diagnosis and classification of different types of lung cancer, or diagnosis of stages of cancers. Pemetrexed plus platinum doublet chemotherapy is the standard treatment for lung adenocarcinoma patients who are not eligible for target therapy. However, the response rates in these patients are only ~30–40% [99]. Gong et al. [42] analyzed serum metabolites of 130 patients before and after treatment to predict responses to pemetrexed plus platinum-based chemotherapy in lung adenocarcinoma. The study found 157 metabolites that could differentiate response and non-response groups [42]. Pathway analysis of the 157 metabolites identified enrichment of pathways related to phospholipid biosynthesis and glycerolipid metabolism [42]. Fifteen lipid metabolites were also identified as potentially predictive biomarkers [42]. In addition, this study also found 76 metabolites associated with hematological toxicity related to pemetrexed plus platinum chemotherapy [42]. Those metabolites were annotated to alanine metabolism and glutathione metabolism pathways [42]. Another 54 metabolites, primarily involved in galactose metabolism, lactose degradation, and ketone body metabolism, were also associated with hepatotoxicity, which is a common adverse side effect of pemetrexed plus platinum chemotherapy [42].

4.2 Cardiovascular Disease

4.2.1 Statins

Cardiovascular disease is a leading cause of death worldwide, with the annual deaths due to atherothrombosis expected to increase to 24 million by 2030 [14, 73]. Statins are the largest class of drugs prescribed for cardiovascular disease. They act by competitive inhibition of 3-hydroxy-3-methyl-glytaryl-CoA reductase, resulting in reduced cellular cholesterol synthesis and upregulation of low-density lipoprotein receptors (LDL), thus lowering plasma LDL levels [40]. However, clinical trials have demonstrated a residual CVD risk of approximately 50–80% in statin-treated patients. There are considerable variations in LDL response due to genetic, demographic, and environmental factors [67]. Additionally, statins have a variety of pleiotropic effects, and treatment is also associated with the occurrence of adverse events. Pharmacometabolomics can provide a broad map of the effects of statin treatment. A targeted lipidomics study assessed changes in lipid profiles during a 6-week trial of simvastatin in a group of 944 men and women of European and African American ancestries [67]. There were consistent changes in "good" vs. "poor" responders, for arachidonic acid and linoleic acid. "Good" responders had a significant increase in the ratio of arachidonic acid to its precursor. The concentrations of the purine metabolite xanthine were the most significantly different between "good" and "poor" simvastatin responders. Xanthine is a substrate of xanthine oxidase, which produces hydrogen peroxide and hence is implicated in mechanisms of oxidative stress. Because free radicals can decouple nitric oxide synthase enzymatic activity, the lower basal level of xanthine in "good" responders might be expected to yield more robust nitric oxide synthase signaling, further supporting a link between the benefits of statins on lipids and endothelial function. Baseline levels of 2-hydroxyvaleric acid (2-hydroxypentanoic acid) also strongly discriminated between "good" and "poor" statin responders, with lower levels associated with greater response [67]. Taken together, the studies demonstrate that simvastatin has a wide range of metabolic effects beyond those directly involved in cholesterol metabolism that may contribute to its efficacy in reducing risk of CVD, as well as to its pleiotropic effects and the risk of adverse events. A follow-up study was conducted by Trupp et al. [113], using a non-targeted gas chromatography time-of-flight MS-based metabolomics to evaluate the global metabolic effects of simvastatin from plasma samples of LDL-C responders before and after treatment. Pathway analysis revealed that the metabolites generated from simvastatin exposure were enriched for amino acid degradation. Using orthogonal PLS-DA, baseline metabolic profiling could discriminate between "good" and "poor" responders based on significant metabolites including xanthine, 2-hydroxyvaleric acid, succinic acid, stearic acid, and fructose. These results may help clinicians to discern the beneficial effect vs. adverse event risk of statin therapy within individual patients [113].

4.2.2 Aspirin

Aspirin is the most commonly used drug worldwide for both primary prevention of cardiovascular disease and secondary prevention of recurrent cardiovascular events. However, about 25% of high-risk patients show persistent platelet reactivity while using aspirin. Using mass spectrometry–based metabolomics, Ellero-Simatos et al. [28] found that aspirin induced strong changes in the serum metabolomic profiles of 745 patients; in particular, serotonin levels were higher in those from the fourth quartile of aspirin response. The authors suggest that measuring serotonin could prevent hemorrhages and increase antiplatelet efficacy in high-risk patients [28].

4.3 Type 2 Diabetes Mellitus

Type 2 diabetes mellitus (T2DM) is considered a global pandemic, and its incidence is steadily increasing [137]. Metformin, a biguanide antihyperglycemic agent used as the first-line therapeutic agent for T2DM, has numerous benefits in lowering lipid levels and decreasing diabetes-related mortality [117]. High inter-individual variation is reported among patients using metformin. Previous studies showed metformin might fail to control glucose levels adequately in approximately 50% of patients [19]. Several studies reported the correlation between the risk of type 2 diabetes mellitus progression and certain metabolites or changes in metabolites following antidiabetic drug treatment. Rotroff et al. [95] showed that 12 metabolites were significantly associated with metformin-induced changes in glucose in non-diabetic participants following an oral glucose tolerance test. Of the 12, 2-hydroxybutanoic acid was a potential predictor of metformin response [95]. Another study identified urinary biomarkers associated with inter-individual variation in metformin response of patients newly diagnosed with T2DM. Urine samples from 22 patients taking metformin for 6 months were collected before treatment and also 3 and 6 months after starting metformin, and profiled using an untargeted pharmacometabolomics approach [86]. Multivariate analysis identified seven metabolites, including citric acid, myoinositol, pseudouridine, p-hydroxyphenyl acetic acid, hippuric acid, hypoxanthine, and 3-(3-hydroxulphenyl)-3-hydroxypropanoic acid, as prominent discriminators between responders and non-responders to metformin [86]. Of the six, the relative intensity values of citric acid, myoinositol, and hippuric acid at baseline showed significant differences between the two groups [86]. The level of citric acid in the non-response group was 54.6% lower than that of the responders at baseline, then increased at 3 and 6 months of treatment [86]. Myoinositol levels were increased in the non-responder group and decreased in both groups at 3 and 6 months of treatment [86]. Responders with lower levels of myoinositol at baseline also had a small decrease in this metabolite at 3 and 6 months [86]. Therefore, urinary citric acid and myoinositol may represent biomarkers of metformin responsiveness in patients with T2DM.

4.4 Neurological Diseases

Several pharmacometabolomics studies have implicated potential markers for treatment response in neurological diseases including schizophrenia. Using a specialized lipidomics platform, 300 polar and non-polar lipid metabolites across seven lipid classes were measured in patients with schizophrenia before and after treatment with three commonly prescribed atypical antipsychotics: olanzapine, risperidone, and aripiprazole [58]. Before treatment, major changes were noted for two phospholipid classes, phosphatidylethanolamine and phosphatidylcholine [58]. Olanzapine and risperidone affected a much broader range of lipid classes than did aripiprazole, and approximately 50 lipids increased after exposure to olanzapine and risperidone but not aripiprazole [58]. This study showed the potential of lipids to serve as biomarkers for schizophrenia medications. In another study of bipolar affective disorder, 246 metabolites were profiled in plasma samples from patients before and after treatment with lithium [56]. There was a clear separation of groups by treatment response using PCA and PLS, and 20 metabolites demonstrated significant fluctuations related to lithium treatment [56].

5 Conclusion

5.1 Limitations and Future Directions of Pharmacometabolomics Research

Despite its promise, the application of pharmacometabolomics faces multiple potential challenges, primarily related to study design, sample selection and analytical approaches, and metabolite annotation. We discuss each of these limitations, and their implications, below.

5.1.1 Study Design

Samples and data needed for pharmacometabolomic studies of clinical outcomes are increasingly sourced from databases and sample repositories such as electronic medical records databases, previously published datasets, and biobanks. These resources may be prospective, with ongoing collections (such as biobanks) or retrospectively collected, with no new subjects enrolled. Prospective studies such as randomized controlled trials and cohort studies are costly in terms of time and resources, and prospective data from ongoing sample collections (e.g., through biobanks) is limited by slow accrual rates. In addition, the sample sizes sufficient to confer high statistical power may be required to be much larger for pharmacometabolomics studies. There is often limited ability for investigators to obtain sufficient numbers of samples with the appropriate drug response/effect phenotypes for different patient populations. Patient selection and recruitment for prospective clinical studies can also be challenging, as these studies investigate highly specific phenotypes that may not be readily available within the general patient population, and which also makes replication of findings in comparable populations more difficult. Analyzing retrospective data can avoid many of the challenges involved in recruiting new pharmacometabolomic cohorts, but it is also limited in terms of the type, number, completeness, and appropriateness of the variables or other data collected to address a particular study question or hypothesis. Meta-analysis of multiple studies or cohorts as a pharmacometabolomic strategy is also analytically challenging, and may be limited by a dearth of relevant studies or the inability to generalize the outcome across cohorts. As with other pharmaco-"omics" studies, the ability to generalize findings from pharmacometabolomics studies is challenging in different populations.

Another concern related to the design of pharmacometabolomic studies is related to time and resources, as the large-scale profiling and quality control of data from thousands of metabolites, for even a few hundred samples, can be prohibitively expensive. Repositories may also be few in number, and sample or data availability may be very limited. Finally, many pharmacometabolomics study protocols suffer from a lack of standardization, which would help facilitate replication and generalizability of the findings.

5.1.2 Sample Selection and Measurement of Metabolites

Often, pharmacometabolomics studies are limited to sampling readily accessible biofluids, such as saliva, blood, plasma, or urine. Different types of samples demonstrate different metabolomic profiles, which vary significantly both at baseline and during drug treatment. The relationship between metabolites in circulation and within specific organs and tissues is not well understood, limiting the ability to validate results in different biofluids. Moreover, metabolomic changes in different samples over time due to other external factors such as health status, age, sex, nutrition, and the presence of concomitant medications also frustrate reproducibility of results. There is need for studies that investigate alterations in metabolic pathways correlated with a given state and its progression over time. Furthermore, an available biofluid may not be the best source for profiling metabolites related to a particular outcome, in which case the ability to correlate metabolites across different biofluids (e.g., plasma and urine) is relevant.

Chromatographic separation coupled to mass spectrometry has become the preferred technique for metabolomics; however, the sensitivity of analytical methods requires improvement. A limitation of interpretation of MS-based metabolomics is the limited fraction of signals assigned to a known metabolite. The development of more sensitive, comprehensive, rapid, and specific analytical methods would greatly increase the number of known and novel metabolites available for profiling in a pharmacometabolomics study.

5.1.3 Annotation of Metabolites

Metabolite annotation is one of the major limitations for untargeted metabolomics pipelines, and successful pharmacometabolomics studies require accurate identification of metabolites. While over 30,000 endogenous metabolites have been annotated in public databases, many metabolites in untargeted assays escape characterization, rendering it difficult to further apply biologically informative methods such as pathway or network analysis. Inaccuracies in identification and overlap in metabolite annotations can introduce errors in the analysis. For example, although PubChem identifiers are more widely represented across many platforms, duplicated metabolite entries may result in misidentified metabolites. Missing, ambiguous, or redundant entries have also been found in multiple databases including KEGG, HMDB, and ChEBI. Inaccurate and missing annotation is a critical problem for investigators attempting to connect observed metabolite-phenotype associations with accurate biochemical pathway information.

6 Summary

Pharmacometabolomics is an emerging field based on the practical measurement and comparison of metabolite profiles in patients with drug response phenotypes. With recent advancements in instrumentation and computational methods, pharmacometabolomics confers the ability to predict the effectiveness of a drug prior to dosing, and to improve efficacy and outcomes while avoiding ADRs. The use of pharmacometabolomics to repurpose existing drugs, identify novel drug candidates, and also to identify and validate biomarkers for complex diseases such as cancer, cardiovascular disease, and asthma, reveals new potential opportunities in treatment strategies. Through identifying relationships between the gut microbiome and specific drug classes, it is possible to circumvent poor responsiveness related to drug-microbe interactions. Particularly through integration with other "omics" data types, pharmacometabolomics represents a powerful approach to achieve the goal of personalized medicine: tailoring the right drug, at the optimal dose, to the right patient.

References

1. Abdul-Hamid, E., Kacper, S., Ryan, T. M., Hiba, A., Chung-ke, C., Joanna, L., & Mariusz, J. (2021). Pharmacometabolomics: A New Horizon in Personalized Medicine. In Z. Xianquan (Ed.), *Metabolomics* (pp. Ch. 3). IntechOpen. https://doi.org/10.5772/intechopen.98911
2. Adams, J. C., Keiser, M. J., Basuino, L., Chambers, H. F., Lee, D.-S., Wiest, O. G., & Babbitt, P. C. (2009). A Mapping of Drug Space from the Viewpoint of Small Molecule Metabolism. *PLOS Computational Biology*, *5*(8), e1000474. https://doi.org/10.1371/journal.pcbi.1000474
3. Alarcon-Barrera, J. C., Kostidis, S., Ondo-Mendez, A., & Giera, M. (2022). Recent advances in metabolomics analysis for early drug development. *Drug Discovery Today*, *27*(6), 1763–1773. https://doi.org/10.1016/j.drudis.2022.02.018
4. Asthma, G. I. f. (2019). *2019 GINA Main Report*. www.ginasthma.org/reports
5. Attia, S. M. (2010). Deleterious effects of reactive metabolites. *Oxid Med Cell Longev*, *3*(4), 238–253. https://doi.org/10.4161/oxim.3.4.13246

6. Baptista, R., Fazakerley, D. M., Beckmann, M., Baillie, L., & Mur, L. A. J. (2018). Untargeted metabolomics reveals a new mode of action of pretomanid (PA-824). *Sci Rep*, *8*(1), 5084. https://doi.org/10.1038/s41598-018-23110-1
7. Barupal, D. K., & Fiehn, O. (2017). Chemical Similarity Enrichment Analysis (ChemRICH) as alternative to biochemical pathway mapping for metabolomic datasets. *Scientific Reports*, *7*(1), 14567. https://doi.org/10.1038/s41598-017-15231-w
8. Bateman, E. D., Hurd, S. S., Barnes, P. J., Bousquet, J., Drazen, J. M., FitzGerald, J. M., Gibson, P., Ohta, K., O'Byrne, P., Pedersen, S. E., Pizzichini, E., Sullivan, S. D., Wenzel, S. E., & Zar, H. J. (2008). Global strategy for asthma management and prevention: GINA executive summary. *Eur Respir J*, *31*(1), 143–178. https://doi.org/10.1183/09031936.00138707
9. Beger, R. D., Schmidt, M. A., & Kaddurah-Daouk, R. (2020). Current Concepts in Pharmacometabolomics, Biomarker Discovery, and Precision Medicine. *Metabolites*, *10*(4). https://doi.org/10.3390/metabo10040129
10. Beretov, J., Wasinger, V. C., Graham, P. H., Millar, E. K., Kearsley, J. H., & Li, Y. (2014). Chapter Four - Proteomics for Breast Cancer Urine Biomarkers. In G. S. Makowski (Ed.), *Advances in Clinical Chemistry* (Vol. 63, pp. 123–167). Elsevier. https://doi.org/10.1016/B978-0-12-800094-6.00004-2
11. Boizard, F., Brunchault, V., Moulos, P., Breuil, B., Klein, J., Lounis, N., Caubet, C., Tellier, S., Bascands, J.-L., Decramer, S., Schanstra, J. P., & Buffin-Meyer, B. (2016). A capillary electrophoresis coupled to mass spectrometry pipeline for long term comparable assessment of the urinary metabolome. *Scientific Reports*, *6*(1), 34453. https://doi.org/10.1038/srep34453
12. Campos, A. I., & Zampieri, M. (2019). Metabolomics-Driven Exploration of the Chemical Drug Space to Predict Combination Antimicrobial Therapies. *Molecular Cell*, *74*(6), 1291-1303.e1296. https://doi.org/10.1016/j.molcel.2019.04.001
13. Chan, M. T. S., Leung, D. Y. M., Szefler, S. J., & Spahn, J. D. (1998). Difficult-to-control asthma: Clinical characteristics of steroid-insensitive asthma. *Journal of Allergy and Clinical Immunology*, *101*(5), 594–601. https://doi.org/10.1016/S0091-6749(98)70165-4
14. Chaurasia, V. (2013). Early Prediction of Heart Diseases Using Data Mining Techniques. *Caribbean Journal of Science and Technology*, *Vol.1*, 208–217.
15. Chen, W. H., Lee, P. Y., Ng, W., Tse, H. F., & Lau, C. P. (2004). Aspirin resistance is associated with a high incidence of myonecrosis after non-urgent percutaneous coronary intervention despite clopidogrel pretreatment. *J Am Coll Cardiol*, *43*(6), 1122–1126. https://doi.org/10.1016/j.jacc.2003.12.034
16. Clayton, T. A., Baker, D., Lindon, J. C., Everett, J. R., & Nicholson, J. K. (2009). Pharmacometabonomic identification of a significant host-microbiome metabolic interaction affecting human drug metabolism. *Proceedings of the National Academy of Sciences*, *106*(34), 14728–14733. https://doi.org/10.1073/pnas.0904489106
17. Clayton, T. A., Lindon, J. C., Cloarec, O., Antti, H., Charuel, C., Hanton, G., Provost, J. P., Le Net, J. L., Baker, D., Walley, R. J., Everett, J. R., & Nicholson, J. K. (2006). Pharmacometabonomic phenotyping and personalized drug treatment. *Nature*, *440*(7087), 1073–1077. https://doi.org/10.1038/nature04648
18. Clendinen, C. S., Lee-McMullen, B., Williams, C. M., Stupp, G. S., Vandenborne, K., Hahn, D. A., Walter, G. A., & Edison, A. S. (2014). 13C NMR Metabolomics: Applications at Natural Abundance. *Analytical Chemistry*, *86*(18), 9242–9250. https://doi.org/10.1021/ac502346h
19. Coetzee, E. J., & Jackson, W. P. U. (1979). Metformin in management of pregnant insulin-independent diabetics. *Diabetologia*, *16*(4), 241–245. https://doi.org/10.1007/BF01221950
20. Connor, S. C., Wu, W., Sweatman, B. C., Manini, J., Haselden, J. N., Crowther, D. J., & Waterfield, C. J. (2004). Effects of feeding and body weight loss on the 1H-NMR-based urine metabolic profiles of male Wistar Han rats: implications for biomarker discovery. *Biomarkers*, *9*(2), 156-179. https://doi.org/10.1080/13547500410001720767
21. Coskun, O. (2016). Separation techniques: Chromatography. *North Clin Istanb*, *3*(2), 156–160. https://doi.org/10.14744/nci.2016.32757

22. Dahlin, A., Sordillo, J., McGeachie, M., Kelly, R., Tantisira, K., Lutz, S., Lasky-Su, J., & Wu, A. (2020). Genome-wide interaction study reveals age-dependent determinants of responsiveness to inhaled corticosteroids in individuals with asthma. *PLOS ONE*, *15*, e0229241. https://doi.org/10.1371/journal.pone.0229241
23. Debik, J., Euceda, L. R., Lundgren, S., Gythfeldt, H. v. d. L., Garred, Ø., Borgen, E., Engebraaten, O., Bathen, T. F., & Giskeødegård, G. F. (2019). Assessing Treatment Response and Prognosis by Serum and Tissue Metabolomics in Breast Cancer Patients. *Journal of Proteome Research*, *18*(10), 3649–3660. https://doi.org/10.1021/acs.jproteome.9b00316
24. Denoroy, L., Zimmer, L., Renaud, B., & Parrot, S. (2013). Ultra high performance liquid chromatography as a tool for the discovery and the analysis of biomarkers of diseases: A review. *Journal of chromatography. B, Analytical technologies in the biomedical and life sciences*, *927*, 37–53. https://doi.org/10.1016/j.jchromb.2012.12.005
25. Dowden, H., & Munro, J. (2019). Trends in clinical success rates and therapeutic focus. *Nat. Rev. Drug Discov*, *18*(7), 495–496.
26. Drazen, J. M., Silverman, E. K., & Lee, T. H. (2000). Heterogeneity of therapeutic responses in asthma. *Br Med Bull*, *56*(4), 1054–1070. https://doi.org/10.1258/0007142001903535
27. Edwards, I. R., & Aronson, J. K. (2000). Adverse drug reactions: definitions, diagnosis, and management. *The Lancet*, *356*(9237), 1255–1259. https://doi.org/10.1016/S0140-6736(00)02799-9
28. Ellero-Simatos, S., Lewis, J., Georgiades, A., Yerges-Armstrong, L., Beitelshees, A., Horenstein, R., Dane, A., Harms, A., Ramaker, R., Vreeken, R., Perry, C., Zhu, H., Sànchez, C., Kuhn, C., Ortel, T., Shuldiner, A., Hankemeier, T., & Kaddurah-Daouk, R. (2014). Pharmacometabolomics Reveals That Serotonin Is Implicated in Aspirin Response Variability. *CPT: Pharmacometrics & Systems Pharmacology*, *3*(7), 125. https://doi.org/10.1038/psp.2014.22
29. Emwas, A. H. (2015). The strengths and weaknesses of NMR spectroscopy and mass spectrometry with particular focus on metabolomics research. *Methods Mol Biol*, *1277*, 161–193. https://doi.org/10.1007/978-1-4939-2377-9_13
30. Emwas, A. H., Roy, R., McKay, R. T., Tenori, L., Saccenti, E., Gowda, G. A. N., Raftery, D., Alahmari, F., Jaremko, L., Jaremko, M., & Wishart, D. S. (2019). NMR Spectroscopy for Metabolomics Research. *Metabolites*, *9*(7). https://doi.org/10.3390/metabo9070123
31. Ernst, M., Silva, D., Silva, R., Vêncio, R., & Lopes, N. (2014). Mass spectrometry in plant metabolomics strategies: From analytical platforms to data acquisition and processing. *Natural product reports*, *31*. https://doi.org/10.1039/c3np70086k
32. Everett, J. R. (2016). From Metabonomics to Pharmacometabonomics: The Role of Metabolic Profiling in Personalized Medicine. *Frontiers in Pharmacology*, *7*.
33. Fiehn, O. (2016). Metabolomics by Gas Chromatography-Mass Spectrometry: Combined Targeted and Untargeted Profiling. *Curr Protoc Mol Biol*, *114*, 30.34.31–30.34.32. https://doi.org/10.1002/0471142727.mb3004s114
34. FitzGerald, J. M., Tavakoli, H., Lynd, L. D., Al Efraij, K., & Sadatsafavi, M. (2017). The impact of inappropriate use of short acting beta agonists in asthma. *Respiratory Medicine*, *131*, 135–140. https://doi.org/10.1016/j.rmed.2017.08.014
35. Frolkis, A., Knox, C., Lim, E., Jewison, T., Law, V., Hau, D. D., Liu, P., Gautam, B., Ly, S., Guo, A. C., Xia, J., Liang, Y., Shrivastava, S., & Wishart, D. S. (2009). SMPDB: The Small Molecule Pathway Database. *Nucleic Acids Research*, *38*(suppl_1), D480–D487. https://doi.org/10.1093/nar/gkp1002
36. Garcia-Perez, I., Posma, J. M., Serrano-Contreras, J. I., Boulangé, C. L., Chan, Q., Frost, G., Stamler, J., Elliott, P., Lindon, J. C., & Holmes, E. (2020). Identifying unknown metabolites using NMR-based metabolic profiling techniques. *Nature Protocols*, *15*(8), 2538-2567.
37. Gargano, A. F. G., Roca, L. S., Fellers, R. T., Bocxe, M., Domínguez-Vega, E., & Somsen, G. W. (2018). Capillary HILIC-MS: A New Tool for Sensitive Top-Down Proteomics. *Analytical Chemistry*, *90*(11), 6601–6609. https://doi.org/10.1021/acs.analchem.8b00382

38. Geeraerts, S. L., Kampen, K. R., Rinaldi, G., Gupta, P., Planque, M., Louros, N., Heylen, E., De Cremer, K., De Brucker, K., Vereecke, S., Verbelen, B., Vermeersch, P., Schymkowitz, J., Rousseau, F., Cassiman, D., Fendt, S.-M., Voet, A., Cammue, B. P. A., Thevissen, K., & De Keersmaecker, K. (2021). Repurposing the Antidepressant Sertraline as SHMT Inhibitor to Suppress Serine/Glycine Synthesis–Addicted Breast Tumor Growth. *Molecular Cancer Therapeutics*, *20*(1), 50–63. https://doi.org/10.1158/1535-7163.Mct-20-0480
39. Godzien, J., López-Gonzálvez, Á., García, A., & Barbas, C. (2019). Chapter 5 - Metabolic Phenotyping Using Capillary Electrophoresis Mass Spectrometry. In J. C. Lindon, J. K. Nicholson, & E. Holmes (Eds.), *The Handbook of Metabolic Phenotyping* (pp. 171–204). Elsevier. https://doi.org/10.1016/B978-0-12-812293-8.00005-0
40. Goldstein, Joseph L., & Brown, Michael S. (2015). A Century of Cholesterol and Coronaries: From Plaques to Genes to Statins. *Cell*, *161*(1), 161–172. https://doi.org/10.1016/j.cell.2015.01.036
41. Gómez-Cebrián, N., Vázquez Ferreiro, P., Carrera Hueso, F. J., Poveda Andrés, J. L., Puchades-Carrasco, L., & Pineda-Lucena, A. (2021). Pharmacometabolomics by NMR in Oncology: A Systematic Review. *Pharmaceuticals*, *14*(10), 1015. https://www.mdpi.com/1424-8247/14/10/1015
42. Gong, W.-J., Cao, P., Zhang, Q.-L., Han, X.-Y., Wang, S.-W., Huang, Y.-F., Wu, S.-L., Li, Q., Zhang, R., Xu, S.-B., Liu, Y.-N., Shi, S.-J., & Zhang, Y. (2022). Prediction of response and adverse drug reaction of pemetrexed plus platinum-based chemotherapy in lung adenocarcinoma by serum metabolomic profiling. *Translational Oncology*, *19*, 101393. https://doi.org/10.1016/j.tranon.2022.101393
43. Gum, P. A., Kottke-Marchant, K., Welsh, P. A., White, J., & Topol, E. J. (2003). A prospective, blinded determination of the natural history of aspirin resistance among stable patients with cardiovascular disease. *J Am Coll Cardiol*, *41*(6), 961–965. https://doi.org/10.1016/s0735-1097(02)03014-0
44. Hameedat, F., Hawamdeh, S., Alnabulsi, S., & Zayed, A. (2022). High Performance Liquid Chromatography (HPLC) with Fluorescence Detection for Quantification of Steroids in Clinical, Pharmaceutical, and Environmental Samples: A Review. *Molecules*, *27*(6), 1807. https://www.mdpi.com/1420-3049/27/6/1807
45. Harada, S., Hirayama, A., Chan, Q., Kurihara, A., Fukai, K., Iida, M., Kato, S., Sugiyama, D., Kuwabara, K., Takeuchi, A., Akiyama, M., Okamura, T., Ebbels, T. M. D., Elliott, P., Tomita, M., Sato, A., Suzuki, C., Sugimoto, M., Soga, T., & Takebayashi, T. (2018). Reliability of plasma polar metabolite concentrations in a large-scale cohort study using capillary electrophoresis-mass spectrometry. *PLOS ONE*, *13*(1), e0191230. https://doi.org/10.1371/journal.pone.0191230
46. Harrison, R. K. (2016). Phase II and phase III failures: 2013–2015. *Nat Rev Drug Discov*, *15*(12), 817–818.
47. Holbrook-Smith, D., Durot, S., & Sauer, U. (2022). High-throughput metabolomics predicts drug–target relationships for eukaryotic proteins. *Molecular Systems Biology*, *18*(2), e10767. https://doi.org/10.15252/msb.202110767
48. Hughes, J. P., Rees, S., Kalindjian, S. B., & Philpott, K. L. (2011). Principles of early drug discovery. *Br J Pharmacol*, *162*(6), 1239–1249. https://doi.org/10.1111/j.1476-5381.2010.01127.x
49. Ishikawa, S., Sugimoto, M., Kitabatake, K., Tu, M., Sugano, A., Yamamori, I., Iba, A., Yusa, K., Kaneko, M., Ota, S., Hiwatari, K., Enomoto, A., Masaru, T., & Iino, M. (2017). Effect of timing of collection of salivary metabolomic biomarkers on oral cancer detection. *Amino Acids*, *49*(4), 761–770. https://doi.org/10.1007/s00726-017-2378-5
50. Iwasaki, Y., Sawada, T., Hatayama, K., Ohyagi, A., Tsukuda, Y., Namekawa, K., Ito, R., Saito, K., & Nakazawa, H. (2012). Separation technique for the determination of highly polar metabolites in biological samples. *Metabolites*, *2*(3), 496–515. https://doi.org/10.3390/metabo2030496

51. Jewison, T., Su, Y., Disfany, F. M., Liang, Y., Knox, C., Maciejewski, A., Poelzer, J., Huynh, J., Zhou, Y., Arndt, D., Djoumbou, Y., Liu, Y., Deng, L., Guo, A. C., Han, B., Pon, A., Wilson, M., Rafatnia, S., Liu, P., & Wishart, D. S. (2013). SMPDB 2.0: Big Improvements to the Small Molecule Pathway Database. *Nucleic Acids Research*, *42*(D1), D478–D484. https://doi.org/10.1093/nar/gkt1067
52. Jiang, L., Lee, S. C., & Ng, T. C. (2018). Pharmacometabonomics Analysis Reveals Serum Formate and Acetate Potentially Associated with Varying Response to Gemcitabine-Carboplatin Chemotherapy in Metastatic Breast Cancer Patients. *Journal of Proteome Research*, *17*(3), 1248–1257. https://doi.org/10.1021/acs.jproteome.7b00859
53. Kachroo, P., Sordillo, J. E., Lutz, S. M., Weiss, S. T., Kelly, R. S., McGeachie, M. J., Wu, A. C., & Lasky-Su, J. A. (2021). Pharmaco-Metabolomics of Inhaled Corticosteroid Response in Individuals with Asthma. *J Pers Med*, *11*(11). https://doi.org/10.3390/jpm11111148
54. Kachroo, P., Stewart, I. D., Kelly, R. S., Stav, M., Mendez, K., Dahlin, A., Soeteman, D. I., Chu, S. H., Huang, M., Cote, M., Knihtilä, H. M., Lee-Sarwar, K., McGeachie, M., Wang, A., Wu, A. C., Virkud, Y., Zhang, P., Wareham, N. J., Karlson, E. W., . . . Lasky-Su, J. A. (2022). Metabolomic profiling reveals extensive adrenal suppression due to inhaled corticosteroid therapy in asthma. *Nature Medicine*, *28*(4), 814–822. https://doi.org/10.1038/s41591-022-01714-5
55. Kaddurah-Daouk, R., Baillie, R. A., Zhu, H., Zeng, Z. B., Wiest, M. M., Nguyen, U. T., Watkins, S. M., & Krauss, R. M. (2010). Lipidomic analysis of variation in response to simvastatin in the Cholesterol and Pharmacogenetics Study. *Metabolomics*, *6*(2), 191–201. https://doi.org/10.1007/s11306-010-0207-x
56. Kaddurah-Daouk, R., & Krishnan, K. R. R. (2009). Metabolomics: A Global Biochemical Approach to the Study of Central Nervous System Diseases. *Neuropsychopharmacology*, *34*(1), 173–186. https://doi.org/10.1038/npp.2008.174
57. Kaddurah-Daouk, R., Kristal, B. S., & Weinshilboum, R. M. (2008). Metabolomics: a global biochemical approach to drug response and disease. *Annu Rev Pharmacol Toxicol*, *48*, 653–683. https://doi.org/10.1146/annurev.pharmtox.48.113006.094715
58. Kaddurah-Daouk, R., McEvoy, J., Baillie, R. A., Lee, D., Yao, J. K., Doraiswamy, P. M., & Krishnan, K. R. R. (2007). Metabolomic mapping of atypical antipsychotic effects in schizophrenia. *Molecular Psychiatry*, *12*(10), 934–945. https://doi.org/10.1038/sj.mp.4002000
59. Kałużna-Czaplińska, J. (2011). Current medical research with the application of coupled techniques with mass spectrometry. *Med Sci Monit*, *17*(5), Ra117–123. https://doi.org/10.12659/msm.881756
60. Kapoor, S. R., Filer, A., Fitzpatrick, M. A., Fisher, B. A., Taylor, P. C., Buckley, C. D., McInnes, I. B., Raza, K., & Young, S. P. (2013). Metabolic Profiling Predicts Response to Anti–Tumor Necrosis Factor α Therapy in Patients With Rheumatoid Arthritis. *Arthritis & Rheumatism*, *65*(6), 1448–1456. https://doi.org/10.1002/art.37921
61. Khalil, H., & Huang, C. (2020). Adverse drug reactions in primary care: a scoping review. *BMC Health Serv Res*, *20*(1), 5. https://doi.org/10.1186/s12913-019-4651-7
62. Khan, V., Putluri, N. r., Sreekumar, A., & Mindikoglu, A. (2018). Current Applications of Metabolomics in Cirrhosis. *Metabolites*, *8*, 67. https://doi.org/10.3390/metabo8040067
63. Khosravi, A., Jayaram, B., Goliaei, B., & Masoudi-Nejad, A. (2019). Active repurposing of drug candidates for melanoma based on GWAS, PheWAS and a wide range of omics data. *Molecular Medicine*, *25*(1), 30. https://doi.org/10.1186/s10020-019-0098-x
64. Kohler, I., Sun, M., Groeneveld, G., & Gargano, A. (2020). Not (Only) reversed-phase lc–ms: Alternative lc–ms approaches. *LC-GC North America*, *38*, 507–518.
65. Kohler, I., Verhoeven, M., Haselberg, R., & Gargano, A. F. G. (2022). Hydrophilic interaction chromatography – mass spectrometry for metabolomics and proteomics: state-of-the-art and current trends. *Microchemical Journal*, *175*, 106986. https://doi.org/10.1016/j.microc.2021.106986

66. Koppel, N., Maini Rekdal, V., & Balskus, E. P. (2017). Chemical transformation of xenobiotics by the human gut microbiota. *Science*, *356*(6344), eaag2770. https://doi.org/10.1126/science.aag2770
67. Krauss, R. M., Zhu, H., & Kaddurah-Daouk, R. (2013). Pharmacometabolomics of Statin Response. *Clinical Pharmacology & Therapeutics*, *94*(5), 562–565. https://doi.org/10.1038/clpt.2013.164
68. Kupfer, D. J., Frank, E., & Phillips, M. L. (2012). Major depressive disorder: new clinical, neurobiological, and treatment perspectives. *Lancet*, *379*(9820), 1045–1055. https://doi.org/10.1016/s0140-6736(11)60602-8
69. Lee, J. R., Muthukumar, T., Dadhania, D., Taur, Y., Jenq, R. R., Toussaint, N. C., Ling, L., Pamer, E., & Suthanthiran, M. (2015). Gut microbiota and tacrolimus dosing in kidney transplantation. *PLOS ONE*, *10*(3), e0122399. https://doi.org/10.1371/journal.pone.0122399
70. Li, F., Lu, J., & Ma, X. (2011). Profiling the reactive metabolites of xenobiotics using metabolomic technologies. *Chem Res Toxicol*, *24*(5), 744–751. https://doi.org/10.1021/tx200033v
71. Lindon, J. C., Keun, H. C., Ebbels, T. M., Pearce, J. M., Holmes, E., & Nicholson, J. K. (2005). The Consortium for Metabonomic Toxicology (COMET): aims, activities and achievements. *Pharmacogenomics*, *6*(7), 691–699. https://doi.org/10.2217/14622416.6.7.691
72. Lísa, M., Cífková, E., Khalikova, M., Ovčačíková, M., & Holčapek, M. (2017). Lipidomic analysis of biological samples: Comparison of liquid chromatography, supercritical fluid chromatography and direct infusion mass spectrometry methods. *Journal of Chromatography A*, *1525*, 96–108. https://doi.org/10.1016/j.chroma.2017.10.022
73. Lüscher, T. F. (2016). Prevention: some important steps forward, but many unmet needs in a world with cardiovascular disease as the leading cause of death. *European Heart Journal*, *37*(42), 3179–3181. https://doi.org/10.1093/eurheartj/ehw566
74. Ma, S., & Subramanian, R. (2006). Detecting and characterizing reactive metabolites by liquid chromatography/tandem mass spectrometry. *Journal of Mass Spectrometry*, *41*(9), 1121–1139. https://doi.org/10.1002/jms.1098
75. Maini Rekdal, V., Bess, E. N., Bisanz, J. E., Turnbaugh, P. J., & Balskus, E. P. (2019). Discovery and inhibition of an interspecies gut bacterial pathway for Levodopa metabolism. *Science*, *364*(6445). https://doi.org/10.1126/science.aau6323
76. Martin, R. J., Szefler, S. J., King, T. S., Kraft, M., Boushey, H. A., Chinchilli, V. M., Craig, T. J., Dimango, E. A., Deykin, A., Fahy, J. V., Israel, E., Lazarus, S. C., Lemanske, R. F., Jr., Leone, F. T., Pesola, G. R., Peters, S. P., Sorkness, C. A., Szwejbka, L. A., & Wechsler, M. E. (2007). The Predicting Response to Inhaled Corticosteroid Efficacy (PRICE) trial. *J Allergy Clin Immunol*, *119*(1), 73–80. https://doi.org/10.1016/j.jaci.2006.10.035
77. Mohs, R., & Greig, N. (2017). Drug discovery and development: Role of basic biological research. *Alzheimer's & Dementia: Translational Research & Clinical Interventions*, *3*. https://doi.org/10.1016/j.trci.2017.10.005
78. Monge, M. E., Dodds, J. N., Baker, E. S., Edison, A. S., & Fernández, F. M. (2019). Challenges in Identifying the Dark Molecules of Life. *Annu Rev Anal Chem (Palo Alto Calif)*, *12*(1), 177–199. https://doi.org/10.1146/annurev-anchem-061318-114959
79. Montastruc, J. L., Lafaurie, M., de Canecaude, C., Durrieu, G., Sommet, A., Montastruc, F., & Bagheri, H. (2021). Fatal adverse drug reactions: A worldwide perspective in the World Health Organization pharmacovigilance database. *Br J Clin Pharmacol*, *87*(11), 4334–4340. https://doi.org/10.1111/bcp.14851
80. Nam, H. W., Karpyak, V. M., Hinton, D. J., Geske, J. R., Ho, A. M. C., Prieto, M. L., Biernacka, J. M., Frye, M. A., Weinshilboum, R. M., & Choi, D. S. (2015). Elevated baseline serum glutamate as a pharmacometabolomic biomarker for acamprosate treatment outcome in alcohol-dependent subjects. *Translational Psychiatry*, *5*(8), e621–e621. https://doi.org/10.1038/tp.2015.120
81. Nishijo, N., Hayama, T., Tomita, R., Yamaguchi, M., & Fujioka, T. (2020). Application of a fluorous derivatization method for characterization of glutathione-trapped reactive

metabolites with liquid chromatography-tandem mass spectrometry analysis. *Journal of Chromatography A*, *1622*, 461160. https://doi.org/10.1016/j.chroma.2020.461160

82. Nordström, A., O'Maille, G., Qin, C., & Siuzdak, G. (2006). Nonlinear data alignment for UPLC-MS and HPLC-MS based metabolomics: quantitative analysis of endogenous and exogenous metabolites in human serum. *Anal Chem*, *78*(10), 3289–3295. https://doi.org/10.1021/ac060245f
83. Pang, H., Jia, W., & Hu, Z. (2019). Emerging Applications of Metabolomics in Clinical Pharmacology. *Clinical pharmacology and therapeutics*, *106*(3), 544–556. https://doi.org/10.1002/cpt.1538
84. Pang, Z., Chong, J., Zhou, G., de Lima Morais, D. A., Chang, L., Barrette, M., Gauthier, C., Jacques, P.-É., Li, S., & Xia, J. (2021). MetaboAnalyst 5.0: narrowing the gap between raw spectra and functional insights. *Nucleic Acids Research*, *49*(W1), W388–W396. https://doi.org/10.1093/nar/gkab382
85. Paris, D., Maniscalco, M., & Motta, A. (2018). Nuclear magnetic resonance-based metabolomics in respiratory medicine. *European Respiratory Journal*, *52*(4), 1801107. https://doi.org/10.1183/13993003.01107-2018
86. Park, J. E., Jeong, G. H., Lee, I. K., Yoon, Y. R., Liu, K. H., Gu, N., & Shin, K. H. (2018). A Pharmacometabolomic Approach to Predict Response to Metformin in Early-Phase Type 2 Diabetes Mellitus Patients. *Molecules*, *23*(7). https://doi.org/10.3390/molecules23071579
87. Park, T. J., Park, J. S., Cheong, H. S., Park, B. L., Kim, L. H., Heo, J. S., Kim, Y. K., Kim, K. U., Uh, S. T., Lee, H. S., Na, J. O., Seo, K. H., Choi, J. S., Kim, Y. H., Kim, M. S., Park, C. S., & Shin, H. D. (2014). Genome-wide association study identifies ALLC polymorphisms correlated with FEV_1 change by corticosteroid. *Clin Chim Acta*, *436*, 20–26. https://doi.org/10.1016/j.cca.2014.04.023
88. Pelkonen, O., Pasanen, M., Tolonen, A., Koskinen, M., Hakkola, J., Abass, K., Laine, J., Hakkinen, M., Juvonen, R., Auriola, S., Storvik, M., Huuskonen, P., Rousu, T., & Rahikkala, M. (2015). Reactive metabolites in early drug development: predictive in vitro tools. *Curr Med Chem*, *22*(4), 538–550. https://doi.org/10.2174/0929867321666141012175543
89. Plumb, R., Castro-Perez, J., Granger, J., Beattie, I., Joncour, K., & Wright, A. (2004). Ultra-performance liquid chromatography coupled to quadrupole-orthogonal time-of-flight mass spectrometry. *Rapid Commun Mass Spectrom*, *18*(19), 2331–2337. https://doi.org/10.1002/rcm.1627
90. Puig-Castellví, F., Pérez, Y., Piña, B., Tauler, R., & Alfonso, I. (2018). Comparative analysis of 1H NMR and 1H–13C HSQC NMR metabolomics to understand the effects of medium composition in yeast growth. *Analytical Chemistry*, *90*(21), 12422–12430. https://doi.org/10.1021/acs.analchem.8b01196
91. Pushpakom, S., Iorio, F., Eyers, P. A., Escott, K. J., Hopper, S., Wells, A., Doig, A., Guilliams, T., Latimer, J., McNamee, C., Norris, A., Sanseau, P., Cavalla, D., & Pirmohamed, M. (2019). Drug repurposing: progress, challenges and recommendations. *Nature Reviews Drug Discovery*, *18*(1), 41–58. https://doi.org/10.1038/nrd.2018.168
92. Ramautar, R., Somsen, G. W., & de Jong, G. J. (2019). CE-MS for metabolomics: Developments and applications in the period 2016-2018. *Electrophoresis*, *40*(1), 165–179. https://doi.org/10.1002/elps.201800323
93. Robuck, P. R., & Wurzelmann, J. I. (2005). Understanding the Drug Development Process. *Inflamm Bowel Dis*, *11*.
94. Rotroff, D., Shahin, M., Gurley, S., Zhu, H., Motsinger-Reif, A., Meisner, M., Beitelshees, A., Fiehn, O., Johnson, J., Elbadawi-Sidhu, M., Frye, R., Gong, Y., Weng, L., Cooper-DeHoff, R., & Kaddurah-Daouk, R. (2015). Pharmacometabolomic Assessments of Atenolol and Hydrochlorothiazide Treatment Reveal Novel Drug Response Phenotypes. *CPT: Pharmacometrics & Systems Pharmacology*, *4*(11), 669–679. https://doi.org/10.1002/psp4.12017
95. Rotroff, D. M., Oki, N. O., Liang, X., Yee, S. W., Stocker, S. L., Corum, D. G., Meisner, M., Fiehn, O., Motsinger-Reif, A. A., Giacomini, K. M., & Kaddurah-Daouk, R. (2016).

Pharmacometabolomic Assessment of Metformin in Non-diabetic, African Americans. *Front Pharmacol*, *7*, 135. https://doi.org/10.3389/fphar.2016.00135
96. Saigusa, D., Matsukawa, N., Hishinuma, E., & Koshiba, S. (2021). Identification of biomarkers to diagnose diseases and find adverse drug reactions by metabolomics. *Drug Metabolism and Pharmacokinetics*, *37*, 100373. https://doi.org/10.1016/j.dmpk.2020.11.008
97. Saito, Y., Sato, T., Nomoto, K., & Tsuji, H. (2018). Identification of phenol- and p-cresol-producing intestinal bacteria by using media supplemented with tyrosine and its metabolites. *FEMS Microbiology Ecology*, *94*(9). https://doi.org/10.1093/femsec/fiy125
98. Schenone, M., Dančík, V., Wagner, B. K., & Clemons, P. A. (2013). Target identification and mechanism of action in chemical biology and drug discovery. *Nat Chem Biol*, *9*(4), 232–240. https://doi.org/10.1038/nchembio.1199
99. Sellers, K., Allen, T. D., Bousamra, M., Tan, J., Méndez-Lucas, A., Lin, W., Bah, N., Chernyavskaya, Y., MacRae, J. I., Higashi, R. M., Lane, A. N., Fan, T. W. M., & Yuneva, M. O. (2019). Metabolic reprogramming and Notch activity distinguish between non-small cell lung cancer subtypes. *British Journal of Cancer*, *121*(1), 51–64. https://doi.org/10.1038/s41416-019-0464-z
100. Sévin, D. C., Fuhrer, T., Zamboni, N., & Sauer, U. (2017). Nontargeted in vitro metabolomics for high-throughput identification of novel enzymes in Escherichia coli. *Nature Methods*, *14*(2), 187–194. https://doi.org/10.1038/nmeth.4103
101. Soga, T., Ohashi, Y., Ueno, Y., Naraoka, H., Tomita, M., & Nishioka, T. (2003). Quantitative Metabolome Analysis Using Capillary Electrophoresis Mass Spectrometry. *Journal of Proteome Research*, *2*(5), 488–494. https://doi.org/10.1021/pr034020m
102. Speck, R. M., Sammel, M. D., Farrar, J. T., Hennessy, S., Mao, J. J., Stineman, M. G., & DeMichele, A. (2013). Impact of Chemotherapy-Induced Peripheral Neuropathy on Treatment Delivery in Nonmetastatic Breast Cancer. *Journal of Oncology Practice*, *9*(5), e234–e240. https://doi.org/10.1200/jop.2012.000863
103. Sud, M., Fahy, E., Cotter, D., Azam, K., Vadivelu, I., Burant, C., Edison, A., Fiehn, O., Higashi, R., Nair, K. S., Sumner, S., & Subramaniam, S. (2016). Metabolomics Workbench: An international repository for metabolomics data and metadata, metabolite standards, protocols, tutorials and training, and analysis tools. *Nucleic Acids Res*, *44*(D1), D463–470. https://doi.org/10.1093/nar/gkv1042
104. Sun, D., Gao, W., Hu, H., & Zhou, S. (2022). Why 90% of clinical drug development fails and how to improve it? *Acta Pharmaceutica Sinica B*, *12*(7), 3049–3062. https://doi.org/10.1016/j.apsb.2022.02.002
105. Sun, Y., Kim, J. H., Vangipuram, K., Hayes, D. F., Smith, E. M. L., Yeomans, L., Henry, N. L., Stringer, K. A., & Hertz, D. L. (2018). Pharmacometabolomics reveals a role for histidine, phenylalanine, and threonine in the development of paclitaxel-induced peripheral neuropathy. *Breast Cancer Research and Treatment*, *171*(3), 657–666. https://doi.org/10.1007/s10549-018-4862-3
106. Szefler, S. J., Martin, R. J., King, T. S., Boushey, H. A., Cherniack, R. M., Chinchilli, V. M., Craig, T. J., Dolovich, M., Drazen, J. M., Fagan, J. K., Fahy, J. V., Fish, J. E., Ford, J. G., Israel, E., Kiley, J., Kraft, M., Lazarus, S. C., Lemanske, R. F., Mauger, E., . . . Sorkness, C. A. (2002). Significant variability in response to inhaled corticosteroids for persistent asthma. *Journal of Allergy and Clinical Immunology*, *109*(3), 410–418. https://doi.org/10.1067/mai.2002.122635
107. Tang, Y., Wang, Y., Kiani, M. F., & Wang, B. (2016). Classification, Treatment Strategy, and Associated Drug Resistance in Breast Cancer. *Clinical Breast Cancer*, *16*(5), 335–343. https://doi.org/10.1016/j.clbc.2016.05.012
108. Tantisira, K. G., Damask, A., Szefler, S. J., Schuemann, B., Markezich, A., Su, J., Klanderman, B., Sylvia, J., Wu, R., Martinez, F., Boushey, H. A., Chinchilli, V. M., Mauger, D., Weiss, S. T., & Israel, E. (2012). Genome-wide association identifies the T gene as a novel asthma pharmacogenetic locus. *Am J Respir Crit Care Med*, *185*(12), 1286–1291. https://doi.org/10.1164/rccm.201111-2061OC

109. Tantisira, K. G., Lasky-Su, J., Harada, M., Murphy, A., Litonjua, A. A., Himes, B. E., Lange, C., Lazarus, R., Sylvia, J., Klanderman, B., Duan, Q. L., Qiu, W., Hirota, T., Martinez, F. D., Mauger, D., Sorkness, C., Szefler, S., Lazarus, S. C., Lemanske, R. F., Jr., . . . Weiss, S. T. (2011). Genomewide association between GLCCI1 and response to glucocorticoid therapy in asthma. *N Engl J Med*, *365*(13), 1173–1183. https://doi.org/10.1056/NEJMoa0911353
110. Tee, K. B., Ibrahim, L., Hashim, N. M., Saiman, M. Z., Zakaria, Z. H., & Huri, H. Z. (2022). Pharmacokinetic-Pharmacometabolomic Approach in Early-Phase Clinical Trials: A Way Forward for Targeted Therapy in Type 2 Diabetes. *Pharmaceutics*, *14*(6). https://doi.org/10.3390/pharmaceutics14061268
111. Tiziani, S., Kang, Y., Choi, J. S., Roberts, W., & Paternostro, G. (2011). Metabolomic high-content nuclear magnetic resonance-based drug screening of a kinase inhibitor library. *Nature Communications*, *2*(1), 545. https://doi.org/10.1038/ncomms1562
112. Trapotsi, M. A., Hosseini-Gerami, L., & Bender, A. (2022). Computational analyses of mechanism of action (MoA): data, methods and integration. *RSC Chem Biol*, *3*(2), 170–200. https://doi.org/10.1039/d1cb00069a
113. Trupp, M., Zhu, H., Wikoff, W. R., Baillie, R. A., Zeng, Z.-B., Karp, P. D., Fiehn, O., Krauss, R. M., & Kaddurah-Daouk, R. (2012). Metabolomics Reveals Amino Acids Contribute to Variation in Response to Simvastatin Treatment. *PLOS ONE*, *7*(7), e38386. https://doi.org/10.1371/journal.pone.0038386
114. Twycross, R., Ross, J., Kotlinska-Lemicszek, A., Charlesworth, S., Mihalyo, M., & Wilcock, A. (2015). Variability in Response to Drugs. *Journal of Pain and Symptom Management*, *49*(2), 293–306. https://doi.org/10.1016/j.jpainsymman.2014.10.003
115. van Kessel, S. P., de Jong, H. R., Winkel, S. L., van Leeuwen, S. S., Nelemans, S. A., Permentier, H., Keshavarzian, A., & El Aidy, S. (2020). Gut bacterial deamination of residual levodopa medication for Parkinson's disease. *BMC Biol*, *18*(1), 137. https://doi.org/10.1186/s12915-020-00876-3
116. Ventola, C. L. (2013). The role of pharmacogenomic biomarkers in predicting and improving drug response: part 2: challenges impeding clinical implementation. *P t*, *38*(10), 624–627.
117. Vigneri, R., & Goldfine, I. D. (1987). Role of metformin in treatment of diabetes mellitus. *Diabetes Care*, *10*(1), 118–122. https://doi.org/10.2337/diacare.10.1.118
118. Vignoli, A., Ghini, V., Meoni, G., Licari, C., Takis, P. G., Tenori, L., Turano, P., & Luchinat, C. (2019). High-Throughput Metabolomics by 1D NMR. *Angewandte Chemie International Edition*, *58*(4), 968–994. https://doi.org/10.1002/anie.201804736
119. Walker, D. I., Perry-Walker, K., Finnell, R. H., Pennell, K. D., Tran, V., May, R. C., McElrath, T. F., Meador, K. J., Pennell, P. B., & Jones, D. P. (2019). Metabolome-wide association study of anti-epileptic drug treatment during pregnancy. *Toxicology and Applied Pharmacology*, *363*, 122–130. https://doi.org/10.1016/j.taap.2018.12.001
120. Wang, A. L., Lahousse, L., Dahlin, A., Edris, A., McGeachie, M., Lutz, S. M., Sordillo, J. E., Brusselle, G., Lasky-Su, J., Weiss, S. T., Iribarren, C., Lu, M. X., Tantisira, K. G., & Wu, A. C. (2022). Novel genetic variants associated with inhaled corticosteroid treatment response in older adults with asthma. *Thorax*. https://doi.org/10.1136/thoraxjnl-2021-217674
121. Wang, M., Wang, H., Zheng, H., Uhrin, D., Dewhurst, R. J., & Roehe, R. (2021). Comparison of HPLC and NMR for quantification of the main volatile fatty acids in rumen digesta. *Scientific Reports*, *11*(1), 24337. https://doi.org/10.1038/s41598-021-03553-9
122. Wang, Y., Ai, F., Ng, S. C., & Tan, T. T. (2012). Sub-2 μm porous silica materials for enhanced separation performance in liquid chromatography. *J Chromatogr A*, *1228*, 99–109. https://doi.org/10.1016/j.chroma.2011.08.085
123. Wilson, I. D., Nicholson, J. K., Castro-Perez, J., Granger, J. H., Johnson, K. A., Smith, B. W., & Plumb, R. S. (2005). High Resolution "Ultra Performance" Liquid Chromatography Coupled to oa-TOF Mass Spectrometry as a Tool for Differential Metabolic Pathway Profiling in Functional Genomic Studies. *Journal of Proteome Research*, *4*(2), 591–598. https://doi.org/10.1021/pr049769r

124. Winnike, J. H., Li, Z., Wright, F. A., Macdonald, J. M., O'Connell, T. M., & Watkins, P. B. (2010). Use of Pharmaco-Metabonomics for Early Prediction of Acetaminophen-Induced Hepatotoxicity in Humans. *Clinical Pharmacology & Therapeutics*, *88*(1), 45–51. https://doi.org/10.1038/clpt.2009.240
125. Wishart, D. S., Guo, A., Oler, E., Wang, F., Anjum, A., Peters, H., Dizon, R., Sayeeda, Z., Tian, S., Lee, B. L., Berjanskii, M., Mah, R., Yamamoto, M., Jovel, J., Torres-Calzada, C., Hiebert-Giesbrecht, M., Lui, V. W., Varshavi, D., Varshavi, D., . . . Gautam, V. (2022). HMDB 5.0: the Human Metabolome Database for 2022. *Nucleic Acids Res*, *50*(D1), D622–d631. https://doi.org/10.1093/nar/gkab1062
126. Wishart, D. S., Tzur, D., Knox, C., Eisner, R., Guo, A. C., Young, N., Cheng, D., Jewell, K., Arndt, D., Sawhney, S., Fung, C., Nikolai, L., Lewis, M., Coutouly, M. A., Forsythe, I., Tang, P., Shrivastava, S., Jeroncic, K., Stothard, P., . . . Querengesser, L. (2007). HMDB: the Human Metabolome Database. *Nucleic Acids Res*, *35*(Database issue), D521–526. https://doi.org/10.1093/nar/gkl923
127. Wu, H., & Tang, K. (2020). Highly Sensitive and Robust Capillary Electrophoresis-Electrospray Ionization-Mass Spectrometry: Interfaces, Preconcentration Techniques and Applications. *Reviews in Analytical Chemistry*, *39*(1), 45–55. https://doi.org/10.1515/revac-2020-0112
128. Xiao, J. F., Zhou, B., & Ressom, H. W. (2012). Metabolite identification and quantitation in LC-MS/MS-based metabolomics. *Trends Analyt Chem*, *32*, 1–14. https://doi.org/10.1016/j.trac.2011.08.009
129. Xu, T., Zhao, H., Wang, M., Chow, A., & Fang, M. (2021). Metabolomics and In Silico Docking-Directed Discovery of Small-Molecule Enzyme Targets. *Analytical Chemistry*, *93*(6), 3072–3081. https://doi.org/10.1021/acs.analchem.0c03684
130. Yao, C.-H., Wang, L., Stancliffe, E., Sindelar, M., Cho, K., Yin, W., Wang, Y., & Patti, G. J. (2020). Dose-Response Metabolomics To Understand Biochemical Mechanisms and Off-Target Drug Effects with the TOXcms Software. *Analytical Chemistry*, *92*(2), 1856–1864. https://doi.org/10.1021/acs.analchem.9b03811
131. Yuan, H. J., & Lee, W. M. (2008). Nonresponse to treatment for hepatitis C: current management strategies. *Drugs*, *68*(1), 27–42. https://doi.org/10.2165/00003495-200868010-00003
132. Yukta Munjal, Rajiv Kumar Tonk, & Sharma, R. (2022). Analytical Techniques Used in Metabolomics: A Review. *Systematic Reviews in Pharmacy*, *13*(5). https://doi.org/10.31858/0975-8453.13.5.550-556
133. Zampieri, M., Szappanos, B., Buchieri, M. V., Trauner, A., Piazza, I., Picotti, P., Gagneux, S., Borrell, S., Gicquel, B., Lelievre, J., Papp, B., & Sauer, U. (2018). High-throughput metabolomic analysis predicts mode of action of uncharacterized antimicrobial compounds. *Sci Transl Med*, *10*(429). https://doi.org/10.1126/scitranslmed.aal3973
134. Zampieri, M., Zimmermann, M., Claassen, M., & Sauer, U. (2017). Nontargeted Metabolomics Reveals the Multilevel Response to Antibiotic Perturbations. *Cell Reports*, *19*(6), 1214–1228. https://doi.org/10.1016/j.celrep.2017.04.002
135. Zeki Ö. C., Eylem, C. C., Reçber, T., Kır, S., & Nemutlu, E. (2020). Integration of GC-MS and LC-MS for untargeted metabolomics profiling. *J Pharm Biomed Anal*, *190*, 113509. https://doi.org/10.1016/j.jpba.2020.113509
136. Zhang, X. W., Li, Q. H., Xu, Z. D., & Dou, J. J. (2020). Mass spectrometry-based metabolomics in health and medical science: a systematic review. *RSC Adv*, *10*(6), 3092–3104. https://doi.org/10.1039/c9ra08985c
137. Zheng, Y., Ley, S. H., & Hu, F. B. (2018). Global aetiology and epidemiology of type 2 diabetes mellitus and its complications. *Nature Reviews Endocrinology*, *14*(2), 88–98. https://doi.org/10.1038/nrendo.2017.151
138. Zimmermann, M., Zimmermann-Kogadeeva, M., Wegmann, R., & Goodman, A. L. (2019). Mapping human microbiome drug metabolism by gut bacteria and their genes. *Nature*, *570*(7762), 462–467. https://doi.org/10.1038/s41586-019-1291-3

Microbial Metabolomics: An Overview of Applications

Pieter M. M. van der Velden and Robert S. Jansen

Abbreviations

13CMFA	^{13}C-based metabolic flux analysis
ADP	Adenosine diphosphate
AMP	Adenosine monophosphate
ASD	Autism spectrum disorder
ATP	Adenosine triphosphate
BGC	Biosynthetic gene clusters
CoA	Coenzyme A
DAP-AT	Succinyldiaminopimelate aminotransferase
DI-FT-ICR-MS	Direct-infusion ion-cyclotron-resonance Fourier-transform mass spectrometry
DNA	Deoxyribonucleic acid
FISH	Fluorescent in situ hybridization
FTICR-MS	Fourier-transform ion-cyclotron-resonance mass spectrometry
GABA	Gamma aminobutyric acid
GCF	Gene cluster family
GC-MS	Gas chromatography-mass spectrometry
GMP	Guanosine monophosphate
GNPS	Global Natural Products Social Molecular Networking
GSMM	Genome-scale metabolic model
IBD	Inflammatory bowel disease

P. M. M. van der Velden · R. S. Jansen (✉)
Radboud University, Department of Microbiology, Radboud Institute for Biological and Environmental Sciences, Heyendaalseweg, Nijmegen, The Netherlands
e-mail: Robert.jansen@ru.nl

V. Soni, T. E. Hartman (eds.), *Metabolomics*,
https://doi.org/10.1007/978-3-031-39094-4_6

KEGG	Kyoto Encyclopedia of Genes and Genomes
LC-MS	Liquid chromatography-mass spectrometry
MALDI-MSI	Matrix-assisted laser desorption/ionization-mass spectrometry imaging
MAMBO	Metabolomic analysis of metagenomes using flux balance analysis and optimization
MS	Mass spectrometry
NAD(P)H	Nicotinamide adenine dinucleotide (phosphate)
NanoDESI MS	Nanospray desorption electrospray ionization mass spectrometry
nanoSIMS	Nanoscale secondary ion mass spectrometry
NMR	Nuclear magnetic resonance
RNA	Ribonucleic acid
SeMeCos	Self-establishing metabolically cooperating yeast communities
TCA	Tricarboxylic acid
VMH	Virtual Metabolic Human
WoM	Web of Microbes

1 Introduction

Microbes are everywhere, from hypersaline polar lakes, to the superheated walls of hydrothermal vents, our guts, and sites of infection teeming with hostile immune cells [25, 159]. To achieve this, they have developed impressive adaptations to not only survive but even thrive in environments that seem physically inhospitable and devoid of nutrients. For humans, microbes can be friend or foe. The human gut microbiome, for example, is essential for nutrient uptake and health, but disbalances or infections can cause life-threatening illnesses. Similarly, some environmental microorganisms help mitigating environmental pollution by devouring harmful chemicals, like heavy metals and pharmaceuticals, while others produce lethal agents like botulinum toxin. Microbial metabolism is, therefore, relevant to fields ranging from biotechnology to medicine and environmental sciences. In this chapter we describe the various ways in which metabolomics can be used to study microbial metabolism.

Microbial metabolism enables microorganisms to use nutrients and energy sources required for essential cell functions but also shapes the interactions with other organisms and the environment. Microbial metabolomics can therefore provide insights into fundamental aspects of life, like the link between metabolism and cell replication, as well as specialized functions like the role of a gut community member in health.

Microbial metabolomics aims to study all aspects of microbial metabolism by measuring the metabolites inside and outside microbes. Consequently, microbial metabolomics is a very diverse field with diverse techniques and applications. Many of the applications we describe can also be applied to multicellular organisms, but relatively easy and controllable cultivation of many microbes, in combination with genetic tractability, makes microbes popular organisms for most metabolomic applications.

Like the microbes under study, the field of microbial metabolomics is extremely diverse (reviewed in [6, 11, 50, 180]). Therefore, it is impossible to provide an all-encompassing overview in a single chapter. Instead, we aim to highlight important applications and refer to key literature for further reading. Microbial metabolomics can be subdivided into application areas like medical microbiology, environmental microbiology, and biotechnology, but since the experimental approaches are similar across these application areas, this chapter provides an overview organized by experimental approach. The first part of this chapter focuses on applications related to microbial monocultures, progressing from relatively simple techniques to assess nutrient consumption to complex metabolic models of single microorganisms. The second part focuses on microbial communities, progressing from studies on community isolates to studies on complex microbial communities. Throughout the chapter, microbial metabolomics approaches are highlighted using selected examples from environmental, medical, and biotechnological microbiology.

2 Metabolomics on Microbial Monocultures

Historically, pure monocultures have been essential to study microorganisms. Although technological advances now allow studying complete microbial communities (see Sect. 3), pure cultures still enable the most in-depth biochemical and physiological studies today. Applying current metabolomics techniques to pure microbial cultures provides insight into properties ranging from basic metabolic pathways and nutrient use to specialized metabolism, stress responses, and metabolic regulation. Integrating these properties with other omics data ultimately allows constructing genome-scale metabolic and kinetic models.

Bioinformatics tools are increasingly adept at distilling metabolic pathway potential from microbial genomes [147], but their reliance on correct gene annotations renders them far from flawless. Moreover, genomes harbor information on metabolic potential, not on actual metabolic activity. Unlike genomics, metabolomics does not depend on gene annotation and measures actual metabolic activity. Therefore, metabolomics is a powerful technique to explore known and unknown metabolic pathways involved in core and secondary metabolism.

In this section we will provide an overview the various ways in which metabolomics can be applied to study microbial monocultures, moving from simple questions to complex models. After describing extraction methods (Sect. 2.1), Sects. 2.2, 2.3, and 2.4 focus on metabolism that is essential for growth and replication, also known as primary metabolism. Sections 2.4 and 2.5 then progress into secondary metabolism – metabolism that is inessential but that confers a selective advantage in environmental interactions. In Sects. 2.6 and 2.7, we describe how metabolomics can be used to improve culture media and optimize strain engineering. Finally, in Sect. 2.8, we will discuss how metabolomics can be used to infer system-level information on microbial metabolism.

2.1 Extraction Methods

Metabolomic analysis almost always requires the extraction of metabolites. Various techniques can be used to extract metabolites from microbes, but the most commonly techniques are based on rapid filter-based sampling followed by quenching in a water-organic solvent mixture [16, 83, 99, 119, 123, 125, 128, 131, 136, 182]. The mere presence or consumption of pathway-specific metabolites is often sufficient to confirm activity of a pathway, but tracing of stable isotopes enables detailed analysis of the metabolic fluxes carried by various pathways.

2.2 Nutrient Use

Every single organism across the microbial world requires nutrients for energy conservation and as building blocks for growth. Therefore, microbes have evolved a wide diversity of metabolic pathways to dissimilate or assimilate and distribute a plethora of carbon, nitrogen, and other element sources. Various metabolomics approaches have proven useful for the characterization of these pathways (Fig. 1).

2.2.1 Exometabolomics to Measure the Consumption of Nutrients

Exometabolomics is the subfield of metabolomics that profiles extracellular metabolites, for example, in culture media [74, 96, 124]. The exometabolome is linked to intracellular metabolism through metabolite consumption and secretion and is particularly well suited to profile nutrient consumption (Fig. 1a). Antunes et al., for example, applied exometabolomics to identify host-derived nutrients that the

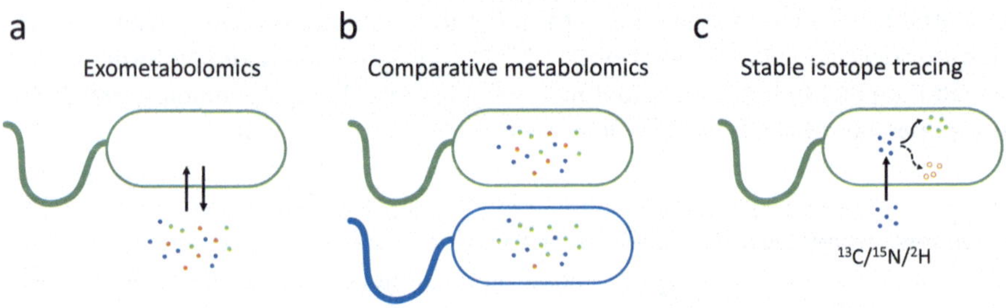

Fig. 1 Metabolomic approaches to characterize microbial nutrient use. Exometabolomics focuses on the extracellular metabolome, which is linked to intracellular metabolism (panel ***a***). In comparative metabolomics, metabolite levels are compared between different conditions, for example, two types of microorganisms, or the same microorganism grown on different nutrients (panel ***b***). Stable isotope tracing uses the resolving power of mass spectrometry to follow the metabolic faith of stable isotope-labeled nutrients (panel ***c***). Metabolites are represented as circles. In panel **c**, labeled metabolites are represented as filled circles, and non-labeled metabolites as open circles

pathogen *Salmonella enterica* serovar Typhimurium – the bacterium that causes typhoid fever – consumes during infection [3]. Despite the antibacterial action of bile, *Salmonella* thrives in the gastrointestinal tract, in the liver, and even in the gallbladder, suggesting that it can feed on bile-derived nutrients. To find these nutrients, Antunes et al. incubated extracted mouse bile with *Salmonella* and profiled the consumption of its components using direct-infusion ion-cyclotron-resonance Fourier-transform mass spectrometry (DI-FT-ICR-MS) [3]. A reduction of several glycerophospholipids suggested that these metabolites were used as nutrients. Similar changes were found when the bile of uninfected mice was compared to that of infected mice. Finally, in vitro growth experiments confirmed that the glycerophospholipid lyso-phosphocholine could serve as sole carbon source in a minimal medium [3].

To link metabolite consumption with specific genes, Baran et al. expanded the exometabolomics approach by applying high-throughput exometabolomics to mutant libraries of *Escherichia coli* and *Shewanella oneidensis* [7]. From an initial screen for consumed metabolites, ten metabolites were selected for high-throughput profiling using a rapid liquid chromatography-mass spectrometry (LC-MS) method. In a technical tour de force, the researchers profiled spent media of 3901 *E. coli* and 4141 *S. oneidensis* mutants and found several mutants that were less proficient in taking up or using the metabolites. Subsequent validation confirmed the role of previously characterized genes in nutrient use and also identified the function of genes of unknown function [7].

2.2.2 Comparative Metabolomics to Identify Nutrient-Related Metabolic Pathways

While exometabolomics is a powerful tool to screen for nutrient consumption, comparative intracellular metabolomics is a powerful tool to identify pathways associated with the use of known nutrients. In comparative metabolomics, metabolite levels are compared between different conditions, for example, two types of microorganisms, or the same microorganism grown on different nutrients (Fig. 1b), as exemplified by work from Griffin et al. [54]. *Mycobacterium tuberculosis*, the causative agent of tuberculosis, has evolved from a soil bacterium into one of the most successful obligate pathogens. As a consequence, it has metabolically adapted from a soil-derived diet to a host-derived one. Because *M. tuberculosis* harbors a gene cluster encoding cholesterol catabolism that is essential for growth on cholesterol and infection, cholesterol has been considered an important host-derived carbon source [120, 165]. The final products of this pathway were unclear, however, leaving the metabolic adaptations required for growth on cholesterol unknown. To identify key metabolic pathways involved in cholesterol catabolism, Griffin et al. performed LC-MS and gas chromatography-mass spectrometry (GC-MS) metabolomics on *M. tuberculosis* grown on various carbon sources, including cholesterol [54]. Growth on cholesterol resulted in the accumulation of several intermediates of the methylcitrate cycle, a variant of the TCA cycle that detoxifies propionyl-CoA. Subsequent experiments showed that methylcitrate cycle-deficient mutants were unable to grow on cholesterol, confirming that propionyl-CoA detoxification through the methylcitrate cycle is essential for cholesterol catabolism.

2.2.3 Stable Isotope Tracing to Track the Metabolic Fate of Nutrients

Although the presence or accumulation of pathway-specific metabolites is often sufficient to show pathway activity, stable isotope tracing provides information about metabolic fluxes through pathways, even when pathways share the same metabolites. Stable isotope tracing builds on the capacity of mass spectrometers to distinguish light isotopes (e.g., ^{12}C, ^{14}N, and ^{1}H) that are dominant in nature from heavy isotopes (e.g., ^{13}C, ^{15}N, and ^{2}H) that can artificially be incorporated into metabolites. Compounds enriched in heavy isotopes, along with their "heavy" metabolic products, can thus be traced over time (Fig. 1c).

Anaerobic ammonium-oxidizing (anammox) bacteria are microorganisms capable of converting ammonium and nitrite into dinitrogen gas [156]. Anammox bacteria are estimated to be the source of up to 50% of N_2 production in the ocean and are successfully applied in wastewater treatment plants [4, 73]. Based on genomic predictions, anammox bacteria use the Wood-Ljungdahl pathway to fix carbon from CO_2, but the absence of an annotated citrate synthase has led to the hypothesis that the TCA cycle runs in the reductive direction [82, 157]. Lawson et al. exposed a highly enriched (~97%) bioreactor culture of the anammox bacterium *Candidatus* "Kuenenia stuttgartiensis" to ^{13}C-bicabonate and traced the incorporation of $^{13}CO_2$ into metabolites using ion-pairing LC-MS [82]. Notwithstanding the lack of an annotated citrate synthase, ^{13}C-labeling of α -ketoglutarate was faster than labeling of succinate, suggesting the presence of a functional citrate synthase operating in the oxidative TCA cycle. To confirm this, the authors performed a labeling experiment with ^{13}C-formate which would lead to single- or double-labeled α-ketoglutarate with a reductive or oxidative TCA cycle, respectively. The dominant presence of M + 2 α-ketoglutarate confirmed the ^{13}C-bicarbonate results and confirmed the presence of an oxidative TCA cycle. Using stable isotope tracing metabolomics, Lawson et al. thus corrected an erroneous genome prediction and functionally identified a non-canonical citrate synthase. A non-canonical pathway for methionine biosynthesis was similarly identified [82]. Both enzymes remain to be identified. Concluding, stable isotope tracing can be used to elucidate the metabolic pathways used to metabolize a single carbon source.

With a related approach, isotope tracing has been used to define the direction of metabolic pathways during growth on multiple carbon sources. Bacteria are classically considered to use nutrients one by one, resulting in diauxic growth. *M. tuberculosis*, however, can use multiple carbon sources at the same time [31]. De Carvalho et al. used ^{13}C-isotope tracing to individually define the metabolic fate of dextrose, acetate, and glycerol catabolism during growth on a mixture of these substrates. Each carbon source had a distinct metabolic fate. Dextrose-derived carbon, for example, was enriched in the glycolytic and pentose phosphate pathway, while acetate-derived carbon was enriched in the TCA cycle. In some cases, *M. tuberculosis* metabolized the carbon sources in contradictory directions, such as simultaneous glycolysis and gluconeogenesis. Since the same pathway cannot operate in two directions at the same time, these results imply that carbon metabolism in *M. tuberculosis* is compartmentalized [31].

Because nitrogen is rapidly exchanged between amino acids, ^{15}N-isotope tracing is less frequently applied than ^{13}C-isotope tracing. Still, ^{15}N-isotope tracing has been instrumental in elucidating specific metabolic pathways. Kurczy et al., for example, applied metabolomics to trace the fate of $^{15}NO_3$ in three nitrate-reducing *Pseudomonas* strains [81]. To assimilate nitrogen, soil bacteria commonly prefer ammonia over other nitrogen sources [102]. In one of the three strains, Kurczy et al. indeed observed that the level of nitrogen-containing metabolites changed when ammonia was added to a nitrate-containing medium. However, two of the three strains showed much less metabolic perturbations, indicating that these strains did not prefer ammonia over nitrate. To confirm this hypothesis, Kurczy et al. traced the fate of $^{15}NO_3$ in the presence and absence of unlabeled ammonia. The single strain did not assimilate ^{15}N from nitrate in the presence of ammonia, while the two other strains did, confirming that these strains had the unexpected capability of using nitrate and ammonia as nitrogen sources at the same time.

To characterize nitrogen metabolism in *M. tuberculosis*, Agapova et al. traced the metabolic fate of a panel of ^{15}N-labeled amino acids and $^{15}NH_4$ [1]. In contrast to common views, the pool size and labeling patterns of alanine suggested a role for alanine dehydrogenase in nitrogen liberation from alanine. Growth experiments subsequently confirmed that an alanine dehydrogenase-deficient strain was unable to grow on alanine as a sole nitrogen source. In similar work, Jansen et al. combined ^{15}N- and ^{13}C-isotope tracing experiments to demonstrate the essential role of the main aspartate aminotransferase in *M. tuberculosis* [67]. In vitro experiments indicated that the uncharacterized protein Rv3722c is an aspartate aminotransferase, an enzyme that transfers the amino group between aspartate and glutamate [67]. To explore the function of this enzyme in vivo, Jansen et al. incubated Rv3722c-proficient and Rv3722c-deficient *M. tuberculosis* with ^{13}C-aspartate, ^{13}C-glutamate, ^{15}N-aspartate, and ^{15}N-glutamate. Isotope tracing showed that Rv3722c is required to channel the carbon from aspartate into the TCA cycle and also to channel assimilated nitrogen from glutamate into essential amino acids, cofactors, and nucleotides.

As exemplified in the work by Jansen et al., carbon metabolism and nitrogen metabolism are closely linked. Diazobacteria are biotechnologically interesting microorganisms because they can fix atmospheric dinitrogen gas to ammonia using the enzyme nitrogenase. To characterize the adaptations in central carbon metabolism that allow diazobacteria to fix nitrogen, Wu et al. applied ^{13}C-metabolic flux analyses on *Azotobacter vinelandii* [175]. Although the activity of the glycolytic Entner-Doudoroff pathway positively correlated with the growth rate, the activity of the respiratory TCA cycle correlated with increased nitrogenase activity, leading the authors to conclude that the TCA cycle supplies ATP and reducing equivalents to sustain nitrogenase activity.

Taken together, exometabolomics, comparative, and isotope-tracing metabolomics approaches have provided invaluable insights into microbial nutrient use.

2.3 Energy Conservation

In addition to nutrients that serve as building blocks (Sect. 2.1), microorganisms also need energy to fuel metabolism. To conserve energy and generate ATP, microorganisms can perform fermentation and respiration. Metabolomics can distinguish between these modes of energy conservation and helps identifying metabolites involved in redox balance (Fig. 2). In addition, metabolomics can define the energy status of microbes.

2.3.1 Measuring Energy Charge

In its conceptually simplest form, metabolomics can determine the status of a cell by measuring key intermediates like AMP, ADP and ATP (adenylate energy charge), or NAD^+ and NADH (redox balance). Though conceptually simple, accurate determination of these metabolites is technically challenging due to their enzymatic and chemical instability. Comparing several extraction strategies for mammalian cells, Lu et al. concluded that extraction with a mixture of acetonitrile, methanol, and water with 0.1 M formic acid minimized interconversion of NADPH/$NADP^+$ and NADH/NAD^+ [90]. After extraction, rapid pH neutralization furthermore prevented acid-catalyzed degradation. For microorganisms, rapid sampling and quenching techniques have been indispensable to accurately determine energy and redox status [13, 131, 137]. Using these techniques in combination with ^{13}C-based absolute

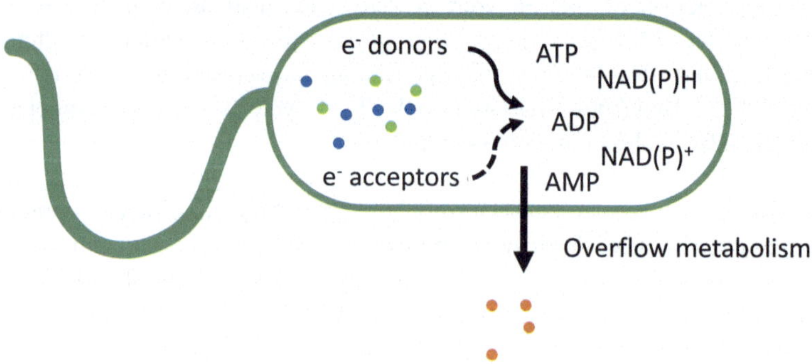

Fig. 2 Microbial energy conservation. Microbes can perform respiration and fermentation to generate ATP and reducing equivalents. Metabolites (represented as circles) can directly serve as electron (e-) donor or acceptor but can also indirectly support energy conservation. Overflow metabolism is characterized by the excretion of fermentation products

quantification, Radoš et al. showed that the energy charge in *E. coli* is almost constant across 19 different growth conditions, demonstrating highly regulated energy homeostasis [132].

2.3.2 Respiration

Shewanella and *Geobacter* spp. are bacteria capable of using extracellular metals as terminal electron acceptors for respiration through extracellular electron transfer [149]. Since they can also transfer electrons to electrodes, these microorganisms hold great promise for biotechnology. To explore the metabolic adaptations to different electron donor-acceptor ratios, Wang et al. profiled the metabolome of *S. oneidensis* using GC-MS [167]. Most extracellular reduction occurred at limiting electron acceptor concentrations at which the reducing power was stored in the form of lactate and pyruvate. Under conditions with excess electron acceptor, metabolites from non-energy-yielding pathways like putrescine were more abundant. Based on these metabolomics results, the authors conclude that the central metabolism of *Shewanella* is streamlined to yield energy under conditions with low electron acceptors. Shi et al. similarly show that the metabolome of *Geobacter sulfurreducens* is dependent on the type of electron acceptor supplied [150].

^{13}C-based metabolic flux analysis (13CMFA) is a technique to assess metabolic fluxes that is widely used in combination with metabolomics. Using 13CMFA, Yang et al. assessed the effect of various electron donor (acetate and hydrogen) and acceptor (Fe^{III} and fumarate) conditions on the metabolism of *G. sulfurreducens* [178]. In contrast to previous conceptions, the flux analysis suggested that fumarate can serve as carbon source and even electron donor in addition to its role as an electron acceptor.

When respiring via an electrode in a microbial fuel cell, microorganisms like *G. sulfurreducens* can couple chemical conversions to electric currents [15]. Song et al. used targeted LC-MS and GC-MS metabolomics to explore the effect of electrode potential on intracellular metabolites [154]. When comparing cells grown on −0.2 V with those grown on +0.2 V, Song found that levels of TCA cycle intermediates and the ATP/ADP ratio were increased in the +0.2 V condition, while metabolites involved in gluconeogenesis and the $NAD(P)H/NAD(P)^+$ ratio were decreased. These results suggest that a larger electric current is associated with defined changes in intracellular metabolism and show that the electrode potential can be used to steer microbial metabolism in microbial fuel cells [154]. Similar work with *Alcanivorax xenomutans* linked the presence of oxygen to metabolism and power output [93].

2.3.3 Fermentation and Overflow Metabolism

In the absence of substrates to respire, organisms can switch to less energy-efficient fermentative metabolism to generate ATP. Many organisms, however, also use fermentative metabolism when conditions would allow more energy-efficient respiration [106]. This phenomenon, which is known as the Crabtree effect in microorganisms, is characterized by overflow metabolism in which metabolites like ethanol, lactate, and acetate are excreted into the extracellular environment. The Crabtree effect has been known for almost a century, but the underlying mechanisms and evolutionary advantages are still incompletely defined [9, 115]. To identify defining metabolic features for the Crabtree effect, Christen and Sauer applied intracellular ^{13}C-flux analysis on seven different yeast strains [24]. As expected, Crabtree-negative strains showed higher fluxes through respiratory pathways, but the metabolite concentrations were mainly species-specific and independent of the Crabtree effect. Only fructose-1,6-bisphosphate and dihydroxyacetone phosphate correlated with the Crabtree effect across all strains, indicating a key role in the switch between respiratory and fermentative metabolism. In a similar work, Yasid et al. used 1H NMR to monitor the intracellular and extracellular metabolites in an *E. coli* culture while shifting it from anaerobic to aerobic growth [179]. Most metabolites gradually changed upon air exposure, but pyruvate only changed extracellularly and not intracellularly. Subsequent electron microscopy studies revealed that pyruvate kinase, a key enzyme in pyruvate metabolism, relocates to the cell membrane upon aeration and potentially facilitates pyruvate excretion while maintaining constant intracellular levels.

Owing to recent advances in metabolomics, it has become apparent that excreted metabolites are more diverse than previously thought. Paczia et al., for example, profiled the exometabolome of *E. coli*, *Corynebacterium glutamicum*, *Bacillus licheniformis*, and *Saccharomyces cerevisiae* and detected a surprisingly wide variety of central metabolic intermediates and amino acids [119]. After carefully ruling out sampling artifacts, they termed this phenomenon "extended" overflow metabolism.

Although ATP and electron carriers are common energy intermediates of all microbes, the pathways generating them vary widely. *Clostridium sporogenes* is a mutualistic bacterium and common component of our gut microbiome. *C. sporogenes* belongs to a group of anaerobic bacteria that obtain their energy by coupling the oxidation of one amino acid to the reduction of another through the Stickland fermentation [86]. The oxidative pathways are assumed to produce ATP, while the reductive pathways are thought to balance redox. Liu et al. used metabolomics to profile metabolites excreted by *C. sporogenes* and identified four of them in plasma of gnotobiotic mice colonized with *C. sporogenes*. In vitro incubation with deuterium-labeled amino acids confirmed that these four metabolites were end products of reductive Stickland metabolism, confirming activity of the reductive Stickland pathways in vivo. To explore whether the reductive pathways can also produce ATP, Liu et al. grew *C. sporogenes* on substrates specific for the reductive Strickland pathways. Subsequent formation of ATP confirmed that reductive Strickland pathways can produce ATP, a process in which the *Rhodobacter* nitrogenase-like (Rnf) complex was previously implied. Metabolomics analysis of Rnf mutants revealed an increase in oxidative products and a block in reductive metabolism, confirming the role of Rnf in ATP generation though the reductive Stickland reaction [86].

Taken together, these examples show that the determination of energy charge, comparative metabolomics, and isotope tracing are valuable techniques to determine the interplay between respiration, fermentation, and metabolism.

2.4 *Metabolism and Cell Replication*

Beyond its role in nutrients and energy homeostasis, metabolism supports and often even regulates many secondary processes that take place in microorganisms. In this section, we will highlight how metabolomics can be applied to discover the intimate links between metabolism, cell structure, and replication (Fig. 3).

Cell shape and size are closely linked with cellular metabolism [155]. In a recent study, Irnov et al. applied metabolomics to assess the effect of deleting *hfq* on the metabolite profile of *Caulobacter crescentus*, a model microorganism for cell division [64]. Hfq is an RNA chaperone that affects the expression of up to one in five genes and causes morphological defects in various bacteria through an uncharacterized mechanism. Metabolomic analysis revealed a 35-fold increase in α-ketoglutarate, pointing towards a role of this TCA cycle intermediate in cell morphology. The authors reasoned that α-ketoglutarate could be linked with cell morphology via succinyldiaminopimelate aminotransferase (DAP-AT), an enzyme that produces the peptidoglycan precursor *meso*-diaminopimelate via an α-ketoglutarate-forming reaction. The accumulation of UDP-N-acetylmuramoyl-L-alanyl-D-glutamate, the intermediate in the peptidoglycan biosynthesis pathway immediately before

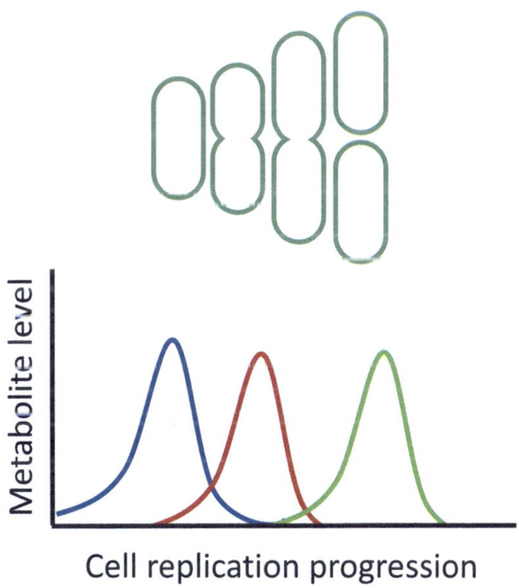

Fig. 3 Metabolism and cell replication. Cell division is a highly organized series of events which is reflected by specific cell cycle-dependent metabolic changes. Metabolomics has been used to discover the intimate links between metabolism and cell structure

meso-diaminopimelate, corroborated this hypothesis. Furthermore, the addition of diaminopimelate, a precursor of *meso*-diaminopimelate, rescued the morphological defect. Thus, by applying metabolomics to a microbe with a morphological defect, this study revealed that α-ketoglutarate is a regulator of peptidoglycan biosynthesis [64].

Cell division is a highly organized series of events in which metabolism takes an equally important role [71]. To discover key cell cycle-related metabolites, Hartl et al. performed untargeted metabolomics on synchronized *C. crescentus* cells progressing through the cell cycle [59]. By mixing experimental extracts from cells grown on ^{12}C-glucose with a standard extract from cells grown on ^{13}C-glucose followed by extensive bioinformatic filtering for $^{12}C/^{13}C$ pairs and removing MS artifacts like adducts and multimers, the authors were able to reduce over 5000 LC-MS features to approximately 400 high-confidence metabolites. Of these features about 50% could be annotated, 17 of which showed cell cycle-related changes. Among these metabolites were well-known cell cycle-dependent metabolites like cyclic-di-GMP and precursors for de novo purine biosynthesis but also metabolites involved in sulfur metabolism like methionine, *S*-adenosylmethionine, glutathione, *O*-acetylserine, and *O*-acetyl-homoserine. Flux analysis with ^{13}C-glucose confirmed cell cycle-dependent changes in the biosynthesis of glutathione, a major antioxidant in many bacteria. To establish a driving role of glutathione in cell cycle progression, the authors generated a *C. crescentus* strain lacking the glutathione synthetase gene (*gshB*). This *gshB* mutant displayed uncoordinated cell division, confirming the previously unknown role of glutathione in bacterial cell cycle progression. The work by Hartl et al. showcases that following metabolite profiles during the cell cycle can elucidate the role of metabolism therein. Moreover, the work includes an elegant workflow to extract high-confidence metabolites from untargeted LC-MS data [59].

2.5 *Metabolic Flexibility and Adaptation*

Microorganisms often live in rapidly changing and sometimes even extreme environments. To survive these conditions, microorganisms display impressive metabolic flexibility towards the environmental changes and adapted to almost all habitats on Earth (Fig. 4). Metabolomics has provided invaluable insights into the flexibility and adaptation of microbial metabolism [151].

2.5.1 Metabolic Adaptation to Nutrient Availability

For environmental microorganisms, key nutrients like carbon and nitrogen can change seasonally but also from hour to hour. To elucidate how cyanobacteria respond to nitrogen fluctuations, Zhang et al. exposed the cyanobacterium *Synechocystis* sp. PCC 6803 to an upshift in its nitrogen source nitrate and monitored the metabolic response using LC-MS metabolomics [187]. In addition to

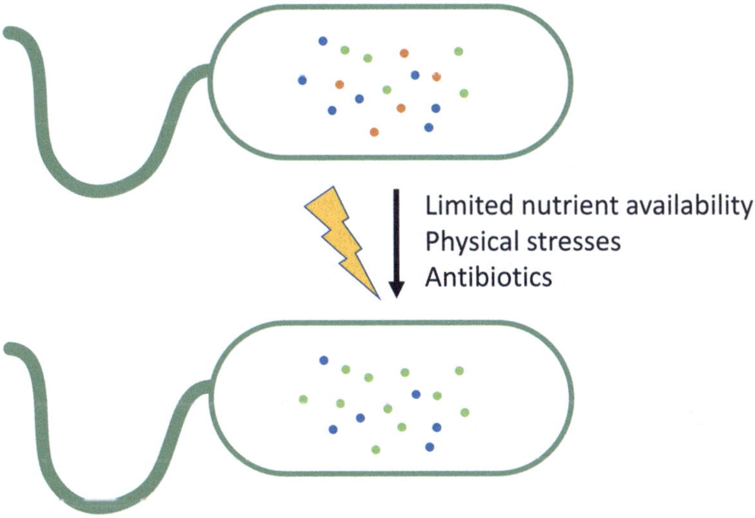

Fig. 4 Metabolic flexibility and adaptation. Microorganisms have evolved metabolic adaptation strategies to cope with environmental changes. Metabolomics has been instrumental in understanding these changes

expected changes in α-ketoglutarate and glutamine, unexpected increases were observed in arginine-biosynthetic-pathway intermediates like N-acetylornithine, ornithine, citrulline, and argininosuccinate. Next, ^{15}N- and ^{13}C-isotope tracing and modeling revealed that arginine was converted into ornithine by an unknown enzyme. Knocking out the uncharacterized gene *sll1336* resulted in arginine accumulation and ornithine depletion, suggesting it functions as an arginine dihydrolase. Finally, growth experiments revealed that *sll1336*-deficient bacteria grew slower than wild-type bacteria under nitrogen-oscillating culture conditions. By measuring the metabolic response to nitrogen availability, Zhang et al. thus discovered a new mechanism to cope with rapidly fluctuating nitrogen concentrations [187].

Pathogenic microorganisms are similarly exposed to changes in nutrient availability in their host. Metabolic adaptations to these changing conditions have been found to play a key role in virulence [41, 42, 60, 135].

2.5.2 Metabolic Adaptation to Stress

Metabolism also plays a major role in the protection of other stresses, like oxidative stress, hypoxia, and osmotic stress. Using ^{13}C-isotope tracing with glucose, Eoh et al. showed that metabolites accumulating in *M. tuberculosis* after hypoxia were mainly unlabeled [44]. This indicated that the carbons in these metabolites originated from a preexisting carbon source, not from ^{13}C-glucose. Subsequent experiments revealed that *M. tuberculosis* liberates carbons from its cell surface trehalose mycolates and preferentially uses them to synthesize peptidoglycan precursors. The authors reason that these peptidoglycan precursors are pre-made to quickly initiate the cell cycle when oxygen reappears and named the effect "metabolic anticipation" [44].

Sévin et al. applied untargeted metabolomics to define the salt stress response of 12 bacteria and 2 yeasts strains and found species-specific accumulation of osmolytes like trehalose, amino acids, and N-acetylated amino acids but also changes in pathways unrelated to osmoprotection [145]. For exometabolomics of salt-tolerant microorganisms, a special LC-MS method has been developed [127]. Metabolomics has also been used to characterize the microbial metabolic adaptations to extreme temperatures [45, 70, 77, 95, 160], acidity [108], sun light [26] salt concentrations [79], and oxidative stress [172].

2.5.3 Metabolism and Antibiotics

It is increasingly clear that metabolism and metabolic enzymes play an essential role in microbial resistance against antibiotics [88]. Metabolomics can detect endogenous metabolites, while at the same time detecting antibiotics and their metabolites, making it an attractive technique to study the microbial defense against antibiotics.

Using a screen for the isolation of *M. tuberculosis* mutants with increased antibiotics survival, Schrader et al. discovered that perturbations in arginine biosynthesis result in resistance to diverse antibiotics [142]. Metabolomic analyses revealed that these perturbations resulted in the accumulation of several arginine biosynthesis intermediates. Next, transcriptomic analyses showed that perturbed arginine biosynthesis led to increased transcription of the transcription factor *whiB7*, which was subsequently shown to be responsible for the high-survival phenotype. How perturbations in arginine biosynthesis lead to increased *whiB7* expression remains to be investigated.

While this example demonstrates the power of metabolomics to understand the effect of metabolism on the effect of antibiotics, metabolomics can also be used to profile the effect of an antibiotic on metabolism [68]. Metabolomics can, for example, identify the mechanism of action of an uncharacterized antimicrobial compound by revealing specific metabolic changes that point to metabolic enzyme inhibition [169]. On a more global scale, metabolomics analyses have revealed that antibiotic efficacy is linked to cellular respiration [87] and that distinct antibiotics all lead to oxidative stress [110]. Finally, metabolic profiling of antibiotics with known and unknown mechanisms of action enables the classification of metabolic responses. This classification allows predicting the mechanism of action of uncharacterized antibiotics [56, 183, 184]. Similarly, classifying the metabolic profiles of genetically silenced mutants with drug-treated bacteria enables functional annotation of uncharacterized antimicrobial compounds [2].

2.6 *Discovery and Biosynthesis of Secondary Metabolites*

Microorganisms are proliferous producers of secondary metabolites. Secondary metabolites are inessential for microbial core metabolism but fulfill important roles in ecological interactions such as inhibiting the growth of competitors (antibiotics) acquiring metals (siderophores) and modulating the host immune system (diverse metabolites). Discovering these bioactive molecules and understanding their biosynthesis and function increase our fundamental understanding of the physiology of microorganisms and pave the way for their biotechnological and medical application. Many secondary metabolites are encoded by biosynthetic gene clusters (BGCs) consisting of neighboring biosynthetic genes [101]. Although bioinformatic tools like antiSMASH can identify BGCs, the exact chemical structure of the encoded secondary metabolites can be ill-defined. Moreover, many secondary metabolites are not encoded in BGCs and therefore difficult to predict. Metabolomics fills these shortcomings of genomic predictions and plays a pivotal role in secondary metabolite discovery [139].

From a medical and societal perspective, antibiotics are extremely important secondary metabolites. Globally, antibiotic resistance is increasing resulting in reduced treatment options for infectious diseases [46]. Although most of the currently used antibiotics are of microbial origin, it is estimated that only a fraction of microbial secondary metabolites have been discovered. Classical phenotypic screens often result in the rediscovery of known antibiotics, so approaches to discover completely new antibiotics classes are highly desired [65]. As reviewed elsewhere, metabolomics represents such a tool to identify new antibiotics, along with their biosynthetic pathways [173, 176].

2.6.1 Monitoring the Products of Biosynthetic Gene Clusters

One hurdle in secondary metabolite discovery is that many BGCs are not expressed (cryptic) under standard lab culture conditions. Metabolomics can be used to monitor the appearance of novel secondary metabolites while changing culture conditions or genetically inducing cryptic BGC expression [152]. Wu et al., for example, induced cryptic BGC expression by exposing *Streptomyces* sp. MBT28 to streptomycin and used NMR metabolomics to monitor the biosynthesis of novel secondary metabolites [174]. Using multivariate data analysis, they were able to link a specific NMR signal to the antibiotic activity of several streptomycin-resistant strains. Chromatographic purification of this signal led to the identification of 7-prenylisatin, a previously unknown isatin-type antibiotic. In a high-throughput version of this approach, Covington et al. challenged *Burkholderia gladioli* with 750 potential elicitors – chemicals that can induce the expression of cryptic BGCs [28]. Using a bioinformatic application named Metabolomics Explorer, they discovered several cryptic metabolites that were induced by previously unknown elicitors. Though powerful, approaches like these still have a high chance of rediscovering (replicating) previously known antibiotics or analogs thereof.

2.6.2 Dereplicating Secondary Metabolites

Recent advances in molecular networking now enable dereplicating secondary metabolites by comparing their MS2 fragmentation spectra with a growing library of microbial metabolites, thereby selecting for new families of metabolites [105, 112, 130, 168]. Duncan et al., for example, generated a molecular network of the metabolites present in 35 strains of *Salinispora*, a genus of marine actinomycetes known to produce a wide variety of secondary metabolites [40]. The network consisting of 1137 MS2 spectra was compared with a spectral database containing authentic standards, which revealed the presence of 7 metabolite classes known to be produced by *Salinispora*. Associated with these known secondary products were several putative demethylated, methylated, and hydrated analogs. Though similar to known secondary products, such analogs could still represent interesting antibiotics. The network also contained completely novel metabolites which, by screening for co-occurrence of BCGs and metabolites in the strains, could be linked to uncharacterized BCGs. One unknown metabolite was structurally elucidated and named "retimycin A" after the Latin word "reticulum" (network). Molecular networking can, thus, recognize completely novel metabolites and is now widely used to mine secondary metabolites and link them to BGCs [91, 92, 161, 166, 177].

2.6.3 Linking Secondary Metabolites to Biosynthetic Gene Clusters

Similar to the examples above, linking the presence of secondary metabolites with the presence of BGC across hundreds of microorganisms has also been the basis of recent large-scale data acquisition and mining approaches [20, 35, 103]. McCaughey et al. developed a general approach to link natural products with biosynthetic gene clusters by comparing measured isotope labeling profiles of secondary metabolites with the expected labeling profiles of predicted products of uncharacterized biosynthetic gene clusters [98]. First, microorganisms are grown in the presence of a single-labeled precursor like 1-^{13}C-acetate, 1-^{13}C-propionate, [methyl-^{13}C]methionine, or ^{15}N-glutamate, along with parallel unlabeled cultures. Next a bioinformatics pipeline called IsoAnalyst compares the labeling of detected secondary products to the theoretical labeling of annotated BGCs. The approach successfully predicted previously known BGCs for erythromycins and erythrochelin in *Saccharopolyspora erythraea* but also identified known and unknown desferrioxamine derivatives and a new lobosamide macrolactam in an environmental *Micromonospora* sp. [98].

2.7 *Metabolic Engineering*

Microbes can perform an amazing variety of chemical reactions. Many of these reactions are difficult and expensive to perform using industrial organic chemistry, making microbes attractive "factories" to efficiently carry out complex chemistry. Metabolic engineering aims to optimize the production of desired metabolites by altering the metabolic properties of a cell, often by genetic engineering [23]. Classically, only a subset of key metabolites has been measured to steer a targeted engineering process and increase metabolite production. Metabolomics has been used in a similar semi-targeted way to map the metabolic effect of rational engineering steps in, for example, CO_2 fixation by cyanobacteria [72] and tryptophan production in *E. coli* [140].

2.7.1 Reverse Metabolic Engineering

By measuring a multitude of metabolites, untargeted metabolomics can be a powerful tool to identify unexpected metabolic bottlenecks and steer metabolic engineering [37]. Omics techniques like genomics, transcriptomics, and metabolomics now also allow "reverse metabolic engineering." Reverse engineering is a process in which high producers are first selected through non-targeted evolution or mutagenesis screens. Next, the mechanisms underlying the high production are investigated and used for further optimization [117].

Hong et al., for example, performed metabolomics on three yeast mutants that evolved to grow fast on galactose [63]. Although the intermediates of galactose metabolism were similar to rationally engineered strains, reserve carbohydrates were higher in the evolved strains. Genomic analysis revealed mutations in Ras/PKA pathway which is important for global carbon sensing. Genetically inhibiting this pathway in wild-type yeast increased the growth rate on galactose, confirming that mutations in the Ras/PKA pathway drive increased galactose use. In this example, Hong et al. thus successfully applied metabolomics to identify unexpected drivers of galactose use.

Valle et al. similarly used metabolomics to guide the metabolic engineering of succinic acid production in *E. coli* [164]. Succinate production is increased in *E. coli* strains with an activated glyoxylate shunt or inactivated pentose phosphate pathway, but since the underlying mechanism is unknown, steps for further improvement are unknown. Using metabolomics, Valle et al. found that such overproducing strains contained increased levels of fructose 1,6-biphosphate, trehalose, isovaleric acid, and mannitol, suggesting a loss of succinate production capacity through these metabolites. To further increase succinate production, mannitol biosynthesis was reduced by inactivating mannitol dehydrogenase. Mannitol dehydrogenase was never implicated in succinate production, but deleting it resulted in a 20% increase in succinate production. This example thus demonstrates how metabolomics can reveal unexpected ways to optimize metabolite production.

Finally, metabolomics also provides valuable information to build and test metabolic models that can be used for metabolic engineering [30, 76]. We will not discuss this approach in detail, but metabolic models are discussed in Sect 2.8.

2.8 *Improving Culture Media*

Since the birth of microbiology, the development of new culture media has been a major driver of discoveries [5]. However, despite intensive research, the majority of microorganisms remains unculturable [85]. Moreover, many culturable microorganisms can only be grown in complex, non-defined media which hampers accurate metabolic modeling of nutrient use and often misrepresents natural habitats. Metabolomic analyses of natural habitats now allow rational design of defined media that closely mimic environmental conditions in vitro.

To mimic saprolite soil, for example, Jenkins et al. applied LC-MS and GC-MS to profile water-soluble metabolites in a soil extract [69]. Combining these techniques, a total of 96 metabolites were identified of which 25 were quantified. Based on these results, the authors formulated two "soil defined media" which, at a 10x concentration, supported the growth of approximately half of 30 bacteria of a range of taxa. An improved formulation of these defined media was later found to support the growth of 108 out of 110 phylogenetically diverse soil bacteria and allowed quantifying metabolite use by various soil bacteria [32].

In similar work with *Trypanosoma brucei*, Creek et al. used metabolomics to assess metabolite consumption in the standard artificial culture medium HMI11 and found that over 30 metabolites were present at levels that were irrelevant for growth [29]. Removing these metabolites from the medium simplified the formulation and improved its resemblance to blood and thereby increased the sensitivity towards some drugs by over 100-fold.

2.9 *Systems Microbiology: Towards Complete Metabolic Models*

Up to now we have mainly discussed how metabolomics can help to answer specific question like "What nutrient does a microorganism use?" or "How does metabolism support cell growth?." Ultimately, however, we would like to describe the complete metabolism of a microbe in terms of metabolic fluxes through all of its metabolic pathways. In practice, our current understanding of metabolism is too incomplete to allow accurate kinetic models of metabolism under varying conditions. However, models describing specific metabolic pathways have generated valuable insights into metabolism. Metabolomics is a key technique to build, test, and refine such metabolic models, often in combination with genomics, transcriptomics, and proteomics [133].

2.9.1 Functional Gene Annotation

One major hurdle to construct accurate metabolic models is that even in extensively studied model organisms like *E. coli*, the function of approximately one in three genes is unknown [43]. Although several approaches successfully use metabolomics to annotate enzymes, most have a limited throughput and have been applied gene per gene [22, 43, 129]. Technical advances in genomics and metabolomics now also allow linking genes to metabolites on a global scale. Fuhrer et al., for example, profiled the metabolomes of 3800 *E. coli* mutants with single-gene deletions. Linking these mutations with over 7000 LC-MS features revealed unexpected metabolic influences of annotated genes and predicted the function of 72 unannotated genes [49]. A similar work in yeast found that 1 in 3 genes affects metabolism and clustered 3923 genes according to function [109]. In a related in vitro approach, Sévin et al. screened 1275 uncharacterized *E. coli* genes for metabolic activity by incubating purified protein with a metabolite extract and monitoring the effect on metabolite pools over time [144]. This high-throughput approach led to the putative annotation of 241 enzymes, of which 12 were validated.

2.9.2 Allosteric Regulation

Even when the function of all genes would be known, accurate prediction would still be challenging because the kinetics, genetic regulation, and allosteric regulation of many enzymes are ill-defined. To characterize allosteric binding of *E. coli* proteins, Diether et al. incubated 29 enzymes with a set of 55 metabolites and used NMR to assess protein binding. Of the 98 detected interactions, 76 were new, indicating that protein-metabolite binding is prevalent but understudied [34].

Using well-controlled bioreactors, enzyme kinetics, genetic regulation, and allosteric regulation can also be studied in vivo. To explore the effect of metabolites on gene transcription, Lemmp et al. switched an *E. coli* culture between starvation and growth monitored metabolite levels and activation of transcriptional regulators across ~30 timepoints [84]. By applying advanced statistical tools, they identified 513 putative metabolite-transcription factor interactions. Similar approaches have provided key insight into the allosteric control of nitrogen assimilation and amino acid biosynthesis in *E. coli* [36, 58, 138] but also showed that overflow metabolism regulates the levels of pyrimidines [134].

Systems-level analysis of metabolomics data collected under various culture conditions further demonstrated that substrate levels in yeast and *E. coli* are important determinants of fluxes in general, while enzymes and allosteric regulators are important for irreversible reactions specifically [55, 78].

The insights gained from these and similar studies [19, 121, 143] serve as input for more refined models that might ultimately be able to accurately describe and predict the metabolism of a single organism under various conditions.

3 Metabolomics on Microbial Communities

Although monocultures are extremely useful to study microorganisms in a lab, microbes in the environment usually live in mixed communities (Fig. 5). Natural microbial communities are incredibly diverse and can harbor thousands of species spanning the three domains of life. The interactions between these microorganisms and between them and their environments are equally diverse. The interactions in these communities can be cooperative or competitive and are often mediated through metabolites. Because these systems are genetically, physiologically, and metabolically extremely complex, studying them often requires a combination of scientific disciplines with microbiology at the center. To understand the dynamics of microbial communities and the role of metabolism therein, we must uncover what microorganisms are present and what they are doing. In other words, the question "who is doing what?" must be answered.

Historically, investigating microbial communities has been challenging due to a lack of systems-wide approaches. Functional properties of community members had to be investigated by isolating them, which is slow and labor intensive. Moreover, many microorganisms are currently unculturable, making isolation of a pure culture impossible. Technical advances and reduced costs of sequencing now enable routine sequencing of the DNA of complete communities. As a result, metagenomics has become a cornerstone of research on microbial communities. Currently, the DNA of millions of species has been sequenced in myriad environments. This data not only provides information on who is present in the communities but, using bioinformatic tools, also provides information on what they might be doing. Based on genomes, these bioinformatic tools predict the metabolic potential of individual community

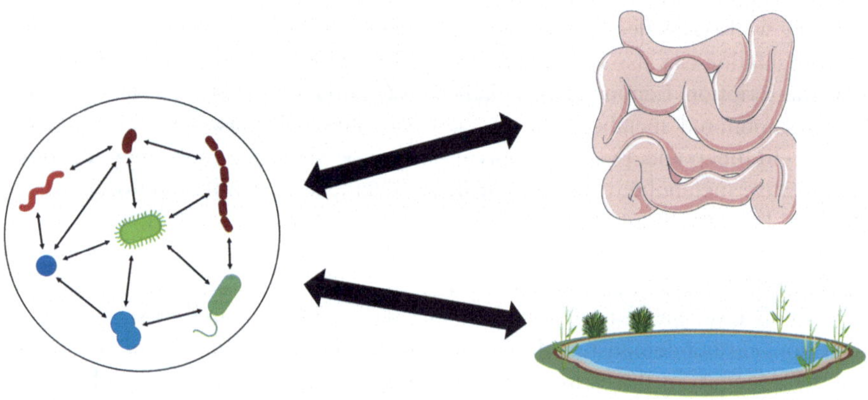

Fig. 5 Metabolic interactions between microorganisms and their environments. Microbes of, for example, the gut microbiome or soil microbiome can engage in a multitude of metabolic interactions with other microorganisms and their environment. These interactions can be positive, for example, in a mutualistic or syntrophic relation, or detrimental, for example, in the case of parasitism or competition. (The figure was partly generated using Servier Medical Art, provided by Servier, licensed under a Creative Commons Attribution 3.0 unported license)

members, which can then be used to infer community interactions [147]. However, as discussed in Sect. 2, these tools only distill metabolic potential and not actual metabolic activity. Though closer to metabolic activity, metatranscriptomics and metaproteomics similarly reveal metabolic potential and not actual activity. By measuring actual metabolites, metabolomics overcomes the limitations of other omics, especially when they are combined [10].

In this section we explore the different ways in which metabolomics-based approaches can be applied to investigate microbial communities, starting with the relatively simple study of community isolates and ending with tools to predict community metabolomes.

3.1 Studying Natural Isolates to Understand Complex Communities

Microbial communities can consist of thousands of species. These species all have their own intracellular metabolome but also continuously take up and release extracellular metabolites. These extracellular metabolites can be nutrients that serve cooperative interactions but also toxins or siderophores that confer competitive advantages over others. As a result, the cellular metabolomes and extracellular exometabolomes of microbial communities are incredibly complex. Therefore, many microbial interactions can currently not be studied in complete communities. Studying representative isolates and synthetic communities, however, can provide detailed mechanistic insights into the dynamics of a microbial community (Fig. 6).

Soil is home to some of the most diverse microbial communities [14]. Baran et al. investigated the metabolic interplay between *Microcoleus vaginatus*, a primary producer isolated from desert soil, and six isolated sympatric heterotrophs [8]. Using metabolomics, they showed that many metabolites present in the *M. vaginatus* exometabolome were used by the heterotrophs. Limited overlap in the consumed metabolites demonstrated a high level of niche partitioning among the heterotrophs. While *M. vaginatus* was the largest producer of metabolites taken up by the heterotrophs, it was also the largest consumer, demonstrating that community interactions can be more complex than the classical primary producer-heterotroph paradigm. In a follow-up study, Swenson et al. tested how well the isolate study by Baran et al. recapitulated the complete natural community metabolome and found that 69% of the metabolite dynamics in desert soil could be explained by the tested isolates [158].

In a similar study, Brisson et al. investigated metabolic interplay between four isolated microalgae [17]. Untargeted metabolomics demonstrated that these algae excreted a wide variety of metabolites, of which eight were tested for their effect on growth. Most of these excreted metabolites showed stimulated growth in one species while inhibiting the growth of others. These results thus revealed a mechanism by which microorganisms can compete in a community.

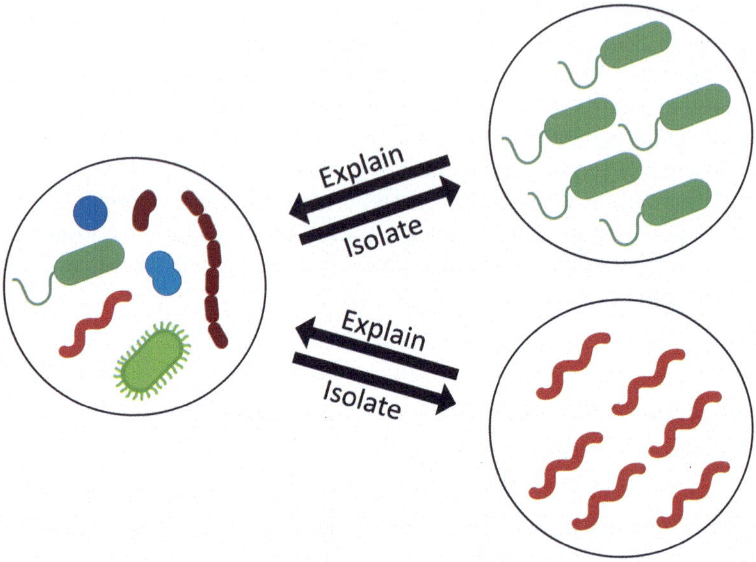

Fig. 6 Deconvoluting the metabolome of complex communities by studying isolated community members. The contribution of community members to the community metabolome can be studied by studying the metabolome of isolated community members

Our gut microbiome has been implicated in almost every disease, and metabolites are considered important mediators. To link metabolites to specific strains, Han et al. profiled the levels of 833 metabolites across 178 gut microorganism strains in vitro [57]. In vitro metabolite profiles were recapitulated in mice colonized with the same strain, which allowed linking specific species to metabolites. In a similar approach, Kešnerová et al. colonized bees with single members of the wild-type bee microbiome [75]. Using untargeted metabolomics, they found that the consumption and production of many metabolites were highly specialized across the tested species, indicating a high level of niche partitioning in the bee gut microbiome.

Zimmerman et al. applied metabolomics to 76 human gut microbiome members to screen for their involvement in metabolizing 271 orally administered drugs [188]. A total of 176 drugs were metabolized, and each species was found to metabolize 11–95 different drugs. When mice where colonized with drug-metabolizing bacteria, their pharmacokinetics changes accordingly, demonstrating an important role of the gut microbiome in drug metabolism.

Taken together, these examples show that metabolomics analyses on individual community members represent a powerful bottom-up approach to understand complex microbial communities.

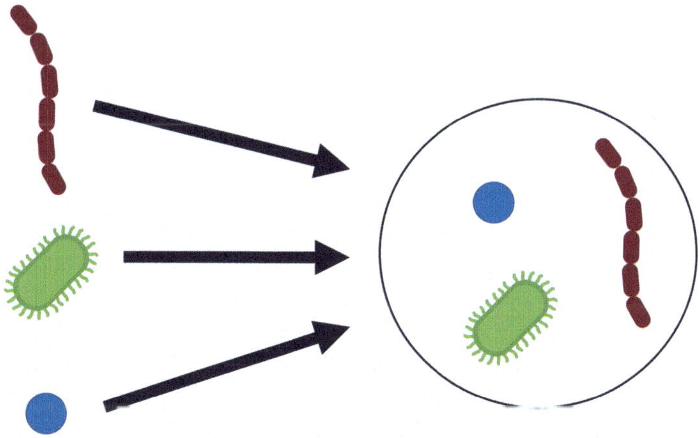

Fig. 7 Synthetic microbial community with a defined set of microorganisms. By establishing a synthetic community, metabolic interactions between community members can be investigated while controlling or applying conditions that are beyond control in natural communities

3.2 Synthetic Microbial Communities

Like natural isolates, synthetic communities grown under well-defined conditions allow studying mechanisms that are hard to study in complex undefined communities (Fig. 7). Yu et al., for example, used metabolomics to explore the effects of auxotrophs in a microbial community [181]. Auxotrophs are unable to synthesize essential metabolites and therefore rely on other community members. Despite this parasitic behavior, auxotrophs are commonly present in natural communities. To explore whether auxotrophs are beneficial to microbial communities, Yu et al. generated self-establishing metabolically cooperating yeast communities (SeMeCos) [18, 181]. SeMeCos are grown from a *his3Δ leu2Δ met15Δ ura3Δ S. cerevisiae* parental strain with corresponding rescuing vectors. As the *his3Δ leu2Δ met15Δ ura3Δ S. cerevisiae* proliferates, the rescuing vectors dilute throughout the population, resulting in a mix of different histidine, leucine, methionine, and uracil auxotrophs. With a targeted metabolomics approach, Yu et al. found that SeMeCos *S. cerevisiae* started sharing not only large amounts of histidine, leucine, methionine, and uracil but also many other metabolites. This increased sharing conferred resistance to antifungal azoles, potentially by increased efflux transporters. Using metabolomics, Yu et al. thus showed that auxotrophs alter the metabolic interactions in a microbial community of similar species.

Metabolic modeling of natural communities suggests that cross-feeding is also common between different species [185]. Ponomarova et al. used exometabolomics to explore interspecies cross-feeding between *S. cerevisiae* and two lactic acid bacteria and found it to be mediated by an overflow of yeast amino acids [126]. Like isolates, synthetic communities thus allow studying metabolic community interactions by a bottom-up approach. Both approaches are culture-dependent, however, limiting their application to the subset of culturable microbes. In the next sections, we will discuss culture-independent techniques that can be applied to natural communities.

3.3 Gnotobiotic Animals

Gnotobiotic animals have a defined microbiome and represent excellent models to study the causal effect of microbiomes on host metabolomes and health (Fig. 8) [27].

In an early metabolomic study with gnotobiotic mice, Wikoff et al. found that hundreds of plasma metabolites were significantly altered in the absence of a gut microbiome, demonstrating the importance of microbiomes to their hosts [171]. More recently, gnotobiotic mice have been "humanized" by colonization with human gut microbiomes, leading to metabolomic profiles that reflected those of the human donors [97].

Sharon et al. applied untargeted metabolomics to the colon content of mice humanized with the gut microbiomes of typically developing individuals and individuals with autism spectrum disorder (ASD) [148]. They found that two $GABA_A$ receptor agonists, 5-aminovaleric acid and taurine, were significantly less abundant in ASD mice. Dietary supplementation of these metabolites alleviated ASD-like behavior in these mice, suggesting a causal link. However, it remained unclear what microorganisms contributed to ASD-like phenotypes.

Metabolomics on gnotobiotic animals thus represents an approach to study complex microbial communities under relatively controlled conditions.

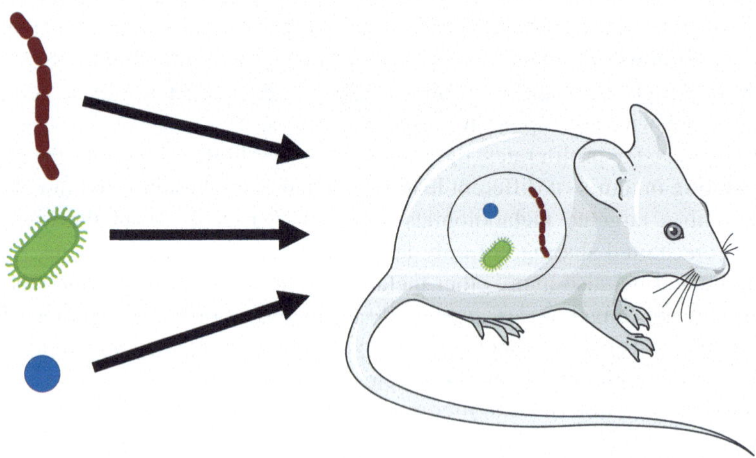

Fig. 8 Gnotobiotic animal with a defined microbiome. Gnotobiotic animals allow studying the metabolic interplay between specific microbiome members and their host. (The figure was partly generated using Servier Medical Art, provided by Servier, licensed under a Creative Commons Attribution 3.0 unported license)

3.4 Molecular Networking to Characterize Community Metabolomes

Microbial communities and their metabolomes are incredibly complex and contain thousands of features that cannot be identified [10]. These high numbers of unknowns have long hampered the application of untargeted metabolomics to microbial communities, but this has changed with the rise of metabolic networking. Molecular networks reveal chemical relatedness between known and unknown metabolites and thereby show the presence of chemical classes rather than specific metabolites [10, 168].

In a landmark study, Nguyen et al. applied molecular networking to over 60 microorganisms and identified several molecular families of peptide natural products produced by various microbes. The presence of these families could subsequently be linked to the presence of biosynthetic gene clusters, which led to the discovery of a biosynthetic gene cluster for a molecular family of bromoalterochromides [112].

Molecular networking is a breakthrough in the representation of complex metabolomes and is applied in a wide variety of ways, as highlighted throughout this manuscript.

3.5 Stable Isotope Tracing in Natural Communities

To gain insight into the metabolic dynamics in natural communities, researchers have successfully applied stable isotope tracing (Fig. 9).

Jang et al., for example, fed mice ^{13}C-labeled fructose and glucose and followed their fate by metabolomics on intestinal content, blood, and various tissues [66]. In contrast to the wide belief that fructose is mainly metabolized in the liver [104], Jang et al. demonstrated that fructose is mainly metabolized in the small intestine

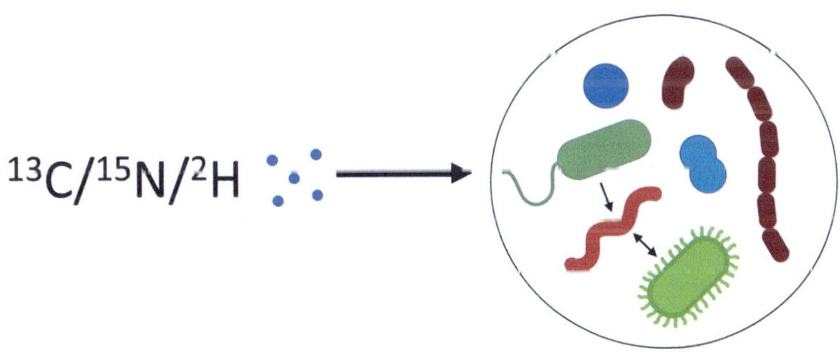

Fig. 9 Stable isotope tracing in microbial communities. Stable isotope tracing can be used to trace the fate of metabolites in complex microbial communities, revealing microbial interactions

and only reaches the liver at high intake levels. Glucose metabolites, however, did accumulate in the muscles and liver. Using isotope labeling, Jan et al. thus revealed that the small intestine plays an important role in fructose metabolism.

Zeng et al. similarly fed mice stable isotope-labeled nutrients but combined metabolomic analyses with metaproteomics to reveal which community members consume which metabolites [186]. Metabolic analyses first showed that inulin, a plant-derived dietary fiber, mostly remained undigested, while a small fraction was broken down into short-chain fatty acids and subsequently labeled microbiome glycolytic and TCA cycle intermediates. The ^{13}C-label of starch, however, was mostly recovered in portal blood glucose, lactate, and alanine. Similarly, the carbons of ^{13}C-labeled algal proteins accumulated in the microbiome, while the carbon of amino acids quickly appeared in the portal blood. These metabolomics data thus suggest that the microbiome preferably consumes nutrients from sources that are hard to digest. Subsequent proteomic analysis revealed that the ^{15}N-label of dietary proteins accumulated in *Acetatifactor muris*, while their ^{13}C-label accumulated in *Bacteroides uniformis*, a well-known commensal in the human microbiome.

3.6 Mass Spectrometry Imaging of Microbial Communities

Natural microbial communities are not only phylogenetically complex but also spatially. This spatial organization is essential for metabolic interactions but is often lost during sample preparation. Mass spectrometry imaging preserves spatial organization and can reveal its role in metabolic interactions. Garg et al., for instance, sectioned a human lung into slices and analyzed the metabolome and microbiome at multiple sites using conventional techniques [51]. Mapping the metabolites and species data onto a computer model of a human lung visualized the spatial organization of microbes and metabolites in the lung of a cystic fibrosis patient, albeit at a low spatial resolution. A higher spatial resolution can be achieved with matrix-assisted laser desorption/ionization-mass spectrometry imaging (MALDI-MSI). Carrell et al. applied MALDI-MSI to a coculture of *Sphagnum angustifolium*, a peat moss, and *Nostoc muscorum*, a cyanobacterium species, to visualize their metabolic interactions [21]. This approach showed that xanthosine, choline O-sulfate, and adenine accumulated at the interface between the *S. angustifolium* and *N. muscorum* cultures, which confirmed cross-feeding experiments.

To explore how *Streptomyces coelicolor* responds to the presence of other actinomycetes, Traxler et al. combined MALDI-MSI with nanospray desorption electrospray ionization mass spectrometry (NanoDESI MS) imaging [162]. In this study, NanoDESI MS was used to obtain fragmentation spectra which can be used to build a molecular network [170]. This network revealed a family of unknown metabolites which were subsequently identified as desferrioxamines, siderophores known to play a role in species interactions.

Combining MS imaging with other imaging techniques allows linking metabolites to species at a high spatial resolution. With an approach called metaFISH,

Geier et al. combined MS imaging with fluorescent in situ hybridization (FISH) to spatially link metabolites to species [53]. Applying metaFISH to a deep-sea mussel and its intracellular symbiotic bacteria led to the discovery of a family of specialized metabolites at the host-microbe interface that remain to be identified. A similar approach combined FISH with nanoscale secondary ion mass spectrometry (nanoSIMS), which allowed linking species information with ^{13}C and ^{15}N abundance at a high spatial resolution [12].

The examples above demonstrate the unique ability of MS imaging to explore the spatial distribution of community metabolomes.

3.7 *Integrating Omics Within Microbial Communities*

As already shown in previous examples, combining metabolomics with other omics techniques provides a much deeper insight into metabolic community interactions than metabolomics alone. In this section we highlight studies applying additional omics combinations to study metabolism within microbial communities.

3.7.1 Metabolomics and Metagenomics

The combination of metabolomics with metagenomics allows linking the presence of metabolites with the abundance of species. Franzosa et al., for example, combined untargeted metabolomics and shotgun metagenomics to link the changes in the gut microbiome of patient with inflammatory bowel disease (IBD) with changes in the gut metabolome [48]. Applying untargeted metabolomics to stool samples from patients and controls revealed IBD-specific changes in metabolite classes like sphingolipids, bile acids, and triacylglycerols. Shotgun metagenomics similarly revealed IBD-specific changes in species composition. Combining these data revealed 122 strong associations between species and metabolites, which could therefore be drivers of the IBD phenotype. For example, caprylic acid, a medium-chain fatty acid enriched in non-IBD patients, correlated positively with several health-associated microbiome species and negatively with *Ruminococcus gnavus*, an IBD-associated bacterium. To test for a causative relation, Franzosa et al. grew *R. gnavus* in the presence of caprylic acid and found reduced growth.

Dekkers et al. combined metabolomics and metagenomics on a large cohort of 8583 participants and found that the gut microbiota explain up to 58% of the variance in human plasma metabolites [33]. Large studies like these enable the identification of subtle, but potentially important, microbiome-metabolite associations.

In a similar environmental study, Shaffer et al. analyzed 880 microbial community samples from the Earth Microbiome Project by metagenomics and untargeted metabolomics, revealing intricate relations between sample sites, microbial community composition, and metabolites [146].

By combining metabolomics and metagenomics, these studies thus link metabolites to microbiota and phenotypes, identifying potential underlying mechanisms.

3.7.2 Metabolomics and Metatranscriptomics

By combining metabolomics with metatranscriptomics, changes in metabolite levels can be linked to species-specific gene transcription and vice versa. Malik et al., for example, integrated metabolomics and metatranscriptomics to study the response to drought of microbial communities living on plant litter [94]. Metatranscriptomics revealed increased transcription of bacterial genes associated with housekeeping, carbohydrate metabolism, extracellular polysaccharides metabolism, and inorganic ion transport, along with a vastly diminished number of eukaryotic transcripts. At the same time, metabolomics showed reduced amounts of free amino acids and sugars, along with an increase in compatible solutes like ectoine, 5-oxo-proline, and trehalose. Together, these results indicate a coordinated metabolic adaptation to drought.

3.7.3 Metabolomics and Metaproteomics

Similar to metatranscriptomics, metaproteomics can be combined with metabolomics to link metabolites with species-specific proteins.

McGivern et al. combined metabolomics with metagenomics and metaproteomics to study how polyphenols can be degraded in anoxic peatlands [100]. Peatlands are widely recognized as some of the most important global carbon sinks. Only the Northern peatlands, for example, are estimated to contain over a 1000 Gt of carbon, which is over 100 times the total global yearly anthropogenic carbon emissions [114]. Peatlands are anoxic and rich in polyphenols, plant metabolites that are thought to inhibit microbial activity at high concentrations. In soil, it is generally believed that microbial degradation of polyphenols requires oxygen and does not occur in anoxic environments [47]. To test whether polyphenols could be degraded in anoxic environments, McGivern et al. performed anoxic batch incubations with wetland subsurface soil and condensed tannin, a model polyphenolic compound [100]. High-resolution Fourier-transform ion-cyclotron-resonance MS (FTICR-MS) in combination with Kendrick mass defect analyses demonstrated biological breakdown of the added condensed tannin. Concomitantly, LC-MS and NMR analyses revealed the appearance of tannin-derived C_{15} flavonoids and (mono) phenolic compounds like caffeic acid, 3,4-dihydroxy phenylalanine, and 3,4-dihydroxy phenylethanol. To gain mechanistic insights into tannin degradation, the batch incubations were analyzed with metagenomics and genome-resolved metaproteomics, which led to the discovery of metabolic pathways that could be responsible for anoxic tannin breakdown. By integrating several omics, McGivern et al. thus showed that anoxic polyphenol degradation is possible, contrary to previous beliefs. Using the same techniques, Solden et al. studied the rumen microbiome of moose and discovered new enzymes involved in carbohydrate degradation [153].

Taken together, these examples demonstrate that integrating metabolomics with other omics can lead to mechanistic insights that were impossible to discover using metabolomics alone.

3.8 Integrating Data Across Microbial Communities and Studies

As we have seen in the previous section, integrating metabolomics with other omics can link metabolites to specific species and proteins within microbial communities. Recent approaches aim to integrate datasets on a much bigger scale and across different communities and studies [89]. Ultimately, these data-mining approaches could link known and unknown metabolites to species and proteins on a scale that is currently impossible.

Web of Microbes (WoM) is a repository for exometabolomes and metadata like community composition and environment [80]. By analyzing the co-occurrence of metabolites, species, and environmental information, WoM aims to, for example, identify metabolite-producing species and predict competition and metabolite exchange between microbes.

Schorn et al. similarly developed the Paired Omics Data Platform, a platform that stores metadata, like growth media, extraction solvent, and ionization mode, to link metabolomics and genomics data, thereby identifying natural product biosynthetic origins and metabolite structures [141]. Focusing on host-microbe interactions, the Virtual Metabolic Human (VMH) seeks to integrate human genetics, microbiome genetics, biochemical pathways, metabolomes with data on diseases, and nutrition [116].

Other platforms do not store data but provide data analysis pipelines. Microbiome and metabolome integrative analysis (M^{2}IA), for example, is an analysis pipeline specialized in integrating metabolomics data and 16S rRNA sequences or metagenomes of microbiomes [113]. M^{2}IA implements a variety of statistical analyses to correlate metabolites in the metabolome with species detected in the metagenome or 16S rRNA gene library. It extracts predicted reaction pathways from the Kyoto Encyclopedia of Genes and Genomes (KEGG), returning probability scores of microbes participating in a specific metabolic reaction or pathway. M^{2}IA can process metabolomics data from sources including feces, plasma, serum, and urine. To the best of our knowledge, M^{2}IA has not been tested on environmental communities.

NPLinker similarly integrates genomic end metabolomic data, but with a focus on finding biochemical relations [62]. NPLinker collapses annotated biosynthetic gene clusters (BGCs) of a given sample into gene cluster families (GCFs) based on similarity-based distances inferred with BiG-SCAPE [111] and links these to metabolites or metabolic networks, thereby relating GCFs to molecules or molecular families, respectively.

3.9 Predicting Community Metabolomes

Ultimately, advanced knowledge on species-metabolite links should allow accurate prediction of community metabolomes based on other omics data.

Currently, several tools integrate omics data to predict community metabolomes. Metabolomic analysis of metagenomes using flux balance analysis and optimization (MAMBO), for example, is metagenomic-based predictor of metabolic microbial community dynamics [52]. MAMBO uses the ModelSEED pipeline to generate genome-scale metabolic models (GSMMs) that encompass the metabolic reactions annotated within reference genomes [61]. A quantitative metabolic model is then constructed from metagenomics-derived abundance data and the reference GSMMs. Similar to NPLinker, Biosynthetic Gene cluster Meta'omics Abundance Profiler (BiG-MAP) uses the BiG-SCAPE tool to collapse BGCs into GCFs [122]. BiG-MAP then calculates GCF expression levels or abundances to model metabolite dynamics in microbial communities. Microbe-metabolite vectors (mmvec) are a neural network that specializes in predicting whole metabolomes from genomic data [107]. The mmvec neural network is iteratively trained with (meta)genomic and metabolomic data and learns co-occurrence probabilities. With several benchmarking tests and in four microbial communities, mmvec outperformed commonly used correlation estimation methods, showcasing the great promise machine learning holds in predicting community metabolomes.

4 Conclusions

In this chapter we provided a broad overview of microbial metabolomics. Metabolomics has helped understanding how single microbes consume and process nutrients and how they adapt their metabolism to cope with stress and antibiotics. Similarly, metabolomics has been instrumental in uncovering the complex interactions in microbial communities in our gut and the environment. Metabolomics has thus become a core technique in modern microbiology.

In the past decades, technical advances in mass spectrometry have pushed metabolomics to the forefront of many life sciences, including microbiology. We expect that in the next decade, advances in data analysis tools will push the boundaries of microbial metabolomics further. Since the field is extremely broad in terms of microbes, environments, and analyses, we focus on overarching trends and challenges here.

Some of the most promising new trends are, in our view, new data analysis tools. Molecular networking, for example, already represents a major development in microbial metabolomics, but given its power to characterize complex metabolomes, its future in the study of microbial communities appears particularly bright [10, 168].

Characterizing complex metabolomes using molecular networking does not allow linking metabolites to specific members of microbial communities though. However, as more microbial metabolomics and metagenomics datasets become available, data mining strategies will become more powerful. Ultimately, big data approaches like the Earth Microbiome Project will enable linking groups of microbes to groups of metabolites, whether annotated or not.

Annotating metabolites represents another major hurdle in microbial metabolomics for which we foresee a bright future for advanced data analysis. The huge annotation gap in metabolites and genes strongly limits the level of insight we gain from metabolomics studies. Recent advances in in silico structure prediction are promising, however, and will be of pivotal importance for microbial metabolomics [38, 39, 163]. Awaiting better in silico structure prediction tools and sharing of unannotated features through platforms like Global Natural Products Social Molecular Networking (GNPS) allow recognizing interesting unknowns across datasets.

Integrating omics data within and across studies thus represents a next frontier in microbial metabolomics but will allow leveraging technical and data analysis advances across omics fields. To apply these big data mining approaches at their full potential, it will be essential to share datasets in formats that allow interoperability [118].

With these exciting developments ahead, we expect a bright future for microbial metabolomics.

References

1. Agapova, A., Serafini, A., Petridis, M., Hunt, D.M., Garza-Garcia, A., Sohaskey, C.D., de Carvalho, L.P.S., 2019. Flexible nitrogen utilisation by the metabolic generalist pathogen Mycobacterium tuberculosis. eLife 8, e41129. https://doi.org/10.7554/eLife.41129
2. Anglada-Girotto, M., Handschin, G., Ortmayr, K., Campos, A.I., Gillet, L., Manfredi, P., Mulholland, C.V., Berney, M., Jenal, U., Picotti, P., Zampieri, M., 2022. Combining CRISPRi and metabolomics for functional annotation of compound libraries. Nat. Chem. Biol. 18, 482–491. https://doi.org/10.1038/s41589-022-00970-3
3. Antunes, L.C.M., Andersen, S.K., Menendez, A., Arena, E.T., Han, J., Ferreira, R.B.R., Borchers, C.H., Finlay, B.B., 2011. Metabolomics reveals phospholipids as important nutrient sources during Salmonella growth in bile in vitro and in vivo. J. Bacteriol. 193, 4719–4725. https://doi.org/10.1128/JB.05132-11
4. Arrigo, K.R., 2005. Marine microorganisms and global nutrient cycles. Nature 437, 349–355. https://doi.org/10.1038/nature04159
5. Atlas, R.M., Atlas, R.M., 2004. Handbook of Microbiological Media, 3rd ed. CRC Press, Boca Raton. https://doi.org/10.1201/9781420039726

6. Baidoo, E.E.K. (Ed.), 2019. Microbial Metabolomics: Methods and Protocols, Methods in Molecular Biology. Springer, New York, NY. https://doi.org/10.1007/978-1-4939-8757-3
7. Baran, R., Bowen, B.P., Price, M.N., Arkin, A.P., Deutschbauer, A.M., Northen, T.R., 2013. Metabolic footprinting of mutant libraries to map metabolite utilization to genotype. ACS Chem. Biol. 8, 189–199. https://doi.org/10.1021/cb300477w
8. Baran, R., Brodie, E.L., Mayberry-Lewis, J., Hummel, E., Da Rocha, U.N., Chakraborty, R., Bowen, B.P., Karaoz, U., Cadillo-Quiroz, H., Garcia-Pichel, F., Northen, T.R., 2015. Exometabolite niche partitioning among sympatric soil bacteria. Nat. Commun. 6, 8289. https://doi.org/10.1038/ncomms9289
9. Basan, M., Hui, S., Okano, H., Zhang, Z., Shen, Y., Williamson, J.R., Hwa, T., 2015. Overflow metabolism in Escherichia coli results from efficient proteome allocation. Nature 528, 99–104. https://doi.org/10.1038/nature15765
10. Bauermeister, A., Mannochio-Russo, H., Costa-Lotufo, L.V., Jarmusch, A.K., Dorrestein, P.C., 2022. Mass spectrometry-based metabolomics in microbiome investigations. Nat. Rev. Microbiol. 20, 143–160. https://doi.org/10.1038/s41579-021-00621-9
11. Beale, D.J., Kouremenos, K.A., Palombo, E.A. (Eds.), 2016. Microbial Metabolomics. Springer International Publishing, Cham. https://doi.org/10.1007/978-3-319-46326-1
12. Behrens, S., Lösekann, T., Pett-Ridge, J., Weber, P.K., Ng, W.-O., Stevenson, B.S., Hutcheon, I.D., Relman, D.A., Spormann, A.M., 2008. Linking microbial phylogeny to metabolic activity at the single-cell level by using enhanced element labeling-catalyzed reporter deposition fluorescence in situ hybridization (EL-FISH) and NanoSIMS. Appl. Environ. Microbiol. 74, 3143–3150. https://doi.org/10.1128/AEM.00191-08
13. Bennett, B.D., Yuan, J., Kimball, E.H., Rabinowitz, J.D., 2008. Absolute quantitation of intracellular metabolite concentrations by an isotope ratio-based approach. Nat. Protoc. 3, 1299–1311. https://doi.org/10.1038/nprot.2008.107
14. Bickel, S., Or, D., 2020. Soil bacterial diversity mediated by microscale aqueous-phase processes across biomes. Nat. Commun. 11, 116. https://doi.org/10.1038/s41467-019-13966-w
15. Bond, D.R., Lovley, D.R., 2003. Electricity production by Geobacter sulfurreducens attached to electrodes. Appl. Environ. Microbiol. 69, 1548–1555. https://doi.org/10.1128/AEM.69.3.1548-1555.2003
16. Brauer, M.J., Yuan, J., Bennett, B.D., Lu, W., Kimball, E., Botstein, D., Rabinowitz, J.D., 2006. Conservation of the metabolomic response to starvation across two divergent microbes. Proc. Natl. Acad. Sci. U. S. A. 103, 19302–19307. https://doi.org/10.1073/pnas.0609508103
17. Brisson, V., Mayali, X., Bowen, B., Golini, A., Thelen, M., Stuart, R.K., Northen, T.R., 2021. Identification of Effector Metabolites Using Exometabolite Profiling of Diverse Microalgae. mSystems 6, e0083521. https://doi.org/10.1128/mSystems.00835-21
18. Campbell, K., Vowinckel, J., Mülleder, M., Malmsheimer, S., Lawrence, N., Calvani, E., Miller-Fleming, L., Alam, M.T., Christen, S., Keller, M.A., Ralser, M., 2015. Self-establishing communities enable cooperative metabolite exchange in a eukaryote. eLife 4, e09943. https://doi.org/10.7554/eLife.09943
19. Canelas, A.B., Ras, C., ten Pierick, A., van Gulik, W.M., Heijnen, J.J., 2011. An in vivo data-driven framework for classification and quantification of enzyme kinetics and determination of apparent thermodynamic data. Metab. Eng. 13, 294–306. https://doi.org/10.1016/j.ymben.2011.02.005
20. Cao, L., Gurevich, A., Alexander, K.L., Naman, C.B., Leão, T., Glukhov, E., Luzzatto-Knaan, T., Vargas, F., Quinn, R., Bouslimani, A., Nothias, L.F., Singh, N.K., Sanders, J.G., Benitez, R.A.S., Thompson, L.R., Hamid, M.-N., Morton, J.T., Mikheenko, A., Shlemov, A., Korobeynikov, A., Friedberg, I., Knight, R., Venkateswaran, K., Gerwick, W.H., Gerwick, L., Dorrestein, P.C., Pevzner, P.A., Mohimani, H., 2019. MetaMiner: A Scalable Peptidogenomics Approach for Discovery of Ribosomal Peptide Natural Products with Blind Modifications from Microbial Communities. Cell Syst. 9, 600–608.e4. https://doi.org/10.1016/j.cels.2019.09.004

21. Carrell, A.A., Veličković, D., Lawrence, T.J., Bowen, B.P., Louie, K.B., Carper, D.L., Chu, R.K., Mitchell, H.D., Orr, G., Markillie, L.M., Jawdy, S.S., Grimwood, J., Shaw, A.J., Schmutz, J., Northen, T.R., Anderton, C.R., Pelletier, D.A., Weston, D.J., 2022. Novel metabolic interactions and environmental conditions mediate the boreal peatmoss-cyanobacteria mutualism. ISME J. 16, 1074–1085. https://doi.org/10.1038/s41396-021-01136-0
22. Caudy, A.A., Hanchard, J.A., Hsieh, A., Shaan, S., Rosebrock, A.P., 2019. Functional genetic discovery of enzymes using full-scan mass spectrometry metabolomics. Biochem. Cell Biol. 97, 73–84. https://doi.org/10.1139/bcb-2018-0058
23. Cheng, Q. (Ed.), 2012. Microbial Metabolic Engineering: Methods and Protocols, Methods in Molecular Biology. Springer, New York, NY. https://doi.org/10.1007/978-1-61779-483-4
24. Christen, S., Sauer, U., 2011. Intracellular characterization of aerobic glucose metabolism in seven yeast species by 13C flux analysis and metabolomics. FEMS Yeast Res. 11, 263–272. https://doi.org/10.1111/j.1567-1364.2010.00713.x
25. Clarke, A., Morris, G.J., Fonseca, F., Murray, B.J., Acton, E., Price, H.C., 2013. A Low Temperature Limit for Life on Earth. PLOS ONE 8, e66207. https://doi.org/10.1371/journal.pone.0066207
26. Coleine, C., Gevi, F., Fanelli, G., Onofri, S., Timperio, A.M., Selbmann, L., 2020. Specific adaptations are selected in opposite sun exposed Antarctic cryptoendolithic communities as revealed by untargeted metabolomics. PloS One 15, e0233805. https://doi.org/10.1371/journal.pone.0233805
27. Corander, J., Hanage, W.P., Pensar, J., 2022. Causal discovery for the microbiome. Lancet Microbe S2666-5247(22)00186-0. https://doi.org/10.1016/S2666-5247(22)00186-0
28. Covington, B.C., Seyedsayamdost, M.R., 2021. MetEx, a Metabolomics Explorer Application for Natural Product Discovery. ACS Chem. Biol. 16, 2825–2833. https://doi.org/10.1021/acschembio.1c00737
29. Creek, D.J., Nijagal, B., Kim, D.-H., Rojas, F., Matthews, K.R., Barrett, M.P., 2013. Metabolomics guides rational development of a simplified cell culture medium for drug screening against Trypanosoma brucei. Antimicrob. Agents Chemother. 57, 2768–2779. https://doi.org/10.1128/AAC.00044-13
30. Curran, K.A., Crook, N.C., Alper, H.S., 2012. Using Flux Balance Analysis to Guide Microbial Metabolic Engineering, in: Cheng, Q. (Ed.), Microbial Metabolic Engineering: Methods and Protocols, Methods in Molecular Biology. Springer, New York, NY, pp. 197–216. https://doi.org/10.1007/978-1-61779-483-4_13
31. de Carvalho, L.P.S., Fischer, S.M., Marrero, J., Nathan, C., Ehrt, S., Rhee, K.Y., 2010. Metabolomics of Mycobacterium tuberculosis reveals compartmentalized co-catabolism of carbon substrates. Chem. Biol. 17, 1122–1131. https://doi.org/10.1016/j.chembiol.2010.08.009
32. de Raad, M., Li, Y.V., Kuehl, J.V., Andeer, P.F., Kosina, S.M., Hendrickson, A., Saichek, N.R., Golini, A.N., Han, L.Z., Wang, Y., Bowen, B.P., Deutschbauer, A.M., Arkin, A.P., Chakraborty, R., Northen, T.R., 2022. A Defined Medium for Cultivation and Exometabolite Profiling of Soil Bacteria. Front. Microbiol. 13, 855331. https://doi.org/10.3389/fmicb.2022.855331
33. Dekkers, K.F., Sayols-Baixeras, S., Baldanzi, G., Nowak, C., Hammar, U., Nguyen, D., Varotsis, G., Brunkwall, L., Nielsen, N., Eklund, A.C., Bak Holm, J., Nielsen, H.B., Ottosson, F., Lin, Y. T., Ahmad, S., Lind, L., Sundström, J., Engström, G., Smith, J.G., Ärnlöv, J., Orho-Melander, M., Fall, T., 2022. An online atlas of human plasma metabolite signatures of gut microbiome composition. Nat. Commun. 13, 5370. https://doi.org/10.1038/s41467-022-33050-0
34. Diether, M., Nikolaev, Y., Allain, F.H., Sauer, U., 2019. Systematic mapping of protein-metabolite interactions in central metabolism of Escherichia coli. Mol. Syst. Biol. 15, e9008. https://doi.org/10.15252/msb.20199008
35. Doroghazi, J.R., Albright, J.C., Goering, A.W., Ju, K.-S., Haines, R.R., Tchalukov, K.A., Labeda, D.P., Kelleher, N.L., Metcalf, W.W., 2014. A roadmap for natural product discovery

based on large-scale genomics and metabolomics. Nat. Chem. Biol. 10, 963–968. https://doi.org/10.1038/nchembio.1659

36. Doucette, C.D., Schwab, D.J., Wingreen, N.S., Rabinowitz, J.D., 2011. α-Ketoglutarate coordinates carbon and nitrogen utilization via enzyme I inhibition. Nat. Chem. Biol. 7, 894–901. https://doi.org/10.1038/nchembio.685
37. Dromms, R.A., Styczynski, M.P., 2012. Systematic applications of metabolomics in metabolic engineering. Metabolites 2, 1090–1122. https://doi.org/10.3390/metabo2041090
38. Dührkop, K., Fleischauer, M., Ludwig, M., Aksenov, A.A., Melnik, A.V., Meusel, M., Dorrestein, P.C., Rousu, J., Böcker, S., 2019. SIRIUS 4: a rapid tool for turning tandem mass spectra into metabolite structure information. Nat. Methods 16, 299–302. https://doi.org/10.1038/s41592-019-0344-8
39. Dührkop, K., Nothias, L.-F., Fleischauer, M., Reher, R., Ludwig, M., Hoffmann, M.A., Petras, D., Gerwick, W.H., Rousu, J., Dorrestein, P.C., Böcker, S., 2021. Systematic classification of unknown metabolites using high-resolution fragmentation mass spectra. Nat. Biotechnol. 39, 462–471. https://doi.org/10.1038/s41587-020-0740-8
40. Duncan, K.R., Crüsemann, M., Lechner, A., Sarkar, A., Li, J., Ziemert, N., Wang, M., Bandeira, N., Moore, B.S., Dorrestein, P.C., Jensen, P.R., 2015. Molecular networking and pattern-based genome mining improves discovery of biosynthetic gene clusters and their products from Salinispora species. Chem. Biol. 22, 460–471. https://doi.org/10.1016/j.chembiol.2015.03.010
41. Ehrt, S., Schnappinger, D., Rhee, K.Y., 2018. Metabolic principles of persistence and pathogenicity in Mycobacterium tuberculosis. Nat. Rev. Microbiol. 16, 496–507. https://doi.org/10.1038/s41579-018-0013-4
42. Eisenreich, W., Dandekar, T., Heesemann, J., Goebel, W., 2010. Carbon metabolism of intracellular bacterial pathogens and possible links to virulence. Nat. Rev. Microbiol. 8, 401–412. https://doi.org/10.1038/nrmicro2351
43. Ellens, K.W., Christian, N., Singh, C., Satagopam, V.P., May, P., Linster, C.L., 2017. Confronting the catalytic dark matter encoded by sequenced genomes. Nucleic Acids Res. 45, 11495–11514. https://doi.org/10.1093/nar/gkx937
44. Eoh, H., Wang, Z., Layre, E., Rath, P., Morris, R., Branch Moody, D., Rhee, K.Y., 2017. Metabolic anticipation in Mycobacterium tuberculosis. Nat. Microbiol. 2, 17084. https://doi.org/10.1038/nmicrobiol.2017.84
45. Esteves, A.M., Graça, G., Peyriga, L., Torcato, I.M., Borges, N., Portais, J.-C., Santos, H., 2019. Combined transcriptomics-metabolomics profiling of the heat shock response in the hyperthermophilic archaeon Pyrococcus furiosus. Extrem. Life Extreme Cond. 23, 101–118. https://doi.org/10.1007/s00792-018-1065-0
46. Fang, F.C., Schooley, R.T., 2020. Antimicrobial Resistance — The Glass Is Half Full. N. Engl. J. Med. 382, 1363–1365. https://doi.org/10.1056/NEJMe2002375
47. Fenner, N., Freeman, C., 2011. Drought-induced carbon loss in peatlands. Nat. Geosci. 4, 895–900. https://doi.org/10.1038/ngeo1323
48. Franzosa, E.A., Sirota-Madi, A., Avila-Pacheco, J., Fornelos, N., Haiser, H.J., Reinker, S., Vatanen, T., Hall, A.B., Mallick, H., McIver, L.J., Sauk, J.S., Wilson, R.G., Stevens, B.W., Scott, J.M., Pierce, K., Deik, A.A., Bullock, K., Imhann, F., Porter, J.A., Zhernakova, A., Fu, J., Weersma, R.K., Wijmenga, C., Clish, C.B., Vlamakis, H., Huttenhower, C., Xavier, R.J., 2019. Gut microbiome structure and metabolic activity in inflammatory bowel disease. Nat. Microbiol. 4, 293–305. https://doi.org/10.1038/s41564-018-0306-4
49. Fuhrer, T., Zampieri, M., Sévin, D.C., Sauer, U., Zamboni, N., 2017. Genomewide landscape of gene-metabolome associations in Escherichia coli. Mol. Syst. Biol. 13, 907. https://doi.org/10.15252/msb.20167150
50. Gao, P., Xu, G., 2015. Mass-spectrometry-based microbial metabolomics: recent developments and applications. Anal. Bioanal. Chem. 407, 669–680. https://doi.org/10.1007/s00216-014-8127-7

51. Garg, N., Wang, M., Hyde, E., da Silva, R.R., Melnik, A.V., Protsyuk, I., Bouslimani, A., Lim, Y.W., Wong, R., Humphrey, G., Ackermann, G., Spivey, T., Brouha, S.S., Bandeira, N., Lin, G.Y., Rohwer, F., Conrad, D.J., Alexandrov, T., Knight, R., Dorrestein, P.C., 2017. Three-Dimensional Microbiome and Metabolome Cartography of a Diseased Human Lung. Cell Host Microbe 22, 705-716.e4. https://doi.org/10.1016/j.chom.2017.10.001
52. Garza, D.R., van Verk, M.C., Huynen, M.A., Dutilh, B.E., 2018. Towards predicting the environmental metabolome from metagenomics with a mechanistic model. Nat. Microbiol. 3, 456–460. https://doi.org/10.1038/s41564-018-0124-8
53. Geier, B., Sogin, E.M., Michellod, D., Janda, M., Kompauer, M., Spengler, B., Dubilier, N., Liebeke, M., 2020. Spatial metabolomics of in situ host-microbe interactions at the micrometre scale. Nat. Microbiol. 5, 498–510. https://doi.org/10.1038/s41564-019-0664-6
54. Griffin, J.E., Pandey, A.K., Gilmore, S.A., Mizrahi, V., McKinney, J.D., Bertozzi, C.R., Sassetti, C.M., 2012. Cholesterol catabolism by Mycobacterium tuberculosis requires transcriptional and metabolic adaptations. Chem. Biol. 19, 218–227. https://doi.org/10.1016/j.chembiol.2011.12.016
55. Hackett, S.R., Zanotelli, V.R.T., Xu, W., Goya, J., Park, J.O., Perlman, D.H., Gibney, P.A., Botstein, D., Storey, J.D., Rabinowitz, J.D., 2016. Systems-level analysis of mechanisms regulating yeast metabolic flux. Science 354, aaf2786. https://doi.org/10.1126/science.aaf2786
56. Halouska, S., Fenton, R.J., Barletta, R.G., Powers, R., 2012. Predicting the in vivo mechanism of action for drug leads using NMR metabolomics. ACS Chem. Biol. 7, 166–171. https://doi.org/10.1021/cb200348m
57. Han, S., Van Treuren, W., Fischer, C.R., Merrill, B.D., DeFelice, B.C., Sanchez, J.M., Higginbottom, S.K., Guthrie, L., Fall, L.A., Dodd, D., Fischbach, M.A., Sonnenburg, J.L., 2021. A metabolomics pipeline for the mechanistic interrogation of the gut microbiome. Nature 595, 415–420. https://doi.org/10.1038/s41586-021-03707-9
58. Hart, Y., Madar, D., Yuan, J., Bren, A., Mayo, A.E., Rabinowitz, J.D., Alon, U., 2011. Robust control of nitrogen assimilation by a bifunctional enzyme in E. coli. Mol. Cell 41, 117–127. https://doi.org/10.1016/j.molcel.2010.12.023
59. Hartl, J., Kiefer, P., Kaczmarczyk, A., Mittelviefhaus, M., Meyer, F., Vonderach, T., Hattendorf, B., Jenal, U., Vorholt, J.A., 2020. Untargeted metabolomics links glutathione to bacterial cell cycle progression. Nat. Metab. 2, 153–166. https://doi.org/10.1038/s42255-019-0166-0
60. Hartman, T.E., Wang, Z., Jansen, R.S., Gardete, S., Rhee, K.Y., 2017. Metabolic Perspectives on Persistence. Microbiol. Spectr. 5. https://doi.org/10.1128/microbiolspec.TBTB2-0026-2016
61. Henry, C.S., DeJongh, M., Best, A.A., Frybarger, P.M., Linsay, B., Stevens, R.L., 2010. High-throughput generation, optimization and analysis of genome-scale metabolic models. Nat. Biotechnol. 28, 977–982. https://doi.org/10.1038/nbt.1672
62. Hjörleifsson Eldjárn, G., Ramsay, A., van der Hooft, J.J.J., Duncan, K.R., Soldatou, S., Rousu, J., Daly, R., Wandy, J., Rogers, S., 2021. Ranking microbial metabolomic and genomic links in the NPLinker framework using complementary scoring functions. PLoS Comput. Biol. 17, e1008920. https://doi.org/10.1371/journal.pcbi.1008920
63. Hong, K.-K., Vongsangnak, W., Vemuri, G.N., Nielsen, J., 2011. Unravelling evolutionary strategies of yeast for improving galactose utilization through integrated systems level analysis. Proc. Natl. Acad. Sci. U. S. A. 108, 12179–12184. https://doi.org/10.1073/pnas.1103219108
64. Irnov, I., Wang, Z., Jannetty, N.D., Bustamante, J.A., Rhee, K.Y., Jacobs-Wagner, C., 2017. Crosstalk between the tricarboxylic acid cycle and peptidoglycan synthesis in Caulobacter crescentus through the homeostatic control of α-ketoglutarate. PLoS Genet. 13, e1006978. https://doi.org/10.1371/journal.pgen.1006978
65. Ito, T., Masubuchi, M., 2014. Dereplication of microbial extracts and related analytical technologies. J. Antibiot. (Tokyo) 67, 353–360. https://doi.org/10.1038/ja.2014.12

66. Jang, C., Hui, S., Lu, W., Cowan, A.J., Morscher, R.J., Lee, G., Liu, W., Tesz, G.J., Birnbaum, M.J., Rabinowitz, J.D., 2018. The Small Intestine Converts Dietary Fructose into Glucose and Organic Acids. Cell Metab. 27, 351–361.e3. https://doi.org/10.1016/j.cmet.2017.12.016
67. Jansen, R.S., Mandyoli, L., Hughes, R., Wakabayashi, S., Pinkham, J.T., Selbach, B., Guinn, K.M., Rubin, E.J., Sacchettini, J.C., Rhee, K.Y., 2020. Aspartate aminotransferase Rv3722c governs aspartate-dependent nitrogen metabolism in Mycobacterium tuberculosis. Nat. Commun. 11, 1960. https://doi.org/10.1038/s41467-020-15876-8
68. Jansen, R.S., Rhee, K.Y., 2017. Emerging Approaches to Tuberculosis Drug Development: At Home in the Metabolome. Trends Pharmacol. Sci. 38, 393–405. https://doi.org/10.1016/j.tips.2017.01.005
69. Jenkins, S., Swenson, T.L., Lau, R., Rocha, A.M., Aaring, A., Hazen, T.C., Chakraborty, R., Northen, T.R., 2017. Construction of Viable Soil Defined Media Using Quantitative Metabolomics Analysis of Soil Metabolites. Front. Microbiol. 8, 2618. https://doi.org/10.3389/fmicb.2017.02618
70. Jousse, C., Dalle, C., Canet, I., Lagrée, M., Traïkia, M., Lyan, B., Mendes, C., Sancelme, M., Amato, P., Delort, A.-M., 2017. Metabolomic study of the response to cold shock in a strain of Pseudomonas syringae isolated from cloud water. Metabolomics Off. J. Metabolomic Soc. 14, 11. https://doi.org/10.1007/s11306-017-1295-7
71. Kalucka, J., Missiaen, R., Georgiadou, M., Schoors, S., Lange, C., De Bock, K., Dewerchin, M., Carmeliet, P., 2015. Metabolic control of the cell cycle. Cell Cycle Georget. Tex 14, 3379–3388. https://doi.org/10.1080/15384101.2015.1090068
72. Kanno, M., Carroll, A.L., Atsumi, S., 2017. Global metabolic rewiring for improved CO2 fixation and chemical production in cyanobacteria. Nat. Commun. 8, 14724. https://doi.org/10.1038/ncomms14724
73. Kartal, B., Kuenen, J.G., van Loosdrecht, M.C.M., 2010. Sewage Treatment with Anammox. Science 328, 702–703. https://doi.org/10.1126/science.1185941
74. Kell, D.B., Brown, M., Davey, H.M., Dunn, W.B., Spasic, I., Oliver, S.G., 2005. Metabolic footprinting and systems biology: the medium is the message. Nat. Rev. Microbiol. 3, 557–565. https://doi.org/10.1038/nrmicro1177
75. Kešnerová, L., Mars, R.A.T., Ellegaard, K.M., Troilo, M., Sauer, U., Engel, P., 2017. Disentangling metabolic functions of bacteria in the honey bee gut. PLoS Biol. 15, e2003467. https://doi.org/10.1371/journal.pbio.2003467
76. Khanijou, J.K., Kulyk, H., Bergès, C., Khoo, L.W., Ng, P., Yeo, H.C., Helmy, M., Bellvert, F., Chew, W., Selvarajoo, K., 2022. Metabolomics and modelling approaches for systems metabolic engineering. Metab. Eng. Commun. 15, e00209. https://doi.org/10.1016/j.mec.2022.e00209
77. Kim, Y.-M., Nowack, S., Olsen, M.T., Becraft, E.D., Wood, J.M., Thiel, V., Klapper, I., Kühl, M., Fredrickson, J.K., Bryant, D.A., Ward, D.M., Metz, T.O., 2015. Diel metabolomics analysis of a hot spring chlorophototrophic microbial mat leads to new hypotheses of community member metabolisms. Front. Microbiol. 6, 209. https://doi.org/10.3389/fmicb.2015.00209
78. Kochanowski, K., Okano, H., Patsalo, V., Williamson, J., Sauer, U., Hwa, T., 2021. Global coordination of metabolic pathways in Escherichia coli by active and passive regulation. Mol. Syst. Biol. 17, e10064. https://doi.org/10.15252/msb.202010064
79. Kol, S., Merlo, M.E., Scheltema, R.A., de Vries, M., Vonk, R.J., Kikkert, N.A., Dijkhuizen, L., Breitling, R., Takano, E., 2010. Metabolomic characterization of the salt stress response in Streptomyces coelicolor. Appl. Environ. Microbiol. 76, 2574–2581. https://doi.org/10.1128/AEM.01992-09
80. Kosina, S.M., Greiner, A.M., Lau, R.K., Jenkins, S., Baran, R., Bowen, B.P., Northen, T.R., 2018. Web of microbes (WoM): a curated microbial exometabolomics database for linking chemistry and microbes. BMC Microbiol. 18, 115. https://doi.org/10.1186/s12866-018-1256-y
81. Kurczy, M.E., Forsberg, E.M., Thorgersen, M.P., Poole, F.L., Benton, H.P., Ivanisevic, J., Tran, M.L., Wall, J.D., Elias, D.A., Adams, M.W.W., Siuzdak, G., 2016. Global Isotope

Metabolomics Reveals Adaptive Strategies for Nitrogen Assimilation. ACS Chem. Biol. 11, 1677–1685. https://doi.org/10.1021/acschembio.6b00082
82. Lawson, C.E., Nuijten, G.H.L., de Graaf, R.M., Jacobson, T.B., Pabst, M., Stevenson, D.M., Jetten, M.S.M., Noguera, D.R., McMahon, K.D., Amador-Noguez, D., Lücker, S., 2021. Autotrophic and mixotrophic metabolism of an anammox bacterium revealed by in vivo 13C and 2H metabolic network mapping. ISME J. 15, 673–687. https://doi.org/10.1038/s41396-020-00805-w
83. Lei, T., Mao, Q., Chen, C., Ji, Y., 2020. Metabolomic Profiling of Staphylococcus aureus. Methods Mol. Biol. Clifton NJ 2069, 177–186. https://doi.org/10.1007/978-1-4939-9849-4_13
84. Lempp, M., Farke, N., Kuntz, M., Freibert, S.A., Lill, R., Link, H., 2019. Systematic identification of metabolites controlling gene expression in E. coli. Nat. Commun. 10, 4463. https://doi.org/10.1038/s41467-019-12474-1
85. Lewis, W.H., Tahon, G., Geesink, P., Sousa, D.Z., Ettema, T.J.G., 2021. Innovations to culturing the uncultured microbial majority. Nat. Rev. Microbiol. 19, 225–240. https://doi.org/10.1038/s41579-020-00458-8
86. Liu, Y., Chen, H., Van Treuren, W., Hou, B.-H., Higginbottom, S.K., Dodd, D., 2022. Clostridium sporogenes uses reductive Stickland metabolism in the gut to generate ATP and produce circulating metabolites. Nat. Microbiol. 7, 695–706. https://doi.org/10.1038/s41564-022-01109-9
87. Lobritz, M.A., Belenky, P., Porter, C.B.M., Gutierrez, A., Yang, J.H., Schwarz, E.G., Dwyer, D.J., Khalil, A.S., Collins, J.J., 2015. Antibiotic efficacy is linked to bacterial cellular respiration. Proc. Natl. Acad. Sci. U. S. A. 112, 8173–8180. https://doi.org/10.1073/pnas.1509743112
88. Lopatkin, A.J., Stokes, J.M., Zheng, E.J., Yang, J.H., Takahashi, M.K., You, L., Collins, J.J., 2019. Bacterial metabolic state more accurately predicts antibiotic lethality than growth rate. Nat. Microbiol. 4, 2109–2117. https://doi.org/10.1038/s41564-019-0536-0
89. Louwen, J.J.R., van der Hooft, J.J.J., 2021. Comprehensive Large-Scale Integrative Analysis of Omics Data To Accelerate Specialized Metabolite Discovery. mSystems 6, e0072621. https://doi.org/10.1128/mSystems.00726-21
90. Lu, W., Wang, L., Chen, L., Hui, S., Rabinowitz, J.D., 2018. Extraction and Quantitation of Nicotinamide Adenine Dinucleotide Redox Cofactors. Antioxid. Redox Signal. 28, 167–179. https://doi.org/10.1089/ars.2017.7014
91. Machushynets, N.V., Elsayed, S.S., Du, C., Siegler, M.A., de la Cruz, M., Genilloud, O., Hankemeier, T., van Wezel, G.P., 2022. Discovery of actinomycin L, a new member of the actinomycin family of antibiotics. Sci. Rep. 12, 2813. https://doi.org/10.1038/s41598-022-06736-0
92. Machushynets, N.V., Wu, C., Elsayed, S.S., Hankemeier, T., van Wezel, G.P., 2019. Discovery of novel glycerolated quinazolinones from Streptomyces sp. MBT27. J. Ind. Microbiol. Biotechnol. 46, 483–492. https://doi.org/10.1007/s10295-019-02140-2
93. Mahidhara, G., Ch, S., Ch, V.R., 2017. Comparative metabolomic studies of Alkanivorax xenomutans showing differential power output in a three chambered microbial fuel cell. World J. Microbiol. Biotechnol. 33, 102. https://doi.org/10.1007/s11274-017-2268-8
94. Malik, A.A., Swenson, T., Weihe, C., Morrison, E.W., Martiny, J.B.H., Brodie, E.L., Northen, T.R., Allison, S.D., 2020. Drought and plant litter chemistry alter microbial gene expression and metabolite production. ISME J. 14, 2236–2247. https://doi.org/10.1038/s41396-020-0683-6
95. Mandelli, F., Couger, M.B., Paixao, D. A. A., Machado, C.B., Carnielli, C.M., Aricetti, J.A., Polikarpov, I., Prade, R., Caldana, C., Paes Leme, A.F., Mercadante, A.Z., Riaño-Pachón, D.M., Squina, F.M., 2017. Thermal adaptation strategies of the extremophile bacterium Thermus filiformis based on multi-omics analysis. Extrem. Life Extreme Cond. 21, 775–788. https://doi.org/10.1007/s00792-017-0942-2

96. Mapelli, V., Olsson, L., Nielsen, J., 2008. Metabolic footprinting in microbiology: methods and applications in functional genomics and biotechnology. Trends Biotechnol. 26, 490–497. https://doi.org/10.1016/j.tibtech.2008.05.008
97. Marcobal, A., Kashyap, P.C., Nelson, T.A., Aronov, P.A., Donia, M.S., Spormann, A., Fischbach, M.A., Sonnenburg, J.L., 2013. A metabolomic view of how the human gut microbiota impacts the host metabolome using humanized and gnotobiotic mice. ISME J. 7, 1933–1943. https://doi.org/10.1038/ismej.2013.89
98. McCaughey, C.S., van Santen, J.A., van der Hooft, J.J.J., Medema, M.H., Linington, R.G., 2022. An isotopic labeling approach linking natural products with biosynthetic gene clusters. Nat. Chem. Biol. 18, 295–304. https://doi.org/10.1038/s41589-021-00949-6
99. McCloskey, D., Utrilla, J., Naviaux, R.K., Palsson, B.O., Feist, A.M., 2015. Fast Swinnex filtration (FSF): a fast and robust sampling and extraction method suitable for metabolomics analysis of cultures grown in complex media. Metabolomics 11, 198–209. https://doi.org/10.1007/s11306-014-0686-2
100. McGivern, B.B., Tfaily, M.M., Borton, M.A., Kosina, S.M., Daly, R.A., Nicora, C.D., Purvine, S.O., Wong, A.R., Lipton, M.S., Hoyt, D.W., Northen, T.R., Hagerman, A.E., Wrighton, K.C., 2021. Decrypting bacterial polyphenol metabolism in an anoxic wetland soil. Nat. Commun. 12, 2466. https://doi.org/10.1038/s41467-021-22765-1
101. Medema, M.H., Blin, K., Cimermancic, P., de Jager, V., Zakrzewski, P., Fischbach, M.A., Weber, T., Takano, E., Breitling, R., 2011. antiSMASH: rapid identification, annotation and analysis of secondary metabolite biosynthesis gene clusters in bacterial and fungal genome sequences. Nucleic Acids Res. 39, W339–346. https://doi.org/10.1093/nar/gkr466
102. Merrick, M.J., Edwards, R.A., 1995. Nitrogen control in bacteria. Microbiol. Rev. 59, 604–622. https://doi.org/10.1128/mr.59.4.604-622.1995
103. Merwin, N.J., Mousa, W.K., Dejong, C.A., Skinnider, M.A., Cannon, M.J., Li, H., Dial, K., Gunabalasingam, M., Johnston, C., Magarvey, N.A., 2020. DeepRiPP integrates multiomics data to automate discovery of novel ribosomally synthesized natural products. Proc. Natl. Acad. Sci. U. S. A. 117, 371–380. https://doi.org/10.1073/pnas.1901493116
104. Mirtschink, P., Krishnan, J., Grimm, F., Sarre, A., Hörl, M., Kayikci, M., Fankhauser, N., Christinat, Y., Cortijo, C., Feehan, O., Vukolic, A., Sossalla, S., Stehr, S.N., Ule, J., Zamboni, N., Pedrazzini, T., Krek, W., 2015. HIF-driven SF3B1 induces KHK-C to enforce fructolysis and heart disease. Nature 522, 444–449. https://doi.org/10.1038/nature14508
105. Mohimani, H., Gurevich, A., Shlemov, A., Mikheenko, A., Korobeynikov, A., Cao, L., Shcherbin, E., Nothias, L.-F., Dorrestein, P.C., Pevzner, P.A., 2018. Dereplication of microbial metabolites through database search of mass spectra. Nat. Commun. 9, 4035. https://doi.org/10.1038/s41467-018-06082-8
106. Molenaar, D., van Berlo, R., de Ridder, D., Teusink, B., 2009. Shifts in growth strategies reflect tradeoffs in cellular economics. Mol. Syst. Biol. 5, 323. https://doi.org/10.1038/msb.2009.82
107. Morton, J.T., Aksenov, A.A., Nothias, L.F., Foulds, J.R., Quinn, R.A., Badri, M.H., Swenson, T.L., Van Goethem, M.W., Northen, T.R., Vazquez-Baeza, Y., Wang, M., Bokulich, N.A., Watters, A., Song, S.J., Bonneau, R., Dorrestein, P.C., Knight, R., 2019. Learning representations of microbe-metabolite interactions. Nat. Methods 16, 1306–1314. https://doi.org/10.1038/s41592-019-0616-3
108. Mosier, A.C., Justice, N.B., Bowen, B.P., Baran, R., Thomas, B.C., Northen, T.R., Banfield, J.F., 2013. Metabolites associated with adaptation of microorganisms to an acidophilic, metal-rich environment identified by stable-isotope-enabled metabolomics. mBio 4, e00484–00412. https://doi.org/10.1128/mBio.00484-12
109. Mülleder, M., Calvani, E., Alam, M.T., Wang, R.K., Eckerstorfer, F., Zelezniak, A., Ralser, M., 2016. Functional Metabolomics Describes the Yeast Biosynthetic Regulome. Cell 167, 553–565.e12. https://doi.org/10.1016/j.cell.2016.09.007
110. Nandakumar, M., Nathan, C., Rhee, K.Y., 2014. Isocitrate lyase mediates broad antibiotic tolerance in Mycobacterium tuberculosis. Nat. Commun. 5, 4306. https://doi.org/10.1038/ncomms5306

111. Navarro-Muñoz, J.C., Selem-Mojica, N., Mullowney, M.W., Kautsar, S.A., Tryon, J.H., Parkinson, E.I., De Los Santos, E.L.C., Yeong, M., Cruz-Morales, P., Abubucker, S., Roeters, A., Lokhorst, W., Fernandez-Guerra, A., Cappelini, L.T.D., Goering, A.W., Thomson, R.J., Metcalf, W.W., Kelleher, N.L., Barona-Gomez, F., Medema, M.H., 2020. A computational framework to explore large-scale biosynthetic diversity. Nat. Chem. Biol. 16, 60–68. https://doi.org/10.1038/s41589-019-0400-9
112. Nguyen, D.D., Wu, C.-H., Moree, W.J., Lamsa, A., Medema, M.H., Zhao, X., Gavilan, R.G., Aparicio, M., Atencio, L., Jackson, C., Ballesteros, J., Sanchez, J., Watrous, J.D., Phelan, V.V., van de Wiel, C., Kersten, R.D., Mehnaz, S., De Mot, R., Shank, E.A., Charusanti, P., Nagarajan, H., Duggan, B.M., Moore, B.S., Bandeira, N., Palsson, B.Ø., Pogliano, K., Gutiérrez, M., Dorrestein, P.C., 2013. MS/MS networking guided analysis of molecule and gene cluster families. Proc. Natl. Acad. Sci. U. S. A. 110, E2611–2620. https://doi.org/10.1073/pnas.1303471110
113. Ni, Y., Yu, G., Chen, H., Deng, Y., Wells, P.M., Steves, C.J., Ju, F., Fu, J., 2020. M2IA: a web server for microbiome and metabolome integrative analysis. Bioinforma. Oxf. Engl. 36, 3493–3498. https://doi.org/10.1093/bioinformatics/btaa188
114. Nichols, J.E., Peteet, D.M., 2019. Rapid expansion of northern peatlands and doubled estimate of carbon storage. Nat. Geosci. 12, 917–921. https://doi.org/10.1038/s41561-019-0454-z
115. Niebel, B., Leupold, S., Heinemann, M., 2019. An upper limit on Gibbs energy dissipation governs cellular metabolism. Nat. Metab. 1, 125–132. https://doi.org/10.1038/s42255-018-0006-7
116. Noronha, A., Modamio, J., Jarosz, Y., Guerard, E., Sompairac, N., Preciat, G., Daníelsdóttir, A.D., Krecke, M., Merten, D., Haraldsdóttir, H.S., Heinken, A., Heirendt, L., Magnúsdóttir, S., Ravcheev, D.A., Sahoo, S., Gawron, P., Friscioni, L., Garcia, B., Prendergast, M., Puente, A., Rodrigues, M., Roy, A., Rouquaya, M., Wiltgen, L., Žagare, A., John, E., Krueger, M., Kuperstein, I., Zinovyev, A., Schneider, R., Fleming, R.M.T., Thiele, I., 2019. The Virtual Metabolic Human database: integrating human and gut microbiome metabolism with nutrition and disease. Nucleic Acids Res. 47, D614–D624. https://doi.org/10.1093/nar/gky992
117. Oud, B., van Maris, A.J.A., Daran, J.-M., Pronk, J.T., 2012. Genome-wide analytical approaches for reverse metabolic engineering of industrially relevant phenotypes in yeast. FEMS Yeast Res. 12, 183–196. https://doi.org/10.1111/j.1567-1364.2011.00776.x
118. Pacheco, A.R., Pauvert, C., Kishore, D., Segrè, D., 2022. Toward FAIR Representations of Microbial Interactions. mSystems 7, e0065922. https://doi.org/10.1128/msystems.00659-22
119. Paczia, N., Nilgen, A., Lehmann, T., Gätgens, J., Wiechert, W., Noack, S., 2012. Extensive exometabolome analysis reveals extended overflow metabolism in various microorganisms. Microb. Cell Factories 11, 122. https://doi.org/10.1186/1475-2859-11-122
120. Pandey, A.K., Sassetti, C.M., 2008. Mycobacterial persistence requires the utilization of host cholesterol. Proc. Natl. Acad. Sci. U. S. A. 105, 4376–4380. https://doi.org/10.1073/pnas.0711159105
121. Park, J.O., Rubin, S.A., Xu, Y.-F., Amador-Noguez, D., Fan, J., Shlomi, T., Rabinowitz, J.D., 2016. Metabolite concentrations, fluxes and free energies imply efficient enzyme usage. Nat. Chem. Biol. 12, 482–489. https://doi.org/10.1038/nchembio.2077
122. Pascal Andreu, V., Augustijn, H.E., van den Berg, K., van der Hooft, J.J.J., Fischbach, M.A., Medema, M.H., 2021. BiG-MAP: an Automated Pipeline To Profile Metabolic Gene Cluster Abundance and Expression in Microbiomes. mSystems 6, e0093721. https://doi.org/10.1128/mSystems.00937-21
123. Pezzatti, J., Bergé, M., Boccard, J., Codesido, S., Gagnebin, Y., Viollier, H. P., González-Ruiz, V., Rudaz, S. 2019. Choosing an Optimal Sample Preparation in Caulobacter crescentus for Untargeted Metabolomics Approaches. Metabolites 9, E193. https://doi.org/10.3390/metabo9100193
124. Pinu, F.R., Villas-Boas, S.G., 2017. Extracellular Microbial Metabolomics: The State of the Art. Metabolites 7, E43. https://doi.org/10.3390/metabo7030043
125. Planck, K.A., Rhee, K., 2021. Metabolomics of Mycobacterium tuberculosis. Methods Mol. Biol. Clifton NJ 2314, 579–593. https://doi.org/10.1007/978-1-0716-1460-0_25

126. Ponomarova, O., Gabrielli, N., Sévin, D.C., Mülleder, M., Zirngibl, K., Bulyha, K., Andrejev, S., Kafkia, E., Typas, A., Sauer, U., Ralser, M., Patil, K.R., 2017. Yeast Creates a Niche for Symbiotic Lactic Acid Bacteria through Nitrogen Overflow. Cell Syst. 5, 345–357.e6. https://doi.org/10.1016/j.cels.2017.09.002
127. Pontrelli, S., Sauer, U., 2021. Salt-Tolerant Metabolomics for Exometabolomic Measurements of Marine Bacterial Isolates. Anal. Chem. 93, 7164–7171. https://doi.org/10.1021/acs.analchem.0c04795
128. Prasannan, C.B., Jaiswal, D., Davis, R., Wangikar, P.P., 2018. An improved method for extraction of polar and charged metabolites from cyanobacteria. PloS One 13, e0204273. https://doi.org/10.1371/journal.pone.0204273
129. Prosser, G.A., Larrouy-Maumus, G., de Carvalho, L.P.S., 2014. Metabolomic strategies for the identification of new enzyme functions and metabolic pathways. EMBO Rep. 15, 657–669. https://doi.org/10.15252/embr.201338283
130. Purves, K., Macintyre, L., Brennan, D., Hreggviðsson, G.Ó., Kuttner, E., Ásgeirsdóttir, M.E., Young, L.C., Green, D.H., Edrada-Ebel, R., Duncan, K.R., 2016. Using Molecular Networking for Microbial Secondary Metabolite Bioprospecting. Metabolites 6, E2. https://doi.org/10.3390/metabo6010002
131. Rabinowitz, J.D., Kimball, E., 2007. Acidic acetonitrile for cellular metabolome extraction from Escherichia coli. Anal. Chem. 79, 6167–6173. https://doi.org/10.1021/ac070470c
132. Radoš, D., Donati, S., Lempp, M., Rapp, J., Link, H., 2022. Homeostasis of the biosynthetic E. coli metabolome. iScience 25, 104503. https://doi.org/10.1016/j.isci.2022.104503
133. Reaves, M.L., Rabinowitz, J.D., 2011. Metabolomics in systems microbiology. Curr. Opin. Biotechnol. 22, 17–25. https://doi.org/10.1016/j.copbio.2010.10.001
134. Reaves, M.L., Young, B.D., Hosios, A.M., Xu, Y.-F., Rabinowitz, J.D., 2013. Pyrimidine homeostasis is accomplished by directed overflow metabolism. Nature 500, 237–241. https://doi.org/10.1038/nature12445
135. Rhee, K.Y., de Carvalho, L.P.S., Bryk, R., Ehrt, S., Marrero, J., Park, S.W., Schnappinger, D., Venugopal, A., Nathan, C., 2011. Central carbon metabolism in Mycobacterium tuberculosis: an unexpected frontier. Trends Microbiol. 19, 307–314. https://doi.org/10.1016/j.tim.2011.03.008
136. Rosebrock, A.P., Caudy, A.A., 2017. Metabolite Extraction from Saccharomyces cerevisiae for Liquid Chromatography-Mass Spectrometry. Cold Spring Harb. Protoc. 2017, pdb.prot089086. https://doi.org/10.1101/pdb.prot089086
137. Sander, K., Asano, K.G., Bhandari, D., Van Berkel, G.J., Brown, S.D., Davison, B., Tschaplinski, T.J., 2017. Targeted redox and energy cofactor metabolomics in Clostridium thermocellum and Thermoanaerobacterium saccharolyticum. Biotechnol. Biofuels 10, 270. https://doi.org/10.1186/s13068-017-0960-4
138. Sander, T., Farke, N., Diehl, C., Kuntz, M., Glatter, T., Link, H., 2019. Allosteric Feedback Inhibition Enables Robust Amino Acid Biosynthesis in E. coli by Enforcing Enzyme Overabundance. Cell Syst. 8, 66–75.e8. https://doi.org/10.1016/j.cels.2018.12.005
139. Scherlach, K., Hertweck, C., 2021. Mining and unearthing hidden biosynthetic potential. Nat. Commun. 12, 3864. https://doi.org/10.1038/s41467-021-24133-5
140. Schoppel, K., Trachtmann, N., Korzin, E.J., Tzanavari, A., Sprenger, G.A., Weuster-Botz, D., 2022. Metabolic control analysis enables rational improvement of E. coli L-tryptophan producers but methylglyoxal formation limits glycerol-based production. Microb. Cell Factories 21, 201. https://doi.org/10.1186/s12934-022-01930-1
141. Schorn, M.A., Verhoeven, S., Ridder, L., Huber, F., Acharya, D.D., Aksenov, A.A., Aleti, G., Moghaddam, J.A., Aron, A.T., Aziz, S., Bauermeister, A., Bauman, K.D., Baunach, M., Beemelmanns, C., Beman, J.M., Berlanga-Clavero, M.V., Blacutt, A.A., Bode, H.B., Boullie, A., Brejnrod, A., Bugni, T.S., Calteau, A., Cao, L., Carrión, V.J., Castelo-Branco, R., Chanana, S., Chase, A.B., Chevrette, M.G., Costa-Lotufo, L.V., Crawford, J.M., Currie, C.R., Cuypers, B., Dang, T., de Rond, T., Demko, A.M., Dittmann, E., Du, C., Drozd, C., Dujardin, J.-C., Dutton, R.J., Edlund, A., Fewer, D.P., Garg, N., Gauglitz, J.M., Gentry, E.C.,

Gerwick, L., Glukhov, E., Gross, H., Gugger, M., Guillén Matus, D.G., Helfrich, E.J.N., Hempel, B.-F., Hur, J.-S., Iorio, M., Jensen, P.R., Kang, K.B., Kaysser, L., Kelleher, N.L., Kim, C.S., Kim, K.H., Koester, I., König, G.M., Leao, T., Lee, S.R., Lee, Y.-Y., Li, X., Little, J.C., Maloney, K.N., Männle, D., Martin H. C., McAvoy, A.C., Metcalf, W.W., Mohimani, H., Molina-Santiago, C., Moore, B.S., Mullowney, M.W., Muskat, M., Nothias, L.-F., O'Neill, E.C., Parkinson, E.I., Petras, D., Piel, J., Pierce, E.C., Pires, K., Reher, R., Romero, D., Roper, M.C., Rust, M., Saad, H., Saenz, C., Sanchez, L.M., Sørensen, S.J., Sosio, M., Süssmuth, R.D., Sweeney, D., Tahlan, K., Thomson, R.J., Tobias, N.J., Trindade-Silva, A.E., van Wezel, G.P., Wang, M., Weldon, K.C., Zhang, F., Ziemert, N., Duncan, K.R., Crüsemann, M., Rogers, S., Dorrestein, P.C., Medema, M.H., van der Hooft, J.J.J., 2021. A community resource for paired genomic and metabolomic data mining. Nat. Chem. Biol. 17, 363–368. https://doi.org/10.1038/s41589-020-00724-z

142. Schrader, S.M., Botella, H., Jansen, R., Ehrt, S., Rhee, K., Nathan, C., Vaubourgeix, J., 2021. Multiform antimicrobial resistance from a metabolic mutation. Sci. Adv. 7, eabh2037. https://doi.org/10.1126/sciadv.abh2037
143. Schuetz, R., Zamboni, N., Zampieri, M., Heinemann, M., Sauer, U., 2012. Multidimensional optimality of microbial metabolism. Science 336, 601–604. https://doi.org/10.1126/science.1216882
144. Sévin, D.C., Fuhrer, T., Zamboni, N., Sauer, U., 2017. Nontargeted in vitro metabolomics for high-throughput identification of novel enzymes in Escherichia coli. Nat. Methods 14, 187–194. https://doi.org/10.1038/nmeth.4103
145. Sévin, D.C., Stählin, J.N., Pollak, G.R., Kuehne, A., Sauer, U., 2016. Global Metabolic Responses to Salt Stress in Fifteen Species. PloS One 11, e0148888. https://doi.org/10.1371/journal.pone.0148888
146. Shaffer, J.P., Nothias, L.-F., Thompson, L.R., Sanders, J.G., Salido, R.A., Couvillion, S.P., Brejnrod, A.D., Lejzerowicz, F., Haiminen, N., Huang, S., Lutz, H.L., Zhu, Q., Martino, C., Morton, J.T., Karthikeyan, S., Nothias-Esposito, M., Dührkop, K., Böcker, S., Kim, H.W., Aksenov, A.A., Bittremieux, W., Minich, J.J., Marotz, C., Bryant, M.M., Sanders, K., Schwartz, T., Humphrey, G., Vásquez-Baeza, Y., Tripathi, A., Parida, L., Carrieri, A.P., Beck, K.L., Das, P., González, A., McDonald, D., Ladau, J., Karst, S.M., Albertsen, M., Ackermann, G., DeReus, J., Thomas, T., Petras, D., Shade, A., Stegen, J., Song, S.J., Metz, T.O., Swafford, A.D., Dorrestein, P.C., Jansson, J.K., Gilbert, J.A., Knight, R., Earth Microbiome Project 500 (EMP500) Consortium, 2022. Standardized multi-omics of Earth's microbiomes reveals microbial and metabolite diversity. Nat. Microbiol. 7, 2128–2150. https://doi.org/10.1038/s41564-022-01266-x
147. Shaffer, M., Borton, M.A., McGivern, B.B., Zayed, A.A., La Rosa, S.L., Solden, L.M., Liu, P., Narrowe, A.B., Rodríguez-Ramos, J., Bolduc, B., Gazitúa, M.C., Daly, R.A., Smith, G.J., Vik, D.R., Pope, P.B., Sullivan, M.B., Roux, S., Wrighton, K.C., 2020. DRAM for distilling microbial metabolism to automate the curation of microbiome function. Nucleic Acids Res. 48, 8883–8900. https://doi.org/10.1093/nar/gkaa621
148. Sharon, G., Cruz, N.J., Kang, D.-W., Gandal, M.J., Wang, B., Kim, Y.-M., Zink, E.M., Casey, C.P., Taylor, B.C., Lane, C.J., Bramer, L.M., Isern, N.G., Hoyt, D.W., Noecker, C., Sweredoski, M.J., Moradian, A., Borenstein, E., Jansson, J.K., Knight, R., Metz, T.O., Lois, C., Geschwind, D.H., Krajmalnik-Brown, R., Mazmanian, S.K., 2019. Human Gut Microbiota from Autism Spectrum Disorder Promote Behavioral Symptoms in Mice. Cell 177, 1600–1618.e17. https://doi.org/10.1016/j.cell.2019.05.004
149. Shi, L., Dong, H., Reguera, G., Beyenal, H., Lu, A., Liu, J., Yu, H.-Q., Fredrickson, J.K., 2016. Extracellular electron transfer mechanisms between microorganisms and minerals. Nat. Rev. Microbiol. 14, 651–662. https://doi.org/10.1038/nrmicro.2016.93
150. Shi, Z., 2019. Methylome and Metabolome Analyses Reveal Adaptive Mechanisms in Geobacter sulfurreducens Grown on Different Terminal Electron Acceptors. J. Proteome Res. 18, 1494–1502. https://doi.org/10.1021/acs.jproteome.8b00763

151. Shimizu, K., 2013. Regulation Systems of Bacteria such as Escherichia coli in Response to Nutrient Limitation and Environmental Stresses. Metabolites 4, 1–35. https://doi.org/10.3390/metabo4010001
152. Sinha, R., Sharma, B., Dangi, A.K., Shukla, P., 2019. Recent metabolomics and gene editing approaches for synthesis of microbial secondary metabolites for drug discovery and development. World J. Microbiol. Biotechnol. 35, 166. https://doi.org/10.1007/s11274-019-2746-2
153. Solden, L.M., Naas, A.E., Roux, S., Daly, R.A., Collins, W.B., Nicora, C.D., Purvine, S.O., Hoyt, D.W., Schückel, J., Jørgensen, B., Willats, W., Spalinger, D.E., Firkins, J.L., Lipton, M.S., Sullivan, M.B., Pope, P.B., Wrighton, K.C., 2018. Interspecies cross-feeding orchestrates carbon degradation in the rumen ecosystem. Nat. Microbiol. 3, 1274–1284. https://doi.org/10.1038/s41564-018-0225-4
154. Song, J., Sasaki, D., Sasaki, K., Kato, S., Kondo, A., Hashimoto, K., Nakanishi, S., 2016. Comprehensive metabolomic analyses of anode-respiring Geobacter sulfurreducens cells: The impact of anode-respiration activity on intracellular metabolite levels. Process Biochem. 51, 34–38. https://doi.org/10.1016/j.procbio.2015.11.012
155. Sperber, A.M., Herman, J.K., 2017. Metabolism Shapes the Cell. J. Bacteriol. 199, e00039–17. https://doi.org/10.1128/JB.00039-17
156. Strous, M., Fuerst, J.A., Kramer, E.H., Logemann, S., Muyzer, G., van de Pas-Schoonen, K.T., Webb, R., Kuenen, J.G., Jetten, M.S., 1999. Missing lithotroph identified as new planctomycete. Nature 400, 446–449. https://doi.org/10.1038/22749
157. Strous, M., Pelletier, E., Mangenot, S., Rattei, T., Lehner, A., Taylor, M.W., Horn, M., Daims, H., Bartol-Mavel, D., Wincker, P., Barbe, V., Fonknechten, N., Vallenet, D., Segurens, B., Schenowitz-Truong, C., Médigue, C., Collingro, A., Snel, B., Dutilh, B.E., Op den Camp, H.J.M., van der Drift, C., Cirpus, I., van de Pas-Schoonen, K.T., Harhangi, H.R., van Niftrik, L., Schmid, M., Keltjens, J., van de Vossenberg, J., Kartal, B., Meier, H., Frishman, D., Huynen, M.A., Mewes, H.-W., Weissenbach, J., Jetten, M.S.M., Wagner, M., Le Paslier, D., 2006. Deciphering the evolution and metabolism of an anammox bacterium from a community genome. Nature 440, 790–794. https://doi.org/10.1038/nature04647
158. Swenson, T.L., Karaoz, U., Swenson, J.M., Bowen, B.P., Northen, T.R., 2018. Linking soil biology and chemistry in biological soil crust using isolate exometabolomics. Nat. Commun. 9, 19. https://doi.org/10.1038/s41467-017-02356-9
159. Takai, K., Nakamura, K., Toki, T., Tsunogai, U., Miyazaki, M., Miyazaki, J., Hirayama, H., Nakagawa, S., Nunoura, T., Horikoshi, K., 2008. Cell proliferation at 122°C and isotopically heavy CH4 production by a hyperthermophilic methanogen under high-pressure cultivation. Proc. Natl. Acad. Sci. 105, 10949–10954. https://doi.org/10.1073/pnas.0712334105
160. Trauger, S.A., Kalisak, E., Kalisiak, J., Morita, H., Weinberg, M.V., Menon, A.L., Poole, F.L., Adams, M.W.W., Siuzdak, G., 2008. Correlating the transcriptome, proteome, and metabolome in the environmental adaptation of a hyperthermophile. J. Proteome Res. 7, 1027–1035. https://doi.org/10.1021/pr700609j
161. Trautman, E.P., Crawford, J.M., 2016. Linking Biosynthetic Gene Clusters to their Metabolites via Pathway- Targeted Molecular Networking. Curr. Top. Med. Chem. 16, 1705–1716. https://doi.org/10.2174/1568026616666151012111046
162. Traxler, M.F., Watrous, J.D., Alexandrov, T., Dorrestein, P.C., Kolter, R., 2013. Interspecies interactions stimulate diversification of the Streptomyces coelicolor secreted metabolome. mBio 4, e00459–13. https://doi.org/10.1128/mBio.00459-13
163. Tripathi, A., Vázquez-Baeza, Y., Gauglitz, J.M., Wang, M., Dührkop, K., Nothias-Esposito, M., Acharya, D.D., Ernst, M., van der Hooft, J.J.J., Zhu, Q., McDonald, D., Brejnrod, A.D., Gonzalez, A., Handelsman, J., Fleischauer, M., Ludwig, M., Böcker, S., Nothias, L.-F., Knight, R., Dorrestein, P.C., 2021. Chemically informed analyses of metabolomics mass spectrometry data with Qemistree. Nat. Chem. Biol. 17, 146–151. https://doi.org/10.1038/s41589-020-00677-3
164. Valle, A., Soto, Z., Muhamadali, H., Hollywood, K.A., Xu, Y., Lloyd, J.R., Goodacre, R., Cantero, D., Cabrera, G., Bolivar, J., 2022. Metabolomics for the design of new metabolic

engineering strategies for improving aerobic succinic acid production in Escherichia coli. Metabolomics Off. J. Metabolomic Soc. 18, 56. https://doi.org/10.1007/s11306-022-01912-9

165. Van der Geize, R., Yam, K., Heuser, T., Wilbrink, M.H., Hara, H., Anderton, M.C., Sim, E., Dijkhuizen, L., Davies, J.E., Mohn, W.W., Eltis, L.D., 2007. A gene cluster encoding cholesterol catabolism in a soil actinomycete provides insight into Mycobacterium tuberculosis survival in macrophages. Proc. Natl. Acad. Sci. U. S. A. 104, 1947–1952. https://doi.org/10.1073/pnas.0605728104
166. Vizcaino, M.I., Engel, P., Trautman, E., Crawford, J.M., 2014. Comparative metabolomics and structural characterizations illuminate colibactin pathway-dependent small molecules. J. Am. Chem. Soc. 136, 9244–9247. https://doi.org/10.1021/ja503450q
167. Wang, H., Correa, E., Dunn, W.B., Winder, C.L., Goodacre, R., Lloyd, J.R., 2013. Metabolomic analyses show that electron donor and acceptor ratios control anaerobic electron transfer pathways in Shewanella oneidensis. Metabolomics 9, 642–656.https://doi.org/10.1007/s11306-012-0488-3
168. Wang, M., Carver, J.J., Phelan, V.V., Sanchez, L.M., Garg, N., Peng, Y., Nguyen, D.D., Watrous, J., Kapono, C.A., Luzzatto-Knaan, T., Porto, C., Bouslimani, A., Melnik, A.V., Meehan, M.J., Liu, W.-T., Crüsemann, M., Boudreau, P.D., Esquenazi, E., Sandoval-Calderón, M., Kersten, R.D., Pace, L.A., Quinn, R.A., Duncan, K.R., Hsu, C.-C., Floros, D.J., Gavilan, R.G., Kleigrewe, K., Northen, T., Dutton, R.J., Parrot, D., Carlson, E.E., Aigle, B., Michelsen, C.F., Jelsbak, L., Sohlenkamp, C., Pevzner, P., Edlund, A., McLean, J., Piel, J., Murphy, B.T., Gerwick, L., Liaw, C.-C., Yang, Y.-L., Humpf, H.-U., Maansson, M., Keyzers, R.A., Sims, A.C., Johnson, A.R., Sidebottom, A.M., Sedio, B.E., Klitgaard, A., Larson, C.B., Cristopher A Boya, P., Torres-Mendoza, D., Gonzalez, D.J., Silva, D.B., Marques, L.M., Demarque, D.P., Pociute, E., O'Neill, E.C., Briand, E., Helfrich, E.J.N., Granatosky, E.A., Glukhov, E., Ryffel, F., Houson, H., Mohimani, H., Kharbush, J.J., Zeng, Y., Vorholt, J.A., Kurita, K.L., Charusanti, P., McPhail, K.L., Nielsen, K.F., Vuong, L., Elfeki, M., Traxler, M.F., Engene, N., Koyama, N., Vining, O.B., Baric, R., Silva, R.R., Mascuch, S.J., Tomasi, S., Jenkins, S., Macherla, V., Hoffman, T., Agarwal, V., Williams, P.G., Dai, J., Neupane, R., Gurr, J., Rodríguez, A.M.C., Lamsa, A., Zhang, C., Dorrestein, K., Duggan, B.M., Almaliti, J., Allard, P.-M., Phapale, P., Nothias, L.-F., Alexandrov, T., Litaudon, M., Wolfender, J.-L., Kyle, J.E., Metz, T.O., Peryea, T., Nguyen, D.-T., VanLeer, D., Shinn, P., Jadhav, A., Müller, R., Waters, K.M., Shi, W., Liu, X., Zhang, L., Knight, R., Jensen, P.R., Palsson, B.O., Pogliano, K., Linington, R.G., Gutiérrez, M., Lopes, N.P., Gerwick, W.H., Moore, B.S., Dorrestein, P.C., Bandeira, N., 2016. Sharing and community curation of mass spectrometry data with Global Natural Products Social Molecular Networking. Nat. Biotechnol. 34, 828–837. https://doi.org/10.1038/nbt.3597
169. Wang, Z., Soni, V., Marriner, G., Kaneko, T., Boshoff, H.I.M., Barry, C.E., Rhee, K.Y., 2019. Mode-of-action profiling reveals glutamine synthetase as a collateral metabolic vulnerability of M. tuberculosis to bedaquiline. Proc. Natl. Acad. Sci. U. S. A. 116, 19646–19651. https://doi.org/10.1073/pnas.1907946116
170. Watrous, J., Roach, P., Alexandrov, T., Heath, B.S., Yang, J.Y., Kersten, R.D., van der Voort, M., Pogliano, K., Gross, H., Raaijmakers, J.M., Moore, B.S., Laskin, J., Bandeira, N., Dorrestein, P.C., 2012. Mass spectral molecular networking of living microbial colonies. Proc. Natl. Acad. Sci. U. S. A. 109, E1743–1752. https://doi.org/10.1073/pnas.1203689109
171. Wikoff, W.R., Anfora, A.T., Liu, J., Schultz, P.G., Lesley, S.A., Peters, E.C., Siuzdak, G., 2009. Metabolomics analysis reveals large effects of gut microflora on mammalian blood metabolites. Proc. Natl. Acad. Sci. U. S. A. 106, 3698–3703. https://doi.org/10.1073/pnas.0812874106
172. Wirgot, N., Lagrée, M., Traïkia, M., Besaury, L., Amato, P., Canet, I., Sancelme, M., Jousse, C., Diémé, B., Lyan, B., Delort, A.-M., 2019. Metabolic modulations of Pseudomonas graminis in response to H2O2 in cloud water. Sci. Rep. 9, 12799. https://doi.org/10.1038/s41598-019-49319-2

173. Wu, C., Choi, Y.H., van Wezel, G.P., 2016. Metabolic profiling as a tool for prioritizing antimicrobial compounds. J. Ind. Microbiol. Biotechnol. 43, 299–312. https://doi.org/10.1007/s10295-015-1666-x
174. Wu, C., Du, C., Gubbens, J., Choi, Y.H., van Wezel, G.P., 2015a. Metabolomics-Driven Discovery of a Prenylated Isatin Antibiotic Produced by Streptomyces Species MBT28. J. Nat. Prod. 78, 2355–2363. https://doi.org/10.1021/acs.jnatprod.5b00276
175. Wu, Chao, Herold, R.A., Knoshaug, E.P., Wang, B., Xiong, W., Laurens, L.M.L., 2019a. Fluxomic Analysis Reveals Central Carbon Metabolism Adaptation for Diazotroph Azotobacter vinelandii Ammonium Excretion. Sci. Rep. 9, 13209. https://doi.org/10.1038/s41598-019-49717-6
176. Wu, C., Kim, H.K., van Wezel, G.P., Choi, Y.H., 2015b. Metabolomics in the natural products field--a gateway to novel antibiotics. Drug Discov. Today Technol. 13, 11–17. https://doi.org/10.1016/j.ddtec.2015.01.004
177. Wu, Changsheng, van der Heul, H.U., Melnik, A.V., Lübben, J., Dorrestein, P.C., Minnaard, A.J., Choi, Y.H., van Wezel, G.P., 2019b. Lugdunomycin, an Angucycline-Derived Molecule with Unprecedented Chemical Architecture. Angew. Chem. Int. Ed Engl. 58, 2809–2814. https://doi.org/10.1002/anie.201814581
178. Yang, T.H., Coppi, M.V., Lovley, D.R., Sun, J., 2010. Metabolic response of Geobacter sulfurreducens towards electron donor/acceptor variation. Microb. Cell Factories 9, 90. https://doi.org/10.1186/1475-2859-9-90
179. Yasid, N.A., Rolfe, M.D., Green, J., Williamson, M.P., 2016. Homeostasis of metabolites in Escherichia coli on transition from anaerobic to aerobic conditions and the transient secretion of pyruvate. R. Soc. Open Sci. 3, 160187. https://doi.org/10.1098/rsos.160187
180. Ye, D., Li, X., Shen, J., Xia, X., 2022. Microbial metabolomics: From novel technologies to diversified applications. TrAC Trends Anal. Chem. 148, 116540. https://doi.org/10.1016/j.trac.2022.116540
181. Yu, J.S.L., Correia-Melo, C., Zorrilla, F., Herrera-Dominguez, L., Wu, M.Y., Hartl, J., Campbell, K., Blasche, S., Kreidl, M., Egger, A.-S., Messner, C.B., Demichev, V., Freiwald, A., Mülleder, M., Howell, M., Berman, J., Patil, K.R., Alam, M.T., Ralser, M., 2022. Microbial communities form rich extracellular metabolomes that foster metabolic interactions and promote drug tolerance. Nat. Microbiol. 7, 542–555. https://doi.org/10.1038/s41564-022-01072-5
182. Yuan, J., Bennett, B.D., Rabinowitz, J.D., 2008. Kinetic flux profiling for quantitation of cellular metabolic fluxes. Nat. Protoc. 3, 1328–1340. https://doi.org/10.1038/nprot.2008.131
183. Zampieri, M., Szappanos, B., Buchieri, M.V., Trauner, A., Piazza, I., Picotti, P., Gagneux, S., Borrell, S., Gicquel, B., Lelievre, J., Papp, B., Sauer, U., 2018. High-throughput metabolomic analysis predicts mode of action of uncharacterized antimicrobial compounds. Sci. Transl. Med. 10, eaal3973. https://doi.org/10.1126/scitranslmed.aal3973
184. Zampieri, M., Zimmermann, M., Claassen, M., Sauer, U., 2017. Nontargeted Metabolomics Reveals the Multilevel Response to Antibiotic Perturbations. Cell Rep. 19, 1214–1228. https://doi.org/10.1016/j.celrep.2017.04.002
185. Zelezniak, A., Andrejev, S., Ponomarova, O., Mende, D.R., Bork, P., Patil, K.R., 2015. Metabolic dependencies drive species co-occurrence in diverse microbial communities. Proc. Natl. Acad. Sci. U. S. A. 112, 6449–6454. https://doi.org/10.1073/pnas.1421834112
186. Zeng, X., Xing, X., Gupta, M., Keber, F.C., Lopez, J.G., Lee, Y.-C.J., Roichman, A., Wang, L., Neinast, M.D., Donia, M.S., Wühr, M., Jang, C., Rabinowitz, J.D., 2022. Gut bacterial nutrient preferences quantified in vivo. Cell 185, 3441–3456.e19. https://doi.org/10.1016/j.cell.2022.07.020
187. Zhang, H., Liu, Y., Nie, X., Liu, L., Hua, Q., Zhao, G.-P., Yang, C., 2018. The cyanobacterial ornithine-ammonia cycle involves an arginine dihydrolase. Nat. Chem. Biol. 14, 575–581. https://doi.org/10.1038/s41589-018-0038-z
188. Zimmermann, M., Zimmermann-Kogadeeva, M., Wegmann, R., Goodman, A.L., 2019. Mapping human microbiome drug metabolism by gut bacteria and their genes. Nature 570, 462–467. https://doi.org/10.1038/s41586-019-1291-3

Metabolomics in Autoimmunity, Infections, and Physiological Diseases

Suyasha Roy

Abbreviations

AD	Alzheimer's disease
AIDS	Acquired immunodeficiency syndrome
ALS	Amyotrophic lateral sclerosis
ATP	Adenosine triphosphate
BCAAs	Branched-chain amino acids
CAD	Coronary artery disease
CD	Crohn's disease
CE-TOFS	Capillary electrophoresis time-of-flight mass spectrometry
CNS	Central nervous system
COVID-19	Coronavirus disease 2019
CSF	Cerebrospinal fluid
CVD	Cardiovascular disease
DAGs	Diacylglycerol
dNTP	Deoxynucleotide triphosphate
FBP	Fructose 1,6-bisphosphate
G3P	Glyceraldehyde-3-phosphate
GC	Gas chromatography
GC-TOFMS	Gas chromatography coupled to time of flight mass spectrometry
GSH	Glutathione
HD	Huntington's disease
HDL	High-density lipoprotein
HIV	Human immunodeficiency virus

S. Roy (✉)
Laboratory of Molecular Immunology, National Heart, Lung and Blood Institute,
National Institutes of Health, Bethesda, MD, USA
e-mail: suyasharoy@gmail.com

V. Soni, T. E. Hartman (eds.), *Metabolomics*,
https://doi.org/10.1007/978-3-031-39094-4_7

HPLC	High performance liquid chromatography
IBD	Inflammatory bowel disease
LC	Liquid chromatography
LDL	Low-density lipoprotein
MI	Myocardial infarction
MS	Mass spectrometry
MS	Multiple sclerosis
Mtb	*Mycobacterium tuberculosis*
NAC	N-acetylcysteine
NAD	Nicotinamide adenine dinucleotide
NADPH	Nicotinamide adenine dinucleotide phosphate
NMR	Nuclear magnetic resonance
NO	Nitric oxide
PC	Phosphatidylcholine
PD	Parkinson's disease
PE	Phosphatidylethanolamine
PPP	Pentose phosphate pathway
PUFA	Polyunsaturated fatty acids
RA	Rheumatoid arthritis
ROS	Reactive oxygen species
SARS-COV-2	Severe acute respiratory syndrome coronavirus 2
SLE	Systemic lupus erythematosus
SM	Sphingomyelin
TB	Tuberculosis
TCA	Tricarboxylic acid
TMAO	Trimethylamine-N-oxide
Tregs	Regulatory T cells
UC	Ulcerative colitis

1 Introduction

Metabolites are the derived products from cellular metabolism and executes essential cellular functions such as signal transduction, cell growth and survival, and energy production. Metabolites are also derived from microorganisms, dietary, exogeneous, and xenobiotic sources [1]. Since metabolites have a wide range of cellular functions and physiological role, it is imperative to identify metabolites and comprehend different metabolic pathways associated with a unique biological phenotype. Metabolomics is the profiling of metabolites in the biological samples such as cells, tissues, and biofluids including blood serum, plasma, urine, saliva, etc. which could be implemented as a biomarker for providing insights into the mechanisms underlying diseases [2].

There are various methodologies used for carrying out metabolomics to understand the normal physiology and pathophysiology in diseases. Mass spectrometry (MS) is the main analytical platform which provides a highly sensitive, versatile, and reproducible metabolomic analysis. Targeted and untargeted (global) mass spectrometry-based metabolomics are the major approaches for performing quantitative metabolite analysis in the biological samples. Untargeted (global) metabolomics measures a broad range of metabolites without a prior information of the metabolome; in contrast, targeted metabolomics relies on the prior knowledge of the metabolome and have a greater sensitivity and specificity [2].

Recent advancements in innovative informatics have made it possible for more accurate quantification and broader coverage of metabolites with the identification of novel or previously unknown functions of metabolites in health and diseases. In this chapter, applications of metabolomics in the context of autoimmunity, infections, and physiological diseases have been discussed to unveil new frontiers in the diagnosis, prognosis, treatment, and drug discovery for the amelioration of diseases.

2 Metabolomics in Autoimmune Diseases

2.1 Metabolomics in Rheumatoid Arthritis (RA)

Rheumatoid arthritis (RA) is a systemic autoimmune disorder characterized by inflammation of synovial joints which progresses to disability [3]. Since RA still remains uncured, metabolomics can have a paramount importance in the early diagnosis and better understanding of the pathogenesis of RA. Biological samples such as synovial fluid, plasma, or serum from RA patients have been exploited for LC or GC coupled MS-based metabolomics studies. Serum-based metabolomics showed an increase in pyruvate and decrease in glucose in RA patients [4]. Elevated levels of ribose, fumarate, and citrate were found in the serum of RA patients indicative of active pentose phosphate pathway (PPP) and TCA cycle [5]. Amino acids such as glutamate and tryptophan were higher, while threonine, isoleucine, methionine, valine, histidine, serine, and alanine were lower in RA patients as compared to healthy individuals [6].

Plasma lipidomic profiling in RA patients identified variations in different classes of lipids with remarkable differences in lysophosphatidylinositol, dihydroceramides, phosphatidylserine, alkyl phosphatidylethanolamine, and alkenyl phosphatidylethanolamine, in comparison to healthy controls [7]. Further, significant rise in the concentrations of monounsaturated and polyunsaturated fatty acids and decline in saturated fatty acids were evident in RA patients [5, 6]. Metabolomic studies employing synovial fluid from RA patients observed increased lactate concentration marking enhanced glycolytic activity [8, 9]. These studies also reported a down-regulation of TCA cycle with reduction in citrate and ATP production and a decrease in acylcarnitine and LDL-lipids in RA patients [8, 9].

Macrophages and T lymphocytes are the most relevant disease-specific cells in RA and possess immunometabolic abnormalities [10]. In RA, these cells display unusual metabolic demands due to abnormal proliferation and pro-inflammatory effector functions. Increased glucose uptake and production of reactive oxygen species (ROS) and ATP in macrophages and lower levels of ATP and high NADPH and glucose shunting into the PPP in T cells are striking features in RA (Fig. 1) [10, 11]. Different metabolomic studies have been conducted to predict the response of RA patients to various treatment regimens exploiting biological agents such as TNF inhibitors, rituximab, and abatacept [12, 13]. These studies profiled serum samples to distinguish RA patients who were responders from non-responders before and after the treatment [12, 13]. Lastly, NMR-based serum metabolomics could be exploited for identifying pathogenic pathways in the RA patients [14].

2.2 *Metabolomics in Systemic Lupus Erythematosus (SLE)*

Systemic lupus erythematosus (SLE) is a chronic autoimmune inflammatory disease manifested by the production of autoantibodies with the activation of different types of immune cells affecting multi-organ system [15]. The pathogenesis of SLE is not completely understood, highlighting the need to fully understand the factors contributing to the development and progression of the disease. Metabolites have been ascribed critical roles in providing valuable insights on the pathogenesis of SLE owing to the modulation of immune responses mounted by different immune cells. In addition to genomics, transcriptomics, and proteomics, metabolomics has shed light on the pathophysiology underlying SLE which has substantially improved our understanding of SLE pathogenesis [16].

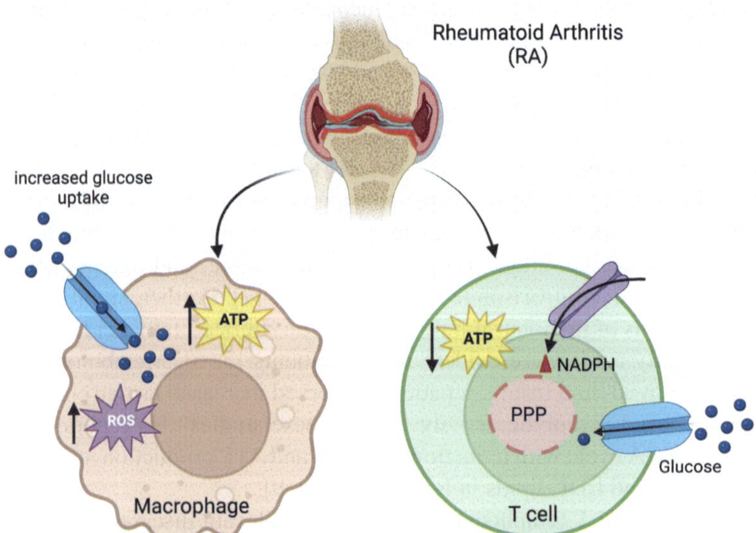

Fig. 1 Immunometabolic perturbations in rheumatoid arthritis (RA): Metabolomics showed increased glucose uptake with higher ATP and ROS production in macrophages while lower levels of ATP in T cells of RA patients

Metabolomics in SLE patients have reported perturbations in glycolysis, TCA cycle, fatty acid oxidation, and amino acid metabolism [17]. Evidences suggests lowering levels of lactate and pyruvate in SLE patients [17]. In the context of TCA cycle metabolites, there was decline in malate, alpha-ketoglutarate, fumarate, and citrate in patients suffering from SLE. With regard to metabolites of the amino acid metabolism, there was a diminution of amino acids in SLE patients including tyrosine, tryptophan, histidine, alanine, phenylalanine, valine, leucine, isoleucine, glutamine, lysine, glycine, asparagine, proline, serine, and aspartate [18–20]. Lipidomic profiling demonstrated fluctuations in fatty acid metabolism with surge in LDL and VLDL, whereas levels of cholesterol and HDL were diminished in SLE patients [20]. Lipidomic studies revealed lower levels of glycerol, polyunsaturated fatty acids (PUFA), and oleic and arachidonic acids and increased levels of monoacylglycerols [17, 18]. Li et al. showed elevation in sphingomyelin, ceramide, phosphatidylethanolamine, and diacylglycerol with concomitant decrease in acylcarnitine in SLE patients [17, 21]. Yan et al. also reported the differences in metabolites as biomarkers for distinguishing SLE patients with inactive and active disease, among them are linoleic acid, glutamate, citrate, propylparaben, 2-hydroxyisobutyrate, and glycerol [18].

Bengtsson et al. illustrated a reduction in amino acid, arginine which induces nitric oxide (NO) production leading to oxidative stress, and aggravated nitro-oxidative stress could modulate disease severity further regulating the SLE pathogenesis [22, 23]. Thus, oxidative stress promotes SLE by modulating signal transduction and cytokine secretion, and the hallmark of oxidative stress includes accumulation of kynurenine, cystine, and threonate and lower levels of cysteine and glutathione (GSH) in the serum of SLE patients [18]. N-acetylcysteine (NAC) is a precursor of GSH which improves disease severity in SLE by overcoming GSH depletion and blocking mTOR activation [24]. mTOR is known to inhibit the development of regulatory T cells (Tregs) which are deficient in SLE patients, and so mTOR blockade would restrain this inhibitory effect on Tregs resulting in disease remission [25]. Xanthine is another metabolite related to oxidative stress which generates ROS and was found in higher levels in SLE patients [21]. Lastly, increase in lipid metabolites in the serum and lower citrate levels in the urine after the treatment with the drug cyclophosphamide has been considered as a non-invasive biomarker for evaluating the response to treatment in SLE patients [26, 27].

2.3 *Metabolomics in Inflammatory Bowel Disease (IBD)*

Inflammatory bowel disease (IBD) is an autoimmune condition including Crohn's disease (CD) and ulcerative colitis (UC) and involves inflammation of the colon and small intestine in the digestive tract, respectively [28]. IBD is marked by disrupted immune regulation, perturbed intestinal permeability, and altered gut microbiota [28]. At present, there is no cure for IBD, and the etiology is not completely known. The pathogenesis of IBD is complex due to the interplay of the immune system, gut microbiota, environmental factors, and the genome. Thus, understanding the pathology of IBD is critical; in lieu of this, metabolomics has explicitly contributed to identify metabolic biomarkers associated with disease pathogenesis providing insights into the mechanisms underlying IBD.

IBD metabolome is quite diverse with serum and plasma metabolic profile attributing to changes in the host metabolism; on the other hand, fecal and urine metabolic profile corresponds to the gut microbial activity or diet [29]. In IBD, alterations in metabolic signatures of energy metabolism were quite evident with significant decrease in TCA cycle metabolites, including malate, succinate, aconitate, fumarate, citrate, and alpha-ketoglutarate in biological samples from IBD patients as compared to healthy individuals [30]. Concerning the lipid profile in IBD, ketone bodies were found in higher levels in serum on account of increased fatty acid oxidation with concomitant accumulation of glucose due to the body's inability to utilize glucose resulting in the elevation of ketone bodies in the serum of IBD patients [31]. Lower levels of amino acids in the blood and urine while increased levels of amino acids in the feces of IBD patients were remarkable due to malabsorption of amino acids which results from compromised gut epithelial barrier by inflammation [32].

Gut dysbiosis is a common phenomenon in IBD pathogenesis and is characterized by diminution in *Faecalibacterium prausnitzii* and *Clostridium* cluster IV which are important for maintaining intestinal homeostasis [33]. Hippurate is a microbial product of dietary metabolism and was found in diminishing levels in urine of patients with UC and CD [34]. Williams et al. also showed that *p*-cresol sulphate, a metabolic product derived from tyrosine and produced by *Clostridium* spp., was lower in the urine of CD patients [34]. Further, studies have shown increasing concentrations of tyrosine in the feces and 3-methylindole in serum and declining levels of indole-3-acrylic acid and indole-3-propionic acid in the serum samples of CD patients due to impaired microbiota as compared to healthy subjects [35, 36]. Lastly, metabolomics has been valuable in distinguishing CD and UC patients upon the diagnosis of IBD by identifying distinct metabolic profile including alterations in TCA cycle metabolites, fatty acids, and amino acids [37].

3 Metabolomics in Infectious Diseases

3.1 Metabolomics in COVID-19

Coronavirus disease 2019 (COVID-19) is a respiratory pandemic caused by severe acute respiratory syndrome coronavirus 2 (SARS-CoV-2). The virus infects lungs and respiratory tract of humans resulting in cough, fever, myalgia and in some instances culminates in acute respiratory distress syndrome (ARDS) [38]. While most patients have mild-to-moderate symptoms, in some patients the infection leads to severe respiratory distress, multi-organ dysfunction, and failure eventually resulting in death. The sudden outbreak of this pandemic in December 2019 led to unprecedented million deaths worldwide, intensifying the need to understand the pathobiology of SARS-CoV-2 in humans. In the fight against the COVID-19, metabolomics has been employed to dissect the metabolic fluctuations and predict the disease progression caused by SARS-CoV-2 infection by recognizing the diagnostic and prognostic biomarker metabolites [39].

A recently published study done in a large cohort of COVID-19 patients and healthy individuals elucidated metabolic perturbations associated with the peripheral immune responses with SARS-CoV-2 infection which could be helpful in predicting disease severity and survival probability [40]. GC-MS, LC-MS/MS, and NMR techniques have been performed for metabolite detection using biofluids such as serum, plasma, or urine of COVID-19 patients. Plasma metabolomics of COVID-19 patients revealed differences in the proportion of lipids such as decreased diacylglycerols (DAGs) and higher levels of monosialodihexosylgangliosides (GM3s) and sphingomyelins (SMs) suggesting the utilization of host-derived lipid membranes by SARS-CoV-2 [41]. Further, decreasing concentrations of phosphatidylcholines (PCs) and higher levels of free fatty acids, diglycerides, and triglycerides were observed in the plasma lipidomic profile of patients with severe COVID-19 symptoms [42]. Reduction in carbohydrate metabolites such as glycerol-3-phosphate and malic acid was a metabolic signature in symptomatic COVID-19 patients [42]. Blasco et al. showed changes in the metabolism of pyrimidine, nicotinamide, and tryptophan with two key metabolites, indole-3-acetic acid, and cytosine as major distinguishing feature for COVID-19 patients [43]. The indole-3-acetic acid is derived from tryptophan degradation, and nicotinamide is derived from nicotinic acid which serves as a precursor for NAD^+ and $NADP^+$, coenzymes important for various biochemical processes. Danlos et al. carried out the plasma metabolomics in COVID-19 patients at different stages of SARS-CoV-2 infection [44]. Among the 77 metabolites that were found to have changed in patients with severe COVID-19, remarkably tyrosine, phenylalanine, arginine, glutamate, leucylproline, aspartate, and S-adenosylmethionine were higher [44]. Moreover, changes in tryptophan metabolism were evident in critical COVID-19 patients with increase in kynurenine metabolite since tryptophan converts into NAD (nicotinamide adenine dinucleotide) through kynurenine pathway [44].

Urine metabolome also showed the dysfunction in purine metabolism, energy metabolism, and NAD^+ synthesis in COVID-19 patients [45]. Similar to plasma, metabolomics of serum from COVID-19 patients indicated alterations in tryptophan metabolism with higher levels of glucose and fatty acids which was linked to the presence of inflammatory markers such as IL-6, C-reactive protein, and renal function markers such as blood urea nitrogen [46]. Further, increased glucuronate, glucose, products derived from bilirubin breakdown, and bile acid derivatives were found in the serum of COVID-19 patients suggesting disrupted liver detoxification functions associated with SARS-CoV2 infection [47]. Consistently, Xiao et al. demonstrated changes in tryptophan metabolism with increased levels of kynurenine and nicotinic acid and inflammatory response accompanied by cytokine release syndrome along with alterations in purine and arginine metabolism, as a

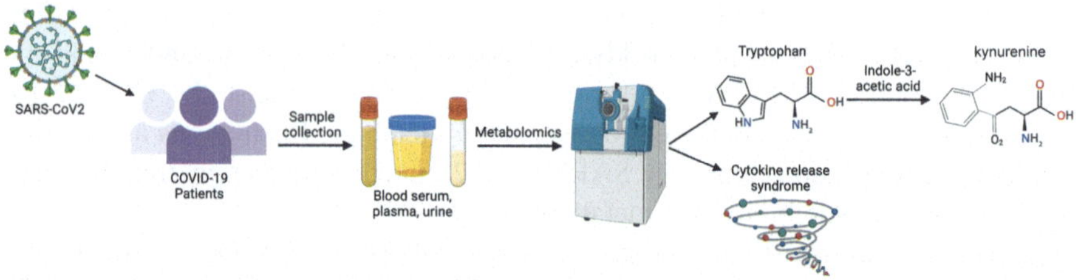

Fig. 2 Metabolic changes in COVID-19 patients: Metabolomics of biological samples from COVID-19 patients showed increased kynurenine concentration as a consequence of enhanced tryptophan metabolism with higher levels of cytokine production

consequence of SARS-CoV-2 infection (Fig. 2) [48]. Dei Cas et al. identified in the sera of severe COVID-19 patients, lower levels of diacylglycerols, sphingomyelins, and higher levels of phosphatidylethanolamines (PE) and acylcarnitine [49]. GC-MS-based serum metabolomics of COVID-19 patients revealed fluctuations in the amino acid catabolism such as glutamate, glutamine, threonine, and alterations in energy metabolism as evident in Warburg effect and Krebs cycle [50]. Further, Shi et al. detected a combination of cholesterol, succinate, 3-hydroxybutyrate, 2-hydroxy-3-methylbutyrate, oleic acid, palmitate, and ornithine in the patient's sera as a prognosis of COVID-19 severity [51]. Finally, Kaur et al. showed elevated concentrations of arachidonic acid, sphingomyelins (SM), phosphatidylcholines (PC), and tryptophan metabolites in the serum of severe COVID-19 patients which is helpful in differentiating severe cases from recovered patients [52].

3.2 *Metabolomics in HIV Infection*

Human immunodeficiency virus (HIV) is a retrovirus causing acquired immunodeficiency syndrome (AIDS) with high morbidity and mortality rates worldwide [53]. Despite the lack of cure, HIV patients can survive with undergoing antiretroviral therapy (ART). HIV binds, infects, and replicates within cells resulting in disease progression which is embraced by gradual decline in $CD4^+$ T cells [54]. Retroviruses such as HIV have shown to alter host cell metabolism [55]. By virtue of this, metabolomics has been applied for the identification of biomarkers from biofluids of HIV patients assisting in better disease diagnosis and development of AIDS vaccines. Initial studies distinguished HIV^+ patients receiving ART from HIV^- individuals exploiting metabolomics-based profiling of serum which reported remarkable fluctuations in the concentrations of glucose and lipids [56, 57]. Consistently, these findings were validated in plasma and cerebrospinal fluid (CSF) from HIV^+ patients using ultrahigh-performance liquid chromatography mass spectrometry (UHLC/MS/MS) and GC/MS [58].

Human immunodeficiency virus type 1 (HIV-1) accounts for most global AIDS cases, while only 30% of individuals infected with human immunodeficiency virus type 2 (HIV-2) eventually manifest AIDS [59]. HIV-1 is more transmissible and virulent ensuing a much faster deficit in $CD4^+$ T cells as compared to HIV-2 infection which is characterized by a slower decrease in $CD4^+$ T cells [54]. Although the glycolytic and TCA metabolic profiles were alike among HIV-1 and HIV-2 infections, metabolomics of HIV-2 infections showed a characteristic higher level of deoxynucleotide triphosphates (dNTPs), which tends to be associated with viral protein x (Vpx) [60]. Vpx degrades SAMHD1, a host antiviral factor which diminishes the availability of dNTPs required for reverse transcription of virus on account of its dNTPase activity [60]. HIV-infection in macrophages is generally long-lasting, whereas $CD4^+$ T cells exhibits acute lytic infection. HIV infection in macrophages resulted in decreasing levels of glycolytic metabolites, while $CD4^+$ T cells demonstrated higher concentrations of glycolytic metabolites upon HIV infection [61]. LC-MS/MS-based metabolomics revealed elevated levels of glyceraldehyde-3-phosphate (G3P) and fructose 1,6-bisphosphate (FBP), key glycolytic intermediates, in HIV-1-infected macrophages while accumulation of quinolinate with HIV-2 infection. Quinolinate is involved in NAD^+ production in kynurenine pathway, which is initiated by tryptophan degradation, and diminishing levels of tryptophan is indicative of T-cell exhaustion and immunosuppression [62].

HIV metabolomic studies have utilized biofluids such as blood, CSF, urine, etc., from patients and control subjects to identify metabolite biomarkers defining the pathogenic signatures of HIV infection and assists in understanding the disease progression and immunological responses to treatment [63]. Metabolomics of blood plasma from HIV-infected patients reported higher levels of L-aspartate, dicarboxylic acylcarnitines, and phosphatidylcholines and lower levels of dopamine and sphingomyelins [64]. HIV-induced oxidative stress such as changes in amino acid metabolism has been studied using both targeted and untargeted metabolomics [65]. A study conducted in HIV^+ sub-Saharan populations revealed soaring levels of indoleamine 2,3-dioxygenase (IDO) after performing GC/MS analysis of plasma samples from HIV^+ patients [66]. This increase led to accelerated degradation of tryptophan attributing to increased concentration of kynurenine pathway intermediate, quinolinate [66], thus, reflecting a correlation between tryptophan levels and the development of HIV in HIV^+ patients in comparison to healthy individuals.

3.3 *Metabolomics in Tuberculosis (TB)*

Tuberculosis (TB) is the leading cause of most of the deaths in the world due to the infection caused by bacterial pathogen *Mycobacterium tuberculosis* (*Mtb*) [67]. *Mtb* causes latent TB infection and only 10% of cases with latent infection progress to active TB disease. In TB, the bacterial pathogen adapts its metabolism to the host environment in order to survive and cause latent infection [68]. Metabolism of both *Mtb* and host plays an important role in the disease progression and is utilized for the development of biomarkers for a better disease diagnosis since the existing diagnostic tests for TB has poor sensitivity. *Mtb* primarily infects macrophages, and thus, macrophage metabolism has been greatly studied in *Mtb* infection [69–71]. Metabolomics in alveolar and interstitial macrophages with *Mtb* infection helped to unravel whether these belong to M1 or M2 phenotypes [72, 73]. Interstitial macrophages displayed M1 phenotype with diminished bacterial replication, while alveolar macrophages exhibited M2 phenotype with optimal bacterial replication [72, 73]. Interstitial macrophages showed increased lactate production as compared to alveolar macrophages and blocking glycolysis using 2-deoxy-D-glucose (2-DG) in bone marrow-derived macrophages (BMDMs) promoted bacterial growth reflecting that increased glycolysis is helpful in restraining *Mtb* growth [73].

Carbon metabolism is particularly essential for *Mtb* growth since deficiency of carbon sources inhibits its replication and survival as demonstrated in animal models [74]. Metabolomics using ^{13}C-labeled carbon substrates fueling *Mtb* growth showed the utilization of different carbon substrates such as glycerol, acetate, and dextrose supporting the growth and survival of *Mtb* [75]. Global metabolite profiling of murine lung tissues infected with *Mtb* identified increased succinate metabolism and alterations in the metabolites involved in the oxidative stress and redox pathways such as xanthine oxidase-related metabolites and inducible nitric oxide synthase (iNOS) [76]. Levels of xanthine and hypoxanthine subsides after 4 weeks of *Mtb* infection in the murine lungs which surges again after 9 weeks of infection. Further, elevated levels of iNOS metabolites, citrulline, and arginine were detected at both weeks 4 and 9 in the lungs of *Mtb*-infected mice [76]. LC/MS and GC/MS studies of biological samples such as sputum, plasma, and serum from TB patients revealed fluctuations in the concentration of metabolites which allowed the discrimination of TB-positive individuals from healthy controls [77–79]. Further, HPLC-MS-based metabolomics of urine samples from active *Mtb*-infected patients showed that metabolites such as N-acetylhexosamine, neopterin, diacetyl-spermine, and sialic acid could be used as potential urinary biomarkers for the diagnosis of TB patients [80].

Amino acids such as aspartate, glutamate, and sulfoxy methionine were found in increasing concentrations, while declining concentrations of asparagine, glutamine, and methionine were reported in the serum of active TB patients in comparison to healthy individuals or individuals with latent TB infection [81]. Tryptophan metabolism is particularly critical for *Mtb* survival, and increased catabolism of tryptophan to kynurenine was observed in the plasma of both active and latent TB patients [82]. Tryptophan catabolism promotes *Mtb* survival and persistence at the site of infection by triggering immune tolerance and regulating $CD4^+$ T-cell responses. Metabolomics has been useful in distinguishing latent and active TB patients by identifying immunometabolic pathways linked to the development of TB [83]. It was also shown that the advancement of TB infection to disease is correlated with the aminoacyl tRNA pathway [83]. Lastly, fatty acids are also important for TB infection as *Mtb* has a selective metabolic preference for fatty acids as a source of nutrients which is evident by enhanced expression of genes associated with the fatty acid metabolism [84, 85].

4 Metabolomics in Physiological Diseases

4.1 Metabolomics in Cancer

Cancer is one of the most dreadful human diseases which causes majority of mortalities worldwide, and despite numerous advancements in cancer research, there is still a challenge for its accurate diagnosis and treatment. There is a rewiring of metabolism which supports the uncontrolled growth and proliferation of cancer cells that is essential for them to adapt to the tumor microenvironment. Cancer cells have selective preference towards anaerobic glycolysis even in the presence of sufficient oxygen and produces higher amounts of lactate with increased glucose consumption which is popularly known as the "Warburg effect" (Fig. 3) [86, 87]. By switching to anaerobic glycolysis, which is faster than oxidative phosphorylation, cancer cells meet their increased demand for energy and biosynthetic molecules in order to adapt to hypoxic tumor microenvironment. Thus, cancer cells have increased influx of glucose for glycolysis and produces metabolites needed to support their abnormal growth [88, 89]. Tumor cells primarily uses glucose and glutamine to produce energy and synthesizes carbohydrates, amino acids, fatty acids, and nucleotides to support their unrestrained proliferation and increased protein synthesis (Fig. 3). Since tumor cells modulate the cellular metabolism, metabolomics is presently being exploited for the detection of altered metabolic pathways and biomarkers. This helps in the early cancer diagnosis and evaluation of effectiveness of the cancer drug therapies that primarily targets altered metabolic pathways [90]. It has

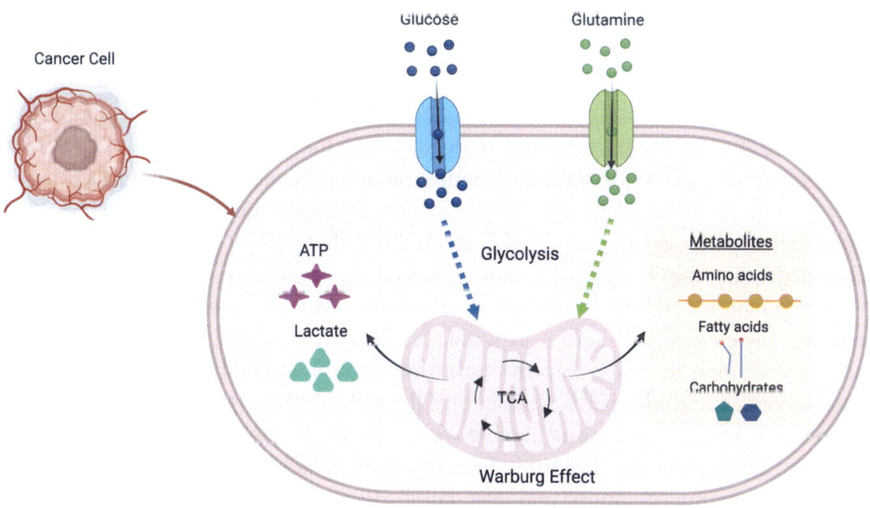

Fig. 3 Warburg effect in cancer cell: Cancer cells produce higher amounts of lactate and ATP through anaerobic glycolysis even in the presence of oxygen primarily utilizing glucose and glutamine as substrates with increased synthesis of carbohydrates, fatty acids, and protein to support their growth and proliferation

been shown that renal carcinoma cells have higher levels of enzymes such as hexokinase-1 pyruvate kinase and lactate dehydrogenase A, which are involved in glycolysis [91]. Perturbations in glucose metabolism has been shown to be associated with lung cancer carcinogenesis, while breast cancer cells have shown to poorly proliferate in low-glucose environment [92, 93]. In human hepatocellular carcinoma, arginine N-methyltransferase 6 (PRMT6) regulates the Warburg effect and glycolysis by controlling the nuclear re-localization of pyruvate kinase M2 (PKM2) [94]. Apart from glycolysis, pentose phosphate pathway (PPP), hexosamine pathway, and glycogenesis also need glucose and are all reprogrammed in cancer cells which endows the possibility to target cancer cells with specificity [95].

In addition to glucose metabolism, tumor cells also display changes in the lipid metabolism which includes enhanced fatty acid β-oxidation and lipid synthesis [96, 97]. Increased lipid synthesis provides surplus energy required for the rapid proliferation and membrane synthesis in the tumor cells. Metabolomics have shown an accelerated utilization of fatty acids by cancer cells as depicted in kidney cancer [98], and altered cholesterol metabolism with amended peroxisome proliferator-activated receptor (PPAR) signaling in tumor cells with similar tissue origin [95]. High-throughput metabolomics demonstrated higher levels of β-hydroxybutyrate, heptanoic acid, hexadecenoic acid, and docosahexaenoic acid in gastric cancer [99]. By virtue of its reliance on fatty acids, tumor cells showed retarded growth in the presence of fatty acid inhibitors such as the PPARα antagonist GW6471 as tested in the renal cell cancer models [100]. It has been shown that within the tumor microenvironment, oxysterols activate transcription factor, liver X receptor (LXR) in macrophages which halts the migration of dendritic cells, and neutrophils recruitment at the tumor site leading to immunosuppression [101]. Alterations in the lipid metabolism activates carcinogenic signaling pathways such as Hippo/YAP and Wnt/β-catenin pathways, along with increased biomass and energy production in cancer stem cells (CSCs) [102].

Amino acids such as glycine, glutamine, and serine, etc. have been valuable for identifying cancer biomarkers and understanding the carcinogenesis of different malignant tumors [103]. Glutamine, being the most abundant free amino acid, has shown to be involved in bioenergy production, biomolecule synthesis, and signal transduction pathways supporting the survival of cancer cells [104, 105]. Glutamine aids in driving the TCA cycle under glucose deprivation with higher levels of malate, fumarate, and citrate in tumor cells [106]. Metabolomic-based assessment revealed lower levels of serum aspartate in the breast cancer patients. Aspartate induces apoptosis in tumor cells through Akt pathway and inhibits their growth and thus, serves as a biomarker for indicating the risk of breast cancer [107]. Further, abnormal levels of arginine and alanine were found to be directly correlated with the occurrence of gastric cancer [108]. Thus, molecules targeting amino acid synthesis are currently being developed and tested clinically for cancer therapy [109].

Since metabolic reprogramming is a hallmark of cancer progression, metabolic biomarkers have been discovered to trace the cancer pathogenesis. There are different types of biological samples such as blood plasma, urine, cerebrospinal fluid (CSF), saliva, etc., which have been used to discover biomarkers and perform metabolomics studies for cancer [95]. NMR spectroscopy identified accumulation of amino acids and reduction in fatty acids in plasma and thyroid tissues from papillary thyroid microcarcinoma patients [110]. For pancreatic cancer, metabolomics of blood plasma showed that metabolites such as glycocholic acid, inosine, sphinganine, β-sitosterol, creatine, phosphatidylcholine, and phosphatidylethanolamine were directly correlated with the clinical outcome and survival of patients [111, 112]. Metabolomics has identified metabolites as potential biomarkers in the plasma of patients suffering from lung, liver, and breast cancer [113, 114]. Further, urine biomarkers are non-invasive, making them apt for understanding tumor pathology and cancer diagnosis. In patients with renal cancer, fatty acid metabolism was altered with increased production of acetyl coenzyme A (acetyl-CoA) resulting in higher levels of acylcarnitine in the urine [115]. Other than kidney cancer, urine biomarkers are also critical in understanding the tumorigenesis and distinguishing among non-urinary tumors such as liver, cervical, stomach, and breast cancers [116]. Omran et al. showed accumulating concentrations of 8-hydroxy-2′-deoxyguanosine, 1-methylguanosine (1-MG), and 1-methyl adenosine (1-MA) in the urine, an indication of early stages of breast cancer [117].

Cerebrospinal fluid (CSF) has been used to assess metabolites to characterize brain tumors particularly glioma. The first metabolomics study using CSF was carried out in patients with malignant glioma which showed the correlation between glioma malignancy and CSF metabolites [118]. The analysis of CSF metabolites in patients with metastatic tumors and glioma showed the presence of amino acids such as methionine, tryptophan, and products of TCA cycle [119]. In addition to CSF, urine and blood, saliva has been valuable for the early diagnosis of specific cancers. Comprehensive metabolomics analysis of saliva samples from patients suffering with oral cancer reported higher levels of polyamines, piperidine and taurine, reflecting specific biomarkers aiding in the screening of oral cancer [120]. Sugimoto et al. also identified cancer-specific biomarkers using capillary electrophoresis time-of-flight mass spectrometry (CE-TOFMS)-based saliva metabolomics in patients with breast and pancreatic cancers [120]. Using CE-TOFMS, S-adenosylmethionine and pipecolate were identified as promising markers for distinguishing oral cancer patients from healthy individuals [121]. There are other studies depicting the identification and validation of the metabolite profiles of saliva from breast and pancreatic cancer patients which is important in discriminating discrete tumors [122, 123]. Therefore, metabolomics is a powerful tool for cancer prognosis, diagnosis, and monitoring the efficacy of cancer drug therapies and interventions by identifying key biomarkers and drivers of tumorigenesis.

4.2 *Metabolomics in Neurological Diseases*

Neurodegenerative diseases are characterized by the progressive destruction of the brain and nerves which leads to irreversible damage and inflammation of the neurons. Despite rigorous efforts, there is no cure for neurological disorders, and their treatment is mainly based on the usage of medications aimed towards reducing the symptoms [124, 125]. Metabolomics has emerged as a valuable tool aiding in the identification of symptoms of neurodegeneration by profiling changes in the brain metabolism (Fig. 4). Perturbations in the metabolic pathways contribute to the pathogenesis of neurodegeneration, and so the assessment of global metabolic profiles of biofluids and biomarker discovery defining the disease status will profoundly enhance the clinical diagnostics for the neurological disorders [126]. Cerebrospinal fluid (CSF) serves as the most promising biomarker for neurodegenerative diseases succoring in evaluating the efficacy of clinical diagnosis and early detection of response to therapy through metabolomic profiling [127].

Amyotrophic lateral sclerosis (ALS) is a devastating motor neuron disease marked by the degradation of both lower and upper motor neurons (Fig. 4) [128]. The pathophysiology involved in ALS are complex, and due to limited reliable diagnostic procedures, high-throughput metabolomics has garnered much interest for the detection of biomarkers [129]. GC-TOFMS (gas chromatography coupled to time-of-flight mass spectrometry) has been exploited to study the CSF metabolome in ALS patients [130]. Hundreds of metabolites were studied in the CSF of ALS patients using GC-TOFMS detecting changes in the metabolite concentrations [131]. H^1 NMR spectroscopy analyzed biomarkers in CSF of ALS patients in the early stages of the disease with increased levels of pyruvate, ascorbate, and acetone and lower levels of acetate which showcased perturbations in brain glucose metabolism [132]. Among different metabolites, glutamate was identified as an important

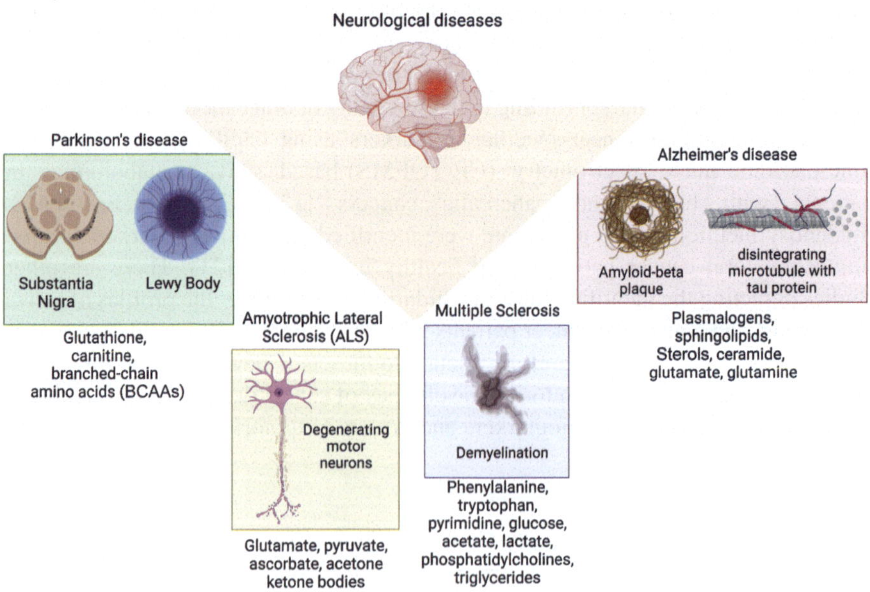

Fig. 4 Key metabolites in neurological diseases: Different neurological diseases such as Parkinson's disease, amyotrophic lateral sclerosis (ALS), multiple sclerosis (MS), Alzheimer's disease with their key metabolites defining their disease status

circulating metabolite in ALS patients [131–134]. An increase in glutamate concentration with a concomitant reduction in glutamine levels in ALS patients was indicative of excitotoxicity which occurs due to an imbalance in the glutamate-glutamine conversion cycle [134]. ALS patients harboring mutations in superoxide dismutase 1 (SOD1) have a more distinct metabolomic profile with strikingly reduced levels of creatinine, glutamate, and glutamine in CSF [131]. Alterations in lipid metabolism have been found to be associated with early phases of disease progression supporting the previous report using animal model showing the metabolic preference for lipids such as cholesteryl ester and phosphatidylcholine during asymptomatic stage of the disease [135, 136]. Higher levels of ketone bodies were evident in symptomatic ALS patients [134]. The limited usage of lipids as a source of energy by motor neurons results in oxidative stress leading to neuroinflammation, mitochondrial dysfunction, excitotoxicity, and cell death [137, 138]. Metabolomic examination of spinal cord samples from ALS patients revealed higher levels of lipid metabolite docosahexaenoic acid (DHA), which is critical for nervous system homeostasis. The accumulation of docosahexaenoic acid-derived resolvin D, and lipid peroxidation-derived molecules leads to exacerbated cell stress resulting in aggregation of TDP-43 and loss of motor neurons [139]. Thus, metabolomics has provided strong assistance to understand the underlying molecular mechanisms of ALS.

Alzheimer's disease (AD) is the most prevalent neurological disease and the leading cause of dementia mostly affecting the elderly people beyond the age of 60. AD is characterized by atrophy, deterioration of the cerebral cortex, and progressive brain degeneration with loss of memory and reasoning (Fig. 4) [140, 141]. AD is a complex disease, and its pathogenesis is poorly understood which embarks the necessity for the development of novel biomarkers for early diagnosis and detection. Advancements in metabolomics has showcased a significant decline in ethanolamine plasmalogens in both the brain and plasma in animal models and human subjects suffering from AD [142–146]. Plasmalogens are antioxidants providing protection to lipid and lipoproteins from oxidative stress [147]. The metabolic imbalance leading to oxidative stress in AD patients was evident in CSF and blood samples as well with reduced levels of plasmalogens, sphingolipids, phosphatidylcholines, phosphatidylinositols, phospholipids, and sterols [148, 149]. Studies have shown the correlation between altered plasma ceramide levels and serum triglycerides to neuropathological perturbations in AD [150, 151]. The findings suggest that polyunsaturated fatty acid-containing triglycerides and apolipoprotein E (APOE) ε4 were linked with entorhinal cortical thickness and CSF β-amyloid1–42 values [151]. In AD patients, elevated levels of glutamate and glutamine and lower levels of branched chain amino acids (BCAAs), creatinine, and taurine were detected [152–155]. Further, increased lactate concentration in CSF reflects mitochondrial alteration due to tau proteins, impeding oxidative phosphorylation in patients suffering from AD [156–158]. A recent study has done comparison of systemic metabolic AD biomarkers in blood and CSF samples and corroborated with previous findings in reporting higher levels of glutamine in plasma [159, 160]. This study proclaimed a positive association between levels of CSF p-Tau181 and t-Tau with apolipoprotein E (APOE) ε4 genotype [159]. Moreover, the study showed diminishing piperine levels attributing to its antioxidant, anti-secretase, and anti-inflammatory functions, which were discordantly correlated with CSF p-Tau18 [161–163]. Therefore, the complexities associated with AD have been well exemplified with the evolution in the field of metabolomics underscoring challenges with the progress of potent therapeutic interventions [164].

Multiple sclerosis (MS) is a chronic autoimmune neurological disorder manifested by inflammation and demyelination of the central nervous system (CNS) (Fig. 4) [165]. MS is characterized by episodes of clinical deterioration and periods of relapsing-remission gradually transforming into irreversible neurodegeneration [165]. The harmful effects of MS could be attenuated if the symptoms are diagnosed at an early stage. However, the diagnosis of MS is often confounded with other neurological diseases with non-specific clinical symptoms emphasizing the need to discover novel biomarkers for early diagnosis and therapeutic interventions [166]. Metabolomic approaches seems attractive for monitoring the pathogenesis of MS and have been gaining attention for the potential use of metabolic biomarkers in the early diagnosis of the disease [166]. Metabolomic studies have identified reduction in the levels of mannose, citrate, acetate, and phenylalanine while increasing levels of glucose, lactate, formate, acetone, choline, threonate, and myoinositol in CSF of MS patients [167–170]. Patients with progressive and relapsing-remitting MS witnessed variations in the levels of phenylalanine, tryptophan, and pyrimidine in CSF as compared to healthy controls [171]. NMR-based metabolomics in plasma and serum of MS patients identified changes in the energy and xenobiotic metabolism with altered metabolites such as glucose, lactate, glutamine, glutamate, acetate, valine, lysine, and scylloinositol [172, 173]. Elevated levels of phospholipids and fatty acids were also observed in the serum of MS patients [174]. Lipidomic studies revealed decreased levels of phosphatidylcholines (PCs) and altered sphingolipid metabolism in MS patients [175–177]. A distinctive lipid profile was observed in CSF samples of MS patients with reduction in triglycerides and upregulation of diglycerides [178]. Further, an increased lipid peroxidation was evident with higher levels of 8-iso-prostaglandin in CSF from MS patients [179]. Thus, metabolomics could be applied to correctly diagnose MS and distinguish it from other neurological diseases having overlapping symptoms.

Parkinson's disease (PD) is a progressive neurological disease manifested by the degeneration of dopaminergic neurons within the substantia nigra and the accumulation of intracellular inclusions known as Lewy bodies containing a protein, α-synuclein in the CNS (Fig. 4) [180, 181]. However, the exact mechanism underlying the pathogenesis of PD remains unclear, emphasizing the need to identify biomarkers assisting in the accurate diagnosis of the disease and evaluating the efficacy of the therapies. Metabolomic-based blood biomarker investigations revealed an increase in glutathione in response to oxidative damage and lower concentrations of uric acid in plasma from PD patients [182]. Oxidative damage was also evident with decreased levels of bilirubin/biliverdin ratio in the blood serum of PD patients [183]. Oxidative damage also leads to accumulation of ROS resulting in mitochondrial dysfunction in PD [184]. Mitochondrial dysfunction has been linked with increased fatty acid oxidation resulting in elevated levels of hexanoylcarnitine, decanoylcarnitine, and myristoleoylcarnitine in the blood serum and hexanoylglycine, malonylcarnitine, and furoylglycine in the urine of PD patients, respectively [185–188]. Further, higher concentration of branched-chain amino acids (BCAAs) such as valine, leucine, and isoleucine was found in the animal model and patients with PD and found to be associated with oxidative damage and mitochondrial dysfunction [187–189]. Thus, metabolomics could be exploited as a powerful tool for discovering drug targets and elucidating molecular mechanisms related to PD [190].

Huntington's disease (HD) is a rare heritable neurological disorder caused by repeated trinucleotide (CAG) of the HTT gene encoding Huntington's protein and is characterized by loss of neurons responsible for voluntary movements [124, 191]. Despite advances in the diagnostic procedures, there is still a need for better clinical management of HD since the exact pathogenesis remains unclear [192]. Since HD affects metabolism, investigating magnitude of metabolites possesses a valuable biomarker strategy for the accurate diagnosis of HD. Metabolomics profiling showed branched-chain amino acids (BCAAs), which is important for neurotransmitter synthesis, act as an important biomarker for HD with reduced levels of tyrosine, valine, phenylalanine, leucine, and isoleucine in the postmortem brain and plasma of HD patients [193–195]. Lower levels of phosphatidylcholine and lysophosphatidylcholine were found in the blood of HD patients as a consequence of downregulated expression of *PCYT1A* and higher phospholipase A2 enzymatic activity [193, 196]. Further, CSF metabolome showed higher concentrations of glucose and lactate in HD patients [197]. Studies have shown that dysregulated C/EBP represses argininosuccinic acid synthetase and argininosuccinase acid lyase resulting in increased accumulation of citrulline due to defective urea and NO cycles in preclinical and clinical models of HD [198–200]. Thus, a defect in energy metabolism substantially directs biomarker discovery for the better treatment of rare neurological diseases such as HD.

4.3 Metabolomics in Cardiovascular Health and Diseases

Cardiovascular diseases (CVDs) are multifactorial diseases accompanied by alterations in the cardiac function and myocardial metabolism. Emerging metabolomic-based fingerprinting of patients provide insight into the pathophysiology of CVDs by highlighting novel biomarkers through the quantification of metabolites in body fluids or biopsies. This aids in a better understanding of the complexities associated with the diagnosis and ramification of CVDs such as atherosclerosis, heart failure, and ischemic and non-ischemic cardiomyopathy.

One of the leading causes of CVDs is "atherosclerosis" which is manifested by the progressive constriction of the arteries. Blood-based metabolomics showcased the circulating trimethylamine-N-oxide (TMAO) as a prognostic marker for atherosclerosis and cardiovascular risk prediction for myocardial infarction and stroke [201, 202]. Hazen and colleagues described in their study that the TMAO levels were higher in CVD patients gradually leading to stroke, heart attack, or

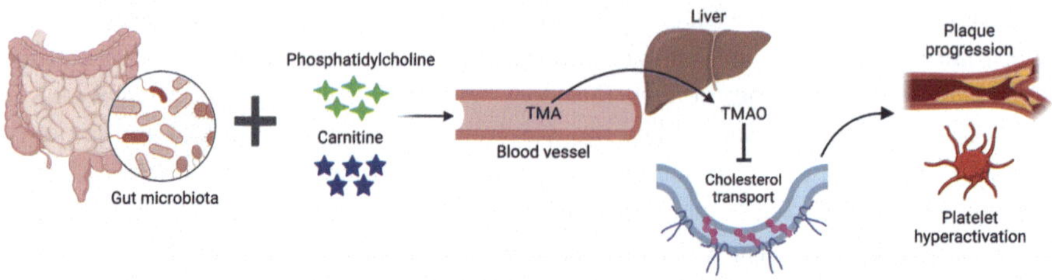

Fig. 5 TMAO as a prognostic biomarker in atherosclerosis: Gut microbiota in combination with dietary metabolites, phosphatidylcholine, and carnitine results in the production of TMAO. TMAO, in turn, inhibits cholesterol transport, which leads to plaque progression and platelet hyperactivity

heart-related death [203]. The gut microbiome plays an integral role in determining the circulating levels of host metabolite, trimethylamine (TMA) production from dietary phosphatidylcholine, carnitine, and choline, which serve as a precursor for the formation of TMAO. The interaction between gut bacteria with dietary phosphatidylcholine and carnitine results in the production of TMAO. After releasing into the bloodstream, TMA is converted to TMAO in the liver where it hinders the cholesterol transport, thereby accelerating the plaque progression resulting in enhanced cardiovascular risks and platelet hyperactivity (Fig. 5) [201, 202]. Moreover, TMAO is a uremic toxin which impairs kidney functions, increasing the likelihood of adverse cardiac events. This phenotype was replicated in apolipoprotein $E^{-/-}$ mice in which mice were fed with high-TMAO and high-choline diets shown to be highly susceptible to the formation of atherosclerotic plaques [2]. Furthermore, in a mouse model of atherosclerosis, reduced levels of circulating TMAO were achieved by compromising the gut microbiome's ability to convert dietary carnitine or choline into TMA, by inhibiting bacterial TMA lyase with 3,3-dimethyl-1-butanol (a structural analogue to choline), mitigating the progression of atherosclerotic lesions [204]. Thus, these studies reflect a strong correlation between levels of TMAO in plasma and atherosclerotic lesion development which could serve as a diagnostic biomarker for atherosclerosis.

Blood-based metabolomic profiling also showed that lipid intermediates such as phosphatidylcholines, sphingomyelins, palmitate, or diacylglycerols are other potential biomarkers shown to be associated with higher risk of myocardial infarction [205]. Decreased levels of 18:2 lysophosphatidylcholine and 28:1 sphingomyelin and increased levels of 18:2 monoglyceride in the plasma were associated with increased risk for myocardial infarction [206]. There was an increased abundance of unsaturated fatty acids comprised lysophosphatidylcholines and lower levels of circulating lysophosphatidylcholines containing saturated fatty acids in patients with angina or myocardial infarction [205].

Coronary artery disease (CAD) remains one of the leading causes of death in both men and women in the world. CAD can be classified into stable and unstable angina, myocardial infarction (MI), and sudden cardiac death. During ischemic cardiomyopathy, there are remarkable changes in the myocardial energy metabolism due to the reduced supply of oxygen and nutrition to the affected myocardium. In the course of ischemic periods, glycolytic rate increases with overall reduction in the oxidative metabolism. The duration and intensity of ischemia is directly proportional to an enhanced myocardial glycolysis [207]. During ischemia, myocardial lactate levels increase consequently due to enhanced glycolytic rates and glycogenolysis with reduced glucose oxidation rates [208–210]. This reflects accelerated anaerobic glycolytic metabolism with higher lactate levels in the circulation as observed in patients with acute ischemia and CAD [211, 212]. Due to reduced levels of oxygen during ischemia, aerobic oxidation gets impaired leading to decreased production of TCA cycle intermediates such as succinate and fumarate [213–215]. In a mouse model of ischemia-reperfusion injury, mitochondrial ROS accumulates because of enhanced oxidation of elevated succinate levels in the ischemic heart tissue [216].

Circulating branched-chain amino acids (BCAAs) such as proline, methionine, glutamine, glutamate, leucine, isoleucine, and valine have shown to be promising biomarkers reflecting the risk of coronary artery disease (CAD) [217]. Amino acids such as tyrosine and lysine break down while levels of alanine, leucine, and isoleucine rise during ischemic myocardium [215]. During ischemia, glutamic oxaloacetic transaminase levels rise in the serum which inhibits the transamination of glutamate into α-ketoglutarate enhancing the glutamate levels in the tissue during myocardial ischemia which could activate ROS production directing toward cardiomyocyte death [218, 219]. Active amino acid biosynthesis is an indicative of acute myocardial infarction with higher levels of N-phenylacetyl-L-glutamine, arginine, leucine, and tryptophan in acute MI patients [212, 220].

Fatty acid oxidation drops as a result of reduced oxygen supply in patients with CAD [221]. Higher levels of short-chain (SC) dicarboxylic acylcarnitines predicted the risk for cardiovascular events in CAD patients [208]. During ST-segment elevation myocardial infarction (STEMI), fatty acids such as palmitic acid, oleic acid, linoleic acid, and stearic acid were found to be increased in the plasma reflecting the ischemia-induced perturbations in the cardiac metabolism [222]. Inflammation induced during myocardial infarction is reflected by the increased levels of eicosatetraenoic acid and eicosatrienoic acid [223]. The impairment of β-oxidation of unsaturated fatty acids during myocardial infarction contributes to myocardial ischemia [224, 225]. Patients with MI were shown to have elevated levels of sphingomyelin and ceramide and perturbed sphingolipid metabolism which is critical for vascular maturation and wound healing [205, 226]. Glycerophospholipids such as linoleamide glycerophosphate choline, phosphatidylserine, lyso-PC (C16:0), lyso-PC (C18:1), and lyso-PC (18:2) appear to lower down during MI [224]. Since altered lipid metabolism is evident during ischemic cardiomyopathy, metabolomics-based assessment of CAD is desirable, particularly when angiography, which is widely used for the diagnosis of CAD, is both expensive and invasive.

5 Future Perspectives

Metabolomics is an evolving and rousing research area, extending from biomarker discovery to providing valuable cognizance toward mechanisms underlying normal physiology and pathophysiology in diseases. Metabolomics have been foremost impactful as a global profiling tool in the discovery of metabolites and metabolic pathways pivotal for maintaining physiological homeostasis in humans. Interestingly, the combination of different "omics" technologies such as genomics and proteomics with metabolomics has furnished novel insights into local and global disease processes. Cumulatively, the amalgamation of metabolomics into basic and biomedical research is immensely upgrading the understanding of diseases through precise and swift diagnosis, prognosis, evaluating response to treatments, and prediction of disease outcomes. Therefore, the future prospect of metabolomics relies on the integration of "multi-omics" strategy into systems biology in view to impart unique biological information in a given physiological milieu.

References

1. Johnson CH, Patterson AD, Idle JR, Gonzalez FJ. Xenobiotic metabolomics: major impact on the metabolome. Annu Rev Pharmacol Toxicol. 2012;52:37–56.
2. Johnson CH, Ivanisevic J, Siuzdak G. Metabolomics: beyond biomarkers and towards mechanisms. Nat Rev Mol Cell Biol. 2016;17(7):451–9.
3. Schoffel N, Mache S, Quarcoo D, Scutaru C, Vitzthum K, Groneberg DA, et al. Rheumatoid arthritis: scientific development from a critical point of view. Rheumatol Int. 2010;30(4):505–13.
4. Sasaki C, Hiraishi T, Oku T, Okuma K, Suzumura K, Hashimoto M, et al. Metabolomic approach to the exploration of biomarkers associated with disease activity in rheumatoid arthritis. PLoS One. 2019;14(7):e0219400.
5. Zhou J, Chen J, Hu C, Xie Z, Li H, Wei S, et al. Exploration of the serum metabolite signature in patients with rheumatoid arthritis using gas chromatography-mass spectrometry. J Pharm Biomed Anal. 2016;127:60–7.
6. Li J, Che N, Xu L, Zhang Q, Wang Q, Tan W, et al. LC-MS-based serum metabolomics reveals a distinctive signature in patients with rheumatoid arthritis. Clin Rheumatol. 2018;37(6):1493–502.
7. Duarte-Delgado NP, Cala MP, Barreto A, Rodriguez CL. Metabolites and metabolic pathways associated with rheumatoid arthritis and systemic lupus erythematosus. J Transl Autoimmun. 2022;5:100150.
8. Yang XY, Zheng KD, Lin K, Zheng G, Zou H, Wang JM, et al. Energy Metabolism Disorder as a Contributing Factor of Rheumatoid Arthritis: A Comparative Proteomic and Metabolomic Study. PLoS One. 2015a;10(7):e0132695.
9. Young SP, Kapoor SR, Viant MR, Byrne JJ, Filer A, Buckley CD, et al. The impact of inflammation on metabolomic profiles in patients with arthritis. Arthritis Rheum. 2013;65(8):2015–23.
10. Weyand CM, Zeisbrich M, Goronzy JJ. Metabolic signatures of T-cells and macrophages in rheumatoid arthritis. Curr Opin Immunol. 2017;46:112–20.
11. Weyand CM, Goronzy JJ. Immunometabolism in early and late stages of rheumatoid arthritis. Nat Rev Rheumatol. 2017;13(5):291–301.

12. Takahashi S, Saegusa J, Onishi A, Morinobu A. Biomarkers identified by serum metabolomic analysis to predict biologic treatment response in rheumatoid arthritis patients. Rheumatology (Oxford). 2019;58(12):2153–61.
13. Sweeney SR, Kavanaugh A, Lodi A, Wang B, Boyle D, Tiziani S, et al. Metabolomic profiling predicts outcome of rituximab therapy in rheumatoid arthritis. RMD Open. 2016;2(2):e000289.
14. Narasimhan R, Coras R, Rosenthal SB, Sweeney SR, Lodi A, Tiziani S, et al. Serum metabolomic profiling predicts synovial gene expression in rheumatoid arthritis. Arthritis Res Ther. 2018;20(1):164.
15. Tsokos GC. Systemic lupus erythematosus. N Engl J Med. 2011;365(22):2110–21.
16. Zhang T, Mohan C. Caution in studying and interpreting the lupus metabolome. Arthritis Res Ther. 2020;22(1):172.
17. Wu T, Xie C, Han J, Ye Y, Weiel J, Li Q, et al. Metabolic disturbances associated with systemic lupus erythematosus. PLoS One. 2012;7(6):e37210.
18. Yan B, Huang J, Zhang C, Hu X, Gao M, Shi A, et al. Serum metabolomic profiling in patients with systemic lupus erythematosus by GC/MS. Mod Rheumatol. 2016;26(6):914–22.
19. Guleria A, Pratap A, Dubey D, Rawat A, Chaurasia S, Sukesh E, et al. NMR based serum metabolomics reveals a distinctive signature in patients with Lupus Nephritis. Sci Rep. 2016;6:35309.
20. Ouyang X, Dai Y, Wen JL, Wang LX. (1)H NMR-based metabolomic study of metabolic profiling for systemic lupus erythematosus. Lupus. 2011;20(13):1411–20.
21. Li Y, Liang L, Deng X, Zhong L. Lipidomic and metabolomic profiling reveals novel candidate biomarkers in active systemic lupus erythematosus. Int J Clin Exp Pathol. 2019;12(3):857–66.
22. Bengtsson AA, Trygg J, Wuttge DM, Sturfelt G, Theander E, Donten M, et al. Metabolic Profiling of Systemic Lupus Erythematosus and Comparison with Primary Sjogren's Syndrome and Systemic Sclerosis. PLoS One. 2016;11(7):e0159384.
23. Scavuzzi BM, Simao ANC, Iriyoda TMV, Lozovoy MAB, Stadtlober NP, Franchi Santos L, et al. Increased lipid and protein oxidation and lowered anti-oxidant defenses in systemic lupus erythematosus are associated with severity of illness, autoimmunity, increased adhesion molecules, and Th1 and Th17 immune shift. Immunol Res. 2018;66(1):158–71.
24. Lai ZW, Hanczko R, Bonilla E, Caza TN, Clair B, Bartos A, et al. N-acetylcysteine reduces disease activity by blocking mammalian target of rapamycin in T cells from systemic lupus erythematosus patients: a randomized, double-blind, placebo-controlled trial. Arthritis Rheum. 2012;64(9):2937–46.
25. Valencia X, Yarboro C, Illei G, Lipsky PE. Deficient CD4+CD25high T regulatory cell function in patients with active systemic lupus erythematosus. J Immunol. 2007;178(4):2579–88.
26. Guleria A, Phatak S, Dubey D, Kumar S, Zanwar A, Chaurasia S, et al. NMR-Based Serum Metabolomics Reveals Reprogramming of Lipid Dysregulation Following Cyclophosphamide-Based Induction Therapy in Lupus Nephritis. J Proteome Res. 2018;17(7):2440–8.
27. Ganguly S, Kumar U, Gupta N, Guleria A, Majumdar S, Phatak S, et al. Nuclear magnetic resonance-based targeted profiling of urinary acetate and citrate following cyclophosphamide therapy in patients with lupus nephritis. Lupus. 2020;29(7):782–6.
28. Burisch J, Jess T, Martinato M, Lakatos PL, EpiCom E. The burden of inflammatory bowel disease in Europe. J Crohns Colitis. 2013;7(4):322–37.
29. Lenz EM, Bright J, Wilson ID, Morgan SR, Nash AF. A 1H NMR-based metabonomic study of urine and plasma samples obtained from healthy human subjects. J Pharm Biomed Anal. 2003;33(5):1103–15.
30. Aldars-Garcia L, Gisbert JP, Chaparro M. Metabolomics Insights into Inflammatory Bowel Disease: A Comprehensive Review. Pharmaceuticals (Basel). 2021a;14(11).

31. Zhang Y, Lin L, Xu Y, Lin Y, Jin Y, Zheng C. 1H NMR-based spectroscopy detects metabolic alterations in serum of patients with early-stage ulcerative colitis. Biochem Biophys Res Commun. 2013a;433(4):547–51.
32. Murgia A, Hinz C, Liggi S, Denes J, Hall Z, West J, et al. Italian cohort of patients affected by inflammatory bowel disease is characterised by variation in glycerophospholipid, free fatty acids and amino acid levels. Metabolomics. 2018;14(10):140.
33. Aldars-Garcia L, Chaparro M, Gisbert JP. Systematic Review: The Gut Microbiome and Its Potential Clinical Application in Inflammatory Bowel Disease. Microorganisms. 2021b;9(5).
34. Williams HR, Cox IJ, Walker DG, North BV, Patel VM, Marshall SE, et al. Characterization of inflammatory bowel disease with urinary metabolic profiling. Am J Gastroenterol. 2009;104(6):1435–44.
35. Bjerrum JT, Wang Y, Hao F, Coskun M, Ludwig C, Gunther U, et al. Metabonomics of human fecal extracts characterize ulcerative colitis, Crohn's disease and healthy individuals. Metabolomics. 2015;11:122–33.
36. Lai Y, Xue J, Liu CW, Gao B, Chi L, Tu P, et al. Serum Metabolomics Identifies Altered Bioenergetics, Signaling Cascades in Parallel with Exposome Markers in Crohn's Disease. Molecules. 2019;24(3).
37. Scoville EA, Allaman MM, Brown CT, Motley AK, Horst SN, Williams CS, et al. Alterations in Lipid, Amino Acid, and Energy Metabolism Distinguish Crohn's Disease from Ulcerative Colitis and Control Subjects by Serum Metabolomic Profiling. Metabolomics. 2018;14(1):17.
38. Valdes A, Moreno LO, Rello SR, Orduna A, Bernardo D, Cifuentes A. Metabolomics study of COVID-19 patients in four different clinical stages. Sci Rep. 2022;12(1):1650.
39. Costanzo M, Caterino M, Fedele R, Cevenini A, Pontillo M, Barra L, et al. COVIDomics: The Proteomic and Metabolomic Signatures of COVID-19. Int J Mol Sci. 2022;23(5).
40. Lee JW, Su Y, Baloni P, Chen D, Pavlovitch-Bedzyk AJ, Yuan D, et al. Integrated analysis of plasma and single immune cells uncovers metabolic changes in individuals with COVID-19. Nat Biotechnol. 2022;40(1):110–20.
41. Song JW, Lam SM, Fan X, Cao WJ, Wang SY, Tian H, et al. Omics-Driven Systems Interrogation of Metabolic Dysregulation in COVID-19 Pathogenesis. Cell Metab. 2020;32(2):188–202 e5.
42. Wu D, Shu T, Yang X, Song JX, Zhang M, Yao C, et al. Plasma metabolomic and lipidomic alterations associated with COVID-19. Natl Sci Rev. 2020;7(7):1157–68.
43. Blasco H, Bessy C, Plantier L, Lefevre A, Piver E, Bernard L, et al. The specific metabolome profiling of patients infected by SARS-COV-2 supports the key role of tryptophan-nicotinamide pathway and cytosine metabolism. Sci Rep. 2020;10(1):16824.
44. Danlos FX, Grajeda-Iglesias C, Durand S, Sauvat A, Roumier M, Cantin D, et al. Metabolomic analyses of COVID-19 patients unravel stage-dependent and prognostic biomarkers. Cell Death Dis. 2021;12(3):258.
45. Raines NH, Cheung MD, Wilson LS, Edberg JC, Erdmann NB, Schmaier AA, et al. Nicotinamide Adenine Dinucleotide Biosynthetic Impairment and Urinary Metabolomic Alterations Observed in Hospitalized Adults With COVID-19-Related Acute Kidney Injury. Kidney Int Rep. 2021;6(12):3002–13.
46. Thomas T, Stefanoni D, Reisz JA, Nemkov T, Bertolone L, Francis RO, et al. COVID-19 infection alters kynurenine and fatty acid metabolism, correlating with IL-6 levels and renal status. JCI Insight. 2020;5(14).
47. Shen B, Yi X, Sun Y, Bi X, Du J, Zhang C, et al. Proteomic and Metabolomic Characterization of COVID-19 Patient Sera. Cell. 2020;182(1):59–72 e15.
48. Xiao N, Nie M, Pang H, Wang B, Hu J, Meng X, et al. Integrated cytokine and metabolite analysis reveals immunometabolic reprogramming in COVID-19 patients with therapeutic implications. Nat Commun. 2021;12(1):1618.
49. Dei Cas M, Ottolenghi S, Morano C, Rinaldo R, Roda G, Chiumello D, et al. Link between serum lipid signature and prognostic factors in COVID-19 patients. Sci Rep. 2021;11(1):21633.

50. Paez-Franco JC, Torres-Ruiz J, Sosa-Hernandez VA, Cervantes-Diaz R, Romero-Ramirez S, Perez-Fragoso A, et al. Metabolomics analysis reveals a modified amino acid metabolism that correlates with altered oxygen homeostasis in COVID-19 patients. Sci Rep. 2021;11(1):6350.
51. Shi D, Yan R, Lv L, Jiang H, Lu Y, Sheng J, et al. The serum metabolome of COVID-19 patients is distinctive and predictive. Metabolism. 2021;118:154739.
52. Kaur G, Ji X, Rahman I. SARS-CoV2 Infection Alters Tryptophan Catabolism and Phospholipid Metabolism. Metabolites. 2021;11(10).
53. Liebenberg C, Luies L, Williams AA. Metabolomics as a Tool to Investigate HIV/TB Co-Infection. Front Mol Biosci. 2021;8:692823.
54. Vidya Vijayan KK, Karthigeyan KP, Tripathi SP, Hanna LE. Pathophysiology of CD4+ T-Cell Depletion in HIV-1 and HIV-2 Infections. Front Immunol. 2017;8:580.
55. Diray-Arce J, Conti MG, Petrova B, Kanarek N, Angelidou A, Levy O. Integrative Metabolomics to Identify Molecular Signatures of Responses to Vaccines and Infections. Metabolites. 2020;10(12).
56. Hewer R, Vorster J, Steffens FE, Meyer D. Applying biofluid 1H NMR-based metabonomic techniques to distinguish between HIV-1 positive/AIDS patients on antiretroviral treatment and HIV-1 negative individuals. J Pharm Biomed Anal. 2006;41(4):1442–6.
57. Philippeos C, Steffens FE, Meyer D. Comparative 1H NMR-based metabonomic analysis of HIV-1 sera. J Biomol NMR. 2009;44(3):127–37.
58. Cassol E, Misra V, Dutta A, Morgello S, Gabuzda D. Cerebrospinal fluid metabolomics reveals altered waste clearance and accelerated aging in HIV patients with neurocognitive impairment. AIDS. 2014;28(11):1579–91.
59. De Cock KM, Jaffe HW, Curran JW. The evolving epidemiology of HIV/ AIDS. AIDS. 2012;26(10):1205–13.
60. Goldstone DC, Ennis-Adeniran V, Hedden JJ, Groom HC, Rice GI, Christodoulou E, et al. HIV-1 restriction factor SAMHD1 is a deoxynucleoside triphosphate triphosphohydrolase. Nature. 2011;480(7377):379–82.
61. Hollenbaugh JA, Munger J, Kim B. Metabolite profiles of human immunodeficiency virus infected CD4+ T cells and macrophages using LC-MS/MS analysis. Virology. 2011;415(2):153–9.
62. Jenabian MA, El-Far M, Vyboh K, Kema I, Costiniuk CT, Thomas R, et al. Immunosuppressive Tryptophan Catabolism and Gut Mucosal Dysfunction Following Early HIV Infection. J Infect Dis. 2015;212(3):355–66.
63. Sitole LJ, Williams AA, Meyer D. Metabonomic analysis of HIV-infected biofluids. Mol Biosyst. 2013;9(1):18–28.
64. Scarpellini B, Zanoni M, Sucupira MC, Truong HM, Janini LM, Segurado ID, et al. Plasma Metabolomics Biosignature According to HIV Stage of Infection, Pace of Disease Progression, Viremia Level and Immunological Response to Treatment. PLoS One. 2016;11(12):e0161920.
65. Williams AA, Sitole LJ, Meyer D. HIV/HAART-associated oxidative stress is detectable by metabonomics. Mol Biosyst. 2017;13(11):2202–17.
66. Bipath P, Levay P, Olorunju S, Viljoen M. A non-specific biomarker of disease activity in HIV/AIDS patients from resource-limited environments. Afr Health Sci. 2015;15(2):334–43.
67. GLOBAL TUBERCULOSIS REPORT 2020: World Health Organization.
68. Tounta V, Liu Y, Cheyne A, Larrouy-Maumus G. Metabolomics in infectious diseases and drug discovery. Mol Omics. 2021;17(3):376–93.
69. Vrieling F, Kostidis S, Spaink HP, Haks MC, Mayboroda OA, Ottenhoff THM, et al. Analyzing the impact of Mycobacterium tuberculosis infection on primary human macrophages by combined exploratory and targeted metabolomics. Sci Rep. 2020;10(1):7085.
70. Cumming BM, Addicott KW, Adamson JH, Steyn AJ. Mycobacterium tuberculosis induces decelerated bioenergetic metabolism in human macrophages. Elife. 2018;7.

71. Borah K, Beyss M, Theorell A, Wu H, Basu P, Mendum TA, et al. Intracellular Mycobacterium tuberculosis Exploits Multiple Host Nitrogen Sources during Growth in Human Macrophages. Cell Rep. 2019;29(11):3580–91 e4.
72. Nazarova EV, Montague CR, Huang L, La T, Russell D, VanderVen BC. The genetic requirements of fatty acid import by Mycobacterium tuberculosis within macrophages. Elife. 2019;8.
73. Huang L, Nazarova EV, Tan S, Liu Y, Russell DG. Growth of Mycobacterium tuberculosis in vivo segregates with host macrophage metabolism and ontogeny. J Exp Med. 2018;215(4):1135–52.
74. Marrero J, Rhee KY, Schnappinger D, Pethe K, Ehrt S. Gluconeogenic carbon flow of tricarboxylic acid cycle intermediates is critical for Mycobacterium tuberculosis to establish and maintain infection. Proc Natl Acad Sci U S A. 2010;107(21):9819–24.
75. de Carvalho LP, Fischer SM, Marrero J, Nathan C, Ehrt S, Rhee KY. Metabolomics of Mycobacterium tuberculosis reveals compartmentalized co-catabolism of carbon substrates. Chem Biol. 2010;17(10):1122–31.
76. Fernandez-Garcia M, Rey-Stolle F, Boccard J, Reddy VP, Garcia A, Cumming BM, et al. Comprehensive Examination of the Mouse Lung Metabolome Following Mycobacterium tuberculosis Infection Using a Multiplatform Mass Spectrometry Approach. J Proteome Res. 2020;19(5):2053–70.
77. du Preez I, Loots DT. New sputum metabolite markers implicating adaptations of the host to Mycobacterium tuberculosis, and vice versa. Tuberculosis (Edinb). 2013;93(3):330–7.
78. Frediani JK, Jones DP, Tukvadze N, Uppal K, Sanikidze E, Kipiani M, et al. Plasma metabolomics in human pulmonary tuberculosis disease: a pilot study. PLoS One. 2014;9(10):e108854.
79. Weiner J, 3rd, Maertzdorf J, Sutherland JS, Duffy FJ, Thompson E, Suliman S, et al. Metabolite changes in blood predict the onset of tuberculosis. Nat Commun. 2018;9(1):5208.
80. Isa F, Collins S, Lee MH, Decome D, Dorvil N, Joseph P, et al. Mass Spectrometric Identification of Urinary Biomarkers of Pulmonary Tuberculosis. EBioMedicine. 2018;31:157–65.
81. Cho Y, Park Y, Sim B, Kim J, Lee H, Cho SN, et al. Identification of serum biomarkers for active pulmonary tuberculosis using a targeted metabolomics approach. Sci Rep. 2020;10(1):3825.
82. Collins JM, Siddiqa A, Jones DP, Liu K, Kempker RR, Nizam A, et al. Tryptophan catabolism reflects disease activity in human tuberculosis. JCI Insight. 2020;5(10).
83. Duffy FJ, Weiner J, 3rd, Hansen S, Tabb DL, Suliman S, Thompson E, et al. Immunometabolic Signatures Predict Risk of Progression to Active Tuberculosis and Disease Outcome. Front Immunol. 2019;10:527.
84. Lee W, VanderVen BC, Fahey RJ, Russell DG. Intracellular Mycobacterium tuberculosis exploits host-derived fatty acids to limit metabolic stress. J Biol Chem. 2013;288(10):6788–800.
85. Kinsella RJ, Fitzpatrick DA, Creevey CJ, McInerney JO. Fatty acid biosynthesis in Mycobacterium tuberculosis: lateral gene transfer, adaptive evolution, and gene duplication. Proc Natl Acad Sci U S A. 2003;100(18):10320–5.
86. Warburg O, Wind F, Negelein E. The Metabolism of Tumors in the Body. J Gen Physiol. 1927;8(6):519–30.
87. Warburg O. On the origin of cancer cells. Science. 1956;123(3191):309–14.
88. Vander Heiden MG. Targeting cancer metabolism: a therapeutic window opens. Nat Rev Drug Discov. 2011;10(9):671–84.
89. Gatenby RA, Gillies RJ. Why do cancers have high aerobic glycolysis? Nat Rev Cancer. 2004;4(11):891–9.
90. Beger RD. A review of applications of metabolomics in cancer. Metabolites. 2013;3(3):552–74.
91. Perroud B, Lee J, Valkova N, Dhirapong A, Lin PY, Fiehn O, et al. Pathway analysis of kidney cancer using proteomics and metabolic profiling. Mol Cancer. 2006;5:64.
92. Ding M, Li F, Wang B, Chi G, Liu H. A comprehensive analysis of WGCNA and serum metabolomics manifests the lung cancer-associated disordered glucose metabolism. J Cell Biochem. 2019;120(6):10855–63.

93. Maldonado R, Talana CA, Song C, Dixon A, Uehara K, Weichhaus M. beta-hydroxybutyrate does not alter the effects of glucose deprivation on breast cancer cells. Oncol Lett. 2021;21(1):65.
94. Wong TL, Ng KY, Tan KV, Chan LH, Zhou L, Che N, et al. CRAF Methylation by PRMT6 Regulates Aerobic Glycolysis-Driven Hepatocarcinogenesis via ERK-Dependent PKM2 Nuclear Relocalization and Activation. Hepatology. 2020;71(4):1279–96.
95. Han J, Li Q, Chen Y, Yang Y. Recent Metabolomics Analysis in Tumor Metabolism Reprogramming. Front Mol Biosci. 2021;8:763902.
96. Currie E, Schulze A, Zechner R, Walther TC, Farese RV, Jr. Cellular fatty acid metabolism and cancer. Cell Metab. 2013;18(2):153–61.
97. Peng X, Chen Z, Farshidfar F, Xu X, Lorenzi PL, Wang Y, et al. Molecular Characterization and Clinical Relevance of Metabolic Expression Subtypes in Human Cancers. Cell Rep. 2018;23(1):255–69 e4.
98. Ganti S, Taylor SL, Abu Aboud O, Yang J, Evans C, Osier MV, et al. Kidney tumor biomarkers revealed by simultaneous multiple matrix metabolomics analysis. Cancer Res. 2012;72(14):3471–9.
99. Stuart E, Buchert M, Putoczki T, Thiem S, Farid R, Elzer J, et al. Therapeutic inhibition of Jak activity inhibits progression of gastrointestinal tumors in mice. Mol Cancer Ther. 2014;13(2):468–74.
100. Abu Aboud O, Donohoe D, Bultman S, Fitch M, Riiff T, Hellerstein M, et al. PPARalpha inhibition modulates multiple reprogrammed metabolic pathways in kidney cancer and attenuates tumor growth. Am J Physiol Cell Physiol. 2015;308(11):C890–8.
101. Traversari C, Sozzani S, Steffensen KR, Russo V. LXR-dependent and -independent effects of oxysterols on immunity and tumor growth. Eur J Immunol. 2014;44(7):1896–903.
102. Yi M, Li J, Chen S, Cai J, Ban Y, Peng Q, et al. Correction to: Emerging role of lipid metabolism alterations in Cancer stem cells. J Exp Clin Cancer Res. 2018;37(1):155.
103. Yip-Schneider MT, Simpson R, Carr RA, Wu H, Fan H, Liu Z, et al. Circulating Leptin and Branched Chain Amino Acids-Correlation with Intraductal Papillary Mucinous Neoplasm Dysplastic Grade. J Gastrointest Surg. 2019;23(5):966–74.
104. Coloff JL, Murphy JP, Braun CR, Harris IS, Shelton LM, Kami K, et al. Differential Glutamate Metabolism in Proliferating and Quiescent Mammary Epithelial Cells. Cell Metab. 2016;23(5):867–80.
105. Wang Y, Bai C, Ruan Y, Liu M, Chu Q, Qiu L, et al. Coordinative metabolism of glutamine carbon and nitrogen in proliferating cancer cells under hypoxia. Nat Commun. 2019;10(1):201.
106. Spinelli JB, Yoon H, Ringel AE, Jeanfavre S, Clish CB, Haigis MC. Metabolic recycling of ammonia via glutamate dehydrogenase supports breast cancer biomass. Science. 2017;358(6365):941–6.
107. Zhang Y, Gao Y, Li Y, Zhang X, Xie H. Characterization of the Relationship Between the Expression of Aspartate beta-Hydroxylase and the Pathological Characteristics of Breast Cancer. Med Sci Monit. 2020;26:e926752.
108. Chen JL, Tang HQ, Hu JD, Fan J, Hong J, Gu JZ. Metabolomics of gastric cancer metastasis detected by gas chromatography and mass spectrometry. World J Gastroenterol. 2010;16(46):5874–80.
109. Tabe Y, Lorenzi PL, Konopleva M. Amino acid metabolism in hematologic malignancies and the era of targeted therapy. Blood. 2019;134(13):1014–23.
110. Lu J, Hu S, Miccoli P, Zeng Q, Liu S, Ran L, et al. Non-invasive diagnosis of papillary thyroid microcarcinoma: a NMR-based metabolomics approach. Oncotarget. 2016;7(49):81768–77.
111. Luo X, Liu J, Wang H, Lu H. Metabolomics identified new biomarkers for the precise diagnosis of pancreatic cancer and associated tissue metastasis. Pharmacol Res. 2020;156:104805.
112. Tao L, Zhou J, Yuan C, Zhang L, Li D, Si D, et al. Metabolomics identifies serum and exosomes metabolite markers of pancreatic cancer. Metabolomics. 2019;15(6):86.

113. Xu H, Liu R, He B, Bi CW, Bi K, Li Q. Polyamine Metabolites Profiling for Characterization of Lung and Liver Cancer Using an LC-Tandem MS Method with Multiple Statistical Data Mining Strategies: Discovering Potential Cancer Biomarkers in Human Plasma and Urine. Molecules. 2016;21(8).
114. His M, Viallon V, Dossus L, Gicquiau A, Achaintre D, Scalbert A, et al. Prospective analysis of circulating metabolites and breast cancer in EPIC. BMC Med. 2019;17(1):178.
115. Huang Z, Lin L, Gao Y, Chen Y, Yan X, Xing J, et al. Bladder cancer determination via two urinary metabolites: a biomarker pattern approach. Mol Cell Proteomics. 2011;10(10):M111 007922.
116. Dinges SS, Hohm A, Vandergrift LA, Nowak J, Habbel P, Kaltashov IA, et al. Cancer metabolomic markers in urine: evidence, techniques and recommendations. Nat Rev Urol. 2019;16(6):339–62.
117. Omran MM, Rashed RE, Darwish H, Belal AA, Mohamed FZ. Development of a gas chromatography-mass spectrometry method for breast cancer diagnosis based on nucleoside metabolomes 1-methyl adenosine, 1-methylguanosine and 8-hydroxy-2'-deoxyguanosine. Biomed Chromatogr. 2020;34(1):e4713.
118. Locasale JW, Melman T, Song S, Yang X, Swanson KD, Cantley LC, et al. Metabolomics of human cerebrospinal fluid identifies signatures of malignant glioma. Mol Cell Proteomics. 2012;11(6):M111 014688.
119. Ballester LY, Lu G, Zorofchian S, Vantaku V, Putluri V, Yan Y, et al. Analysis of cerebrospinal fluid metabolites in patients with primary or metastatic central nervous system tumors. Acta Neuropathol Commun. 2018;6(1):85.
120. Sugimoto M, Wong DT, Hirayama A, Soga T, Tomita M. Capillary electrophoresis mass spectrometry-based saliva metabolomics identified oral, breast and pancreatic cancer-specific profiles. Metabolomics. 2010;6(1):78–95.
121. Ishikawa S, Sugimoto M, Kitabatake K, Sugano A, Nakamura M, Kaneko M, et al. Identification of salivary metabolomic biomarkers for oral cancer screening. Sci Rep. 2016;6:31520.
122. Asai Y, Itoi T, Sugimoto M, Sofuni A, Tsuchiya T, Tanaka R, et al. Elevated Polyamines in Saliva of Pancreatic Cancer. Cancers (Basel). 2018;10(2).
123. Murata T, Yanagisawa T, Kurihara T, Kaneko M, Ota S, Enomoto A, et al. Salivary metabolomics with alternative decision tree-based machine learning methods for breast cancer discrimination. Breast Cancer Res Treat. 2019;177(3):591–601.
124. Botas A, Campbell HM, Han X, Maletic-Savatic M. Metabolomics of neurodegenerative diseases. Int Rev Neurobiol. 2015;122:53–80.
125. Duraes F, Pinto M, Sousa E. Old Drugs as New Treatments for Neurodegenerative Diseases. Pharmaceuticals (Basel). 2018;11(2).
126. Zhang AH, Sun H, Wang XJ. Recent advances in metabolomics in neurological disease, and future perspectives. Anal Bioanal Chem. 2013b;405(25):8143–50.
127. Goodpaster AM, Romick-Rosendale LE, Kennedy MA. Statistical significance analysis of nuclear magnetic resonance-based metabonomics data. Anal Biochem. 2010;401(1):134–43.
128. Hardiman O, Al-Chalabi A, Chio A, Corr EM, Logroscino G, Robberecht W, et al. Amyotrophic lateral sclerosis. Nat Rev Dis Primers. 2017;3:17071.
129. Pradat PF, Dib M. Biomarkers in amyotrophic lateral sclerosis: facts and future horizons. Mol Diagn Ther. 2009;13(2):115–25.
130. Wuolikainen A, Andersen PM, Moritz T, Marklund SL, Antti H. ALS patients with mutations in the SOD1 gene have an unique metabolomic profile in the cerebrospinal fluid compared with ALS patients without mutations. Mol Genet Metab. 2012;105(3):472–8.
131. Wuolikainen A, Moritz T, Marklund SL, Antti H, Andersen PM. Disease-related changes in the cerebrospinal fluid metabolome in amyotrophic lateral sclerosis detected by GC/TOFMS. PLoS One. 2011;6(4):e17947.
132. Blasco H, Corcia P, Moreau C, Veau S, Fournier C, Vourc'h P, et al. 1H-NMR-based metabolomic profiling of CSF in early amyotrophic lateral sclerosis. PLoS One. 2010;5(10):e13223.

133. Blasco H, Corcia P, Pradat PF, Bocca C, Gordon PH, Veyrat-Durebex C, et al. Metabolomics in cerebrospinal fluid of patients with amyotrophic lateral sclerosis: an untargeted approach via high-resolution mass spectrometry. J Proteome Res. 2013;12(8):3746–54.
134. Kumar A, Bala L, Kalita J, Misra UK, Singh RL, Khetrapal CL, et al. Metabolomic analysis of serum by (1) H NMR spectroscopy in amyotrophic lateral sclerosis. Clin Chim Acta. 2010;411(7–8):563–7.
135. Bjornevik K, Zhang Z, O'Reilly EJ, Berry JD, Clish CB, Deik A, et al. Prediagnostic plasma metabolomics and the risk of amyotrophic lateral sclerosis. Neurology. 2019;92(18):e2089–e100.
136. Palamiuc L, Schlagowski A, Ngo ST, Vernay A, Dirrig-Grosch S, Henriques A, et al. A metabolic switch toward lipid use in glycolytic muscle is an early pathologic event in a mouse model of amyotrophic lateral sclerosis. EMBO Mol Med. 2015;7(5):526–46.
137. Schonfeld P, Reiser G. Why does brain metabolism not favor burning of fatty acids to provide energy? Reflections on disadvantages of the use of free fatty acids as fuel for brain. J Cereb Blood Flow Metab. 2013;33(10):1493–9.
138. Tracey TJ, Steyn FJ, Wolvetang EJ, Ngo ST. Neuronal Lipid Metabolism: Multiple Pathways Driving Functional Outcomes in Health and Disease. Front Mol Neurosci. 2018;11:10.
139. Cacabelos D, Ayala V, Granado-Serrano AB, Jove M, Torres P, Boada J, et al. Interplay between TDP-43 and docosahexaenoic acid-related processes in amyotrophic lateral sclerosis. Neurobiol Dis. 2016;88:148–60.
140. Francis PT, Palmer AM, Snape M, Wilcock GK. The cholinergic hypothesis of Alzheimer's disease: a review of progress. J Neurol Neurosurg Psychiatry. 1999;66(2):137–47.
141. Lane CA, Hardy J, Schott JM. Alzheimer's disease. Eur J Neurol. 2018;25(1):59–70.
142. Ginsberg L, Rafique S, Xuereb JH, Rapoport SI, Gershfeld NL. Disease and anatomic specificity of ethanolamine plasmalogen deficiency in Alzheimer's disease brain. Brain Res. 1995;698(1–2):223–6.
143. Han X, Holtzman DM, McKeel DW, Jr. Plasmalogen deficiency in early Alzheimer's disease subjects and in animal models: molecular characterization using electrospray ionization mass spectrometry. J Neurochem. 2001;77(4):1168–80.
144. Han X. Lipid alterations in the earliest clinically recognizable stage of Alzheimer's disease: implication of the role of lipids in the pathogenesis of Alzheimer's disease. Curr Alzheimer Res. 2005;2(1):65–77.
145. Goodenowe DB, Cook LL, Liu J, Lu Y, Jayasinghe DA, Ahiahonu PW, et al. Peripheral ethanolamine plasmalogen deficiency: a logical causative factor in Alzheimer's disease and dementia. J Lipid Res. 2007;48(11):2485–98.
146. Wood PL, Mankidy R, Ritchie S, Heath D, Wood JA, Flax J, et al. Circulating plasmalogen levels and Alzheimer Disease Assessment Scale-Cognitive scores in Alzheimer patients. J Psychiatry Neurosci. 2010;35(1):59–62.
147. Brites P, Waterham HR, Wanders RJ. Functions and biosynthesis of plasmalogens in health and disease. Biochim Biophys Acta. 2004;1636(2–3):219–31.
148. Oresic M, Hyotylainen T, Herukka SK, Sysi-Aho M, Mattila I, Seppanan-Laakso T, et al. Metabolome in progression to Alzheimer's disease. Transl Psychiatry. 2011;1:e57.
149. Xu XH, Huang Y, Wang G, Chen SD. Metabolomics: a novel approach to identify potential diagnostic biomarkers and pathogenesis in Alzheimer's disease. Neurosci Bull. 2012;28(5):641–8.
150. Mielke MM, Bandaru VV, Haughey NJ, Rabins PV, Lyketsos CG, Carlson MC. Serum sphingomyelins and ceramides are early predictors of memory impairment. Neurobiol Aging. 2010;31(1):17–24.
151. Bernath MM, Bhattacharyya S, Nho K, Barupal DK, Fiehn O, Baillie R, et al. Serum triglycerides in Alzheimer disease: Relation to neuroimaging and CSF biomarkers. Neurology. 2020;94(20):e2088–e98.

152. Chouraki V, Preis SR, Yang Q, Beiser A, Li S, Larson MG, et al. Association of amine biomarkers with incident dementia and Alzheimer's disease in the Framingham Study. Alzheimers Dement. 2017;13(12):1327–36.
153. van der Lee SJ, Teunissen CE, Pool R, Shipley MJ, Teumer A, Chouraki V, et al. Circulating metabolites and general cognitive ability and dementia: Evidence from 11 cohort studies. Alzheimers Dement. 2018;14(6):707–22.
154. Toledo JB, Arnold M, Kastenmuller G, Chang R, Baillie RA, Han X, et al. Metabolic network failures in Alzheimer's disease: A biochemical road map. Alzheimers Dement. 2017;13(9):965–84.
155. Tynkkynen J, Chouraki V, van der Lee SJ, Hernesniemi J, Yang Q, Li S, et al. Association of branched-chain amino acids and other circulating metabolites with risk of incident dementia and Alzheimer's disease: A prospective study in eight cohorts. Alzheimers Dement. 2018;14(6):723–33.
156. Liguori C, Stefani A, Sancesario G, Sancesario GM, Marciani MG, Pierantozzi M. CSF lactate levels, tau proteins, cognitive decline: a dynamic relationship in Alzheimer's disease. J Neurol Neurosurg Psychiatry. 2015;86(6):655–9.
157. Cheng Y, Bai F. The Association of Tau With Mitochondrial Dysfunction in Alzheimer's Disease. Front Neurosci. 2018;12:163.
158. Parnetti L, Reboldi GP, Gallai V. Cerebrospinal fluid pyruvate levels in Alzheimer's disease and vascular dementia. Neurology. 2000;54(3):735–7.
159. Niedzwiecki MM, Walker DI, Howell JC, Watts KD, Jones DP, Miller GW, et al. High-resolution metabolomic profiling of Alzheimer's disease in plasma. Ann Clin Transl Neurol. 2020;7(1):36–45.
160. Bonomo R, Cavaletti G, Skene DJ. Metabolomics markers in Neurology: current knowledge and future perspectives for therapeutic targeting. Expert Rev Neurother. 2020;20(7):725–38.
161. Mittal R, Gupta RL. In vitro antioxidant activity of piperine. Methods Find Exp Clin Pharmacol. 2000;22(5):271–4.
162. Mujumdar AM, Dhuley JN, Deshmukh VK, Raman PH, Naik SR. Anti-inflammatory activity of piperine. Jpn J Med Sci Biol. 1990;43(3):95–100.
163. Murata K, Matsumura S, Yoshioka Y, Ueno Y, Matsuda H. Screening of beta-secretase and acetylcholinesterase inhibitors from plant resources. J Nat Med. 2015;69(1):123–9.
164. Wilkins JM, Trushina E. Application of Metabolomics in Alzheimer's Disease. Front Neurol. 2017;8:719.
165. Bhargava P, Anthony DC. Metabolomics in multiple sclerosis disease course and progression. Mult Scler. 2020;26(5):591–8.
166. Donatti A, Canto AM, Godoi AB, da Rosa DC, Lopes-Cendes I. Circulating Metabolites as Potential Biomarkers for Neurological Disorders-Metabolites in Neurological Disorders. Metabolites. 2020;10(10).
167. Lynch J, Peeling J, Auty A, Sutherland GR. Nuclear magnetic resonance study of cerebrospinal fluid from patients with multiple sclerosis. Can J Neurol Sci. 1993;20(3):194–8.
168. Simone IL, Federico F, Trojano M, Tortorella C, Liguori M, Giannini P, et al. High resolution proton MR spectroscopy of cerebrospinal fluid in MS patients. Comparison with biochemical changes in demyelinating plaques. J Neurol Sci. 1996;144(1–2):182–90.
169. Reinke SN, Broadhurst DL, Sykes BD, Baker GB, Catz I, Warren KG, et al. Metabolomic profiling in multiple sclerosis: insights into biomarkers and pathogenesis. Mult Scler. 2014;20(10):1396–400.
170. Kim HH, Jeong IH, Hyun JS, Kong BS, Kim HJ, Park SJ. Metabolomic profiling of CSF in multiple sclerosis and neuromyelitis optica spectrum disorder by nuclear magnetic resonance. PLoS One. 2017;12(7):e0181758.
171. Herman S, Akerfeldt T, Spjuth O, Burman J, Kultima K. Biochemical Differences in Cerebrospinal Fluid between Secondary Progressive and Relapsing(-)Remitting Multiple Sclerosis. Cells. 2019a;8(2).

172. Mehrpour M, Kyani A, Tafazzoli M, Fathi F, Joghataie MT. A metabonomics investigation of multiple sclerosis by nuclear magnetic resonance. Magn Reson Chem. 2013;51(2):102–9.
173. Moussallieh FM, Elbayed K, Chanson JB, Rudolf G, Piotto M, De Seze J, et al. Serum analysis by 1H nuclear magnetic resonance spectroscopy: a new tool for distinguishing neuromyelitis optica from multiple sclerosis. Mult Scler. 2014;20(5):558–65.
174. Villoslada P, Alonso C, Agirrezabal I, Kotelnikova E, Zubizarreta I, Pulido-Valdeolivas I, et al. Metabolomic signatures associated with disease severity in multiple sclerosis. Neurol Neuroimmunol Neuroinflamm. 2017;4(2):e321.
175. Stoessel D, Stellmann JP, Willing A, Behrens B, Rosenkranz SC, Hodecker SC, et al. Metabolomic Profiles for Primary Progressive Multiple Sclerosis Stratification and Disease Course Monitoring. Front Hum Neurosci. 2018;12:226.
176. Del Boccio P, Pieragostino D, Di Ioia M, Petrucci F, Lugaresi A, De Luca G, et al. Lipidomic investigations for the characterization of circulating serum lipids in multiple sclerosis. J Proteomics. 2011;74(12):2826–36.
177. Miller LG, Jr., Young JA, Ray SK, Wang G, Purohit S, Banik NL, et al. Sphingosine Toxicity in EAE and MS: Evidence for Ceramide Generation via Serine-Palmitoyltransferase Activation. Neurochem Res. 2017;42(10):2755–68.
178. Nogueras L, Gonzalo H, Jove M, Sol J, Gil-Sanchez A, Hervas JV, et al. Lipid profile of cerebrospinal fluid in multiple sclerosis patients: a potential tool for diagnosis. Sci Rep 2019;9(1):11313.
179. Gonzalo H, Brieva L, Tatzber F, Jove M, Cacabelos D, Cassanye A, et al. Lipidome analysis in multiple sclerosis reveals protein lipoxidative damage as a potential pathogenic mechanism. J Neurochem. 2012;123(4):622–34.
180. Mhyre TR, Boyd JT, Hamill RW, Maguire-Zeiss KA. Parkinson's disease. Subcell Biochem. 2012;65:389–455.
181. Beitz JM. Parkinson's disease: a review. Front Biosci (Schol Ed). 2014;6(1):65–74.
182. Bogdanov M, Matson WR, Wang L, Matson T, Saunders-Pullman R, Bressman SS, et al. Metabolomic profiling to develop blood biomarkers for Parkinson's disease. Brain. 2008;131(Pt 2):389–96.
183. Hatano T, Saiki S, Okuzumi A, Mohney RP, Hattori N. Identification of novel biomarkers for Parkinson's disease by metabolomic technologies. J Neurol Neurosurg Psychiatry 2016;87(3):295–301.
184. Shoffner JM, Watts RL, Juncos JL, Torroni A, Wallace DC. Mitochondrial oxidative phosphorylation defects in Parkinson's disease. Ann Neurol. 1991;30(3):332–9.
185. Farmer K, Smith CA, Hayley S, Smith J. Major Alterations of Phosphatidylcholine and Lysophosphotidylcholine Lipids in the Substantia Nigra Using an Early Stage Model of Parkinson's Disease. Int J Mol Sci. 2015;16(8):18865–77.
186. Burte F, Houghton D, Lowes H, Pyle A, Nesbitt S, Yarnall A, et al. metabolic profiling of Parkinson's disease and mild cognitive impairment. Mov Disord. 2017;32(6):927–32.
187. Luan H, Liu LF, Meng N, Tang Z, Chua KK, Chen LL, et al. LC-MS-based urinary metabolite signatures in idiopathic Parkinson's disease. J Proteome Res. 2015a;14(1):467–78.
188. Luan H, Liu LF, Tang Z, Zhang M, Chua KK, Song JX, et al. Comprehensive urinary metabolomic profiling and identification of potential noninvasive marker for idiopathic Parkinson's disease. Sci Rep. 2015b;5:13888.
189. Valerio A, D'Antona G, Nisoli E. Branched-chain amino acids, mitochondrial biogenesis, and healthspan: an evolutionary perspective. Aging (Albany NY). 2011;3(5):464–78.
190. Shao Y, Le W. Recent advances and perspectives of metabolomics-based investigations in Parkinson's disease. Mol Neurodegener. 2019;14(1):3.
191. Saudou F, Humbert S. The Biology of Huntingtin. Neuron. 2016;89(5):910–26.
192. Bax BE. Biomarkers in Rare Diseases. Int J Mol Sci. 2021;22(2).
193. Mochel F, Charles P, Seguin F, Barritault J, Coussieu C, Perin L, et al. Early energy deficit in Huntington disease: identification of a plasma biomarker traceable during disease progression. PLoS One. 2007;2(7):e647.

194. Graham SF, Kumar PK, Bjorndahl T, Han B, Yilmaz A, Sherman E, et al. Metabolic signatures of Huntington's disease (HD): (1)H NMR analysis of the polar metabolome in post-mortem human brain. Biochim Biophys Acta. 2016;1862(9):1675–84.
195. Herman S, Niemela V, Emami Khoonsari P, Sundblom J, Burman J, Landtblom AM, et al. Alterations in the tyrosine and phenylalanine pathways revealed by biochemical profiling in cerebrospinal fluid of Huntington's disease subjects. Sci Rep. 2019b;9(1):4129.
196. Cheng ML, Chang KH, Wu YR, Chen CM. Metabolic disturbances in plasma as biomarkers for Huntington's disease. J Nutr Biochem. 2016;31:38–44.
197. Verwaest KA, Vu TN, Laukens K, Clemens LE, Nguyen HP, Van Gasse B, et al. (1)H NMR based metabolomics of CSF and blood serum: a metabolic profile for a transgenic rat model of Huntington disease. Biochim Biophys Acta. 2011;1812(11):1371–9.
198. Chiang MC, Chen HM, Lee YH, Chang HH, Wu YC, Soong BW, et al. Dysregulation of C/EBPalpha by mutant Huntingtin causes the urea cycle deficiency in Huntington's disease. Hum Mol Genet. 2007;16(5):483–98.
199. Chen CM, Lin YS, Wu YR, Chen P, Tsai FJ, Yang CL, et al. High Protein Diet and Huntington's Disease. PLoS One. 2015;10(5):e0127654.
200. Skene DJ, Middleton B, Fraser CK, Pennings JL, Kuchel TR, Rudiger SR, et al. Metabolic profiling of presymptomatic Huntington's disease sheep reveals novel biomarkers. Sci Rep. 2017;7:43030.
201. Tang WH, Hazen SL. The contributory role of gut microbiota in cardiovascular disease. J Clin Invest. 2014;124(10):4204–11.
202. Zhu W, Gregory JC, Org E, Buffa JA, Gupta N, Wang Z, et al. Gut Microbial Metabolite TMAO Enhances Platelet Hyperreactivity and Thrombosis Risk. Cell. 2016;165(1):111–24.
203. Wang Z, Klipfell E, Bennett BJ, Koeth R, Levison BS, Dugar B, et al. Gut flora metabolism of phosphatidylcholine promotes cardiovascular disease. Nature. 2011;472(7341):57–63.
204. Wang Z, Roberts AB, Buffa JA, Levison BS, Zhu W, Org E, et al. Non-lethal Inhibition of Gut Microbial Trimethylamine Production for the Treatment of Atherosclerosis. Cell. 2015;163(7):1585–95.
205. Park JY, Lee SH, Shin MJ, Hwang GS. Alteration in metabolic signature and lipid metabolism in patients with angina pectoris and myocardial infarction. PLoS One. 2015;10(8):e0135228.
206. Ganna A, Salihovic S, Sundstrom J, Broeckling CD, Hedman AK, Magnusson PK, et al. Large-scale metabolomic profiling identifies novel biomarkers for incident coronary heart disease. PLoS Genet. 2014;10(12):e1004801.
207. Stanley WC, Lopaschuk GD, Hall JL, McCormack JG. Regulation of myocardial carbohydrate metabolism under normal and ischaemic conditions. Potential for pharmacological interventions. Cardiovasc Res. 1997;33(2):243–57.
208. Wisneski JA, Gertz EW, Neese RA, Mayr M. Myocardial metabolism of free fatty acids. Studies with 14C-labeled substrates in humans. J Clin Invest. 1987;79(2):359–66.
209. Vary TC, Reibel DK, Neely JR. Control of energy metabolism of heart muscle. Annu Rev Physiol. 1981;43:419–30.
210. Neely JR, Rovetto MJ, Whitmer JT, Morgan HE. Effects of ischemia on function and metabolism of the isolated working rat heart. Am J Physiol. 1973;225(3):651–8.
211. Turer AT, Stevens RD, Bain JR, Muehlbauer MJ, van der Westhuizen J, Mathew JP, et al. Metabolomic profiling reveals distinct patterns of myocardial substrate use in humans with coronary artery disease or left ventricular dysfunction during surgical ischemia/reperfusion. Circulation. 2009;119(13):1736–46.
212. Lewis GD, Wei R, Liu E, Yang E, Shi X, Martinovic M, et al. Metabolite profiling of blood from individuals undergoing planned myocardial infarction reveals early markers of myocardial injury. J Clin Invest. 2008;118(10):3503–12.
213. Li F, Xu Q, Zheng T, Huang F, Han L. Metabonomic analysis of Allium macrostemon Bunge as a treatment for acute myocardial ischemia in rats. J Pharm Biomed Anal. 2014;88:225–34.
214. Mohamad N, Ismet RI, Rofiee M, Bannur Z, Hennessy T, Selvaraj M, et al. Metabolomics and partial least square discriminant analysis to predict history of myocardial infarction of

self-claimed healthy subjects: validity and feasibility for clinical practice. J Clin Bioinforma. 2015;5:3.
215. Wang X, Wang D, Wu J, Yu X, Lv J, Kong J, et al. Metabolic Characterization of Myocardial Infarction Using GC-MS-Based Tissue Metabolomics. Int Heart J. 2017;58(3):441–6.
216. Chouchani ET, Pell VR, Gaude E, Aksentijevic D, Sundier SY, Robb EL, et al. Ischaemic accumulation of succinate controls reperfusion injury through mitochondrial ROS. Nature. 2014;515(7527):431–5.
217. Yang RY, Wang SM, Sun L, Liu JM, Li HX, Sui XF, et al. Association of branched-chain amino acids with coronary artery disease: A matched-pair case-control study. Nutr Metab Cardiovasc Dis. 2015b;25(10):937–42.
218. Camici P, Marraccini P, Lorenzoni R, Ferrannini E, Buzzigoli G, Marzilli M, et al. Metabolic markers of stress-induced myocardial ischemia. Circulation. 1991;83(5 Suppl):III8–13.
219. Minamisawa M, Motoki H, Izawa A, Kashima Y, Hioki H, Abe N, et al. Comparison of Inflammatory Biomarkers in Outpatients With Prior Myocardial Infarction. Int Heart J. 2016;57(1):11–7.
220. Shah SH, Sun JL, Stevens RD, Bain JR, Muehlbauer MJ, Pieper KS, et al. Baseline metabolomic profiles predict cardiovascular events in patients at risk for coronary artery disease. Am Heart J. 2012;163(5):844–50 e1.
221. Lloyd SG, Wang P, Zeng H, Chatham JC. Impact of low-flow ischemia on substrate oxidation and glycolysis in the isolated perfused rat heart. Am J Physiol Heart Circ Physiol. 2004;287(1):H351–62.
222. Ali SE, Farag MA, Holvoet P, Hanafi RS, Gad MZ. A Comparative Metabolomics Approach Reveals Early Biomarkers for Metabolic Response to Acute Myocardial Infarction. Sci Rep. 2016;6:36359.
223. Fan Y, Li Y, Chen Y, Zhao YJ, Liu LW, Li J, et al. Comprehensive Metabolomic Characterization of Coronary Artery Diseases. J Am Coll Cardiol. 2016;68(12):1281–93.
224. Zhu M, Han Y, Zhang Y, Zhang S, Wei C, Cong Z, et al. Metabolomics Study of the Biochemical Changes in the Plasma of Myocardial Infarction Patients. Front Physiol. 2018;9:1017.
225. Hjelte LE, Nilsson A. Arachidonic acid and ischemic heart disease. J Nutr. 2005;135(9):2271–3.
226. Watterson K, Sankala H, Milstien S, Spiegel S. Pleiotropic actions of sphingosine-1-phosphate. Prog Lipid Res. 2003;42(4):344–57.

Nutrimetabolomics: Metabolomics in Nutrition Research

Upasna Srivastava, Swarna Kanchan, Minu Kesheri, and Satendra Singh

Abbreviations

FAO	Food and Agriculture Organization
FFQs	Food frequency questionnaires
GNPS	Global Natural Products Social Molecular Networking
IBRD	International Bank for Reconstruction and Development
MS-DIAL	Multiple-Stage Mass Spectrometry-Based Data Independent Acquisition Library
OPLS	Orthogonal partial least squares
PLS-DA	Partial least squares discriminant analysis
SIM	Selected ion monitoring
WHO	World Health Organization
XCMS	XCentric mass spectrometry
XCMS	X-chromatography/mass spectrometry

U. Srivastava (✉)
University of California, San Diego, CA, USA

School of Medicine, Yale University, New Haven, CT, USA
e-mail: upasna.srivastava@yale.edu

S. Kanchan
INRIA, Bordeaux Sud-Ouest, Talence, France

Boise State University, Boise, ID, USA

M. Kesheri
Boise State University, Boise, ID, USA

S. Singh
Department of Computational Biology and Bioinformatics JSSB, Sam Higginbottom Institute of Agriculture, Technology and Sciences (Formerly Allahabad Agriculture Institute), Allahabad, Uttar Pradesh, India

V. Soni, T. E. Hartman (eds.), *Metabolomics*,
https://doi.org/10.1007/978-3-031-39094-4_8

1 Introduction

Life sciences are currently transformed by the developments in molecular biology-based analysis [55, 56], which include important advances in the instrumental analysis as well bioinformatics analysis (including machine learning approaches) of genes, RNA, proteins, metabolites, etc. [57–61, 63, 64] and drug development [54, 62, 66], revolutionizing the pharma [65]; and life sciences industries. Now the focus has broadened unlike to drug molecules, called as metabolites, because the molecular profile of the metabolites represents the actual healthy/unhealthy condition of a human. The modern age employs diverse strategies to handle various forms of single-cell omics methodologies, including computational pipelines for transcriptomics such as sc-RNA sequencing, metabolomics, and epigenomics like scATAC. Additionally, it focuses on the integration of omics techniques, methodologies, and the associated challenges within the context of crop plants. According to Srivastava and Singh [67], nutrimetabolomics is a further extension of metabolomics which represents the application of metabolomics in nutrition research and investigates the effects of whole diets, specific foods, and food components on the human metabolome. Nutrimetabolomics is the study of the metabolic responses of biological systems to nutritional interventions. It combines the fields of nutrition, metabolomics, and systems biology to investigate the impact of nutrients on biochemical pathways, gene expression, and overall physiological responses. It can help identify the specific metabolic pathways that are influenced by different nutrients, providing insights into how nutrients are processed by the body and how they contribute to the overall health. Besides this by analyzing the metabolites produced in response to different nutrients, nutrimetabolomics can identify biomarkers that can be used to monitor nutrient status and predict health outcomes. With recent advancements, nutrimetabolomics is found to be very helpful in identifying individual differences in nutrient metabolism and in defining personalized nutrition strategies that are tailored to an individual's unique metabolic profile. Interestingly, experts are not using this methodology to identify the bioactive compounds in foods and supplements that have specific health benefits and to develop new functional foods and supplements that target specific metabolic pathways. Undoubtedly, nutrimetabolomics is playing an important role in advancing our understanding of the complex interactions between nutrition, metabolism, and health. Therefore, in this chapter we have discussed various methodologies of nutrimetabolomics and their applications to identify important dietary biomarkers. We have also discussed the role of machine learning-based approach to develop novel strategies towards precision medicine. Overall, we believe that this chapter would serve as a good starting point to understand the revolutionizing area of the study of nutrition and metabolism, thus called nutrimetabolomics.

2 Metabolomics: A Multidisciplinary Approach

The metabolome is defined as qualitative and quantitative agglomeration of all low-molecular-weight molecules, which are called metabolites, present in the cell which are participants in general metabolic reactions. These metabolites are required for the maintenance, growth, and normal function of a cell. Collectively, metabolites and their interactions within a biological system are termed as the metabolome.

Metabolomics is the large-scale study of all low-molecular-weight molecules, commonly known as metabolites, which are present within cells. Metabolomics has yielded many important insights into biological processes and diseases [1–5]. In other words, metabolomics is the "systematic study of the unique small-molecule metabolite-based chemical fingerprints generated after every cellular process. Moreover, metabolomics can give a blueprint of the physiology of that cell, and thus, metabolomics provides a direct actual condition of the physiological state" of an organism.

There are three different strategies applied to study metabolomics – untargeted, semi-targeted, and targeted. Both untargeted and semi-targeted approaches are applied in hypothesis-generating-based metabolomics studies, while targeted assays are usually applied to validate the novel hypothesis-generating metabolomics studies. The major differences between untargeted, semi-targeted, and targeted studies are based on:

- The level of sample preparation required
- The number of metabolites detected
- The level of quantification of the metabolites

Based on the objectives of the metabolomics studies, one of the above three analytical strategies are applied.

2.1 *Technical Approaches to Nutrimetabolomics and Analysis*

There are generally two main approaches to metabolomic analysis as illustrated in Fig. 1.

2.1.1 Targeted Metabolomics

Targeted metabolomics is applied to study a relatively small and specific number of metabolites, typically up to 20 metabolites. These metabolites are chemically characterized first, followed by biochemical annotation with established biological importance. Targeted metabolomics-based approaches have a greater selectivity and sensitivity compared to untargeted methods. Targeted metabolomics is applied when authenticated chemical standard of the metabolite is available which are used

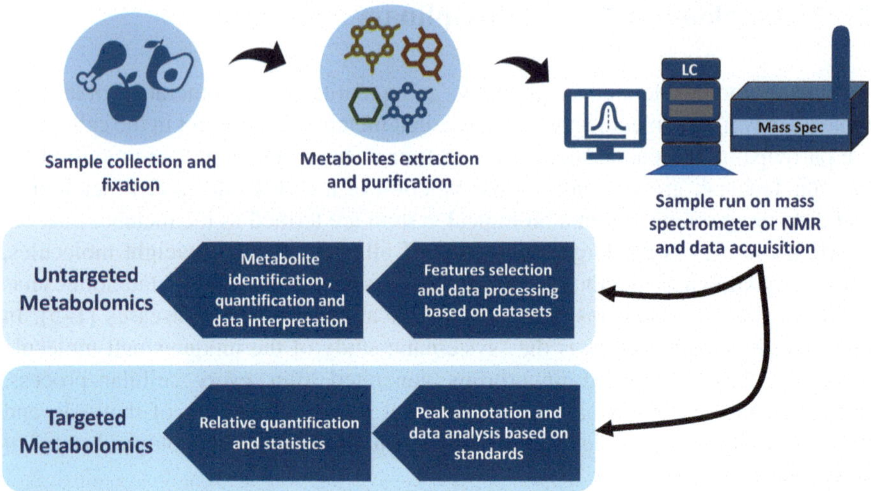

Fig. 1 Approaches to nutrimetabolomics as untargeted and targeted metabolomics strategy pipeline

to construct calibration curves for each of the metabolite under study. In targeted metabolomics-based studies, sample preparation methods are optimized to retain the metabolites of interest and to discard other biological species and analytical artifacts. Here, only the metabolites of interest are carried through to the downstream analysis. Since the number of metabolites under study is less and limited in numbers, therefore, the analysis of the data and interpretation of biological significance are much simpler in targeted metabolomics-based studies.

2.1.2 Untargeted Metabolomics

These approaches provide the most appropriate route to detect unexpected changes in metabolite concentrations; the aim is to maximize the number of metabolites detected and therefore provide the opportunity to observe unexpected changes. In an untargeted metabolomics, hundreds to thousands of metabolites are measured together. However, for measuring a large number of metabolites, a single analytical method is not considered appropriate to detect all of the metabolites in a biological system. It is therefore desirable to combine multiple analytical approaches (such as complementary UPLC methods) to maximize the number of metabolites detected and increase the coverage of the metabolome.

In untargeted metabolomics-based studies, sample preparation involves the extraction of the metabolites from the biological sample using a suitable solvent. The extracted sample is analyzed by an appropriate analytical method, for example, reversed-phase liquid chromatography, mass spectrometry, etc. Liquid chromatography-mass spectrometry-based peaks are used as the parameter in the statistical analysis to signify the concentration differences between the different biological samples measured. This is referred to as relative quantification as there is no comparison with calibration curves that are required for full quantification, constructed with chemical standards. The biological importance of each metabolite is determined through metabolite identification followed by biological interpretation which is performed at the end of the experimental pipeline. The chemical identity of each metabolite in the study is performed after data acquisition.

One of the major limitations in untargeted approaches is the identification of metabolites. Due to the presence of a large number of metabolites considered in untargeted metabolomics, it becomes impossible to identify all the metabolites highlighted in the statistical analysis. Metabolite identification is a major area of concern in metabolomics study-based projects.

2.2 Mass Spectrometry LC-MS Methods

LC-MS analysis and data acquisition strategies are defined by the analytical process which needs to first choose between a targeted and an untargeted approach. In the targeted metabolomics data analysis, the compounds of interest are known, and the entire pipeline is focused towards the detection of these metabolites. However, the range of molecules of interest must also be restricted to something reasonable in the untargeted strategy (e.g., according to polarity, *m/z* range, etc.). On the basis of the properties of the expected molecules and the complexity of the sample, an extraction method (disintegration method, solvents, centrifugation, filtration, etc.) and a separation method need to be defined, after which the ionization method, polarity, and a suitable mass analyzer need to be chosen appropriately.

An automated precursor-ion fragmentation process, also known as data-dependent acquisition (DDA), provides structural information for the identification of compounds. However, conventional DDA methods with MS and alternating MS^2 scans suffer from low efficiencies due to scan-cycle time limitations. Therefore, separate target-directed DDA experiments for the acquisition of fragmentation data provide superior results [6]. In addition, the additional data provided by MS full scan mode (i.e., without intermittent fragmentation scans) deliver a higher data density for quantitative analyses. For the same reason, the sensitivities towards ions can be improved by selecting the ions of interest. In selected ion monitoring (SIM), only the defined ions of interest are recorded, thereby maximizing the sensitivities for those signals. In addition, only MS ions of a defined *m/z* precursor can be acquired in selected reaction monitoring (SRM). Consecutive Reaction Monitoring (CRM) pertains to a multi-stage involving three or more stages of m/z separation. In this approach, the focus is on detecting products resulting from sequential fragmentation or bimolecular reactions. Finally, SRM can be applied to multiple product ions from one or more precursor ions, resulting in multiple reaction monitoring (MRM). Data are collected in either profile or centroid mode. As explained above, centroid spectra contain fewer data points, but are less bulky, and data are acquired faster by the mass analyzer. The acquisition of centroid data is often preferable, especially in time-critical applications such as LC-MS (proteomics and metabolomics) and imaging. For producing spectra of the highest possible quality, the collection of profile spectra is recommended. Multiple sample types (organs, tissues, liquids, etc.), extraction methods (polar, nonpolar, pH, etc.), and analytical strategies (MS, NMR, etc.) need to be combined to obtain a comprehensive picture of the physiology of a complex organism, such as a human being or a plant [7, 8].

Footnote

NMR spectroscopy provides qualitative and quantitative information on different small molecules which could be present in a biological sample, without a prior selection of specific biochemical pathways, thus enabling a broad unbiased approach. Additionally, NMR allows a high-throughput analysis and high reproducibility, and it is an intrinsically quantitative technique over a wide dynamic range, thanks to the linear response of NMR signals within concentration and with the use of just one internal reference compound.

3 Analysis of Nutrimetabolomics Data

The analysis of the acquired data has been divided in various sections. Interestingly, there are various machine learning and statistical analysis-based methods which could be used in the nutrimetabolomics approach such as (1) data analysis and compound identification; (2) transformation, normalization, centering, and scaling; and (3) imputation of missing values. Here, we categorized them as follows in Fig. 2.

3.1 Clustering Analysis

Clustering analysis is an effective method to visualize the metabolomics data where grouping of the metabolites and samples based on similarities and abundance profile in their metabolite is performed. Different types of methods can be used for clustering, such as k-means, hierarchical clustering, fuzzy clustering, density-based clustering, and model-based clustering. It also provides visualization in the form of a heat map. We utilize k-means clustering and hierarchical clustering analysis

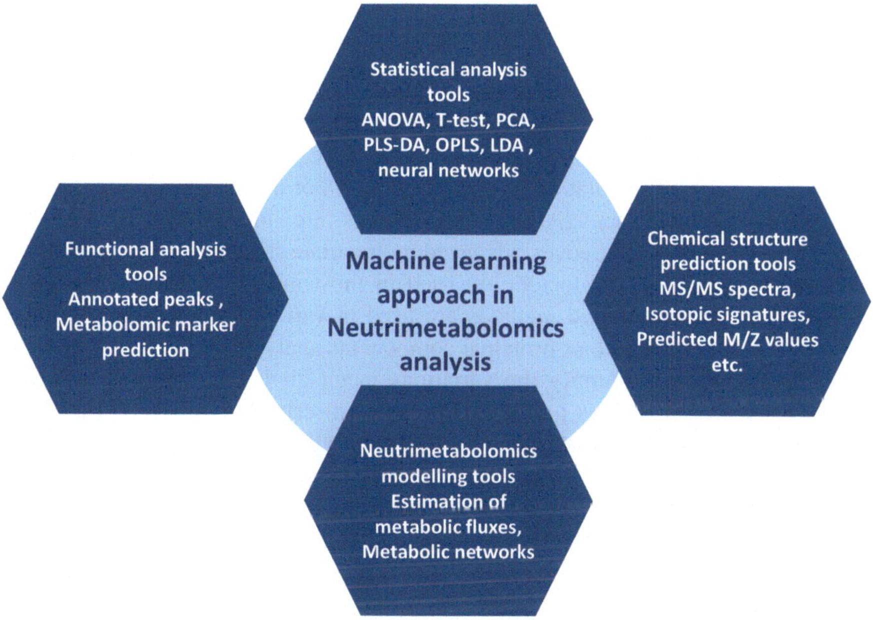

Fig. 2 Machine learning approaches to enhance the dynamics of nutrimetabolomics analysis

methods, which is helpful to group the metabolites into separate clusters and into a hierarchy tree, respectively. k-means clustering is most important where the optimal number of clusters can be estimated using the NbClust package. Also, it uses the fviznbclust function for visualizing the results. In case of hierarchical clustering, hclust package creates clusters with sets of data that are similar internally but different from each other externally, and here dendrogram serves as useful graphical representation of molecular clusters. After clustering analysis, the heat maps generated from the analyses can be used to assess the impact of the intervention and the number and proportion of metabolites behaving in a certain manner. There are several R packages that generate heat maps between the identified metabolites and their associations with clinical phenotypes, which shows additional information that may be added to each cell, such as the statistical significance with circles, where a larger circle represents a lower p-value. In case of the hierarchical clustering, we can select whether to cluster only the features or samples as well. Pearson correlation and average linkage clustering are also important. In another clustering method – k-means clustering – we choose cluster genes, based on Pearson correlation, calculate k-means, and repeat the procedure by increasing the number of clusters until no more clusters with a unique pattern emerge, and finally the highest number of clusters is chosen based on this visual optimization.

3.2 Statistical Analysis Tools

Statistical analysis is performed to identify important features in the datasets that vary significantly in any metabolic experiment. Additional data processing steps are also required sometimes prior to statistical analysis, e.g., batch correction, normalization, filtering, and imputation of missing values. For univariate statistical analysis, analysis of variance (ANOVA) or t-tests are implemented in smaller metabolomics studies. Moreover, advanced multivariate statistical methods that utilize multiple experimental factors are used in most metabolomics studies. Both unsupervised and supervised methods are used in various metabolomics studies. Unsupervised method such as principal components analysis (PCA) is mainly used as a first approach to identify the variations in metabolomics datasets, whereas supervised methods such as partial least squares discriminant analysis (PLS-DA) and orthogonal partial least squares (OPLS) are used to identify different features in the metabolomics data. Moreover, clustering, linear discriminant analysis, random forest, and neural networks are also implemented in the development of software tools dedicated for statistical analysis of metabolomics datasets.

3.3 Functional Analysis Tools

Functional analysis mainly utilizes information such as annotated peak intensities and their corresponding properties to infer changes in metabolic reaction modules – pathways to facilitate biological interpretation related to experiment. The functional annotation of metabolomics datasets acts as metabolomic markers which are also associated with disease and phenotype and not just an altered pathway. Mostly, functional analysis utilizes various known metabolites to integrate into the species/ context-dependent metabolic network of the biological system. Reconstruction of the metabolic network is also required sometimes based on prior knowledge of enzymes, reactions, and pathways using automated text mining or manual curation of the metabolomics datasets.

3.4 Nutrimetabolomics Modeling Tools

These tools represent enzymes or biochemical reactions using appropriate kinetic and thermodynamic parameters as a mathematical model. This facilitates the development of kinetic or flux models of metabolic networks as well as provides a dynamical assessment of metabolic model networks and enables prediction or estimation of metabolic fluxes from metabolomics measurements.

3.5 *Chemical Property Prediction Tools*

It is designed to build or augment libraries of chemical properties that can be measured in metabolomics experiments (i.e., relative intensities, m/z, retention times, collision cross sections, chemical shifts, MS/MS spectra, isotopic signatures, etc.). Libraries can consist of properties derived from laboratory analysis of authentic reference standards (the traditional gold standard approach) or from the use of property prediction software to create in silico libraries. Historically, predicted m/z values have been routinely used in MS-based metabolomics analysis, as this property is readily calculated at higher accuracy than is achievable experimentally. Recently, various software tools based on molecular dynamics, quantum chemical calculations, and machine learning are being used to predict NMR chemical shifts, chromatographic retention times, isotopic signatures, ion-mobility collisions, MS peak shifts, and other chemical properties.

4 Computational Pipeline for Nutrimetabolomics

There are various metabolomics software which could be helpful to analyze the metabolomics data such as MetaboAnalyst, MZmine2, MS-DIAL, XCMS, GNPS, XCMS online, Mummichog, chemical property prediction tools, etc. The metabolomics pipeline is basically composed of five steps. In general, these steps involve:

(a) **Experimental design**: This step shows how one could prepare the sample for analysis.
(b) **Sample preparation** and (c) **data acquisition**: The metabolomics workflow starts with detecting mass spectrometry peaks. Peak 60 identification is commonly performed using open-source tools like XCMS [9], or we need to do data cleaning after the sample preparation. The objective of data cleaning was to remove the least informative metabolites and samples, using a number of subjective criteria. First, the pipeline excluded metabolites and samples exceeding a certain threshold of missingness in each sample at least 50%. After that for (d) **data processing**, we can use the data-dependent spectra files for further statistical analysis.
(e) **Statistical analysis**: The statistical analysis approach generally involves applying a multivariate statistical strategy such as univariate and multivariate statistical methods, principal component analysis (PCA), and k-means cluster analysis to dietary data to identify dietary patterns and then through the use of regression statistical method, linking these to metabolomic profiles in order to identify dietary biomarkers. Shapiro-Wilk test and Leven test are used to test normal distribution of the data and homogeneity of variances, respectively. One-way

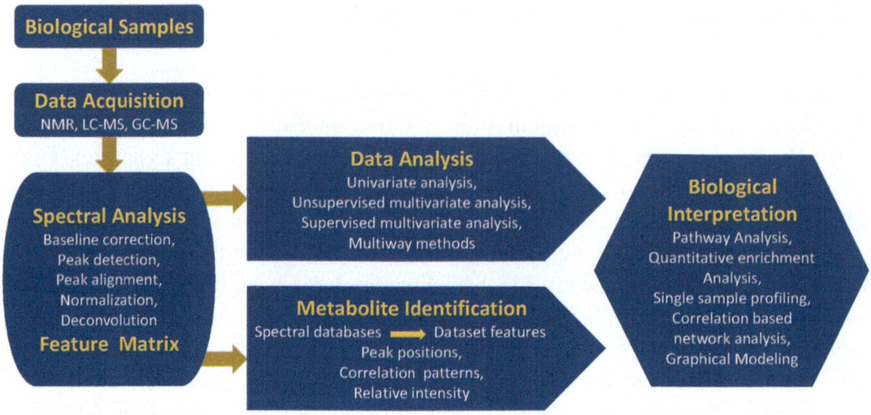

Fig. 3 Computational nutrimetabolomics pipeline

ANOVA is conducted to compare metabolic data as phenolic metabolites, immune markers, SCFAs, and MCFAs. The least significant differences are calculated by Tukeýs test ($p < 0.05$). A paired-samples t-test is also conducted to assess differences in metabolic data. Additionally, the relationships between metabolic parameters could be investigated by using the Pearson correlation coefficient (Fig. 3).

4.1 R Packages and Software Tools for Nutrimetabolomics Studies

Many untargeted metabolomics analysis tools such as MetaboAnalyst, CDK-R, mzR, and some other as MALDIquant package (provides a complete analysis pipeline for MALDITOF and other 2D mass spectrometry data) supporting packages are found for processing and analyzing of such metabolomics data, which are based on R platform. These R-based packages are used to perform mass spectral peak annotation and identify metabolites and molecular pathways associated to various metabolites. These R-based packages are also helpful to identify, design, and perform chemometric analysis that helps in designing drugs against potential targets. Additionally, MetDNA is useful for metabolite annotation and dysregulated network analysis of untargeted metabolomics. MetaboAnalyst is a popular platform dedicated for metabolomics data analysis via user-friendly, web-based interface which is freely accessible for which R code is freely available at GitHub as the MetaboAnalystR (https://github.com/xia-lab/MetaboAnalystR).

4.2 R Codes for Importing and Visualization of Chemical Molecules Using R Packages

```
#Downloading R packages
source("https://bioconductor.org/biocLite.R")
biocLite("ChemmineR")
library(rcdk)#
library(chemometrics)#
library(rJava)
library (ChemmineR)
library(cluster)
library(rgl)
library(ggplot2)
library(vegan)
library(factoextra)
library(fingerprint)
library(fmcsR)
source("https://bioconductor.org/biocLite.R")
biocLite("fmcsR")
library (NbClust)
library(iqspr)
library(ggplot2)
library(gridExtra)
library(fpc)
#################################################
#Reading and visualizing the methotrexate molecule in SMILES format
mol       <-        parse.smiles('CCOC(=O)C1=C(CN=C1C)\\N=N\\C1=C(O)
C=CC2=C1C=C(O)C=C2',kekulise=TRUE)[[1]]
mol
view.molecule.2d(mol)
#Reading and visualizing the methotrexate (CMP1) in sdf format
CMP1 <- load.molecules( c('CMP1.sdf') )
 view.molecule.2d(CMP1[[1]])
#########################################################
#General properties of methotrexate molecule
cat('No. of atoms =', length(atoms), '\n')
cat('No. of bonds =', length(bonds), '\n')
atoms <- get.atoms((CMP1[[1]]))
atoms
bonds <- get.bonds((CMP1[[1]]))
bonds
coords <- get.point2d(atoms[[1]])
coords
```

```
coords <- do.call('rbind' , lapply(atoms, get.point2d))
coords
##############################################################
#Descriptors categories in rcdk package:
descNames  <-  unique(unlist(sapply(get.desc.categories(),  get.
desc.names)))
descNames
dc <- get.desc.categories()
dc
descriptors = get.desc.names(type="all")
descriptors
#"Constitutional" descriptors
dn <- get.desc.names(dc[2])
dn
#Calculus of a descriptor - 14 "AlogP"
aDesc <- eval.desc(CMP1, dn[14])
aDesc

# Topological Polar Surface Area, xlogp, alogp, total charge
get.tpsa(mol)
get.xlogp(mol)
get.alogp(mol)
get.total.charge(mol)

#Calculus of all the descriptors -1
allDescs <- eval.desc(CMP1, dn)
allDescs
require(rcdk)
#Calculus of all the descriptors -2
drug.mols <- load.molecules(molfiles="CMP1.sdf")
descNames  <-  unique(unlist(sapply(get.desc.categories(),  get.
desc.names)))
drug.descs <- eval.desc(drug.mols, descNames, verbose=T)
drug.descs
descs <- eval.desc(mols, descNames)

###########################################################
#Calculs of the molecular fingerprint for methotrexate (CMP1)-maccs
fps <- get.fingerprint(CMP1[[1]], type='maccs')
fps
#Calculs of the molecular fingerprint for methotrexate (CMP1)-extended
fps <- get.fingerprint(CMP1[[1]], type='extended')
fps
## Reading and visualizing the set of 24 molecules
mols <- load.molecules(c('CMP1.sdf', 'CMP2.sdf', 'CMP3.sdf','CMP4.
```

```
sdf', 'CMP5.sdf', 'CMP6.sdf', 'CMP7.sdf', 'CMP8.sdf', 'CMP9.sdf',
'CMP10.sdf', 'CMP11.sdf', 'CMP12.sdf', 'CMP13.sdf', 'CMP14.sdf',
'CMP15.sdf','CMP16.sdf',    'CMP17.sdf','CMP18.sdf',    'CMP19.
sdf','CMP20.sdf','CMP21.sdf','CMP22.sdf',              'CMP23.
sdf','CMP24.sdf'))

view.molecule.2d(mols, ncol = 4,width = 200, height = 200, depic-
tor = NULL, type="isomeric")
view.molecule.2d(mols)
mols

#The molecular fingerprints for the set of molecules
fps <- lapply(mols, get.fingerprint, type='extended')
fps
fp.sim <- fp.sim.matrix(fps, method='tanimoto')
fp.dist <- 1 - fp.sim
fp.dist
```

#Identification of molecules located at a certain distance from the target

```
query.fp<-get.fingerprint(CMP1[[1]], type = 'maccs')
target. mols <-mols
target.fps <- lapply(target.mols, get.fingerprint, type='maccs')
target.fps
sims <- data.frame(sim=do.call(rbind, lapply(target.fps,
 fingerprint::distance,
 fp2=query.fp, method='tanimoto')))
subset(sims, sim >= 0.5)
hits <- which(sims >= 0.5)
hits

#Distances between CMP1 and the rest of molecules
query.mol<-load.molecules( c('CMP1.sdf') )
query.mol
target. Mols <- mols
target.mols
fp.sim <- fingerprint::fp.sim.matrix(fps, method='tanimoto')
fp.dist <- 1 - fp.sim
fp.dist
##############################################################
##########
```

#Assessment of the optimal number of clusters using NbClust R package

```
fviz_nbclust(fp.dist, kmeans, method = "wss") +
 geom_vline(xintercept = 3, linetype = 2)
```

```
#Hierarchical clustering with hclust using Ward's method
d <- dist(fp.dist, method = "euclidean")
res.hc <- hclust(d, method = "ward.D2" )
grp <- cutree(res.hc, k = 3)
plot(res.hc, cex = 0.6)
rect.hclust(res.hc, k = 3, border = 2:5)

#K-means clustering
fviz_nbclust(fp.dist, method = "gap_stat")
km.res <- kmeans(fp.dist, 3, nstart = 5)
km.res
fviz_cluster(km.res, data = fp.dist, ellipse.type = "convex")
theme_minimal()

####################################
#Clustering analysis
#Clusters statistics for K-mean
silinfo <- km.res$silinfo
names(silinfo)
km_stats <- cluster.stats(fp.dist, km.res$cluster)
km_stats

# Silhouette coefficient of observations
library("cluster")
sil <- silhouette(km.res$cluster, dist(fp.dist))
head (sil[, 1:3], 10)
plot (sil, main ="Silhouette plot - K-means")

#####################################
#Visualisation of the molecules from cluster
molsc <- load.molecules(c('CMP1.sdf', 'CMP2.sdf', 'CMP3.sdf',
'CMP5.sdf', 'CMP6.sdf', 'CMP7.sdf', 'CMP8.sdf'))
view.molecule.2d(molsc)
```

4.3 R-Code for Preprocessing of Mass Spectrometry Imaging (MSI) Data Using MALDIquant

[10] where he used the publicly available dataset submitted by Brittney Gorman(PNNL) published on 28 November, 2022.

```
library (MALDIquant)
library (MALDIquantForeign)
library(irlba)
library(viridis)
# Importing ImzML
#Importing of publicly available datasets on METASPACE2020
(https://metaspace2020.eu/)
We call the function importImzML from MALDIquant to load the MSI
dataset imzML file. The dataset is in centroided mode, so we will
set the option (centroided = TRUE). This operation usually takes
few minutes, depending on the dataset size.
peaks <- importImzMl ('Exampledataset.imzML', centroided = TRUE,
verbose = FALSE)
# Total number of pixels
print(length(peaks))
# Spatial dimensions (same metadata for all pixels)
print(peaks[[1]]@metaData$imaging$size)
orderPixels <- function(peaks) {
 require(MALDIquant)
 coords <- coordinates(peaks)
 ord <- c()
 for (y in sort(unique(coords[, 2]))) {
 curr.y <- which(coords[, 2] == y)
 ord <- c(ord, curr.y[order(coords[curr.y, 1])])
 }
return(ord)
}
px.ord <- orderPixels(peaks)

# To reorder the peaks
peaks <- peaks[px.ord]
# Let's have a look at the distribution of the number of detected
peaks and pixels mean intensities (in the log-space):
n.peaks <- unlist(lapply(peaks, function(x) length(mass(x))))
#histogram plot
hist(n.peaks)
mu.peaks <- unlist(lapply(peaks, function(x) mean(intensity(x)))
hist(log1p(mu.peaks))
# We can check that we have a set of MS peaks (class MassPeaks):
print(class(peaks[[1]]))
plot(peaks[[3000]])
# A quick way to check the spatial properties of an MSI dataset is
to plot the TIC image. In this image, each pixel represents its
total peaks intensity:
tic <- unlist(lapply(peaks, function(x) sum(intensity(x))))
```

```
tic <- matrix(tic, peaks[[1]]@metaData$imaging$size)
image(tic, col=viridis(64))
title('TIC')
# if we calculate the TIC of the log-transformed intensities:
tic.log        <-        unlist(lapply(peaks,        function(x)
sum(log1p(intensity(x)))))
tic.log <- matrix(tic.log, peaks[[1]]@metaData$imaging$size)
#Peak binning (or peak matching)
This can be done by calling the function binPeaks from MALDIquant.
We save the new masses in the same list of MassPeaks.
image(tic.log, col=viridis(64))
title('TIC (log)')
# Normalization and log-transformation
scaling.factor <- apply(X, 1, sum)
X <- X / scaling.factor
# Be sure that the empty pixels are set to 0
X[scaling.factor == 0, ] <- 0
X <- log1p(X)
results   <-   irlba::prcomp_irlba(X,   center   =   TRUE,   scale.   =
TRUE, n = 3)
# First rescale the scores in [0, 1]
p                                                                c
<- apply(results$x, 2, function(x) (x - min(x)) / (max(x) - min(x)))
# Create the RGB
im <- rgb(pc[, 1], pc[, 2], pc[, 3])
im <- matrix(im, peaks[[1]]@metaData$imaging$size)
# Plot the image as a raster (transpose before)
plot(as.raster(t(im)), interpolate = FALSE)
```

5 Applications and Importance of Nutrimetabolomics

Nutrimetabolomics is an integrative action for metabolomic analyses in human nutritional studies. Target-Approaches could be used for finding the effects of a drug on a specific enzyme, where the therapeutic and genetic modifications that occur because of drug metabolism are studied. Metabolomics therefore has a great potential for improving diagnosis, therapeutic treatment, and care of disease. In the past decade, metabolomics has already proved to be useful for the characterization of several pathological conditions and offers promises as a clinical tool. In future work, the potential biomarkers should be further validated with a large enough patient cohort to achieve earlier diagnosis.

5.1 *Food Metabolome and Dietary Biomarkers*

Food-Metabolomics is defined as the composition of human metabolism which is derived from the digestion and biotransformation of food composition and these complete set of food metabolites in a biological cell, tissue, organ, or organism, which are the end products of cellular processes. So, nutrimetabolomics is a powerful approach because food metabolites and their concentrations, unlike other "omics" measures, directly reflect the underlying biochemical activity and the state of cells or tissues. The most promising application of nutrimetabolomics studies in the nutrition sciences is the identification of new dietary biomarkers. New dietary biomarkers are usually determined using supervised analysis models since they are capable to aggregate the evidence of multiple nutrimetabolomics data, analyzing dietary patterns in conjunction with metabolomic profiles to identify biomarkers and nutrition types.

5.2 *In Physiological Monitoring of Diet and Nutrition Studies*

5.2.1 Requirement for Measuring Dietary Intake

Traditional dietary intake methods are mainly divided into three distinct categories which are based on the information reported by the subjects. These methods include food records, food frequency questionnaires (FFQs), and 24-h recalls. Food records are commonly documented for 3–7 days, but 7-day records are considered as "gold standard" for validating other methods. Food records are mainly based on self-reported information by the subjects. In this method, biomarkers measured in body fluids or tissues are variables which reflect the intake of various food components due to variability in biomarker content; therefore, this method is not considered as a good method.

FFQs are most commonly used for groups of people usually for a period of 6 months to 1 year to provide estimates of usual dietary intake. FFQs are mainly used in large cohort studies to place individuals into broad categories. FFQs provide specific foods sometimes based on culture (e.g., Japanese, Korean, Indian, etc.) to eat. The major drawback of FFQs is that the FFQs method is not designed to assess current energy intake. The measurement of current energy should be considered because it is considered an important component of diet therapy for obesity treatment. Modified FFQs such as quick screening questionnaires have been developed to identify people with a high intake of fat and a low intake of fiber/fruit/vegetable [11]. FFQs method was developed to categorize groups of people by their long-term intake of various nutrients for epidemiological studies.

The 24-hour recall could be conducted in person or by telephone, and it is mainly focused on quantitative assessment of current nutrient intake [12]. The 24-hour recall method is popular in low-income countries, and it is also popular in low-literacy populations because the subjects do not need to read or write to complete the recall [13]. The major drawback of the 24-hour recall includes the inability of a single day's intake to describe the usual diet [14]. Food records and 24-hour recalls are designed to estimate current nutrient intake.

Nutrition monitoring in the United States is a group of coordinated activities that provides information about the dietary, nutritional, and related health status of Americans. Nutrition monitoring also includes the relationships between diet and health and the factors affecting dietary and nutritional status.

5.2.2 Requirements for Measuring Physical Activity and Sedentary Behavior

Physical inactivity is strongly associated with bad health, increasing an individual's risk for developing cardiovascular disease. On the other hand, physical activity (PA) broadly includes all activities above the resting stage including all physical movement performed during purposeful activities (i.e., aerobic/resistance exercise), occupational, household, active transportation, and/or recreation.

The degree of physical activities to which health benefits are acquired depends on the frequency, intensity, time, and the volume of PA performed. Regular routine exercises are very advantageous to us because they improve cardiorespiratory fitness, body composition, resting blood pressure, blood glucose, and circulating lipoprotein levels in blood.

Greater volumes of physical activity contribute to the protection against various diseases/conditions such as premature cardiovascular disease (CVD), certain types of cancers, metabolic disorders, obesity, dementia, and morbidity [15]. Nowadays, adopting a physical activity in daily life is highly recommended by many health organizations of repute to combat the development of noncommunicable diseases (NCDs) [16–18]. Increasing physical activity and exercise levels from low to nationally recommended volumes is known to contribute to increases in cardiovascular fitness of the people of a country. Recent studies have therefore emphasized reporting the quantity of participants favorably and meaningfully "responding" to varying amounts and intensities of exercise. Therefore, exercise prescription is promoted in the context of precision medicine [19, 20] in recent times.

5.3 *In the Study of Diet-Related Diseases*

5.3.1 From Health to Disease Status

Good health can be described as a state of complete physical, mental, and social well-being. Tracking trends in general health status can help us in identifying various interventions which have improved the health of a population or where interventions may be required (e.g., by exploring causative factors and preventive measures). The health status of a population can be measured by a wide range of factors such as birth and death rates, life expectancy, quality of life, morbidity from specific diseases and conditions, environmental risk factors, use of ambulatory care and inpatient care, financial and geographical accessibility of health personnel and facilities, health insurance coverage, and many other factors. Any impairment of normal physiological function affecting all or part of an organism or any pathological change caused by infection, stress, etc., producing characteristic symptoms, illness, or sickness, in general is called disease.

5.3.2 Discovering Disease-Related Dietary Factors

Approximately 11 million deaths worldwide were estimated in 2017 attributed to dietary risk factors. Diet-related noncommunicable diseases (NCDs) include obesity, cardiovascular diseases (such as arterial hypertension, myocardial infarction, stroke), diabetes mellitus, cancers, osteoporosis, etc. An unhealthy diet which is associated with numerous dietary risk factors contributes to the development of diseases/disorders commonly known as metabolic syndrome [21, 22]. However, an excessive sodium intake and a low intake of whole grains as well as fruits and vegetables are considered the most important dietary risk factors. Sugar intake is also considered as an important dietary risk factor, which increases the risk for tooth decay, obesity, and cardiovascular diseases. Moreover, the excessive consumption of saturated and trans fats also contributes to cardiovascular diseases [23].

A healthy eating plan includes eating of green vegetables, salads, fruits, whole grains, and fat-free or low-fat dairy products. It also includes eating of lean meats, poultry, fish, beans, eggs, and nuts. In a healthy eating plan, we should limit the intake of saturated and trans fats, sodium, and added sugars. A regular healthy eating plan can lower the risk for heart disease, stroke, diabetes, and other health conditions. However, many natural resources such as fishes (good source of animal protein) are excessively exploited, which needs our attention. The rapid urbanization and industrialization have forced us to change our lifestyles as well as our dietary and nutritional patterns, especially in developing countries [24–26]. Other factors such as a large amount of food intake and eating out, as well as an increase in food portion sizes, were also observed [26] as contributors of various diseases. For example, Poland (a high-income country in Central and Eastern Europe (CEE)) has undergone a significant transition in terms of food intake (based on change in food market) and especially after joining the European Union (EU) over the past few decades [27, 28]. The dietary patterns of the people of Poland have changed a lot due to an increase in the gross domestic product (GDP), urbanization, and agricultural sector [28]. Changes in nutritional behaviors led to an increase in obesity among adults in Poland [28, 29]. The portion of obese adults in Poland (58%) is higher than the EU average (53%) [29] leading to high cancer and myocardial infarction reported in Poland [30].

To better understand how different dietary components affect the risk for various diseases such as heart disease, stroke, and type 2 diabetes, a research was conducted by a team of researchers led by Dr. Dariush Mozaffarian at Tufts University in the United States. Dr. Dariush Mozaffarian conducted a research and analyzed the data from the Centers for Disease Control and Prevention (CDC) in the United States and elucidated the relationships of ten different foods/nutrients with deaths attributed to heart disease, stroke, and type 2 diabetes. In their research they also included various other factors such as participants' age, sex, ethnicity, and education. In this research, they observed that the cardiometabolic diseases were responsible for around 50% deaths in the United Sates in 2012 which were mainly associated with suboptimal eating habits. Deaths due to heart disease, stroke, and type 2 diabetes were linked with inadequate consumption of certain foods and nutrients. Around 10% of cardiometabolic disease-related deaths was due to the consumption of excess amount of sodium. Eating a low amount of nuts and seeds, seafood omega-3 fats, vegetables, and fruits also increased the risk of deaths due to various metabolic disorders. The lower consumption of whole grains or polyunsaturated fats also increased the risk of death due to certain metabolic disorders compared with people having an adequate amount of these foods/nutrients. Eating too much processed meat (8.2%), sugar-sweetened beverages (7.4%), and unprocessed red meat (0.4%) also increased the risk of heart disease-, stroke-, and type 2 diabetes-related deaths. The results showed that a suboptimal diet which varied in men (higher) and women (lower) was the main reason behind the large proportion of deaths. This proportion was observed to be higher among blacks and Hispanics compared to whites and among those with lower education levels.

6 Role of Nutrimetabolomics in Precision Nutrition

"Precision nutrition" is an emerging area of nutrition research that focuses on understanding metabolic variability within and between individuals. Precision nutrition helps us in developing customized dietary plans and interventions to maintain optimal individual health. One of the ultimate goals of the precision nutrition is the design of tailored nutritional recommendations to treat or prevent metabolic disorders. Precision nutrition includes various disciplines such as nutritional genomic (gene-nutrient interactions), epigenetic, microbiome, and environmental factors.

6.1 Deep Phenotyping: High-Quality Phenotypes to Stratify Obesity

Various factors such as obesity, body composition, resting energy expenditure, satiety, satiation, eating behavior, affect, and physical activity were used to classify obesity phenotypes measured by validated methods and questionnaires [31]. In a study of 450 patients, 4 phenotypes of obesity were identified in 382 of 450 participants (85%) which includes hungry brain (mainly controlled by the brain-gut axis and abnormal calories needed to reach fullness), emotional hunger (desire to eat to cope with positive or negative emotions), hungry gut (abnormal duration of fullness), and slow burn (decreased metabolic rate) [31]. In many cases, biological and behavioral phenotypes are associated with human obesity and can be targeted with medications to promote weight loss.

6.2 *Metabolomics: Towards a Better Characterization of Eating*

A healthy condition of a living being works properly when it is properly backed by optimal nutritional inputs. Inadequate or poor nutritional intakes are linked with a number of metabolic diseases, such as diabetes, obesity, atherosclerosis, hypertension, and osteoporosis. Metabolomics science deals with genomics, transcriptomics, and proteomics, in order to understand the changes in the profiles of low-molecular-weight metabolites. Metabolomics qualitatively and quantitatively defines metabolites (small molecules) present in biological samples, which has become popular and important in nutritional research. Metabolomics approaches enable monitoring of metabolites in people, taking into consideration their age, gender, drug toxicology, lifestyle, health status, and most notably nutrition intake, in correlation with genetic and environmental components [32, 33, 40]. The changes in the metabolite profiles in a person could be due to genetics, the environment, and dietary intake [34]. In brief, metabolomics is "the measurement of metabolite concentrations in cells and tissues". Metabolomics employs various analytical platforms including high-performance liquid chromatography (HPLC) [35, 36], Fourier transform infrared (FTIR) spectroscopy [37], mass spectrometry (MS) [38, 39], and nuclear magnetic resonance (NMR) spectroscopy [41].

Not only the variety of diet is directly responsible for the metabolites but also the gut microbial population and our own metabolism [34, 42] are responsible for various metabolites. Approximately 10^{14} bacterial cells are present in the gastrointestinal tract of a human, which is about 10 times the total number of human cells in the body, with a total biomass of 2 kg [43, 44]. Approximately, 1000 different bacterial species [44], some of which are crucial for our well-being, such as lactic acid bacteria and bifidobacteria, are present in the human gut. These bacterial species synthesize vitamin K as well as many B vitamins, such as biotin, cobalamin, folate, nicotinic acid, pantothenic acid, pyridoxine, riboflavin, and thiamine [45, 46]. The microbiota population and various metabolic pathways are influenced by the genetics, geographical regions, diets, antibiotics, and other therapies of the host [47]. Metabolomics also helps us to obtain information on an individual's diet from food-induced shifts in metabolites. The analysis of dietary patterns allows researchers to gain a broader insight into dietary intake, applying metabolomics to achieve this. In a study by Posma and colleagues [48], urine was collected from 1848 Americans, and ^{1}H NMR spectroscopy was used to measure the urinary metabolome, producing a wide range of chemical profiles of metabolites. It was observed that 46 metabolites can differentiate people with healthy and unhealthy dietary patterns. These 46 metabolites in association between sodium and calcium with citrate and format have effect on blood pressure, adiposity, and renal function, and a correlation between fructose, glucose, and vitamin C with biomarkers of citrus fruit consumption [48]. In another research study, intake of a Western diet (WD), which is rich in saturated fats and simple sugars, is associated not only with weight gain, diabetes, and metabolic diseases but also with impaired hippocampal-dependent memory and hippocampal pathologies. Nowadays, various therapeutic diets have become an integral part of the clinical treatment for obesity, dyslipidemias, diabetes, cardiovascular disease, and hypertension [49]. Metabolomics also helps us not only in the analysis of nutrients in single food products but also in the analysis of complete diet. A study conducted by Jin and co-workers showed the importance of metabolomics approach in investigating the effects of a Mediterranean diet and the role of the microbiome [50]. In brief, metabolomics is helping researchers to understand the mechanisms following dietary interventions, which in turn advances our knowledge of the relationships between diet and health/disease. Chromatography and NMR spectroscopy, separately or combined, deliver qualitative and quantitative data on the molecular contents of all food types, which can be collected and disseminated by databases to the scientific community. Each metabolomics workflow starts with a biological sample preparation, metabolite extraction, separation and analysis, data processing, statistical analysis and classification, identification, and biological pathway mapping. Each metabolomics workflow needs to satisfy standards for the chosen analytical method. The numerous analytical data are analyzed with statistical methods and then gathered in the databases which provide various information about the product of interest from NMR/MS reference spectra of a product's metabolites, gene structures and expression profiles of genes, etc. For example, the Tea Metabolome Database (TMDB) [51] provides users with information about small chemical compounds found in *Camellia sinensis*. MS/MS data can be obtained from this database for the purpose of identification of metabolites. At the time of the database creation in 2014, it contained more than 1473 compound (713 compounds in green tea, 497 in black tea, 140 in oolong tea, and 445 for dark tea) entries [51]. The entries included in TMDB were collected based on the information obtained in 364 published books, journal articles, and electronic databases. Out of 1473 compounds, most of the compounds (74%) collected in the database have a molecular weight of less than 500 Da.

FooDB (https://foodb.ca/, accessed on 24 November 2022) is another world's largest open-access database that provides information on macronutrients and micronutrients as well as compound nomenclature, descriptions, information on structure, chemical class, physicochemical data, food source(s), color, aroma, taste, physiological effects, presumptive health effects (from published studies), and concentrations in various foods. Currently in 2022, approximately 800 types of products are listed in this database, with more than 70,000 compounds characterized.

PhytoHUB (http://phytohub.eu/, accessed on 24 November 2022) is a freely available electronic database containing detailed information about dietary phytochemicals and their human and animal metabolites [52]. The content of PhytoHUB consists of about 1850 entries of which around 1200 are polyphenols, terpenoids, alkaloids, and other plant secondary metabolites, with 560 human or animal metabolites. The total number of plant-based foods featured in PhytoHUB, as of 2022, is 381 [52].

7 Challenges and Future Prospects for Nutrimetabolomics

Since the last 7–8 years, nutrimetabolomics approach has produced a number of robust biomarkers of dietary intake. This approach has resulted in the identification of a number of putative biomarkers of specific foods and drinks such as citrus fruit, cruciferous vegetables, red meat, coffee, tea, sugar-sweetened beverages, and wine, and the search for biomarkers of specific foods can also be carried out through the use of cohort studies for which the dietary data are collected using a traditional method to identify consumers and nonconsumers of a specific food. After that, nutrimetabolomics profiles are then compared between these groups in order to identify potential dietary biomarkers. These cohort studies rely on self-reported dietary assessment methods which are prone to error, due to which it also needs to be highlighted that the biomarkers identified in cohort studies do not assess the direct relationships of food amounts consumed, and on the basis of statistical test; they are simply correlations between the food and the metabolites, and therefore the relationship only shows an association (Brennan et al. 2015).

Precision nutrition integrates genetic, metagenomic, metabolomic, physiopathological, behavioral, and sociocultural cues to understand metabolism and human well-being and implement health actions. Personalized nutrition considers the differential response to dietary intake due to individual endogenous aspects that influence nutrient intake and uptake, metabolism, assimilation, and excretion [53]. Personalized nutrition focuses on the assessment of diet and health along with the use of "omics" technologies (nutrigenomics, metagenomics, and metabolomics) to develop optimal and customized dietary support. Personalized nutrition also promotes health maintenance and disease prevention for each individual.

8 Conclusion

In conclusion, nutrimetabolomics is a rapidly evolving field that aims to understand the complex interactions between nutrition and metabolism. It combines analytical techniques such as mass spectrometry and nuclear magnetic resonance spectroscopy with bioinformatics tools to identify and quantify metabolites in biological samples. Nutrimetabolomics has the potential to reveal biomarkers of dietary intake and to elucidate the mechanisms underlying the health benefits of certain foods and nutrients. This chapter has demonstrated the utility of nutrimetabolomics in the identification of dietary biomarkers, the characterization of metabolic pathways, and the investigation of the effects of dietary interventions on metabolism. However, challenges such as data integration and standardization, as well as the need for larger sample sizes and diverse populations, still need to be addressed. We have also discussed various computational and machine learning pipelines for the data analysis and interpretation. Despite many challenges, nutrimetabolomics holds great promise for advancing our understanding of the intricate relationship between diet and health. It has the potential to inform the development of personalized nutrition recommendations, aid in the prevention and management of chronic diseases, and ultimately improve human health and well-being. Further research in this field will undoubtedly yield valuable insights and innovations in the years to come.

References

1. Carlson, A.K., Rawle, R.A., Adams, E., Greenwood, M.C., Bothner, B. and June, R.K. (2018) Application of global metabolomic profiling of synovial fluid for osteoarthritis biomarkers. Biochem Biophys Res Commun 499, 182–188.
2. Carlson, A.K., Rawle, R.A., Wallace, C.W., Adams, E., Greenwood, M.C., Bothner, B. and June, R.K. (2019a) Global metabolomic profiling of human synovial fluid for rheumatoid arthritis biomarkers. Clin Exp Rheumatol 37, 393–399.
3. Carlson, A.K., Rawle, R.A., Wallace, C.W., Brooks, E.G., Adams, E., Greenwood, M.C., Olmer, M., Lotz, M.K., Bothner, B. and June, R.K. (2019b) Characterization of synovial fluid metabolomic phenotypes of cartilage morphological changes associated with osteoarthritis. Osteoarthritis Cartilage 27, 1174–1184.
4. Hahn, A.K., Batushansky, A., Rawle, R.A., Prado Lopes, E.B., June, R.K. and Griffin, T.M. (2021) Effects of long-term exercise and a high-fat diet on synovial fluid metabolomics and joint structural phenotypes in mice: an integrated network analysis. Osteoarthritis Cartilage 29, 1549–1563.
5. Hislop, B.D., Devine, C., June, R.K. and Heveran, C.M. (2022) Subchondral bone structure and synovial fluid metabolism are altered in injured and contralateral limbs 7 days after non-invasive joint injury in skeletally-mature C57BL/6 mice. Osteoarthritis Cartilage.
6. Wang, Yang et al. "Enhanced MS/MS coverage for metabolite identification in LC-MS-based untargeted metabolomics by target-directed data dependent acquisition with time-staggered precursor ion list." *Analytica chimica acta* vol. 992 (2017): 67–75. https://doi.org/10.1016/j.aca.2017.08.044
7. Ulaszewska, Marynka M et al. "Nutrimetabolomics: An Integrative Action for Metabolomic Analyses in Human Nutritional Studies." *Molecular nutrition & food research* vol. 63,1 (2019): e1800384. https://doi.org/10.1002/mnfr.201800384
8. Hegeman, Adrian D. "Plant metabolomics--meeting the analytical challenges of comprehensive metabolite analysis." *Briefings in functional genomics* vol. 9,2 (2010): 139–48. https://doi.org/10.1093/bfgp/elp053
9. Smith, Colin A et al. "XCMS: processing mass spectrometry data for metabolite profiling using nonlinear peak alignment, matching, and identification." *Analytical chemistry* vol. 78,3 (2006): 779–87. https://doi.org/10.1021/ac051437y

10. Gibb S, Strimmer K (2012). "MALDIquant: a versatile R package for the analysis of mass spectrometry data." Bioinformatics, 28(17), 2270–2271. https://doi.org/10.1093/bioinformatics/bts447
11. Block G, Clifford C, Naughton MD, Henderson M, McAdam's M.A brief dietary screen for high fat intake. J Nutr Edu.1989;21:199–207.
12. Tran KM, Johnson RK, Soultanakis RB, Mathhews DE. In person vs telephone-administered multiple-pass 24-hour recalls in women: validation with doubly labeled water. JAmDiet Assoc.2000; 100:777–83.11.
13. Johnson RK, Soultanakis RP, Matthews DE. Literacy and body fatness are associated with underreporting of energy intake in U.S. low-income women using the multiple-pass24-hour recall: a doubly labeled water study. J Am Diet Assoc.1998; 98:1136–40.12
14. Basiotis PP, Walsh SO, Cronin RJ, Kelsay JL, Mertz W. Number of days of food intake records required to estimate individual and group nutrient intakes with defined confidence. J Nutr.1987;117:1638–41.
15. G.F. Fletcher, C. Landolfo, J. Niebauer, C. Ozemek, R. Arena, C.J. Lavie Promoting physical activity and exercise: JACC health promotion series J Am Coll Cardiol, 72 (14) (2018), pp. 1622–1639
16. M.D. Jensen, D.H. Ryan, C.M. Apovian, et al. (2013) AHA/ACC/TOS'S guideline for the management of overweight and obesity in adults: a report of the American College of Cardiology/American Heart Association Task Force on Practice Guidelines and The Obesity Society J Am Coll Cardiol, 63, (2013), pp. 2985–3023.
17. M.F. Piepoli, A.W. Hoes, S. Agewall, et al.2016 European guidelines on cardiovascular disease prevention in clinical practice Rev Esp Cardiol (Engl Ed), 69 (10) (2016), p. 939
18. C.W. Yancy, M. Jessup, B. Bozkurt, et al. (2013) ACCF/AHA guideline for the management of heart failure: executive summary: a report of the American College of Cardiology Foundation/American Heart Association Task Force on practice guidelines. Circulation, 128 (16) (2013), pp. 1810–1852
19. R. Ross, L. de Lannoy, P.J. Stotz. Separate effects of intensity and amount of exercise on inter-individual cardiorespiratory fitness response. Mayo Clin Proc, 90 (11) (2015), pp. 1506–1514.
20. Ozemek C, Arena R. Precision in Promoting Physical Activity and Exercise with the Overarching Goal of Moving More. Prog Cardiovasc Dis. 2019 Jan-Feb;62(1):3–8. Epub 2018 Dec 7. PMID: 30529578. https://doi.org/10.1016/j.pcad.2018.12.001
21. Alberti, K.G.; Zimmet, P.; Shaw, J.; IDF Epidemiology Task Force Consensus Group. The metabolic syndrome-a new worldwide definition. Lancet 2005, 366, 1059–1062.
22. Micha, R.; Peñalvo, J.L.; Cudhea, F.; Imamura, F.; Rehm, C.D.; Mozaffarian, D. Association Between Dietary Factors and Mortality from Heart Disease, Stroke, and Type 2 Diabetes in the United States. JAMA 2017, 317, 912–924.
23. Wang, Q.; Afshin, A.; Yakoob, M.Y.; Singh, G.M.; Rehm, C.D.; Khatibzadeh, S.; Micha, R.; Shi, P.; Mozaffarian, D.; Global Burden of Diseases Nutrition and Chronic Diseases Expert Group (NutriCoDE). Impact of Nonoptimal Intakes of Saturated, Polyunsaturated, and Trans Fat on Global Burdens of Coronary Heart Disease. J. Am. Heart Assoc. 2016, 5, e002891.
24. Olatona, F.A.; Onabanjo, O.O.; Ugbaja, R.N.; Nnoaham, K.E.; Adelekan, D.A. Dietary habits and metabolic risk factors for non-communicable diseases in a university undergraduate population. J. Health Popul Nutr. 2018, 37, 21.
25. Popkin, B.M.; Adair, L.S.; Ng, S.W. Global nutrition transition and the pandemic of obesity in developing countries. Nutr. Rev. 2012, 70, 3–21.
26. World Health Organization. Healthy Diet. Available online: https://www.who.int/news-room/fact-sheets/detail/healthy-diet (accessed on 21 July 2022).
27. Grzelak-Kostulska, E.; Sypion-Dutkowska, N.; Michalski, T. Changes in the health situation of the population of Poland following the accession to the European Union (compared to Central and Eastern European countries). J. Geogr. Politics Soc. 2017, 7, 24–38.
28. Stos, K.; Rychlik, E.; Woźniak, A.; Ołtarzewski, M.; Jankowski, M.; Gujski, M.; Juszczyk, G. Prevalence and Sociodemographic Factors Associated with Overweight and Obesity among

Adults in Poland: A 2019/2020 Nationwide Cross-Sectional Survey. Int. J. Environ. Res. Public Health 2022, 19, 1502.
29. Eurostat. Overweight and Obesity-BMI Statistics. Available online: https://ec.europa.eu/eurostat/statistics-explained/index.php?title=Overweight_and_obesity_-_BMI_statistics (accessed on 24 November 2022).
30. Wojtyniak, B.; Gorynski, P. Health Status of Polish Population and Its Determinants 2020. Available online: https://www.pzh.gov.pl/sytuacja-zdrowotna-ludnosci-polski-i-jej-uwarunkowania-raport-za-2020-rok/ (accessed on 24 November 2022).
31. Acosta A, Camilleri M, Abu Dayyeh B, et al. Selection of anti-obesity medications based on phenotypes enhances weight loss: a pragmatic trial in an obesity clinic. Obesity. 2021; 29:662–671.
32. Emwas A.-H.M., Salek R.M., Griffin J.L., Merzaban J.S. NMR-Based Metabolomics in Human Disease Diagnosis: Applications, Limitations, and Recommendations. Metabolomics. 2013; 9:1048–1072. https://doi.org/10.1007/s11306-013-0524-y.
33. Gowda G.A.N., Zhang S., Gu H., Asiago V., Shanaiah N., Raftery D. Metabolomics-Based Methods for Early Disease Diagnostics. Expert Rev. Mol. Diagn. 2008; 8:617–633. https://doi.org/10.1586/14737159.8.5.617.
34. Guo, Y.; Bian, X.; Liu, J.; Zhu, M.; Li, L.; Yao, T.; Tang, C.; Ravichandran, V.; Liao, P.; Papadimitriou, K.; et al. Dietary Components, Microbial Metabolites and Human Health: Reading between the Lines. Foods 2020, 9, 1045.
35. Defernez, M.; Gunning, Y.M.; Parr, A.J.; Shepherd, L.V.T.; Davies, H.V.; Colquhoun, I.J. NMR and HPLC-UV Profiling of Potatoes with Genetic Modifications to Metabolic Pathways. J. Agric. Food Chem. 2004, 52, 6075–6085.
36. Al-Talla, Z.; Akrawi, S.H.; Emwas, A.-H.M. Solid State NMR and Bioequivalence Comparison of the Pharmacokinetic Parameters of Two Formulations of Clindamycin. Int. J. Clin. Pharmacol. Ther. 2011, 49, 469–476.
37. Ahmed, A.E.-S.I.; Wardell, J.N.; Thumser, A.E.; Avignone-Rossa, C.A.; Cavalli, G.; Hay, J.N.; Bushell, M.E. Metabolomic Profiling can differentiate between Bactericidal Effects of Free and Polymer Bound Halogen. J. Appl. Polym. Sci. 2011, 119, 709–718.
38. Lin, S.; Liu, N.; Yang, Z.; Song, W.; Wang, P.; Chen, H.; Lucio, M.; Schmitt-Kopplin, P.; Chen, G.; Cai, Z. GC/MS-Based Metabolomics Reveals Fatty Acid Biosynthesis and Cholesterol Metabolism in Cell Lines Infected with Influenza A Virus. Talanta 2010, 83, 262–268.
39. Riccio, M.F.; Saraiva, S.A.; Marques, L.A.; Alberici, R.; Haddad, R.; Möller, J.C.; Eberlin, M.N.; Catharino, R.R. Easy Mass Spectrometry for Metabolomics and Quality Control of Vegetable and Animal Fats. Eur. J. Lipid Sci. Technol. 2010, 112, 434–438.
40. Emwas, A.-H.M. The Strengths and Weaknesses of NMR Spectroscopy and Mass Spectrometry with Particular Focus on Metabolomics Research. In Metabonomics: Methods and Protocols; Bjerrum, J.T., Ed.; Springer: New York, NY, USA, 2015; pp. 161–193.
41. Emwas, A.-H.; Roy, R.; McKay, R.T.; Tenori, L.; Saccenti, E.; Gowda, G.A.N.; Raftery, D.; AlAhmari, F.; Jaremko, L.; Jaremko, M.; et al. NMR Spectroscopy for Metabolomics Research. Metabolites 2019, 9, 123.
42. Roager, H.M.; Dragsted, L.O. Diet-Derived Microbial Metabolites in Health and Disease. Nutr. Bull. 2019, 44, 216–227.
43. Qin, J.; Li, R.; Raes, J.; Arumugam, M.; Burgdorf, K.S.; Manichanh, C.; Nielsen, T.; Pons, N.; Levenez, F.; Yamada, T.; et al. A Human Gut Microbial Gene Catalogue Established by Metagenomic Sequencing. Nature 2010, 464, 59–65.
44. Farag, M.A.; Abdelwareth, A.; Sallam, I.E.; Shorbagi, M.; Jehmlich, N.; Fritz-Wallace, K.; Schäpe, S.S.; Rolle-Kampczyk, U.; Ehrlich, A.; Wessjohann, L.A.; et al. Metabolomics Reveals Impact of Seven Functional Foods on Metabolic Pathways in a Gut Microbiota Model. J. Adv. Res. 2020, 23, 47–59.
45. LeBlanc, J.G.; Milani, C.; de Giori, G.S.; Sesma, F.; van Sinderen, D.; Ventura, M. Bacteria as Vitamin Suppliers to Their Host: A Gut Microbiota Perspective. Curr. Opin. Biotechnol. 2013, 24, 160–168.

46. Gu, Q.; Li, P. Biosynthesis of Vitamins by Probiotic Bacteria. In Probiotics and Prebiotics in Human Nutrition and Health; Rao, V., Rao, L., Eds.; Intech Open: London, UK, 2016
47. Hasan, N.; Yang, H. Factors Affecting the Composition of the Gut Microbiota, and Its Modulation. PeerJ 2019, 7, e7502.
48. Posma, J.M.; Garcia-Perez, I.; Frost, G.; Aljuraiban, G.S.; Chan, Q.; Van Horn, L.; Daviglus, M.; Stamler, J.; Holmes, E.; Elliott, P.; et al. Nutriome–Metabolome Relationships Provide Insights into Dietary Intake and Metabolism. Nat. Food 2020, 1, 426–436.
49. Berkow, S.E.; Barnard, N.; Eckart, J.; Katcher, H. Four Therapeutic Diets: Adherence and Acceptability. Can. J. Diet. Pr. Res. 2010,71, 199–204.
50. Jin, Q., Black, A., Kales, S. N., Vattem, D., Ruiz-Canela, M., & Sotos-Prieto, M. (2019). Metabolomics and Microbiomes as Potential Tools to Evaluate the Effects of the Mediterranean Diet. *Nutrients*, *11*(1), 207. https://doi.org/10.3390/nu11010207.
51. Yue, Y., Chu, GX., Liu, XS. et al. TMDB: A literature-curated database for small molecular compounds found from tea. *BMC Plant Biol* **14**, 243 (2014). https://doi.org/10.1186/s12870-014-0243-1.
52. Forte, G.; Bocca, B.; Senofonte, O.; Petrucci, F.; Brusa, L.; Stanzione, P.; Zannino, S.; Violante, N.; Alimonti, A.; Sancesario, G. Trace and Major Elements in Whole Blood, Serum, Cerebrospinal Fluid and Urine of Patients with Parkinson's Disease. J. Neural Transm. 2004, 111, 1031–1040.
53. Ferguson, D.J., Gonnermann, H.M., Ruprecht, P. *et al.* Magma decompression rates during explosive eruptions of Kīlauea volcano, Hawaii, recorded by melt embayments. *Bull Volcanol* **78**, 71 (2016). https://doi.org/10.1007/s00445-016-1064-x
54. Sahu N, Mishra S., Minu K, Kanchan S, Sinha RP. 2022. Identification of cyanobacteria-based natural inhibitors against SARS-CoV-2 druggable target ACE2 using molecular docking study, ADME and toxicity analysis. Indian Journal of Clinical Biochemistry. https://doi.org/10.1007/s12291-022-01056-6.
55. Kesheri M, Kanchan S, Sinha RP. 2022. Responses of antioxidants for resilience to temporal variations in the cyanobacterium Microcystis aeruginosa. South African Journal of Botany 148:190–199. https://doi.org/10.1016/j.sajb.2022.04.017
56. Kesheri M, Kanchan S, Sinha RP. 2021. Isolation and in-silico analysis of antioxidants in response to temporal variations in the cyanobacterium Oscillatoria sp. Gene Reports. 23:101023. https://doi.org/10.1016/j.genrep.2021.101023.
57. Kanchan S., Sharma P., and Chowdhury, S. 2019. Evolution of endonuclease IV protein family: an in-silico analysis. 3 Biotech. 9:168. https://doi.org/10.1007/s13205-019-1696-6 PMID: 30997305
58. Singla S. Kesheri M., Kanchan S., Aswath S. 2019. Current Status and Data Analysis of Diabetes in India. Internal Journal of Innovative Technology and Exploring Engineering (IJITEE). 8(9):1920–1934. https://doi.org/10.35940/ijitee.I8403.078919
59. Kumari A., Kesheri M., Sinha R.P., Kanchan S. 2018. Integration of soft computing approach in plant biology and its applications in agriculture. In: Soft computing in biological sciences (Eds. H.J. Purohit, D.V. Raje, R. P. More, V. C. Kalia), pp. 265–278, Springer Nature Singapore. https://doi.org/10.1007/978-981-10-7455-4_16.
60. Kanchan S, Mehrotra R, Chowdhury S. 2015. In-silico study of endonuclease III protein family identifies key residues and processes during evolution. Journal of molecular evolution. 81:54–67. https://doi.org/10.1007/s00239-015-9689-5.
61. Kesheri M, Kanchan S, Richa, and Sinha RP. 2014. Isolation and in-silico analysis of Fe-superoxide dismutase in Nostoc commune. Gene, 553(2), pp. 117–125. https://doi.org/10.1016/j.gene.2014.10.010.
62. Priya P, Kesheri M, Sinha RP, Kanchan S. 2017. Molecular dynamics simulations for Biological Systems. In: Biomedical Image Analysis and Mining Techniques for Improved Health Outcomes (Karâa W. B. A., Dey N. (eds.), 286–313. IGI Global, USA. https://doi.org/10.4018/978-1-4666-8811-7.ch014.

63. Kumari A, Kanchan S, Kesheri M. 2016. Applications of Bio-Molecular Databases in Bioinformatics. In: Advancements in Bio-Medical Sensing, Imaging, Measurements and Instrumentation (N. Dey, V. Bhateja, A.E. Hassanien (eds.)), Vol. 651, Ch 15, (pp. 329–351). Springer-Verlag, Germany. https://doi.org/10.1007/978-3-319-33793-7_15.
64. Kesheri M, Kanchan S*, Chowdhury S, Sinha RP. 2015. Secondary and Tertiary Structure Prediction of Proteins: A Bioinformatic Approach. In: Complex system modelling and control through intelligent soft computations (Q. Zhu, A.T Azar (eds.)), 319:541–569, Springer-Verlag, Germany. https://doi.org/10.1007/978-3-319-12883-2_19. (Scopus indexed)
65. Ghai R., Nagarajan K., Singh J., Swarup S., Kesheri M. 2016. Evaluation of anti-oxidant status in-vitro and in-vivo in hydro-alcoholic extract of Eugenia caryophyllus. International Journal of Pharmacology and Toxicology, 4(1), 19–24. https://doi.org/10.14419/ijpt.v4i1.5880
66. Gahoi S, Mandal RS, Ivanisenko N, Shrivastava P, Jain S, Singh AK, Raghunandanan MV, Kanchan S, Taneja B, Mandal C, Ivanisenko VA, Kumar A, Kumar R, Open Source Drug Discovery Consortium, Ramachandran S. 2013. Computational screening for new inhibitors of M. tuberculosis mycolyltransferases antigen 85 group of proteins as potential drug targets. Journal of Biomolecular Structure and Dynamics. 31(1):30–43. https://doi.org/10.1080/07391102.2012.691343.
67. Srivastava, U., Singh, S. (2022). Approaches of Single-Cell Analysis in Crop Improvement. In: Wani, S.H., Kumar, A. (eds) Genomics of Cereal Crops. Springer Protocols Handbooks. Humana, New York, NY. https://doi.org/10.1007/978-1-0716-2533-0_14

Metabolomics in Natural Product Discovery and Their Applications

Seema Nath and Láisa Gomes Dias

Abbreviations

CAWG	Chemical Analysis Working Group
CE	Capillary electrophoresis
EAE	Enzyme-assisted extraction
EIC	Extracted ion chromatogram
ESI	Electrospray ionization
FDA	Food and Drug Administration
FID	Flame ionization detector
GC	Gas chromatography
HILIC	Hydrophilic interaction liquid chromatography
HMDB	Human metabolome database
HPLC	High-performance liquid chromatography
HPLC-ESI-IT-TOF-MS	High-performance liquid chromatography electrospray ionization ion trap time-of-flight multistage mass spectrometry
HR-ESI-MS	High-resolution electrospray ionization mass spectrometry
HTS	High-throughput screening
ICP-MS	Inductively coupled plasma mass spectrometry
IM-MS	Ion mobility-mass spectrometry
KEGG	Kyoto Encyclopedia of Genes and Genomes
LC	Liquid chromatography
LLF	Liquid-liquid fractionation
m/z	Mass-to-charge ratio
MAE	Microwave-assisted extraction

S. Nath (✉)
University of Texas Health Science Center at San Antonio (UTHSCSA),
San Antonio, TX, USA
e-mail: naths@uthscsa.edu

L. G. Dias
School of Agronomy, Federal University of Goiás (UFG), Goiás, Brazil
e-mail: laisa_dias@ufg.br

V. Soni, T. E. Hartman (eds.), *Metabolomics*,
https://doi.org/10.1007/978-3-031-39094-4_9

MS	Mass spectrometry
MSI	Metabolomics standards initiative
MTBE	Methyl tert-butyl ether
NIST	National Institute of Standards and Technology
NMR	Nuclear magnetic resonance
NP	Natural product
OPLS-DA	Orthogonal partial least squares discriminant analysis
OSMAC	One strain many compound
OTC	Over the counter
PCA	Principal component analysis
PDA	Photodiode array
PEFE	Pulsed electric field extraction
PLE	Pressurized liquid extraction
PLS-DA	Partial least squares discriminant analysis
PTFE	Polytetrafluoroethylene
QC	Quality control
RI	Retention indices
SFE	Supercritical fluid extraction
SM	Secondary metabolite
SPE	Solid-phase extraction
TIC	Total ion chromatogram
TOF	Time-of-flight
UAE	Ultrasound-assisted extraction
UHPLC	Ultrahigh-performance liquid chromatography
UPLC-QqQ MS	Ultrahigh-performance liquid chromatography-triple quadrupole mass spectrometry
UPLC-QTOFMS	Ultrahigh-performance liquid chromatography-quadrupole time-of-flight mass spectrometry

1 Introduction

Natural products (NPs) are defined as organic compounds produced by primary and/or secondary metabolisms of any living creatures, and the systematic study of these metabolic processes is broadly termed as metabolomics [1]. NPs could be broadly divided into three categories – (i) growth factors, energy producing units, nucleotides, amino acids, etc. which are produced due to primary metabolism function as the building blocks of living beings [1–3]; (ii) secondary metabolites (SMs), which are characteristic features of plants and few microorganisms are produced as response factors to external stimuli by plants, fungi, bacteria, and marine species and are involved in the survival and adaptation processes; and (iii) high-molecular-weight compounds such as cellulose, lignin, proteins, etc. The bioactive constituents of NPs studied so far are mostly SMs, and more than 40% medicines have been derived from NPs [2]. SMs are also believed to play a pivotal role in the defense mechanisms in plants against pathogens and predators [4, 5]. They are produced by different biosynthetic pathways and are classified according to their pathways. Different classes of compounds belong to SMs – phenolics and phenylpropanoids, terpenes and steroids, polyketides and fatty acids, alkaloids, specialized amino acids and peptides, and specialized carbohydrates [2, 6]. Due to their diverse chemical nature and biological activities, SMs are endowed with diverse range of functions and have profound effects in the well-being of human health [7–9]. The wealth and strength of technologies have contributed to the existing knowledge of the structural and functional aspects of NPs and due to their economical values, especially SMs have been found to possess immense usages in different industries and hold a wide range of possibilities for their valorization and reuse in the circular economy as well [10, 11].

The description of obtaining oil from *Cupressus sempervirens* (cypress), *Commiphora species* (myrrh), *Glycyrrhiza glabra* (licorice), and *Papaver somniferum* (poppy juice) as documented on clay tablets in cuneiform from Mesopotamia (2600 BC) is believed to be the first record of NP extraction from natural sources. Till date, these extracts are still used for treating cold and cough, parasitic infections, inflammation, and other ailments. The ancient records of Egyptian medicines, traditional Chinese medicines, Indian Ayurveda, Tibetan medicine, African Tribal medicine, and Greek compilation of medicinal herbs recognized the importance of NPs, extracted mostly from plant sources, and are being used in treating different diseases. The Arabs were the pioneer in establishing privately owned drug stores harnessing the strength and knowledge of Indo-Western world of medicinal herbs. The discovery of the antimalarial drug, quinine, from the bark of *Cinchona* species by French pharmacists Caventou and Pelletier followed by the isolation of the analgesic morphine from the opium poppy, *Papaver somniferum*, by the German pharmacist Serturner strengthens the premise of using NPs as potential drug targets. In 1929, the discovery and broad therapeutic use of penicillin from the filamentous fungus, *Penicillium notatum*, by Fleming opened the door of exploring other living beings as potential sources for extracting NPs [12].

Some other examples of NPs obtained from different sources are as follows: natural rubber or latex which is formed by condensation of one of the SMs called isoprene obtained from different species of rubber trees (*Hevea brasiliensis*, Russian dandelion) [13]; curcumin (diferuloylmethane), a low-molecular-weight polyphenol isolated from the rhizome of *Curcuma longa* L. (turmeric) that has been shown to possess wide range of biological and pharmacological properties [14, 15]; and hyrtiomanzamine 4, a β-carboline alkaloid, isolated from the marine sponge *H. erectus* that exhibits in vitro immunosuppressive activity [16, 17]. Gradually the bioactive compounds present in NPs have widely been recognized as valuable products in drug discovery, food additives, cosmetics, and pigments and have been used in different industries. Table 1 summarizes the importance of some bioactive compounds known as NPs and extracted by different methods from a wide range of sources.

Table 1 Different extraction methods employed for bioactive compounds obtained from diverse sources

Bioactive compound	Source	Method	Importance	References
Pectin	Dried mango peel	Hot acid extraction followed by purification steps	Used in food industry	[18]
Cellulose and microcrystalline cellulose	Rice straw and banana plant waste	Chemical treatment followed by pulping and bleaching technique	Used as raw material for fuel and other chemical production	[19]
Bromelain (protease), cellulose and hemicellulose, polyphenol,	Pineapple waste (peel, pomace, core, and crown)	Buffer extraction followed by purification	Used as food additive, therapeutic agent, and polymer industry	[20]
succinic acid, and essential oil	Citrus peel waste	Extraction and acid and enzyme hydrolysis	Energy-intensive bioprocess of succinic acid production	[21]
Steviol glycosides	Extract of candy leaf (*Stevia rebaudiana*)	Extracted in water: acetonitrile (80:20, v/v) followed by UHPLC purification	Used as a natural sweetener	[22].
Taxol	Stem bark of the western yew (*Taxus brevifolia*)	Enrichment of alcoholic extract of the stem bark followed by partitioning between water and chloroform	The most well-known natural source of cancer drug	[23]
Azadirachtin	Seed kernels of neem (*Azadirachta indica*)	Pressurized liquid extraction followed by HPLC analyses	Natural insecticide	[24]
Eugenol	Abundantly found in clove buds; also found in cinnamon bark and leaves, tulsi leaves, turmeric, pepper, ginger, oregano, and thyme	Solvent extraction, hydrodistillation/ microwave-assisted extraction/ supercritical carbon dioxide extraction/ ultrasound-based extraction	Used as essential oil	[25]

(continued)

Table 1 (continued)

Bioactive compound	Source	Method	Importance	References
Resin	Poplar shoots	Extracted in dichloromethane followed by evaporation under nitrogen and resuspension in methanol and HPLC/MS-TOF analyses	Important compound in pharmacy and chemotaxonomy	[26]
Pyrethrin	Petals of *Chrysanthemum cinerariaefolium*	Solvent extraction (petroleum ether and methanol) followed by HPLC analyses	Natural insecticide	[27]
Anthocyanin	Blueberry fruits	Microwave extraction of blueberry powder followed by ethanol wash	Medicinal and nutritional value	[28]
Glycyrrhizic acid (triterpene glycoside) and glabridin (isoflavonoid)	Licorice roots	Solvent extraction followed by reverse-phase HPLC	Used in food and pharma industry	[29]

2 Metabolomics in NP Discovery

The term metabolome was first coined in 1998 by Steven Oliver in a review article on yeast functional genomics adding the newest member of the triad – genome, transcriptome, and proteome [30, 31]. This article reviews the quantitative analyses of gene function in the context of analyzing metabolic control and explores the effect of deletion or overexpression of genes on changing concentrations of metabolites. Followed by this, several articles explored the possibilities of using metabolite profiling to study metabolic regulation and extend it to cellular functions. A brief summary of those articles is provided in Table 2.

The aim of metabolomics is comprehensive qualitative and/or quantitative analyses of all metabolites present in a living system at a specific time and under specific influencing factors [37–40]. This approach also addresses the relationship between biochemical status and gene functions of the sourced organism and can further be used as biological fingerprinting of natural extracts [41]. Before the metabolomics approaches, well-characterized compounds were used as references to identify the functional roles of metabolites. The development of metabolomics methodology has been proved to be useful in investigating the metabolite variation unraveling the multiple pathways affected by external factors [42]. For example, an MS-based metabolomic study of the red algae *Gracilaria vermiculophylla* in response to

Table 2 Metabolite profiling of different classes of compounds from diverse sources

Metabolites	Source	Method	References
326 compounds: fatty acids, fatty alcohols, sterols, and aliphatics, hydroxy and amino acids, sugars, sugar alcohols, organic monophosphates, (poly)amines, and aromatic acids	Arabidopsis thaliana leaf extract	Solvent extraction, derivatization, and GC-MS	[32]
Isoprenoids: carotenes, xanthophylls, ubiquinones, tocopherols, and plastoquinones	Arabidopsis thaliana and tomato fruits	Solvent extraction and HPLC-PDA	[33]
Homoserine	Sycamore (*Acer pseudoplatanus*) and weed (*Echinochloa colonum*)	Perchloric acid extraction and NMR	[34]
Compounds generated in sucrose metabolism	Potato tuber tissue (*Solanum tuberosum*)	Solvent extraction, derivatization and GC-MS	[35]
Bioactive natural products such as barettin (brominated alkaloid) and 8,9-dihydrobarettin	Extracts from the marine sponge *Geodia macandrewii*	Lyophilization of frozen samples followed by aqueous and organic solvent extraction and UHPLC-ESI-HR-MS analysis	[36]

herbivory showed upregulated metabolites with some compounds increasing more than 100-fold in concentration, illustrating the biochemical pathway involved in the defense mechanism [43]. Modern approaches to identifying, analyzing, and quantifying NPs are heavily based on traditional methodologies of sample collection, harvesting, and extraction which with time have sophisticated in terms of employing cutting-edge technologies. Nuclear magnetic resonance (NMR) spectroscopy and mass spectrometry (MS) are two common analytical approaches to generate metabolomic data [44]. Spectral data obtained in both methods are used to characterize the chemical nature and quantification of metabolites [45, 46]. Unlike the MS approach which requires a rigorous sample preparation step involving the separation of the analyte of interest from complex biological matrices, NMR needs a less stringent sample process [45, 46]. Liquid chromatography (LC) and gas chromatography (GC) are generally used for separating the analytes before MS and together termed as LC-MS and GC-MS techniques [47]. A generalized workflow is described in Fig. 1.

The improved methodologies in extracting and characterizing NPs as potential components for different industries, especially as drug targets, have led to the rise of more than several thousands of NPs sourced from plants, fungi, and bacteria. Table 3 summarizes the list of NPs derived from different microbial sources and their potential use as therapeutic agents [48].

Fig. 1 Steps involved in metabolomic workflow start with sample preparation. Sample purification or enrichment is an optional step applicable to selected samples followed by detection and analyses by different analytical tools. The acquired data is thus processed and undergoes statistical analyses before submitting to the database

Table 3 Usages of different classes of NPs as potent therapeutic agents [48]

Name of NP	Class of NP	Name of microbial source	Potential usage	References
Erythromycin A	Polyketide	*Saccharopolyspora erythraea*	Macrolide group of antibiotics	[49–51]
Tetracycline	Polyketide	*Streptomyces* genus of *Actinobacteria*	Polyketide antibiotic	[52, 53]
Vancomycin	Amphoteric glycopeptide	*Amycolatopsis orientalis*	Antibiotic	[54, 55]
Streptomycin	Aminoglycoside	*Streptomyces griseus*	Antibiotics	[56, 57]
Nisin A	Polycyclic peptide	*Lactococcus* and *Streptococcus* species	Antibiotic	[58, 59]
Amphotericin B	Polyene	*Streptomyces nodosus*	Antifungal antibiotic	[60, 61]
Bleomycin	Glycopeptide	*Streptomyces verticillus*	Chemotherapy agent	[62, 63]
Rapamycin	Macrolide	*Streptomyces hygroscopicus*	Immunosuppressive agent	[64–68]

In some cases, NP derivatives and mimics serve the purpose as well. The Food and Drug Administration (FDA) has approved a significant number of modern drugs that are unaltered NPs and NP derivatives and mimetics along with synthetic drugs [69]. The growing number of reports of discovery and probable effects of NPs have led to the creation of databases [70]. The list of NP databases available to date is given in Table 4.

This chapter discusses the role of metabolomics in the qualitative and quantitative estimation of NPs elucidating the steps – sample preparation techniques and analytical methods. In addition, computational tools used in metabolomics data processing and metabolite identification are also discussed. Finally, the applications and challenges of metabolomics in the identification and quality control of NPs are outlined

Table 4 List of various NP databases available to date [70]

Databases	About	Number of NPS	References
Commercial databases			
Reaxys https://www.reaxys.com/	Database for substances, reactions and documents compiled	>200,000	n/a
SciFinder	The biggest collection of curated chemicals, and, subsequently, of NPs	>300,000	[71]
National Institute of Standards and Technology-NIST (version 17) https://www.nist.gov/	Standard reference databases for mass spectra (MS)	>250,000 molecules of natural origin (separation of NPs is not clearly marked)	n/a
Natural Product Discovery System (NADI)	Natural compounds from Malaysian plant species	>3000 natural compounds	[72]
Open-access databases			
ChEBI	Main focus is chemical ontologies	>15,000	[73]
ChEMBL	Considered as a repository for experimentally elucidated molecular structures and, in particular, drugs and drug-like chemical	>1800	[74]
ChemSpider	Chemical database offering very rich metadata, cross-references to a lot of other chemical sources, and advanced search	>9700	[75]
PubChem	Is an integrated platform of small molecules and biological activities	>3500	[76]
KEGG	Metabolites, chemicals that are produced by living organisms and that are involved in primary and secondary metabolisms	No estimate	[77]
MetaCyc	Metabolites, chemicals that are produced by living organisms and that are involved in primary and secondary metabolisms	No estimate	[78]
Databases for dereplication			
METLIN	Characterization of known metabolites and a technology platform for the identification of known and unknown metabolites and other chemical entities	>1 million molecules including primary metabolites, toxins, small peptides, and NPs	[79]
Databases for dereplication for NMR data			
NMRdata http://www.nmrdata.com/	Chinese initiative for the storage and elucidation of NP structures from NMR data	About 1,167,468 spectra	n/a

(continued)

Table 4 (continued)

Databases	About	Number of NPS	References
Generalistic databases of natural products			
SuperNatural II	Is a database that contains NPs together with their two-dimensional structures, computed physicochemical properties and predicted toxicity.	>300,000	[80]
NP from microorganisms			
Natural Products Atlas (NP Atlas) https://www.npatlas.org/	Cover NPs from microbes (bacteria, fungi, lichens, and cyanobacteria)	>33,000	n/a
Databases of drug-like natural compounds			
ChemIDplus	Database part of the TOXicology DataNETwork and chemicals that have a relationship with diseases, environment, environmental health, and poisoning	>9000 entries	[81]
Food			
FooDB https://foodb.ca/	Reference database on chemical food constituents associated with extremely rich and diverse metadata	>22,000	n/a

3 Metabolomic Workflow

When a whole set of metabolites present in a given biological matrix (plant, bacteria, fungi, animal, human, etc.) becomes the subject of experiment, then it is termed as untargeted or global analysis. In the case of targeted analysis, only selected metabolites treated in the similar physicochemical properties or involved in the same biochemical pathway are analyzed and quantified [82]. In both analyses, sample size, sample preparation, and statistical analysis of the data obtained are crucial to validate the hypothesis [83]. Randomization and biological replicates are important while selecting the samples [84, 85]. In this chapter, we will limit the workflow within the sample preparation and analyses of NPs.

The diverse chemical nature, wide concentration range, and influences of external stimuli complicate the extraction and quantitation process of NPs and give rise to different metabolomic strategies such as metabolite profiling, metabolic fingerprinting, metabolite target analysis, metabonomic, etc. Since, metabolomic analyses represent the snapshot of metabolites present in a certain time frame, minor changes in any step(s) of the workflow may highly affect the data interpretation and final outcome of the investigation. For plant sources, cultivation parameters, type of tissue collected, season of collection, developmental stage of the plant, harvesting time, etc., should be reported as metabolites are greatly affected by such parameters [86, 87]. For bacterial sources, the type of media used, bacterial strains, and conditions of growing bacterial cultures significantly affect the production of metabolites

[88]. External factors such as environmental conditions may induce both qualitative and quantitative variations in the metabolite composition of both primary and secondary metabolites [89]. Hence each step in metabolomic experiment starting from sample procuring to data analysis require meticulous attention to details [40, 90, 91].

3.1 Sample Preparation

(i) Identification of source, harvesting and storage: The metabolomic workflow starts with the identification of appropriate source to procure the sample followed by proper harvesting and storage [37, 85]. Minor changes in the sample collection, harvesting, or storage steps may highly impact the final analyses of metabolites leading to major changes in the observed metabolome [37]. The samples should be collected in a fast and uniform way to minimize technical bias and changes due to external factors, especially in the case of volatile compounds [92, 93]. In order to minimize enzyme-induced metabolic changes, sample should be frozen using dry ice or liquid nitrogen while harvesting fresh samples. In case of plant sources, unwanted components such as soil particles and dried foliage are also recommended to be removed before collection. Long-term storage should be avoided, and for short-term storage, samples are better stored using liquid nitrogen or kept in a freezer at −80°C [94].

Depending on the nature of metabolites obtained from plant samples, drying is an extra step to minimize decomposition due to enzymatic activity in the presence of moisture. Also, varying levels of water in the sample may affect the quantitation of metabolites as well [87]. Drying can be carried out by airdrying, oven-drying, freeze-drying, and trap-drying [95]. Among these, freeze-drying (lyophilization or cryodesiccation) is the most suitable method to dry plant material [87]. In this method, the tissue is rapidly frozen and dehydrated at a temperature below −30°C under vacuum with a desiccant. The dried tissue is then infiltrated with paraffin under a vacuum. Leaf tissue and root tips are the easiest and the most difficult to freeze and dry, respectively [96]. The microwave heating method can also be used for drying samples as metabolomic changes due to enzyme activities such as peroxidase and glycosidase are almost destroyed in this method [97]. For example, classical extraction showed the presence of multiple glycosides as found in fresh flowers compared to only one glycoside while the sample was given a short microwave treatment [98]. The Metabolomics Standards Initiative (MSI) is by far the best approach to determining the minimum parameters related to experimental design, sample extraction to data analysis for validating plant metabolomics studies [94].

The bacterial strains used for metabolomics study can significantly affect the sample preparation, e.g., the presence/absence of cell wall, chemical composition, and permeability of cell membranes of gram-positive and gram-negative bacteria require different sample preparation protocols. Bacterial cells growing in culture flasks/bioreactors are harvested by centrifugation or fast vacuum filtration [88]. This step separates the bacterial cells from the culture media followed by the extraction of intracellular metabolites from bacterial cell pellets and extracellular metabolites from the supernatant secreted by the bacteria during its different growth stages [99]. In this context, it is important to note the time point when the cells are harvested as bacterial cultures which produce different types and levels of metabolites at its different growth phases which indicate the dynamics of physiological processes [88].

Marine organisms, for example, coral is recommended to be harvested from a sheltered lagoon or reef flat in shallow water, whereas the collection of coral from deeper water follows a completely different technique [100]. Again the production of peloruside A from aquaculture of the sponge *Mycale hentscheli* is affected by parameters such as light and fouling intensity in the farm setting [101]. Different biotic and abiotic conditions greatly affect the production of distinct metabolites indicating that factors responsible for metabolite production are important in the discovery of novel compounds [102, 103]. This has led to the development of "OSMAC" (one strain many compounds) and co-culture methodologies to produce differential compounds by modifying the abiotic or biotic culture [104, 105].

(ii) Sample quenching and extraction: Followed by the previous steps, the quenching of the samples at a specific time is necessary to stop the physiological processes of plant or microbial source and "freeze" the metabolic snapshot [106, 107]. Although quenching with cold methanol is regarded as gold standard for plant and microbial samples, leakage of intracellular metabolites from microbial sources were reported [108]. Alternative solvent systems and the fingerprinting study was used to minimize the leakage and/or compensate for the error induced by leakage [109]. The use of liquid nitrogen compared to ethanol/methanol/ice-cold buffer was reported to have less influence on cell viability [110]. However, the ratio of quenching solvent to sample, solvent concentration, presence of additives, processing time, and growth phase of microbial culture play important roles in minimizing the metabolite leakage [108].

Due to the diverse chemical nature of metabolites and complex biological matrices, different extraction protocols are employed for obtaining the bioactive compound(s) [111]. Depending on the biological material, sometimes the harvested and quenched samples may require an extra processing step by lyophilization, cell lysis, and/or grinding prior to extraction. Based on the physiochemical nature of the metabolites, different methods such as solvent extraction, distillation, pressing, and sublimation could be used. Generally, organic solvents are selected and combined depending on respective solvent polarity index and **miscibility** (when more than one solvent is used) according to HPLC Solvent Guide, solvent miscibility and viscosity chart [112]. In the solvent extraction method, factors such as chemical properties of the extracting solvent and analytes, particle size of the sample, solvent-to-solid sample ratio, solubility, dissolution rate, extraction temperature, processing time and method (ultrasonic treatment), and miscibility of solvents affect the extraction efficiency [87, 112]. Solvent polarity, selectivity, toxicity, and inertness plays a major role in the extraction step affecting both the quality and quantity of the metabolites. Generally, hydrophilic solvents are used to extract polar and semipolar metabolites, whereas lipids can be extracted with more hydrophobic

Table 5 The use of different solvent systems and extraction methods applied to diverse classes of metabolites

Class of metabolites	Solvent(s) used	Extraction method used	References
Untargeted large-scale plant metabolomics	Ice-cold 99.875% methanol acidified with 0.125% formic acid	Sonication at 40 Hz for 15 min at 20°C followed by centrifugation and filter through 0.2-μM PTFE membrane	[118]
Sinapates, glucosinolates, flavonoids, and anthocyanins and primary metabolites from *Arabidopsis* leaves	Methyl *tert*-butyl ether and methanol (3:1, v/v), solvent 1 Water and methanol (3:1, v/v), solvent 2	Briefly vortexed after adding solvent 1 followed by incubation, sonication, and phase separation with solvent 2	[119]
(i) Primary metabolites (lipids, proteins, cell wall polymers) from *Arabidopsis* seeds (ii) Phytohormones from *Arabidopsis* seeds	(i) Extraction in ice-cold methyl *tert*-butyl ether followed by phase separation in water and methanol (3:1, v/v) (ii) Extracted in pre-cooled buffer (2-propanol:H_2O:HCl, 2:1:0.002) followed by incubation at 4°C and liquid-liquid separation by dichloromethane	Briefly vortexed, incubated, and centrifuged for 5 min at 4°C followed by drying and resuspension of the samples	[120]
Phytohormones from *Arabidopsis* rosettes	Extracted with 80% (v/v) methanol	Frozen ground material was extracted and analyzed	[121]
(i) Lipids and (ii) primary metabolites from soft corals	(i) Methyl *tert*-butyl ether and methanol (3:1, v/v), solvent 1 Water and methanol (3:1, v/v), solvent 2 (ii) Methoxyamine-HCl/pyridine solution	(i) Incubated at 4°C in solvent 1 followed by sonication 15 min in an ice-cooled bath and phase separation with solvent 2 (ii) Resuspended and heated at 37°C for 1.5 h	[122]
Biosurfactants from *Rhodococcus* bacterial cultures	MTBE, dichloromethane, chloroform-methanol (1:2 or 2:1 v/v) and MTBE-chloroform (1:1 v/v)	Extracted followed by sonication, phase separation, and rotary evaporation at 50°C under reduced pressure	[123]

solvents. Metabolomic studies are designed to detect as many metabolites as possible in an organism; thus, solvents capable of extracting diverse groups of metabolites are preferable [87].

As extraction is a preliminary step of purifying or concentrating the analytes from the sourced samples, liquid-liquid fractionation (LLF) method which employs solvent extraction and partitioning is useful in removing contaminants or interfering compounds, e.g., removing complex carbohydrates while extracting polyphenols or deproteinization [37, 87]. Chloroform/methanol mixture in different proportions is the gold standard of LLF [113, 114]. Other alternatives are chloroform/methanol/water and methyl *tert*-butyl ether for the analysis of lipids and polar metabolites from plant, microbes, and mammalian sources [32, 115–117]. Few examples of different solvents and extraction methods used are outlined in Table 5.

Apart from LLF several other methods such as Soxhlet extraction, distillation, reflux extraction, decoction, percolation, maceration, solid-phase extraction (SPE), supercritical fluid extraction (SFE), pulsed electric field extraction (PEFE), microwave-assisted extraction (MAE), pressurized liquid extraction (PLE), enzyme-assisted extraction (EAE), and ultrasound-assisted extraction (UAE) are also used for metabolite extraction. Soxhlet extraction has been a standard technique for over a century where the extract from the sample is continuously condensed at best ambient temperature. Although shortcomings in terms of the final yield of the product have been reported about this technique, it has been improvised in recent years by introducing high pressure, automation,microwave assistance and ultrasound assistance [124, 125].

3.2 *Analytical Methods of Purification and Quantitation of the Analyte of Interest*

Between the metabolite extraction and its analyses, they are further processed/purified for enrichment [87]. Liquid chromatography (LC) and gas chromatography (GC) are the main methods used in this step. The interaction between the metabolites and the materials used in LC and GC columns is responsible for the separation of the analytes of interest. Due to the diversity of physical-chemical properties – polarity, molecular weight, solubility, and volatility – the simultaneous characterization of all metabolites is difficult, requiring the combination of techniques expanding the metabolite profile [40, 126]. The determination of the best techniques to be used is directly related to the experimental objective, the sample matrix, and its properties [126, 127]. Hence GC-MS or LC-MS and (NMR) spectroscopy have been the most commonly used analytical techniques for metabolomics studies; as well as capillary electrophoresis (CE), which combined with mass spectrometry (CE-MS), has become a promising tool [126, 128]. This section discusses these analytical techniques for metabolomics studies, highlighting the advantages, drawbacks, and differences in the applicability of each technique.

(i) **Gas Chromatography-Mass Spectrometry (GC-MS)**

GC is one of the most used and pioneering techniques for metabolic profiling in different research fields coupled with mass spectrometry (MS). It is a suitable tool for identification and quantification of metabolites, especially volatile and semi-volatile organic compounds, as it combines high-separation efficiency with selective and sensitive mass detection [126, 129]. GC-MS has been widely applied in metabolomics studies and is considered as one of the most efficient and reproducible analytical technique for metabolomics research for metabolite profiling and quantification [127]. Other important attributes to GC-MS are its versatility in detecting compounds of a wide range of molecular weights and polarities, robustness, separation capacity, selectivity, sensitivity, and reproducibility [37, 40, 127].

In this technique, the sample analytes obtained in the extraction step is passed through the chromatographic column, composed of a stationary phase and a mobile phase (gaseous), called "carrier gas," which must have a high degree of purity and inert to the sample. The most commonly used carrier gases are helium and nitrogen. Often hydrogen is also used. Two types of columns are used in gas chromatography – packed columns and capillary columns which differ in diameter, length, and material. Capillary columns are the most used due to better separation resolution and are of the liquid-on-solid type, which can be polar (e.g., polyethylene glycol) or nonpolar (e.g. 5%-phenyl-methylpolysiloxane). The separation occurs based on the boiling point of the compound and its interaction with the stationary phase [130]. Gas chromatography can be coupled to different detectors, such as a flame ionization detector (FID) or mass spectrometry (MS), the latter being the most widespread technique for identifying and quantifying metabolites. MS is capable of detecting all ionizable compounds and obtaining mass spectra at each time point, providing information on the molecular structure of compounds, and can be used for automated search and match with that of mass spectrum library [131].

Temperature is one of the most critical parameters that must be strictly controlled, especially in GC-MS analysis, as high temperatures increase the possibility of thermal degradation of analytes leading to multiple peaks [132]. Volatile metabolites, released by plants, for example, during their growth and development function in the defense mechanism against predators, also attract pollinators [127]. The untargeted approach to identify this group of compounds resulted in the creation of the term "volatome" or "volatilome," designated for the comprehensive analysis of volatile compounds in any matrix. Detection of polar, thermolabile, nonvolatile metabolites require chemical derivatization prior to analysis. Silylation, acylation, and alkylation are common types of derivatization methods used prior to injecting samples to the GC-MS [127, 132]. This improves volatility, thermal stability, sensitivity, and detector response. Several classes of volatile and nonvolatile metabolites, such as phenolics, fatty acids (FAs), alkaloids, and terpenoids, have been analyzed and identified by GC-MS [37, 126, 127].

The use of a metabolomics approach based on target and nontarget GC-MS in natural products has been reported to investigate the chemical profile of the lipophilic extract of the bark of *Vitex pinnata* and reveal the presence of several classes of phytochemicals, such as hydrocarbons, terpenes (monoterpenes, diterpenes, sesquiterpenes and triterpenes), and phytosterols (γ-sitosterol and stigmasterol) [133]. The extract of *V. pinnata* bark in n-hexane was analyzed using GC-MS equipped with RTX-5MS fused bonded column (30 m × 0.25 mm i.d., × 0.25 μm film 47 thickness). The initial column temperature was 45°C to final temperature of 300°C. Injector temperature was 250°C; ionization voltage, 70 eV; and ion source,

200°C. The sample was injected at a split ratio 1: 15. To identify the compounds, the retention indices (RI) were calculated by injecting the standard n-alkanes series (C8–C40) under similar conditions. Subsequently, the mass spectra and RI were compared with those of the National Institute of Standards and Technology NIST chemistry webbook library and other literature. Another example of using GC-MS is to discriminate the metabolic profiles of whole sorghum containing high tannin and low tannin and its derivatives obtained by fermentation [134]. GC-MS has also been used for the identification of sugars and phenolic compounds in honey powders [135], as well as the use of two-dimensional GC × GC-MS to investigate volatile compounds in the samples of green tea grown at low and high altitudes, detecting more than 200 sensory and bioactive compounds in each of them [136].

However, this method has limited capacity to separate and identify low-molecular-weight compounds (50–600 Da) with high-vapor pressures (volatile compounds). As complex matrices are generally analyzed, the sample preparation steps are expensive and time-consuming and often require extra steps of derivatization to adequately vaporize the desired analyte [129]. Another challenge is the separation of overlapping peaks in the raw chromatogram. However, this limitation has been overcome with the use of two-dimensional GC × GC together with high-resolution mass spectrometry. In this way, the compounds could be separated into two columns with different properties (nonpolar versus polar), thereby producing enhanced resolution and peak capacity [37].

(ii) **Liquid Chromatography-Mass Spectrometry (LC-MS)**

Liquid chromatography-mass spectrometry (LC-MS) has become a widely used technique due to its applicability to measure a wide range of metabolites. The advantages being (i) direct injection of extracts obtained with (ii) easy sample preparation and (iii) easy extraction, unlike GC-MS where derivatization is prerequisite [137]. GC-MS and LC-MS techniques are often used in a complementary way to improve the identification of metabolites; however, this choice increases analysis time and cost. A correct selection of instrumental parameters and separation mechanisms corroborate to obtain satisfactory results. For LC-MS, chromatographic separation is one of the important factors. According to their chemical properties, the metabolites are separated using specialized columns, such as C8 and C18, reverse-phase column. Due to its ability to separate semipolar compounds such as phenolic acids, flavonoids, glycosylated steroids, alkaloids, and other glycosylated species, C18 is often the popular choice of column [138, 139]. Ion exchange and hydrophobic interaction columns (HILIC) are also used. As every technique has its limitations, the most exhausting step in LC-MS is the identification of the metabolites obtained, making it one of the biggest challenges of this tool for metabolomic analysis [139].

Examples of LC-MS-based NP studies include analyses of extracts obtained from different parts of the Brazilian plant *Annona crassiflora* to discover compounds with larvicidal activity against *Aedes aegypti* [140]. The *Annona crassiflora* extract was partitioned using Diol cartridges using hexane, ethyl acetate, and methanol as mobile phases followed by chromatographic separation in C18 column. The results confirmed the ability of metabolomics to discriminate metabolites between active and inactive samples using LC-MS and LC-MS/MS data. Other authors observed that LC/MS-PCA, known as secondary metabolomics, was effective in the selection of natural bacterial products derived from the sea, helping in the discovery of new drugs, in addition to highlighting the use of the technique as a way to reduce the time of LC/MS analysis of chromatographic data [141]. Furthermore, metabolic profiles and variations among five truffle species using untargeted metabolomics technology based on an ultrahigh-performance liquid chromatography tandem mass spectrometry method (UHPLC-MS/MS) were also reported [142].

(iii) **Nuclear Magnetic Resonance Spectroscopy (NMR)**

Another important analytical technique in metabolomics is NMR. Although NMR and MS can be used as isolated techniques, they are often considered as complementary way to improve analytical performance [143]. Among the main advantages of NMR are the identification of unknown compounds, with the elucidation of structures and recovery of used samples [143, 144]. Low sensitivity and requirement of higher amount of sample compared to MS are the drawbacks. The considerable signal overlap in the NMR spectra could make it difficult to identify and properly integrate the peaks [144, 145]. However, the development of NMR hardware could improve the sensitivity, and the use of two-dimensional NMR may serve as a partial solution of the peak overlapping problem since it presents much better resolution than one-dimensional ^{1}H NMR [144].

The greater applicability of NMR is observed for compounds of moderate to high polarity (Kim et al., 2010). NMR-based metabolomics analysis has been successfully applied to NPs produced by endophytic fungi and isolated from the leaves of medicinal plants *Hypericum perforatum*. Using NMR metabolomics, two natural products, lignicol and isolignicol, have been isolated [146].

(iv) **Capillary Electrophoresis Coupled with Electrospray Mass Spectrometry (CE-MS)**

Another extremely important separation method in the metabolomics approach is electrophoresis. While GC is limited to volatile and thermostable compounds, or compounds that can be derivatized to produce volatile products, LC is more robust with the ability to separate semipolar compounds. Despite these advantages, many metabolites are too polar to be significantly retained by the most used reverse-phase columns. As most metabolites are polar and ionic, a good option is to use capillary electrophoresis (CE-MS). This separation technique is based on their mass-to-charge ratios. This could be regarded as a complementary technique to GC and LC. The advantages of CE technique are the (i) low sample demand, (ii) speed, (iii) separation efficiency, (iv) minimal pre-treatment of the sample, and (v) low cost. As a limitation, there is low sensitivity to concentration, which can be mitigated with the use of MS detector, confirming CE-MS as an important ally for metabolomics investigations [147].

A capillary electrophoresis method using nonaqueous separation solutions in combination with MS and MS/MS for the identification and quantification of glycoalkaloids and their relative aglycones has been reported [148].

3.3 *Data Processing and Curation*

The aim of MDP is to extract biologically relevant information from the acquired data and use it in the subsequent steps to interpret the significance of various metabolic pathways. Analyses of metabolomic data are comprised of four steps: (1) raw data preprocessing including compound identification, (2) data processing including data transformation and data normalization, (3) statistical analysis, and (4) data interpretation [149]. Firstly, the raw data collected as signals and peaks from analytical techniques – nuclear magnetic resonance (NMR) and mass spectrometry (MS), respectively – are preprocessed which include noise reduction, background correction, peak picking, and compound identification [150]. In the data transformation step, the preprocessed or clean data from the previous step is converted into more useful forms either by mathematical operations or by changing formats [149]. This converted format is then normalized to minimize systematic and technical variations before statistical evaluation [150]. Statistical analyses and subsequent data interpretation are used to screen metabolites based on the experimental design and hypothesis followed by a downstream process of the network- and pathway-based data analyses [151].

For untargeted metabolomics study using NMR, the cleaned data excluding the noise region can be used as the input for principal component analysis (PCA) and partial least squares discriminant analysis (PLS-DA) or orthogonal partial least squares discriminant analysis (OPLS-DA) to find out the metabolite features [85]. For untargeted MS data, detection of chromatographic peaks, extracted ion chromatograms (EICs), annotated features, and chromatogram alignment are important aspects for downstream process [152]. The raw MS data contain mass-to-charge ratio (m/z); retention time, i.e., specific time when the mass is acquired; and relative abundance of each m/z ratio. Data preprocessing is done in two steps: Firstly, the m/z ratio and relative abundance for each retention time are processed, and secondly, the retention time and relative abundance for a particular m/z ratio or all mass-to-charge ratios are calculated. The relationship between the retention time and the summation of relative abundance values for all the m/z ratios produces the total ion chromatogram (TIC), whereas that between the retention time and relative abundance values for a particular mass-to-charge ratio gives the EIC [85].

For metabolomics data, some of the available tools for preprocessing are [153] cloud version of XCMS [154]; open-source software MZmine [155], MS-DIAL [156], and MAVEN [157]; SIEVE by Thermo Fisher, Python Package-TidyMS

[158], and NeatMS [159]; and R-package: AutoTuner [160], MetumpX [161], MeTaQuaC [162], Dbnorm [163], and MetaClean [164]. Data obtained from NMR spectroscopy is subjected to chemometric analysis [165]. Then commercially available softwares including ACD (www.acdlabs.com), Chenomx (www.chenomx.com), and MestreNova (http://mestrelab.com/) are used for preprocessing the raw data during which the added internal standard is assigned to 0 ppm [166].

For metabolomics data, some of the available data-based normalization tools are [167] R-based, MetabR [168]; Microsoft Excel based, MetaboDrift and NormalizeMets [169, 170]; and web-based, NormalyzerDE [171], NOREVA [172], MetaboGroupS [173], and PseudoQC [174].

While processing one-dimensional NMR spectra several elements like protons coming from water, urea, proton resonances of noise regions of upfield to sodium trimethylsilylpropanesulfonate or DSS peak and downfield of metabolites are need to be removed prior to statistical analyses [166]. Also, the pH-based chemical shift needs to be taken care of by using NMR peak alignment tools [175] or pH-sensitive NMR libraries such as Chenomx (https://www.chenomx.com/) or using a buffer solution to control pH [166].

Large metabolomics datasets generated by MS contain thousands of features identified falsely or with imperfect integration. Hence filtering methods are crucial to remove noise prior to explaining the metabolome. Programs like Metaboanalyst and Workflow4Metabolomics could be used apart from manual filtering to remove the noise. In manual filtering, the cleaned data is separated into high/low features followed by appropriate cutoff [176].

For metabolomics data, the most popular statistical methods are univariate and multivariate. Principal component analysis (PCA) and partial least squares (PLS) are established methods for multivariate analysis of metabolomics data. Some of the available statistical analyses and data interpretation/visualization tools are: R-package including struct (statistics in R using class-based templates) [177] and rawR [178]; web-based – EpiMetal [179], Metabolite-Investigator [180], and VIIME (VIsualization and Integration of Metabolomics Experiments) [181]; web-based and R-package – NORmalization and EVAluation (NOREVA 2.0) [172], Metabolite AutoPlotter [182]; SAS-based %polynova_2way [179]; standalone softwares – Metaboverse [183], JavaScript mass spectrometry (JS-MS) 2.0 [184].

Finally, the output is used for enrichment and pathway analyses to identify significant expression changes among the functionally related metabolites [150]. MetPA and MSEA are web-based tools used for pathway analysis and data visualization based on the Kyoto Encyclopedia of Genes and Genomes (KEGG) [185, 186].

For broader application-based research, metabolomics works need to be replicated and reproduced in laboratories of diverse backgrounds. Availability of detailed experimental protocols which are prerequisite to generate metabolomics data with highest level accuracy are often not found in the publications due to various reasons. Hence there are specified standards for reporting minimum meta-data metabolomics studies developed over the years by CAWG including sample preparation, extraction protocol, details of analytical techniques used, data processing strategies, annotation, and statistical analyses [94, 99, 187].

4 Applications of Metabolomics in NPs

4.1 Dereplication of NPs

NPs being the source of important bioactive compounds bear wide-spectrum applications, especially in drug discovery. To find the "lead compounds," screening of NPs follows two types of procedures – (i) using previously purified natural compounds obtained from commercially purified NP libraries and (ii) the crude extract of NP subjected to a primary screening followed by the detection and activity-guided purification of the bioactive constituent [80]. Open-access spectral repository like Global Natural Products Social Molecular Networking (GNPS; http://gnps.ucsd.edu) provides raw, processed, or identified tandem mass (MS/MS) spectrometry data useful in the discovery of new NPs [188]. Dereplication uses the chemical information of a known natural product to identify that compound in an experimental sample without repeating the steps of isolation and structure determination [189]. Figure 2 illustrates the steps involved in the dereplication process.

An example of combinatorial approaches to dereplicate NPs from bacteria grown on corals involve both in silico and experimental procedures coupled with genome analyses where metabolomics data is useful in selecting the appropriate candidates involved in symbiotic relationship between the bacteria and coral [190]. Computational approaches of dereplication method includes algorithms like DEREPLICATOR and VarQuest to identify NPs from mass spectra database [189, 191].

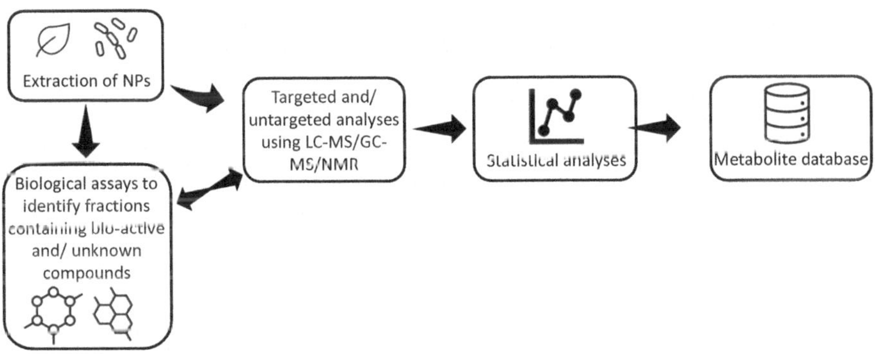

Fig. 2 Steps in dereplication process of NPs – the extracted NP either directly subjected to targeted or untargeted metabolomics followed by biochemical assays for identifying bioactive compounds followed by metabolomics depending on the sample or the reverse way. The acquired data is processed, analyzed, and submitted to database

4.2 Metabolic Map Profiling of NPs

Metabolic profiles of NPs help in understanding their respective pharmacological significance and toxicity, if any, along with other biological activities [37]. For example, the metabolic map of siamenoside I profiled in rat by a high-performance liquid chromatography-electrospray ionization-ion trap time of flight-multistage mass spectrometry (HPLC-ESI-IT-TOF-MS) method identified 86 metabolites, of which 8 were already established as bioactive compounds [192]. Siamenoside I is the sweetest mogroside (glycoside of cucurbitane, a plant triterpene) obtained from a fruit, *Siraitiae* sp. Different morgosides and morgol generated by the metabolism of siamenoside I have been shown to have hypoglycemic, antioxidant, anti-inflammatory, anticarcinogenic, and antitumor targeting effects [193]. Of these, the metabolism of morgoside V in healthy and type 2 diabetic rats has potential for developing new drugs for treating insulin resistance and diabetes [194, 195].

4.3 Quality Control (QC) of NPs

Increased interests in using plant-derived medicines to prevent diseases and/or to improve the overall quality of life have attracted the attention of scientists and pharmacists [196]. They have been shown to have fewer side effects and are convenient and economical to use [197]. As NP-derived products, especially drugs, contain multiple compounds, it is important to use metabolic profiling of multiple constituents to produce quality compounds especially NP-derived drugs [86]. HPLC was used to be the technique for QC, e.g., turmeric (*Curcuma longa rhizoma*), which was shown to have antiparasitic, antimutagenic, and antimicrobial properties, was analyzed using HPLC, and curcumin, desmethyoxycurcuin, and bisdesmethoxycurcuin were found to be the main bioactive substances [198]. Red ginseng (*Ginseng radix rubra*) contains ginsenoside Rg3 and white ginseng (*Ginseng radix alba)* contains ginsenoside Rb1, ginsenoside Rg1, and panaxadiol as bioactive compounds as analyzed by HPLC and GC [199, 200]. Ginsenoside has been shown to possess anticarcinogenic, immunomodulatory, anti-inflammatory, antiallergic, antiatherosclerotic, antihypertensive, antidiabetic, and antistress activity and effects on the central nervous system [201]. Due to the presence of multiple constituents in NP-derived drug, HPLC-based QC methods have now been replaced with other analytical tools including GC-MS, ICP-MS, LC-MS, and multivariate statistical analyses to quantify the ingredients of NP-derived drugs [202, 203]. Figure 3 illustrates the application of plant metabolomics in the quality control of Chinese material medicines derived from various plant sources [204].

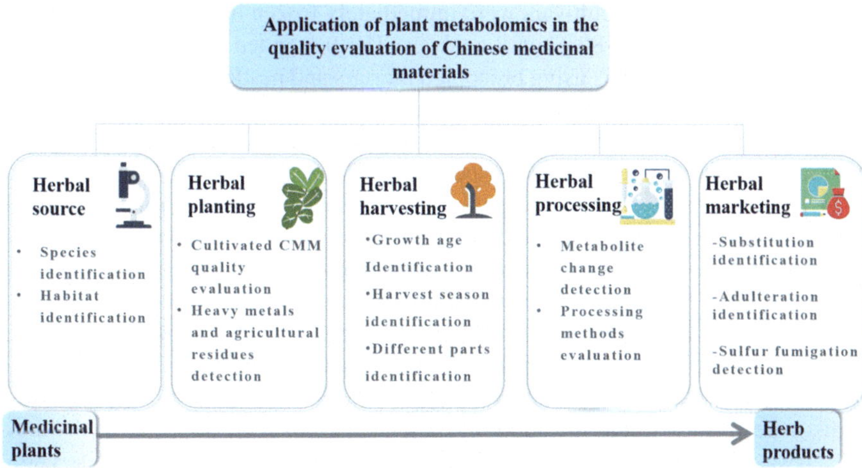

Fig. 3 Application of plant metabolomics in the quality control of Chinese material medicines derived from various plant sources [193]. From identification of medicinal plant sources to commercial production of herbal product, metabolomics is used in qualitative and quantitative analyses

4.4 Metabolomics in Revealing the Medicinal Properties of NPs

Over the past 20 years, more than one-third of FDA-approved therapeutic agents are derived from NPs, and more than 50% small-molecule drugs have been sourced from NPs, semisynthetic NPs, and NP-derived mimetics [205]. Till date, FDA has approved two NPs sourced from plants as prescribed botanical drug products after those have fulfilled the Botanical Guidance definition – Veregen® ointment 15% and Fulyzaq™. Along with these, some other botanical drugs that are included in the over-the-counter (OTC) drug review is cascara, psyllium, and senna (https://www.fda.gov/). Veregen™ ointment 15% is a topical medicine for skin use only for the treatment of warts on the outside of the genitals and around the outside of the anus. The active ingredient of Veregen™ ointment is sinecatechins (15%), a partially purified fraction of the water extract of green tea leaves from *Camellia sinensis* (L.) O Kuntze. A total of 150 mg of sinecatechins is present per gram of the ointment in a water-free ointment base consisting of isopropyl myristate, white petrolatum, white wax, propylene glycol palmitostearate, and oleyl alcohol. It includes epigallocatechin gallate (EGCg), epicatechin (EC), epigallocatechin (EGC), epicatechin gallate (ECg), and some additional minor catechin derivatives, i.e., gallocatechin gallate (GCg), gallocatechin (GC), catechin gallate (Cg), and catechin (C). Apart from that the extract also contains gallic acid, caffeine, and theobromine and undefined botanical constituents derived from green tea leaves.

The protocols used for metabolite analyses of *Camellia sinensis* extract and subsequent biological studies explain how the metabolomic workflow helps in unravelling various usages of available bioactive compounds in NPs [206, 207]. The excised tea leaves from different cultivars were immediately frozen in liquid nitrogen upon collection and stored at −80°C until analysis. The frozen tea leaves were individually ground to fine powders using precooled mortars and pestles followed by lyophilization and cold methanol extraction. In one method, the extracted sample was then centrifuged and run on UPLC-QTOFMS and UPLC-QqQ MS for both nontargeted and targeted metabolite analyses, respectively. The acquired data were processed using Progenesis QI software for peak picking, normalization (normalized to all compounds), signal integration, and initial peak assignments followed by PCA and PLS-DA analyses. Finally, metabolites were identified by comparing accurate masses, MS/MS fragmentation patterns, and isotope patterns with authentic standards as available in different metabolite databases such as Metlin, HMDB, MassBank, ReSpect, and KNApSAcK, and literature references. The individual mass spectrum was manually inspected to verify if the software-predicted fragments were from a single metabolite [208]. In the other method, the methanolic extract was derivatized using BSTFA and subjected to GC–TOF-MS. The chromatograms were processed using ChromaTOF software followed by noise removal, peak picking, normalization, and metabolite annotation with the NIST 05 Standard mass spectral database and Fiehn database. The normalized datasets were analyzed using SIMCA-P+11.5 for multivariate statistical analysis followed by PCA and PLS-DA analyses. Finally, the Kyoto Encyclopedia of Genes and Genomes (KEGG) database was used to explain the biological significance [209]. For ^{1}H NMR method, the collected spectra were analyzed for chemical shift w.r.t. deuterated methanol [210].

Fulyzaq™ is orally consumable tablet administered for noninfectious diarrhea in patients with HIV/AIDS on antiretroviral therapy. It contains 125 mg of crofelemer, NP with clinical significance and derived from the red latex of *Croton lechleri*. Crofelemer is an oligomeric proanthocyanidin and mixture of catechin, epicatechin, gallocatechin, and epigallocatechin monomer units linked in random sequence. The whole plant except root was dried in shade for about 10 days and then grounded using a laboratory mill followed by solvent extraction using solvents of different polarity and maceration. The extracted sample was then dried, resuspended, and subjected to HPLC-MS and one-dimensional and two-dimensional NMR studies [211].

5 Future Directions – Challenges and Prospects

In the past few years, the enormous advancement in technologies like separation science, spectroscopic-based methods, ultrasensitive in vitro assays, and high-throughput screening (HTS) have had a profound impact in NP-based research [212]. It envisaged the increased applications of NP-derived products, especially in the pharma industry. Green chemistry and circular economy are a few of the beneficial by-products of increased use of NP-derived products [213]. However, like other fields, existing methodologies used in NP works have some barriers, and steps involved in metabolomics, one of the major techniques in NP research, have challenges as well. Starting from the sample preparation to data analyses in the metabolomics workflow as represented in Fig. 1, shortcomings have been seen from the beginning and rerouted accordingly to streamline the process leading to the present-day cutting-edge technology.

Many NPs have poor or low solubility, and chemical stability, toxicity, and the concentration of different metabolites are greatly affected by parameters like harvesting time, condition/growth stage of sources, etc. which make the sample preparation steps non-reproducible [42]. NMR and MS are the two important analytical tools for identification and quantitation of NPs so far. Unlike MS spectrometry, NMR spectroscopy is a direct quantitation method involving fewer and simpler sample preparation steps. However, the sensitivity of MS outweighs NMR in many cases [214]. The apparent strength of MS in terms of sensitivity is often compromised due to a phenomenon called ion suppression caused by the matrix effect of diverse chemical compounds present in a biological sample [93]. Another problem caused by matrix effect is frequent occurrence of contamination of the MS source and adduct formation. The SPE (solid-phase extraction) method is often employed to efficiently remove matrix components such as proteins and salts [215]. The use of isotopic standards in quantitative analyses of metabolites is important for understanding flux and biomarker discovery. However, synthesizing stable isotopic standard is often difficult and expensive adding to a bottleneck in accurate quantitation of metabolites [216]. Few of the major challenges in the identification of compounds from natural sources is frequently encountering already-discovered compounds while analyzing the big pool of spectral data and the manual curation of spectral interpretation in MS/MS-based metabolomics data [205, 217]. In order to enable this time-consuming and tedious way of NP screening to become more efficient, dereplication method is employed to analyze the extracts of microbial and plant samples [202]. Again, for untargeted metabolomics, annotation of metabolites is often a bottleneck which is being addressed by ion mobility-mass spectrometry (IM-MS) technology. It provides the ion mobility CCS atlas, known as AllCCS, useful in predicting both known and unknown metabolite annotations from biological samples while combined with in silico MS/MS spectra [218].

Integration of metabolomics with other high-throughput omics technologies like transcriptomics and genomics and bioassays could be more effective in understanding the molecular mechanisms of metabolite production and identification of potent bioactive compounds [219, 220].

Apart from the troubleshooting in the technical steps, there's a major concern about the environmental impact preceded by the exploitation of NPs [205]. For example, natural rubber obtained as latex from *Hevea brasiliensis* has shifted the economic condition of small-holding farmers overnight due to huge demand of rubber-based products, high income, government incentives, surge of investors, etc., impacting natural resources and biodiversity to a great extent. Intercropping, substitutes of rubber tree with native plants producing latex, could be remedial actions to control the ecological damages [221]. Regarding marine organism-derived NPs, coral reefs are experiencing rapid global degradation due to both climate change and exploitation to fulfill the demand for reef-derived natural resources [222]. Another example is 13 tons of the marine bryozoan *Bugula neritina* were harvested producing 18 g of bryostatin 1 for clinical phase 1 studies as anticancer compound [223]. One way of freezing the present decay and restoration of the marine environment could be achieved by controlled and ecologically feasible ways of collecting samples – another way of encouraging marine biotechnology or by chemical synthesis [224].

6 Conclusion

Metabolomics has emerged as one of the powerful tools in the discovery of bioactive compounds available in natural resources. However, the amount or concentration of those compounds vary greatly with time and within different parts of the same source. Also, the NPs are heavily compacted with millions of metabolites, each with varying degree of concentration and produced by continuous physiological processes which in turn are under the direct/indirect influence of environmental stimuli. The interconnected metabolic pathways also impart positive/negative feedback effects on each other with accelerated/deaccelerated rate of generation, conversion, and degradation of different metabolites. In this complex cellular milieu, it is needless to say that sample preparation for metabolomic studies require a great deal of meticulous practices in each step – sample collection, harvesting, and storage with an additional step of sample quenching to freeze the metabolome snapshot. This intact sample is then passed through a pipeline to generate the information of metabolites captured in that snapshot. The pipeline is composed of processing steps to extracting detectable and quantifiable amount, identifying, characterizing, and/quantifying the bioactive compounds present in the crowd of metabolites with minimum perturbation. The presence of a large number of metabolites belonging to different classes of diverse physiochemical natures complicates the steps in the pipeline. The choice of solvent system plays a pivotal role in the dissolution of the sample and acts as a preliminary round of separation of several class of metabolites. The mixture is further processed to remove the impurities by different separation techniques for enrichment of metabolites of interest. The purified sample is then subjected to analytical tools like NMR and MS to generate raw data which goes through processing steps to tidy up the acquired data by removing background noise, bad signals. Different web-based and standalone programs perform the data processing, analyses, and visualization steps. Finally, the detected compounds are assigned as tentative metabolites present in the sample based on matching their mass/spectra with publicly available database or mass/spectral libraries. The data is then deposited in suitable database following the rules of minimum requirements of submission. However, metabolomics tools involved in each step has respective pitfalls, and no generalized protocol is applicable to all samples. The latter part of the workflow as discussed in automated although labor-intensive human interventions are prerequisites up to the point of acquiring raw data. Metabolomics results obtained from plants, microbial sources, have already been shown to contain profound pharmacological effects against hepatitis, cancer, and diabetes, and few of the NPs have already been approved by the FDA as prescribed drugs. The technical advancements and growing number of metabolite libraries and populated databases are catalyzing the discovery of more NPs valuable to mankind.

References

1. J. Li, C. A. Larregieu, and L. Z. Benet, "Classification of natural products as sources of drugs according to the biopharmaceutics drug disposition classification system (BDDCS).," *Chin. J. Nat. Med.*, vol. 14, no. 12, pp. 888–897, Dec. 2016, https://doi.org/10.1016/S1875-5364(17)30013-4.
2. M. P. Puglisi, "Natural Products: the Secondary Metabolites By James R. Hanson (University of Sussex). The Royal Society of Chemistry, Cambridge, UK. 2003. vi + 147 pp. 24.5 × 19 cm. $28.00. ISBN 0-85404-490-6.," *J. Nat. Prod.*, vol. 67, no. 10, pp. 1772–1773, Oct. 2004, https://doi.org/10.1021/np030747d.
3. H. G. Mikail, M. Mohammed, H. D. Umar, and M. M. Suleiman, "Secondary Metabolites: The Natural Remedies," R. Vijayakumar and S. S. S. Raja, Eds. Rijeka: IntechOpen, 2022, p. Ch. 3.
4. A. Ramakrishna and G. A. Ravishankar, "Influence of abiotic stress signals on secondary metabolites in plants.," *Plant Signal. Behav.*, vol. 6, no. 11, pp. 1720–1731, Nov. 2011, https://doi.org/10.4161/psb.6.11.17613.
5. J. L. Berini et al., "Combinations of Abiotic Factors Differentially Alter Production of Plant Secondary Metabolites in Five Woody Plant Species in the Boreal-Temperate Transition Zone," *Front. Plant Sci.*, vol. 9, 2018, https://doi.org/10.3389/fpls.2018.01257.
6. I. F. G. Mera, D. E. G. Falconí, and V. M. Córdova, "Secondary metabolites in plants: main classes, phytochemical analysis and pharmacological activities," *Bionatura*, vol. 4, pp. 1000–1009, 2019.
7. F. Bourgaud, A. Gravot, S. Milesi, and E. Gontier, "Production of plant secondary metabolites: a historical perspective," *Plant Sci.*, vol. 161, pp. 839–851, 2001.
8. J. B. Harborne, "CLASSES AND FUNCTIONS OF SECONDARY PRODUCTS FROM PLANTS," 1999.
9. N. J. Walton and D. E. Brown, "Chemicals from Plants: Perspectives on Plant Secondary Products," 1999.
10. I. Chiocchio, M. Mandrone, P. Tomasi, L. Marincich, and F. Poli, "Plant Secondary Metabolites: An Opportunity for Circular Economy.," *Molecules*, vol. 26, no. 2, Jan. 2021, https://doi.org/10.3390/molecules26020495.
11. A. Ariyawardana, "The natural products industry: A global and African economic perspective," 2009, pp. 7–28.
12. G. M. Cragg and D. J. Newman, "Biodiversity: A continuing source of novel drug leads," vol. 77, no. 1, pp. 7–24, 2005, https://doi.org/10.1351/pac200577010007.
13. M. Salehi, M. Bahmankar, M. R. Naghavi, and K. Cornish, "Rubber and latex extraction processes for Taraxacum kok-saghyz," *Ind. Crops Prod.*, vol. 178, p. 114562, 2022, https://doi.org/10.1016/j.indcrop.2022.114562.
14. K. I. Priyadarsini, "The chemistry of curcumin: from extraction to therapeutic agent.," *Molecules*, vol. 19, no. 12, pp. 20091–20112, Dec. 2014, https://doi.org/10.3390/molecules191220091.
15. D. Stan et al., "Natural Compounds With Antimicrobial and Antiviral Effect and Nanocarriers Used for Their Transportation," *Front. Pharmacol.*, vol. 12, 2021, https://doi.org/10.3389/fphar.2021.723233.

16. M. L. Bourguet-Kondracki, M. T. Martin, and M. Guyot, "A new β-carboline alkaloid isolated from the marine sponge Hyrtios erecta," *Tetrahedron Lett.*, vol. 37, no. 20, pp. 3457–3460, 1996, https://doi.org/10.1016/0040-4039(96)00573-4.
17. N. H. Shady, E. M. El-Hossary, M. A. Fouad, T. A. M. Gulder, M. S. Kamel, and U. R. Abdelmohsen, "Bioactive Natural Products of Marine Sponges from the Genus Hyrtios.," *Molecules*, vol. 22, no. 5, May 2017, https://doi.org/10.3390/molecules22050781.
18. A. Nagel, C. Winkler, R. Carle, H.-U. Endress, C. Rentschler, and S. Neidhart, "Processes involving selective precipitation for the recovery of purified pectins from mango peel," *Carbohydr. Polym.*, vol. 174, pp. 1144–1155, 2017, https://doi.org/10.1016/j.carbpol.2017.07.005.
19. M. M. Ibrahim, W. K. El-Zawawy, Y. Jüttke, A. Koschella, and T. Heinze, "Cellulose and microcrystalline cellulose from rice straw and banana plant waste: preparation and characterization," *Cellulose*, vol. 20, no. 5, pp. 2403–2416, 2013, https://doi.org/10.1007/s10570-013-9992-5.
20. S. Banerjee, V. Ranganathan, A. Patti, and A. Arora, "Valorisation of pineapple wastes for food and therapeutic applications," *Trends Food Sci. Technol.*, vol. 82, pp. 60–70, 2018, https://doi.org/10.1016/j.tifs.2018.09.024.
21. M. Patsalou, K. K. Menikea, E. Makri, M. I. Vasquez, C. Drouza, and M. Koutinas, "Development of a citrus peel-based biorefinery strategy for the production of succinic acid," *J. Clean. Prod.*, vol. 166, pp. 706–716, 2017, https://doi.org/10.1016/j.jclepro.2017.08.039.
22. C. Gardana and P. Simonetti, "Determination of steviol glycosides in commercial extracts of Stevia rebaudiana and sweeteners by ultra-high performance liquid chromatography Orbitrap mass spectrometry.," *J. Chromatogr. A*, vol. 1578, pp. 8–14, Nov. 2018, https://doi.org/10.1016/j.chroma.2018.09.057.
23. M. C. Wani, H. L. Taylor, M. E. Wall, P. Coggon, and A. T. McPhail, "Plant antitumor agents. VI. The isolation and structure of taxol, a novel antileukemic and antitumor agent from Taxus brevifolia.," *J. Am. Chem. Soc.*, vol. 93, no. 9, pp. 2325–2327, May 1971, https://doi.org/10.1021/ja00738a045.
24. G. C. Jadeja, R. C. Maheshwari, and S. N. Naik, "Extraction of natural insecticide azadirachtin from neem (Azadirachta indica A. Juss) seed kernels using pressurized hot solvent," *J. Supercrit. Fluids*, vol. 56, no. 3, pp. 253–258, 2011, https://doi.org/10.1016/j.supflu.2011.01.004.
25. A. A. Khalil, U. ur Rahman, M. R. Khan, A. Sahar, T. Mehmood, and M. Khan, "Essential oil eugenol: sources{,} extraction techniques and nutraceutical perspectives," *RSC Adv.*, vol. 7, no. 52, pp. 32669–32681, 2017, https://doi.org/10.1039/C7RA04803C.
26. B. Trudić et al., "Original Scientific paper: HPLC/MS-TOF Analysis of Surface Resins from Three Poplar Clones Grown in Serbia," *SEEFOR-South-East Eur. For.*, vol. 7, p. early view, Sep. 2016, https://doi.org/10.15177/seefor.16-12.
27. M. Shawkat, A. Khazaal, and, Majid Majeed, *EXTRACTION OF PYRETHRINS FROM Chrysanthemum cinerariaefolium PETALS AND STUDY ITS ACTIVITY AGAINST BEETLE FLOUR Tribolium castaneum*. 2011.
28. C. Liu et al., "Improvement of anthocyanins rate of blueberry powder under variable power of microwave extraction," *Sep. Purif. Technol.*, vol. 226, pp. 286–298, 2019, https://doi.org/10.1016/j.seppur.2019.05.096.
29. M. Tian, H. Yan, and K. H. Row, "Extraction of glycyrrhizic acid and glabridin from licorice.," *Int. J. Mol. Sci.*, vol. 9, no. 4, pp. 571–577, Apr. 2008, https://doi.org/10.3390/ijms9040571.
30. S. Alseekh and A. R. Fernie, "Metabolomics 20 years on: what have we learned and what hurdles remain?," *Plant J.*, vol. 94, no. 6, pp. 933–942, Jun. 2018, https://doi.org/10.1111/tpj.13950.
31. S. G. Oliver, M. K. Winson, D. B. Kell, and F. Baganz, "Systematic functional analysis of the yeast genome," *Trends Biotechnol.*, vol. 16, no. 9, pp. 373–378, 1998, https://doi.org/10.1016/S0167-7799(98)01214-1.

32. Fiehn O, Kopka J, Dörmann P, Altmann T, Trethewey RN, and Willmitzer L., "Metabolite profiling for plant functional genomics," *Nat. Biotechnol.*, vol. 18, no. 11, pp. 1157–61, 2000.
33. P. D. Fraser, M. E. Pinto, D. E. Holloway, and P. M. Bramley, "Technical advance: application of high-performance liquid chromatography with photodiode array detection to the metabolic profiling of plant isoprenoids.," *Plant J.*, vol. 24, no. 4, pp. 551–558, Nov. 2000, https://doi.org/10.1046/j.1365-313x.2000.00896.x.
34. S. Aubert, G. Curien, R. Bligny, E. Gout, and R. Douce, "Transport, Compartmentation, and Metabolism of Homoserine in Higher Plant Cells. Carbon-13- and phosphorus-31-nuclear magnetic resonance studies Carbon-13- and Phosphorus-31-Nuclear Magnetic Resonance Studies.," *Plant Physiol.*, vol. 116, no. 2, pp. 547–557, Feb. 1998, https://doi.org/10.1104/pp.116.2.547.
35. U. Roessner, "Metabolic Profiling Allows Comprehensive Phenotyping of Genetically or Environmentally Modified Plant Systems," *PLANT CELL ONLINE*, vol. 13, pp. 11–29, Jan. 2001, https://doi.org/10.1105/tpc.13.1.11.
36. E. K. Olsen, K. L. Søderholm, J. Isaksson, J. H. Andersen, and E. Hansen, "Metabolomic Profiling Reveals the N-Acyl-Taurine Geodiataurine in Extracts from the Marine Sponge Geodia macandrewii (Bowerbank)," *J. Nat. Prod.*, vol. 79, no. 5, pp. 1285–1291, May 2016, https://doi.org/10.1021/acs.jnatprod.5b00966.
37. M. A. Salem et al., "Metabolomics in the context of plant natural products research: From sample preparation to metabolite analysis," *Metabolites*, vol. 10, no. 1, p. 37, Jan. 2020, https://doi.org/10.3390/metabo10010037.
38. M. Wang et al., "Metabolomics in the context of systems biology: bridging traditional Chinese medicine and molecular pharmacology.," *Phytother. Res.*, vol. 19, no. 3, pp. 173–182, Mar. 2005, https://doi.org/10.1002/ptr.1624.
39. R. Goodacre, S. Vaidyanathan, W. B. Dunn, G. G. Harrigan, and D. B. Kell, "Metabolomics by numbers: acquiring and understanding global metabolite data," *Trends Biotechnol.*, vol. 22, no. 5, pp. 245–252, 2004, https://doi.org/10.1016/j.tibtech.2004.03.007.
40. N. D. Yuliana, A. Khatib, Y. H. Choi, and R. Verpoorte, "Metabolomics for bioactivity assessment of natural products," *Phyther. Res.*, vol. 25, no. 2, pp. 157–169, 2011, https://doi.org/10.1002/ptr.3258.
41. G. Ulrich-Merzenich, H. Zeitler, D. Jobst, D. Panek, H. Vetter, and H. Wagner, "Application of the '-Omic-' technologies in phytomedicine," *Phytomedicine*, vol. 14, no. 1, pp. 70–82, 2007, https://doi.org/10.1016/j.phymed.2006.11.011.
42. M. Reverter, S. Rohde, C. Parchemin, N. Tapissier-Bontemps, and P. J. Schupp, "Metabolomics and Marine Biotechnology: Coupling Metabolite Profiling and Organism Biology for the Discovery of New Compounds," *Front. Mar. Sci.*, vol. 7, no. December, pp. 1–8, 2020, https://doi.org/10.3389/fmars.2020.613471.
43. G. M. Nylund, F. Weinberger, M. Rempt, and G. Pohnert, "Metabolomic assessment of induced and activated chemical defence in the invasive red alga Gracilaria vermiculophylla.," *PLoS One*, vol. 6, no. 12, p. e29359, 2011, https://doi.org/10.1371/journal.pone.0029359.
44. K. N. Turi, L. Romick-Rosendale, K. K. Ryckman, and T. V Hartert, "A review of metabolomics approaches and their application in identifying causal pathways of childhood asthma.," *J. Allergy Clin. Immunol.*, vol. 141, no. 4, pp. 1191–1201, Apr. 2018, https://doi.org/10.1016/j.jaci.2017.04.021.
45. J. Draper et al., "Metabolite signal identification in accurate mass metabolomics data with MZedDB, an interactive m/z annotation tool utilising predicted ionisation behaviour 'rules'.," *BMC Bioinformatics*, vol. 10, p. 227, Jul. 2009, https://doi.org/10.1186/1471-2105-10-227.
46. O. Beckonert et al., "Beckonert O, Keun HC, Ebbels TMD, Bundy JG, Holmes E, Lindon JC, Nicholson JKMetabolic profiling, metabolomic and metabonomic procedures for NMR spectroscopy of urine, plasma, serum and tissue extracts. Nat Protoc 2: 2692-2703," *Nat. Protoc.*, vol. 2, pp. 2692–2703, Feb. 2007, https://doi.org/10.1038/nprot.2007.376.

47. G. Theodoridis, H. G. Gika, and I. D. Wilson, "Mass spectrometry-based holistic analytical approaches for metabolite profiling in systems biology studies.," *Mass Spectrom. Rev.*, vol. 30, no. 5, pp. 884–906, 2011, https://doi.org/10.1002/mas.20306.
48. J. V Pham et al., "A Review of the Microbial Production of Bioactive Natural Products and Biologics," *Front. Microbiol.*, vol. 10, 2019, https://doi.org/10.3389/fmicb.2019.01404.
49. J. M. McGUIRE et al., "Ilotycin, a new antibiotic.," *Antibiot. Chemother. (Northfield, Ill.)*, vol. 2, no. 6, pp. 281–283, Jun. 1952.
50. H. Zhang, Y. Wang, J. Wu, K. Skalina, and B. A. Pfeifer, "Complete Biosynthesis of Erythromycin A and Designed Analogs Using E. coli as a Heterologous Host," *Chem. Biol.*, vol. 17, no. 11, pp. 1232–1240, 2010, https://doi.org/10.1016/j.chembiol.2010.09.013.
51. R. E. Cobb, Y. Luo, T. Freestone, and H. Zhao, "Chapter 10 – Drug Discovery and Development via Synthetic Biology," H. B. T.-S. B. Zhao, Ed. Boston: Academic Press, 2013, pp. 183–206.
52. I. Chopra and M. Roberts, "Tetracycline Antibiotics: Mode of Action, Applications, Molecular Biology, and Epidemiology of Bacterial Resistance," *Microbiol. Mol. Biol. Rev.*, vol. 65, no. 2, pp. 232–260, 2001, https://doi.org/10.1128/MMBR.65.2.232-260.2001.
53. A. L. Demain, "Antibiotics: natural products essential to human health.," *Med. Res. Rev.*, vol. 29, no. 6, pp. 821–842, Nov. 2009, https://doi.org/10.1002/med.20154.
54. J. E. Geraci, F. R. Heilman, D. R. Nichols, E. W. Wellman, and G. T. Ross, "Some laboratory and clinical experiences with a new antibiotic, vancomycin.," *Antibiot. Annu.*, pp. 90–106.
55. A. Dasgupta, "Advances in antibiotic measurement.," *Adv. Clin. Chem.*, vol. 56, pp. 75–104, 2012, https://doi.org/10.1016/b978-0-12-394317-0.00013-3.
56. S. A. Waksman, H. C. Reilly, and D. B. Johnstone, "Isolation of Streptomycin-producing Strains of Streptomyces griseus.," *J. Bacteriol.*, vol. 52, no. 3, pp. 393–397, Sep. 1946.
57. A. Schatz, E. Bugie, and S. A. Waksman, "Streptomycin, a substance exhibiting antibiotic activity against gram-positive and gram-negative bacteria. 1944.," *Clin. Orthop. Relat. Res.*, no. 437, pp. 3–6, Aug. 2005, https://doi.org/10.1097/01.blo.0000175887.98112.fe.
58. J. W.-H. Li and J. C. Vederas, "Drug discovery and natural products: end of an era or an endless frontier?," *Science*, vol. 325, no. 5937, pp. 161–165, Jul. 2009, https://doi.org/10.1126/science.1168243.
59. R. Gyawali and S. A. Ibrahim, "Natural products as antimicrobial agents," *Food Control*, vol. 46, pp. 412–429, 2014, https://doi.org/10.1016/j.foodcont.2014.05.047.
60. K. M. Abu-Salah, "Amphotericin B: an update.," *Br. J. Biomed. Sci.*, vol. 53, no. 2, pp. 122–133, Jun. 1996.
61. A. N. Tevyashova et al., "Structure-antifungal activity relationships of polyene antibiotics of the amphotericin B group.," *Antimicrob. Agents Chemother.*, vol. 57, no. 8, pp. 3815–3822, Aug. 2013, https://doi.org/10.1128/AAC.00270-13.
62. L. H. Einhorn and J. Donohue, "Cis-diamminedichloroplatinum, vinblastine, and bleomycin combination chemotherapy in disseminated testicular cancer.," *Ann. Intern. Med.*, vol. 87, no. 3, pp. 293–298, Sep. 1977, https://doi.org/10.7326/0003-4819-87-3-293.
63. A. L. Demain and P. Vaishnav, "Natural products for cancer chemotherapy.," *Microb. Biotechnol.*, vol. 4, no. 6, pp. 687–699, Nov. 2011, https://doi.org/10.1111/j.1751-7915.2010.00221.x.
64. C. Vézina, A. Kudelski, and S. N. Sehgal, "Rapamycin (AY-22,989), a new antifungal antibiotic. I. Taxonomy of the producing streptomycete and isolation of the active principle.," *J. Antibiot. (Tokyo).*, vol. 28, no. 10, pp. 721–726, Oct. 1975, https://doi.org/10.7164/antibiotics.28.721.
65. J. Mann, "Natural products as immunosuppressive agents.," *Nat. Prod. Rep.*, vol. 18, no. 4, pp. 417–430, Aug. 2001, https://doi.org/10.1039/b001720p.
66. A. L. Demain, "Importance of microbial natural products and the need to revitalize their discovery.," *J. Ind. Microbiol. Biotechnol.*, vol. 41, no. 2, pp. 185–201, Feb. 2014, https://doi.org/10.1007/s10295-013-1325-z.

67. Y. H. Ban, S. R. Park, and Y. J. Yoon, "The biosynthetic pathway of FK506 and its engineering: from past achievements to future prospects.," *J. Ind. Microbiol. Biotechnol.*, vol. 43, no. 2–3, pp. 389–400, Mar. 2016, https://doi.org/10.1007/s10295-015-1677-7.
68. Y. J. Yoo, H. Kim, S. R. Park, and Y. J. Yoon, "An overview of rapamycin: from discovery to future perspectives.," *J. Ind. Microbiol. Biotechnol.*, vol. 44, no. 4–5, pp. 537–553, May 2017, https://doi.org/10.1007/s10295-016-1834-7.
69. D. J. Newman and G. M. Cragg, "Natural Products as Sources of New Drugs over the Nearly Four Decades from 01/1981 to 09/2019," *J. Nat. Prod.*, vol. 83, no. 3, pp. 770–803, Mar. 2020, https://doi.org/10.1021/acs.jnatprod.9b01285.
70. M. Sorokina and C. Steinbeck, "Review on natural products databases: where to find data in 2020.," *J. Cheminform.*, vol. 12, no. 1, p. 20, Apr. 2020, https://doi.org/10.1186/s13321-020-00424-9.
71. S. W. Gabrielson, "SciFinder.," *Journal of the Medical Library Association : JMLA*, vol. 106, no. 4. pp. 588–590, Oct-2018, https://doi.org/10.5195/jmla.2018.515.
72. N. K. K. Ikram et al., "A Virtual Screening Approach For Identifying Plants with Anti H5N1 Neuraminidase Activity," *J. Chem. Inf. Model.*, vol. 55, no. 2, pp. 308–316, Feb. 2015, https://doi.org/10.1021/ci500405g.
73. J. Hastings et al., "The ChEBI reference database and ontology for biologically relevant chemistry: enhancements for 2013.," *Nucleic Acids Res.*, vol. 41, no. Database issue, pp. D456-63, Jan. 2013, https://doi.org/10.1093/nar/gks1146.
74. A. Gaulton et al., "The ChEMBL database in 2017.," *Nucleic Acids Res.*, vol. 45, no. D1, pp. D945–D954, Jan. 2017, https://doi.org/10.1093/nar/gkw1074.
75. H. E. Pence and A. Williams, "ChemSpider: An Online Chemical Information Resource," *J. Chem. Educ.*, vol. 87, no. 11, pp. 1123–1124, Nov. 2010, https://doi.org/10.1021/ed100697w.
76. V. D. Hähnke, S. Kim, and E. E. Bolton, "PubChem chemical structure standardization.," *J. Cheminform.*, vol. 10, no. 1, p. 36, Aug. 2018, https://doi.org/10.1186/s13321-018-0293-8.
77. M. Kanehisa, M. Furumichi, M. Tanabe, Y. Sato, and K. Morishima, "KEGG: new perspectives on genomes, pathways, diseases and drugs.," *Nucleic Acids Res.*, vol. 45, no. D1, pp. D353–D361, Jan. 2017, https://doi.org/10.1093/nar/gkw1092.
78. R. Caspi et al., "The MetaCyc database of metabolic pathways and enzymes – a 2019 update.," *Nucleic Acids Res.*, vol. 48, no. D1, pp. D445–D453, Jan. 2020, https://doi.org/10.1093/nar/gkz862.
79. C. Guijas et al., "METLIN: A Technology Platform for Identifying Knowns and Unknowns," *Anal. Chem.*, vol. 90, no. 5, pp. 3156–3164, Mar. 2018, https://doi.org/10.1021/acs.analchem.7b04424.
80. P. Banerjee, J. Erehman, B.-O. Gohlke, T. Wilhelm, R. Preissner, and M. Dunkel, "Super Natural II--a database of natural products.," *Nucleic Acids Res.*, vol. 43, no. Database issue, pp. D935–9, Jan. 2015, https://doi.org/10.1093/nar/gku886
81. P. Tomasulo, "ChemIDplus-super source for chemical and drug information.," *Med. Ref. Serv. Q.*, vol. 21, no. 1, pp. 53–59, 2002, https://doi.org/10.1300/J115v21n01_04.
82. J. Jacyna, M. Kordalewska, and M. J. Markuszewski, "Design of Experiments in metabolomics-related studies: An overview," *J. Pharm. Biomed. Anal.*, vol. 164, pp. 598–606, 2019, https://doi.org/10.1016/j.jpba.2018.11.027.
83. A. M. Rodrigues, A. I. Ribeiro-Barros, and C. António, "Experimental Design and Sample Preparation in Forest Tree Metabolomics.," *Metabolites*, vol. 9, no. 12, Nov. 2019, https://doi.org/10.3390/metabo9120285
84. N. Zamboni, A. Saghatelian, and G. J. Patti, "Defining the metabolome: size, flux, and regulation.," *Mol. Cell*, vol. 58, no. 4, pp. 699–706, May 2015, https://doi.org/10.1016/j.molcel.2015.04.021.
85. S. Barnes et al., "Training in metabolomics research. I. Designing the experiment, collecting and extracting samples and generating metabolomics data.," *J. Mass Spectrom.*, vol. 51, no. 7, pp. 461–475, Jul. 2016, https://doi.org/10.1002/jms.3782.

86. W. B. Dunn and D. I. Ellis, "Metabolomics: Current analytical platforms and methodologies," *TrAC – Trends Anal. Chem.*, vol. 24, no. 4, pp. 285–294, 2005, https://doi.org/10.1016/j.trac.2004.11.021.
87. H. K. Kim and R. Verpoorte, "Sample preparation for plant metabolomics," *Phytochem. Anal.*, vol. 21, no. 1, pp. 4–13, 2010, https://doi.org/10.1002/pca.1188.
88. K. Mohd Kamal, M. H. Mahamad Maifiah, N. Abdul Rahim, Y. Z. H.-Y. Hashim, M. S. Abdullah Sani, and K. A. Azizan, "Bacterial Metabolomics: Sample Preparation Methods.," *Biochem. Res. Int.*, vol. 2022, p. 9186536, 2022, https://doi.org/10.1155/2022/9186536.
89. M. Ernst, D. Silva, R. Silva, R. Vêncio, and N. Lopes, "Mass spectrometry in plant metabolomics strategies: From analytical platforms to data acquisition and processing," *Nat. Prod. Rep.*, vol. 31, Mar. 2014, https://doi.org/10.1039/c3np70086k.
90. S. Rochfort, "Metabolomics reviewed: a new 'omics' platform technology for systems biology and implications for natural products research.," *J. Nat. Prod.*, vol. 68, no. 12, pp. 1813–1820, Dec. 2005, https://doi.org/10.1021/np050255w.
91. B. Ruiz et al., "Production of microbial secondary metabolites: Regulation by the carbon source," *Crit. Rev. Microbiol.*, vol. 36, no. 2, pp. 146–167, 2010, https://doi.org/10.3109/10408410903489576.
92. E. Riekeberg and R. Powers, "New frontiers in metabolomics: from measurement to insight.," *F1000Research*, vol. 6, p. 1148, 2017, https://doi.org/10.12688/f1000research.11495.1.
93. U. Christians, J. Klawitter, A. Hornberger, and J. Klawitter, "How unbiased is non-targeted metabolomics and is targeted pathway screening the solution?," *Curr. Pharm. Biotechnol.*, vol. 12, no. 7, pp. 1053–1066, Jul. 2011, https://doi.org/10.2174/138920111795909078.
94. L. W. Sumner et al., "Proposed minimum reporting standards for chemical analysis Chemical Analysis Working Group (CAWG) Metabolomics Standards Initiative (MSI).," *Metabolomics*, vol. 3, no. 3, pp. 211–221, Sep. 2007, https://doi.org/10.1007/s11306-007-0082-2.
95. N. Harbourne, E. Marete, J. C. Jacquier, and D. O'Riordan, "Effect of drying methods on the phenolic constituents of meadowsweet (Filipendula ulmaria) and willow (Salix alba)," *LWT – Food Sci. Technol.*, vol. 42, no. 9, pp. 1468–1473, 2009, https://doi.org/10.1016/j.lwt.2009.05.005.
96. W. A. Jensen, "The Application of Freeze-Dry Methods to Plant Material," *Stain Technol.*, vol. 29, no. 3, pp. 143–150, Jan. 1954, https://doi.org/10.3109/10520295409115459.
97. H.-Y. Zhou and C.-Z. Liu, "Microwave-assisted extraction of solanesol from tobacco leaves.," *J. Chromatogr. A*, vol. 1129, no. 1, pp. 135–139, Sep. 2006, https://doi.org/10.1016/j.chroma.2006.07.083.
98. R. Verpoorte, Y. H. Choi, N. R. Mustafa, and H. K. Kim, "Metabolomics: back to basics," *Phytochem. Rev.*, vol. 7, no. 3, pp. 525–537, 2008, https://doi.org/10.1007/s11101-008-9091-7.
99. S. Alseekh et al., "Mass spectrometry-based metabolomics: a guide for annotation, quantification and best reporting practices," *Nat. Methods*, vol. 18, no. 7, pp. 747–756, 2021, https://doi.org/10.1038/s41592-021-01197-1.
100. B. R. Gordon, W. Leggat, and C. A. Motti, "Extraction protocol for nontargeted NMR and LC-MS metabolomics-based analysis of hard coral and their algal symbionts.," *Methods Mol. Biol.*, vol. 1055, pp. 129–147, 2013, https://doi.org/10.1007/978-1-62703-577-4_10.
101. M. Page, L. West, P. Northcote, C. Battershill, and M. Kelly, "Spatial and temporal variability of cytotoxic metabolites in populations of the New Zealand sponge Mycale hentscheli.," *J. Chem. Ecol.*, vol. 31, no. 5, pp. 1161–1174, May 2005, https://doi.org/10.1007/s10886-005-4254-0.
102. B. C. Elias, S. Said, S. de Albuquerque, and M. T. Pupo, "The influence of culture conditions on the biosynthesis of secondary metabolites by Penicillium verrucosum Dierck.," *Microbiol. Res.*, vol. 161, no. 3, pp. 273–280, 2006, https://doi.org/10.1016/j.micres.2005.10.003.
103. S. Romano, S. A. Jackson, S. Patry, and A. D. W. Dobson, "Extending the 'One Strain Many Compounds' (OSMAC) Principle to Marine Microorganisms.," *Mar. Drugs*, vol. 16, no. 7, Jul. 2018, https://doi.org/10.3390/md16070244.

104. A. Sproule et al., "Terrosamycins A and B, Bioactive Polyether Ionophores from Streptomyces sp. RKND004 from Prince Edward Island Sediment.," *Mar. Drugs*, vol. 17, no. 6, Jun. 2019, https://doi.org/10.3390/md17060347.
105. B. Fan, D. Parrot, M. Blümel, A. Labes, and D. Tasdemir, "Influence of OSMAC-Based Cultivation in Metabolome and Anticancer Activity of Fungi Associated with the Brown Alga Fucus vesiculosus.," *Mar. Drugs*, vol. 17, no. 1, Jan. 2019, https://doi.org/10.3390/md17010067.
106. X. Duportet, R. B. M. Aggio, S. Carneiro, and S. G. Villas-Bôas, "The biological interpretation of metabolomic data can be misled by the extraction method used," *Metabolomics*, vol. 8, no. 3, pp. 410–421, 2012, https://doi.org/10.1007/s11306-011-0324-1.
107. X. Liu et al., "Optimized sampling protocol for mass spectrometry-based metabolomics in Streptomyces," *Bioresour. Bioprocess.*, vol. 6, Dec. 2019, https://doi.org/10.1186/s40643-019-0269-1.
108. R. V. Kapoore and S. Vaidyanathan, "Quenching for Microalgal Metabolomics: A Case Study on the Unicellular Eukaryotic Green Alga Chlamydomonas reinhardtii.," *Metabolites*, vol. 8, no. 4, Oct. 2018, https://doi.org/10.3390/metabo8040072.
109. C. L. Winder et al., "Global Metabolic Profiling of Escherichia coli Cultures: an Evaluation of Methods for Quenching and Extraction of Intracellular Metabolites," *Anal. Chem.*, vol. 80, no. 8, pp. 2939–2948, Apr. 2008, https://doi.org/10.1021/ac7023409.
110. I. Bertini, X. Hu, and C. Luchinat, "Global metabolomics characterization of bacteria: Pre-analytical treatments and profiling," *Metabolomics*, vol. 10, Apr. 2014, https://doi.org/10.1007/s11306-013-0571-4.
111. A. R. Fernie, R. N. Trethewey, A. J. Krotzky, and L. Willmitzer, "Metabolite profiling: from diagnostics to systems biology.," *Nature reviews. Molecular cell biology*, vol. 5, no. 9. England, pp. 763–769, Sep-2004, https://doi.org/10.1038/nrm1451.
112. Q. W. Zhang, L. G. Lin, and W. C. Ye, "Techniques for extraction and isolation of natural products: A comprehensive review," *Chinese Med. (United Kingdom)*, vol. 13, no. 1, pp. 1–26, 2018, https://doi.org/10.1186/s13020-018-0177-x.
113. E. G. Bligh and W. J. Dyer, "A rapid method of total lipid extraction and purification.," *Can. J. Biochem. Physiol.*, vol. 37, no. 8, pp. 911–917, Aug. 1959, https://doi.org/10.1139/o59-099.
114. J. Folch, M. Lees, and G. H. S. Stanley, "A SIMPLE METHOD FOR THE ISOLATION AND PURIFICATION OF TOTAL LIPIDES FROM ANIMAL TISSUES," *J. Biol. Chem.*, vol. 226, no. 1, pp. 497–509, 1957, https://doi.org/10.1016/S0021-9258(18)64849-5.
115. O. Fiehn, J. Kopka, R. N. Trethewey, and L. Willmitzer, "Identification of uncommon plant metabolites based on calculation of elemental compositions using gas chromatography and quadrupole mass spectrometry.," *Anal. Chem.*, vol. 72, no. 15, pp. 3573–3580, Aug. 2000, https://doi.org/10.1021/ac991142i.
116. J. Lisec, N. Schauer, J. Kopka, L. Willmitzer, and A. R. Fernie, "Gas chromatography mass spectrometry-based metabolite profiling in plants.," *Nat. Protoc.*, vol. 1, no. 1, pp. 387–396, 2006, https://doi.org/10.1038/nprot.2006.59.
117. P. Giavalisco et al., "Elemental formula annotation of polar and lipophilic metabolites using (13) C, (15) N and (34) S isotope labelling, in combination with high-resolution mass spectrometry.," *Plant J.*, vol. 68, no. 2, pp. 364–376, Oct. 2011, https://doi.org/10.1111/j.1365-313X.2011.04682.x.
118. R. C. H. De Vos, S. Moco, A. Lommen, J. J. B. Keurentjes, R. J. Bino, and R. D. Hall, "Untargeted large-scale plant metabolomics using liquid chromatography coupled to mass spectrometry," *Nat. Protoc.*, vol. 2, no. 4, pp. 778–791, 2007, https://doi.org/10.1038/nprot.2007.95.
119. M. A. Salem, J. Jüppner, K. Bajdzienko, and P. Giavalisco, "Protocol: a fast, comprehensive and reproducible one-step extraction method for the rapid preparation of polar and semi-polar metabolites, lipids, proteins, starch and cell wall polymers from a single sample.," *Plant Methods*, vol. 12, p. 45, 2016, https://doi.org/10.1186/s13007-016-0146-2.

120. M. A. Salem, Y. Li, A. Wiszniewski, and P. Giavalisco, "Regulatory-associated protein of TOR (RAPTOR) alters the hormonal and metabolic composition of Arabidopsis seeds, controlling seed morphology, viability and germination potential.," *Plant J.*, vol. 92, no. 4, pp. 525–545, Nov. 2017, https://doi.org/10.1111/tpj.13667.
121. D. Alhajturki et al., "Dose-dependent interactions between two loci trigger altered shoot growth in BG-5 × Krotzenburg-0 (Kro-0) hybrids of Arabidopsis thaliana.," *New Phytol.*, vol. 217, no. 1, pp. 392–406, Jan. 2018, https://doi.org/10.1111/nph.14781.
122. M. A. Farag et al., "Comparative Metabolomics Approach Detects Stress-Specific Responses during Coral Bleaching in Soft Corals," *J. Proteome Res.*, vol. 17, no. 6, pp. 2060–2071, Jun. 2018, https://doi.org/10.1021/acs.jproteome.7b00929.
123. M. Kuyukina, I. Ivshina, J. Philp, N. Christofi, S. Dunbar, and M. Ritchkova, "Recovery of Rhodococcus biosurfactants using methyl-tertiary butyl ether (MTBE) extraction," *J. Microbiol. Methods*, vol. 46, pp. 149–156, Sep. 2001, https://doi.org/10.1016/S0167-7012(01)00259-7.
124. C. Chin, K. Chong, M. Atong, and N. K. U. I. Wong, "Tea polyphenols and alkaloids content using Soxhlet and direct extraction method," *World J. Agric. Sci.*, vol. 9, pp. 266–270, Jan. 2013.
125. M. D. Luque de Castro and F. Priego-Capote, "Soxhlet extraction: Past and present panacea," *J. Chromatogr. A*, vol. 1217, no. 16, pp. 2383–2389, 2010, https://doi.org/10.1016/j.chroma.2009.11.027.
126. K. Segers, S. Declerck, D. Mangelings, Y. Vander Heyden, and A. Van Eeckhaut, "Analytical techniques for metabolomic studies: a review," *Bioanalysis*, vol. 11, no. 24, pp. 2297–2318, Dec. 2019, https://doi.org/10.4155/bio-2019-0014.
127. D. J. Beale et al., "Review of recent developments in GC–MS approaches to metabolomics-based research," *Metabolomics*, vol. 14, no. 11, p. 152, Nov. 2018, https://doi.org/10.1007/s11306-018-1449-2.
128. W. Zhang, T. Hankemeier, and R. Ramautar, "Next-generation capillary electrophoresis–mass spectrometry approaches in metabolomics," *Curr. Opin. Biotechnol.*, vol. 43, pp. 1–7, Feb. 2017, https://doi.org/10.1016/j.copbio.2016.07.002.
129. S. Naz, D. C. Moreira dos Santos, A. García, and C. Barbas, "Analytical protocols based on LC–MS, GC–MS and CE–MS for nontargeted metabolomics of biological tissues," *Bioanalysis*, vol. 6, no. 12, pp. 1657–1677, Jun. 2014, https://doi.org/10.4155/bio.14.119.
130. C. H. Collins, G. L. Braga, and P. S. Bonato, *Fundamentos de cromatografia*. Campinas: Editora da Unicamp, 2006.
131. X. Hu, L. Lu, Z. Guo, and Z. Zhu, "Volatile compounds, affecting factors and evaluation methods for rice aroma: A review," *Trends Food Sci. Technol.*, vol. 97, pp. 136–146, Mar. 2020, https://doi.org/10.1016/J.TIFS.2020.01.003.
132. D. Dudzik, C. Barbas-Bernardos, A. García, and C. Barbas, "Quality assurance procedures for mass spectrometry untargeted metabolomics. a review," *J. Pharm. Biomed. Anal.*, vol. 147, pp. 149–173, Jan. 2018, https://doi.org/10.1016/j.jpba.2017.07.044.
133. S. Abdelbaset, D. M. El-Kersh, I. M. Ayoub, and O. A. Eldahshan, "GC-MS profiling of Vitex pinnata bark lipophilic extract and screening of its anti-TB and cytotoxic activities," *Nat. Prod. Res.*, pp. 1–7, Sep. 2022, https://doi.org/10.1080/14786419.2022.2124512.
134. O. A. Adebo, E. Kayitesi, F. Tugizimana, and P. B. Njobeh, "Differential metabolic signatures in naturally and lactic acid bacteria (LAB) fermented ting (a Southern African food) with different tannin content, as revealed by gas chromatography mass spectrometry (GC–MS)-based metabolomics," *Food Res. Int.*, vol. 121, pp. 326–335, Jul. 2019, https://doi.org/10.1016/j.foodres.2019.03.050.
135. K. Kozłowicz et al., "Identification of sugars and phenolic compounds in honey powders with the use of GC–MS, FTIR spectroscopy, and X-ray diffraction," *Sci. Rep.*, vol. 10, no. 1, p. 16269, Oct. 2020, https://doi.org/10.1038/s41598-020-73306-7.
136. J. Morimoto, M. C. Rosso, N. Kfoury, C. Bicchi, C. Cordero, and A. Robbat, "Untargeted/targeted 2D gas chromatography/mass spectrometry detection of the total volatile tea

metabolome," *Molecules*, vol. 24, no. 20, p. 3757, Oct. 2019, https://doi.org/10.3390/molecules24203757.
137. Q. Zhao, J. Le Zhang, and F. Li, "Application of Metabolomics in the Study of Natural Products," *Nat. Products Bioprospect.*, vol. 8, no. 4, pp. 321–334, 2018, https://doi.org/10.1007/s13659-018-0175-9.
138. L. P. de Souza, T. Naake, T. Tohge, and A. R. Fernie, "From chromatogram to analyte to metabolite. How to pick horses for courses from the massive web resources for mass spectral plant metabolomics," *Gigascience*, vol. 6, no. 7, Jul. 2017, https://doi.org/10.1093/gigascience/gix037.
139. Ö. C. Zeki, C. C. Eylem, T. Reçber, S. Kır, and E. Nemutlu, "Integration of GC–MS and LC–MS for untargeted metabolomics profiling," *J. Pharm. Biomed. Anal.*, vol. 190, p. 113509, Oct. 2020, https://doi.org/10.1016/j.jpba.2020.113509.
140. D. P. Demarque et al., "Mass spectrometry-based metabolomics approach in the isolation of bioactive natural products," *Sci. Rep.*, vol. 10, no. 1, p. 1051, Jan. 2020, https://doi.org/10.1038/s41598-020-58046-y.
141. Y. Hou et al., "Microbial strain prioritization using metabolomics tools for the discovery of natural products," *Anal. Chem.*, vol. 84, no. 10, pp. 4277–4283, May 2012, https://doi.org/10.1021/ac202623g.
142. X. Li et al., "LC-MS-based metabolomic approach revealed the significantly different metabolic profiles of five commercial truffle species," *Front. Microbiol.*, vol. 10, Sep. 2019, https://doi.org/10.3389/fmicb.2019.02227.
143. A. S. Edison et al., "NMR: Unique Strengths That Enhance Modern Metabolomics Research," *Anal. Chem.*, vol. 93, no. 1, pp. 478–499, Jan. 2021, https://doi.org/10.1021/acs.analchem.0c04414.
144. H. K. Kim, Y. H. Choi, and R. Verpoorte, "NMR-based metabolomic analysis of plants," *Nat. Protoc.*, vol. 5, no. 3, pp. 536–549, Mar. 2010, https://doi.org/10.1038/nprot.2009.237.
145. J. L. Markley et al., "The future of NMR-based metabolomics," *Curr. Opin. Biotechnol.*, vol. 43, pp. 34–40, Feb. 2017, https://doi.org/10.1016/j.copbio.2016.08.001.
146. T. N. Clark, "The application of NMR and LC-HRMS based prioritization strategies for the discovery of natural products produced by endophytic fungi from medicinal plants," University of New Brunswick, 2019.
147. M. R. N. Monton and T. Soga, "Metabolome analysis by capillary electrophoresis–mass spectrometry," *J. Chromatogr. A*, vol. 1168, no. 1–2, pp. 237–246, Oct. 2007, https://doi.org/10.1016/j.chroma.2007.02.065.
148. G. Bianco, P. Schmitt-Kopplin, G. De Benedetto, A. Kettrup, and T. R. I. Cataldi, "Determination of glycoalkaloids and relative aglycones by nonaqueous capillary electrophoresis coupled with electrospray ionization-ion trap mass spectrometry," *Eletrophoresis*, vol. 23, no. 17, pp. 2904–2912, Sep. 2002, https://doi.org/10.1002/1522-2683(200209)23:17<2904::AID-ELPS2904>3.0.CO;2-1.
149. K. Wanichthanarak, S. Fan, D. Grapov, D. K. Barupal, and O. Fiehn, "Metabox: A Toolbox for Metabolomic Data Analysis, Interpretation and Integrative Exploration.," *PLoS One*, vol. 12, no. 1, p. e0171046, 2017, https://doi.org/10.1371/journal.pone.0171046.
150. A. Cambiaghi, M. Ferrario, and M. Masseroli, "Analysis of metabolomic data: tools, current strategies and future challenges for omics data integration.," *Brief. Bioinform.*, vol. 18, no. 3, pp. 498–510, May 2017, https://doi.org/10.1093/bib/bbw031.
151. K. Wanichthanarak, J. F. Fahrmann, and D. Grapov, "Genomic, Proteomic, and Metabolomic Data Integration Strategies.," *Biomark. Insights*, vol. 10, no. Suppl 4, pp. 1–6, 2015, https://doi.org/10.4137/BMI.S29511.
152. S. Hemmer, S. K. Manier, S. Fischmann, F. Westphal, L. Wagmann, and M. R. Meyer, "Comparison of Three Untargeted Data Processing Workflows for Evaluating LC-HRMS Metabolomics Data.," *Metabolites*, vol. 10, no. 9, Sep. 2020, https://doi.org/10.3390/metabo10090378.

153. B. B. Misra, "New software tools, databases, and resources in metabolomics: updates from 2020," *Metabolomics*, vol. 17, no. 5, pp. 1–24, 2021, https://doi.org/10.1007/s11306-021-01796-1.
154. X. Domingo-Almenara et al., "XCMS-MRM and METLIN-MRM: a cloud library and public resource for targeted analysis of small molecules," *Nat. Methods*, vol. 15, no. 9, pp. 681–684, 2018, https://doi.org/10.1038/s41592-018-0110-3.
155. T. Pluskal, S. Castillo, A. Villar-Briones, and M. Orešič, "MZmine 2: Modular framework for processing, visualizing, and analyzing mass spectrometry-based molecular profile data," *BMC Bioinformatics*, vol. 11, no. 1, p. 395, 2010, https://doi.org/10.1186/1471-2105-11-395.
156. H. Tsugawa et al., "MS-DIAL: data-independent MS/MS deconvolution for comprehensive metabolome analysis," *Nat. Methods*, vol. 12, no. 6, pp. 523–526, 2015, https://doi.org/10.1038/nmeth.3393.
157. P. Seitzer, B. Bennett, and E. Melamud, "MAVEN2: An Updated Open-Source Mass Spectrometry Exploration Platform.," *Metabolites*, vol. 12, no. 8, Jul. 2022, https://doi.org/10.3390/metabo12080684.
158. G. Riquelme, N. Zabalegui, P. Marchi, C. M. Jones, and M. E. Monge, "A Python-Based Pipeline for Preprocessing LC-MS Data for Untargeted Metabolomics Workflows.," *Metabolites*, vol. 10, no. 10, Oct. 2020, https://doi.org/10.3390/metabo10100416.
159. Y. Gloaguen, J. A. Kirwan, and D. Beule, "Deep Learning-Assisted Peak Curation for Large-Scale LC-MS Metabolomics," *Anal. Chem.*, vol. 94, no. 12, pp. 4930–4937, Mar. 2022, https://doi.org/10.1021/acs.analchem.1c02220.
160. C. McLean and E. B. Kujawinski, "AutoTuner: High Fidelity and Robust Parameter Selection for Metabolomics Data Processing," *Anal. Chem.*, vol. 92, no. 8, pp. 5724–5732, Apr. 2020, https://doi.org/10.1021/acs.analchem.9b04804.
161. B. Wajid, H. Iqbal, M. Jamil, H. Rafique, and F. Anwar, "MetumpX-a metabolomics support package for untargeted mass spectrometry.," *Bioinformatics*, vol. 36, no. 5, pp. 1647–1648, Mar. 2020, https://doi.org/10.1093/bioinformatics/btz765.
162. M. Kuhring et al., "Concepts and Software Package for Efficient Quality Control in Targeted Metabolomics Studies: MeTaQuaC," *Anal. Chem.*, vol. 92, no. 15, pp. 10241–10245, Aug. 2020, https://doi.org/10.1021/acs.analchem.0c00136.
163. N. Bararpour et al., "DBnorm as an R package for the comparison and selection of appropriate statistical methods for batch effect correction in metabolomic studies," *Sci. Rep.*, vol. 11, no. 1, p. 5657, 2021, https://doi.org/10.1038/s41598-021-84824-3.
164. K. Chetnik, L. Petrick, and G. Pandey, "MetaClean: a machine learning-based classifier for reduced false positive peak detection in untargeted LC-MS metabolomics data.," *Metabolomics*, vol. 16, no. 11, p. 117, Oct. 2020, https://doi.org/10.1007/s11306-020-01738-3.
165. J. Trygg, E. Holmes, and T. Lundstedt, "Chemometrics in Metabonomics," *J. Proteome Res.*, vol. 6, no. 2, pp. 469–479, Feb. 2007, https://doi.org/10.1021/pr060594q.
166. S. Barnes et al., "Training in metabolomics research. II. Processing and statistical analysis of metabolomics data, metabolite identification, pathway analysis, applications of metabolomics and its future.," *J. Mass Spectrom.*, vol. 51, no. 8, pp. 535–548, Aug. 2016, https://doi.org/10.1002/jms.3780.
167. B. B. Misra, "Data normalization strategies in metabolomics: Current challenges, approaches, and tools.," *Eur. J. Mass Spectrom. (Chichester, Eng).*, vol. 26, no. 3, pp. 165–174, Jun. 2020, https://doi.org/10.1177/1469066720918446.
168. B. Ernest, J. R. Gooding, S. R. Campagna, A. M. Saxton, and B. H. Voy, "MetabR: an R script for linear model analysis of quantitative metabolomic data," *BMC Res. Notes*, vol. 5, no. 1, p. 596, 2012, https://doi.org/10.1186/1756-0500-5-596.
169. C. Thonusin, H. B. IglayReger, T. Soni, A. E. Rothberg, C. F. Burant, and C. R. Evans, "Evaluation of intensity drift correction strategies using MetaboDrift, a normalization tool for multi-batch metabolomics data.," *J. Chromatogr. A*, vol. 1523, pp. 265–274, Nov. 2017, https://doi.org/10.1016/j.chroma.2017.09.023.

170. A. M. De Livera, G. Olshansky, J. A. Simpson, and D. J. Creek, "NormalizeMets: assessing, selecting and implementing statistical methods for normalizing metabolomics data.," *Metabolomics*, vol. 14, no. 5, p. 54, Mar. 2018, https://doi.org/10.1007/s11306-018-1347-7.
171. J. Willforss, A. Chawade, and F. Levander, "NormalyzerDE: Online Tool for Improved Normalization of Omics Expression Data and High-Sensitivity Differential Expression Analysis," *J. Proteome Res.*, vol. 18, no. 2, pp. 732–740, Feb. 2019, https://doi.org/10.1021/acs.jproteome.8b00523.
172. Q. Yang et al., "NOREVA: enhanced normalization and evaluation of time-course and multi-class metabolomic data," *Nucleic Acids Res.*, vol. 48, no. W1, pp. W436–W448, 2020, https://doi.org/10.1093/nar/gkaa258.
173. S. Wang et al., "MetaboGroupS: A Group Entropy-Based Web Platform for Evaluating Normalization Methods in Blood Metabolomics Data from Maintenance Hemodialysis Patients," *Anal. Chem.*, vol. 90, no. 18, pp. 11124–11130, Sep. 2018, https://doi.org/10.1021/acs.analchem.8b03065.
174. S. Wang and H. Yang, "pseudoQC: A Regression-Based Simulation Software for Correction and Normalization of Complex Metabolomics and Proteomics Datasets.," *Proteomics*, vol. 19, no. 19, p. e1900264, Oct. 2019, https://doi.org/10.1002/pmic.201900264.
175. F. Savorani, G. Tomasi, and S. B. Engelsen, "icoshift: A versatile tool for the rapid alignment of 1D NMR spectra.," *J. Magn. Reson.*, vol. 202, no. 2, pp. 190–202, Feb. 2010, https://doi.org/10.1016/j.jmr.2009.11.012.
176. C. Schiffman et al., "Filtering procedures for untargeted LC-MS metabolomics data.," *BMC Bioinformatics*, vol. 20, no. 1, p. 334, Jun. 2019, https://doi.org/10.1186/s12859-019-2871-9.
177. G. R. Lloyd, A. Jankevics, and R. J. M. Weber, "Struct: an R/bioconductor-based framework for standardised metabolomics data analysis and beyond.," *Bioinformatics*, vol. 36, no. 22–23, pp. 5551–5552, Dec. 2020, https://doi.org/10.1093/bioinformatics/btaa1031.
178. T. Kockmann and C. Panse, "The rawrr R Package: Direct Access to Orbitrap Data and Beyond," *J. Proteome Res.*, vol. 20, no. 4, pp. 2028–2034, Apr. 2021, https://doi.org/10.1021/acs.jproteome.0c00866.
179. R. Manjarin, M. A. Maj, M. R. La Frano, and H. Glanz, "%polynova_2way: A SAS macro for implementation of mixed models for metabolomics data.," *PLoS One*, vol. 15, no. 12, p. e0244013, 2020, https://doi.org/10.1371/journal.pone.0244013.
180. C. Beuchel, H. Kirsten, U. Ceglarek, and M. Scholz, "Metabolite-Investigator: an integrated user-friendly workflow for metabolomics multi-study analysis.," *Bioinformatics*, vol. 37, no. 15, pp. 2218–2220, Aug. 2021, https://doi.org/10.1093/bioinformatics/btaa967.
181. R. Choudhury et al., "Viime: Visualization and Integration of Metabolomics Experiments.," *J. open source Softw.*, vol. 5, no. 54, 2020, https://doi.org/10.21105/joss.02410.
182. M. Pietzke and A. Vazquez, "Metabolite AutoPlotter – an application to process and visualise metabolite data in the web browser.," *Cancer Metab.*, vol. 8, p. 15, 2020, https://doi.org/10.1186/s40170-020-00220-x.
183. J. Berg et al., *Gazing into the Metaboverse: Automated exploration and contextualization of metabolic data*. 2020.
184. J. Henning and R. Smith, "A web-based system for creating, viewing, and editing precursor mass spectrometry ground truth data.," *BMC Bioinformatics*, vol. 21, no. 1, p. 418, Sep. 2020, https://doi.org/10.1186/s12859-020-03752-7.
185. J. Xia and D. S. Wishart, "MetPA: a web-based metabolomics tool for pathway analysis and visualization.," *BioInformatics*, vol. 26, no. 18, pp. 2342–2344, Sep. 2010, https://doi.org/10.1093/bioinformatics/btq418.
186. J. Xia and D. S. Wishart, "MSEA: a web-based tool to identify biologically meaningful patterns in quantitative metabolomic data.," *Nucleic Acids Res.*, vol. 38, no. Web Server issue, pp. W71–7, Jul. 2010, https://doi.org/10.1093/nar/gkq329.
187. A. R. Fernie et al., "Recommendations for reporting metabolite data.," *The Plant cell*, vol. 23, no. 7. England, pp. 2477–2482, Jul-2011, https://doi.org/10.1105/tpc.111.086272.

188. M. Wang et al., "Sharing and community curation of mass spectrometry data with Global Natural Products Social Molecular Networking," *Nat. Biotechnol.*, vol. 34, no. 8, pp. 828–837, 2016, https://doi.org/10.1038/nbt.3597.
189. H. Mohimani et al., "Dereplication of microbial metabolites through database search of mass spectra," *Nat. Commun.*, vol. 9, no. 1, p. 4035, 2018, https://doi.org/10.1038/s41467-018-06082-8.
190. J. M. Deutsch et al., "Metabolomics Approaches to Dereplicate Natural Products from Coral-Derived Bioactive Bacteria," *J. Nat. Prod.*, vol. 85, no. 3, pp. 462–478, 2022, https://doi.org/10.1021/acs.jnatprod.1c01110.
191. A. Gurevich, A. Mikheenko, A. Shlemov, A. Korobeynikov, H. Mohimani, and P. A. Pevzner, "Increased diversity of peptidic natural products revealed by modification-tolerant database search of mass spectra.," *Nat. Microbiol.*, vol. 3, no. 3, pp. 319–327, Mar. 2018, https://doi.org/10.1038/s41564-017-0094-2.
192. X.-R. Yang et al., "Metabolites of Siamenoside I and Their Distributions in Rats.," *Molecules*, vol. 21, no. 2, p. 176, Jan. 2016, https://doi.org/10.3390/molecules21020176.
193. C. Liu, L. Dai, Y. Liu, D. Dou, Y. Sun, and L. Ma, "Pharmacological activities of mogrosides.," *Future Med. Chem.*, vol. 10, no. 8, pp. 845–850, Apr. 2018, https://doi.org/10.4155/fmc-2017-0255.
194. G. Zhou, Y. Zhang, Y. Li, M. Wang, and X. Li, "The metabolism of a natural product mogroside V, in healthy and type 2 diabetic rats.," *J. Chromatogr. B, Anal. Technol. Biomed. life Sci.*, vol. 1079, pp. 25–33, Mar. 2018, https://doi.org/10.1016/j.jchromb.2018.02.002.
195. H. Yan, L. Tao, X. Qu, L. Zhou, and S. Zhang, "Quantitative determination of mogroside V in rat plasma by LC-MS/MS and its application to a pharmacokinetic study.," *Pak. J. Pharm. Sci.*, vol. 31, no. 3, pp. 867–873, May 2018.
196. A. Santini, G. C. Tenore, and E. Novellino, "Nutraceuticals: A paradigm of proactive medicine.," *Eur. J. Pharm. Sci. Off. J. Eur. Fed. Pharm. Sci.*, vol. 96, pp. 53–61, Jan. 2017, https://doi.org/10.1016/j.ejps.2016.09.003.
197. X. Xu, L. Jia, X. Ma, H. Li, and C. Sun, "Application Potential of Plant-Derived Medicines in Prevention and Treatment of Platinum-Induced Peripheral Neurotoxicity.," *Front. Pharmacol.*, vol. 12, p. 792331, 2021, https://doi.org/10.3389/fphar.2021.792331.
198. W. Wichitnithad, N. Jongaroonngamsang, S. Pummangura, and P. Rojsitthisak, "A simple isocratic HPLC method for the simultaneous determination of curcuminoids in commercial turmeric extracts.," *Phytochem. Anal.*, vol. 20, no. 4, pp. 314–319, 2009, https://doi.org/10.1002/pca.1129.
199. K. Samukawa, H. Yamashita, H. Matsuda, and M. Kubo, "[Simultaneous analysis of ginsenosides of various ginseng radix by HPLC].," *Yakugaku Zasshi*, vol. 115, no. 3, pp. 241–249, Mar. 1995, https://doi.org/10.1248/yakushi1947.115.3_241.
200. A. M. Komishon et al., "The effect of ginseng (genus Panax) on blood pressure: a systematic review and meta-analysis of randomized controlled clinical trials.," *J. Hum. Hypertens.*, vol. 30, no. 10, pp. 619–626, Oct. 2016, https://doi.org/10.1038/jhh.2016.18.
201. L. P. Christensen, "Ginsenosides chemistry, biosynthesis, analysis, and potential health effects.," *Adv. Food Nutr. Res.*, vol. 55, pp. 1–99, 2009, https://doi.org/10.1016/S1043-4526(08)00401-4.
202. K. M. Lee, J. Y. Jeon, B. J. Lee, H. Lee, and H. K. Choi, "Application of metabolomics to quality control of natural product derived medicines," *Biomol. Ther.*, vol. 25, no. 6, pp. 559–568, 2017, https://doi.org/10.4062/biomolther.2016.249.
203. A. Balekundri and V. Mannur, "Quality control of the traditional herbs and herbal products: a review," *Futur. J. Pharm. Sci.*, vol. 6, Dec. 2020, https://doi.org/10.1186/s43094-020-00091-5.
204. Q. Xiao et al., "Plant metabolomics: a new strategy and tool for quality evaluation of Chinese medicinal materials," *Chin. Med.*, vol. 17, Dec. 2022, https://doi.org/10.1186/s13020-022-00601-y.

205. F. Li, Y. Wang, D. Li, Y. Chen, and Q. P. Dou, "Are we seeing a resurgence in the use of natural products for new drug discovery?," *Expert opinion on drug discovery*, vol. 14, no. 5. England, pp. 417–420, May-2019, https://doi.org/10.1080/17460441.2019.1582639.
206. W. C. Reygaert, "Green Tea Catechins: Their Use in Treating and Preventing Infectious Diseases.," *Biomed Res. Int.*, vol. 2018, p. 9105261, 2018, https://doi.org/10.1155/2018/9105261.
207. M. A. Salem, R. A. Radwan, E. S. Mostafa, S. Alseekh, A. R. Fernie, and S. M. Ezzat, "Using an UPLC/MS-based untargeted metabolomics approach for assessing the antioxidant capacity and anti-aging potential of selected herbs.," *RSC Adv.*, vol. 10, no. 52, pp. 31511–31524, Aug. 2020, https://doi.org/10.1039/d0ra06047j.
208. S. Chen et al., "Metabolite Profiling of 14 Wuyi Rock Tea Cultivars Using UPLC-QTOF MS and UPLC-QqQ MS Combined with Chemometrics.," *Molecules*, vol. 23, no. 2, Jan. 2018, https://doi.org/10.3390/molecules23020104.
209. C.-F. Li, M.-Z. Yao, C.-L. Ma, J.-Q. Ma, J.-Q. Jin, and L. Chen, "Differential Metabolic Profiles during the Albescent Stages of 'Anji Baicha' (Camellia sinensis).," *PLoS One*, vol. 10, no. 10, p. e0139996, 2015, https://doi.org/10.1371/journal.pone.0139996.
210. J. J. Kellogg et al., "Comparison of Metabolomics Approaches for Evaluating the Variability of Complex Botanical Preparations: Green Tea (Camellia sinensis) as a Case Study.," *J. Nat. Prod.*, vol. 80, no. 5, pp. 1457–1466, May 2017, https://doi.org/10.1021/acs.jnatprod.6b01156.
211. S. Dall'Acqua et al., "Evaluation of Antioxidant and Enzyme Inhibition Properties of Croton hirtus L'Hér. Extracts Obtained with Different Solvents.," *Molecules*, vol. 26, no. 7, Mar. 2021, https://doi.org/10.3390/molecules26071902.
212. S. D. Sarker and L. Nahar, "An introduction to natural products isolation.," *Methods Mol. Biol.*, vol. 864, pp. 1–25, 2012, https://doi.org/10.1007/978-1-61779-624-1_1.
213. W. R. Stahel, "The circular economy," *Nature*, vol. 531, no. 7595, pp. 435–438, 2016, https://doi.org/10.1038/531435a.
214. A.-H. M. Emwas, "The strengths and weaknesses of NMR spectroscopy and mass spectrometry with particular focus on metabolomics research.," *Methods Mol. Biol.*, vol. 1277, pp. 161–193, 2015, https://doi.org/10.1007/978-1-4939-2377-9_13.
215. A. David and P. Rostkowski, "Analytical techniques in metabolomics," 2020, pp. 35–64.
216. F. A. de Jong and C. Beecher, "Addressing the current bottlenecks of metabolomics: Isotopic Ratio Outlier Analysis™, an isotopic labeling technique for accurate biochemical profiling.," *Bioanalysis*, vol. 4, no. 18, pp. 2303–2314, Sep. 2012, https://doi.org/10.4155/bio.12.202.
217. U. R. Abdelmohsen et al., "Potential of marine natural products against drug-resistant fungal, viral, and parasitic infections.," *Lancet. Infect. Dis.*, vol. 17, no. 2, pp. e30–e41, Feb. 2017, https://doi.org/10.1016/S1473-3099(16)30323-1.
218. Z. Zhou et al., "Ion mobility collision cross-section atlas for known and unknown metabolite annotation in untargeted metabolomics," *Nat. Commun.*, vol. 11, no. 1, p. 4334, 2020, https://doi.org/10.1038/s41467-020-18171-8.
219. J.-L. Wolfender et al., "Metabolomics in Ecology and Bioactive Natural Products Discovery: Challenges and Prospects for a Comprehensive Study of the Specialised Metabolome," *Chimia (Aarau).*, vol. 76, p. 954, Nov. 2022, https://doi.org/10.2533/chimia.2022.954.
220. M. A. Beniddir, K. Bin Kang, G. Genta-Jouve, F. Huber, S. Rogers, and J. J. J. van der Hooft, "Advances in decomposing complex metabolite mixtures using substructure- and network-based computational metabolomics approaches," *Nat. Prod. Rep.*, vol. 38, pp. 1967–1993, 2021.
221. I. Häuser et al., "Environmental and socio-economic impacts of rubber cultivation in the Mekong region: Challenges for sustainable land use," *CAB Rev. Perspect. Agric. Vet. Sci. Nutr. Nat. Resour.*, vol. 10, Sep. 2015, https://doi.org/10.1079/PAVSNNR201510027.
222. A. L. Rhyne, M. F. Tlusty, and L. Kaufman, "Is sustainable exploitation of coral reefs possible? A view from the standpoint of the marine aquarium trade," *Curr. Opin. Environ. Sustain.*, vol. 7, pp. 101–107, 2014, https://doi.org/10.1016/j.cosust.2013.12.001.

223. D. E. Schaufelberger et al., "The large-scale isolation of bryostatin 1 from Bugula neritina following current good manufacturing practices.," *J. Nat. Prod.*, vol. 54, no. 5, pp. 1265–1270, 1991, https://doi.org/10.1021/np50077a004.
224. U. Lindequist, "Marine-Derived Pharmaceuticals – Challenges and Opportunities.," *Biomol. Ther. (Seoul).*, vol. 24, no. 6, pp. 561–571, Nov. 2016, https://doi.org/10.4062/biomolther.2016.181.

Metabolomics Approach in Environmental Studies: Current Progress, Analytical Challenges, and Future Recommendations

Vijay Soni, Nicholas Bartelo, Ramya Venkataraman, and Bhupendra Singh Rawat

Abbreviations

AOPs	Adverse outcome pathways
CE	Capillary electrophoresis
DDT	Dichlorodiphenyltrichloroethane
DEHTP	Di(2-ethylhexyl) terephthalate
EAGMST	Extended Advisory Group for Molecular Screening and Toxicogenomics
FT-ICR-MS	Fourier transform ion cyclotron resonance MS
GC	Gas chromatography
GDM	Gestational diabetes mellitus
HPLC	High-performance liquid chromatography
HR-MAS	High-resolution magic angle spinning
HRMS	High-resolution MS
LC	Liquid chromatography

These authors contributed equally to this work.

V. Soni (✉)
Division of Infectious Diseases, Weill Department of Medicine, Weill Cornell Medicine, New York, NY, USA
e-mail: vis2032@med.cornell.edu

N. Bartelo (✉)
Department of Physiology and Biophysics, Weill Cornell Medicine, Institute for Computational Biomedicine, Englander Institute for Precision Medicine, New York, NY, USA
e-mail: nib4003@med.cornell.edu

R. Venkataraman
Laboratory of Innate Immunity, National Institute of Immunology, New Delhi, India

B. S. Rawat
Center for Immunity and Inflammation, Rutgers New Jersey Medical School, Newark, NJ, USA

V. Soni, T. E. Hartman (eds.), *Metabolomics*,
https://doi.org/10.1007/978-3-031-39094-4_10

LPCs	Lysophosphatidylcholines
MDCs	Metabolism disrupting chemicals
MECPTP	Mono(2-ethyl-5-carboxypentyl) terephthalate
MERIT	Metabolomics Standards Initiative in Toxicology
MRF	Metabolomic Reporting Framework
MS	Mass spectrometry
MTBE	Methyl-tertbutyl ether
NMR	Nuclear magnetic resonance
PAHs	Polycyclic aromatic hydrocarbons
POPs	Persistent organic pollutants
POPs	Persistent organic pollutants
PUFAs	Polyunsaturated fatty acids
QA	Quality assurance
QC	Quality control
SVOCs	Semi volatile organic compounds
T2D	Type 2 diabetes
TCE	Trichloroethylene

1 Introduction

When asked to describe the power of metabolomics, most would immediately mention patient studies in which food, drink, drugs, etc., are consumed and metabolized, and the effect on health through computational analyses is reported. A less-known role of metabolomics is its integral part in our understanding of the environment, including environmental changes in response to manmade products. Many of us have heard the continual news stories about the population struggle of one of the world's most popular pollinators, the honeybee. The honeybee endured a sharp decline in 2006 and many years following [2]. Still recovering from this population reduction, metabolomics is at the forefront of methodologies to discover responses to different stimuli. This area of research attempts to experimentally test different factors of nutrition to create an optimal metabolic response in bees.

Similar to current trends in human healthcare, the scientific approach to developing a desired metabolic response in honeybees provided with different environmental conditions has presented great strides in precision nutrition for bees. Targeted and untargeted metabolomic approaches compared metabolite abundances for bees with different levels of carbohydrates in their diets to identify the impact of low-carbohydrate nutritional stress on metabolism [3]. Machine learning classification models of select metabolites showed perfect discrimination between bees in different treatment groups, and the authors concluded that increased protein catabolism resulted from nutritional stress [3]. Ricigliano et al. performed untargeted metabolomics on caged honeybees fed sugar, pollen, *Chlorella*, and spirulina diets to compare the metabolomes of a natural pollen diet with those of artificial, microalgae-based diets [4]. This work concluded similar nutritional and metabolic impacts of microalgae-based and pollen diets, suggesting microalgae-based additions to feed can be important as health modulators and further experiments can create precise, diet-inducing changes using metabolomic analyses [4]. In addition to nutrition, experiments focused on the effects of pesticides on metabolomes of bees fed pollen or nectar with different phytochemicals have provided recommendations such as the use of flavonoids to increase detoxification systems which reduce the concentration of the pesticide tau-fluvalinate [5]. These studies provide recommendations and assessments for bee health status in different environmental conditions aimed to help bees thrive and multiply. Through the power of metabolomics, we can create an optimal future for honeybees and many other living organisms.

This is just an example to show how the environment is important, and we need better tools to monitor adulteration. Ecometabolomics is an emerging field that combines the techniques of metabolomics and ecological research to understand the metabolic processes of organisms in their natural environments (Fig. 1) [1]. It offers a promising avenue for investigating the complex interactions between living organisms and their surrounding environment, from the molecular to the ecosystem level. This chapter aims to provide a comprehensive overview of the current progress, applications, challenges, and future recommendations of ecometabolomics. We will discuss the analytical techniques and tools used in ecometabolomics research, the ecological and environmental factors influencing metabolite production, and the advancements of ecometabolomics. Moreover, we will address the challenges facing ecometabolomics, including data analysis and interpretation, and standardization of protocols. Finally, we will provide future recommendations for advancing ecometabolomics research, including novel tools for monitoring, multipollutant prospective, different technical improvements, and application of exposome wearables. We believe that this chapter will be a valuable resource for researchers

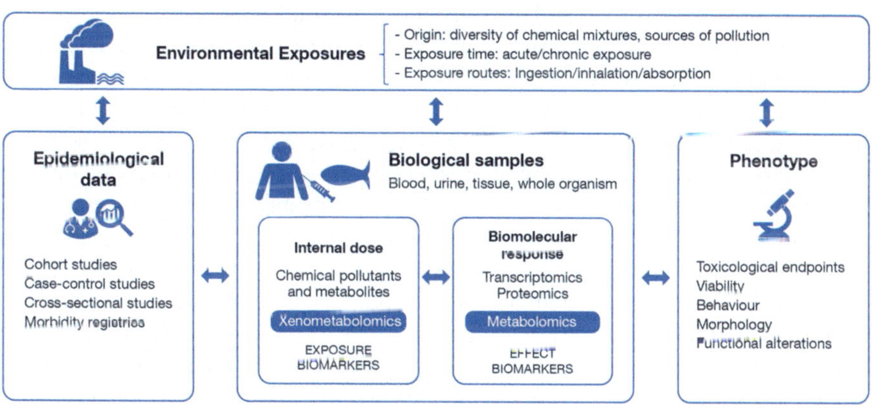

Fig. 1 Schema of the different data associations linking exposures and adverse effect/disease outcomes in exposome studies, with emphasis on the role of metabolomics to investigate the molecular responses to chemical pollutant exposure. (Figure and caption are adopted from [1])

interested in ecometabolomics and its potential for advancing our understanding of the complex interactions between living organisms and their environment.

2 General Techniques in Environmental Metabolomics

Normally, for biomonitoring purposes, targeted analysis is in use, but it cannot be applied to identify the toxic derivatives of that chemical or pollutant. Therefore, the exposomic approach is the key to the development of targeted and untargeted assessment of pollutants and their byproducts [6]. Starting with sample preparation, which involves two primary steps: the first being sample collection and the second being sample extraction. The process of sample extraction from biofluids and low-mass tissue samples has always been rather laborious, inundated by inconsistency, proving to be a bottleneck for high-throughput metabolomics [7]. However, new technologies are changing the landscape of sample preparation methodologies for ecometabolomics. Depending upon the properties of the metabolites, researchers can opt for acid-based extraction or organic solvent-based extraction. The former is employed for polar metabolites and the latter for moderately polar, nonpolar, or hydrophobic metabolites [7]. A biphasic solvent system utilizing chloroform/methanol/water (2/2/1.8, v/v/v) was employed for extracting both polar (methanol/water phase) and nonpolar (chloroform phase) compounds at the same time. But this method had a few drawbacks, one of them being an accumulation of cellular debris between the upper polar and lower nonpolar phases, hindering the clean aspiration of the lower phase [8]. Due to these reasons, the quest for a better method of extraction has been ongoing. This method was succeeded by the Matyash method, which replaced chloroform with methyl tert-butyl ether (MTBE), which is noncarcinogenic [9]. In a recent study by Sostare et al. on *Daphnia magna*, the Matyash method has further been modified leading to an increased yield and reproducibility in samples [8].

Two techniques most commonly employed for the identification of metabolites include nuclear magnetic resonance (NMR) and mass spectrometry (MS), and with technological advancements, the precision and sensitivity of these tools have improved manifold in the past few years [10]. Though NMR is one of the most widely utilized methods for metabolite identification and quantification, its limited sensitivity does not allow the entire sample metabolome to be unraveled. Furthermore, most NMR spectrometers in environmental metabolomics laboratories require large sample volumes. However, a recent study by Poynton et al. reported using just 3-μl hemolymph from six adult water fleas (*Daphnia magna*) for measurement of metabolites by employing a 1-mm TXI microliter NMR probe [11] acquiring spectra from such a small sample volume opens up the possibility of conducting many more chemical risk assessments in important species using metabolomics [12]. Extraction of metabolites from samples, a step preluding performing NMR, is not only time-consuming, but it can also lead to sample loss or degradation. High-resolution magic angle spinning (HR-MAS) NMR eliminates the extraction procedure, directly allowing the use of tissue samples [13]. HR-MAS is being widely employed for analyzing the plant metabolome under the influence of different biotic and abiotic stressors [14–17]. In an interesting study, Sarou-Kanian et al. employed HR-MAS slice-localized spectroscopy and chemical shift imaging to obtain metabolic profiles in localized regions of live *Drosophila* [18]. HR-MAS NMR has further been modified to use a microprobe, wherein high-quality data can be acquired even from samples in microgram quantities and this technique has been successfully used for generating localized metabolic profiles of four different regions of a garlic clove with less than 0.5 mg sampling mass [19]. These techniques provide a unique opportunity for performing localized metabolomic analysis of organ or tissue-specific changes in response to various stressors, even with a limited amount of sample.

MS is suitable for high-sensitivity and high-throughput analysis and depending upon the metabolite properties, samples can be analyzed either with gas chromatography (GC), liquid chromatography (LC), or capillary electrophoresis (CE) equipped with MS [10]. Nowadays, LC coupled with high-resolution MS (HRMS) is being employed in several plant metabolomic studies for further enhanced performance. In one study, a multifunctional mass analyzer, triple quadrupole QTRAP mass spectrometry, has been employed with multiple reaction monitoring for identifying important metabolites responsible for antioxidant activity in *Foeniculum vulgare* [20]. HRMS is ideal when adopting an untargeted approach for comparing and analyzing hundreds of metabolites in an unbiased manner [10].

Both NMR and MS provide an abundant amount of information about the damage caused by metabolism-disrupting chemicals (MDCs), but it needs to be validated by more precise methods to characterize the exact molecule and for potential disease prevention. Most of the exposomic and metabolomics-based data are coming from a single type of sample which may not be a proper reflective of the exposure or damage. Therefore, it is important to confirm the results by taking repeated samples from other possible sources and bio reservoirs [21]. For a broad cumulative risk assessment, it is important to include better technology and advanced computational methods along with adverse outcome pathways (AOPs) research models [22]. In a review, Vermeulen et al. have proposed different exposomic frameworks to manage ecometabolomics and associated risk assessment [23]. For example, a multilayer network-based framework can combine exposomic data with genomics and artificial intelligence-based data analysis. It provides an urban exposome framework to access public health disparities and detection of vulnerable people with better disease surveillance [24–26].

3 Environmental Exposure and Role of Metabolomics

Environmental stressors are categorized into two groups: abiotic and biotic [27]. Abiotic stressors include pollution, other toxic chemicals, and climate change. Biotic stressors include pathogens and predation. Analysis of the metabolome of organisms exposed to environmental stressors provides insights into biological pathways which are being directly altered, allowing for a deeper comprehension of resulting phenotypic changes. In addition to environmental stressors, metabolomics has been used to study agricultural practices, such as fertilizer efficiency and fertilizer effect on animals. The measurement of metabolic changes due to fertilizer in plants and animals will aid in the development of the largest crop yield with lowest negative impact on the surrounding ecosystem. As in most biological domains, model organisms for the study of environmental metabolomics have been determined. These species have the greatest influence on our understanding of metabolic changes due to the ability to directly test variables of interest in controlled experimental setups.

Xenobiotics are human-made, chemical substances, foreign to living organisms [28], which encompass pollutants and other toxic chemicals released into the environment. Examples of xenobiotics include personal care products, pesticides, nanomaterials, and heavy metals. These materials are transferred into the atmosphere (air pollution), water (water pollution), and soil (land pollution). Each type of pollution can affect the metabolism of organisms living in different ecosystems, and metabolomics has been utilized to study the influence of these toxins, demonstrating the negative impacts of pollution on the environment.

3.1 Pollutants Cause Changes in the Metabolome

Air pollution is composed of a combination of gaseous, volatile, semi-volatile, and particulate matter [29]. Air pollution has been shown to change lipid metabolism. Rat studies by Wang et al. [30] and Zhang et al. [31] concluded PM2.5, a fine particulate matter which is a major air pollutant, altered phospholipid and sphingolipid metabolism, and short-term air pollution resulted in an increase of lysophosphatidylcholines (LPCs) due to the activation of phospholipase A2, respectively. Human studies have also demonstrated an association between LPC changes and air pollution using a meet-in-the-middle approach with the SAPALDIA and EPIC cohorts [32] and randomized crossover trials using the Oxford Street II (London) and the TAPAS II (Barcelona) studies [33]. Air pollution also effects inflammatory and oxidative stress pathways [29, 34]. Gaskins et al. conducted an experiment to study human reproduction in which the authors concluded N-methyltryptamine, 1-methylnicotina-mide, and methyl vanillate were found to significantly mediate the association between air pollution and live birth [34]. Ritz et al. [35] studied air pollution exposure in third-trimester pregnancy and found associations between air pollution exposures with lipid-related metabolic pathways including fatty acid activation, de novo fatty acid biosynthesis, fatty acid metabolism, the carnitine shuttle, and glycerophospholipid metabolism. Oxidative stress in newborns can result due to modifications of these pathways. These studies reveal the power of metabolomic analysis on the determination of biomarkers that can be used as a potential screening for a successful pregnancy and healthy newborn, as women exposed to pollution with higher susceptibility to pregnancy termination can be better identified. Nassan et al. studied the short-term [36] and long-term [37] exposure to air pollution in the same human cohort. Short-term exposure to air pollution was associated with sphingolipid and butanoate metabolisms, whereas long-term exposure was associated with glycerophospholipid, propanoate, sphingolipid, and glutathione metabolism. The overlap of sphingolipid alteration due to air pollution again amplifies the impact on lipid metabolism. Both studies identified pathways related to inflammation, oxidative stress, immunity, and nucleic acid damage and repair. Gruzieva et al. published a review containing ten studies related to air pollution influences on the metabolome [38]. Culmination of research involving air pollution exposure led to the conclusion that metabolites with pro-inflammatory effects generally tended to be upregulated, while anti-inflammatory metabolites appeared to be downregulated. Additionally, they stated that air pollution can perturb metabolic pathways in populations defined by asthma status, age, and sex. Analysis of organisms exposed to air pollution using metabolomics has increased our knowledge of specific alterations in metabolite abundance and metabolic pathways. This may lead to the ability to suggest changes to current air pollution regulations and identify those who are most susceptible of negative health impacts in the future.

In addition to air pollution, the effects of water pollution on organisms have been elucidated using metabolomic analyses. Many wastewater treatment plants remove most, but not all, chemicals in contaminated water, including personal care products and pharmaceuticals. To understand the effect on the animal life surrounding two different wastewater treatment plants on Colorado's South Platte River which produce increased estrogen levels downstream of each plant, NMR spectroscopy was performed on caged male fathead minnows [39]. Minnows downstream of the wastewater treatment plants were characteristic of higher levels of vitellogenin, a protein secreted by the liver in the presence of estrogen, and more abundant alanine and glutamate. These two amino acids were likely increased due to the high demand needed to perform biosynthesis of vitellogenin. Koubova et al. studied the metabolic response of common carp previously living in a severely polluted treated wastewater pond, transferred from unpolluted water into a severely polluted treated wastewater pond, or transferred from a severely polluted treated wastewater pond into unpolluted water [40]. Carp moved from unpolluted water to polluted water had detectable pharmaceutically active compounds in their metabolism within 7 days. However, carp moved from polluted water to unpolluted water eliminated pharmaceutically active compounds 2 weeks after removal. This suggests it is possible for fish to expel pollution in a relatively short period of time. However, the metabolomic analysis demonstrated alterations in fish metabolism 60 days after movement to unpolluted water. If testing was only performed to find the presence of pollutants in the fish, it is possible it would be unknown that the metabolome of the fish was different than a fish not effected by pollution. As humans consume fish living in these conditions, it is important to understand how the metabolic makeup of these fish differ from those living in unpolluted water, as there is a potential impact on the health of the person eating a fish with metabolic changes due to toxins. Mercury is an important water pollutant worldwide due to its ability to bioaccumulate as it is consumed through the food chain. Brandao et al. used metabolomics to study mercury toxicity by identifying changes in metabolites between fish in a severely contaminated region and fish from the same region in unpolluted water [41]. The fish in mercury-polluted waters had significantly increased levels of alanine, phosphocholine, glucose, and glutathione and significantly decreased concentrations of tyrosine, phenylalanine, taurine, and hypoxanthine. The most differential metabolite between populations was phosphocholine, which was characterized by a 93% increase in fish living in mercury-polluted waters. It may be possible to use phosphocholine levels as a biomarker to measure mercury concentration. In addition to fish, other aquatic animals have been used to study the effect of water pollution. Damselfly larvae, aquatic invertebrates, were subject to metabolomic analysis before and after exposure to wastewater treatment plant effluents [42]. Metabolomic pathway analysis revealed significant differences in the D-glutamine and D-glutamate metabolism, and a nonsignificant, yet highly impacted, difference in the alanine, aspartate, and glutamate metabolism. This research by Spath et al. was also novel because the authors additionally studied behavioral responses of the damselfly larvae, which revealed an association between negative fitness consequences and pollution exposure, even though larvae were only exposed to pollutants for 7 days. Metabolomics has allowed for a unique insight into changes in aquatic animals due to pollution, allowing for future tools to identify animals highly influenced by contaminants, new data to support updating environmental practices from a political angle, and a deeper understanding of how polluted waterways can affect humans who are unable to obtain clean water.

Nanoparticles range in size from 1 to 100 nm and are waste products from numerous applications such as electronics, healthcare, energy, and consumer products, which are released into the soil and landfills to create land pollution [28]. These materials can affect the metabolome of plants exposed. Metabolomic analysis of pinto beans before and after exposure to CeO_2 by spray or through soil resulted in downregulation of carotene, lycopene, and rhodopsin derivatives and upregulation of keto-echinenone and glucosinolates [43]. Vecerova et al. studied the difference between metabolite concentrations for barley exposed to CdO nanoparticles and barley without exposure to understand the pollutant's effect on fully developed higher plants [44]. The authors measured differences in concentration for primary metabolites such as amino acids and saccharides between the treatment groups. Experiments involving soil application of TiO_2 nanoparticles to rice demonstrated an increase in proline, aspartic acids, glutamic acids, palmitic acid, glycerol, inositol, ribitol, phosphoric acid, and glycerol-3-phosphate and a decrease in organic acids, fatty acids, and sugars [45]. Interestingly, the uptake of nanoparticles was not observed in the crops, yet metabolic changes were present. Therefore, metabolomic analysis uncovered biological alterations in rice that may have been overlooked given the lack of measured contaminant. This is a very important takeaway because in situations where pollution may not be detected, organisms may still incur negative effects at a metabolic level. In addition to nanoparticles, polybrominated diphenyl ethers (PBDEs) have been used as additives to consumer products and have been found in soil [28]. Metabolomic analysis of two strains of rice, YY-9 and LJ-7, upon exposure to the PBDE 2,2′,4,4′-tetrabromodiphenyl ether demonstrated different responses for each rice strain [46]. LJ-7 was characteristic of increased abundance of 13 amino acids and 24 organic acids, whereas measurement of YY-9 metabolites resulted in decreased abundance of 13 amino acids and 19 organic acids. These outcomes suggest LJ-7 is tolerant of 2,2′,4,4′-tetrabromodiphenyl ether, and YY-9 is susceptible. Therefore, this analysis has demonstrated the ability of metabolomics to identify superior species for specific land effected by pollution. Metabolomic analyses have added an extra dimension to our understanding of land pollutants on biological mechanisms of plants, as controlled experiments allow for the precise measurement of changing metabolites, a more sensitive approach to identifying plants affected by pollutants, and the ability to determine optimal species for future farming in polluted areas.

3.2 *Natural Sources Cause Metabolomic Alterations: Climate Change, Pathogens, and Predation*

Climate change is another abiotic environmental stressor established to produce modifications of metabolomic markers. Under stressors due to climate change, such as drought, temperature, and salinity, increased production of amino acids, soluble sugars, the raffinose family of oligosaccharides, polyols, and polyamines has been validated in forest trees [47]. Correia et al. studied drought-tolerant *Eucalyptus globulus* clones under stresses of drought, heat, and a combination of drought and heat [48]. *Eucalyptus globulus* subject to drought only had reductions in the levels of sugar phosphates, α-glycerophosphate, and shikimate, as well as increases in starch and nonstructural carbohydrates. Under heat stress alone, sugar alcohols and several amino acids increased, whereas starch, fructose-6-phosphate, glucose-6-phosphate, and α-glycerophosphate were reduced. Unlike the drought-only condition, heat stress activated the shikimic acid pathway. When drought and heat stress were combined, accumulation of cinnamate was observed. Therefore, it was concluded that the sum of individual stresses does not create the same effect as the combination of the stresses. This has important applications in future work studying ecosystems suffering from multiple influences of climate change. Ottow et al. analyzed metabolites of the tree *Populus euphratica* Olivier exposed to an environmental gradient of increasing salinity [49]. As more salt stress was added to the tree, increases in free amino acids, ammonia, γ-amino butaric acid, and citrulline were observed, demonstrating the effect of salinity levels on plant metabolism. A comprehensive review focused on metabolomic analyses of fruit under heat, drought, and irradiance stress provides an overview of changes in primary metabolites of sugars, organic acids, amino acids, and fatty acids, as well as secondary metabolites including polyphenols and terpenoids [50]. Climate change is driven by the addition of carbon dioxide (CO_2) released into the atmosphere. The effect of multiple CO_2 levels, low to very high, on eelgrass plants from two different thermal environments demonstrated increased photosynthetic energy capture, sucrose formation, and growth [51]. This analysis suggests eelgrass under increased CO_2 enrichment had higher thermal tolerance, an indicator that eelgrass may counteract some effects of global warming. Using metabolomics, humans can determine the biological mechanisms altered due to environmental changes for major food sources and attempt to rectify these alterations to create climate-resistant plants.

Pathogens are biotic environmental stressors, and the understanding of plant-pathogen interactions has been enhanced through metabolomics. Cultured tobacco cells were treated with different concentrations of the fungal sterol, ergosterol, and varying incubation times to uncover the metabolic response to the pathogen [52]. The metabolome alterations due to ergosterol addition, labeled the "defensome," included an increase in terpenoid metabolites and other metabolites such as abscisic acid and phytosterols. Reviews by Heuberger et al. in 2014 [53] and Castro-Moretti et al. in 2020 [54] provide an overview of research using metabolomics to characterize multiple aspects of plant immunity. In addition to plant-pathogen interactions, metabolomics is beginning to be used to identify biomarkers as a diagnostic for pathogens in drinking water. Yu et al. showed the ability to separate between different types of bacteria and *N. fowleri*, a lethal brain-eating amoeba, using principal component analysis of four metabolite profiles: two lipids, a nucleobase, and an amine [55]. This study demonstrates the power metabolomics has to discriminate between pathogens to identify potential safety hazards. As pathogens are abundant, it is imperative to recognize their impact on the organisms they effect, and metabolomics is a powerful tool to assess the exact biological mechanisms altered upon interaction.

A biotic stressor, perhaps not receiving as much spotlight as the other environmental stressors, is predation. Metabolomics allows for the study of physiological response to pressures from predators, an area not well understood. Differential metabolites in stool between cattle raised in low and high wolf pack interaction locations included amino acids and analogues, carbohydrates, carboxylic acids and derivatives, fatty acids, organonitrogen compounds, phenylpropanoic acids, and phenylacetic acids [56]. Zhang et al. measured the metabolome of the water flea *Daphnia magna* subject to different amounts of predators over 16 years [57]. This study concluded that increased predation pressure resulted in shifts in amino acid and sugar metabolism, and the subpopulation of water flea with the most predators had the strongest phenotypic response. To understand how the mud crab reacts to odor released by a predator, blue crabs were fed a diet which consisted of entirely mud crabs, partially mud crabs, and no mud crabs, and metabolomic analysis of the urine was conducted, as well as the recording of mud crab behavior [58]. The ability for mud crabs to detect varying levels of metabolites in the urine of blue crabs fed different diets was evident, as foraging decreased in the presence of blue crab urine produced after consumption of any proportion of mud crabs. Research involving mechanistic changes in response to predation may determine biomarkers that can be utilized to study current food chains, allowing for any necessary intervention if the prey is coming close to extinction, and may reveal new mechanisms used by organisms to communicate the presence of danger.

3.3 *Fertilizer Exposure Disrupts Biological Mechanisms of Plants and Animals*

Fertilizers perturb the metabolome of plants. Ehime Kashi 34, a hybrid cultivar with fruit, was exposed to three levels of potassium to test the change in metabolic profiles involved in fruit splitting [59]. Treatment of potassium resulted in upregulation of various glycoside metabolites and downregulation of levels of gibberellin and glycoside. The authors hypothesize the decrease of gibberellin in the flesh made the fruit easier to split. By measuring the change in metabolites, the authors were able to identify specific markers correlated with improvement of fruit splitting. Therefore, this method could be generalized to study the effect of fertilizers on numerous plants, allowing for the creation of optimal fertilizers and food with specific characteristics. Ciampa et al. compared the effect of mineral and organo-mineral fertilizers on the metabolome of cultivar grape berries [60]. Valine, leucine, isoleucine, proline, and malic acid were significantly different between berries under different fertilizer usage, showing the ability to easily change the biological mechanisms of plants given specific nutrients. In addition, fertilizer combination with another agricultural additive, pesticides, can have a large influence on the metabolome of animals. Metabolites of juvenile leopard frogs exposed to combinations of one fertilizer and two pesticides were compared to frogs without chemical exposure [61]. Treating frogs with each pesticide separately and together resulted in positive correlations with glycine, an important metabolite in glutathione metabolism. Frogs subject to fertilizer alone were characteristic of the most alterations of metabolites compared with controls, totaling 23. The authors concluded the main perturbations from exposure to individual and combinations of agrochemicals had the greatest influence on gluconeogenesis/glycolysis, the glucose-alanine cycle, glutathione metabolism, and the urea cycle. Dichlorodiphenyltrichloroethane (DDT) is a well-known pesticide due to its detrimental environmental impacts. Transgenic injection of human p-tau protein, a biomarker of Alzheimer's disease, in *C. elegans* with and without exposure to DDT was performed to test the hypothesis that metabolomic analysis can determine whether DDT influences p-tau accumulation and, if so, its driving mechanisms [62]. The conclusion from this study was mitochondrial function was inhibited by DDT exposure and p-tau, and DDT exacerbated the mitochondrial inhibitory effects of p-tau accumulation. Therefore, a synergistic effect of DDT and p-tau accumulation was discovered by studying the metabolome of *C. elegans*. This study is important because it portrays further evidence of the harmful effects of DDT and provides a methodology which other researchers can follow to probe the impact of other pesticides on animals. Meng et al. exposed male mice to the fungicide penconazole and its enantiomers ((±) – PEN, (+) – PEN, and (−) – PEN) to understand metabolic changes on liver function in nontarget animals [63]. All penconazole treatments resulted in the development of liver metabolic disorder, and mice exposed to (±) – PEN and (−) – PEN had observed regulation of fatty acid and triglyceride synthesis, as well as regulated fatty acid β-oxidation pathways, leading to the significant accumulation of lipids. Finally, pink shrimp were exposed to the benzoylurea chitin synthesis inhibitor teflubenzuron, an insecticide widely used against sea lice, showing an association between teflubenzuron and reduced N-acetylglucosamine, due to the insecticide's inhibitory effect on chitin synthesis, and energy metabolism [64]. Metabolomics provides a tool to assess the current impacts of fertilizers and pesticides on a multitude of organisms. It may be possible to formulate optimal chemicals which induce the necessary growth additives and/or insect deterrents while also guaranteeing a non-harmful effect on the environment.

3.4 *Model Organisms Allow for the Scientific Study of Environmental Factors*

The use of model organisms is widespread in biological disciplines. Environmental metabolomics has adopted many of these models to study a multitude of stressors in different ecological systems, as each organism provides a unique characteristic that makes it an optimal candidate for studying the effects of the environment. *Danio rerio* (zebrafish) has cost-effective breeding conditions, a genome similar to the human genome, and high productivity [1]. *Caenorhabditis elegans* has a well-characterized genome, easy maintenance, body transparency, and sensitivity to chemical toxicants [1]. *Oryza sativa* (rice) is a part of diets worldwide, and therefore its maintenance and optimization are a necessity for the growing population. Kim and Kang published a review in 2021 of environmental toxicants on model organisms, summarizing the effects of different pollutants on model organisms, shown in the table below (Table 1) [65].

3.5 *Specific Compounds and Their Effects on Human Health*

In addition to the extensive research conducted on plants and animals to understand environmental stressors, the effect of specific compounds, such as heavy metals, organohalogens, and persistent organic pollutants (POPs), on human health has been deduced. Heavy metals have been demonstrated to cause changes to metabolic processes. Volunteers surrounding a closed zinc smelter which increased environmental cadmium levels gave urine samples to test if there were associations between urinary cadmium and metabolite abundance, thus defining the metabolic signatures of exposure to cadmium [89]. The authors reported a negative correlation between urine cadmium and 3-HV, DMG, creatinine, creatine and 4-DEA, and a negative correlation between urine cadmium and citrate. This study demonstrates the ability for metabolomics to identify biomarkers of response to pollutants at true environmental concentrations. A study of 610 pregnant women found an association between aluminum concentration and gestational diabetes mellitus (GDM) [90]. Additional metabolomic profiling conducted found correlations between plasma aluminum and polyunsaturated fatty acids (PUFAs) and led to the conclusion that

Table 1 Summary of perturbation of metabolic pathways in model organisms in response to environmental toxicants

	Organisms	Environmental toxicants	Experimental conditions	Toxic mechanisms	Biochemical assays	References
Pharmaceuticals and personal care products	*Danio rerio*	Clarithromycin Florfenicol Sulfamethazine	Adult fish ($n = 30$) 0.1 mg/L for 72 h Extraction with Bligh-Dyer LC-QToF-MS	Dysregulation of choline, guanosine, and ADP	Impaired swimming behavior	De Sotto et al. [66]
		Triclosan (1, 30, and 300 μg/L) Methyl triclosan (MTCS) (0.5, 10, and 400 μg/L)	50 embryos (n = 6) for 96 h Extraction with acetonitrile:isopropanol:water (3:3:2) GC-MS	Dysregulation of energy metabolism, nitrogen metabolism, and fatty acid synthesis	Dysregulation of eight genes related to energy metabolism, nitrogen metabolism, and fatty acid synthesis	Fu et al. [67]
	Gammarus pulex	Propranolol (100 and 153 mg/L) Triclosan (0.1 and 0.3 mg/L) Nimesulide (0.5 and 1.4 mg/L)	Adult *G. pulex* (2, 6, and 24 h) (Approximately 100 specimens, $n = 4$) Extraction with 90% MeOH LC-Orbitrap-MS	Possible alterations of protein synthesis and oxidative stress	Three pharmaceuticals affected 23 functional pathways	Sheikholeslami et al. [68]

(continued)

Table 1 (continued)

	Organisms	Environmental toxicants	Experimental conditions	Toxic mechanisms	Biochemical assays	References
	Mytilus galloprovincialis	Diclofenac (100 μg/L)	3 mussels ($n = 6$) for 7 days Extraction with water: methanol: dichloromethane LC-Orbitrap-MS	Dysregulation of tyrosine and tryptophan metabolism	Potential risk of osmoregulation and reproduction	Bonnefille et al. [69]
		Sulfamethoxazole	10 mussels exposed 4 days Extraction with methanol:water (1:2), clean up with SPE LC-QTrap-MS	Significant change in four amino acids, benzoic acid, and inosine	Perturbation of osmoregulation, energy metabolism, and organoleptic properties	Serra-Compte et al. [70]
	Caenorhabditis elegans	Triclosan (0.1 and 1 mg/L)	Adult worms exposed 24 h Extraction with 80% MeOH GC-MS after silylation	Significantly affected amino acids, tricyclic acid intermediates, carbohydrates and poly amines	Decreased lifespan, reproduction, and locomotion. Increased oxidative stress	Kim et al. [71]
Pesticides	*Danio rerio*	Dieldrin (16 or 163.5 ng/g)	Adult fish Extraction with acetone:hexane (5:2), reconstitution with acetonitrile (ACN), clean up with SPE, GC-MS/MS	Dieldrin altered composition and function of intestinal microbiome	No change in body mass, growth rate, or histopathology	Hua et al. [72]
		Isocarbophos (50 and 200 μg/L)	Adult fish exposed 4 days Extraction with 20% MeOH 1H-NMR analysis	Significant alteration with energy related metabolism (lactate, alanine, and creatin)	Significant down-regulation of antioxidant enzyme activity. Accumulation of isocarbophos in zebrafish	Jia et al. [73]

	Oryza sativa	Butachlor (3.148 kg/a.i.ha) Chlorpyrifos (1.440 kg/a.i.ha) Tricyclazole (0.607 kg/a.i.ha)	10 mg of dried leaves Extraction with methanol: chloroform:water (5:2:2) GC-MS after silylation	Significantly affected TCA cycle, amino acid, and fatty acid metabolism	Differentially expressed genes starch-sucrose distribution, protein contents, and photosynthesis	Liu et al. [74]
	Eisenia fetida	Sulfoxaflor (0/2 mg/kg)	Earthworms exposed 14 days Extractions with methanol: acetonitrile:water (2:2:1) LC-Orbitrap-MS	Sulfoxaflor altered carbohydrates, TCA cycle, pyrimidine purine, and some amino acids	Oxidative damage by sulfoxaflor was confirmed by SOD, CAT, GST, and MDA assay	Fang et al. [75]
		Imidacloprid Dinotefuran	Earthworms exposed 7, 14, 21, and 28 days Extraction with acetonitrile: methanol:water (1:2:1) LC-QToF-MS	Disturbance of TCA/urea cycle, energy production and oxidative stress	Alteration of activity acetylcholinesterase, superoxide dismutase, and catalase	Zhang et al. [76]
	Caenorhabditis elegans	Atrazine (4 mg/L)	10,000 worms ($n = 5$) for 48 h Extractions with methanol:acetonitrile:water (2:2:1) LC-Orbitrap-MS	Perturbation of glycolysis, gluconeogenesis, and phosphatidylcholine metabolism	Increased oxidative stress and disrupt ATP synthesis. Reduction of reproduction, locomotion, and brood size	Yin et al. [77]
Nanoparticles	*Oreochromis mossambicus*	100 nm polystyrene (20 mg/L)	20 fish for 7 days Extractions with methanol:acetonitrile:water (2:2:1) LC-QToF-MS	Disorder of energy, amino acid, and lipid metabolism	Damage of feeding and sensing behavior and signaling disorder	Pang et al. [78]

(continued)

Table 1 (continued)

	Organisms	Environmental toxicants	Experimental conditions	Toxic mechanisms	Biochemical assays	References
	Danio rerio	Polypropylene fibers (10 and 100 μg/L)	Adult fish for 21 days Intestines were extracted with methanol:water (4:1) UPLC-MS	Upregulation of glycerophospholipids metabolism and downregulation of fatty acyls metabolism	Intestinal damage, nutritional deficiency, and oxidative stress were induced by microplastic fibers	Zhao et al. [79]
	Cyprinus carpio	Silver nanoparticle (0.1, 0.5, 1, and 2 mg/L)	10 adult fish for 24 ~ 96 h Fish gills were extracted with methanol:water (4:1) UPLC-QToF-MS	Inhibition of TCA cycle. Perturbation of lipid metabolism	Induced epithelial hyperplasia of the gill. Perturbation of genes in aspartate metabolism pathways	Xiang et al. [80]
	Poterioochromonas malhamensis	Silver nanoparticle (1 mg/L)	Algae for 2 and 24 h Extractions with methanol:water (4:1) LC-QqQ-MS	Perturbation of amino acids, nucleobases, sugars, and fatty acids metabolism	Increased level of ROS and decrease of the photosynthetic efficiency	Liu et al. [81]
	Eisenia fetida	TiO2 nanoparticles (5, 50, and 500 mg/kg)	Earthworms for 120 days Extractions with chloroform:water:methanol (2:2:5) GC-MS after silylation	Alteration of glutathione and starch, sucrose metabolism	Decreased GSH/GSSG ratio. Slight increase in ROS level. Alteration of genes in TGF-beta singaling pathway	Zhu et al. [82]
	Enchytraeus crypticus	Silver nanoparticles (60–102 mg/kg) Silver ions (45–60 mg/kg)	Worms ($n = 5$) for 7 and 14 days Extraction with methanol LC-Orbitrap-MS	Alteration of phenylalanine, histidine, lipid, and energy metabolism	Activation of cellular iron ion homeostasis, tyrosine catabolism, glycosylation, and stress response	Maria et al. [83]

Consumer products additives	*Drosophila melanogaster*	BDE-47 (2, 10, or 50 μM)	5 flies for 30 days Extraction with methanol LC-Orbitrap-MS Extraction with methanol:water (4:1) GC-MS after silylation	Perturbation of metabolites involved in tryptophan, phenylalanine, and purine metabolism	Decreased ratio of SAM/SAH and GSH/GSSG. Imbalance of kynurenine metabolism and oxidative stress	Ji et al. [84]
	Eisenia fetida	BDE-47 BDE-209 (10, 50, 100, and 200 mg/kg)	3 earthworms ($n = 6$) for 14 days Extraction with water 1H-NMR analysis	Increase of lactate, glutamate, betaine leucine, and lysine. Decrease of fumarate and glycine	Toxic effects by disturbing osmoregulation, energy metabolism, nerve activities, and TCA cycle	Liang et al. [85]
	Danio rerio	Bisphenol A (4.4, 8.8, and 17.5 μM)	20 zebrafish embryos ($n = 6$) Extraction with methanol:water:chloroform LC-QToF-MS	Perturbation of amino acids, prostaglandin, folate, ascorbate, and nucleotide metabolic pathways	Altered gene expression with estrogenic, CYP450 enzyme, tissue development, and cell proliferation	Ortiz-Villanueva et al. [86]
	Rattus norvegicus	Bisphenol A Bisphenol S (50 μg/kg)	Rat plasma ($n = 14$) Extraction with methanol:water (4:1) LC-Orbitrap-MS	BPA exposure decreased citric acid, oxoglutaric acid, and malic acid, while BPS decreased polyunsaturated fatty acid	Toxic effects of endocrine disruption, cytotoxicity, and genotoxicity	Mao et al. [87]
	Mytilus coruscus	Phthalates (0.04, 0.40, and 1.00 mg/L)	4 mussels for 7 days Extraction with methanol LC-QToF-MS	Significant changes in amino acids, lipids, energy storage compounds, osmolytes, and neurotransmitters	Activation of antioxidant defense system	Gu et al. [88]

Table adopted from Kim and Kang [65]

n-6 PUFAs could be potential mediators for the relationship between aluminum exposure and the risk of developing GDM. At multiple time points, arsenic was measured in the water supply of an area in Bangladesh and in 112 participant urine samples from this area, and participant metabolite abundances were taken [91]. Significant associations with water or urine arsenic included 1,2-dithiane-4,5-diol, l-threonine, phosphoric acid, pyroglutamic acid, (R*,S*)-3,4-dihydroxybutanoic acid, and succinic acid. This study determined long-term reproducible urinary metabolites associated with arsenic exposure.

Organohalogens, compounds with at least one halogen bonded to a carbon, are associated with metabolic disorder. Trichloroethylene (TCE) has been linked to different cancers, and therefore 80 exposed workers and 95 controls were included in a metabolomic analysis to discover TCE's mode of action [92]. Response to TCE included disruption in purine catabolism and decreases in sulfur amino acid and bile acid biosynthesis pathways. Many metabolites were associated with TCE exposure including uric acid, glutamine, cystine, methylthioadenosine, taurine, and chenodeoxycholic acid. An enriched level of uric acid has been identified as a marker of kidney disease, thus suggesting TCE has the potential to drive this phenotype. Wang et al. investigated the effects of polycyclic aromatic hydrocarbons (PAHs) on 566 subjects split into a control and exposed group [93]. Significant changes elicited by PAHs were observed in amino acid, purine, lipid, and glucuronic acid metabolism, and 1-hydroxyphenanthrene and dodecadiene-L-carnitine were identified as potential biomarkers for PAH exposure, allowing for the potential to create a future diagnostic through metabolomic screening.

Persistent organic pollutants (POPs) are highly lipophilic compounds resistant to degradation with a tendency to bioaccumulate [29]. Hu et al. analyzed DDT and its degradation products DDE, specifically p,p'-DDT, o,p'-DDT, and p,p'-DDE, in 397 maternal serum samples collected during pregnancy to identify their impact on the metabolome [94]. Results indicated two shared pathways (glycine, threonine, alanine, and serine metabolism and phosphatidylinositol phosphate metabolism) between all exposures, three pathways (urea cycle/amino groups, arginine and proline metabolism, aspartate and asparagine metabolism) between p,p'-DDT and o,p'-DDT but not p,p'-DDE, and multiple pathways including fatty acid metabolism and carnitine shuttle uniquely impacted by p,p'-DDE. Interestingly, the degradation products of DDT initiated a very different response from DDT itself, demonstrating the specificity of each toxicant on the metabolome. Perfluoroalkyl substances (PFASs) are POPs which have been studied in a cohort of 965 individuals in Sweden to determine correlations between metabolites with 6 different PFAS levels [95]. A total of 15 metabolites, most from lipid pathways including glycerolipids, glycerophospholipids, and fatty acids were associated with one or more PFAS. Significant enrichment of glycerophospholipid metabolism, linoleic acid metabolism, and α-linoleic acid metabolism was discovered using metabolic pathway analysis tool MetaboAnalyst 3.0. A type 2 diabetes (T2D) case-control study of 187 matched pairs was conducted under the hypothesis that there is an association between T2D and PFAS [96]. The authors concluded a positive correlation with PFAS and negative correlation with T2D for glycerophospholipids, and a negative correlation with

PFAS and positive correlation with T2D for diacylglycerols, suggesting the influenced lipid species of PFAS have opposite relations with T2D. Polychlorinated biphenyls (PCBs), in the class of POPs, was studied in maternal and umbilical cord serum in 93 patients to determine changes in biological pathways and identify biomarkers of PCB exposure [97]. In a machine learning random forest model of metabolite abundances, citraconic acid level in maternal serum, as well as ethanolamine, p-hydroxybenzoate, and purine in umbilical cord serum, predicted high or low level of PCB with an area under the receiver operating characteristic curve greater than 0.70. This result suggests a relatively high ability for few metabolites to discriminate between high- and low-PCB exposure, potentially to be used as biomarkers in the future to determine the risk of PCB-related health problems for the fetus. A study encompassing multiple POPs across populations from Europe discovered sphingolipid and glycerophospholipid families to be significantly altered upon comparison of subjects exposed to both the class and concentration (high and low) of POPs [98]. The authors conclude, through metabolomic profiling, biomarkers of different types of POP exposure can be discovered. Metabolomics provides the foundation upon which researchers can identify specific physiological changes in humans due to environmental stressors, determine populations at high risk for developing health problems by profiling different geographical locations likely to be impacted by pollutants, and discover specific metabolites to be used as diagnostic tools in healthcare through large studies.

3.6 Precision Environmental Health Monitoring

As metabolomics has been utilized on large-scale studies to uncover mechanisms for human interactions with the environment, a very new focus on precision medicine has influenced the development of research involving the study of a single individual with substantial amounts of data measured multiple times over a specified time period, published in 2022 by Gao et al. [99]. The goal of this experiment was to dissect the complexity and heterogeneity of individualized environmental exposures, i.e., the physical, chemical, and biological components collectively in the human external and internal environment, termed the exposome. Over 52 days, blood and stool samples were taken from the participant to perform tests on the gut microbiome, proteome, metabolome, toxins and carcinogens, cytokines, and blood tests. Exposome-metabolome analysis identified positive metabolite correlations with the biological component fungi, and the chemicals salicylic acid, dinoseb, and dibromoethane. Protein digestion and absorption and aminoacyl-tRNA biosynthesis were metabolic pathways significantly correlated with both chemical and biological exposomes. The authors also discovered 19 highest-degree metabolites related to protein metabolism, inflammation, and kidney and liver functions, indicating a possible link of these metabolites directly to responses of the exposome. This first of a kind study demonstrates the capability of metabolomics to discover personalized metabolic changes, possibly leading to tailored recommendations for optimal health for the individual studied and will serve as a model for future research.

4 Recent Advancements in Ecometabolomics and Biomonitoring

Ecometabolomics is still at a nascent stage of development compared to the application of metabolomics in other fields such as biomedicine and nutrition. Nevertheless, a continuum of developments in analytical techniques allows ecometabolomics applications to increase every year, making it a powerful tool for studying the environment. There have been rapid advancements at every focal point of ecometabolomics and biomonitoring that include technological improvements in sample preparation and identification methods, advancements in conducting field experiments studying stress responses in organisms, and bioassessment strategies being employed for monitoring [12]. The number of applications for which environmental metabolomics is being utilized is also ever-increasing, including but not limited to understanding responses of organisms to biotic pressure, studying the effect of anthropogenic pollutants on aquatic and terrestrial life and assessing organismal responses, characterizing aerosol composition in the environment and its impact on ecosystem [100], etc. In this section, we would be highlighting recent examples and novel developments in the field of environmental metabolomics.

4.1 Advancements in Conducting Field Experiments

While ecometabolomics studies are being performed in the laboratory commonly for some time now, a huge effort is being made to extend effect-based monitoring to the field as well. Such field-based investigations have been conducted in terrestrial as well as aquatic environments. Metabolomic data gathered from on-field experiments are proving to be a powerful tool for complementing laboratory-derived data, to gain deeper insight into the effect of various stressors on the environment [101].

In a study by Skelton et al., metabolomics was employed for in situ monitoring of surface waters impacted by point and nonpoint source water contaminants [101]. Male fathead minnows (*Pimephales promelas*) held in mobile monitoring units were exposed to surface waters upstream and downstream from the effluent release site and at the site of effluent release as well. NMR and GC-MS analysis successfully revealed significant differences in the liver metabolome of minnows exposed to surface water with varying degrees of agricultural and industrial activity [101]. Majority of the terrestrial studies performed focus on the aboveground tissues and belowground tissues are largely ignored. In recent years, there have been exciting developments in the collection and untargeted analysis of plant root exudates in response to biotic and abiotic stressors [102–105]. Changes in the root exudate composition play a vital role on the plant itself, soil properties, soil microbes, and the plants around it. A study by Gargallo-Garriga et al. successfully establish an experimental drought gradient and subsequent recovery for identifying the changes in root exudates in *Quercus ilex* (Holm oak) [102]. In another laboratory bioassay, the allelopathic influence of Alfalfa genotypes on annual rye grass has been analyzed by conducting a metabolomic analysis of the root exudates of Alfalfa genotypes using quadruple time of flight [105].

Aerosols play a pivotal role in controlling the functioning of the ecosystem and atmospheric composition. In a first-of-its-kind study by Guenther and group, an efficient and detailed methodology for characterizing the chemical composition of low-molecular-weight compounds, by employing liquid and gas chromatography-mass spectrometry (MS) and Fourier transform ion cyclotron resonance MS (FT-ICR-MS), has been described to gain a deeper insight into the aerosol-biosphere interface [100]. In the study, aerosols collected in different seasons, at different levels of biological activity, were analyzed, and despite the sample complexity, they could identify clear changes in the composition of aerosols at higher levels of biological activity. This strategy thus demonstrates a novel approach to understanding the biosphere-aerosols dynamics and could improve the identification and quantification of biomarkers [100].

4.2 Advances in Biomonitoring and Bioassessment

Biomonitoring involves supervising an environment or organism based on biological responses. Initially, sensitive species (sentinel species) in an environment such as planktons, lichens, bryophytes, etc. were monitored for assessing environmental health and detecting changes that may impact human society [106]. With advancements in the omics approach especially metabolomics, minute changes in organisms in response to stressors can now be monitored. Since the last decade, environmental metabolomics has been at the helm of monitoring the impact of various biotic and abiotic stressors on the different components of the ecosystem [107]. Using a wide variety of both model and non-model organisms, metabolomics has assisted researchers in discovering and validating biomarkers with accurate diagnosis [108].

Identification of biomarkers in the terrestrial and aquatic environments is useful in assessing changes in signaling at the molecular level and preventing deleterious effects in the population [107, 108]. Changes in metabolome precede pathological changes, and identifying early response biomarkers would enable more efficient testing [109]. Furthermore, identifying early changes in metabolome also has the potential to predict changes at a higher level of organization [110]. Taylor et al. have employed metabolomic analyses to discover early response biomarkers that can predict chronic reproduction fitness in *Daphnia magna*, reducing the testing time from 21 days to 24 h [109]. In another study by Hines et al., a metabolic signature predicting whole-body toxicological stress has been identified. They exposed marine mussels to copper and pentachlorophenol and measured the metabolic fingerprint predictive of overall fitness and scope for growth. The identified signatures could accurately predict decreased fitness in animal samples from a contaminated site in comparison to a rural site [111].

Several metabolomic studies have successfully scrutinized data sets belonging to polluted and nonpolluted areas, identifying even minute changes between the two [107, 112–114]. To employ metabolomics as a diagnostic tool for ecotoxicological research in polluted areas, it is essential to test if biomarker profile analyzed across different field sites with varying physicochemical properties remains relevant. This has been assessed in a study by Bundy et al. where they collected earthworms (*Lumbricus rubellus*) from sites with radically different soil types with metal contamination, and despite the drastic difference in soil properties, the biomarker profiles from all the sites could discern metabolites specific for zinc contamination [115].

Further studies have also proven that metabolic responses in an organism are governed by its ontogeny. Wu et al. demonstrated differential metabolic responses in different life stages of mussels (*Mytilus galloprovincialis*) as a result of cadmium exposure. The contaminant induced osmotic stress in all three stages. The larval and juvenile mussels exhibited higher sensitivity to cadmium, while adult mussels interestingly sustained cadmium exposure by decreasing their energy requirements [116]. Neonate and adult *Daphnia magna* were also found to respond uniquely to sub-lethal level exposure to atrazine and perfluorooctane sulfonic (PFOS) acid, but not so much difference was observed with propranolol exposure [113]. These reports indicate that the metabolic responses of species are controlled by the contaminant and stage of development. Unlike many monitoring techniques, metabolomics has the potential to distinguish between stressors in a multi-stressor environment. This has been demonstrated in a study by Khan et al., where exposure to two different metals, lead, and cadmium resulted in different metabolic profiles for each metal toxicity supporting the applicability of metabolomics as an effective biomonitoring tool in presence of multiple stressors [117]. Additionally, Izral et al. exposed crayfish to varying nutrient and dissolved oxygen concentrations (high, medium, and low) to identify if metabolomic analysis can be effectively employed as a bioassessment tool [118].

Sun et al. published a review in 2022 identifying different biomarkers and associated metabolic diseases as a result of exposure to various metabolism-disrupting chemicals (MDCs) in both animal experiments (Table 2, Fig. 2) and human epidemiological studies (Table 3, Fig. 2) detailed in the tables below [119].

5 Challenges of Eco-Metabolomics

The ecological system is complex and requires expert analysts with an extensive understanding of biology, chemistry, statistics, and data science. Besides this limited detection ranges and narrow molecular resolution of NMR and MS devises also

Table 2 Metabolomic studies reporting disrupting effect biomarkers and biological pathways caused by MDCs in animal studies

Animal strain	Exposure groups	Exposure route	Timing of exposure	Phenotype changes	Sample types	Metabolic biomarkers	References
C57BL/6 female mice	*Arsenic.* 10 ppm	Drinking arsenic water	PND 42–84	disturbance of intestinal microbiota	Urine and feces	↓ Glycocholic acid, guanosine, N6-methyllysine dihydrodaidzein, cuminaldehyde, indole-3-carbinol, daidzein, phenylpyruvic acid, indolelactic acid, 5-methyltetrahydro-folic acid, 6-hydroxy-5-methoxyindole glucuronide, coumarin, phenylacetaldehyde ↑ O-Desmethylangolensin, hydroxyphenylacetic acid, 1b-hydroxycholic acid, tetrahydroxy-5b-cholanoic acid, MG (0:0/18:2 (9Z,12Z)/0:0), cinnamaldehyde, 7α-hydroxy-3-oxo-4--cholestenoate, lysoPC(P-18:0), indoxyl	Lu et al. [148]
Wild-type N2 *C. elegans*	*Atrazine:* 0.0004 mg/L 0.004 mg/L 0.04 mg/L 0.4 mg/L 4 mg/L 40 mg/L	Living in the solution	Larvae	↓ Brood size body, length and width, locomotion behavior ↑ Generation time, oxidative stress, ATP synthesis disruption	Nematodes extract solution	↓ Phosphatidylcholine, CDP-choline, GMP, 7,8-dihydroneopterin 2′,3′-cyclic-p, 2-oxobutanoate, neopterin, 7,8-dihydroneopterin ↑ Salicin 2-hydroxyethyl-ThPP	Yin et al. [77]

(continued)

Table 2 (continued)

Animal strain	Exposure groups	Exposure route	Timing of exposure	Phenotype changes	Sample types	Metabolic biomarkers	References
Manila clam *R. philippinarum*	*B[a]P:* 0.02 μM 0.2 μM	Living in the solution	adult	↑ Carcinogenesis, teratogenesis, mutagenesis	Gill tissues	↓ Phosphocholine and homarine Dimethylamine, betaine, taurine, glucose, and glycogen ↑ Chain amino acids (valine, leucine, and isoleucine), alanine, succinate, malonate	Zhang et al. [149]
Adult zebrafish	*BPA*: 0.01 mg/L 0.1 mg/L 1 mg/L	Living in BPA solutions	72 h	↓ Oxidative stress	Tissue powdery	↓ AMP, glutamine, inosine, lactate, succinate	Yoon et al. [150]
Sprague-Dawley rats	*BPA*: 0.5 μg/kg/day 50 mg/kg/day	Gavage	PND 42–56	↓ Body weight Liver weight	Urine	↓ Riboflavin, biotin ↑ Methylated amino acids	Wang et al. [151]
SPF-grade male Sprague-Dawley rats	*Cd*: 10 mg/kg 50 mg/kg bw/day	Gavage		↑ Renal toxicity, oxidative stress	Blood/kidney	↓ UA, hypoxanthine (Hyp), xanthosine, deoxyinosine, 12-HETE, guanidinosuccinic acid, taurocholic acid, hydroxy phenylacetylglycine, ATP ↑ Formiminoglutamic acid, arachidonic acid	Guan et al. [152]

Female Wistar rats	*DEHP*: 90 mg/kg/day	Gavage	PND 28–58	↑ Oxidative stress, blood Urea nitrogen, urinary creatinine, urea, fatty liver	urine	↓ Kynurenic acid, *N*6-methyladenosine, L-isoleucyl-L-proline, xanthurenic acid, hippuric acid, tetrahydrocortisol, citric acid, phenylpropionylglycine, cPA(18:2/0:0), LysoPC (14:1)	Dong et al. [153]
Kunming mice	*DEHP*: 4.8 mg/kg bw/day	Injection	PND 0–21	↓ Body weight BMI Heart weight Plasma total cholesterol ↑ Liver weight/bw Brain weight/bw Liver TBA Blood glucose	Plasma	↓ PC, PC-(O), PC-(P), LPC18:0, and PIs ↑ PE, PI, PS, and CE	Huang et al. [154]
Kunming mice	*DEHP*: 0.048 mg/kg bw/day	Subcutaneous injection	PND 0–21	↓ Liver weight/bw Brain weight/bw Heart weight/bw Plasma total diglyceride ↑ Body weight BMI Blood glucose Liver TBA	Plasma	↑ PS in female pups CE20:1, TG18:2/18:2/18:2	Huang et al. [154]

(continued)

Table 2 (continued)

Animal strain	Exposure groups	Exposure route	Timing of exposure	Phenotype changes	Sample types	Metabolic biomarkers	References
Kunming mice	*DINP*: 4.8 mg/kg bw/day	Subcutaneous injection	PND 0–21	↑ Liver total bile acid levels	Liver tissue extraction	Female: ↓ PG, GM ↑ PC, LPC, PC(O), PC(P), PE, LPE, PE(O), PE(P), PS, PI, BMP dhCER, CER, MHC, DHC, SM, DG, TG, CE(14:0), CE (15:0), CE(16:0), CE(16:1), CE(16:2), CE(17:1), CE(18:3), CE(20:3), CE(22:5), COH Male: ↓ PG, SM ↑ LPC, PC(O) PC(P), PE,LPE,PE(O), PE(P), PS,PI,BMP, dhCER, CER, MHC, DHC, DG, TG, CE(14:0), CE(15:0), CE(16:2), CE(17:1), CE(18:0), CE(20:1), CE(22:1), CE(22:5), CE(24:0), CE(24:1), CE(24:6), COH	Yang et al. [155]
C57BL/6 J mice	*DEHP*: 625 mg/kg 1250 mg/kg 2500 mg/kg bw	Gavage	PND 56–84	↑ Liver weight Fatty liver	Blood	↑ LTB4 8S-HETE Arachidonic acid Bishomo-γ-linolenic acid γ-Linolenic acid Linolenic acid	Li et al. [156]

Earthworm	*DDT: For ¹H NMR* 0.5 μg/cm² 1.0 μg/cm² 2.0 μg/cm² *For GC-MS* 0.5 μg/cm² 1.0 μg/cm²	Glass filter paper	/	↓ Body weight	Lyophilized earthworm tissues	↑ Leucine, alanine	McKelvie et al. [157]
CD-1 mice	*DDT*: 300 mg/kg	Gavage	PND 17–20	↑ Uterine wet weight, luminal epithelial cell height, uterine proliferation	Brain and uterine extracts	↓ Ethanol, phosphorylcholine, glycerophosphocholine, taurine, glycine, fumarate, hypoxanthine ↑ Glutamine, choline, succinate, citrate, inosine, uridine, tyrosine, phenylalanine	Wang et al. [158]
Adult tropical 5D strain zebrafish	*PAHs*: 4 μM	Living in solution		↑ Oxidative stress, impaired locomotor activity, anxiety, developmental neurotoxicity	Zebrafish larvae extraction solvent	↑ Erine, threonine, cystathionine, *S*-adenosylmethionine, and *S*-adenosylhomocysteine, glutathione, *S*-transferase, glutathione peroxidase, guanosine, hypoxanthine, inosine, xanthine, phosphocreatine, docosahexaenoic acid	Elie et al. [159]

(continued)

Table 2 (continued)

Animal strain	Exposure groups	Exposure route	Timing of exposure	Phenotype changes	Sample types	Metabolic biomarkers	References
C57BL/6 male mice	*PCB126*: 1.53 μmol/kg	Gavage	PND 56–91	↑ Liver injury, oxidative stress, systemic inflammation	Liver	↓ Glucose 6-phosphate, fructose 6-phosphate, fructose 1,6-diphosphate, glucose 1,6-diphosphate, dihydroxyacetone phosphate, maltohexaose, maltopentaose, maltotetraose, maltotriose, maltose, liver cysteine GSH and CoA ↑ 5-phosphate, ribose 1-phosphate, *N*-carbamoylaspartate, orotate, 4-hydroxy-nonenal	Deng et al. [160]

Sprague-Dawley rats	*PBDE*: 10 mg/kg bw	Gavage	PND 35–49	↓ Fetal development ↑ Neurodevelopmental toxicity, abnormal brain development	Urine and blood	↓ Acetyl-CoA ↑ Pyruvate	Jung et al. [161]
The C57BL/6J gpt delta mice	*PBDE (BDE-3)*: 0.0015 mg/kg 1.5 mg/kg 10 mg/kg 30 mg/kg	Gavage	PND 105–174	↓ Sperm count Germ cells ↑ Reproductive toxic Oxidative stress Abnormal sperms	Urine	↓3-Hydroxy-L-proline, acetylpyruvate, alanopine, beta-hydroxyarginine, creatine, D-biotin, L-leucine, L-thyronine, L-tyrosine, Lyso-PAF C-16, PG(22:4/0:0), PG(P-20:0/0:0), PGE2, reduced riboflavin, riboflavin, succinic anhydride, xanthurenic acid, *N*-acetyl-L-phenylalanine ↑ L-proline, L-valine, ketoleucine, L-leucine, 2-phenylacetamide, L-methionine, 4-imidazolone-5-propi-onic acid, L-carnitine, L-phenylalanine, L-tyrosine fragment, phenyl sulfate, L-tyrosine, 2-hendecenoic acid, L-tryptophan fragment, pyrocatechol sulfate, L-tryptophan, indoxyl sulfuric acid, dodecanedioic acid, inosine, LysoPE(0:0/16:0), LysoPE(18:1/0:0), LysoPE(18:2/0:0), LysoPC(16:1), LysoPC(18:3), LysoPC(18:1), LysoPC(20:5/0:0), LysoPC(20:4), LysoPC (20:3), LysoPC(22:5) cholic acid glucuronide	Wei et al. [162]

(continued)

Table 2 (continued)

Animal strain	Exposure groups	Exposure route	Timing of exposure	Phenotype changes	Sample types	Metabolic biomarkers	References
Male Balb/c mice	*PFOA*: 0.5 mg/kg, 2.5 mg/kg bw/day	Gavage	PND 42–70	↓ Brain weight ↑ Hepatotoxicity Neurobehavioral liver weight	Extracts of brain or liver	↓ α-linolenic acid ↑ Linoleic acid Arachidonic acid ACC18, ACC16, ACC8:1, ACC5:1, ACC3:1	Yu et al. [163]
M. edulis specimens	*PFOA*: 20 μg/L 200 μg/L 2000 μg/L	Living in PFOA seawater		↑ Oxidative stress	Gland	↓ LysoPC, taurine, glutamine, oleic acid	Li et al. [164]
Sprague-Dawley rats	*PFOS*: 1 mg/kg/d	Gavage	PND0–21	↑ Hyperactivity Motor function Developmental toxicity	Brain section	↑ Chenodeoxycholate, MDA, pyruvate, serine, glutamate, cystein, glysine ↓ Phosphatidylcholines, ↑ GABA, taurine, Gly, Met, Pro, Ser, and T4-hydroxyproline, N-methyl-daspartate (NMDA)	Reardon et al. [165]
Kunming mice	*PM2.5*: 10 mg/ml 20 mg/ml 30 mg/ml	Instillation		↑ Oxidative damage, liver oxidative stress, inflammatory damage, degraded nervous system, fat deposition, and insufficient energy supply	Liver homogenate	↓ 4-Pyridine acid, succinate, proline, KYNA, spermidine, serotonin ↑ Taurine, bile acid	Wang et al. [166]

Adult zebrafish	*Tributyltin*: 0.1 μM 0.01 μM	Living in tributyltin solution	Adult zebrafish	↓ Lipid of embryo ↑ Yolk sack area, development of the nervous system	Zebrafish embryos extract	↓ alanine, 4-aminobutanoate, L-valine, taurine, creatine, L-methionine, L-carnitine, 6-succinoa-minopurine, 6-pyruvoyltetrahydrop- terin, hypoxanthine, adenosine monophosphate, guanosine monophosphate, 5β-cyprinol sulfate, oxidized glutathione, PS(44:12) ↑ L-glutamine, 5-Oxo-D-proline, L-tyrosine, L-acetylcarnitine, 4-aminohippuric acid, neopterin, inosine, methyl hexadecanoic acid, dehydrodiconiferyl alcohol, LysoPC(18:1), LysoPC(18:0), PC(32:1), PC(36:5)	Ortiz-Villanueva et al. [167]
C57BL/6J and DBA/2J mice	*TCDD*: 0 μg/kg/d 20 μg/kg/d	Gavage		↑ Neurotoxicity, liver damage, insulin resistance, immunotoxicity, reproductive toxicity, carcinogenicity and wasting syndromes	Serum	↓ l-Arginine, l-aspartic acid, EPA, DHA, glucuronide-conjugate ↑ LysoPE(18:3), LysoPC(22:5), LysoPE(18:2), LysoPE(18:1), elaidic carnitine, palmitoylcarnitine, sphingosine-1-phosphate	Lin et al. [167]

(continued)

Table 2 (continued)

Animal strain	Exposure groups	Exposure route	Timing of exposure	Phenotype changes	Sample types	Metabolic biomarkers	References
Wild-type AB strain zebrafish (*Danio rerio*)	*Microplastics*: 5 and 50 μm: 1000 μg/L	Living in the solution	Larval	↑ Oxidative stress, imbalance of antioxidant defense system, neurotoxicity	Larval zebrafish	↓ Sarcosine, 2-monoolein, aconitic acid, zymosterol, putrescine, s-carboxymethylcysteine, 5,6-dimethyl-benzim-idazole, 6-hydroxynicotinic acid, lactobionic acid, ascorbate, tagatose, prostaglandin A2, norvaline, androsterone, and maleic acid ↑ Succinic acid, pyruvic acid, creatine, glucose, 2-monopalmitin, methyl palmitoleate, fructose-6-phosphate, ribose-5-phosphate, nicotinic acid, 3-phosphoglycerate, serine, gallic acid, noradrenaline, aspartic acid, hydroquinone, glycine, ribose, lyxonic acid, 1,4-lactone, uridine, adenosine, uracil, xanthine, hypoxanthine, 5-ami-novaleric acid lactam, mannitol, octanal, phloroglucinol, sophorose, sorbose, d-Talose, *trans*-4-hydroxy-l-proline, pelargonic acid, 2-methylglutaric acid, scopoletin, alpha-ketoisocaproic acid, gly-pro, and urea	Wan et al. [168]

Male ICR mice	*Microplastics* 5 and 20 μm 0.01 mg/day 0.1 mg/day 0.5 mg/day	Gavage	PND 35–77	↓ Relative liver weight ↑ Oxidative stress neurotoxic (AchE)	Liver and serum	↓ Creatine, 2-oxoglutarate, choline, pyruvate, lysine, citrate ↑ Taurine, ethanol, lipids, threonine, aspartate	Deng et al. [169].

Adopted from Kim and Kang [119]

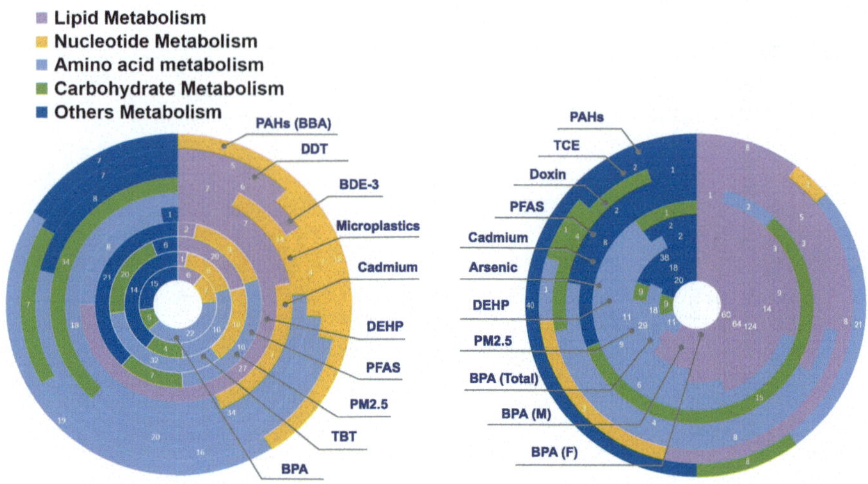

Fig. 2 Summaries of disrupted metabolic pathways within which biomarkers were identified responsive to metabolism disrupting chemicals (MDCs) in animal (left) and human (right) studies. The circular chart shows the number/proportion of major metabolic biomarkers identified after exposure to each type of environmental pollutants found through literature search. We labeled the four metabolic pathways and other metabolic pathways with different colors. (Figure and caption are adopted from reference [119])

require attention. Due to the multi-subject nature of environmental samples and narrow range of signals, it is important to have a single reliable analytical platform for consistent identification of all metabolites, chemicals, and pollutants within a single sample.

An additional challenge in the study of environmental metabolomics is the lack of agreement on a standardized pipeline for analysis. One fact important to understand about the concentration of chemicals in organisms is that their exposure pattern, uptake, metabolism, and half-life result in large fluctuations over time [120]. Therefore, it may be difficult to compare the effects of chemicals on different organisms, or even on the same organism with multiple measurements taken. It is a necessity to carefully design each experiment such that confounders have a minimal impact on the conclusions and potentially create guidelines for what confounders to consider for specific situations. As a result, metabolomic deductions will have a greater chance to be reproducible. The opinions of experts have been culminated to determine the best procedures for metabolomic studies. Evans et al. conducted a survey using a six-page questionnaire to collect the quality assurance (QA) and quality control (QC) practices used in the laboratories of 23 volunteers [121]. Commonly used methods included system suitability assessments, sample run-order randomization, balancing, authentic chemical standards, various quality control samples such as pooled-QC/intra-study QC samples and blanks, compound identification practices, assessment of integration accuracy, and PCA for quality tthe frequency of injections utilized by practitioners, the specific acceptance criteria

Table 3 Metabolomic studies reporting the disrupting effects of MDCs on biomarkers and biological pathways in humans

MDCs	Sampled year	Sample size	Participants	Countries and regions	Sample types	Chemical exposure concentrations	Metabolic biomarkers	References
Aluminum	2013.10–2016.10	610	Pregnant women	China	Plasma	Median (IQR) Al: 69.4 (33.2, 132.8) μg/L	↓ Linoleic acid (18:2n-6) and eicosadienoic acid (20:2n-6) total n-6, and total PUFAs ↑ Total n-3 and n-6/n-3	Huang et al. [90]
Arsenic (μg/g creatinine)	2013.11–2014.5	285	Pregnant women	China	Urine	Mean ± SD Low arsenic exposure group 15.6 ± 2.9 Middle arsenic exposure group 23.6 ± 2.4 High arsenic exposure group 49.9 ± 32.7	↓ Glutathione, thiocysteine, glucuronide ↑ LysoPC (14:0), 18-carboxy-dinor-LTE4, 20-COOH-LTE4, cystathionine ketimine, 1-(beta-d-ribofuranosyl)-1,4-dihydronicotinamide, p-cresol, vanillactic acid	Li et al. [170]
Arsenic (μg/g creatinine)	2000–2006	112	Nonsmoking participants (58 men and 54 women)	Bangladesh	Urine	Mean ± SD Urinary arsenic Original cohort 197.6 ± 162.4 Expansion cohort 174.8 ± 133.4	↓ Glycine, l-threonine, and serine, aminoethanol, β-amino-isobutyric acid, citric acid, 1,2-dithiane-4,5-diol, 3-hydroxyisovaleric acid, indole-3-acetic acid, phosphoric acid, pyroglutamic acid, (*R,S*)-3,4-dihydroxybutanoic acid, succinic acid, uracil ↑ Ethanedioic acid, uric acid Male:	Wu et al. [91]

(continued)

Table 3 (continued)

MDCs	Sampled year	Sample size	Participants	Countries and regions	Sample types	Chemical exposure concentrations	Metabolic biomarkers	References
BPA Urine BPA (μg/g creatinine)	2015	194	Male and female, aged 30–50 years	Korea	Urine	Mean ± SD *Male* Low group 0.89 ± 0.30 *Male* High group 4.82 ± 10.45 *Female* Low group 0.85 ± 0.38 *Female* High group 3.32 ± 2.37	↓ (2)-Octenoyl-CoA, *N*-acylsphingosine sphingomyelin, (*S*)-hydroxyoctanoyl-CoA, palmitic acid, (2*E*)-tetradecanoyl-CoA ↑ (2)-Octenoyl-CoA glucosylceramide, 3-oxododecanoyl-CoA Female: ↓ *N*-Acylsphingosine sphingomyelin, glucosylceramide, tetradecanoyl-CoA, sulfatide ↑ (2E)-octenoyl-CoA, (2E)-tetradecenoyl-CoA; digalactosylceramide; 6-ketoprosta-glandin E1, 3-oxododecanoyl-CoA	Cho et al. [171]
Cadmium (μg/g creatinine)	2013.10–2016.10	610	Pregnant women	China	Urine	Cd: Median (IQR) Control group: 0.59 (0.34–1.06) Case group: 0.69 (0.46–1.07) μg/L	↓ Stearic acid, eicosapentaenoic acid, total odd-chain saturated fatty acids, total n-3 polyunsaturated fatty acids (PUFAs), and n-3 PUFAs/n-6 PUFAs ratio ↑ Eicosadienoic acid and arachidonic acid/ eicosapentaenoic acid ratio	Li et al. [172]

DEHP (μg/g creatinine)	2016	364	Male, 19–44 years	China	Urine	Median (IQR) DEHP: 42.0 (8.5, 166.8)	↓ Alanine, taurine, tryptophan, ornithine, and methylglutaconic acid prostaglandins – hydroxyl-PGE2 and keto-PGE2, diacetylspermine ↑ Cystine, phenylglycine, phenylpyruvic acid, and glutamylphenylalanine Carnitine C8:1 carnitine C18:0acetylneuraminic acid	Zhang et al. [173]
Dioxin	2015	48	Adult males	Switzerland	Urine	Exposure level	↓ Dehydroepiandrosterone 3β-sulfate, androsterone 3α-glucuronide, androsterone 3α-sulfate, pregnanediol 3α-glucuronide, 11-ketoetiocholano-lone 3α-glucuronide, glucuronide conjugates of 11β-hydroxy androsterone, glycochenodeoxycholic acid, glycocholic acid, and glycoursodeoxycholic acid sulfate	Jeanneret et al. [174]
PM2.5 volunteers	2015 11–2016 9	78	College students	China	Blood	Mean (0–max) PM2.5: 53 (0–191) μg/m^3	↓ LysoPE (18:0/0:0) Glutamine Citric acid/isocitric acid ↑ LysoPC (P-20:0), LysoPC (P-18:1 (9z)), LysoPC (20:1), LysoPC (O-16:0), choline, and 1,3-diphenylprop- 2-en-1-one	Chu et al. [175]

(continued)

Table 3 (continued)

MDCs	Sampled year	Sample size	Participants	Countries and regions	Sample types	Chemical exposure concentrations	Metabolic biomarkers	References
Heavy metals (μg/g creatinine)	2015	550	Women	China	Urine	95% CI (median) *High-exposure group* Urine Cr greater than 15 Urine Cd 15.01–40.15 (19.05) Urine Pb 3.508–12.48 (5.454) Urine Zn 197.7–618.9 (329.8) *Low-exposure group* Urine Cr 5–15 Urine Cd 4.102–9.504 (6.314) Urine Pb 0.0929–10.79 (2.694) Urine Zn 272.4–1227 (446.9)	↓ d-Glucose, phosphate, glycine, l-threonine, azelaic acid, 3-α-Mannobiose, d-(+)-Cellobiose ↑ Butanoic acid, α-d-glucopyranosiduronic acid, ribonic acid	Xu et al. [176]
Neonicotinoid	2020	120	9- to 80–year-old residents	China	Serum	Median (IQR) ΣNeonicotinoid insecticide 26 (3.25,120) pg/ mL	↑ FA(22:6), FA(14:0), LysoPC(18:1), LysoPE(18:0), SM(d18:1/12:0), SM (d18:1/18:0), DG (18:1/18:1), S1P	Chen et al. [177]

PAH (μg/g creatinine)	2009–2012	3230	Residents (aged 5–88 yo)	Taiwan	Urine	$Mean_{low} - Mean_{high}$ *Pyrene* Children: 0.023–0.026 Elderly: 0.022–0.030 *Fluoranthene* Children: 0.026–0.028 Elderly: 0.024–0.027 *Dibenzo[a,h] anthracene* Children: 0.011–0.013 Elderly: 0.011–0.014 *Benzo[k] fluoranthene* Children: 0.011–0.017 Elderly: 0.018–0.019 *Benzo[a] anthracene* Children: 0.017–0.019 Elderly: 0.017–0.020	↓ 3-Hydroxyhippuric acid, 4-hydroxybenzoic acid, hippuric acid, azelaic acid, myristic acid, L-aspartic acid, L-histidine, N-acetylglutamic acid, succinic acid, tiglylglycin γ-Aminobutyric acid ↑ 2-Phenylpropanal, o-cymene, decane, dodecane, tridecane, glycerol 3-phosphate, stearic acid Tiglic acid, 3-hydroxyoctanoic acid, aminomalonic acid, fumaric acid, hydroxypyruvic acid, L-alpha-aminobutyric acid	Chen et al. [178]
PFAS (ng/mL)	2001–2012	40	Hispanic children (8–14 years)	USA	Plasma	Geometric mean (SD) PFHxS: 1.65(2) PFOS:12.22(1.91) PFOA:2.78(1.29)	↑ Sphingomyelin, palmitic acid, HPODE, linoleic, tyrosine, phenylalanine, arginine, and aspartate	Alderete et al. [179]
Smoking (μg/g creatinine)	2015	556	238 elderly nonsmokers, 114 elderly smokers, and 214 children	China	Urine	Mean ± SD ΣOH-PAHs *Elderly non-smoker:* Control group	↑ 3-Methylhistidine, dodecadienylcarnitine, nonanoylcarnitine, 3-hydroxydeca-noylcarnitine, and hydroxydodecenoylcarnitine	Wang et al. [93]

(continued)

Table 3 (continued)

MDCs	Sampled year	Sample size	Participants	Countries and regions	Sample types	Chemical exposure concentrations	Metabolic biomarkers	References
						14.4 ± 13.5 Exposure group 21.9 ± 16.1 *Elderly smokers* Control group 21.5 ± 11.1 Exposure group 31.7 ± 26.9	Octenedioylcarnitine, decenedioic acid, and 3-hydroxydecanoylcarnitine	
Trichloroethylene	2016	175	80 exposed workers and 95 controls	China	Plasma	Mean (SD) TCE air level: Control less than 0.03 Exposed 22.19 (35.9) ppm	↓7α-Hydroxycholest-4-en-3-one, chenodeoxycholic acid, creatine, cysteine, glutamine, homocysteine, indolelactic acid, methylthioadenosine, palmitocarnitine, phenylacetic acid, taurine, tryptophan, tyrosine, uric acid, α-linolenic acid	Walker et al. [92]
Vanadium (μmol/mol-creatinine)	2009–2012	3230	Residents who lived for more than 5 years in the study area	Taiwan	Urine	Mean ± SD Vanadium 2.07 ± 2.48 1-OHP 0.22 ± 0.40	↓ Valine, alanine, lysine, glutamine, tyrosine, histidine, phenylalanine, pyruvate, α- & β-glucose α-glucose, *N*-acetyl glycoprotein ↑ VLDL & LDL, lipids, isoleucine, lactate	Yuan et al. [180]

Adapted from Sun et al. [119]

for the process of filtering/removing peaks based on the reproducibility of aligned peaks in the technical replicates of a pooled QC/intra-study QC sample, and some quality assurance practices. The difficulties with QC in metabolomics are mostly attributable to untargeted analyses, as targeted analyses have clear guidelines for analytical validation and the parameters to be reported are available [120]. Viant et al. name the ordered steps in the metabolomics pipeline, all of which need to be standardized to produce reproducible and reliable research: experimental design, quality assurance, quality control, sampling and extraction, data acquisition, data processing, statistical analysis, metabolite annotation and identification if an untargeted study, and data management (Fig. 3) [122]. Once there is consensus about this pipeline, there may be a large increase in metabolomics-based usage, as the benefits of using an unstandardized pipeline have already been shown to be extremely impactful on the study of the environment.

Data reporting is a necessity for all scientific studies, yet there is no standardized protocol of what is important to report for environmental metabolomics. One reason for the lack of a common data reporting standard is the many different applications metabolomics can be used for. The metabolomics standards initiative in toxicology

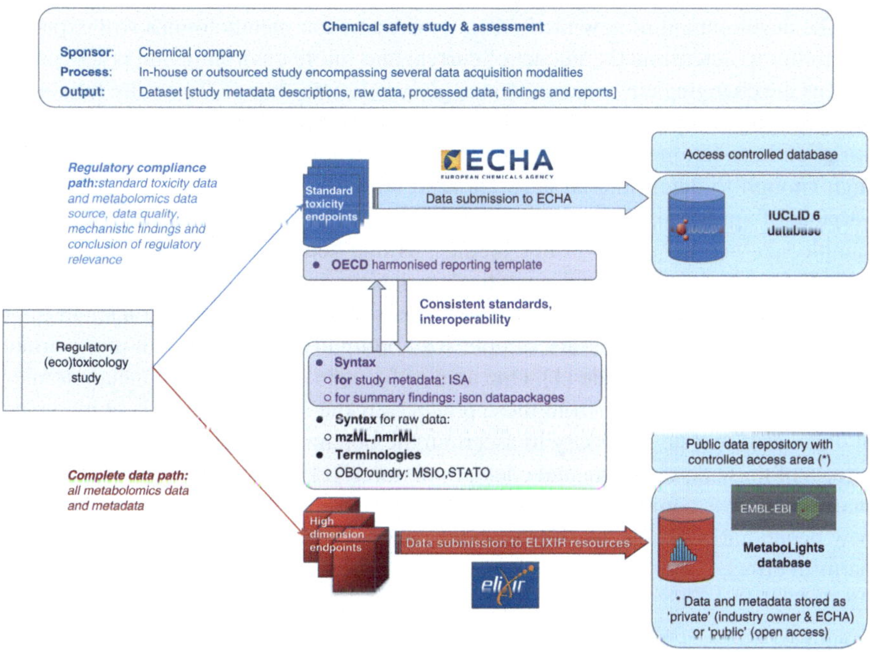

Fig. 3 Proposed management strategy for metabolomics data from a regulatory toxicology study. The strategy benefits from several existing access-controlled and public resources and would allow compliance standards to be checked by the chemical regulator (illustrated here for Europe) as well as open the potential for (metabol)omics data reuse subject to approval by the industry owner. One area of development is to align metadata standards between the regulatory compliance and complete data paths. (Figure and caption are taken from [122])

(MERIT) project has attempted to bridge this gap by presenting multiple scenarios using metabolomic analysis and identifying components important from the purpose, methods, data processing and analysis, and conclusions to report in scientific writing (Fig. 3) [122]. Numerous factors affect the metabolome and environmental chemical concentrations, including age, sex, ethnicity, health status, diet, and lifestyle in general [120]. Therefore, it is important to report these attributes in both human and other organism studies when applicable, as these variables may have a larger effect size on the metabolome than the environmental stressor. If publications do not report these variables, different populations may be used, and results may not be reproducible. The OECD Extended Advisory Group for Molecular Screening and Toxicogenomics (EAGMST) created a metabolomic reporting framework (MRF) which suggests reporting extraction method details, internal reference standards, quality control samples, blank sample processing, details of mass spectrometry and NMR acquisition data (instrument method, untargeted/targeted approach, chromatography details), data reduction approaches (feature detection/peak picking, retention time alignment, grouping), methods of metabolite quantification and identification, and the univariate/multivariate data analysis methods used [1]. These considerations are under review, as multiple trials are testing these guidelines to develop an optimal, standardized pipeline for metabolomics.

The development of new methods in environmental metabolomics will expedite the ability to determine the impact of current and future environmental practices, as well as the changing environment due to global climate change. To ensure reliability and reproducibility in a field characterized by the study of very large biological variation between measurements, the number of samples in an experiment must be high enough to determine an accurate representation of the health impacts of the variable of study [1]. However, it has been noted that it is difficult to estimate the sample size needed in exposomic studies, as this depends on multiple parameters including experiment design, analytical variability, and minimum measured concentration difference expected [1]. Therefore, suggestions to conduct a small pilot study to estimate the necessary sample size, optimally combined with multivariate simulation, have been made [1]. One area of focus in environmental metabolomics that would benefit greatly from these practices is the study of multiple compounds at once, as it will be necessary to ascertain the potential synergistic or antagonistic effects of these co-exposure and cocktail mixtures [119]. The determination of how many chemicals combine to produce a final impact will allow researchers to create new beneficial cocktails, as well as reduce current chemical mixtures which cause harmful effects.

6 Future Perspective of Environmental Metabolomic

Environmental metabolomics is an effective approach to studying ecotoxicity and the mode of action (MoA) of environmental contaminants. The last two decades have witnessed exciting advancements in metabolomics and have made it a routine approach in environmental science. With all the innovation in environmental metabolomics, it has a fundamental problem of not having a simple and universally accepted approach for understanding the mechanistic details with reverse engineering from the metabolomic data. This problem persists in model species and becomes even more complicated in non-model organisms. Despite the tremendous progress, there are some critical challenges in metabolomics such as designing standardized methods, compatible software, and well-recognized databases to validate and interpret the results. The progress in these aspects will primarily determine the future of environmental metabolomics. Keeping this in mind, here we are discussing some fundamental areas which need more effort for the rapid advancement of metabolomics in environmental science.

6.1 The Need for Standardizing Protocols

The progress of ecometabolomics was directly impacted by the advancement in analytical chemistry. New analytical techniques, such as gas chromatography (GC) and high-performance liquid chromatography (HPLC) in combination with high-resolution mass spectrometry (MS), help in the detection, quantification, separation, and characterization of novel metabolites. In general, the current metabolomics research is driven mainly by NMR and chromatography-coupled MS approaches, which need to be improved to better identify and quantify critical metabolites in a sample. Despite the tremendous advancement, one primary challenge is the lack of standardized methods. The standard methodologies for the collection and pretreatment of samples, data acquisition, processing, and analysis are still not well defined. To overcome these aspects, in addition to information on mass spectra and ionization sources, well-defined analytical conditions and parameters, such as column type used in chromatography, could greatly help in metabolite identification. Including the information about the fragments and collision energy in the database can also be helpful. Inefficient metabolite recognition is one of the major limitations of metabolomic studies since many metabolites detected by NMR and MS have an unknown chemical nature.

Metabolomics can also be applied to detect molecular changes in organisms that are exposed to real mixtures of contaminants at an environmentally relevant concentration. The combined use of NMR spectroscopy and MS can be more insightful in terms of providing a holistic picture of the altered metabolic pathways and MoA of contaminants. A single metabolomic profiling approach is not capable of detecting all the metabolites in a biological sample due to the huge complexity of the metabolome. Therefore, it demands a combined usage of advanced analytical approaches to improve the coverage and propose useful biomarkers for environmental monitoring. Consequently, a deep understanding of biomarkers will help in keeping track of the ecosystem in an effective, rapid, and routine manner. However, using metabolomics to determine early threat signals for environmental biomonitoring and assessing ecological risk is one of the primary challenges [123]. The environmental samples are very complex mixtures of organic material and toxic metals. Profiling of the metal ions that take part in various biological processes in combination with proteins and metabolites is known as metallomics [124]. Combined application of metabolomics and metallomics can provide a deeper understanding of the detoxification processes of living organisms after contamination.

Majority of the research done on environmental metabolomics has been focused on understanding aquatic environments. Accordingly, the research was aimed toward identifying biomarkers to assess pollution and have been based on whole organismal metabolomics using typical small fish models. Importantly, now biofluid is expected to be examined thoroughly since it represents the organism's health status at a given time and it is easy to collect and needs less sample preparation. For example, urine (a low-protein biofluid) can be directly used for analysis after dilution. Consequently, such approaches might facilitate the automation in metabolomics, which is a crucial challenge in near future. More importantly, this will help in examining many samples in relatively less time for environmental monitoring. In addition, application of environmental metabolomics on plants is providing useful information about the soil contamination and helping the development of novel phytosanitary compounds. Altogether, this is facilitating a comprehensive environmental assessment by involving test organisms with less ethical issues.

6.2 Developing Reliable Databases

Metabolomic studies are broadly grouped into two categories: targeted and untargeted metabolomics. Targeted metabolomics examines the qualitative data collected from a pre-defined set of compounds, whereas untargeted metabolomics investigates a broader range of metabolites to identify new compounds [125]. The datasets in untargeted metabolomics are large and multidimensional. Even though tremendous advancements have been done in developing computational programs to process mass spectrometry data, better tools are still required for organizing metabolites and their associated metadata to compare different experiments. The information about organic compounds such as metabolites can be found in two types of databases. The first type of databases includes PubChem [126], ChemSpider [127], and METLIN [128] simply store the chemical information of a compound without considering its source. These databases are useful in getting information of metabolites, but they do not provide the environmental context as they do not have metadata. Contrary to this, the second type of database that includes MetaboLights [129, 130] contains the experimental metadata. However, MetaboLights provides information about only known metabolites highlighting the need for a database giving information about the unknown compounds along with the associated experimental metadata. In comparison to other omics databases, a metabolomic database has different complexities as the chemical nature of genes and proteins is simpler than metabolites due to a fewer diversity of building blocks. For example, the genomic sequences are constituted by a combination of four or five nucleotides (A, G, C, T, or U). Therefore, a nucleic acid database like GenBank [131] has less chemical complexity, and the errors are primarily due to the interpretation of the data such as annotation of genes and assessment homology. Opposite to this, metabolites are extremely diverse as they do not have common building blocks like nucleic acids even though they share common elements such as C, H, N, O, S, and P. In addition, interpreting mass spectrometry-based metabolomic data becomes more complicated since a metabolite can be present in one or more adducts having different mass-to-charge value, e.g., $[M+Na]^+$ or $[M+H]^+$. Furthermore, there are errors in measuring mass-to-charge value depending upon the type of instrument used. Currently, available metabolomic reference databases have information about a huge number of metabolites (in hundreds of thousands), but only 5% of them have experimental data from pure standards [132]. Furthermore, in toxicological research, distinguishing the metabolic changes happening due to the toxicant or other damage, and variability among individuals, and for an individual under different situations is often technically tricky. Altogether, these limitations are highlighting the need for well-designed databases that are recognized by the scientific community for validating and interpreting the results.

6.3 *Advance Applications of Metabolomics in Assessing Occupational Health Exposure*

The health concerns associated with chemical exposure at the workplace are continuously increasing [133–135]. Metabolomics has emerged as a novel method for measuring diverse chemicals in biological samples that can provide information about new biomarkers, toxicological mechanisms, and biological effects associated with occupational exposure [136, 137]. Appropriate measurement of harmful effects of occupational exposure to chemicals is a high priority for health workers. This is done mainly in two ways: air monitoring and biological monitoring [138]. In the traditional air monitoring process, samples are collected from the breathing zone of professionals for examination, but it has various limitations considering that many unknown chemicals have no streamlined pipeline for sampling and analysis [139, 140]. Generally, with an established method, only a single chemical is measured even though exposures at workplaces are a complex mixture of chemicals. Variables such as diet, drinking, and smoking habits further complicate the analysis of these exposures [141, 142]. Consequently, these challenges can result in an incorrect assessment of occupational exposure studies. In addition to air monitoring, biological monitoring also has challenges that include a lack of proper Biological Exposure Indices (BEI) for many frequently used chemicals in the industry limiting assessment of their adverse health effects [143]. Together, these challenges highlight that assessment of occupational exposure is a relatively difficult task and therefore demands an urgent need to conduct experiments to get accurate and relevant information regarding unexpected exposures. Metabolomics is used to characterize metabolomic modulations and biochemical responses in professionals with diverse chemical exposures [144]. To address these challenges, metabolomics can serve as an important tool by using it for biomarker discovery, early diagnosis of occupational disease, investigating low-level exposure, interaction of mixed exposure, and dose-response relationship.

Metabolomic studies on occupational exposures highlight that metabolomic profiling of professionals has great potential in biomarker discovery for exposure and disease, and provides deeper insight into the underlying mechanisms of these diseases. Using metabolomics in their studies, the occupational health researcher can provide critical insights into exposure with very limited or no toxicological information. Metabolomics has also empowered research directed toward examining the effects of chemical or nonchemical exposures and discovering key biomarkers to determine potential toxicity. Taken together, metabolomics serves as a key approach for facilitating occupational research and should be included in evaluating health of professionals suspected to encounter harmful occupational exposures.

6.4 *Exposomes and Wearables*

As consumer health monitoring has become increasingly common, driven by wearables such as Fitbit, the emergence of an exposome tracking device has been created as a silicon wristband. This technology uses passive sampling to capture the personal exposure of individuals to chemicals encountered daily, including PAHs, oxygenated PAHs, flame retardants, and pesticides [145]. Studies have combined exposome readings of silicon wristbands with metabolomics to develop additional PAH exposure assessment tools, such as the wristband, to improve public health research pertaining to PAH exposure [145]. Identification of correlation between wristband PAH level and OH-PAH urine metabolites demonstrated the ability for the accurate discovery of some environmental toxins. This is important because PAHs metabolize quickly, and therefore exposure to PAHs even up to a few hours before urine collection may not be quantified using metabolomics. Research focusing on the silicon wristband's ability to capture semivolatile organic compounds (SVOCs) in nail salons discovered exposure to certain phthalates, phthalate alternatives, and OPEs, with metabolites of DEHTP showing the largest increase across a workday [146]. Pre- and post-shift urine were sampled to measure changes in metabolites during this period, which led to the conclusion that phthalate alternative di(2-ethylhexyl) terephthalate (DEHTP): mono(2-ethyl-5-carboxypentyl) terephthalate (MECPTP) more than tripled from 11.7 to 36.6 μg/g creatinine, the greatest shift in the concentration of all metabolites. Detection of SVOCs was found using both wristbands and urine samples, making them a viable tool for examining potential occupational exposures to SVOC. The company MyExposome is the distributor of silicon wristbands designed to record the exposure of the individual to a wide variety of important chemical compounds, thereby informing that person, and society-at-large, about the presence or absence of critical chemical exposures in every tested individual's environment [147]. Currently, wristbands can measure endocrine-disrupting compounds, consumer and personal care products, PAHs,

PCBs, pesticides, and flame retardants. As additional studies on environmental exposure are reported and the many effects of chemicals are realized first by scientists and then the public, it is possible that an increase in consumer demand for the MyExposome silicon wristband will occur.

7 Conclusion

With the development of next-generation omics and analysis instruments (NMR and MS), applications of metabolomics are constantly expanding. Advanced developments in computational tools for feature determination, it is now easy to detect and characterize many unknown molecules and chemicals. This is opening new avenues to better understand living organisms and the environment. Ecometabolomics is arising as a unique branch of metabolomics and helping ecologists to find out the unexplored role of various environmental factors, chemicals, and pollutants in the ecosystem's functioning.

This chapter has described expanding landscapes of ecometabolomics and described its applications, challenges, technicalities, and future perspectives. Due to the variable nature of the samples and models, it is difficult to develop a unified protocol with greater reproducibility. Besides this, a robust broad-spectrum metabolite library would also be useful to reduce the complexities associated with variable results. It will enable global collaborations and data exchange among researchers from different disciplines and expertise. We have described several environmental stressors (abiotic and biotic) and their impacts and relationship with pollution and environmental health. The inclusion of metabolomic analyses has expanded our comprehension of the impact of land pollutants on plant biology. Controlled experiments provide a precise measurement of fluctuating metabolites, offering a more discerning approach to identifying plants that are influenced by pollutants, and facilitating the determination of optimal plant species for future farming in polluted regions. We have learned that extensive research is being conducted using metabolomics to understand the biological mechanisms perturbed by environmental stressors on both animals and plants.

As humans continue to pollute extensively and use potentially harmful fertilizers and pesticides, it is imperative to continue researching the overall impacts on the metabolomes of organisms, as this may allow for the creation of novel methods which improve animal and plant health. These impacts would reach humans as ecosystems would return to balance, food sources would become more stable, and possibly even increase production. Identification of metabolite biomarkers to specific chemicals may act as a diagnostic tool to identify organisms at risk of developing health problems due to exposure. While there is great potential for metabolomics to enhance the study of environmental impacts, there are still challenges that need to be overcome. Looking ahead, we also recommend some future directions for ecometabolomics such as method standardization, different efforts, and projects to develop reliable databases, applications in occupational health monitoring, and futuristic wearable detecting exposome. In conclusion, ecometabolomics has the potential to understand the chemical ecology of our environment. However, challenges still exist, and more efforts are required toward collaborative research and data exchange. Ecometabolomics offers a multidisciplinary approach to guide our efforts to mitigate the detrimental effects of pollution and climate change.

References

1. Bedia, C., *Metabolomics in environmental toxicology: Applications and challenges.* Trends in Environmental Analytical Chemistry, 2022. **34**.
2. vanEngelsdorp, D., et al., *Colony Collapse Disorder (CCD) and bee age impact honey bee pathophysiology.* PLoS One, 2017. **12**(7): p. e0179535.
3. Wang, L., et al., *Metabolomics-based biomarker discovery for bee health monitoring: A proof of concept study concerning nutritional stress in Bombus terrestris.* Sci Rep, 2019. **9**(1): p. 11423.
4. Ricigliano, V.A., et al., *Metabolomics-Guided Comparison of Pollen and Microalgae-Based Artificial Diets in Honey Bees.* J Agric Food Chem, 2022. **70**(31): p. 9790–9801.
5. Ardalani, H., et al., *Metabolomics unveils the influence of dietary phytochemicals on residual pesticide concentrations in honey bees.* Environ Int, 2021. **152**: p. 106503.
6. Dennis, K.K., et al., *Biomonitoring in the Era of the Exposome.* Environ Health Perspect, 2017. **125**(4): p. 502–510.
7. Mushtaq, M.Y., et al., *Extraction for Metabolomics: Access to The Metabolome.* Phytochemical Analysis, 2014. **25**(4): p. 291–306.
8. Sostare, J., et al., *Comparison of modified Matyash method to conventional solvent systems for polar metabolite and lipid extractions.* Analytica chimica acta, 2018. **1037**: p. 301–315.
9. Matyash, V., et al., *Lipid extraction by methyl-tert-butyl ether for high-throughput lipidomics*s.* Journal of Lipid Research, 2008. **49**(5): p. 1137–1146.
10. Kim, H.M. and J.S. Kang, *Metabolomic Studies for the Evaluation of Toxicity Induced by Environmental Toxicants on Model Organisms.* Metabolites, 2021. **11**(8): p. 485.
11. Poynton, H.C., et al., *Metabolomics of Microliter Hemolymph Samples Enables an Improved Understanding of the Combined Metabolic and Transcriptional Responses of Daphnia magna to Cadmium.* Environmental Science & Technology, 2011. **45**(8): p. 3710–3717.
12. Viant, M., *Recent developments in environmental metabolomics.* Molecular Biosystems, 2008(4): p. 980–86.
13. Augustijn, D., H.J.M. de Groot, and A. Alia, *HR-MAS NMR Applications in Plant Metabolomics.* Molecules, 2021. **26**(4).
14. Pereira, S.I., et al., *Changes in the metabolome of lettuce leaves due to exposure to mancozeb pesticide.* Food Chem, 2014. **154**: p. 291–8.
15. Blondel, C., et al., *Effect of organochlorine pesticides exposure on the maize root metabolome assessed using high-resolution magic-angle spinning (1)H NMR spectroscopy.* Environ Pollut, 2016. **214**: p. 539–548.
16. Coutinho, I.D., et al., *Integrating High-Resolution and Solid-State Magic Angle Spinning NMR Spectroscopy and a Transcriptomic Analysis of Soybean Tissues in Response to Water Deficiency.* Phytochem Anal, 2017. **28**(6): p. 529–540.
17. Mazzei, P., et al., *Metabolomics by Proton High-Resolution Magic-Angle-Spinning Nuclear Magnetic Resonance of Tomato Plants Treated with Two Secondary Metabolites Isolated from Trichoderma.* J Agric Food Chem, 2016. **64**(18): p. 3538–45.
18. Sarou-Kanian, V., et al., *Metabolite localization in living drosophila using High Resolution Magic Angle Spinning NMR.* Scientific Reports, 2015. **5**(1): p. 9872.

19. Lucas-Torres, C., et al., *HR-μMAS NMR-Based Metabolomics: Localized Metabolic Profiling of a Garlic Clove with μg Tissues.* Anal Chem, 2018. **90**(22): p. 13736–13743.
20. Crescenzi, M.A., et al., *A Comparative UHPLC-Q-Trap-MS/MS-Based Metabolomics Analysis to Distinguish Foeniculum vulgare Cultivars’ Antioxidant Extracts.* Molecules, 2023. **28**(2): p. 900.
21. Laine, J.E. and O. Robinson, *Framing Fetal and Early Life Exposome Within Epidemiology*, in *Unraveling the Exposome: A Practical View*, S. Dagnino and A. Macherone, Editors. 2019, Springer International Publishing: Cham. p. 87–123.
22. Smith, M.T., C.M. McHale, and R. de la Rosa, *Using Exposomics to Assess Cumulative Risks from Multiple Environmental Stressors*, in *Unraveling the Exposome: A Practical View*, S. Dagnino and A. Macherone, Editors. 2019, Springer International Publishing: Cham. p. 3–22.
23. Vermeulen, R., et al., *The exposome and health: Where chemistry meets biology.* Science, 2020. **367**(6476): p. 392–396.
24. Sille, F.C.M., et al., *The exposome - a new approach for risk assessment.* ALTEX, 2020. **37**(1): p. 3–23.
25. Andrianou, X.D. and K.C. Makris, *The framework of urban exposome: Application of the exposome concept in urban health studies.* Sci Total Environ, 2018. **636**: p. 963–967.
26. Andrianou, X.D., et al., *Exposome-based public health interventions for infectious diseases in urban settings.* Environ Int, 2021. **146**: p. 106246.
27. Lankadurai, B.P., E.G. Nagato, and M.J. Simpson, *Environmental metabolomics: an emerging approach to study organism responses to environmental stressors.* Environmental Reviews, 2013. **21**(3): p. 180–205.
28. Matich, E.K., et al., *Applications of metabolomics in assessing ecological effects of emerging contaminants and pollutants on plants.* J Hazard Mater, 2019. **373**: p. 527–535.
29. Deng, P., et al., *Application of metabolomics to characterize environmental pollutant toxicity and disease risks.* Rev Environ Health, 2019. **34**(3): p. 251–259.
30. Wang, X., et al., *Comprehensive pulmonary metabolome responses to intratracheal instillation of airborne fine particulate matter in rats.* Sci Total Environ, 2017. **592**: p. 41–50.
31. Zhang, S.Y., et al., *Metabolomics analysis reveals that benzo[a]pyrene, a component of PM(2.5), promotes pulmonary injury by modifying lipid metabolism in a phospholipase A2-dependent manner in vivo and in vitro.* Redox Biol, 2017. **13**: p. 459–469.
32. Jeong, A., et al., *Perturbation of metabolic pathways mediates the association of air pollutants with asthma and cardiovascular diseases.* Environ Int, 2018. **119**: p. 334–345.
33. van Veldhoven, K., et al., *Impact of short-term traffic-related air pollution on the metabolome - Results from two metabolome-wide experimental studies.* Environ Int, 2019. **123**: p. 124–131.
34. Gaskins, A.J., et al., *Periconception air pollution, metabolomic biomarkers, and fertility among women undergoing assisted reproduction.* Environ Int, 2021. **155**: p. 106666.
35. Ritz, B., et al., *Child serum metabolome and traffic-related air pollution exposure in pregnancy.* Environ Res, 2022. **203**: p. 111907.
36. Nassan, F.L., et al., *Metabolomic signatures of the short-term exposure to air pollution and temperature.* Environ Res, 2021. **201**: p. 111553.
37. Nassan, F.L., et al., *Metabolomic signatures of the long-term exposure to air pollution and temperature.* Environ Health, 2021. **20**(1): p. 3.
38. Gruzieva, O., et al., *Air pollution, metabolites and respiratory health across the life-course.* Eur Respir Rev, 2022. **31**(165).
39. Ekman, D.R., et al., *Evaluation of targeted and untargeted effects-based monitoring tools to assess impacts of contaminants of emerging concern on fish in the South Platte River, CO.* Environ Pollut, 2018. **239**: p. 706–713.
40. Koubova, A., et al., *Metabolome adaptation and oxidative stress response of common carp (Cyprinus carpio) to altered water pollution levels.* Environ Pollut, 2022. **303**: p. 119117.

41. Brandao, F., et al., *Unravelling the mechanisms of mercury hepatotoxicity in wild fish (Liza aurata) through a triad approach: bioaccumulation, metabolomic profiles and oxidative stress.* Metallomics, 2015. **7**(9): p. 1352–63.
42. Spath, J., et al., *Wastewater effluent affects behaviour and metabolomic endpoints in damselfly larvae.* Sci Rep, 2022. **12**(1): p. 6830.
43. Salehi, H., et al., *Morphological, proteomic and metabolomic insight into the effect of cerium dioxide nanoparticles to Phaseolus vulgaris L. under soil or foliar application.* Sci Total Environ, 2018. **616–617**: p. 1540–1551.
44. Vecerova, K., et al., *Changes of primary and secondary metabolites in barley plants exposed to CdO nanoparticles.* Environ Pollut, 2016. **218**: p. 207–218.
45. Zahra, Z., et al., *Growth and Metabolic Responses of Rice (Oryza sativa L.) Cultivated in Phosphorus-Deficient Soil Amended with TiO(2) Nanoparticles.* J Agric Food Chem, 2017. **65**(28): p. 5598–5606.
46. Chen, J., et al., *Metabolomic analysis of two rice (Oryza sativa) varieties exposed to 2, 2', 4, 4'-tetrabromodiphenyl ether.* Environ Pollut, 2018. **237**: p. 308–317.
47. Rodrigues, A.M., et al., *Mass spectrometry-based forest tree metabolomics.* Mass Spectrom Rev, 2021. **40**(2): p. 126–157.
48. Correia, B., et al., *Combined Drought and Heat Activates Protective Responses in Eucalyptus globulus That Are Not Activated When Subjected to Drought or Heat Stress Alone.* Front Plant Sci, 2018. **9**: p. 819.
49. Ottow, E.A., et al., *Populus euphratica displays apoplastic sodium accumulation, osmotic adjustment by decreases in calcium and soluble carbohydrates, and develops leaf succulence under salt stress.* Plant Physiol, 2005. **139**(4): p. 1762–72.
50. Romero, H., et al., *Metabolomics-Based Evaluation of Crop Quality Changes as a Consequence of Climate Change.* Metabolites, 2021. **11**(7).
51. Zayas-Santiago, C.C., et al., *Metabolic Profiling Reveals Biochemical Pathways Responsible for Eelgrass Response to Elevated CO(2) and Temperature.* Sci Rep, 2020. **10**(1): p. 4693.
52. Tugizimana, F., et al., *Multi-platform metabolomic analyses of ergosterol-induced dynamic changes in Nicotiana tabacum cells.* PLoS One, 2014. **9**(1): p. e87846.
53. Heuberger, A.L., et al., *Evaluating plant immunity using mass spectrometry-based metabolomics workflows.* Front Plant Sci, 2014. **5**: p. 291.
54. Castro-Moretti, F.R., et al., *Metabolomics as an Emerging Tool for the Study of Plant-Pathogen Interactions.* Metabolites, 2020. **10**(2).
55. Yu, Z., et al., *Development of Untargeted Metabolomics Methods for the Rapid Detection of Pathogenic Naegleria fowleri.* Environ Sci Technol, 2017. **51**(8): p. 4210–4219.
56. Valerio, A., et al., *Detecting the effects of predator-induced stress on the global metabolism of an ungulate prey using fecal metabolomic fingerprinting.* Sci Rep, 2021. **11**(1): p. 6129.
57. Zhang, C., et al., *Resurrecting the metabolome: Rapid evolution magnifies the metabolomic plasticity to predation in a natural Daphnia population.* Mol Ecol, 2021. **30**(10): p. 2285–2297.
58. Weissburg, M., R.X. Poulin, and J. Kubanek, *You Are What you Eat: a Metabolomics Approach to Understanding Prey Responses to Diet-Dependent Chemical Cues Released by Predators.* J Chem Ecol, 2016. **42**(10): p. 1037–1046.
59. Jiao, Y., C. Sha, and Q. Shu, *Integrated Physiological and Metabolomic Analyses of the Effect of Potassium Fertilizer on Citrus Fruit Splitting.* Plants (Basel), 2022. **11**(4).
60. Ciampa, A., et al., *Combined magnetic resonance imaging and high resolution spectroscopy approaches to study the fertilization effects on metabolome, morphology and yeast community of wine grape berries, cultivar Nero di Troia.* Food Chem, 2019. **274**: p. 831–839.
61. Van Meter, R.J., et al., *Induced Hepatic Glutathione and Metabolomic Alterations Following Mixed Pesticide and Fertilizer Exposures in Juvenile Leopard Frogs (Lithobates sphenocephala).* Environ Toxicol Chem, 2022. **41**(1): p. 122–133.
62. Kalia, V., et al., *Cross-species metabolomic analysis of tau- and DDT-related toxicity.* PNAS Nexus, 2022. **1**(2): p. pgac050.

63. Meng, Z., et al., *Different effects of exposure to penconazole and its enantiomers on hepatic glycolipid metabolism of male mice.* Environ Pollut, 2020. **257**: p. 113555.
64. Olsvik, P.A., et al., *Whole-animal accumulation, oxidative stress, transcriptomic and metabolomic responses in the pink shrimp (Pandalus montagui) exposed to teflubenzuron.* J Appl Toxicol, 2019. **39**(3): p. 485–497.
65. Kim, H.M. and J.S. Kang, *Metabolomic Studies for the Evaluation of Toxicity Induced by Environmental Toxicants on Model Organisms.* Metabolites, 2021. **11**(8).
66. Sotto, R.B., et al., *Sub-lethal pharmaceutical hazard tracking in adult zebrafish using untargeted LC-MS environmental metabolomics.* J Hazard Mater, 2017. **339**: p. 63–72.
67. Fu, J., et al., *The toxic effect of triclosan and methyl-triclosan on biological pathways revealed by metabolomics and gene expression in zebrafish embryos.* Ecotoxicol Environ Saf, 2020. **189**: p. 110039.
68. Sheikholeslami, M.N., et al., *Untargeted metabolomics changes on Gammarus pulex induced by propranolol, triclosan, and nimesulide pharmaceutical drugs.* Chemosphere, 2020. **260**: p. 127479.
69. Bonnefille, B., et al., *Metabolomics assessment of the effects of diclofenac exposure on Mytilus galloprovincialis: Potential effects on osmoregulation and reproduction.* Sci Total Environ, 2018. **613–614**: p. 611–618.
70. Serra-Compte, A., et al., *Comprehensive study of sulfamethoxazole effects in marine mussels: Bioconcentration, enzymatic activities and metabolomics.* Environ Res, 2019. **173**: p. 12–22.
71. Kim, H.M., et al., *Metabolomics and phenotype assessment reveal cellular toxicity of triclosan in Caenorhabditis elegans.* Chemosphere, 2019. **236**: p. 124306.
72. Hua, Q., et al., *Microbiome analysis and predicted relative metabolomic turnover suggest bacterial heme and selenium metabolism are altered in the gastrointestinal system of zebrafish (Danio rerio) exposed to the organochlorine dieldrin.* Environ Pollut, 2021. **268**(Pt B): p. 115715.
73. Jia, M., et al., *Toxicity and metabolomics study of isocarbophos in adult zebrafish (Danio rerio).* Ecotoxicol Environ Saf, 2018. **163**: p. 1–6.
74. Liu, N. and L. Zhu, *Metabolomic and Transcriptomic Investigation of Metabolic Perturbations in Oryza sativa L. Triggered by Three Pesticides.* Environ Sci Technol, 2020. **54**(10): p. 6115–6124.
75. Fang, S., et al., *Lethal Toxicity and Sublethal Metabolic Interference Effects of Sulfoxaflor on the Earthworm (Eisenia fetida).* J Agric Food Chem, 2018. **66**(45): p. 11902–11908.
76. Zhang, H., et al., *Differences in kinetic metabolomics in Eisenia fetida under single and dual exposure of imidacloprid and dinotefuran at environmentally relevant concentrations.* J Hazard Mater, 2021. **417**: p. 126001.
77. Yin, J., et al., *Non-targeted metabolomic profiling of atrazine in Caenorhabditis elegans using UHPLC-QE Orbitrap/MS.* Ecotoxicol Environ Saf, 2020. **206**: p. 111170.
78. Pang, M., et al., *Transcriptome sequencing and metabolite analysis reveal the toxic effects of nanoplastics on tilapia after exposure to polystyrene.* Environ Pollut, 2021. **277**: p. 116860.
79. Zhao, Y., et al., *Metabolomic profiling reveals the intestinal toxicity of different length of microplastic fibers on zebrafish (Danio rerio).* J Hazard Mater, 2021. **403**: p. 123663.
80. Xiang, Q.Q., et al., *Integration of transcriptomics and metabolomics reveals damage and recovery mechanisms of fish gills in response to nanosilver exposure.* Aquat Toxicol, 2021. **237**: p. 105895.
81. Liu, W., et al., *Metabolomics for early detection of stress in freshwater alga Poterioochromonas malhamensis exposed to silver nanoparticles.* Sci Rep, 2020. **10**(1): p. 20563.
82. Zhu, Y., et al., *Integration of transcriptomics and metabolomics reveals the responses of earthworms to the long-term exposure of TiO(2) nanoparticles in soil.* Sci Total Environ, 2020. **719**: p. 137492.
83. Maria, V.L., et al., *Multiomics assessment in Enchytraeus crypticus exposed to Ag nanomaterials (Ag NM300K) and ions (AgNO(3)) - Metabolomics, proteomics (& transcriptomics).* Environ Pollut, 2021. **286**: p. 117571.

84. Ji, F., et al., *Study of metabolic disorders associated with BDE-47 exposure in Drosophila model by MS-based metabolomics.* Ecotoxicol Environ Saf, 2019. **184**: p. 109606.
85. Liang, R., et al., *Toxicological effects on earthworms (Eisenia fetida) exposed to sub-lethal concentrations of BDE-47 and BDE-209 from a metabolic point.* Environ Pollut, 2018. **240**: p. 653–660.
86. Ortiz-Villanueva, E., et al., *Metabolic disruption of zebrafish (Danio rerio) embryos by bisphenol A. An integrated metabolomic and transcriptomic approach.* Environ Pollut, 2017. **231**(Pt 1): p. 22–36.
87. Mao, L., et al., *Effects of Bisphenol A and Bisphenol S Exposure at Low Doses on the Metabolome of Adolescent Male Sprague-Dawley Rats.* Chem Res Toxicol, 2021. **34**(6): p. 1578–1587.
88. Gu, Y.Y., et al., *A comprehensive study of the effects of phthalates on marine mussels: Bioconcentration, enzymatic activities and metabolomics.* Mar Pollut Bull, 2021. **168**: p. 112393.
89. Ellis, J.K., et al., *Metabolic profiling detects early effects of environmental and lifestyle exposure to cadmium in a human population.* BMC Med, 2012. **10**: p. 61.
90. Huang, Y., et al., *Aluminum Exposure and Gestational Diabetes Mellitus: Associations and Potential Mediation by n-6 Polyunsaturated Fatty Acids.* Environ Sci Technol, 2020. **54**(8): p. 5031–5040.
91. Wu, F., et al., *Arsenic Exposure from Drinking Water and Urinary Metabolomics: Associations and Long-Term Reproducibility in Bangladesh Adults.* Environ Health Perspect, 2018. **126**(1): p. 017005.
92. Walker, D.I., et al., *High-resolution metabolomics of occupational exposure to trichloroethylene.* Int J Epidemiol, 2016. **45**(5): p. 1517–1527.
93. Wang, Z., et al., *Human metabolic responses to chronic environmental polycyclic aromatic hydrocarbon exposure by a metabolomic approach.* J Proteome Res, 2015. **14**(6): p. 2583–93.
94. Hu, X., et al., *Metabolome Wide Association Study of serum DDT and DDE in Pregnancy and Early Postpartum.* Reprod Toxicol, 2020. **92**: p. 129–137.
95. Salihovic, S., et al., *Identification of metabolic profiles associated with human exposure to perfluoroalkyl substances.* J Expo Sci Environ Epidemiol, 2019. **29**(2): p. 196–205.
96. Schillemans, T., et al., *Plasma metabolites associated with exposure to perfluoroalkyl substances and risk of type 2 diabetes - A nested case-control study.* Environ Int, 2021. **146**: p. 106180.
97. Eguchi, A., et al., *Exploration of potential biomarkers and related biological pathways for PCB exposure in maternal and cord serum: A pilot birth cohort study in Chiba, Japan.* Environ Int, 2017. **102**: p. 157–164.
98. Carrizo, D., et al., *Untargeted metabolomic analysis of human serum samples associated with exposure levels of Persistent organic pollutants indicate important perturbations in Sphingolipids and Glycerophospholipids levels.* Chemosphere, 2017. **168**: p. 731–738.
99. Gao, P., et al., *Precision environmental health monitoring by longitudinal exposome and multi-omics profiling.* Genome Res, 2022. **32**(6): p. 1199–1214.
100. Rivas-Ubach, A., et al., *Atmo-ecometabolomics: a novel atmospheric particle chemical characterization methodology for ecological research.* Environmental Monitoring and Assessment, 2019. **191**(2).
101. Skelton, D.M., et al., *Metabolomics for in Situ Environmental Monitoring of Surface Waters Impacted by Contaminants from Both Point and Nonpoint Sources.* Environmental Science & Technology, 2014. **48**(4): p. 2395–2403.
102. Gargallo-Garriga, A., et al., *Root exudate metabolomes change under drought and show limited capacity for recovery.* Sci Rep, 2018. **8**(1): p. 12696.
103. Valentinuzzi, F., et al., *Phosphorus and iron deficiencies induce a metabolic reprogramming and affect the exudation traits of the woody plant Fragaria×ananassa.* J Exp Bot, 2015. **66**(20): p. 6483–95.

104. van Dam, N.M. and H.J. Bouwmeester, *Metabolomics in the Rhizosphere: Tapping into Belowground Chemical Communication.* Trends Plant Sci, 2016. **21**(3): p. 256–265.
105. Zubair, H.M., et al., *Allelopathic interference of alfalfa (Medicago sativa L.) genotypes to annual ryegrass (Lolium rigidum).* J Plant Res, 2017. **130**(4): p. 647–658.
106. Parmar, T.K., D. Rawtani, and Y.K. Agrawal, *Bioindicators: the natural indicator of environmental pollution.* Frontiers in Life Science, 2016. **9**(2): p. 110–118.
107. Santos, L.H.M.L.M., et al., *Linking chemical exposure and fish metabolome: Discovering new biomarkers of environmental exposure of Argyrosomus regius to the antidepressant venlafaxine.* Environmental Toxicology and Pharmacology, 2023. **98**: p. 104063.
108. Porto, V.A., et al., *NMR-based metabolomics applied to ecotoxicology with zebrafish (Danio rerio) as a prominent model for metabolic profiling and biomarker discovery: Overviewing the most recent approaches.* Sci Total Environ, 2023: p. 161737.
109. Taylor, N.S., A. Gavin, and M.R. Viant, *Metabolomics Discovers Early-Response Metabolic Biomarkers that Can Predict Chronic Reproductive Fitness in Individual Daphnia magna.* Metabolites, 2018. **8**(3): p. 42.
110. Pomfret, S.M., et al., *Metabolomics for biomonitoring: an evaluation of the metabolome as an indicator of aquatic ecosystem health.* Environmental Reviews, 2020. **28**(1): p. 89–98.
111. Hines, A., et al., *Discovery of metabolic signatures for predicting whole organism toxicology.* Toxicol Sci, 2010. **115**(2): p. 369–78.
112. Gago-Tinoco, A., et al., *Metabolic signatures associated with environmental pollution by metals in Doñana National Park using P. clarkii as bioindicator.* Environ Sci Pollut Res Int, 2014. **21**(23): p. 13315–23.
113. Wagner, N.D., A.J. Simpson, and M.J. Simpson, *Metabolomic responses to sublethal contaminant exposure in neonate and adult Daphnia magna.* Environ Toxicol Chem, 2017. **36**(4): p. 938–946.
114. Wu, H. and W.X. Wang, *NMR-based metabolomic studies on the toxicological effects of cadmium and copper on green mussels Perna viridis.* Aquat Toxicol, 2010. **100**(4): p. 339–45.
115. Bundy, J.G., et al., *Metabolic Profile Biomarkers of Metal Contamination in a Sentinel Terrestrial Species Are Applicable Across Multiple Sites.* Environmental Science & Technology, 2007. **41**(12): p. 4458–4464.
116. Wu, H., et al., *Differential metabolic responses in three life stages of mussels Mytilus galloprovincialis exposed to cadmium.* Ecotoxicology, 2017. **26**(1): p. 74–80.
117. Khan, S.A., et al., *1H NMR-based serum metabolic profiling of Carassius auratus gibelio under the toxicity of Pb2+ and Cd2+.* International Journal of Environmental Science and Technology, 2016. **13**(11): p. 2597–2608.
118. Izral, N.M., et al., *Developing metabolomics-based bioassessment: crayfish metabolome sensitivity to food and dissolved oxygen stress.* Environmental Science and Pollution Research, 2018. **25**(36): p. 36184–36193.
119. Sun, J., et al., *A review of environmental metabolism disrupting chemicals and effect biomarkers associating disease risks: Where exposomics meets metabolomics.* Environ Int, 2022. **158**: p. 106941
120. Hyotylainen, T., *Analytical challenges in human exposome analysis with focus on environmental analysis combined with metabolomics.* J Sep Sci, 2021. **44**(8): p. 1769–1787.
121. Evans, A.M., et al., *Dissemination and analysis of the quality assurance (QA) and quality control (QC) practices of LC-MS based untargeted metabolomics practitioners.* Metabolomics, 2020. **16**(10): p. 113.
122. Viant, M.R., et al., *Use cases, best practice and reporting standards for metabolomics in regulatory toxicology.* Nat Commun, 2019. **10**(1): p. 3041.
123. Bahamonde, P.A., et al., *Defining the role of omics in assessing ecosystem health: Perspectives from the Canadian environmental monitoring program.* Environ Toxicol Chem, 2016. **35**(1): p. 20–35.
124. Haraguchi, H., *Metallomics: the history over the last decade and a future outlook.* Metallomics, 2017. **9**(8): p. 1001–1013.

125. Patti, G.J., O. Yanes, and G. Siuzdak, *Metabolomics: the apogee of the omics trilogy.* Nature Reviews Molecular Cell Biology, 2012. **13**(4): p. 263–269.
126. Bolton, E.E., et al., *Chapter 12 - PubChem: Integrated Platform of Small Molecules and Biological Activities*, in *Annual Reports in Computational Chemistry*, R.A. Wheeler and D.C. Spellmeyer, Editors. 2008, Elsevier. p. 217–241.
127. Pence, H.E. and A. Williams, *ChemSpider: An Online Chemical Information Resource.* Journal of Chemical Education, 2010. **87**(11): p. 1123–1124.
128. Zhu, Z.-J., et al., *Liquid chromatography quadrupole time-of-flight mass spectrometry characterization of metabolites guided by the METLIN database.* Nature Protocols, 2013. **8**(3): p. 451–460.
129. Steinbeck, C., et al., *MetaboLights: towards a new COSMOS of metabolomics data management.* Metabolomics, 2012. **8**(5): p. 757–760.
130. Haug, K., et al., *MetaboLights—an open-access general-purpose repository for metabolomics studies and associated meta-data.* Nucleic Acids Research, 2012. **41**(D1): p. D781–D786.
131. Benson, D.A., et al., *GenBank.* Nucleic Acids Res, 2013. **41**(Database issue): p. D36–42.
132. Frainay, C., et al., *Mind the Gap: Mapping Mass Spectral Databases in Genome-Scale Metabolic Networks Reveals Poorly Covered Areas.* Metabolites, 2018. **8**(3).
133. Paraskevaidou, K., et al., *Asthma and rhinitis in Greek furniture workers.* J Asthma, 2021. **58**(2): p. 170–179.
134. Pałaszewska-Tkacz, A., et al., *Cytostatics as hazardous chemicals in healthcare workers' environment.* Int J Occup Med Environ Health, 2019. **32**(2): p. 141–159.
135. Kim, K.H., et al., *A review of airborne polycyclic aromatic hydrocarbons (PAHs) and their human health effects.* Environ Int, 2013. **60**: p. 71–80.
136. Roux, A., et al., *Applications of liquid chromatography coupled to mass spectrometry-based metabolomics in clinical chemistry and toxicology: A review.* Clin Biochem, 2011. **44**(1): p. 119–35.
137. Ladva, C.N., et al., *Metabolomic profiles of plasma, exhaled breath condensate, and saliva are correlated with potential for air toxics detection.* J Breath Res, 2017. **12**(1): p. 016008.
138. Fostinelli, J., et al., *Environmental and biological monitoring of occupational exposure to polynuclear aromatic hydrocarbons during highway pavement construction in Italy.* Toxicol Lett, 2018. **298**: p. 134–140.
139. Ramachandran, G., et al., *A strategy for assessing workplace exposures to nanomaterials.* J Occup Environ Hyg, 2011. **8**(11): p. 673–85.
140. Patton, A.P., et al., *Transferability and generalizability of regression models of ultrafine particles in urban neighborhoods in the Boston area.* Environ Sci Technol, 2015. **49**(10): p. 6051–60.
141. Rauh, V.A., et al., *Developmental effects of exposure to environmental tobacco smoke and material hardship among inner-city children.* Neurotoxicol Teratol, 2004. **26**(3): p. 373–85.
142. Fortes, C., et al., *The relation between dietary habits and urinary levels of 3-phenoxybenzoic acid, a pyrethroid metabolite.* Food Chem Toxicol, 2013. **52**: p. 91–6.
143. Paustenbach, D.J., D.M. Cowan, and J. Sahmel, *The History and Biological Basis of Occupational Exposure Limits for Chemical Agents*, in *Patty's Industrial Hygiene*. p. 865–955.
144. Lin, C.Y., M.R. Viant, and R.S. Tjeerdema, *Metabolomics: Methodologies and applications in the environmental sciences.* Journal of Pesticide Science, 2006. **31**(3): p. 245–251.
145. Dixon, H.M., et al., *Silicone wristbands compared with traditional polycyclic aromatic hydrocarbon exposure assessment methods.* Anal Bioanal Chem, 2018. **410**(13): p. 3059–3071.
146. Craig, J.A., et al., *Exposure of Nail Salon Workers to Phthalates, Di(2-ethylhexyl) Terephthalate, and Organophosphate Esters: A Pilot Study.* Environ Sci Technol, 2019. **53**(24): p. 14630–14637.
147. MyExposome, I.; Available from: https://www.myexposome.com/approach.
148. K. Lu *et al.*, Arsenic exposure perturbs the gut microbiome and its metabolic profile in mice: an integrated metagenomics and metabolomics analysis. *Environmental health perspectives* **122**, 284–291 (2014).

149. L. Zhang *et al.*, Benzo (a) pyrene-induced metabolic responses in Manila clam Ruditapes philippinarum by proton nuclear magnetic resonance (1H NMR) based metabolomics. *environmental toxicology and pharmacology* **32**, 218–225 (2011).
150. C. Yoon *et al.*, 1H-NMR-based metabolomic studies of bisphenol A in zebrafish (Danio rerio). *Journal of Environmental Science and Health, Part B* **52**, 282–289 (2017).
151. X. Wang *et al.*, Hepatoprotective effects of oridonin against bisphenol A induced liver injury in rats via inhibiting the activity of xanthione oxidase. *Science of the Total Environment* **770**, 145301 (2021).
152. T. Guan *et al.*, Metabolomics analysis of the effects of quercetin on renal toxicity induced by cadmium exposure in rats. *Biometals* **34**, 33–48 (2021).
153. X. Dong *et al.*, Urinary metabolomic profiling in rats exposed to dietary di (2-ethylhexyl) phthalate (DEHP) using ultra-performance liquid chromatography quadrupole time-of-flight tandem mass spectrometry (UPLC/Q-TOF-MS). *Environmental Science and Pollution Research* **24**, 16659–16672 (2017).
154. Y. Huang *et al.*, DEHP and DINP induce tissue-and gender-specific disturbances in fatty acid and lipidomic profiles in neonatal mice: a comparative study. *Environmental science & technology* **53**, 12812–12822 (2019).
155. Y. Yang *et al.*, Postnatal exposure to DINP was associated with greater alterations of lipidomic markers for hepatic steatosis than DEHP in postweaning mice. *Science of the Total Environment* **758**, 143631 (2021).
156. Y. Li *et al.*, Hepatotoxicity study of combined exposure of DEHP and ethanol: A comprehensive analysis of transcriptomics and metabolomics. *Food and Chemical Toxicology* **141**, 111370 (2020).
157. J. R. McKelvie, J. Yuk, Y. Xu, A. J. Simpson, M. J. Simpson, 1 H NMR and GC/MS metabolomics of earthworm responses to sub-lethal DDT and endosulfan exposure. *Metabolomics* **5**, 84–94 (2009).
158. D. Wang *et al.*, Metabolomics approach to investigate estrogen receptor-dependent and independent effects of o, p′-DDT in the uterus and brain of immature mice. *Journal of Agricultural and Food Chemistry* **65**, 3609–3616 (2017).
159. M. R. Elie *et al.*, Metabolomic analysis to define and compare the effects of PAHs and oxygenated PAHs in developing zebrafish. *Environmental research* **140**, 502–510 (2015).
160. P. Deng *et al.*, Hepatic metabolomics reveals that liver injury increases PCB 126-induced oxidative stress and metabolic dysfunction. *Chemosphere* **217**, 140–149 (2019).
161. Y. S. Jung, J. Lee, J. Seo, G. S. Hwang, Metabolite profiling study on the toxicological effects of polybrominated diphenyl ether in a rat model. *Environmental toxicology* **32**, 1262–1272 (2017).
162. Z. Wei *et al.*, Metabolomics coupled with pathway analysis characterizes metabolic changes in response to BDE-3 induced reproductive toxicity in mice. *Scientific reports* **8**, 1–16 (2018).
163. N. Yu *et al.*, Effects of perfluorooctanoic acid on metabolic profiles in brain and liver of mouse revealed by a high-throughput targeted metabolomics approach. *Scientific reports* **6**, 23963 (2016).
164. F. Li *et al.*, Integrated analysis of physiological, transcriptomics and metabolomics provides insights into detoxication disruption of PFOA exposure in Mytilus edulis. *Ecotoxicology and Environmental Safety* **214**, 112081 (2021).
165. A. J. Reardon *et al.*, Neurodevelopmental and Metabolomic Responses from Prenatal Coexposure to Perfluorooctanesulfonate (PFOS) and Methylmercury (MeHg) in Sprague–Dawley Rats. *Chemical Research in Toxicology* **32**, 1656–1669 (2019).
166. R. Wang, X. Han, H. Pang, Z. Hu, C. Shi, Illuminating a time-response mechanism in mice liver after PM2. 5 exposure using metabolomics analysis. *Science of The Total Environment* **767**, 144485 (2021).
167. E. Ortiz-Villanueva *et al.*, Assessment of endocrine disruptors effects on zebrafish (Danio rerio) embryos by untargeted LC-HRMS metabolomic analysis. *Science of the Total Environment* **635**, 156–166 (2018).

168. Z. Wan *et al.*, Effects of polystyrene microplastics on the composition of the microbiome and metabolism in larval zebrafish. *Chemosphere* **217**, 646–658 (2019).
169. Y. Deng, Y. Zhang, B. Lemos, H. Ren, Tissue accumulation of microplastics in mice and biomarker responses suggest widespread health risks of exposure. *Scientific reports* **7**, 46687 (2017).
170. H. Li *et al.*, Urinary metabolomics revealed arsenic exposure related to metabolic alterations in general Chinese pregnant women. *Journal of Chromatography A* **1479**, 145–152 (2017).
171. S. Cho *et al.*, High resolution metabolomics to determines the risk associated with bisphenol A exposure in humans. *Environmental Toxicology and Pharmacology* **58**, 1–10 (2018).
172. X. Li *et al.*, Association of urinary cadmium, circulating fatty acids, and risk of gestational diabetes mellitus: a nested case-control study in China. *Environment international* **137**, 105527 (2020).
173. J. Zhang *et al.*, Low-level environmental phthalate exposure associates with urine metabolome alteration in a Chinese male cohort. *Environmental science & technology* **50**, 5953–5960 (2016).
174. F. Jeanneret *et al.*, Evaluation and identification of dioxin exposure biomarkers in human urine by high-resolution metabolomics, multivariate analysis and in vitro synthesis. *Toxicology letters* **240**, 22–31 (2016).
175. H. Chu *et al.*, Metabolomics identifying biomarkers of PM 2.5 exposure for vulnerable population: based on a prospective cohort study. *Environmental Science and Pollution Research* **28**, 14586–14596 (2021).
176. Y. Xu *et al.*, Urine metabolomics of women from small villages exposed to high environmental cadmium levels. *Environmental toxicology and chemistry* **35**, 1268–1275 (2016).
177. Q. Chen *et al.*, Serum concentrations of neonicotinoids, and their associations with lipid molecules of the general residents in Wuxi City, Eastern China. *Journal of hazardous materials* **413**, 125235 (2021).
178. C.-H. S. Chen, T.-H. Yuan, R.-H. Shie, K.-Y. Wu, C.-C. Chan, Linking sources to early effects by profiling urine metabolome of residents living near oil refineries and coal-fired power plants. *Environment international* **102**, 87–96 (2017).
179. T. L. Alderete *et al.*, Perfluoroalkyl substances, metabolomic profiling, and alterations in glucose homeostasis among overweight and obese Hispanic children: A proof-of-concept analysis. *Environment international* **126**, 445–453 (2019).
180. T.-H. Yuan *et al.*, Metabolic profiling of residents in the vicinity of a petrochemical complex. *Science of the Total Environment* **548**, 260–269 (2016).

Deciphering Plant-Pathogen Interactions Through Plant Metabolomics: From Technical Advances to Applied Research

Marisa Maia, Stefania Savoi, and Andreia Figueiredo

Abbreviation

ASCA	ANOVA-simultaneous component analysis
DAMPs	Damage-associated molecular patterns
DoE	Design of experiments
ETI	Effector-triggered immunity
F.A.I.R.	Findable, Accessible, Interoperable, Reusable
FT-ICR MS	Fourier-transform ion cyclotron resonance mass spectrometry
GC	Gas chromatography
GWAS	Genome-wide association studies
JA	Jasmonic acid
LC	Liquid chromatography
LDI	Laser desorption/ionization
m/z	Mass-to-charge ratio
MAMPs	Microbe-associated molecular patterns
MAS	Marker-assisted selection
mQTL	Metabolic quantitative trait loci
MS	Mass spectrometry
MSI	Mass spectrometry imaging
NMR	Nuclear magnetic resonance

Authors: "Marisa Maia", "Stefania Savoi" and "Andreia Figueiredo" have equally contributed to this chapter.

M. Maia · A. Figueiredo (✉)
Grapevine Pathogen Systems Lab, Biosystems and Integrative Sciences Institute (BioISI), Plant Biology Department, Science Faculty of Lisbon University, Lisbon, Portugal
e-mail: mrmaia@fc.ul.pt; aafigueiredo@fc.ul.pt

S. Savoi
Department of Agricultural, Forest and Food Sciences, University of Turin, Turin, Italy
e-mail: stefania.savoi@unito.it

V. Soni, T. E. Hartman (eds.), *Metabolomics*,
https://doi.org/10.1007/978-3-031-39094-4_11

PAMPs	Pathogen-associated molecular patterns
PCA	Principal component analysis (PCA)
PLS-DA	Partial least squares discriminant analysis
PTI	Triggered immunity
QC	Quality control
QTL	Quantitative trait loci
QTOF	Quadrupole time of flight
SA	Salicylic acid
UPLC	Ultrahigh-performance liquid chromatography

1 Introduction

Plant life plays an essential role in the world. The food, water, natural compounds for pharmaceuticals, air, habitats for a huge number of species, soil quality, and climate are only a few examples of resources that plants provide.

Pathogens can have a devastating impact in all of the above mentioned points. They can affect all plant tissues, cause damage, and reduce quality and yield, which lead to significant production losses and consequently affect production and commercialization. Plant-pathogen interactions occur inherently to both host and pathogen evolution. Plants activate several defense layers, including the constitutive and inducible defenses [114], to restrain pathogen growth, while pathogens secrete molecules to deceive the host defense arsenal and establish disease. The success of the primary recognition of pathogen molecules by the plants may dictate the outcome of the interaction [30, 60], as it is crucial for the activation of the different layers of defense. When pathogen molecules, pathogen-/microbe-associated molecular patterns (PAMPs/MAMPs), or damage-associated molecular patterns (DAMPs) are successfully recognized, pathogen-associated molecular pattern (PAMP)-triggered immunity (PTI) is established [60]. Pathogens may then secrete effector proteins to overcome PTI. If plants recognize the secreted pathogen effectors through resistance proteins, effector-triggered immunity (ETI) is established. ETI is often accompanied by a broader modulation of host's defenses and the establishment of a hypersensitive response leading to rapid cell death at the infection sites [59]. Although ETI is generally associated with stronger local responses than PTI, they were suggested to co-occur and to present synergistic effects [7, 13, 136] (Fig. 1).

The establishment and modulation of host's defense mechanisms, as well as the establishment of PTI and ETI, have been thoroughly studied over the years from both partners' sides, but several gaps remain in our understanding of plant-pathogen interactions.

In that sense, it has been widely known that plants are able to synthesize a myriad of chemical compounds. It is also known that perturbations of plant metabolism play a central role in determining the outcome of an infection. In the last decade, advances in the metabolomics field have enabled us to deepen our understanding of the key roles of metabolites in the host signaling (e.g., oxylipins) and defense events.

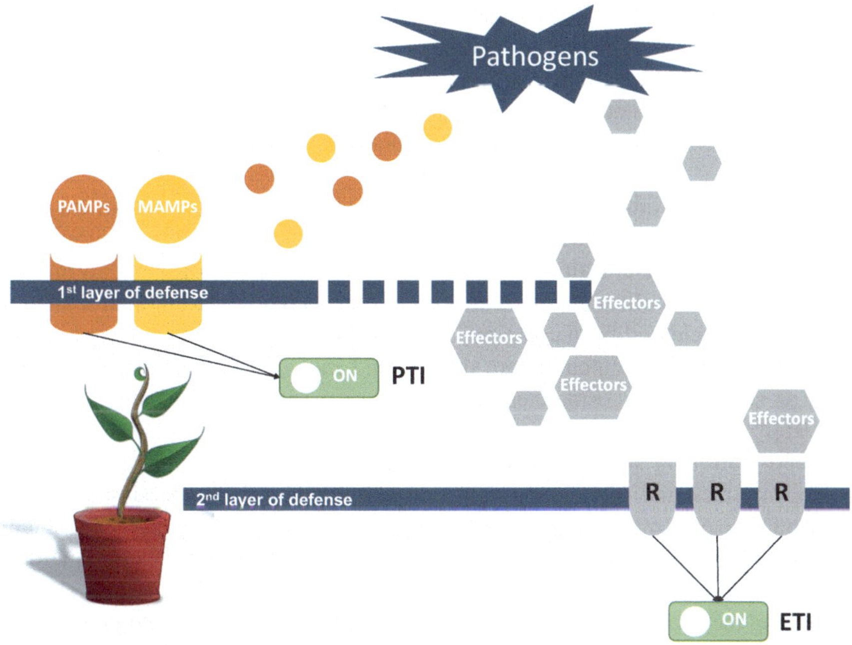

Fig. 1 Pattern-triggered immunity (PTI) and effector-triggered immunity (ETI) following pathogen attack. *PAMP* pathogen-associated molecular pattern, *MAMP* microbe-associated molecular pattern

But the application of this omics to plant pathology is still lagging in comparison with other omics approaches.

This chapter discusses the application of metabolomics to the study of plant-pathogen interactions, highlighting the technical advances in the field. Moreover, recent visualization techniques such as matrix-assisted laser desorption/ionization mass spectrometry imaging (MALDI-MSI) are exploited in the interaction context. Finally, metabolomics data analysis and data integration with other omics are discussed, and metabolomics impact in the development of new approaches for breeding programs is highlighted.

2 Looking into Plant-Pathogen Interactions from a Metabolic Point of View

Different omics approaches have been widely applied to study plant biotic interactions, making metabolomics the newest and one of the most promising approaches [132]. Metabolites play a crucial role in plant-pathogen interactions, providing a snapshot that reflects the regulation at different cell layers (genetic, epigenetic, transcriptomic, and proteomic).

Metabolomic studies of plant-pathogen interactions are quite challenging due to the difficulties underlying the annotation of an entire cell metabolome, the chemical and structural diversity of metabolites, and the complex nature of metabolic regulation and interaction networks applied to both partners. Plants are a metabolically complex system, with over 200,000 metabolites estimated in the plant kingdom [37, 38].

Plant-pathogen interactions represent an even more complex biochemical matrix as plant metabolites include an array of structurally and functionally diverse compounds that may be constitutively present or synthesized in response to different stimuli [91]. Metabolites that present antimicrobial and perform antimicrobial properties and are released from constitutively stored precursors were named phytoanticipins [91]. These compounds are diverse and include several metabolic groups such as saponins, glucosinolates, and cyanogenic and benzoxazinone glucosides [89, 91].

Reprogramming of host primary and secondary metabolisms occurs after a pathogen attack and often starts with the mobilization of photoassimilates to the infected tissues to meet the intensive demand for energy and carbon to fuel defense responses [11, 62, 74, 83]. These sugars play important roles in the regulation of host defense-related metabolic pathways, namely, strengthening the cell wall, enhancing the accumulation of secondary metabolites (e.g., isoflavonoids), and biosynthesis of defense hormones as jasmonic acid (JA) [31, 111], or in signaling mechanisms [43]. In addition, as part of the host defense strategy, several low-molecular-weight secondary metabolites, often with antimicrobial properties, are produced after MAMP recognition (e.g., phytoalexins) [2, 57, 58]. These secondary metabolites contribute to strengthening mechanical barriers, directly killing the invading pathogens, and attracting biocontrol species or beneficial symbionts (reviewed in [120]).

Looking into the pathogens, small molecules are produced and secreted to manipulate the host metabolism (e.g., soluble sugars in the apoplast are often used by pathogens as a source of carbon for their own development, leading to an increase in virulence [92]) and to subvert plant immunity. A well-studied example is the bacterial JA analog, coronatine, which is secreted to interfere with the JA/salicylic acid (SA) balance leading to a lowering of host SA levels [45, 46].

Although thousands of metabolomics studies devoted to plant-pathogen interactions have been published in the past years, some points need to be taken into consideration, and several questions remain to be answered. One of the most pertinent is how may we define the origin of a determined metabolite in a plant-pathogen interaction. This is quite challenging, particularly for biotrophic pathogens that cannot be cultured *in vitro*. Some approaches have been proposed so far, such as dual metabolomics [132] or metabolite fingerprinting of both partners by mass spectrometry imaging (MSI) [14, 17], but a long way is still ahead.

3 Technical Advances in Plant Metabolomics

3.1 Sample Preparation

Multiple factors must be considered upon sample preparation for a metabolomics study. It goes from planning and carefully performing the experiments and collecting the material to preparing the samples to be analyzed (Table 1).

When one intends to extract metabolites from plant tissues, it is important to understand plant organ diversity, morphology, and physiology to adapt the protocol according to it [29, 85, 126]. Also, due to tissue complexity, the selection of the solvent's composition is essential to cover as many chemically diverse structures as possible [84]. When planning an experimental assay, the amount of sample needed for the analysis afterwards is another point to have in consideration. To ensure a proper extraction, the amount of sample should not be limited, being more difficult to control the test conditions leading to more errors, nor too much, which can affect the dissolution of metabolites in the solvents [65].

After these conditions are optimized, sample collection is normally performed by stopping all the metabolic machinery by using low temperatures, such as freezing samples with liquid nitrogen, fast heating, addition of acid, or lyophilization [127].

Table 1 Summary of the different points mentioned in this book section to consider when performing a plant-pathogen interaction experimental assay and metabolite extraction and analysis

Checklist to perform	
Experimental assay of plant-pathogen interaction	Metabolite extraction protocol and analysis
Selection of a plant organ Roots Stems Leaves Flowers Fruits	Selection of an extraction protocol adequate to the plant organ in study
Study the morphology and physiology of the plant organ	Selection of solvent(s) and proportions to extract the desired metabolites
Conditions of the experiment Control vs inoculated samples Time-points to be collected	Extraction time Selection of the equipment needed for the extraction
Follow the extension of the damage of the pathogen	Optimization of the analysis parameters in the analytical technique according to the sample
Plant material collection Enough material for the analysis afterwards Number of replicates Selection of a method to stop all the metabolic machinery	Use of internal and external standards Technical repetitions

In metabolic extractions, although the main goal is to cover the highest number of metabolites with differing chemical properties, the extraction time and equipment required as well as laboratory supplies are also important points to have in consideration to avoid wasting resources, manpower, and time [19, 22, 90, 95]. Nowadays, it is still difficult to combine all these aspects, so depending on the study some compromises need to be made.

Sample number is also worth mentioning. The number of samples and biological replicates for each condition used for analysis is important to avoid repeating differences, increase efficiency, and rule out experimental bias. This ensures that the data obtained is robust and scientifically significant. Finally, extraction protocols should be user-friendly to ensure reproducibility by any other operator.

Through the years, some metabolite extraction protocols for plant tissues and cellular compartments upon pathogenic challenges have been optimized, considering all the above mentioned settings, and published [29, 40, 80, 90, 98].

All these methods are focused on the total accumulation of metabolites in a sample pool of the tissues infected with the pathogens. Although the separation of metabolites in the different subclasses prior to analysis could be helpful to diminish the complexity of the samples, it is still unclear, upon interaction, which metabolites are in fact from the pathogen and/or from the plant.

3.2 *Advances in Analytical Methods and Techniques*

In the past years, different metabolomics techniques have been widely applied to study different pathosystems. Mass spectrometry (MS) and NMR (nuclear magnetic resonance) are the most commonly used techniques.

NMR is a nondestructive, high-throughput, and extremely reproducible technique [6, 24, 27, 125]. In addition, it has the advantage that the signal intensity detected for each metabolite in the spectrum is related to the concentration, allowing the quantification of the metabolite as well as the undeniable identification of compounds through structure determination [82]. However, a lot of initial material is needed, and it lacks sensitivity, which is a bottleneck when it comes to analyzing highly complex samples [34, 95].

MS tends to be the most commonly used method for metabolomics studies [19, 22, 44, 47, 78]. It is selective and highly sensitive and has high-resolution power, which allow the detection and identification of an enormous number of metabolites in complex samples [38, 104, 116]. Furthermore, the different ionization sources available nowadays allow researchers to cover almost all the different chemical structures present at different samples [49, 54, 88, 122]. Moreover, besides the detection of a mass-to-charge ratio of the different metabolites, another dimension can be added by coupling separative steps, such as gas (GC) or liquid chromatography (LC) [8, 88, 122, 137]. All the analytical advances in MS equipment, applications, and improved protocols for the analysis of plant tissues and analysis for MS data have already been extensively reviewed in the past years.

From the plethora of techniques, an interesting fairly new worth mentioning is leaf-spray mass spectrometry. It is an ambient ionization method that permits the analysis of living plants and provides real-time information on small molecules [41, 73]. Also, it is suitable for examining various plant tissues and a wide variety of species. The determination of the different metabolites, such as sugars, fatty acids, lipids, alkaloids, and others, present at the surface of any plant tissue is highly relevant to understand the compounds involved in the first moments of pathogen contact as well as which compounds are constitutively produced by the plant acting as pathogen chemoattractants.

Albeit recent advances in metabolomics studies, the diversity of metabolites is still far from being overcome. Also, in plant metabolomics analysis challenged with any pathogen, another layer of complexity is added as the combined metabolomes are being analyzed simultaneously, being still extremely difficult to distinguish them.

The modulation of metabolites involved in plant-pathogen interaction remains a black box, and it is clear that there is still a lot of progress to be done.

To overcome these difficulties, combining different analytical techniques, and generating different datasets, is an approach commonly used to characterize the regulation happening in plants in defense against pathogens. For example, the combination of MS and NMR studies or coupling LC or GC to MS is a type of approach frequently used.

In recent years, several works have been published with the combination of these techniques to study plant-pathogen interactions. In rice (*Oryza sativa*) infected with the blast fungus (*Magnaporthe oryzae*, Magnaporthaceae), an ultrahigh-performance liquid chromatography quadrupole time-of-flight mass spectrometry (UPLC-QTOF-MS) system was used to characterize this interaction using an untargeted metabolomics approach [87]. To metabolically investigate the interaction of tomato (*Solanum lycopersicum*) with *Phytophthora infestans*, liquid chromatography mass spectrometry (LC-MS) and matrix-assisted laser desorption/ionization mass spectrometry (MALDI-MS) were used [42]. The combination of GC-MS, LC-TOF-MS, and NMR allowed better understanding of the metabolic reprogramming and resistance of barley against the biotrophic fungus *Blumeria hordei* [69].

Another example was the use of LC-MS/MS and GC-MS to identify and quantify lipids, phenols, and primary compounds as well as GC-MS for the semi-quantification of volatile compounds in the interaction of grapevine (*Vitis vinifera* L.) with *Plasmopara viticola* [23].

Moreover, combining direct (allow the undeniable identification of the disease/pathogen) and indirect methods (identification of the disease/pathogen through physiological changes in the plant) to study plant pathogen interactions will provide an overall characterization of the biologically important pathways [36]. This will allow the research community to integrate the information obtained, reveal plant-pathogen metabolomics dynamics, and unveil the inherent resistance/susceptibility of plants as well as their defenses upon attack and a possibility to also characterize pathogen metabolome.

As an example of the combination of a direct method with metabolomics analysis, after a primary metabolomics approach, discriminant metabolites can be identified and mapped in metabolic pathways, and genes coding for key enzymes involved in their biosynthesis/catalysis can be selected [71, 75, 103]. A combined transcriptomics and metabolomics approach was used in rose plant (*Rosa chinensis*) after the infection with powdery mildew (*Podosphaera pannosa*) and after the treatment with exogenous salicylic acid in order to explore the resistance mechanisms of rose plant [134]. Another study analyzed a resistant and a susceptible black pepper species (*Piper flaviflorum* and *Piper nigrum*, respectively) upon infection by the oomycete *Phytophthora capsici*. In this work, the authors highlight differences in the expression profiles of certain genes which are correlated to specific metabolic pathways [35].

On the other hand, upon pathogen interaction, it is known that several metabolic classes are modulated (e.g., flavonoids, lipids, and sugars) and that some compounds are even directly related to photosynthesis modulation (sucrose and chloroplast lipids).

The combination of chlorophyll fluorescence imaging measurements (an indirect detection method) with metabolomics qualitative and quantitative data can show the state of the photosynthesis apparatus, upon pathogen infection, to understand how these pathogens modulate plant photosynthesis and how it affects the plant [55, 96, 133]. Bonfig et al. [16] used this approach to examine the effects of virulent and avirulent strains of *P. syringae* in *Arabidopsis thaliana* to investigate the effect of the pathogen on the photosynthetic activity and carbohydrate metabolism [16]. Also with *Arabidopsis thaliana*, the chlorophyll content and carbohydrates were measured to examine the effect of *Albugo candida* (white blister rust) on the regulation of photosynthesis, carbohydrate metabolism, and gene expression [25].

Metabolomics has a high potential for trait screening. Ultimately, technological improvements could lead to portable devices that could analyze directly, in a user-friendly way, *in vivo* plant-pathogen data, detecting early pathogenic infections through metabolic biomarkers and controlling them in a more precise and sustainable way.

3.3 Metabolic Visualization Techniques: The Case of Mass Spectrometry Imaging

The selection of the most thorough extraction protocol and the choice of the analytical technique with the best resolution for the detection and identification of metabolites have been helping researchers to slowly answer several questions, such as:

- Which metabolites are constitutively present in plants that are used to counteract pathogen attack?
- Which plant metabolites are produced after stress?
- Is there any interplay between these metabolites?
- Are these metabolites host- or pathogen-specific?
- Where are the metabolites accumulated upon pathogen interaction?

Metabolomics has the potential to tackle these questions, allowing a better understanding of a biochemically complex system as plant-pathogen interactions. Nevertheless, metabolome analyses are based on extracted metabolites from a pool of samples from a plant tissue (e.g., roots, stems, leaves, fruits), from a specific plant compartment (e.g., cytoplasm, apoplast, chloroplast, mitochondria), or from the pathogens. However, this information obtained is not locally specific. Therefore, the real location of these metabolites and their accumulation patterns through time could be crucial for plant survival upon pathogen interaction.

Several techniques have been used to allow the visualization of biomolecules in different cellular structures, such as immunolocalization [64]; microscopy techniques [26], techniques that rely on the interaction between the molecule of interest and an external probe [28, 135]; and others [67].

These visualization techniques allow the identification of molecules. However, they require personnel experience to obtain good results reproducibly. In addition, sample preparation is sample-specific, the materials for analysis are costly, and most importantly, it is necessary to know the molecules of interest. This last point is quite challenging for any technique as it is estimated that plants possess an enormous number of unknown metabolites with different properties and structures [50, 104].

Taking this into consideration, mass spectrometry imaging (MSI) has been pointed out as a promising technique for studying plant metabolites, particularly in the context of plant-pathogen interactions [77, 120]. This technique has the ability to identify and reveal the spatial distribution of specific biomolecules with a high mass accuracy and a high resolution across a sample in a label-free and non-targeted mode without having to consider the complexity of the sample and without prior knowledge of their content. Moreover, it has the advantage of allowing the analysis of all the different plant organs (roots, stems, leaves, flowers, and fruits) (Fig. 2).

In recent years, MSI has been widely applied to visualize differences in the distribution of specific metabolites at different maturity stages [129], determine the location of metabolites, correlate this information with tissue architecture [12], and study the plant contaminations with metals [3], the metabolites of interest for food analysis [66], and the migration and distribution behavior of pesticide residues [130]. But its application in plant-pathogen interactions is scarce [100].

Concerning MSI analytical techniques, currently, there are several approaches with different ionization sources, mass range detection, spatial resolution, and modes of analysis that can be applied to study plant metabolites. In the past years, different reviews have been published describing in detail how these techniques work and highlighting their advantages and analytical differences and capabilities [5, 14, 15, 17, 32, 44, 93, 108].

To perform any MSI analysis in plant tissues, with or without pathogen challenge, one needs to take into consideration the different challenges for the preparation of the tissue sections. Sample preparation and handling are crucial for not disrupting or causing any delocalization of the molecules, lateral displacement of the ions, achieving high-quality signals, correct ionization, and confirming the spatial distribution of the molecules [14, 15, 17, 33, 48, 63]. Several protocols have

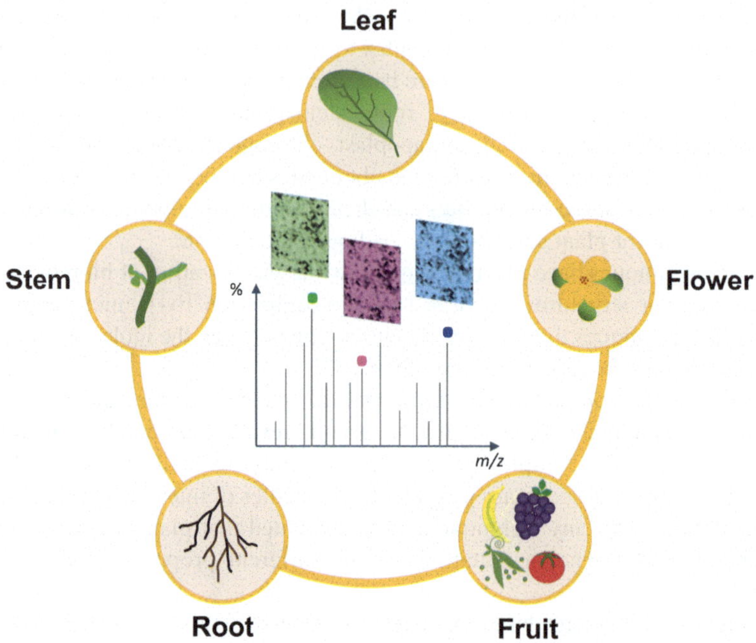

Fig. 2 Schematic representation of the different plant tissues which could be used for MSI experiments

already been proposed for different plant tissues; however, it is always necessary to perform some experiments and systematic observations and analyze some data to ensure to obtain good results. Plant tissues are overall more difficult to handle than animal tissues [17, 53]. Due to their physical properties upon sample preparation, several points need to be considered, such as water content, sample shrinkage, flaking, dehydration, and fragility after freezing [20, 33, 117]. Moreover, plants are sessile organisms and rely on their physical structures to restrain pathogen attacks, penetration, and development. The first barriers of defense are the wax cuticle and trichomes. Although these structures act as active shields against different pathogens, their analysis is not straightforward as some MSI ionization techniques may not be able to penetrate through them and analyze their metabolites [33, 79, 121]. Also, cell walls, another barrier of defense, are highly complex due to their being actively remodeled and reinforced upon any stimuli, hampering the analysis of internal cell metabolites.

Moreover, another layer of concern is added when preparing plant tissues after pathogen interaction. Plant tissues become more fragile and lose structure, and the ion signal could be difficult to obtain depending on the pathogen life cycle. Also, some pathogens create structures on top of the tissues, becoming another physical barrier for the MSI technique to penetrate. Also, it can hamper the analysis of the tissue itself and can lead to inaccurate ion ionizations creating misleading results.

To date, the literature on the application of MSI techniques to study plant-pathogen interaction is limited to the comparison between challenged and non-challenged plant tissues. The combination of high-performance liquid chromatography-ultraviolet (HPLC-UV) method with matrix-assisted laser desorption/ionization mass spectrometry imaging (MALDI-MSI) was used to quantify and investigate the localization of hesperidin in orange (*Citrus sinensis*) grafted onto mandarin lime (*C. limonia*) upon challenge with *Xylella fastidiosa* [110]. Also, imaging desorption electrospray ionization mass spectrometry (DESI-MS) was used to investigate fluctuations in glycoalkaloids present in sprouted potatoes (*Solanum tuberosum*) infected by the phytopathogen *Pythium ultimum* [118]. In *Vitis vinifera* L. (grapevine), different MSI techniques have been applied to study the leaf surface during the interaction with the biotroph oomycete *Plasmopara viticola*. With laser desorption/ionization (LDI), it was reported the accumulation of stilbene phytoalexins (e.g., resveratrol, piceid, pterostilbene) around infection sites [9, 51], and with MALDI, it was shown a pattern of accumulation of resveratrol, piceid, and viniferins close to *P. viticola* lesions [10]. With MALDI, the localization of sucrose was linked to the development structures of *P. viticola* [79]. Germinating soybeans inoculated with Aspergillus oryzae [1] as well as pea (Pisum sativum) pod tissues exposed to Fusarium solani f. sp. phaseoli spores [107] were visually investigated using MALDI-imaging.

Despite the modest applicability of MSI to complex pathosystems, it has proven its value. The different techniques mentioned will certainly tackle the challenges of deciphering the tight host-pathogen communication and improve our understanding of pathogen infection and development in plant tissues.

4 Metabolomics Data Analysis and Omics Data Integration

As an omics technology, metabolomics, officially born at the beginning of the twenty-first century [39], outputs large datasets. These require a high data mining capacity to fulfill two objectives in the direction of linking a genotype to a phenotype [38] or from a relevant biological question to arrive at its biological interpretation (Fig. 3). The first is to identify and quantify as many organ- or plant-specific metabolites as possible and then to measure their modulation and association under different conditions or in time series. However, the metabolomics and data analysis challenges lie in the complexity of the metabolome, which entails highly different compounds that vary in sizes, chemical properties, and concentrations and where most remain unknown.

A precondition to guarantee reliable results is the thoughtful planning of a high-quality design of experiments (DoE), an imminent gold standard in metabolomics [56]. Among all, it requires a representative array of randomized and unbiased samples adequately collected, stored, prepared, measured, and analyzed in order to guarantee enough statistical power for answering and interpreting the research questions [105, 123].

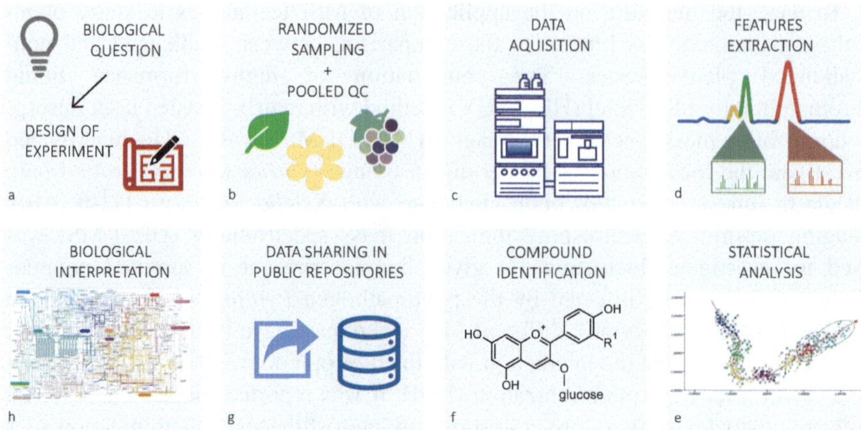

Fig. 3 Eight significant steps in metabolomics data analysis (panels **a**–**h**)

Metabolomic analyses can be tackled with a targeted or an untargeted approach. The choice is based on metabolite coverage and prior knowledge. The targeted method quantitatively analyzes a predefined array of known compounds extracted with an ad hoc analytical method. At the same time, the untargeted holistically aims to comprehensively measure the totality of the metabolites of a biological matrix. In any case, the repeated analysis of pooled quality control (QC) samples, intended to represent the sample variability, is needed for monitoring and adjusting the performance of the analytical instrument in terms of retention time shift, signal intensity stability, and mass calibration, removing systematic trends and batch effects. These corrections can also be controlled by measuring samples spiked with one or multiple internal standards or by analyzing several reference standards [18, 123].

Once the spectra have been acquired, data analysis is a significant element of the metabolomics workflow, where compound identification can be extremely time-consuming as well as the most critical rate-limiting bottleneck. In a nutshell, metabolomics data analysis involves feature extraction, statistical analysis, compound identification, and marker discovery and interpretation. In every analytical technique utilized (e.g., liquid/gas chromatography often coupled with mass spectrometry (LC-MS or GC-MS), Fourier-transform ion cyclotron resonance mass spectrometry (FT-ICR MS), or nuclear magnetic resonance spectroscopy (NMR)), each metabolite is recognized and characterized by a set of highly correlating different signals such as a suite of mass-to-charge ratio (*m/z*) ions or proton peaks. These signals or features (possibly converted in open-source formats) need to be extracted within a processing method by performing actions such as peak picking, retention time correction, noise reduction, baseline correction, spectra deconvolution, etc., to finally get a raw data matrix containing the corresponding list of peaks and associated areas across all samples analyzed. Several tools are available for processing metabolomics data [81]. The users can opt for commercial software (e.g., Analyst, Chromeleon CDS, Compound Discoverer, MassLynx, MarkerView, Mass Profiler

Professional, Progenesis QI, etc.), open-source packages that can be downloaded and executed offline (e.g., MS-DIAL, MZmine, XCMS, etc.), or web-based solutions to run online (e.g., MetaboAnalyst, XCMS online) [21, 70]. Often, this data matrix needs to be transformed, scaled, and normalized before performing any other analysis by applying, for example, a logarithmic or square root transformation, a Pareto, or unit variance scaling [123, 124]. Statistical data analysis for modeling, marker discovery, and data visualization relies mainly on multivariate analysis, such as principal component analysis (PCA), partial least squares discriminant analysis (PLS-DA), and ANOVA-simultaneous component analysis (ASCA); on deep learning methods; and, to some extent, on univariate approaches, based on t-test, linear mixed models, and analysis of variance [97, 123]. Finally, the last step in a metabolomics workflow data analysis is the identification of the key compounds that drive differences among groups of samples. Again, if a targeted scenario has been used, this task is facilitated by comparing and identifying (often manually) the spectra with reference standards or with known metabolite libraries available online (mass spectral library of NIST, METLIN, MassBank, mzCloud, etc.) or by using personal libraries built in-house. More challenging is the identification of unknown features extracted from untargeted analysis. Currently, three methods guide the users in finding the appropriate annotation confidence levels to adapt when reporting their findings. The first one published was based on four levels of annotation summarized in decreasing order as (i) identified compounds, (ii) putatively annotated compounds, (iii) putatively characterized compound classes, and (iv) unknown compounds [115]. It was refined by the work of Schymanski et al. [106], which proposed the identification of metabolites in five confidence levels described as (i) confirmed structure by reference standards, (ii) probable structure by library spectrum match, (iii) tentative candidates, (iv) unequivocal molecular formula, and (v) exact mass of interest. Lastly, in 2019 the Metabolomics Society's Metabolite Identification Task Group released a document with a more detailed definition and information for metabolite annotation based on seven confidence levels, from A to G (document available at https://drive.google.com/file/d/1PJLdPCkz8ymX8SgZ4Wl5Sw4ZG-dlyWWU/view, accessed in January 2023).

The demand for transparency in omics sciences and open data research has become increasingly urgent [113]. The publication of workflows, codes, analytical procedures, and data and metadata is necessary to evaluate and make the metabolomics works reproducible according to the F.A.I.R. (Findable, Accessible, Interoperable, Reusable) principles [131]. To this scope, a step-by-step guideline for data and metadata management has been recently published addressing the entire metabolomics community [4] or specifically the grapevine and wine science [101], the latter one sponsored by the work of the EU-funded Cost Action CA17111 INTEGRAPE. There are several benefits to data sharing. Since data are the basis of scientific research, the availability of complex datasets allows for deeper analysis and scientific progress (e.g., meta-analysis); complying with guidelines for minimum reporting standards (strongly promoted by the Metabolomics Standards Initiative) allows for the reproducibility of the experiments by other users or the simple reanalysis of datasets with different inspection angles; the accurate metadata description sustains plant phenotyping and grants biological significance to the data. This is possible only by ensuring data sharing and, therefore, by uploading experiments data and metadata to one of the existing public and freely accessible repositories, such as MetaboLights, the Metabolomics Workbench, etc. The more and more availability of curated open-access data can prompt more sophisticated analysis aimed at integrating metabolomics with other omics science, paving the way for a more comprehensive understanding of an organ/plant system and, ultimately, attaining systems biology studies [94]. However, further progress is needed in providing different datasets from different analytical techniques and omics sciences compatible [102].

5 Breeding Programs: The Impact of Metabolome-Based Knowledge

Disease management in crops is mainly achieved through chemical control of pathogens, which is neither sustainable nor environmentally friendly. Also, the pathogen evolution rate is relatively high, and many develop pesticide resistance [76]. Therefore, the development of disease-resistant varieties appears as one of the more sustainable approaches for crop disease management. It may ensure global food security and become an environmentally friendly measure to reduce chemical input.

Plant disease resistance is often under genetic control and of quantitative nature. Quantitative trait loci (QTL) identification is one of the most used approaches to identifying and associating genetic factors with complex trait control. The identification of QTLs comes from genome-wide association studies (GWAS) that determine the genomic regions associated with a given phenotype [99]. These regions might be used to establish molecular marker-assisted selection (MAS) of a particular trait in breeding programs [52], shortening the time needed to obtain new commercially suitable varieties.

Many studies have been conducted to identify resistance-associated QTLs for different crops to be used in breeding programs (recently reviewed in [61, 72, 99, 109]). Recently, metabolomics has also contributed to the definition of new QTLs, metabolite QTLs (mQTLs) [86]. In plant-pathogen interactions, the identification of mQTLs is still poorly represented but present an enourmous potential for breeding programs.

In *Brassica napus* roots, mQTLs involved in resistance and in metabolic adjustments were established for *Plasmodiophora brassicae* resistance [128], namely, the association between gluconasturtiin and two unknown metabolites in the resistance conferred by two QTLs and the association between glycine and glutathione and three main resistance QTLs [128] and also for glucosinolate content [112]. Also, for grapevine, mQTLs associated with flavonoid biosynthetic pathways that are likely involved in the production of secondary metabolites, including phytoalexins and stilbenoids, were established on chromosome 18 [119]. Functional mapping of these mQTLs on the VitisNet network database revealed chromosome 18 as a major hotspot of disease-resistance motifs [119]. In carrot, mQTLs for *Alternaria dauci* resistance were also established, highlighting camphene, α-pinene, α-bisabolene, β-cubebene, caryophyllene, germacrene D, and α-humulene as the terpenes potentially involved in carrot resistance [68]. Functional analyses revealed that α-humulene and caryophyllene exhibited fungitoxic properties, consistent with a direct role of these compounds in disease resistance [68].

These studies show the potential of metabolomics application in plant breeding. The definition of metabolic markers to evaluate performance and select plants with a desired trait will also improve MAS application in crop improvement.

Acknowledgments This work was financially supported by national funds through the FCT (Foundation for Science and Technology) through Marisa Maia research contract 2022.07433. CEECIND and the BioISI funding (UIDP/04046/2020).

References

1. Abe, C., Zhang, Y., Takao, K., Sasaki, K., Ochiai, K., and Matsui, T. (2021). Visualization Analysis of Glyceollin Production in Germinating Soybeans by Matrix-Assisted Laser Desorption/Ionization Mass Spectrometric Imaging Technique. *J. Agric. Food Chem.* 69, 7057–7063. doi: https://doi.org/10.1021/acs.jafc.1c02261.
2. Ahuja, I., Kissen, R., and Bones, A. M. (2012). Phytoalexins in defense against pathogens. *Trends Plant Sci* 17, 73–90. doi: https://doi.org/10.1016/j.tplants.2011.11.002.
3. Alcantara, H. J. P., Jativa, F., Doronila, A. I., Anderson, C. W. N., Siegele, R., Spassov, T. G., et al. (2020). Localization of mercury and gold in cassava (Manihot esculenta Crantz). *Environ Sci Pollut Res Int* 27, 18498–18509. doi: https://doi.org/10.1007/s11356-020-08285-3.
4. Alseekh, S., Aharoni, A., Brotman, Y., Contrepois, K., D'Auria, J., Ewald, J., et al. (2021). Mass spectrometry-based metabolomics: a guide for annotation, quantification and best reporting practices. *Nat Methods* 18, 747–756. doi: https://doi.org/10.1038/s41592-021-01197-1.
5. Amstalden van Hove, E. R., Smith, D. F., and Heeren, R. M. A. (2010). A concise review of mass spectrometry imaging. *Journal of Chromatography A* 1217, 3946–3954. doi: https://doi.org/10.1016/j.chroma.2010.01.033.
6. Aranibar, N., Singh, B. K., Stockton, G. W., and Ott, K. H. (2001). Automated mode-of-action detection by metabolic profiling. *Biochem Biophys Res Commun* 286, 150–155. doi: https://doi.org/10.1006/bbrc.2001.5350.
7. B. Santos, R., Nascimento, R., V. Coelho, A., and Figueiredo, A. (2020). Grapevine–Downy Mildew Rendezvous: Proteome Analysis of the First Hours of an Incompatible Interaction. *Plants* 9, 1498. doi: https://doi.org/10.3390/plants9111498.
8. Beale, D. J., Pinu, F. R., Kouremenos, K. A., Poojary, M. M., Narayana, V. K., Boughton, B. A., et al. (2018). Review of recent developments in GC–MS approaches to metabolomics-based research. *Metabolomics* 14, 152. doi: https://doi.org/10.1007/s11306-018-1449-2.
9. Becker, L., Bellow, S., Carré, V., Latouche, G., Poutaraud, A., Merdinoglu, D., et al. (2017). Correlative Analysis of Fluorescent Phytoalexins by Mass Spectrometry Imaging and Fluorescence Microscopy in Grapevine Leaves. *Anal Chem* 89, 7099–7106. doi: https://doi.org/10.1021/acs.analchem.7b01002.
10. Becker, L., Carré, V., Poutaraud, A., Merdinoglu, D., and Chaimbault, P. (2014). MALDI Mass Spectrometry Imaging for the Simultaneous Location of Resveratrol, Pterostilbene and

Viniferins on Grapevine Leaves. *Molecules* 19, 10587–10600. doi: https://doi.org/10.3390/molecules190710587.

11. Bezrutczyk, M., Yang, J., Eom, J.-S., Prior, M., Sosso, D., Hartwig, T., et al. (2018). Sugar flux and signaling in plant–microbe interactions. *The Plant Journal* 93, 675–685. doi: https://doi.org/https://doi.org/10.1111/tpj.13775.
12. Bhandari, D. R., Wang, Q., Friedt, W., Spengler, B., Gottwald, S., and Römpp, A. (2015). High resolution mass spectrometry imaging of plant tissues: towards a plant metabolite atlas. *Analyst* 140, 7696–7709. doi: https://doi.org/10.1039/C5AN01065A.
13. Bhattacharjee, S., Halane, M. K., Kim, S. H., and Gassmann, W. (2011). Pathogen effectors target Arabidopsis EDS1 and alter its interactions with immune regulators. *Science* 334, 1405–1408. doi: https://doi.org/10.1126/science.1211592.
14. Bjarnholt, N., Li, B., D'Alvise, J., and Janfelt, C. (2014). Mass spectrometry imaging of plant metabolites--principles and possibilities. *Nat Prod Rep* 31, 818–837. doi: https://doi.org/10.1039/c3np70100j.
15. Bodzon-Kulakowska, A., and Suder, P. (2016). Imaging mass spectrometry: Instrumentation, applications, and combination with other visualization techniques. *Mass Spectrom Rev* 35, 147–169. doi: https://doi.org/10.1002/mas.21468.
16. Bonfig, K. B., Schreiber, U., Gabler, A., Roitsch, T., and Berger, S. (2006). Infection with virulent and avirulent P. syringae strains differentially affects photosynthesis and sink metabolism in Arabidopsis leaves. *Planta* 225, 1–12. doi: https://doi.org/10.1007/s00425-006-0303-3.
17. Boughton, B. A., Thinagaran, D., Sarabia, D., Bacic, A., and Roessner, U. (2016). Mass spectrometry imaging for plant biology: a review. *Phytochem Rev* 15, 445–488. doi: https://doi.org/10.1007/s11101-015-9440-2.
18. Broadhurst, D., Goodacre, R., Reinke, S. N., Kuligowski, J., Wilson, I. D., Lewis, M. R., et al. (2018). Guidelines and considerations for the use of system suitability and quality control samples in mass spectrometry assays applied in untargeted clinical metabolomic studies. *Metabolomics* 14, 72. doi: https://doi.org/10.1007/s11306-018-1367-3.
19. Castro-Moretti, F. R., Gentzel, I. N., Mackey, D., and Alonso, A. P. (2020). Metabolomics as an Emerging Tool for the Study of Plant-Pathogen Interactions. *Metabolites* 10. doi: https://doi.org/10.3390/metabo10020052.
20. Cha, S., Zhang, H., Ilarslan, H. I., Wurtele, E. S., Brachova, L., Nikolau, B. J., et al. (2008). Direct profiling and imaging of plant metabolites in intact tissues by using colloidal graphite-assisted laser desorption ionization mass spectrometry. *Plant J* 55, 348–360. doi: https://doi.org/10.1111/j.1365-313X.2008.03507.x.
21. Chang, H.-Y., Colby, S. M., Du, X., Gomez, J. D., Helf, M. J., Kechris, K., et al. (2021). A Practical Guide to Metabolomics Software Development. *Anal. Chem.* 93, 1912–1923. doi: https://doi.org/10.1021/acs.analchem.0c03581.
22. Chen, F., Ma, R., and Chen, X.-L. (2019). Advances of Metabolomics in Fungal Pathogen-Plant Interactions. *Metabolites* 9. doi: https://doi.org/10.3390/metabo9080169.
23. Chitarrini, G., Soini, E., Riccadonna, S., Franceschi, P., Zulini, L., Masuero, D., et al. (2017). Identification of Biomarkers for Defense Response to Plasmopara viticola in a Resistant Grape Variety. *Front. Plant Sci.* 8. doi: https://doi.org/10.3389/fpls.2017.01524.
24. Choi, Y. H., Tapias, E. C., Kim, H. K., Lefeber, A. W. M., Erkelens, C., Verhoeven, J. T. J., et al. (2004). Metabolic Discrimination of Catharanthus roseus Leaves Infected by Phytoplasma Using 1H-NMR Spectroscopy and Multivariate Data Analysis. *Plant Physiology* 135, 2398–2410. doi: https://doi.org/10.1104/pp.104.041012.
25. Chou, H.-M., Bundock, N., Rolfe, S. A., and Scholes, J. D. (2000). Infection of Arabidopsis thaliana leaves with Albugo candida (white blister rust) causes a reprogramming of host metabolism. *Molecular Plant Pathology* 1, 99–113. doi: https://doi.org/10.1046/j.1364-3703.2000.00013.x.
26. Colin, L., Martin-Arevalillo, R., Bovio, S., Bauer, A., Vernoux, T., Caillaud, M.-C., et al. (2022). Imaging the living plant cell: From probes to quantification. *Plant Cell* 34, 247–272. doi: https://doi.org/10.1093/plcell/koab237.

27. Crockford, D. J., Holmes, E., Lindon, J. C., Plumb, R. S., Zirah, S., Bruce, S. J., et al. (2006). Statistical heterospectroscopy, an approach to the integrated analysis of NMR and UPLC-MS data sets: application in metabonomic toxicology studies. *Anal Chem* 78, 363–371. doi: https://doi.org/10.1021/ac051444m.
28. DeVree, B. T., Steiner, L. M., Głazowska, S., Ruhnow, F., Herburger, K., Persson, S., et al. (2021). Current and future advances in fluorescence-based visualization of plant cell wall components and cell wall biosynthetic machineries. *Biotechnology for Biofuels* 14, 78. doi: https://doi.org/10.1186/s13068-021-01922-0.
29. Dietz, K.-J. (2017). Subcellular metabolomics: the choice of method depends on the aim of the study. *J Exp Bot* 68, 5695–5698. doi: https://doi.org/10.1093/jxb/erx406.
30. Diez-Navajas, A. M., Wiedemann-Merdinoglu, S., Greif, C., and Merdinoglu, D. (2008). Nonhost versus host resistance to the grapevine downy mildew, Plasmopara viticola, studied at the tissue level. *Phytopathology* 98, 776–780.
31. Dixon, R., and Paiva, N. (1995). Stress-Induced Phenylpropanoid Metabolism. *Plant Cell* 7, 1085–1097.
32. Dong, Y., and Aharoni, A. (2022). Image to insight: exploring natural products through mass spectrometry imaging. *Nat Prod Rep* 39, 1510–1530. doi: https://doi.org/10.1039/d2np00011c.
33. Dong, Y., Li, B., Malitsky, S., Rogachev, I., Aharoni, A., Kaftan, F., et al. (2016). Sample Preparation for Mass Spectrometry Imaging of Plant Tissues: A Review. *Front. Plant Sci.* 7. doi: https://doi.org/10.3389/fpls.2016.00060.
34. Emwas, A.-H., Roy, R., McKay, R. T., Tenori, L., Saccenti, E., Gowda, G. A. N., et al. (2019). NMR Spectroscopy for Metabolomics Research. *Metabolites* 9, 123. doi: https://doi.org/10.3390/metabo9070123.
35. Fan, R., Tao, X., Xia, Z., Sim, S., Hu, L., Wu, B., et al. (2022). Comparative Transcriptome and Metabolome Analysis of Resistant and Susceptible Piper Species Upon Infection by the Oomycete Phytophthora Capsici. *Frontiers in Plant Science* 13. Available at: https://www.frontiersin.org/articles/10.3389/fpls.2022.864927 [Accessed January 15, 2023].
36. Fang, Y., and Ramasamy, R. P. (2015). Current and Prospective Methods for Plant Disease Detection. *Biosensors* 5, 537–561. doi: https://doi.org/10.3390/bios5030537.
37. Fernie, A. R., and Tohge, T. (2017). The Genetics of Plant Metabolism. *Annual Review of Genetics* 51, 287–310. doi: https://doi.org/10.1146/annurev-genet-120116-024640.
38. Fiehn, O. (2002). Metabolomics – the link between genotypes and phenotypes. *Plant Mol Biol* 48, 155–171. doi: https://doi.org/10.1023/A:1013713905833.
39. Fiehn, O., Kopka, J., Dörmann, P., Altmann, T., Trethewey, R. N., and Willmitzer, L. (2000). Metabolite profiling for plant functional genomics. *Nat Biotechnol* 18, 1157–1161. doi: https://doi.org/10.1038/81137.
40. Figueiredo, J., Cavaco, A. R., Guerra-Guimarães, L., Leclercq, C., Renaut, J., Cunha, J., et al. (2021). An apoplastic fluid extraction method for the characterization of grapevine leaves proteome and metabolome from a single sample. *Physiol Plantarum* 171, 343–357. doi: https://doi.org/10.1111/ppl.13198.
41. Freund, D. M., Sammons, K. A., Makunga, N. P., Cohen, J. D., and Hegeman, A. D. (2018). Leaf Spray Mass Spectrometry: A Rapid Ambient Ionization Technique to Directly Assess Metabolites from Plant Tissues. *J Vis Exp*, 57949. doi: https://doi.org/10.3791/57949.
42. Galeano Garcia, P., Neves dos Santos, F., Zanotta, S., Eberlin, M. N., and Carazzone, C. (2018). Metabolomics of Solanum lycopersicum Infected with Phytophthora infestans Leads to Early Detection of Late Blight in Asymptomatic Plants. *Molecules* 23, 3330. doi: https://doi.org/10.3390/molecules23123330.
43. Gebauer, P., Korn, M., Engelsdorf, T., Sonnewald, U., Koch, C., and Voll, L. M. (2017). Sugar Accumulation in Leaves of Arabidopsis sweet11/sweet12 Double Mutants Enhances Priming of the Salicylic Acid-Mediated Defense Response. *Front Plant Sci* 8. doi: https://doi.org/10.3389/fpls.2017.01378.

44. Gemperline, E., Keller, C., and Li, L. (2016). Mass Spectrometry in Plant-omics. *Anal. Chem.* 88, 3422–3434. doi: https://doi.org/10.1021/acs.analchem.5b02938.
45. Geng, X., Cheng, J., Gangadharan, A., and Mackey, D. (2012). The Coronatine Toxin of Pseudomonas syringae Is a Multifunctional Suppressor of Arabidopsis Defense. *The Plant Cell* 24, 4763–4774. doi: https://doi.org/10.1105/tpc.112.105312.
46. Geng, X., Jin, L., Shimada, M., Kim, M. G., and Mackey, D. (2014). The phytotoxin coronatine is a multifunctional component of the virulence armament of Pseudomonas syringae. *Planta* 240, 1149–1165. doi: https://doi.org/10.1007/s00425-014-2151-x.
47. Gowda, G. A. N., and Djukovic, D. (2014). Overview of Mass Spectrometry-Based Metabolomics: Opportunities and Challenges. *Methods Mol Biol* 1198, 3–12. doi: https://doi.org/10.1007/978-1-4939-1258-2_1.
48. Grassl, J., Taylor, N. L., and Millar, A. H. (2011). Matrix-assisted laser desorption/ionisation mass spectrometry imaging and its development for plant protein imaging. *Plant Methods* 7, 21. doi: https://doi.org/10.1186/1746-4811-7-21.
49. Gross, J. H. (2017). *Mass Spectrometry: A Textbook*. 3rd ed. Springer International Publishing doi: https://doi.org/10.1007/978-3-319-54398-7.
50. Hall, R., Beale, M., Fiehn, O., Hardy, N., Sumner, L., and Bino, R. (2002). Plant Metabolomics: The Missing Link in Functional Genomics Strategies. *Plant Cell* 14, 1437–1440. doi: https://doi.org/10.1105/tpc.140720.
51. Hamm, G., Carré, V., Poutaraud, A., Maunit, B., Frache, G., Merdinoglu, D., et al. (2010). Determination and imaging of metabolites from *Vitis vinifera* leaves by laser desorption/ionisation time-of-flight mass spectrometry: Imaging of metabolites from *Vitis vinifera* leaves. *Rapid Commun. Mass Spectrom.* 24, 335–342. doi: https://doi.org/10.1002/rcm.4395.
52. Hasan, N., Choudhary, S., Naaz, N., Sharma, N., and Laskar, R. A. (2021). Recent advancements in molecular marker-assisted selection and applications in plant breeding programmes. *Journal of Genetic Engineering and Biotechnology* 19, 128. doi: https://doi.org/10.1186/s43141-021-00231-1.
53. Heyman, H. M., and Dubery, I. A. (2016). The potential of mass spectrometry imaging in plant metabolomics: a review. *Phytochem Rev* 15, 297–316. doi: https://doi.org/10.1007/s11101-015-9416-2.
54. Hiraoka, K. (2013). *Fundamentals of Mass Spectrometry*. Springer Science & Business Media.
55. Hupp, S., Rosenkranz, M., Bonfig, K., Pandey, C., and Roitsch, T. (2019). Noninvasive Phenotyping of Plant–Pathogen Interaction: Consecutive In Situ Imaging of Fluorescing Pseudomonas syringae, Plant Phenolic Fluorescence, and Chlorophyll Fluorescence in Arabidopsis Leaves. *Frontiers in Plant Science* 10. Available at: https://www.frontiersin.org/article/10.3389/fpls.2019.01239 [Accessed March 8, 2022].
56. Jacyna, J., Kordalewska, M., and Markuszewski, M. J. (2019). Design of Experiments in metabolomics-related studies: An overview. *Journal of Pharmaceutical and Biomedical Analysis* 164, 598–606. doi: https://doi.org/10.1016/j.jpba.2018.11.027.
57. Jeandet, P., Hébrard, C., Deville, M.-A., Cordelier, S., Dorey, S., Aziz, A., et al. (2014). Deciphering the Role of Phytoalexins in Plant-Microorganism Interactions and Human Health. *Molecules* 19, 18033–18056. doi: https://doi.org/10.3390/molecules191118033.
58. Jeandet, P., Vannozzi, A., Sobarzo-Sánchez, E., Uddin, M. S., Bru, R., Martínez-Márquez, A., et al. (2021). Phytostilbenes as agrochemicals: biosynthesis, bioactivity, metabolic engineering and biotechnology. *Nat. Prod. Rep.* 38, 1282–1329. doi: https://doi.org/10.1039/D0NP00030B.
59. Johansson, O. N., Nilsson, A. K., Gustavsson, M. B., Backhaus, T., Andersson, M. X., and Ellerström, M. (2015). A quick and robust method for quantification of the hypersensitive response in plants. *PeerJ* 3, e1469. doi: https://doi.org/10.7717/peerj.1469.
60. Jones, J. D. G., and Dangl, J. L. (2006). The plant immune system. *Nature* 444, 323–329. doi: https://doi.org/10.1038/nature05286.

61. Kang, Y., Zhou, M., Merry, A., and Barry, K. (2020). Mechanisms of powdery mildew resistance of wheat – a review of molecular breeding. *Plant Pathology* 69, 601–617. doi: https://doi.org/10.1111/ppa.13166.
62. Kanwar, P., and Jha, G. (2019). Alterations in plant sugar metabolism: signatory of pathogen attack. *Planta* 249, 305–318. doi: https://doi.org/10.1007/s00425-018-3018-3.
63. Kaspar, S., Peukert, M., Svatos, A., Matros, A., and Mock, H.-P. (2011). MALDI-imaging mass spectrometry – An emerging technique in plant biology. *Proteomics* 11, 1840–1850. doi: https://doi.org/10.1002/pmic.201000756.
64. Kemen, E., Mendgen, K., and Voegele, R. T. (2011). "Immunolocalization of Pathogen Effectors," in *Plant Immunity: Methods and Protocols* Methods in Molecular Biology., ed. J. M. McDowell (Totowa, NJ: Humana Press), 211–225. doi: https://doi.org/10.1007/978-1-61737-998-7_17.
65. Kim, H. K., and Verpoorte, R. (2010). Sample preparation for plant metabolomics. *Phytochem. Anal.* 21, 4–13. doi: https://doi.org/10.1002/pca.1188.
66. Kokesch-Himmelreich, J., Wittek, O., Race, A. M., Rakete, S., Schlicht, C., Busch, U., et al. (2022). MALDI mass spectrometry imaging: From constituents in fresh food to ingredients, contaminants and additives in processed food. *Food Chemistry* 385, 132529. doi: https://doi.org/10.1016/j.foodchem.2022.132529.
67. Kopittke, P. M., Lombi, E., van der Ent, A., Wang, P., Laird, J. S., Moore, K. L., et al. (2020). Methods to Visualize Elements in Plants1[OPEN]. *Plant Physiology* 182, 1869–1882. doi: https://doi.org/10.1104/pp.19.01306.
68. Koutouan, C. E., Le Clerc, V., Suel, A., Hamama, L., Claudel, P., Halter, D., et al. (2023). Co-Localization of Resistance and Metabolic Quantitative Trait Loci on Carrot Genome Reveals Fungitoxic Terpenes and Related Candidate Genes Associated with the Resistance to Alternaria dauci. *Metabolites* 13, 71. doi: https://doi.org/10.3390/metabo13010071.
69. Laupheimer, S., Kurzweil, L., Proels, R., Unsicker, S. B., Stark, T. D., Dawid, C., et al. (2023). Volatile-mediated signalling in barley induces metabolic reprogramming and resistance against the biotrophic fungus Blumeria hordei. *Plant Biology* 25, 72–84. doi: https://doi.org/10.1111/plb.13487.
70. Li, Z., Lu, Y., Guo, Y., Cao, H., Wang, Q., and Shui, W. (2018). Comprehensive evaluation of untargeted metabolomics data processing software in feature detection, quantification and discriminating marker selection. *Analytica Chimica Acta* 1029, 50–57. doi: https://doi.org/10.1016/j.aca.2018.05.001.
71. Lievens, B., Brouwer, M., Vanachter, A. C. R. C., Cammue, B. P. A., and Thomma, B. P. H. J. (2006). Real-time PCR for detection and quantification of fungal and oomycete tomato pathogens in plant and soil samples. *Plant Science* 171, 155–165. doi: https://doi.org/10.1016/j.plantsci.2006.03.009.
72. Lin, F., Chhapekar, S. S., Vieira, C. C., Da Silva, M. P., Rojas, A., Lee, D., et al. (2022). Breeding for disease resistance in soybean: a global perspective. *Theor Appl Genet* 135, 3773–3872. doi: https://doi.org/10.1007/s00122-022-04101-3.
73. Liu, J., Wang, H., Cooks, R. G., and Ouyang, Z. (2011). Leaf Spray: Direct Chemical Analysis of Plant Material and Living Plants by Mass Spectrometry. *Anal. Chem.* 83, 7608–7613. doi: https://doi.org/10.1021/ac2020273.
74. Liu, Y.-H., Song, Y.-H., and Ruan, Y.-L. (2022). Sugar conundrum in plant–pathogen interactions: roles of invertase and sugar transporters depend on pathosystems. *Journal of Experimental Botany* 73, 1910–1925. doi: https://doi.org/10.1093/jxb/erab562.
75. López, M. M., Bertolini, E., Olmos, A., Caruso, P., Gorris, M. T., Llop, P., et al. (2003). Innovative tools for detection of plant pathogenic viruses and bacteria. *Int Microbiol* 6, 233–243. doi: https://doi.org/10.1007/s10123-003-0143-y.
76. Lucas, J. A., Hawkins, N. J., and Fraaije, B. A. (2015). "Chapter Two – The Evolution of Fungicide Resistance," in *Advances in Applied Microbiology*, eds. S. Sariaslani and G. M. Gadd (Academic Press), 29–92. doi: https://doi.org/10.1016/bs.aambs.2014.09.001.

77. Maia, M., Carré, V., Aziz, A., and Jeandet, P. (2022a). Molecular Localization of Phytoalexins at the Micron Scale: Toward a Better Understanding of Plant-Phytoalexin-Pathogen Dynamics. *J. Agric. Food Chem.* 70, 9243–9245. doi: https://doi.org/10.1021/acs.jafc.2c04208.
78. Maia, M., Figueiredo, A., Cordeiro, C., and Sousa Silva, M. (2021). FT-ICR-MS-based metabolomics: A deep dive into plant metabolism. *Mass Spectrometry Reviews* n/a, 1–22. doi: https://doi.org/10.1002/mas.21731.
79. Maia, M., McCann, A., Malherbe, C., Far, J., Cunha, J., Eiras-Dias, J., et al. (2022b). Grapevine leaf MALDI-MS imaging reveals the localisation of a putatively identified sucrose metabolite associated to Plasmopara viticola development. *Frontiers in Plant Science* 13. Available at: https://www.frontiersin.org/articles/10.3389/fpls.2022.1012636 [Accessed January 15, 2023].
80. Maia, M., Monteiro, F., Sebastiana, M., Marques, A. P., Ferreira, A. E. N., Freire, A. P., et al. (2016). Metabolite extraction for high-throughput FTICR-MS-based metabolomics of grapevine leaves. *EuPA Open Proteom* 12, 4–9. doi: https://doi.org/10.1016/j.euprot.2016.03.002.
81. Misra, B. B. (2021). New software tools, databases, and resources in metabolomics: updates from 2020. *Metabolomics* 17, 49. doi: https://doi.org/10.1007/s11306-021-01796-1.
82. Moco, S. (2022). Studying Metabolism by NMR-Based Metabolomics. *Frontiers in Molecular Biosciences* 9. Available at: https://www.frontiersin.org/articles/10.3389/fmolb.2022.882487 [Accessed January 15, 2023].
83. Morkunas, I., and Ratajczak, L. (2014). The role of sugar signaling in plant defense responses against fungal pathogens. *Acta Physiol Plant* 36, 1607–1619. doi: https://doi.org/10.1007/s11738-014-1559-z.
84. Mushtaq, M. Y., Choi, Y. H., Verpoorte, R., and Wilson, E. G. (2014). Extraction for Metabolomics: Access to The Metabolome. *Phytochemical Analysis* 25, 291–306. doi: https://doi.org/10.1002/pca.2505.
85. Nelissen, H., and Gonzalez, N. (2020). Understanding plant organ growth: a multidisciplinary field. *Journal of Experimental Botany* 71, 7–10. doi: https://doi.org/10.1093/jxb/erz448.
86. Nunes-Nesi, A., Alseekh, S., de Oliveira Silva, F. M., Omranian, N., Lichtenstein, G., Mirnezhad, M., et al. (2019). Identification and characterization of metabolite quantitative trait loci in tomato leaves and comparison with those reported for fruits and seeds. *Metabolomics* 15, 46. doi: https://doi.org/10.1007/s11306-019-1503-8.
87. Oh, M., Park, S., Kim, H., Choi, G. J., and Kim, S. H. (2021). Application of UPLC-QTOF-MS Based Untargeted Metabolomics in Identification of Metabolites Induced in Pathogen-Infected Rice. *Plants* 10, 213. doi: https://doi.org/10.3390/plants10020213.
88. Olshina, M. A., and Sharon, M. (2016). Mass Spectrometry: A Technique of Many Faces. *Q Rev Biophys* 49, e18. doi: https://doi.org/10.1017/S0033583516000160.
89. Pedras, M. S. C., and Yaya, E. E. (2015). Plant chemical defenses: are all constitutive antimicrobial metabolites phytoanticipins? *Nat Prod Commun* 10, 209–218.
90. Perez de Souza, L., Alseekh, S., Naake, T., and Fernie, A. (2019). Mass Spectrometry-Based Untargeted Plant Metabolomics. *Current Protocols in Plant Biology* 4, e20100. doi: https://doi.org/10.1002/cppb.20100.
91. Piasecka, A., Jedrzejczak-Rey, N., and Bednarek, P. (2015). Secondary metabolites in plant innate immunity: conserved function of divergent chemicals. *New Phytologist* 206, 948–964. doi: https://doi.org/10.1111/nph.13325.
92. Pommerrenig, B., Müdsam, C., Kischka, D., and Neuhaus, H. E. (2020). Treat and trick: common regulation and manipulation of sugar transporters during sink establishment by the plant and the pathogen. *Journal of Experimental Botany* 71, 3930–3940. doi: https://doi.org/10.1093/jxb/eraa168.
93. Qi, K., Wu, L., Liu, C., and Pan, Y. (2021). Recent Advances of Ambient Mass Spectrometry Imaging and Its Applications in Lipid and Metabolite Analysis. *Metabolites* 11, 780. doi: https://doi.org/10.3390/metabo11110780.

94. Rajasundaram, D., and Selbig, J. (2016). More effort — more results: recent advances in integrative 'omics' data analysis. *Current Opinion in Plant Biology* 30, 57–61. doi: https://doi.org/10.1016/j.pbi.2015.12.010.
95. Razzaq, A., Sadia, B., Raza, A., Khalid Hameed, M., and Saleem, F. (2019). Metabolomics: A Way Forward for Crop Improvement. *Metabolites* 9. doi: https://doi.org/10.3390/metabo9120303.
96. Rolfe, S. A., and Scholes, J. D. (2010). Chlorophyll fluorescence imaging of plant–pathogen interactions. *Protoplasma* 247, 163–175. doi: https://doi.org/10.1007/s00709-010-0203-z.
97. Saccenti, E., Hoefsloot, H. C. J., Smilde, A. K., Westerhuis, J. A., and Hendriks, M. M. W. B. (2014). Reflections on univariate and multivariate analysis of metabolomics data. *Metabolomics* 10, 361–374. doi: https://doi.org/10.1007/s11306-013-0598-6.
98. Salem, M. A., Jüppner, J., Bajdzienko, K., and Giavalisco, P. (2016). Protocol: a fast, comprehensive and reproducible one-step extraction method for the rapid preparation of polar and semi-polar metabolites, lipids, proteins, starch and cell wall polymers from a single sample. *Plant Methods* 12, 45. doi: https://doi.org/10.1186/s13007-016-0146-2.
99. Sánchez-Martín, J., and Keller, B. (2019). Contribution of recent technological advances to future resistance breeding. *Theor Appl Genet* 132, 713–732. doi: https://doi.org/10.1007/s00122-019-03297-1.
100. Sarabia, L. D., Boughton, B. A., Rupasinghe, T., van de Meene, A. M. L., Callahan, D. L., Hill, C. B., et al. (2018). High-mass-resolution MALDI mass spectrometry imaging reveals detailed spatial distribution of metabolites and lipids in roots of barley seedlings in response to salinity stress. *Metabolomics* 14, 63. doi: https://doi.org/10.1007/s11306-018-1359-3.
101. Savoi, S., Arapitsas, P., Duchêne, É., Nikolantonaki, M., Ontañón, I., Carlin, S., et al. (2021). Grapevine and Wine Metabolomics-Based Guidelines for FAIR Data and Metadata Management. *Metabolites* 11, 757. doi: https://doi.org/10.3390/metabo11110757.
102. Savoi, S., Santiago, A., Orduña, L., and Matus, J. T. (2022). Transcriptomic and metabolomic integration as a resource in grapevine to study fruit metabolite quality traits. *Frontiers in Plant Science* 13. Available at: https://www.frontiersin.org/articles/10.3389/fpls.2022.937927 [Accessed January 15, 2023].
103. Schaad, N. W., and Frederick, R. D. (2002). Real-time PCR and its application for rapid plant disease diagnostics. *Canadian Journal of Plant Pathology* 24, 250–258. doi: https://doi.org/10.1080/07060660209507006.
104. Schauer, N., and Fernie, A. R. (2006). Plant metabolomics: towards biological function and mechanism. *Trends Plant Sci* 11, 508–516. doi: https://doi.org/10.1016/j.tplants.2006.08.007.
105. Schripsema, J., and Dagnino, D. (2019). "Metabolomics: Experimental Design, Methodology, and Data Analysis," in *Encyclopedia of Analytical Chemistry* (John Wiley & Sons, Ltd), 1–22. doi: https://doi.org/10.1002/9780470027318.a9939.pub2.
106. Schymanski, E. L., Jeon, J., Gulde, R., Fenner, K., Ruff, M., Singer, H. P., et al. (2014). Identifying Small Molecules via High Resolution Mass Spectrometry: Communicating Confidence. *Environ. Sci. Technol.* 48, 2097–2098. doi: https://doi.org/10.1021/es5002105.
107. Seneviratne, H. K., Dalisay, D. S., Kim, K.-W., Moinuddin, S. G. A., Yang, H., Hartshorn, C. M., et al. (2015). Non-host disease resistance response in pea (Pisum sativum) pods: Biochemical function of DRR206 and phytoalexin pathway localization. *Phytochemistry* 113, 140–148. doi: https://doi.org/10.1016/j.phytochem.2014.10.013.
108. Sgobba, E., Daguerre, Y., and Giampà, M. (2021). Unravel the Local Complexity of Biological Environments by MALDI Mass Spectrometry Imaging. *Int J Mol Sci* 22, 12393. doi: https://doi.org/10.3390/ijms222212393.
109. Simko, I., Jia, M., Venkatesh, J., Kang, B.-C., Weng, Y., Barcaccia, G., et al. (2021). Genomics and Marker-Assisted Improvement of Vegetable Crops. *Critical Reviews in Plant Sciences* 40, 303–365. doi: https://doi.org/10.1080/07352689.2021.1941605.
110. Soares, M. S., da Silva, D. F., Forim, M. R., da Silva, M. F. das G. F., Fernandes, J. B., Vieira, P. C., et al. (2015). Quantification and localization of hesperidin and rutin in Citrus sinensis grafted on C. limonia after Xylella fastidiosa infection by HPLC-UV and MALDI

imaging mass spectrometry. *Phytochemistry* 115, 161–170. doi: https://doi.org/10.1016/j.phytochem.2015.02.011.

111. Solfanelli, C., Poggi, A., Loreti, E., Alpi, A., and Perata, P. (2006). Sucrose-Specific Induction of the Anthocyanin Biosynthetic Pathway in Arabidopsis. *Plant Physiology* 140, 637–646. doi: https://doi.org/10.1104/pp.105.072579.
112. Sotelo, T., Soengas, P., Velasco, P., Rodríguez, V. M., and Cartea, M. E. (2014). Identification of Metabolic QTLs and Candidate Genes for Glucosinolate Synthesis in Brassica oleracea Leaves, Seeds and Flower Buds. *PLOS ONE* 9, e91428. doi: https://doi.org/10.1371/journal.pone.0091428.
113. Spicer, R. A., and Steinbeck, C. (2017). A lost opportunity for science: journals promote data sharing in metabolomics but do not enforce it. *Metabolomics* 14, 16. doi: https://doi.org/10.1007/s11306-017-1309-5.
114. Spoel, S. H., and Dong, X. (2012). How do plants achieve immunity? Defence without specialized immune cells. *Nature Reviews Immunology* 12, 89–100. doi: https://doi.org/10.1038/nri3141.
115. Sumner, L. W., Amberg, A., Barrett, D., Beale, M. H., Beger, R., Daykin, C. A., et al. (2007). Proposed minimum reporting standards for chemical analysis. *Metabolomics* 3, 211–221. doi: https://doi.org/10.1007/s11306-007-0082-2.
116. Sumner, L. W., Mendes, P., and Dixon, R. A. (2003). Plant metabolomics: large-scale phytochemistry in the functional genomics era. *Phytochemistry* 62, 817–836. doi: https://doi.org/10.1016/s0031-9422(02)00708-2.
117. Takahashi, K., Kozuka, T., Anegawa, A., Nagatani, A., and Mimura, T. (2015). Development and Application of a High-Resolution Imaging Mass Spectrometer for the Study of Plant Tissues. *Plant Cell Physiol* 56, 1329–1338. doi: https://doi.org/10.1093/pcp/pcv083.
118. Tata, A., Perez, C. J., Hamid, T. S., Bayfield, M. A., and Ifa, D. R. (2015). Analysis of Metabolic Changes in Plant Pathosystems by Imprint Imaging DESI-MS. *J. Am. Soc. Mass Spectrom.* 26, 641–648. doi: https://doi.org/10.1007/s13361-014-1039-0.
119. Teh, S. L., Rostandy, B., Awale, M., Luby, J. J., Fennell, A., and Hegeman, A. D. (2019). Genetic analysis of stilbenoid profiles in grapevine stems reveals a major mQTL hotspot on chromosome 18 associated with disease-resistance motifs. *Horticulture Research* 6, 1–11. doi: https://doi.org/10.1038/s41438-019-0203-x.
120. Tenenboim, H., and Brotman, Y. (2016). Omic Relief for the Biotically Stressed: Metabolomics of Plant Biotic Interactions. *Trends in Plant Science* 21, 781–791. doi: https://doi.org/10.1016/j.tplants.2016.04.009.
121. Thunig, J., Hansen, S. H., and Janfelt, C. (2011). Analysis of Secondary Plant Metabolites by Indirect Desorption Electrospray Ionization Imaging Mass Spectrometry. *Anal. Chem.* 83, 3256–3259. doi: https://doi.org/10.1021/ac2004967.
122. Tomita, M., and Nishioka, T. eds. (2005). *Metabolomics: The Frontier of Systems Biology*. Springer Japan doi: https://doi.org/10.1007/4-431-28055-3.
123. Ulaszewska, M. M., Weinert, C. H., Trimigno, A., Portmann, R., Andres Lacueva, C., Badertscher, R., et al. (2019). Nutrimetabolomics: An Integrative Action for Metabolomic Analyses in Human Nutritional Studies. *Molecular Nutrition & Food Research* 63, 1800384. doi: https://doi.org/10.1002/mnfr.201800384.
124. van den Berg, R. A., Hoefsloot, H. C., Westerhuis, J. A., Smilde, A. K., and van der Werf, M. J. (2006). Centering, scaling, and transformations: improving the biological information content of metabolomics data. *BMC genomics* 7, 142. doi: https://doi.org/10.1186/1471-2164-7-142.
125. Viant, M. R., Rosenblum, E. S., and Tieerdema, R. S. (2003). NMR-based metabolomics: a powerful approach for characterizing the effects of environmental stressors on organism health. *Environ Sci Technol* 37, 4982–4989. doi: https://doi.org/10.1021/es034281x.
126. Vissenberg, K., and Gonzalez, N. (2020). Plant organ and tip growth. *J Exp Bot* 71, 2363–2364. doi: https://doi.org/10.1093/jxb/eraa163.

127. Vuckovic, D. (2012). Current trends and challenges in sample preparation for global metabolomics using liquid chromatography-mass spectrometry. *Anal Bioanal Chem* 403, 1523–1548. doi: https://doi.org/10.1007/s00216-012-6039-y.
128. Wagner, G., Laperche, A., Lariagon, C., Marnet, N., Renault, D., Guitton, Y., et al. (2019). Resolution of quantitative resistance to clubroot into QTL-specific metabolic modules. *Journal of Experimental Botany* 70, 5375–5390. doi: https://doi.org/10.1093/jxb/erz265.
129. Wang, J., Yang, E., Chaurand, P., and Raghavan, V. (2021). Visualizing the distribution of strawberry plant metabolites at different maturity stages by MALDI-TOF imaging mass spectrometry. *Food Chemistry* 345, 128838. doi: https://doi.org/10.1016/j.foodchem.2020.128838.
130. Wang, M., Tian, Q., Li, H., Dai, L., Wan, Y., Wang, M., et al. (2023). Visualization and metabolome for the migration and distribution behavior of pesticides residue in after-ripening of banana. *Journal of Hazardous Materials* 446, 130665. doi: https://doi.org/10.1016/j.jhazmat.2022.130665.
131. Wilkinson, M. D., Dumontier, M., Aalbersberg, Ij. J., Appleton, G., Axton, M., Baak, A., et al. (2016). The FAIR Guiding Principles for scientific data management and stewardship. *Sci Data* 3, 160018. doi: https://doi.org/10.1038/sdata.2016.18.
132. William Allwood, J., Clarke, A., Goodacre, R., and Mur, L. A. J. (2010). Dual metabolomics: A novel approach to understanding plant–pathogen interactions. *Phytochemistry* 71, 590–597. doi: https://doi.org/10.1016/j.phytochem.2010.01.006.
133. Xue, C., Liu, Z., Dai, L., Bu, J., Liu, M., Zhao, Z., et al. (2018). Changing Host Photosynthetic, Carbohydrate, and Energy Metabolisms Play Important Roles in Phytoplasma Infection. *Phytopathology®* 108, 1067–1077. doi: https://doi.org/10.1094/PHYTO-02-18-0058-R.
134. Yang, F., Wu, C., Zhu, G., Yang, Q., Wang, K., and Li, Y. (2022a). An integrated transcriptomic and metabolomic analysis for changes in rose plant induced by rose powdery mildew and exogenous salicylic acid. *Genomics* 114, 110516. doi: https://doi.org/10.1016/j.ygeno.2022.110516.
135. Yang, J.-F., Chen, W.-J., Zhou, L.-M., Hewage, K. A. H., Fu, Y.-X., Chen, M.-X., et al. (2022b). Real-Time Fluorescence Imaging of the Abscisic Acid Receptor Allows Nondestructive Visualization of Plant Stress. *ACS Appl. Mater. Interfaces* 14, 28489–28500. doi: https://doi.org/10.1021/acsami.2c02156.
136. Yuan, M., Jiang, Z., Bi, G., Nomura, K., Liu, M., Wang, Y., et al. (2021). Pattern-recognition receptors are required for NLR-mediated plant immunity. *Nature* 592, 105–109. doi: https://doi.org/10.1038/s41586-021-03316-6.
137. Zhou, B., Xiao, J. F., Tuli, L., and Ressom, H. W. (2012). LC-MS-based metabolomics. *Mol Biosyst* 8, 470–481. doi: https://doi.org/10.1039/c1mb05350g.

Metabolomics in Fundamental Plant Research

José Juan Ordaz-Ortiz, Anita Arroyo-Silva, and Moisés Guerrero-Esperanza

Abbreviations

$[M + H]^+$	Protonated ion
^1H NMR	Proton nuclear magnetic resonance
2D-NMR	Bidimensional nuclear magnetic resonance
CE-MS	Capillary electrophoresis-mass spectrometry
cmQTL	Canalization metabolite quantitative trait loci
Da	Dalton
DART-MS	Direct analysis in real time-mass spectrometry
DEGs	Differentially expressed genes
DESI	Desorption electrospray ionization
FACS	Fluorescence-activated cell sorting
FT-ICR-MS	Fourier transform ion cyclotron resonance-mass spectrometry
GC	Gas chromatography
GC-MS	Gas chromatography-mass spectrometry
GC-QqQ-MS	Gas chromatography-triple quadrupole-mass spectrometry
GC-TOF-MS	Gas chromatography-time-of-flight mass spectrometry
GLC	Gas-liquid chromatography
HPLC	High-performance liquid chromatography
LC	Liquid chromatography
LC-ESI-MS	Liquid chromatography-electrospray-mass spectrometry
LC-HRMS	Liquid chromatography-high-resolution mass spectrometry
LCM	Laser-capture microdissection
LC-MS	Liquid chromatography-mass spectrometry
LMD	Laser microdissection
LMPC	Laser microdissection and pressure catapulting
m/z	Mass-to-charge ratio

J. J. Ordaz-Ortiz (✉) · A. Arroyo-Silva · M. Guerrero-Esperanza
Metabolomics and Mass Spectrometry Laboratory, Unidad de Genómica Avanzada, Irapuato, Gto, Mexico
e-mail: jose.ordaz.ortiz@cinvestav.mx

V. Soni, T. E. Hartman (eds.), *Metabolomics*,
https://doi.org/10.1007/978-3-031-39094-4_12

MALDI	Matrix-assisted laser desorption/ionization
mGWAS	Metabolite genome-wide association studies
mQTL	Metabolite quantitative trait loci
MS	Mass spectrometry
MS/MS	Tandem mass spectrometry, fragmentation
MSI	Mass spectrometry imaging
NMR	Nuclear magnetic resonance
PCA	Principal component analysis
PCR	Polymerase chain reaction
PLS-DA	Partial least squares discriminant analysis
SIMS	Secondary ion mass spectrometry
SNPs	Single-nucleotide polymorphisms
UHPLC-MS	Ultrahigh-pressure liquid chromatography-mass spectrometry
UPLC	Ultra-performance liquid chromatography
VOC	Volatile organic compounds
WGCNA	Correlation and weighted gene coexpression network analysis

1 Introduction

Plant metabolomics is an emerging research field focusing on the comprehensive analysis of metabolites, small molecules (<1500 Da) that play a critical role in plant growth, development, and response to environmental changes. By applying advanced analytical techniques, such as mass spectrometry and nuclear magnetic resonance spectroscopy, researchers can identify and quantify thousands of metabolites in a single sample, providing a holistic view of the metabolic pathways and processes that occur in plants. The insights gained from plant metabolomics research have broad implications for plant biology, agriculture, and biotechnology, as well as for human health and the environment.

Metabolomics is crucial to studying abiotic stress tolerance, pathogen resistance, robust ecotypes, and metabolic-assisted breeding of crops. The plant kingdom contains a considerable diversity of metabolites of approximately 200,000 compounds; the majority are still unknown. It is estimated that around 10,000 secondary metabolites have been discovered in different plant species [1], since metabolites have a wide range of physicochemical properties and functions. This makes it challenging for metabolomics techniques to study their diversity and gain insights into plant biology.

Since the word "metabolomics" was mentioned for the first time in the literature, it has evolved and been applied to many disciplines, such as plant biology. Plant metabolomics has become hugely modernized in the last decade. This chapter describes novel applications of mass spectrometry-based metabolomics approaches in recent fundamental plant research. We first commented on different metabolomics techniques, mass spectrometry, and nuclear magnetic resonance-based metabolomics. Mass spectrometry-based metabolomics is the widest in use; the main approaches are targeted or quantitative metabolomics and untargeted, nontargeted, or discovery metabolomics, also known as global metabolomics. We then explored contemporary literature on gene identification and their functional characterization for crop improvement enabled by applying nontargeted and targeted metabolomic analysis in combination with genome-wide association studies, metabolite quantitative trait loci, and transcriptomics.

From there, we explored the current application of metabolomics methods for plant species identification, a trendy topic that is enabling researchers to support many aspects of food authentication, food quality control, and traceability; plant species identification is an essential factor in understanding biodiversity, the discovery of bioactives from herbal medicines and correlating chemical components from plants with chemical markers of patients who intakes herbal medication, in the same manner monitoring food intake in foodomics studies.

We exemplified by commenting on the use of molecular networking analysis and its application to classify plant species; by providing an example of the Malpighiaceae family, chemotaxonomic studies guided by metabolomics methods become hugely in use as it allows rapid classification of plant samples based on the endogenous chemical content. We also discussed pioneering work on classifying and discriminating cinnamon, vanilla, and coffee plant species using different metabolomics techniques.

We then explored the impact of climate change on the root metabolome and the differences in root abiotic and metabolic changes associated with other biotic factors interactions, highlighting key metabolites involved in root exudates when exposed to these types of stresses. In addition, we reviewed recent literature on plant biomarker discovery outlining different application areas, such as the food industry, where the identification of biomarkers has also worked in quality processes for food authenticity and food traceability matrices of plant origin. We outlined a list of key metabolites identified in various plant species using different analytical techniques including metabolites detected in transgenic plants.

Finally, we briefly explored the emerging field of single-cell metabolomics methods. We describe the latest development in mass spectrometry imaging, including different approaches for collecting single-cell from plant tissues and a revision of some essential techniques on mass spectrometry imaging. Mention has also been made on the challenging and future needs for plant metabolomics research.

2 Novel Gene Identifications and Their Functional Characterization for Crop Improvement

Metabolomics studies all small molecules – metabolites – content of a cell or whole organism. Plant metabolomics refers to comprehensive, nonbiased, high-throughput analyses of complex metabolite mixtures typical of plant extracts. The role of metabolomics in such studies is twofold: (1) to identify the spatial and temporal

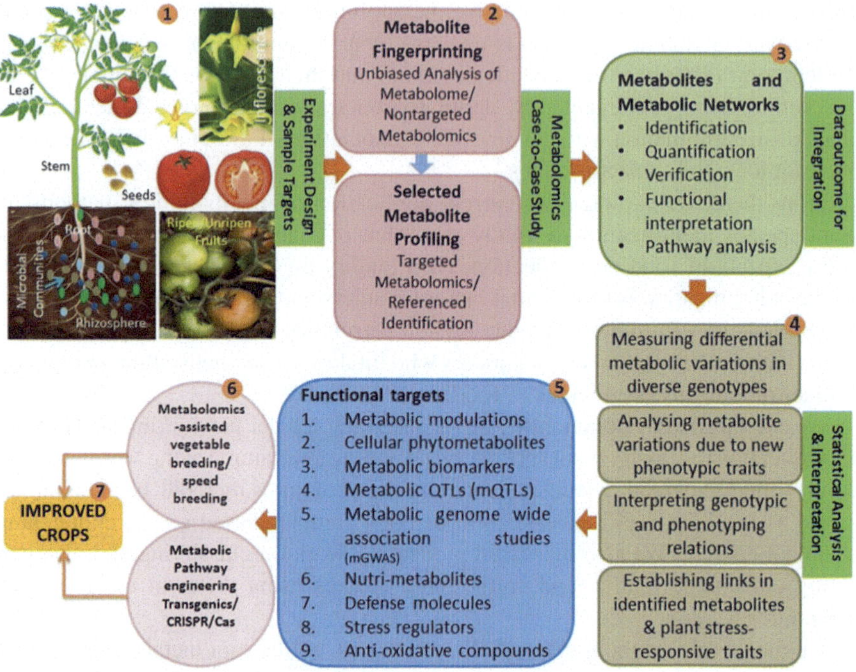

Fig. 1 Schematic presentation of metabolomics applications in crop improvement programs: (1) A representative sampling source of vegetable crop plants (tomato) as a biological source from which cellular metabolome can be extracted from almost all the plant parts and the rhizosphere under varying experimental environmental conditions; (2) data acquisition approach in metabolomics to be applied whether unbiased nontargeted fingerprinting is required or the analysis and quantification of a few selected target molecules is the need of the experiment; (3) study outcomes which needs biological interpretation for hypothesis questions; (4) possible answers to the hypothesis questions in the cellular chemistry and its entwining relations with the environmental impacts; (5) functional targets that could be achieved through metabolomics analysis of the vegetable plants; (6) result-oriented applications of the data outcomes in crop improvement practices; and (7) the "end product" of the experimental metabolomics exercise in vegetable crops. (Reproduced from Ref. [2])

distribution of the target compounds as influenced by plant development and environmental cues and (2) to identify related phytochemicals, which may be considered as either intermediate of biosynthesis or alternative or alternative products of promiscuous enzymes that support the biosynthesis of the target phytochemical [1].

This chapter describes the latest development and application of plant metabolomics in combination with metabolite genome-wide association studies (mGWAS), metabolite quantitative trait loci (mQTL), and transcriptomics for the discovery and characterization of genes and enzymes associated with the biosynthesis of specialized metabolites in significant crops such as maize, rice, and tomato (Fig. 1).

Different metabolomics techniques have been developed in the last two decades, including mass spectrometry-based metabolomics (MS); nuclear magnetic

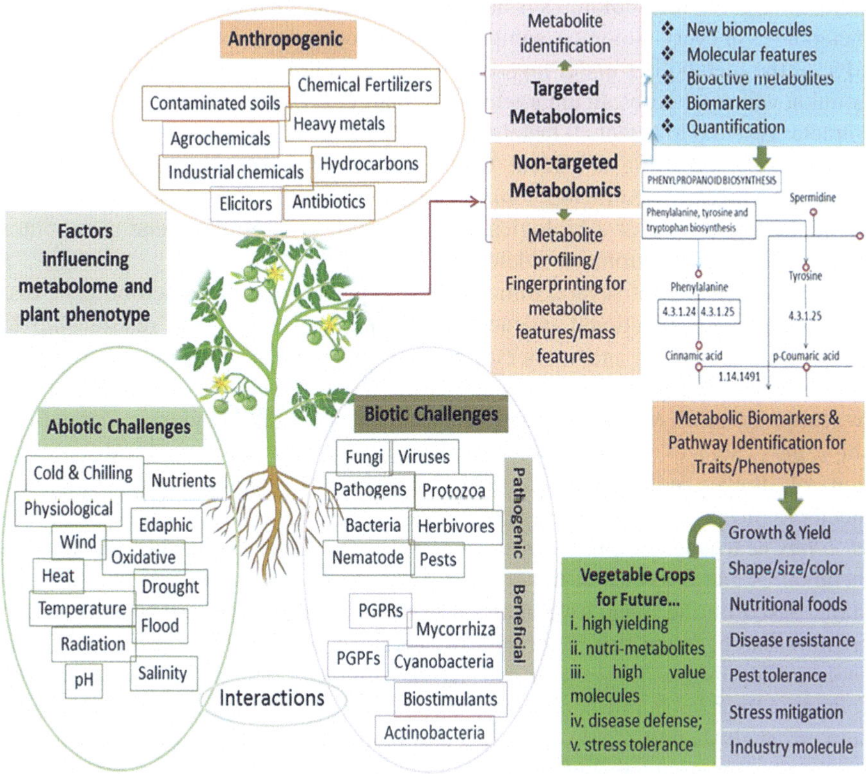

Fig. 2 Illustrative diagram of possible plant environmental interactions, which are supposed to influence the metabolic status of the crop plants. Analyzing the metabolome of plants exposed to such challenges using metabolomics approaches can yield competitive vegetables crops with better yield, high level of defense and stress-mitigating capabilities. (Reproduced from Ref. [2])

resonance spectroscopy (NMR); gas-chromatography-mass spectrometry (GC-MS); capillary electrophoresis-mass spectrometry (CE-MS); liquid chromatography-mass spectrometry (LC-MS); and more recently the implementation of high-resolution metabolomics with the aid of Fourier transform ion cyclotron resonance mass spectrometry (FT-ICR MS). Mass spectrometry-based metabolomics is one of the most used in plant metabolomics, where researchers can undertake two main approaches: targeted and untargeted metabolomics (Fig. 2). Targeted metabolomics is a hypothesis-driven approach focusing on a specific set of metabolites. This approach is often used when the researcher has prior knowledge about the metabolites of interest or when a specific metabolic pathway is under investigation. In targeted metabolomics, a set of known metabolites is selected, and the mass spectrometer is set up to detect these specific metabolites. Targeted metabolomics can provide more accurate and quantitative information about specific metabolites, but it requires prior knowledge of the metabolites under investigation; on the other hand, untargeted metabolomics is an exploratory approach that aims to identify all metabolites in every biological sample. This approach does not require prior knowledge of the metabolites under investigation, and it allows for detecting unexpected or unknown metabolites. In untargeted or nontargeted metabolomics, the mass spectrometer is set up to detect all metabolites within a certain mass range. The resulting data is analyzed using statistical methods and bioinformatics tools to identify metabolites. Untargeted metabolomics can provide a more comprehensive view of the metabolic profile of a biological sample. Still, it may miss some metabolites outside the mass spectrometer's detection range.

The combination of several omics has been recently implemented on gene discovery and their functional characterization to investigate gene relationship to metabolites supporting and accelerating crop improvements. For example, Wu et al. established a high-quality chromosome-level genome assembly of *Melilotus albus*, by resequencing 94 Melilotus accessions to characterize their phylogenetic relationships, the genetic exchange between *M. albus* and *M. officinalis*, and the differentiation of flower color and coumarin content. In addition, transcriptomics, metabolomics, and bulked segregant analysis (BSA) have been used to investigate *M. albus* near-isogenic lines segregating at the coumarin level to identify the key metabolites and enzymes in the coumarin biosynthesis pathway [3]. Similarly, Li et al., based on the integrative analysis of the transcriptomics and targeted carotenoid research, found that differentially expressed genes (DEGs) related to carotenoid metabolism had a stronger correlation with the critical carotenoid metabolite content in the panicle of foxtail millet. Correlation and weighted gene coexpression network analysis (WGCNA) identified and predicted the gene regulation network related to carotenoid metabolism [4].

Zheng et al. used a combination of metabolome and transcriptome of 11 tea cultivars and then a WGCNA-based biological system strategy to interpret metabolomic flux, predicted gene functions, and mined critical regulators involved in the flavonoid biosynthesis pathway; in this manner, they revealed new uncharacterized transcription factors (TFs) such as MADS, WRKYs, and SBP; and microRNAs (including 17 conserved and 15 novel microRNAs) that are potentially implicated in different steps of the catechin biosynthesis. In addition, they applied the metabolic-signature-based association method to capture additional critical regulators involved in the catechin pathway. This provides important clues for the functional characterization of five SCPL1A acyltransferase family members, which might be implicated in the production balance of anthocyanins, galloylated catechins, and proanthocyanins in tea cultivars [5].

Another approach has been implementing metabolite genome-wide association studies (mGWAS) and metabolite quantitative trait loci (mQTL), especially in cereal grains such as wheat. Chen et al. developed an approach that has also been applied in other major crops such as maize, barley, tomato, and blueberry, while the mQTL approach has been used in crops sunch as soybean, rice, and carrot [6]. Traditionally the output of mQTL/mGWAS was merely linkages/associations between chromosomal locations and metabolite contents, and this was basically due to the lack of genomic information for some crops until very recently, a wheat mGWAS successfully identified 26 candidate genes with high confidence, among them two were validated to be involved in flavonoid metabolism pathway of wheat [6, 7].

This way, by combining interval mapping and genome-wide association studies (GWAS), the genetic determinants of tocochromanol accumulation in tomato (*Solanum lycopersicum*) fruits have been unveiled. Specifically, the content of vitamin E has been enhanced in tomato plants by identifying the genes involved in the chorismate-tyrosine pathway [8]. With this approach, Alseekh et al. reported a large-scale metabolic quantitative trait loci (mQTL) analysis on the well-characterized *Solanum pennellii* introgression lines to investigate the genomic regions associated with secondary metabolism in tomato fruit pericarp [9]. In total, 679 mQTLs were detected across the 76 introgression lines. Heritability analyses revealed that the environment affected mQTLs of secondary metabolism less than mQTLs of primary metabolism. Network analysis allowed to assess the interconnectivity of primary and secondary metabolism and compare their respective associations with morphological traits. Additionally, a real-time quantitative PCR platform demonstrated a transcriptional control mechanism of a subset of the mQTLs, including those for hydroxycinnamates, acyl-sugar, naringenin chalcone, and a range of glycoalkaloids. Intriguingly, many of these compounds displayed a dominant-negative mode of inheritance, contrary to the conventional wisdom that secondary metabolite contents decreased on domestication. Additionally, two candidate genes for glycoalkaloid mQTLs via virus-induced gene silencing were also validated [9].

More recently, Alseekh et al. identified several metabolite quantitative trait loci that reduce variability for both primary and secondary metabolites (phenylalanine, glucose-6-P, fructose-6-P, and maltose), which they named canalization metabolite quantitative trait loci (cmQTL); on their study nine cmQTL were validated using an independent population of backcross inbred lines, derived from the same parents, which allows increased resolution in mapping the QTL previously identified in the introgression lines. These cmQTL showed little overlap with QTL for the metabolite levels themselves. Moreover, the intervals they mapped harbor few metabolism-associated genes, suggesting that regulatory genes largely control the canalization of metabolism [10].

Maize has also been favored by the combination of GWAS and metabolomics profiling to highlight genes involved in the biosynthesis of several metabolites. Among these compounds, Liang et al. reported the identification of metabolites biomarkers for the tolerance to salt-induced osmotic stress; a citrate synthase, a glucosyltransferase, and a cytochrome P450 were found to be responsible for controlling the associations between the genotype and metabolites that induced the tolerance [11]. Owens et al. reported essential genes controlling maize grain carotenoid composition by using GWAS of quantified seed carotenoids across a panel of maize inbreeds ranging from light yellow to dark orange in grain color; significant associations at the genome-wide level were detected within the coding region of zep1 and lut1, carotenoid biosynthetic genes not previously shown to impact grain carotenoid composition in association studies, as well as within previously associated lcyE and crtRB1 genes [12]. Similarly, Li et al. examined the genetic architecture of maize oil biosynthesis in a genome-wide association study using 1.03 million SNPs characterized in 368 inbred maize lines, including "high-oil" lines. Seventy-four loci were significantly associated with kernel oil concentration and fatty acid composition [4]. Another combination of maize metabolic profiling with GWAS has been reported by Riedelsheimer et al. who outlined an association of genetic variants and concentration of 118 metabolites in leaves of 289 diverse maize inbred lines from worldwide sources; genome-wide association mapping with correction for population structure and cryptic relatedness identified for 26 distinct metabolites with strong associations with SNPs explains up to 32.0% of the observed genetic variance [13]. Similarly, Chen et al. described a comprehensive profiling of 840 metabolites and a further metabolic genome-wide association study based on ~6.4 million SNPs obtained from 529 diverse accessions of *Oryza sativa*. They identified hundreds of common variants influencing numerous secondary metabolites with significant effects at high resolution. They observed substantial heterogeneity in the natural variation of metabolites and their underlying genetic architectures among different rice subspecies [13, 14]. Data mining identified 36 candidate genes modulating metabolite levels that are potentially physiological and nutritionally important. As a proof of concept, they functionally identified (annotated) five candidate genes influencing metabolic traits; the study provides first-time insights into the genetic and biochemical bases of rice metabolome variation and can be used as a powerful complementary tool to classical phenotypic trait mapping for rice improvement. Besides, Dong et al. reported a comprehensive metabolic profiling and natural variation analysis of phenolamides in rice using a liquid chromatography (LC)-mass spectrometry (MS)-based targeted metabolomics method; spatiotemporal controlled accumulations were observed for most phenolamides, together with their differential accumulations between the two major subspecies of rice.

Further GWAS on rice leaves identified *Os12g27220* and *Os12g27254* as spermidine hydroxycinnamoyl transferases that might underlie the natural variation of spermidine conjugate levels in rice [15]. Likewise, Chen et al. identified 32 candidate genes underlying metabolic traits in rice grains; 8 candidate genes were involved in the biosynthesis and transportation of amino acids and their derivatives. Three candidates were assigned to the choline levels and its lysophosphatidyl derivatives. Precise signals for trigonelline, a bioactive compound implicated in cell cycle control, resulted in the assignment of seven candidate genes for this metabolite. Furthermore, mGWAS in rice grains revealed 40 candidates (both regulatory and structural genes) involved in the biosynthesis, modification, and transportation of phenylpropanoids, including the C-glycosyl flavones, the primary class of flavonoids in cereals [7]. More recently, Yang et al. reported the identification of a gene s07g32020 (UGT707A3) that encodes a glucosyltransferase that converts naringenin and uridine diphosphate-glucose to naringenin-7-O-β-D-glucoside; the function of Os07g32020 was verified with CRISPR/Cas9 mutant lines, which accumulated more naringenin and less naringenin-7-O-β-D-glucoside and apigenin-7-O-β-D-glucoside than wild-type Nipponbare [16].

It is evident that when working with a crop with a reference genome sequence, such as maize, rice, and wheat, to mention some of the global agricultural importance, enormous advances have been made in elucidating genes and their functional annotation for their improvements; sadly this is not the case for minor crops, medicinal plants, and other regional staple foods. Therefore, considerable effort must be made to genome sequence minor crops and develop a collection of segregating crop populations. On the other hand, plant metabolome lacks a free online accessible metabolite database for support in the annotation of unknown metabolites, and most of the studies carried out under a nontargeted metabolomic approach remain with several novel metabolites, making association studies such as transcriptomics, mQTL, and GWAS difficult when searching for gene annotation and their functional characterization. A summary of mQTL and mGWAS is presented in Table 1.

Table 1 Summary of mQTL and mGWAS studies in plants [17]

Species	Tissue	Population type	Method	Metabolic traits	Ref.
mQTL study					
Arabidopsis	Harvested seed	Recombinant inbred lines	HPLC	Tocopherol	[18]
Arabidopsis	Leaf	Recombinant inbred lines	GC-TOF-MS	Metabolome	[19]
Arabidopsis	Seed	Recombinant inbred lines	LC-MS	Flavonoids	[20]
Arabidopsis	Seedling	RILs and introgression lines	GC-TOF-MS	Metabolome	[21]
Brassica napus	Leaf/seed	Double haploid lines	HPLC	Glucosinolates	[22]
Maize	Leaf	RILS and natural accessions	GC-TOF-MS	Primary metabolites	[23]
Rice	Seed	Chromosomal segment substitution lines	LC-Q-TOF-MS	Metabolome	[24]
Rice	Seed	F2, F2-derived lines	GC-MS	Lipids	[25]
Rice	Flag leaf/ germinating seed	RILs	LC-EI-MS	Metabolome	[26]
Tomato	Fruit	Introgression lines	GC-MS	Metabolome	[27]
Tomato	Fruit	Introgression lines	GC-MS	Metabolome	[28]
Tomato	Fruit	Introgression lines	LC-MS	Metabolome	[29]
Tomato	Fruit	Introgression lines	GC-MS	Primary metabolites	[30]
Tomato	Fruit	Introgression lines	UPLC	Secondary metabolites	[9]
Wheat	Flag leaf	Doubled haploid lines	LC-ESI-MS	Metabolome	[31]
Wheat	Flag leaf	Doubled haploid lines	GC-MS	Metabolome	[32]

(continued)

Table 1 (continued)

Species	Tissue	Population type	Method	Metabolic traits	Ref.
mGWAS study					
Arabidopsis	Seed	Natural accessions	LC-MS	Branched-chain amino acids	[33]
Arabidopsis	Leaf/seedling	Natural accessions	LC-MS	Glucosinolates	[34]
Arabidopsis	Leaf	Natural accessions	GC-TOF-MS	Metabolome	[35]
Maize	Kernel	Natural accessions	UPLC-MS	Metabolome	[36]
Maize	Grain	Natural accessions	HPLC	Carotenoid	[12]
Maize	Grain	Natural accessions	HPLC	Tocochromanol	[37]
Maize	Leaf	Natural accessions	GC-MS	Metabolome	[38]
Maize	Leaf	Natural accessions	GC-MS	Metabolome	[13]
Maize	Kernel	Natural accessions	LC-MS	Metabolome	[39]
Potato	Tuber	Natural accessions	GC-MS	Primary metabolites	[40]
Rice	Leaf	Natural accessions	LC-QTOF-MS	Secondary metabolites	[41]
Rice	Leaf	Natural accessions	LC-MS	Metabolome	[14]
Rice	Leaf	Natural accessions	LC-MS	Phenolamides	[15]
Tomato	Fruit	Natural accessions	GC-MS	Metabolome	[42]

3 Plant Species Identification

The plant metabolome constitutes an enormous reservoir of bioactive compounds; many of these are products of secondary or specialized metabolism. Their taxonomic distribution is in relatively narrow phylogenetic clades within *Plantae* [1]. While the identification of plant species based on morphological characteristics is a well-established practice in botany, being able to identify plant species is an essential factor in understanding biodiversity, the discovery of bioactives from herbal medicines, and correlating chemical components from plants with chemical markers of patients who intake herbal medication, as well as monitoring food intake in foodomics studies, food authentication, and fraud detection, among other applications. Since the small-molecule profile of an organism ultimately reflects the genes that distinguish it, the information content of the metabolome might be just as well suited to genomic fingerprinting and assessment of genetic relatedness between species as the genomes themselves [43].

Despite its practical importance, the establishment of phylogenetic diversification and distribution patterns of secondary plant metabolites is still in its early steps, and several plant families have not been deeply explored to date in this context; Mannochio-Russo et al. described a strategy for chemotaxonomic investigations using the Malpighiaceae botanical family as a model; their workflow (Fig. 3) was based on MS/MS untargeted metabolomics, spectral searches, and recently described in silico classification tools, which were mapped into the latest molecular phylogeny accepted for this family, the workflow combines several approaches to

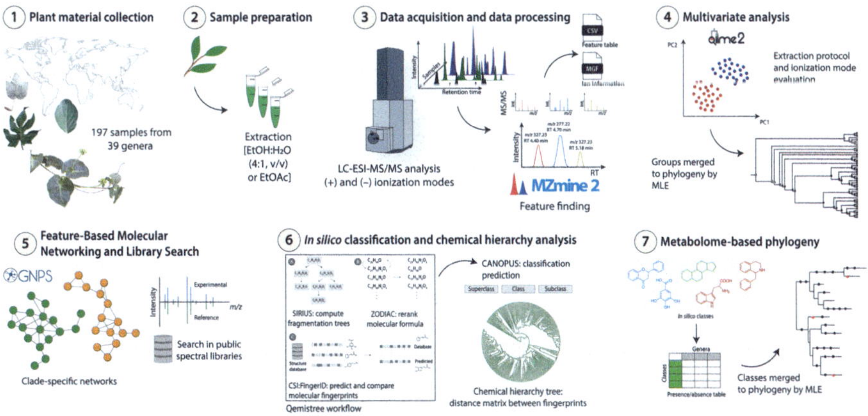

Fig. 3 Experimental workflow followed for the metabolomics and chemosystematics analyses of Malpighiaceae samples. (1) The samples were initially collected. (2) The extracts were prepared with different solvents ($EtOH:H_2O$ (4:1, v/v) or EtOAc) and then (3) subjected to LC-ESI-MS/MS analysis in positive and negative ionization modes in an untargeted method. (4) The data acquired were processed for feature finding, and the exported data were used for multivariate analysis. The clustering groups observed were merged to the phylogeny using the maximum likelihood estimation (MLE) for preliminary chemotaxonomic investigations. (5) The data were also used for feature-based molecular networking and library searches workflows to observe clade-specific molecular families. (6) A chemical hierarchy analysis and in silico classifications were obtained and finally (7) merged to the currently accepted Malpighiaceae phylogeny to determine the ubiquitous and the taxa-specific in silico classes. (Reproduced from Ref. [44])

perform a comprehensive evolutionary chemical study. It is expected to be used in further chemotaxonomic investigations [44]

The metabolomic analysis revealed that different ionization modes and extraction protocols significantly impacted the chemical profiles, influencing the chemotaxonomic results. In addition to the library searches for metabolite annotation, the MS/MS data generated were visualized by molecular networking analysis (Fig. 4). Molecular families constructed by such analysis represent the similarity of fragmentation patterns obtained by tandem mass spectrometry (MS/MS) analysis. These molecular families consisted of nodes (representing MS/MS spectra) and edges of connecting these nodes (representing the cosine similarity between two nodes, which measure the relatedness in MS/MS spectra). The library matches retrieved from the analysis obtained in the positive ionization mode showed the presence of a high diversity of compound classes, including *C*-glycosylated and *O*-glycosylated flavonoids, lipids, alkaloids, quinic acid derivatives, amides, triterpenes, iridoids, and lignans [44].

Cinnamon is one of the oldest spices used in the world. A growing number of studies have illustrated varied phytochemical compositions among cinnamon species. Primary cinnamon metabolites, such as coumarin, cinnamaldehyde, cinnamic acid, cinnamyl alcohols, and proanthocyanidins, are shown to be differentially produced among various species; in this context, Zhang et al. developed a metabolomic

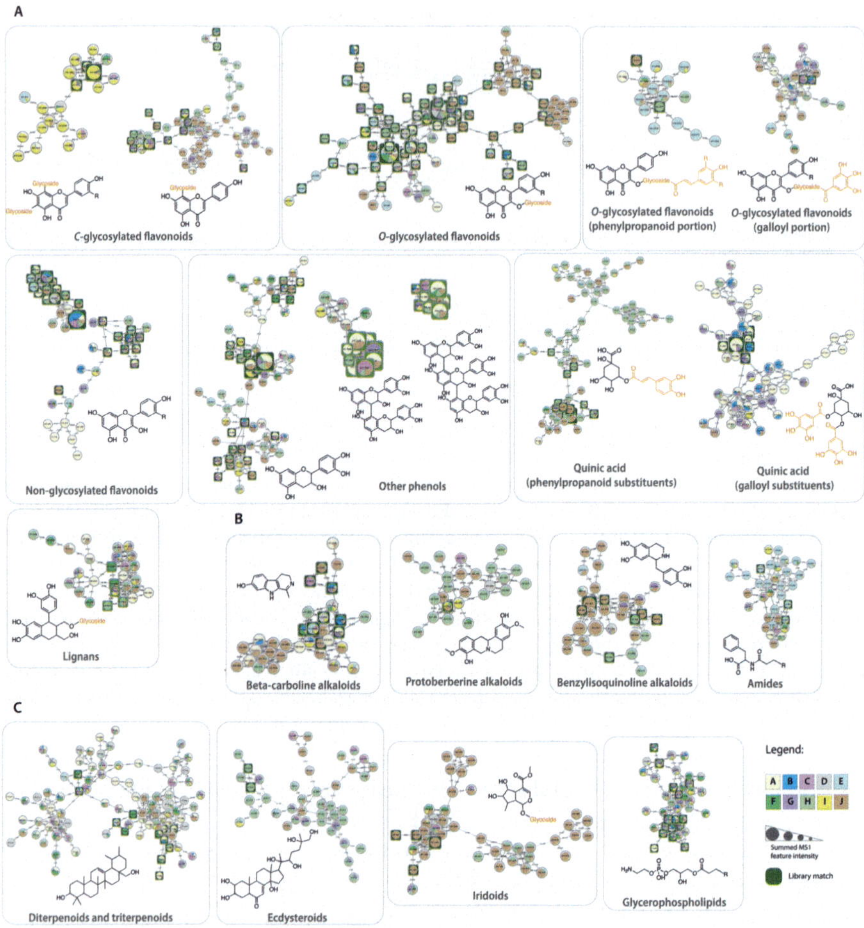

Fig. 4 Molecular families obtained from the feature-based molecular networking workflow and annotated based on spectral matches within the GNPS platform: (**a**) phenolic compounds, (**b**) alkaloids, and (**c**) lipids and terpenoids. Each node represents a tandem mass spectrometry spectra (MS/MS), while the edges that connect them represent the MS/MS fragmentation similarity (cosine >0.7). Pie charts indicate the relative abundance of ion features in each Malpighiaceae phylogenetic clade (A–J). Node sizes are relative to the summed peak areas of the precursor ion in MS1 scans. These are level 2 or 3 annotations according to the 2007 metabolomics standards initiative [45]. (Reproduced from Ref. [44])

ratio rule-based classification method for the automated metabolite profiling and differentiation of four cinnamon species using ultra-performance liquid chromatography-high-resolution mass spectrometry. Among the species studied were *Cinnamomum cassia* (Chinese cinnamon), *C. loureiroi* (Vietnamese cinnamon), *C. verum* (Ceylon cinnamon), and *C. burmannii* (Korintje cinnamon); proanthocyanidins, coumarin, and cinnamaldehyde were the preselected metabolites allowing the classification [46].

The genus *Vanilla*, a source of the most appreciated flavor worldwide, comprise over 110 species. Currently, only three species have commercial relevance, *Vanilla planifolia* Andrews, *V. tahitensis* J. W. Moore, and *V. pompona* Schiede. *V. planifolia* are preferred by industry due to its higher content of vanillin, the main flavor component; more attention needs to be made to other *Vanilla* species. Leyva et al. developed a nuclear magnetic resonance (NMR) metabolomic platform to profile for the first-time leaves that are known to accumulate putative vanillin precursors of *V. planifolia* and those of Peruvian *V. pompona*, *V. palmarum*, and *V. ribeiro* to determine metabolite difference among them. Their NMR analysis identified 36 metabolites, and multivariate analysis identified malic and homocitric acids, together with 2 vanillin precursors (glucoside A and B), as relevant markers for species identification [47].

Coffee is appreciated worldwide for its aroma, flavor, and stimulant properties. Souard et al. examined leaves of nine *Coffea* species grown in the same environmental conditions by an untargeted liquid chromatography high-resolution mass spectrometry (LC-HRMS) approach, with the primary objective of identifying metabolites that significantly contribute to the classification between *Coffea* species. Based on their results of multivariate analyses, 1637 variables (metabolites) were analyzed, from which 92% (1505 metabolites) were significantly different overall taxa. Among the species studied, when two well-known *C. arabica* and *C. canephora* were compared, a feature with an $m/z = 195.0870$ corresponding to $[M + H]^+$ of caffeine came out as the main discriminant compound. Caffeine concentration was approximately 800 times higher in *C. arabica* leaves than in *C. canephora*. Another feature observed at an m/z value of 247.0598 had much higher intensities in C. *arabica* than in *C. canephora*, but unfortunately, this feature was not identified. This metabolic fingerprinting study aimed to determine the specific differences between the metabolomes. All nine clusters of each species studied were observed on both PCA and PLS-DA score plots, with good discrimination between the eight *Coffea* species [48].

Several studies have described the use of metabolomics to distinguish herbal medicinal plants. For example, Lesiak et al., using seeds of the genus *Datura* plants, outlined direct analysis real-time mass spectrometry (DART-MS), which could provide diagnostic fingerprint profiles of nine *Datura* species seeds and whether chemometric processing of the observed profiles could enable species-level identification and differentiation. They confirmed that the seeds could be analyzed by DART-MS directly in a high-throughput manner without using a solvent extract. Each species exhibits a distinct chemical signature, and the processing of this data by multivariate statistical methods enables species-level differentiation. In addition, they observed that while intraspecies chemical signatures are similar, interspecies fingerprints are distinct enough to be discriminated against using multivariate statistical analysis tools [49]. Another example using seed samples of *Polygonatum* species was reported by Qi et al.; *Polygonatum* plant species have properties that make them sound like medicine and food in China. There were almost no differences in the contents of the metabolites in the amino acids and derivatives, nucleotides and products, and others (e.g., saccharides, alcohols, and vitamins) classes among the seed samples. In contrast, the seed samples had some diversity in the contents of lipids, phenolic acids, lignans and coumarins, tannins, and quinones. The flavonoid, steroid, and terpenoid classes and contents varied among the *Polygonatum* seed samples; these compounds have relatively strong pharmacological effects; their findings indicate that different *Polygonatum* seeds differ in terms of their medicinal and nutritional value [50].

Other plant species that have been classified and identified using metabolomic approaches are *Mentha* species [51]; *Acorus [52]* – plants in *Acorus* have been used as herbal medicine by various linguistic groups for thousands of years; and *Phyllanthus* species [53]; among other medicinal plant species.

4 Plant Root Metabolome and Climate Change

Climate change is a relevant issue due to its adverse and high-impact consequences it can cause directly and indirectly at social, ecological, biological, and health levels [54, 55]. The main factors promoting climate change comprise natural and anthropogenic activity; agriculture can produce 30–40% of the total greenhouse gas emissions [56]. In addition, due to these issues, an increase in the use of pesticides is expected, which will significantly affect global crop production, as well as pathogen diseases, abiotic stress, and the decrease in the production of the major crops worldwide [57]. Quality and crop productivity are negatively affected by global warming. It is expected that dramatic increase in the following years due to the increasing annual temperature, solar radiation, changes in precipitation, and high CO_2 levels [54, 56, 57]. Some other factors which reduce crop production quality are floods and droughts [58, 59] and are affected by alterations in rainy seasons, pest invasions, crop disease, water supplies, price of products for agricultural processes, and premature consumption of fertilizers [54].

Humanity's well-being and economics depend strongly on the agricultural sector, which simultaneously depends on the ability to adapt crops to environmental conditions and is therefore considered a climate-dependent industry [58]. To improve productivity, nutrient quality, and crop resilience, it is necessary to adapt strategies and design technologies to help mitigate climate change's effects [55, 60]. The leading crop breeding technologies for adaptation comprise biotechnology techniques, such as next-generation sequencing and RNA-mediated silencing [58, 61, 62]. Another strategy to improve crops and their resistance to biotic and abiotic stress is the engineering of the root microbiota, which represents a promising technology in the future to face climate change [63]. This strategy arises from analyzing the plant microbiome in the rhizosphere and endosphere, in which several interactions occur between the plant, microorganisms, their metabolites, and the metabolites of the surrounding plants. Once the relevance of these interactions on plant development is understood, the rhizosphere microbiome should be manipulated and thus increase production and their well-being; reduce the need for farmland, pesticides, and fertilizers; and thus reduce the intrinsic carbon footprint associated with these activities [64–66].

Plant evolution gave way to the adaptation through metabolite excretion to the soil (exudates) to interact with rhizosphere composition. The exudates may alter the composition and activity of the microbiome around, changing the pH, soil structure, and availability of oxygen or supplying organic compounds as a source of energy [67]. Some of the compounds exudated can also act as chemotrophic signals, which can attract pathogenic microorganisms to the plant, nematodes, or herbivorous arthropods [68]; however, these compounds can also work in the recruitment of beneficial microorganisms which can aid the plant in defense against pathogens, diseases, biotic stress conditions or enhance nutrient absorption [63, 64]. Under stress conditions, plants trigger the production of many secondary metabolites, with defense signals that promote the cope against pathogen organisms. Synthesis of secondary metabolites can relieve stress by modifying root microbiota to further degradation of different types of pollutants to carry out bioremediation [63, 69].

Among the strategies proposed for manipulating the rhizosphere microbiome is the direct inoculation of microorganisms in the soil. However, one of the biggest challenges to achieve is the determination of the species that act on the mechanisms involved, the competitive behavior with the native microbiota, and the effects it has on agricultural conditions. Alternatively, the metabolites observed in rhizosphere exudates should be used and applied in specific areas to stimulate native microorganisms [67]. Therefore, it is necessary to determine qualitatively and quantitatively the composition of the exudates under particular conditions, as well as the metabolomic analysis of the prominent participants in the interactions described and reasonably link the production of metabolites and their primary function in the face of possible types of stress [65, 70].

A metabolomic analysis is an excellent choice to study plant-rhizosphere interactions due to the complexity and quantity of compounds involved in the metabolic relation between roots and microorganisms. Instrumentation such as LC-MS, GC-MS, and NMR are the most widespread techniques and powerful tools used for identifying the compounds present in the rhizosphere [68, 71]. On the other hand, the exometabolomics concept consists of analyzing the metabolic traces of microorganisms present in certain conditions to understand the underlayer mechanisms that exist in the rhizosphere and the determination of substrate consumption by microorganisms [64]. In this context, we reviewed recent literature in metabolomics and exometabolomics, the findings in primary metabolites, and their function on specific rhizosphere interactions.

4.1 Metabolome Changes in Roots by Abiotic Stress

Metabolome alterations in the rhizosphere due to interactions result in significant interest in setting the defense mechanisms that play plants against external agents and expanding the landscape about key metabolites to degrade specific contaminants. Therefore, Wang et al. explored changes in chemical composition in root exudates of the urban greening trees exposed to phenanthrene. In *Loropetalum chinense*, *Gardenia ellis*, and *Rhododendron simsii*, carbohydrate levels increased in the presence of phenanthrene, indicating a regulatory function mediated by the biopolymer's degradation, whereas in *Osmanthus fragrans* levels decreased, suggesting different behavior between species. Phenolic compounds increased in the presence of phenanthrene in *Ligustrum japonicum*, *R. simsii*, *O. fragrans*, *Gardenia jasminoides*, and *Camellia sasanqua*, implying an adaptation to attract rhizobia bacteria, with the aim of cope with exposure to phenanthrene [72]. Regarding heavy metals stress, Lu et al. analyzed two wheat genotypes with different tolerance to Cd: Aikang 58 with low accumulation and Zhenmai 10 with high accumulation. Both phenotypes showed an increase in phenylalanine and tyrosine in the presence of Cd, relating these changes in the shikimate-phenylpropanoid pathway. The rise in acetylglycine and histidine indicates a chelating activity to chelate Cd in vacuoles. At the same time, glutamate, glutamine, aspartic acid, asparagine, and lysine perform an osmotic balance to detoxify heavy metals. In the presence of Cd, an increase in maltose, isomaltose, sorbose, tagatose, and polyols assists the cell wall's structure. In contrast, the addition of glyceric acid, cis-aconitate, malic acid, salicylic acid, and citrate indicates a deterioration in the activity of the tricarboxylic acid cycle activity to assimilate carbon under stress conditions. These alterations in the metabolism promote a high ability to take out and defend the plant against reactive oxygen species, inducing molecular signaling and antioxidant enzymes [73].

A report studying the effect of acid drainage contamination on *Phragmites australis* by Kalu et al. analyzed root and rhizosphere metabolome. The main compounds found in roots at contaminated sites were adenosine, inosine, methionine, carnitine, and dimethylglycine. On the other hand, uridine, dopa, asymmetric dimethylglycine, adenosine, and phenylalanine had a lower abundance in contaminated sites. This alteration has the purpose of recruiting microorganisms that promote the growth of the plant while at the same time attracting microorganisms specialized in heavy metal detoxification. As for bacterial communities, the main phylum in samples grown at contaminated sites was proteobacteria, β-proteobacteria, and the *Methylocystis*, *Rhizobium*, and *Delftias* genera [74]. In salinity stress, Wang et al. analyzed the canola roots metabolome in the presence of NaCl. The abundance of proline and soluble sugars increased in canola roots under saline stress. However, the metabolites with the most significant difference between groups were lipids, primarily fatty acids, which increased compared to controls. In saline stress, lipids affect membrane permeability, fluidity, integrity, and protein transport activity; therefore, the reconstitution of lipids in cells becomes transcendental. This stress triggers the production of polyunsaturated fatty acids, which can help the activation of membrane ATP-loop activity, which is responsible for maintaining homeostasis, facilitating the pumping of Na^+ from the cytosol to the external medium, and blocking K^+ channels [70].

Finally, it is essential also to consider UV radiation stress. Mannucci et al. analyzed the effect of UV exposure on tomato plants and their metabolomic changes in roots and leaves. In the roots of plants exposed to UV radiation, terpenoids and phenylpropanoid derivatives pathway synthesis increased compared to controls. For carbohydrates, degradation processes increased in radiation treatments, suggesting a wide degree of mobilization of reserve compounds to produce necessary precursors for secondary metabolites synthesis, such as flavonoids. As an effect of a rearrangement of lipid membrane composition, monogalactosylglycerol was found in the plants exposed, while 4α-carboxy-5α-cholesta-7,24-dien-3β-ol levels decreased. Finally, phenolic compounds and p-coumaroyl glycolic acid decreased in the UV treatments, a compound with anti-inflammatory properties [75].

4.2 Metabolic Changes Associated with Root and Other Biotic Factor Interactions

The most relevant biotic factors that transcend the metabolome analysis include plant pathogenic microorganisms, it is essential to decipher the compounds produced by plants to counteract the conditions and even the mutual organisms that help defend against these agents. Another biotic stress to consider is the neighborhood of other plant species competing for nutrients. In addition, the importance of the interaction of mycorrhizal fungi and the benefits they have in association with root plants is known. Both metabolisms are relevant to develop strategies that can help to improve crops.

To find out how rye competition affects *Vicia villosa* Roth, Hazrati et al. analyzed the metabolic effect on the roots of these plants. Kaempferol-Rha-Xyl-Gal was the main compound found. This compound decreased in *V. villosa* when it was grown in the presence of rye. Thus, it is estimated that competition produces a deficit of several nutrients in hairy vetch, decreasing the production of flavonoids [76]. *Phytophthora sojae* is known to cause phytophthora root rot disease in soybean; Zhang et al. analyzed the rhizosphere of *Glycine max* inoculated with *P. sojae*. The post-inoculated rhizosphere of a resistant species had a greater abundance of metabolites related to cutin biosynthesis, suberin, wax, arginine, ansamycins, pyrimidine, galactose, linoleic acid, ABC transporters metabolism, and lysine degradation. Most of the metabolites in the post-inoculated rhizosphere include antibiotics, which are responsible for conferring plant resistance to pathogens. On the other hand, some compounds in the control rhizosphere contained compounds to attract possible pathogenic microorganisms to the plant, such as daidzein and genistein. Although some flavones and isoflavones repel zoospores, others have the opposite effect, each specific to the conditions the plants were exposed to. Besides, cutin, suberin, and wax biosynthesis are inferred to provide drought tolerance by preventing water loss and insect tolerance [77].

Finally, interaction with symbiotic microorganisms focuses on knowing the benefits they bring and how they achieve them. Therefore, a view with a metabolomic approach is interesting to know the main compounds in the rhizosphere used to improve crop capacity. In this context, Zhang et al. analyzed the interaction between *Medicago truncatula* and rhizobium bacteria to get the metabolites in the formed nodules. Oxylipin-9-HODE decreased during the application of rhizobia bacteria, indicating decreased jasmonic acid precursors and inhibition due to the interaction [78]. Oxylipins are critical signaling molecules in the defensive response of plants to protect their tissues from attacks by herbivores or pathogens, and some contain antimicrobial properties.

Similarly, Sebastiana et al. analyzed cork oak roots colonized with *Pisolithus tinctorius*, an ectomycorrhizal fungus. In the study, the inoculated roots showed higher levels of γ-aminobutyric acid, alanine, β-glucose, and citrate. It was also demonstrated that the inoculated samples decreased quercitol, glycine-betaine, α-glucose, fructose, malate, and lactate levels. Inoculated samples influenced alkaloids, terpenoids, oxylipins, lipids, carbohydrates, amino acids, nucleic acids, and vitamins. In addition, a decrease in isomers of glucose, sucrose, sorbitol, and mannosyl glycerate was observed, as well as a reduction of isomers of fatty acids and compounds involved in the metabolism of tyrosine and histidine. The decrease of organic acids and glycine-betaine is related to apoplastic protective barriers, indicating the transfer of these metabolites to fungi, in the case of lipids, including monoacylglycerols, which are the main components of suberin and bark. This layer accumulates on the most exposed face of tree stems and roots and protects against drying and pathogen attack [79].

Another example was shown by Hernández et al. who analyzed the beneficial activity of *Rhizobium tropici* in *Phaseolus vulgaris* growth under phosphorous deficiency. It was observed that some organic acids, polyhydroxy acids, sugars, and polyols increased significantly in nodules with phosphorus (P) deficiency. In contrast, some amino acids and nitrogenous compounds decreased, reducing N fixation in P-deficient plants. In addition, they presented sugar accumulation, indicating demand for root photosynthate due to the decrease in the photosynthesis network. On the other hand, changes in carbohydrate content mean glycolysis/C binding pathways are induced in nodules under P deficiency stress [80]. P-deficient roots showed a decrease in the organic acid concentration, suggesting their exudation toward the rhizosphere; this has also been recently demonstrated in a study by Gomez-Zepeda et al. when using mass spectrometry imaging to locate organic acid exudates in P-deficient *Arabidopsis* plants. Organic acid exudation by roots is considered a core response to different types of abiotic stress and the interaction of roots with soil microbes. For decades it has been a target trait to produce plant varieties with increased capacities of inorganic orthophosphate uptake and aluminum tolerance [81].

Microbiota may vary according to the plant growing zone. In this way, Li et al. analyzed metabolome and microbiota in the roots of *Aconitum vilmorinianum* grown in two different sites in China (Luquan and Weixi). The difference observed in the metabolites was an increase in yunnaconitin and vilmorranine A in Weixi and a decrease in amino acids and some derivatives in Luquan. A correlation was found between 137 bacteria and 17 fungi with 75 differential metabolites in the 2 regions, among which the fungus on Cladosporium stands out, with a high probability conditional on aconitine, demonstrating the appearance of this metabolite in Weixi samples. Regarding Luquan, three bacterial and six fungal biomarkers were found, while Weixi showed the presence of five bacterial and five fungal biomarkers. This finding in the microbiome may be due to the environmental temperature, while in Weixi usually snows, and Luquan rarely occurs [82].

Knowledge about metabolome interaction between roots and their environment is crucial to identify the relevant metabolites produced in this medium. With this, it is intended to know which metabolic pathways are altered in plants to produce detoxifying compounds, antibiotics, or those that recruit beneficial microorganisms for plants. On the other hand, the compounds produced by microorganisms and their identification are also convincing to analyze the possibilities of the inoculation of different bacterial species in scenarios of various types of stress. However, it is important to specifically identify these conditions since not all species will behave similarly under the same state or stress. Herewith, it is essential to trace how crops can be treated to face biotic and abiotic stress caused mainly by climate change and improve their production and development.

5 Plant Biomarker Identification

Metabolomics is one of the most recent powerful tools for studying plants and other organisms and is becoming a complementary technique to genomics, transcriptomics, and proteomics [83]. Metabolomics addresses the activity of small molecules (<1500 Da) produced by cells during their life cycles, that is, products of primary or secondary – specialized – metabolism, found in various biological systems, studying how metabolic profiles change within an organism in response to some situations, such as disease or stress [84].

Therefore, unlike other "omics," metabolomics best describes phenotypes, can give instantaneous information on the physiological state of cells and thus provide a broader view of the biochemical state of an organism, and can track the metabolic network of a biological system and its perturbation in response to stimuli. Metabolomics aims not to identify every metabolite observed but to compare patterns of metabolites that change in each biological system. When these analyses are performed on enough biological replicate samples, it allows researchers to discriminate and classify samples and gain insight into changes in metabolome composition related to a particular physiological state, influence of stress or stimulus, genetic modification, or interaction with other organisms.

Metabolites provide a "fingerprint" of the complex interaction between the genome and the environment. They can generally be divided into two groups: primary metabolites, essential for maintaining processes directly involved in plant survival, growth, and reproduction, and secondary metabolites, which contribute to specialized processes in each organism synthesized to fulfill a nonessential function in the plant.

Due to the structural heterogeneity of metabolites and their different ranges of magnitude and concentration, their identification and measurement present a considerable challenge. For the plant kingdom alone, researchers have reported more than 400,000 plant species worldwide [85]. As for structurally distinct secondary metabolites, there are approximately 200,000 to 1,000,000 [86], which is why the field of plant metabolomics is the most advanced with a wide range of applications [1, 85].

All this information and understanding of the metabolome as it is affected by factors including environmental changes, physical changes, biotic stresses, abiotic stresses, and even internal changes in the plant as a function of its developmental stage can be used to monitor significant variations in metabolites and in the search for metabolites that can act as biomarkers.

The study of metabolomic biomarkers is one of the least explored areas in metabolomics. By 2022 only 16% of the publications examined the discovery or discrimination of biomarkers, while 46% of the publications refer to metabolic mechanisms and 33% examined metabolic profiles. However, many of these publications on metabolic profiles include preliminary and descriptive findings for more detailed analysis of the machine tool and discovery of biomarkers [87].

A metabolomic biomarker differs greatly from a protein biomarker and transcriptomic biomarkers because of the close relationship between individual metabolites. Factors measured in other "omics" technologies are independent, although there may be patterns of abundance that reflect a disease state. A metabolomic biomarker is not just a chain of changes in individual metabolites. Instead, it is composed of co-related metabolites that change together [88].

For the discovery and characterization of a metabolomic biomarker, validation based on the environment and research design is necessary to determine whether the proposed biomarker can distinguish between the changes to which plants are subjected [88], that is, for a biomarker to be classified as such, it must meet specific characteristics: be measurable, reflect the qualitative or quantitative interaction of the plant with the chemical of interest, be precise and sensitive, and be commonly shared among individuals in a population and plant species.

In this way, and through preliminary findings of metabolic profiling, some biomarkers have been identified in plants in response to exposure to stress factors: For example, a study on biomarker discovery [89] demonstrated that volatile organic compound (VOC) profiles could be used as diagnostic markers of stress in grapevine; this study shows that VOC emission can be considered as a universal response of grapevine to plant defense elicitors, given that the elicitors evaluated induced the emission of a standard set of VOCs encompassing chemically different compounds, including the sesquiterpenes α-farnesene and β-caryophyllene and that such a response is analogous with the induction of stilbene phytoalexins.

Similarly, plant metabolomics can help to identify resistant metabolites in plants that are subjected to stress conditions [17]; the selected biomarker can be used as a diagnostic metabolite for plant stress, as in the case of the study of wheat and barley resistance against *F. graminearum* infection where they point to various plant hormones that respond to this infection [90]. Such is also the case of phenylpropanoids and organic acids, metabolites identified as biomarkers of nitrogen deficiency in leaves and roots of tea plants (*Camellia sinensis*) that are elevated when there is a nitrogen deficit [91]. For example, hexadecanoic acid and dotriacontane, highly expressed metabolites, were identified as potential biomarkers in rice seeds infected with *Rhizoctonia solani* toxin, metabolites involved in several important rice biosynthetic pathways, such as the biosynthesis of saturated fatty acids and the unsaturated fatty acids cutin, suberin, and wax [92].

However, the identification of metabolites not only corresponds to stress responses, but the detection of metabolic changes at different developmental stages also contributes to finding metabolites characteristic of each stage (biomarkers), as in the case of metabolomic analysis of rice where developmentally controlled phenolamide accumulation patterns are observed [15] or in *Arabidopsis* where patterns of glucosinolate, raffinose, and galactinol accumulation are present at the base of leaves during the senescence stage [90]. Analysis of the spatiotemporal metabolic profile of plant development also allows the identification of potential biomarkers to capture intrinsic genetic features of plant development, as in the study of rice tillering (branching), in which 21 metabolites captured nearly 83% of the metabolic variation [93], and the developmental phase of soybean during the transition from vegetative to reproductive stage, in which eight flavonoid kaempferol glycosides were identified as potential growth markers [94].

In the food industry, the identification of biomarkers has also worked in quality processes for food authenticity and food traceability matrices of plant origin, especially in the field of aromatic herbs and spices, which are very susceptible to food fraud, as in the case of thyme, an aromatic herb traditionally used for food purposes due to its organoleptic characteristics and medicinal properties. In this particular case, it was possible to determine the geographical traceability of thyme based on different origins (Spain, Poland, and Morocco), as well as to evaluate its processing by comparing sterilized thyme with non-sterilized thyme, where 24 differential markers belonging to different classes were identified: among monoterpenoids, diterpenoids, sesquiterpenoids, alkylbenzenes, and other diverse compounds for its authenticity [95].

Another example of this application that helps to detect adulterants in plants that are used commercially is observed in the study by Ivanovic et al. using wild garlic (*Allium ursinum*) and poisonous adulterants *Convallaria majalis* and *Arum maculatum* as a model for the detection of adulterants in edible plants; the metabolites isovitexin, vicenin II, azetidine-2-carboxylic acid, and trigonelline were elucidated as biomarkers of adulteration [96].

On the other hand, metabolomics approaches have also been used to characterize and diagnose plant diseases and thus crop improvement, for example, during the study of the interaction maize-Fusarium *graminearum-Bacillus amyloliquefaciens* or soybean-*Rhizoctonia-B. amyloliquefaciens*, a better understanding of the metabolic regulation of all interacting systems has been achieved, providing valuable insights potentially useful in plant breeding and metabolic bioengineering [97].

Metabolite markers against drought stress (malonate, leucine, 5-oxo-L-proline, saccharic acid, trans-cinnamate, succinate, and glyceric acid) have been reported by Khan et al. who identified biomarkers in the metabolome of chickpea (*Cicer arietinum* L.) when treated with plant growth regulators (salicylic acid and putrescine) and the PGPR growth consortium (*B. thuringiensis*, *Bacillus subtilis*, and *Bacillus megaterium*). Deliberative metabolic reprogramming of chickpea targeting biomarker synthesis pathways resulted in drought-tolerant chickpea varieties [98].

Biomarker identification can also be applied to predict phenotypic traits and provide early detection tools to identify and use them in plant breeding development [99]. In China, for example, hybrid rice combinations have been created using sterile lines and restorer lines to reduce seed deterioration during storage and establish galactose and gluconic acid as metabolic biomarkers that reflect the degree of seed aging [100]. Also, in understanding the functioning of plants growing under extreme conditions, the identification of biomarkers in these plants could provide information that would benefit crop improvement; for example, it was possible to identify associated metabolic biomarkers in an alpine medicinal plant (*Rhodiola crenulata*) that can survive in extreme altitude conditions where the shikimic acid-phenylalanine-phenylpropanoid flavonoid pathway was enhanced with phenylpropanoids upregulating much more than flavonoids [101].

Surveillance for potential pathogens is critical to plant innate immunity, so plants depend on the perception of pathogen-derived molecules to activate defense-related signaling cascades and specialized metabolites in response; in studies of the tomato plant (*Solanum lycopersicum*), by monitoring metabolic profiles of signaling cascades in response to pathogens, significant biomarkers were noted for several classes of metabolites including amino acid derivatives, lipid species, steroidal glycoalkaloids, hydroxybenzoic acids, hydroxycinnamic acids, and products, as well as flavonoids [102].

Other metabolites identified as biomarkers in the plant defense response to pathogens are hydroxycinnamic acids; the conjugation of these acids with amide groups contributes to the regulation of the dynamic metabolic pool of hydroxycinnamates; a wide range of biogenic amine compounds found in most plant cells and these conjugates can scavenge radicals, confer antimicrobial activity, and can be deposited in the cell wall; so finding the activity of these metabolites is indicative of the plant-pathogen response [103]. A summary of primary and secondary or specialized metabolites identified in various plant species is presented in Table 2.

The identification of biomarkers in plants can have diverse applications, as described above; however, to reach the validation of these metabolites, metabolic profiling studies are necessary; metabolomics has been widely applied in the study of plants showing a breakthrough in understanding how the phenotype is related to the metabolome and therefore the function of metabolites under normal conditions, stress, and during their development.

Table 2 Identifying key metabolites in various plant species using different analytical methods [104]

Plant species	Class	Analytical tools	Key metabolites	Ref.
Primary metabolites				
Plantago ovata	Fatty acids	GC-MS	α-Linolenic acid, linoleic acid, and palmitic acid	[105]
P. ovata	Fatty acids	GC-MS	Pentadecanoic acid, palmitic acid, heptadecanoic acid, stearic acid, oleic acid, linoleic acid, γ-linoleic acid, and arachidic acid	[106]
Jatropha curcas	Fatty acids	GC	Oleic acid, palmitic acid, and linoleic acid	[107]
Paeonia rockii, *P. potaninii*, and *P. lutea*	Fatty acids	GC-MS	α-Linolenic acid, oleic acid, and linoleic acid	[108]
Cicer arietinum	Fatty acids	GC-MS	Pentadecanoic acid, palmitic acid, palmitoleic acid, stearic acid, oleic acid, linoleic acid, α-linolenic acid, and arachidic acid	[109]
P. ovata	Amino acids	HPLC	Isoleucine, threonine, leucine, histidine, and lysine	[105]
P. ovata	Amino acids	HPLC	Aspartate, glutamine, glycine, alanine, arginine, serine, proline, isoleucine, and methionine	[106]
Fritillaria thunbergii	Amino acids	GC-MS	Tryptophan, phenylalanine, and histidine	[110]
C. arietinum	Amino acids	GC-MS	L-glutamic acid, L-tryptophan, phenylalanine, glycine, serine, L-threonine, L-valine, L-ornithine, and L proline	[109]
C. arietinum	Sugars and sugar alcohols	GC-MS	Sucrose, cellobiose, galactose, methyl galactoside, and *myo*-inositol	[109]
C. arietinum	Sugar alcohols	GC-QqQ-MS	Galactitol, erythritol, arabitol, xylitol, mannitol, and inositol	[111]
Secondary metabolites				
Beta vulgaris	Terpenes	HPLC-MS	Oleanolic acid, hederagenin, akebonoic acid, and gypsogenin	[112]
Ocimum gratissimum	Terpenes	GC MS	m-Chavicol, t-anethole, germacrene-D, naphthalene, ledene, eucalyptol, azulene, and camphor	[113]
Mentha piperita	Terpenes	GC-MS	Menthone, menthol, pulegone, and menthofuran	[114]
M. arvensis	Terpenes	GLC	Menthol, isomenthone, L-methone, and menthyl acetate	[115]
Achyranthes bidentata	Terpenes	HPLC	Oleanolic acid and ecdysterone	[116]

(continued)

Table 2 (continued)

Plant species	Class	Analytical tools	Key metabolites	Ref.
Arabidopsis thaliana	Phenolics	UHPLC-MS	Scopoletin, unbelliferone and esculetin, scopolin, skimming, and esculin	[117]
P. ovata	Phenolics	LC-MS	Luteolin, quercetagetin, syringetin, kaempferol, limocitrin, helilupolone, and catechin	[105]
P. ovata	Phenolics	LC-MS	Kaempferol, 3-(2″,3″-diacetylrhamnoside)-7-rhamnoside and apigenin 7-rhamnoside	[106]
P. ovata	Alkaloids	LC-MS	Lunamarine, hordatine B, and pinidine	[105]
Dendrobium snowflake "red star"	Alkaloids	^{1}H and 2D NMR	Dendrobine and nobilone	[118]

Understanding the adaptative physiology and biochemistry of plants, as well as the underlying metabolic events, is relevant to have a global perception of the metabolomic status of plants, with the identification of biomarkers providing helpful information on metabolites involved in resistance responses, stress, growth, a better understanding of intra- and interspecific microbial interactions occurring at different heterogeneous levels within the plant habitat, identification of systemic responses of various crops to pathogen stress, and pathogens and their biological control, would allow crop scientists to identify unique metabolic markers that can be applied to early detection of a plant pathogen as well as to the development of bio fungicides, for example, for use during pre-harvest, post-harvest, and harvest storage and large-scale storage of crops. Identifying and applying metabolic biomarkers could favor controlled and semi-controlled planting systems shortly, and if properly integrated into crop protection strategies, food security could be mitigated. However, the applications of these biomarkers could be helpful in various areas, such as the food and pharmaceutical industry in food quality and safety processes, diagnosis and treatment of plant diseases, crop improvement, and analyzing genetic-modified crops. Still, the work done so far is relatively new, and efforts should continue to cover the tremendous potential presented by identifying metabolic biomarkers. Table 3 summarizes the identification of some metabolites in transgenic plants using different analytical techniques.

Table 3 Identification of important metabolites in transgenic plants using different analytical tools [104]

Transgenic plant	Analytical techniques	Key metabolites	Ref.
Artemisia annua	GC-TOF-MS	Borneol, phytol, (3-farnesene, germacrene-D, artemisinic acid, dihydroartemisinic acid, and artemisinin	[119]
Lactuca sativa	NMR	Asparagine, glutamine, valine, isoleucine, cx-ketoglutarate, succinate, fumarate, malate, sucrose, and fructose	[120]
Lycopersicum esculentum	GC-MS	y-Aminobutyric acid, histidine, proline, pyrrol-2-carboxylate, galactitiol/sorbitol, glycerol, maltitol, 3-phosphoglyceric acid, allantoin, homocysteine, caffeate, gluconate, ribonate, lysine, threonine, homoserine, tyrosine, tryptophan, leucine, arginine, and valine	[121]
Nicotiana tabacum	NMR	Chlorogenic acid, 4-O-caffeoylquinic acid, malic acid, threonine, alanine, glycine, fructose, (3-glucose, cx-glucose, sucrose, fumaric acid, and salicylic acid	[121]
N. tabacum	GC-MS	4-Aminobutanoic acid, asparagine, glutamine, glycine, leucine, phenylalanine, proline, serine, threonine, tryptophan, chlorogenic acid, quininic acid, threonic acid, citric acid, malic acid, and ethanolamine	[122]
Oryza sativa	GC-MS	Glycerol-3-phosphate, citric acid, linoleic acid, oleic acid, hexadecanoic acid, 2,3-dihydroxypropyl ester, sucrose, 9-octadecenoic acid, 2,3-dihydroxypropyl ester, sucrose, mannitol, and glutamic acid	[123])
O. sativa	LC-MS	Tryptophan, phytosphingosine, palmitic acid, 5-hydroxy-2-octadenoic acid 9,10,13-trihydroxyoctadec-ll-enoic acid, and ethanolamine	[124]
Populus	GC-MS, HPLC	Caffeoyl and feruloyl conjugates, syringyl-to-guaiacyl ratio, asparagine, glutamine, aspartic acid, y-amino-butyric acid, 5-oxo-proline, salicylic acid-2-O-glucoside, 2, 5-dihydroxybenzoic acid-5-O-glucoside, 2-methoxyhydroquinone-l-O-glucoside, 2-methoxyhydroquinone 4 0-glucoside, salicin, gallic acid, and dihydroxybenzoic acid	[125]
Solanum tuberosum	LC-TOF-MS	Glutathione, y-aminobutyric acid, 3-cyanoalanine, 5-oxoproline, sucrose, glucose-1-phosphate, glucose-6-phosphate, fructose-6-phosphate, ethanolamine, adenosine, and guanosine	[126]
Triticum aestivum	GC-MS	Guanine and 4-hydroxycinnamic acid	[127]

(continued)

Table 3 (continued)

Transgenic plant	Analytical techniques	Key metabolites	Ref.
T. aestivum	LC-MS	Aminoacyl-tRNA biosynthesis, phenylalanine, tyrosine, tryptophan glyoxylic, tartaric acid, oxalic acids, sucrose, galactose, mannitol, leucine, valine, glutamate, proline, pyridoxamine, glutathione, arginine, citrulline, adenosine, hypoxanthine, allantoin, and adenosine monophosphate	[128]
Zea mays	^{1}H NMR	Lactic acid, citric acid, lysine, arginine, glycine-betaine, raffinose, trehalose, galactose, and adenine	[129]

6 Plant Single-Cell Metabolomics

In the past, genomics, transcriptomics, and metabolomics techniques have been applied in bulk plant samples consisting of many cells; in such experiments, the biological process leading to cell heterogeneity is often considered not to be biologically relevant. However, cell heterogeneity has been shown to play important biological roles in many situations for which averaging would mask relevant metabolic processes [130]. Plants contain several cell types and exhibit complex regulatory mechanisms. Studies at the single-cell level have gradually become more common in plant science. Single-cell transcriptomics, spatial transcriptomics, and spatial metabolomics techniques have been combined to analyze plant development. These techniques have been used to study the transcriptomes and metabolomes of plant tissues at the single-cell level, enabling the systematic investigation of gene expression and metabolism in specific tissues and cell types during defined developmental stages [131]. However, single-cell technologies require labor-intensive protocols for plant cell isolation. On that respect several attempts have been developed; these strategies can be classified into three main groups: those that attempt to isolate material of specific cell type to perform the analysis on platforms used for regular metabolomics, which we will refer to as single-cell type metabolomics [132]; those based on micromanipulation of single cells; and those based on mass spectrometry imaging [130].

Several methods for harvesting cells have been developed for single-cell and single-cell type metabolomics, whereby cells can be obtained or extracted in situ. The in situ techniques include micropipetting for isolating the contents of specific cells, laser microdissection (LMD), laser microdissection and pressure catapulting (LMPC), laser capture microdissection (LCM), and fluorescence-activated cell sorting (FACS). Laser microdissection and pressure catapulting and laser capture microdissection use laser to excise single cells or microareas from fixed or frozen intact tissues and are becoming very popular for plant cell and tissue sampling. FACS, on the other hand, is used to obtain specific cell types; for example, those identified from root developmental zones by transgene-labelled nuclei or by immunolabelled-based collection and microfluidic sorting-based methods that exploit intrinsic cell properties [129]. Figure 5 summarizes different approaches for cell-specific metabolomics.

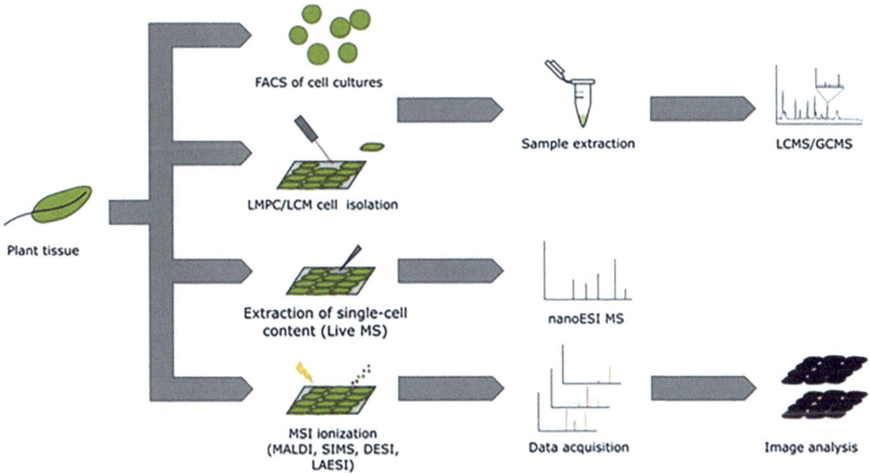

Fig. 5 Overview of experimental steps and data structure from the different approaches for cell-specific metabolomics. (Reproduced from Ref. [130])

However, to obtain single-cell suspension, it is a very challenging activity and deeply laborious. In addition, plant cells are rigid cells when compared to animal cells; rigid cell walls remain the main obstacle for single-cell technologies in plants. Since protoplasts must remain alive and be subjected to a minimal level of disturbance during isolation, for example, for protoplast isolation, the cell wall digestion procedure requires optimization for suitability for the tissue under study [133].

Mass spectrometry imaging (MSI) technique can provide spatially resolved information on the structure and content of metabolites including know and unknow endogenous metabolites, and it thus produces tissue molecular imaging maps. Three MSI techniques have been developed based on different ions sources: secondary ions MS (SIMS), desorption electrospray ionization (DESI), and matrix-assisted laser desorption/ionization (MALDI) [131]. Figure 6 summarizes different ionization techniques used for MSI. Among them, MALDI is the most popular ionization technique for MSI experiments. In MALDI mode a matrix applied to the sample is excited by a laser; this energy is further transferred to the sample resulting in the desorption/ionization event. Preparation for MALDI usually comprises cryosectioning and lyophilizing a frozen sample before applying the matrix by either a sprayer or a special device, as well solvent free sublimation [130]. However, MALDI remains a technique that still lacks significant improvement, for example, matrix selection and the choice of matrix method application, tissue sectioning technique, embedding protocols, sample preparation, and mounting. In other words, MALDI imaging technique requires optimization for every tissue and metabolites of interest; for example, Pérez-López et al. developed a protocol of MALDI imaging by sample imprinting in nylon membranes to locate fructans in stem and rhizome tissues of *Agave tequilana* plants [134]; in addition, the combination of ion mobility

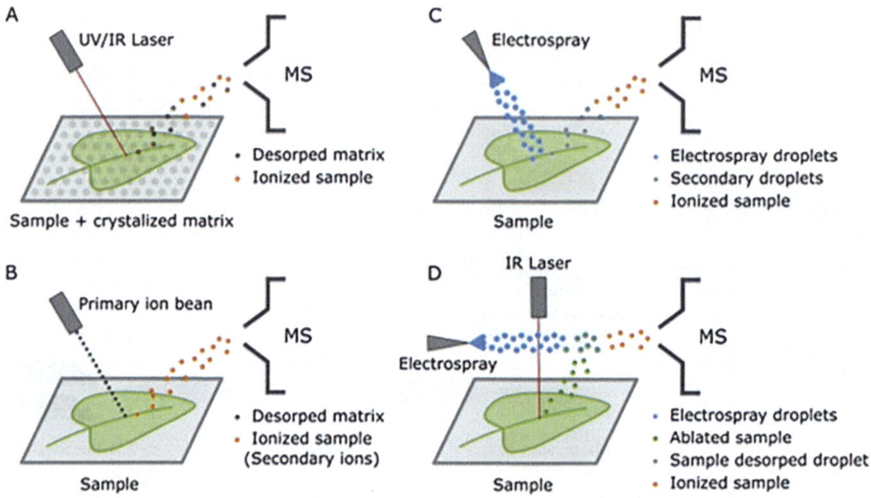

Fig. 6 Schematic representation of the different ionization strategies used for mass spectrometry imaging (MSI). (**a**) MALDI, (**b**) secondary ion mass spectrometry (SIMS), (**c**) desorption electrospray ionization (DESI), (**d**) laser-ablation electrospray ionization (LAESI). (Reproduced from Ref. [130])

spectrometry allowed the detection of fructan isomers even if these have not been mapped on their images obtained; Fig. 7 outlines the protocol developed.

More recently, DESI have become the newest development for mass spectrometry imaging to visualize plant metabolites; DESI offers a great advantage being matrix-free ionization alternative to MALDI. In DESI, a solvent stream originating from an electrospray probe is directed at an angle (most important parameter) toward the sample at ambient pressure, propelling secondary ions to the mass analyzer, enable direct analysis of unprocessed frozen samples sections which simplify samples preparation [135, 136]. Very recently, some metabolites detected with DESI source ranged from monoterpenoid alkaloids, which were localized in several of the major parts of the *Rauvolfia tetraphylla* plant when analyzed by MALDI and DESI-MSI [137], alkaloids in the leaves of *Gelsemium elegans* [138] through cannabinoids and flavonoids in the leaves of *Cannabis sativa* [139], among other recent applications of DESI.

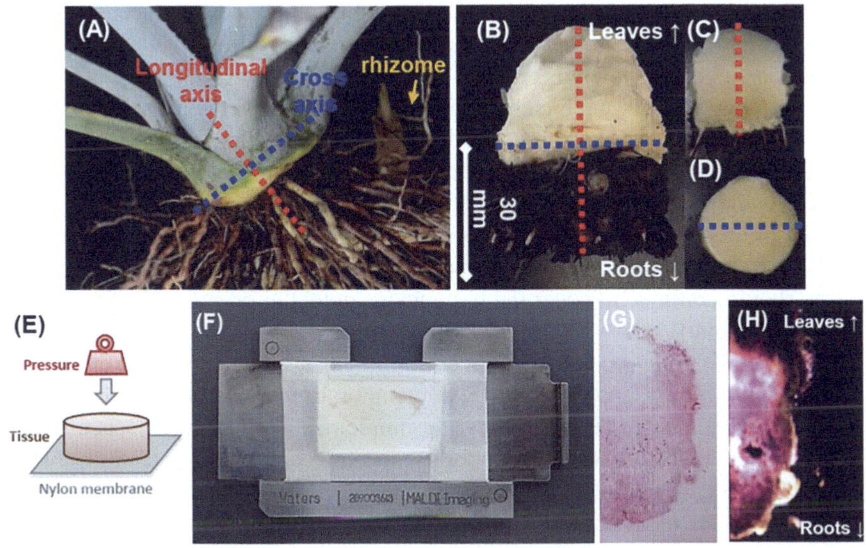

Fig. 7 Outline of tissue printing technique. (**a**) *Agave tequilana* plant showing the crown region (leaf/stem to root transition). Dotted red line indicates the longitudinal axis; dotted blue line indicates the transversal axis. (**b**) *A. tequilana* plant as in (**a**) with leaves and most roots removed and dissected longitudinally. (**c**) Longitudinal stem section. (**d**) Transverse stem section. (**e**) Representation of tissue printing process. (**f**) Tissue-printed transverse section mounted on MALDI Imaging plate. (**g**) PAS staining of a tissue-printed longitudinal section. (**h**) MALDI-ToF-MSI of a tissue-printed longitudinal section obtained, using a sprayer for matrix application and a QTOF SYNAPT G1 spectrometer with a spatial resolution of 100 pm per pixel. (Reproduced from Ref. [134])

7 Concluding Remarks

The utilization of a model plant like *Arabidopsis* or a crop with a reference genome sequence, such as - maize, rice, tomato, and wheat, to mention some of the global agricultural importance - offers a unique opportunity, where approaches such as mQTL, GWAS, mGWAS, and transcriptomics can effectively provide a vast potential to reveal gene annotation and their functional characterization. On the contrary, enormous efforts must be made for other minor crops, medicinal plants, and regional staple foods. However, plant metabolomics research needs to be focused on developing strategies to develop confident metabolite annotation through implementing a free online accessible database for metabolite identification. Plant metabolomics showed, as reviewed here, great potential to assist crop improvements, supporting exploring species identification for diversity and botanical purposes, food authentication, fraud, and traceability.

Without forgetting the knowledge of metabolome interaction between roots and their environment to identify the relevant metabolites produced, on the other hand, metabolomics approaches can play a crucial role in studying the interaction between plants with biotic and abiotic stresses. Knowing the compounds produced by microorganisms and their identification in climate change conditions that enhance crop production and development is crucial.

Identifying and applying metabolic biomarkers could favor controlled and semi-controlled planting systems shortly, and if properly integrated into crop protection strategies, food security could be mitigated. However, the applications of these biomarkers could be helpful in various areas, such as the food and pharmaceutical industry in food quality and safety processes, diagnosis and treatment of plant diseases, and crop improvement. Still, the work done so far is relatively new, and efforts should continue to cover the tremendous potential presented by identifying metabolic biomarkers.

Plant metabolomics could benefit in developing new strategies to face challenges and demands in crop improvement. Spatiotemporal metabolomics can effectively support plant-soil interactions studies; the implementation of mass spectrometry imaging in combination with ion mobility spectroscopy could potentially reveal metabolites' location in plant tissue without the need for extraction, in addition, to providing isomer identification without forgetting the need to develop metabolite databases that can support full plant metabolome coverage.

References

1. Sumner, L.W., et al., *Modern plant metabolomics: advanced natural product gene discoveries, improved technologies, and future prospects.* Natural product reports, 2015. **32**(2): p. 212–229.
2. Singh, D.P., et al., *Metabolomics-Driven Mining of Metabolite Resources: Applications and Prospects for Improving Vegetable Crops.* International Journal of Molecular Sciences, 2022. **23**(20): p. 12062.
3. Wu, F., et al., *Genome and systems biology of Melilotus albus provides insights into coumarins biosynthesis.* Plant Biotechnology Journal, 2022a. **20**(3): p. 592–609.
4. Li, H., et al., *Genome-wide association study dissects the genetic architecture of oil biosynthesis in maize kernels.* Nature genetics, 2013. **45**(1): p. 43–50.
5. Zheng, C., et al., *Gene coexpression networks reveal key drivers of flavonoid variation in eleven tea cultivars (Camellia sinensis).* Journal of agricultural and food chemistry, 2019. **67**(35): p. 9967–9978.
6. Chen, J., et al., *Exploring the genic resources underlying metabolites through mGWAS and mQTL in wheat: From large-scale gene identification and pathway elucidation to crop improvement.* Plant Communications, 2021a. **2**(4): p. 100216.
7. Chen, W., et al., *Comparative and parallel genome-wide association studies for metabolic and agronomic traits in cereals.* Nature communications, 2016. **7**(1): p. 1–10.
8. Burgos, E., et al., *Validated MAGIC and GWAS population mapping reveals the link between vitamin E content and natural variation in chorismate metabolism in tomato.* The Plant Journal, 2021. **105**(4): p. 907–923.
9. Alseekh, S., et al., *Identification and mode of inheritance of quantitative trait loci for secondary metabolite abundance in tomato.* The Plant Cell, 2015. **27**(3): p. 485–512.
10. Alseekh, S., et al., *Canalization of tomato fruit metabolism.* The Plant Cell, 2017. **29**(11): p. 2753–2765.

11. Liang, X., et al., *Metabolomics-driven gene mining and genetic improvement of tolerance to salt-induced osmotic stress in maize.* New Phytologist, 2021. **230**(6): p. 2355–2370.
12. Owens, B.F., et al., *A foundation for provitamin A biofortification of maize: genome-wide association and genomic prediction models of carotenoid levels.* Genetics, 2014. **198**(4): p. 1699–1716.
13. Riedelsheimer, C., et al., *Genome-wide association mapping of leaf metabolic profiles for dissecting complex traits in maize.* Proceedings of the National Academy of Sciences, 2012a. **109**(23): p. 8872–8877.
14. Chen, W., et al., *Genome-wide association analyses provide genetic and biochemical insights into natural variation in rice metabolism.* Nature genetics, 2014. **46**(7): p. 714–721.
15. Dong, X., et al., *Spatiotemporal distribution of phenolamides and the genetics of natural variation of hydroxycinnamoyl spermidine in rice.* Molecular Plant, 2015. **8**(1): p. 111–121.
16. Yang, Z., et al., *Genetic mapping identifies a rice naringenin O-glucosyltransferase that influences insect resistance.* The Plant Journal, 2021. **106**(5): p. 1401–1413.
17. Hong, J., et al., *Plant metabolomics: an indispensable system biology tool for plant science.* International journal of molecular sciences, 2016. **17**(6): p. 767.
18. Gilliland, L.U., et al., *Genetic basis for natural variation in seed vitamin E levels in Arabidopsis thaliana.* Proceedings of the National Academy of Sciences, 2006. **103**(49): p. 18834–18841.
19. Rowe, H.C., et al., *Biochemical networks and epistasis shape the Arabidopsis thaliana metabolome.* The Plant Cell, 2008. **20**(5): p. 1199–1216.
20. Routaboul, J.-M., et al., *Metabolite profiling and quantitative genetics of natural variation for flavonoids in Arabidopsis.* Journal of Experimental Botany, 2012. **63**(10): p. 3749–3764.
21. Lisec, J., et al., *Identification of heterotic metabolite QTL in Arabidopsis thaliana RIL and IL populations.* The Plant Journal, 2009. **59**(5): p. 777–788.
22. Feng, J., et al., *Characterization of metabolite quantitative trait loci and metabolic networks that control glucosinolate concentration in the seeds and leaves of Brassica napus.* New phytologist, 2012. **193**(1): p. 96–108.
23. Wen, W., et al., *Genetic determinants of the network of primary metabolism and their relationships to plant performance in a maize recombinant inbred line population.* The Plant Cell, 2015. **27**(7): p. 1839–1856.
24. Matsuda, F., et al., *Dissection of genotype–phenotype associations in rice grains using metabolome quantitative trait loci analysis.* The Plant Journal, 2012. **70**(4): p. 624–636.
25. Ying, J.-Z., et al., *Identification of quantitative trait loci for lipid metabolism in rice seeds.* Molecular plant, 2012. **5**(4): p. 865–875.
26. Gong, L., et al., *Genetic analysis of the metabolome exemplified using a rice population.* Proceedings of the National Academy of Sciences, 2013. **110**(50): p. 20320–20325.
27. Schauer, N., et al., *Comprehensive metabolic profiling and phenotyping of interspecific introgression lines for tomato improvement.* Nature biotechnology, 2006. **24**(4): p. 447–454.
28. Toubiana, D., et al., *Metabolic profiling of a mapping population exposes new insights in the regulation of seed metabolism and seed, fruit, and plant relations.* PLoS genetics, 2012. **8**(3): p. e1002612.
29. Perez-Fons, L., et al., *A genome-wide metabolomic resource for tomato fruit from Solanum pennellii.* Scientific Reports, 2014. **4**(1): p. 3859.
30. Schauer, N., et al., *Mode of inheritance of primary metabolic traits in tomato.* The Plant Cell, 2008. **20**(3): p. 509–523.
31. Hill, C.B., et al., *Detection of QTL for metabolic and agronomic traits in wheat with adjustments for variation at genetic loci that affect plant phenology.* Plant Science, 2015. **233**: p. 143–154.
32. Hill, C.B., et al., *Whole-genome mapping of agronomic and metabolic traits to identify novel quantitative trait loci in bread wheat grown in a water-limited environment.* Plant Physiology, 2013. **162**(3): p. 1266–1281.

33. Angelovici, R., et al., *Genome-wide analysis of branched-chain amino acid levels in Arabidopsis seeds.* The Plant Cell, 2013. **25**(12): p. 4827–4843.
34. Chan, E.K.F., et al., *Combining genome-wide association mapping and transcriptional networks to identify novel genes controlling glucosinolates in Arabidopsis thaliana.* PLoS biology, 2011. **9**(8): p. e1001125.
35. Chan, E.K.F., et al., *The complex genetic architecture of the metabolome.* PLoS genetics, 2010. **6**(11): p. e1001198.
36. Shen, M., et al., *Leveraging non-targeted metabolite profiling via statistical genomics.* PLoS One, 2013. **8**(2): p. e57667.
37. Lipka, A.E., et al., *Genome-wide association study and pathway-level analysis of tocochromanol levels in maize grain.* G3: Genes, Genomes, Genetics, 2013. **3**(8): p. 1287–1299.
38. Riedelsheimer, C., et al., *Genomic and metabolic prediction of complex heterotic traits in hybrid maize.* Nature genetics, 2012b. **44**(2): p. 217–220.
39. Wen, W., et al., *Metabolome-based genome-wide association study of maize kernel leads to novel biochemical insights.* Nature communications, 2014. **5**(1): p. 3438.
40. Carreno-Quintero, N., et al., *Untargeted metabolic quantitative trait loci analyses reveal a relationship between primary metabolism and potato tuber quality.* Plant physiology, 2012. **158**(3): p. 1306–1318.
41. Matsuda, F., et al., *Metabolome-genome-wide association study dissects genetic architecture for generating natural variation in rice secondary metabolism.* The Plant Journal, 2015. **81**(1): p. 13–23.
42. Sauvage, C., et al., *Genome-wide association in tomato reveals 44 candidate loci for fruit metabolic traits.* Plant physiology, 2014. **165**(3): p. 1120–1132.
43. Musah, R.A., et al., *A high throughput ambient mass spectrometric approach to species identification and classification from chemical fingerprint signatures.* Scientific Reports, 2015. **5**(1): p. 1–16.
44. Mannochio-Russo, H., et al., *Untargeted metabolomics sheds light on the diversity of major classes of secondary metabolites in the Malpighiaceae botanical family.* Frontiers in Plant Science, 2022. **13**.
45. Sumner, L.W., et al., *Proposed minimum reporting standards for chemical analysis: chemical analysis working group (CAWG) metabolomics standards initiative (MSI).* Metabolomics, 2007. **3**: p. 211–221.
46. Zhang, M., et al., *Development of a Metabolite Ratio Rule-Based Method for Automated Metabolite Profiling and Species Differentiation of Four Major Cinnamon Species.* Journal of Agricultural and Food Chemistry, 2022a. **70**(17): p. 5450–5457.
47. Leyva, V.E., et al., *NMR-based leaf metabolic profiling of V. planifolia and three endemic Vanilla species from the Peruvian Amazon.* Food Chemistry, 2021. **358**: p. 129365.
48. Souard, F., et al., *Metabolomics fingerprint of coffee species determined by untargeted-profiling study using LC-HRMS.* Food chemistry, 2018. **245**: p. 603–612.
49. Lesiak, A.D., et al., *Plant seed species identification from chemical fingerprints: a high-throughput application of direct analysis in real time mass spectrometry.* Analytical chemistry, 2015. **87**(17): p. 8748–8757.
50. Qi, J., et al., *Untargeted Metabolomics Analysis Revealed the Major Metabolites in the Seeds of four Polygonatum Species.* Molecules, 2022. **27**(4): p. 1445.
51. Brahmi, F., et al., *Discrimination of Mentha species grown in different geographical areas of Algeria using 1H-NMR-based metabolomics.* Journal of Pharmaceutical and Biomedical Analysis, 2020. **189**: p. 113430.
52. Cheng, Z., et al., *From folk taxonomy to species confirmation of Acorus (Acoraceae): evidences based on phylogenetic and metabolomic analyses.* Frontiers in plant science, 2020: p. 965.
53. Kiran, K.R., et al., *Untargeted metabolomics and DNA barcoding for discrimination of Phyllanthus species.* Journal of Ethnopharmacology, 2021. **273**: p. 113928.

54. Abbass, K., et al., *A review of the global climate change impacts, adaptation, and sustainable mitigation measures.* Environmental Science and Pollution Research, 2022: p. 1–21.
55. García, G.A., et al., *High night temperatures during grain number determination reduce wheat and barley grain yield: a field study.* Global Change Biology, 2015. **21**(11): p. 4153–4164.
56. Reidsma, P., et al., *Regional crop modelling in Europe: the impact of climatic conditions and farm characteristics on maize yields.* 2009. **100**(1-3): p. 51–60.
57. Santoro, A., et al., *A review of the role of forests and agroforestry systems in the FAO Globally Important Agricultural Heritage Systems (GIAHS) programme.* Forests, 2020. **11**(8): p. 860.
58. Munaweera, T., et al., *Modern plant biotechnology as a strategy in addressing climate change and attaining food security.* 2022. **11**(1): p. 1–28.
59. Ortiz, A.M.D., et al., *A review of the interactions between biodiversity, agriculture, climate change, and international trade: Research and policy priorities.* 2021. **4**(1): p. 88–101.
60. Chandio, A.A., et al., *Short and long-run impacts of climate change on agriculture: an empirical evidence from China.* 2020. **12**(2): p. 201–221.
61. Kausch, A.P., et al., *Edit at will: genotype independent plant transformation in the era of advanced genomics and genome editing.* Plant science, 2019. **281**: p. 186–205.
62. Wang, M.-B., et al., *RNA silencing and plant viral diseases.* Molecular Plant-Microbe Interactions, 2012. **25**(10): p. 1275–1285.
63. Hong, Y., et al., *Crafting the plant root metabolome for improved microbe-assisted stress resilience.* New Phytologist, 2022.
64. Chen, L., et al., *Metabolomics in plant-microbe interactions in the roots.* Advances in Botanical Research, 2021b. **98**: p. 133–161.
65. Hacquard, S., et al., *Impact of global change on the plant microbiome.* New Phytologist, 2022. **234**(6): p. 1907–1909.
66. Xun, W., et al., *Rhizosphere microbiome: Functional compensatory assembly for plant fitness.* Computational and Structural Biotechnology Journal, 2021. **19**: p. 5487–5493.
67. Pantigoso, H.A., D. Newberger, and J.M. Vivanco, *The rhizosphere microbiome: Plant–microbial interactions for resource acquisition.* Journal of Applied Microbiology, 2022.
68. van Dam, N.M. and H.J. Bouwmeester, *Metabolomics in the rhizosphere: tapping into belowground chemical communication.* Trends in plant science, 2016. **21**(3): p. 256–265.
69. Saeed, Q., et al., *Rhizosphere bacteria in plant growth promotion, biocontrol, and bioremediation of contaminated sites: a comprehensive review of effects and mechanisms.* International Journal of Molecular Sciences, 2021. **22**(19): p. 10529.
70. Wang, W., et al., *Transcriptomic and metabolomics-based analysis of key biological pathways reveals the role of lipid metabolism in response to salt stress in the root system of Brassica napus.* Plant Growth Regulation, 2022. **97**(1): p. 127–141.
71. Baldwin, M.A., *Mass spectrometers for the analysis of biomolecules.* Methods in enzymology, 2005. **402**: p. 3–48.
72. Wang, J., et al., *Non-targeted metabolomics reveal the impact of phenanthrene stress on root exudates of ten urban greening tree species.* Environmental Research, 2021. **196**: p. 110370.
73. Lu, M., et al., *Physiological and metabolomics responses of two wheat (Triticum aestivum L.) genotypes differing in grain cadmium accumulation.* Science of The Total Environment, 2021. **769**: p. 145345.
74. Kalu, C.M., et al., *Correlations between root metabolomics and bacterial community structures in the phragmites australis under acid mine drainage-Polluted wetland ecosystem.* Current Microbiology, 2022. **79**(1): p. 1–15.
75. Mannucci, A., et al., *Foliar and Root Comparative Metabolomics and Phenolic Profiling of Micro-Tom Tomato (Solanum lycopersicum L.) Plants Associated with a Gene Expression Analysis in Response to Short Daily UV Treatments.* Plants, 2022. **11**(14): p. 1829.
76. Hazrati, H., I.S. Fomsgaard, and P. Kudsk, *Targeted metabolomics unveil alteration in accumulation and root exudation of flavonoids as a response to interspecific competition.* Journal of Plant Interactions, 2021. **16**(1): p. 53–63.

77. Zhang, Z., et al., *Comparative metabolomics reveal the participation of soybean unique rhizosphere metabolites in susceptibility and resistance of host soybean to Phytophthora sojae.* Plant and Soil, 2022b: p. 1–15.
78. Zhang, N., et al., *Metabolomic profiling reveals suppression of oxylipin biosynthesis during the early stages of legume–rhizobia symbiosis.* FEBS letters, 2012. **586**(19): p. 3150–3158.
79. Sebastiana, M., et al., *Metabolomics and transcriptomics to decipher molecular mechanisms underlying ectomycorrhizal root colonization of an oak tree.* Scientific reports, 2021. **11**(1): p. 1–16.
80. Hernández, G., et al., *Global changes in the transcript and metabolic profiles during symbiotic nitrogen fixation in phosphorus-stressed common bean plants.* Plant Physiology, 2009. **151**(3): p. 1221–1238.
81. Gomez-Zepeda, D., et al., *Mass spectrometry-based quantification and spatial localization of small organic acid exudates in plant roots under phosphorus deficiency and aluminum toxicity.* The Plant Journal, 2021. **106**(6): p. 1791–1806.
82. Li, H., et al., *Metabolomics and microbiome reveal potential root microbiota affecting the alkaloidal metabolome in Aconitum vilmorinianum Kom.* BMC microbiology, 2022. **22**(1): p. 1–15.
83. Castro-Santos, P., C.M. Laborde, and R. Díaz-Peña, *Genomics, proteomics and metabolomics: their emerging roles in the discovery and validation of rheumatoid arthritis biomarkers.* Clin Exp Rheumatol, 2015. **33**(2): p. 279–286.
84. Dunn, W.B., N.J. Bailey, and H.E. Johnson, *Measuring the metabolome: current analytical technologies.* Analyst, 2005. **130**(5): p. 606–625.
85. Maroli, A.S., et al., *Omics in weed science: a perspective from genomics, transcriptomics, and metabolomics approaches.* Weed Science, 2018. **66**(6): p. 681–695.
86. Hartmann, T., *From waste products to ecochemicals: fifty years research of plant secondary metabolism.* Phytochemistry, 2007. **68**(22–24): p. 2831–2846.
87. Sakurai, N., *Recent applications of metabolomics in plant breeding.* Breeding Science, 2022: p. 21065.
88. Marchand, C.R., et al., *A framework for development of useful metabolomic biomarkers and their effective knowledge translation.* Metabolites, 2018. **8**(4): p. 59.
89. Lemaitre-Guillier, C., et al., *VOCs are relevant biomarkers of elicitor-induced defences in grapevine.* Molecules, 2021. **26**(14): p. 4258.
90. Watanabe, M., et al., *Comprehensive dissection of spatiotemporal metabolic shifts in primary, secondary, and lipid metabolism during developmental senescence in Arabidopsis.* Plant physiology, 2013. **162**(3): p. 1290–1310.
91. Lin, Z.-H., et al., *The GC-TOF/MS-based Metabolomic analysis reveals altered metabolic profiles in nitrogen-deficient leaves and roots of tea plants (Camellia sinensis).* BMC Plant Biology, 2021. **21**(1): p. 1–13.
92. Dauda, W.P., et al., *Metabolomic analysis of sheath blight disease of rice (Oryza sativa L.) induced by Rhizoctonia solani phytotoxin.* Journal of Applied Microbiology, 2022.
93. Tarpley, L., et al., *Biomarker metabolites capturing the metabolite variance present in a rice plant developmental period.* BMC plant biology, 2005. **5**(1): p. 1–12.
94. Song, H.-H., et al., *Metabolomics investigation of flavonoid synthesis in soybean leaves depending on the growth stage.* Metabolomics, 2014. **10**(5): p. 833–841.
95. Rivera-Pérez, A., R. Romero-González, and A.G. Frenich, *Fingerprinting based on gas chromatography-Orbitrap high-resolution mass spectrometry and chemometrics to reveal geographical origin, processing, and volatile markers for thyme authentication.* Food Chemistry, 2022: p. 133377.
96. Ivanović, S., et al., *Plant Metabolomics as a Tool for Detecting Adulterants in Edible Plant: A Case Study of Allium ursinum.* Metabolites, 2022. **12**(9): p. 849.
97. Adeniji, A.A. and O.O. Babalola, *Metabolomic applications for understanding complex tripartite plant-microbes interactions: Strategies and perspectives.* Biotechnology reports, 2020. **25**: p. e00425.

98. Khan, N., A. Bano, and M.A. Babar, *Metabolic and physiological changes induced by plant growth regulators and plant growth promoting rhizobacteria and their impact on drought tolerance in Cicer arietinum L.* PloS one, 2019. **14**(3): p. e0213040.
99. Razzaq, A., et al., *Metabolomics: A way forward for crop improvement.* Metabolites, 2019. **9**(12): p. 303.
100. Chen, B.-X., et al., *Identification of metabolomic biomarkers of seed vigor and aging in hybrid rice.* Rice, 2022. **15**(1): p. 1–12.
101. Dong, T., et al., *Altitudinal Variation of Metabolites, Mineral Elements and Antioxidant Activities of Rhodiola crenulata (Hook. f. & Thomson) H. Ohba.* Molecules, 2021. **26**(23): p. 7383.
102. Zeiss, D.R., et al., *Altered metabolomic states elicited by Flg22 and FlgII-28 in Solanum lycopersicum: intracellular perturbations and metabolite defenses.* BMC plant biology, 2021a. **21**(1): p. 1–17.
103. Zeiss, D.R., L.A. Piater, and I.A. Dubery, *Hydroxycinnamate amides: Intriguing conjugates of plant protective metabolites.* Trends in plant science, 2021b. **26**(2): p. 184–195.
104. Patel, M.K., et al., *Plants metabolome study: Emerging tools and techniques.* Plants, 2021. **10**(11): p. 2409.
105. Patel, M.K., A. Mishra, and B. Jha, *Non-targeted metabolite profiling and scavenging activity unveil the nutraceutical potential of psyllium (Plantago ovata Forsk).* Frontiers in plant science, 2016. **7**: p. 431.
106. Patel, M.K., et al., *Metabolic profiling and scavenging activities of developing circumscissile fruit of psyllium (Plantago ovata Forsk) reveal variation in primary and secondary metabolites.* BMC plant biology, 2020. **20**(1): p. 1–15.
107. Sinha, P., et al., *Changes in oil content and fatty acid composition in Jatropha curcas during seed development.* Industrial Crops and Products, 2015. **77**: p. 508–510.
108. Zhang, Q.-Y., et al., *Fatty acid and associated gene expression analyses of three tree peony species reveal key genes for α-linolenic acid synthesis in seeds.* Frontiers in plant science, 2018. **9**: p. 106.
109. Pandey, S., et al., *Nitric oxide accelerates germination via the regulation of respiration in chickpea.* Journal of experimental botany, 2019. **70**(17): p. 4539–4555.
110. Cui, M.-C., et al., *Metabolic profiling investigation of Fritillaria thunbergii Miq. by gas chromatography–mass spectrometry.* journal of food and drug analysis, 2018. **26**(1): p. 337–347.
111. Dias, D.A., et al., *Quantitative profiling of polar primary metabolites of two chickpea cultivars with contrasting responses to salinity.* Journal of Chromatography B, 2015. **1000**: p. 1–13.
112. Mikołajczyk-Bator, K., et al., *Characterisation and identification of triterpene saponins in the roots of red beets (Beta vulgaris L.) using two HPLC–MS systems.* Food Chemistry, 2016. **192**: p. 979–990.
113. Hazzoumi, Z., Y. Moustakime, and K. Amrani Joutei, *Effect of gibberellic acid (GA), indole acetic acid (IAA) and benzylaminopurine (BAP) on the synthesis of essential oils and the isomerization of methyl chavicol and trans-anethole in Ocimum gratissimum L.* SpringerPlus, 2014. **3**(1): p. 1–7.
114. Santoro, V.M., et al., *Effects of growth regulators on biomass and the production of secondary metabolites in peppermint (Mentha piperita) micropropagated in vitro.* 2013.
115. Naeem, M., et al., *Triacontanol-mediated regulation of growth and other physiological attributes, active constituents and yield of Mentha arvensis L.* Plant Growth Regulation, 2011. **65**: p. 195–206.
116. Li, J. and Z. Hu, *Accumulation and dynamic trends of triterpenoid saponin in vegetative organs of Achyranthes bidentata.* Journal of integrative plant biology, 2009. **51**(2): p. 122–129.
117. Perkowska, I., et al., *Identification and quantification of coumarins by UHPLC-MS in Arabidopsis thaliana natural populations.* Molecules, 2021. **26**(6): p. 1804.

118. Morita, H., et al., *New picrotoxinin-type and dendrobine-type sesquiterpenoids from Dendrobium snowflake 'Red Star'*. Tetrahedron, 2000. **56**(32): p. 5801–5805.
119. Ma, C., et al., *Terpenoid metabolic profiling analysis of transgenic Artemisia annua L. by comprehensive two-dimensional gas chromatography time-of-flight mass spectrometry.* Metabolomics, 2009. **5**: p. 497–506.
120. Sobolev, A.P., et al., *Quality traits of conventional and transgenic lettuce (Lactuca sativa L.) at harvesting by NMR metabolic profiling.* Journal of agricultural and food chemistry, 2010. **58**(11): p. 6928–6936.
121. Roessner-Tunali, U., et al., *Metabolic profiling of transgenic tomato plants overexpressing hexokinase reveals that the influence of hexose phosphorylation diminishes during fruit development.* Plant Physiology, 2003. **133**(1): p. 84–99.
122. Jha, R.K., et al., *Introgression of a novel cold and drought regulatory-protein encoding CORA-like gene, SbCDR, induced osmotic tolerance in transgenic tobacco.* Physiologia Plantarum, 2021. **172**(2): p. 1170–1188.
123. Zhou, J., et al., *Metabolic profiling of transgenic rice with cryIAc and sck genes: An evaluation of unintended effects at metabolic level by using GC-FID and GC–MS.* Journal of Chromatography B, 2009. **877**(8-9): p. 725–732.
124. Chang, Y., et al., *Metabolic profiling based on LC/MS to evaluate unintended effects of transgenic rice with cry1Ac and sck genes.* Plant Molecular Biology, 2012. **78**: p. 477–487.
125. Payyavula, R.S., et al., *Metabolic profiling reveals altered sugar and secondary metabolism in response to UGPase overexpression in Populus.* BMC Plant Biology, 2014. **14**: p. 1–14.
126. Iwaki, T., et al., *Metabolic profiling of transgenic potato tubers expressing Arabidopsis dehydration response element-binding protein 1A (DREB1A).* Journal of agricultural and food chemistry, 2013. **61**(4): p. 893–900.
127. Stamova, B.S., et al., *Metabolic profiling of transgenic wheat over-expressing the high-molecular-weight Dx5 glutenin subunit.* Metabolomics, 2009. **5**: p. 239–252.
128. Niu, F., et al., *Metabolic profiling of DREB-overexpressing transgenic wheat seeds by liquid chromatography–mass spectrometry.* The Crop Journal, 2020. **8**(6): p. 1025–1036.
129. Piccioni, F., et al., *NMR metabolic profiling of transgenic maize with the Cry1A (b) gene.* Journal of agricultural and food chemistry, 2009. **57**(14): p. 6041–6049.
130. de Souza, L.P., M. Borghi, and A. Fernie, *Plant single-cell metabolomics—challenges and perspectives.* International journal of molecular sciences, 2020. **21**(23): p. 8987.
131. Yu, X., Z. Liu, and X. Sun, *Single-cell and spatial multi-omics in the plant sciences: Technical advances, applications, and perspectives.* Plant Communications, 2022: p. 100508.
132. Misra, B.B., S.M. Assmann, and S. Chen, *Plant single-cell and single-cell-type metabolomics.* Trends in Plant Science, 2014. **19**(10): p. 637–646.
133. Mo, Y. and Y. Jiao, *Advances and applications of single-cell omics technologies in plant research.* The Plant Journal, 2022. **110**(6): p. 1551–1563.
134. Pérez-López, A.V., et al., *Localization and Composition of Fructans in Stem and Rhizome of Agave tequilana Weber var. azul.* Frontiers in plant science, 2021. **11**: p. 2309.
135. Takats, Z., J.M. Wiseman, and R.G. Cooks, *Ambient mass spectrometry using desorption electrospray ionization (DESI): instrumentation, mechanisms and applications in forensics, chemistry, and biology.* Journal of mass spectrometry, 2005. **40**(10): p. 1261–1275.
136. Takats, Z., et al., *Mass spectrometry sampling under ambient conditions with desorption electrospray ionization.* Science, 2004. **306**(5695): p. 471-473.
137. Lorensen, M.D.B.B., et al., *Spatial localization of monoterpenoid indole alkaloids in Rauvolfia tetraphylla by high resolution mass spectrometry imaging.* Phytochemistry, 2023a: p. 113620.
138. Wu, Z., et al., *A mass spectrometry imaging approach on spatiotemporal distribution of multiple alkaloids in Gelsemium elegans.* Frontiers in Plant Science, 2022b. **13**.
139. Lorensen, M.D.B.B., et al., *Leaves of Cannabis sativa and their trichomes studied by DESI and MALDI mass spectrometry imaging for their contents of cannabinoids and flavonoids.* Phytochemical Analysis, 2023b.

Spatial Metabolomics Using Imaging Mass Spectrometry

Kanchustambham Vijaya Lakshmi

Abbreviations

DESI	Desorption electrospray ionization
FT-ICR	Fourier transform ion cyclotron resonance
IBD	Inflammatory bowel disease
LAESI	Laser ablation electrospray ionization
LA-ICP-MS	Laser ablation-inductively coupled plasma mass spectrometry
LASSO	Least absolute selection and shrinkage operator
LC-MS	Liquid chromatography coupled to mass spectrometry
LDI	Laser desorption ionization
LESA	Liquid extraction surface analysis
LMJ-SSP	Liquid microjunction surface sampling
MALDI	Matrix-assisted laser desorption ionization
MHE	Minimal hepatic encephalopathy
NIMS	Nanostructure imaging mass spectrometry
PTMs	Post-translational modifications
QQQ	Orbitrap, triple quadrupole
SIMS	Secondary ion mass spectrometry
ToF	Time-of-flight

K. Vijaya Lakshmi (✉)
Proteomics Center of Excellence, Northwestern University, Evanston, IL, USA
e-mail: Vijayalakshmi.kanchustambh@northwestern.edu

V. Soni, T. E. Hartman (eds.), *Metabolomics*,
https://doi.org/10.1007/978-3-031-39094-4_13

1 Introduction

Complex networks of cells and molecules constitute the biological processes of an organism. The physiology of each tissue depends on the signaling and intercommunication between heterogeneous populations of cells, and gene regulatory networks in a cell. The systematic and hierarchical analysis of the biological networks within each cell, tissue, and organ in the body will greatly benefit the understanding of physiology and of disease pathogenesis at the molecular level [1]. Each single cell has its unique transcriptome with cell-type-specific transcription factors, and cis-regulatory sequences, proteome, and metabolome that ultimately poses cell with functional differences across populations of cells within each single species, and across kingdoms of life [2]. The current technological advances enabled the measurement of molecular-based omics which can be subcategorized into genomics, transcriptomics, proteomics, and metabolomics, all of which provide functional state of biological systems [1, 3–6].

Lyons et al. have successfully applied integration of human transcriptomic data with mouse multi-omics data including RNA microarray, total protein mass spectrometry (MS), and phosphoprotein MS measurements, and has shown implication of elevated p21-activated kinase (Pak) signaling as a driver in chronic colonic inflammation (colitis) [8]. Through these studies, multi-omics integration provides translational insights in to the mechanisms of colitis pathogenesis and identify Pak as a potential therapeutic target for inflammatory bowel disease (IBD) of patients with liver cirrhosis who may develop minimal hepatic encephalopathy (MHE), multi-omics profiling of the transcriptome, metabolome, and a panel of cytokines of blood samples taken from cirrhotic patients with or without MHE suggested a relationship between cytokines CCL20, CX3CL1, CXCL13, IL-15, IL-22, and IL-6 with alteration in chemotaxis and a link between long-chain unsaturated phospholipids and the increased fatty acid transport and prostaglandin production. These findings together with transcriptomic analysis found altered immune pathways that may collectively contribute to the mild cognitive impairment phenotype in MHE which opens new insights to the understanding of the disease [9]. Wozniak et al. has performed metabolomics and multiplexed quantitative proteomics analysis on more than 200 patient serum samples infected with *Staphylococcus aureus* bacteremia (SaB) including uninfected controls, and identified post-translational modifications (PTMs) in carbamylation of albumin and serum transferrin, unmodified fetuin B, and glycosylated fetuin A of cystatin superfamily of proteins, and cytokine signatures such as IL-6, TGF-β1, TNF, IL-1β, and IL-10 in mortality samples of SaB. The depth of analysis was enhanced through computational methods, which accelerated a number of findings focus on leveraging emerging multi-omics analyses to uncover previously unknown mechanisms of disease and to define biomarkers associated with clinical variables as shown in Fig. 1 [7].Imaging mass spectrometry (IMS) allows the investigation of the spatial distribution of molecules coupled with high mass accuracy and chemical specificity. The combination of molecular speciation with MS analysis renders a chemical microscope that can be used for the label-free

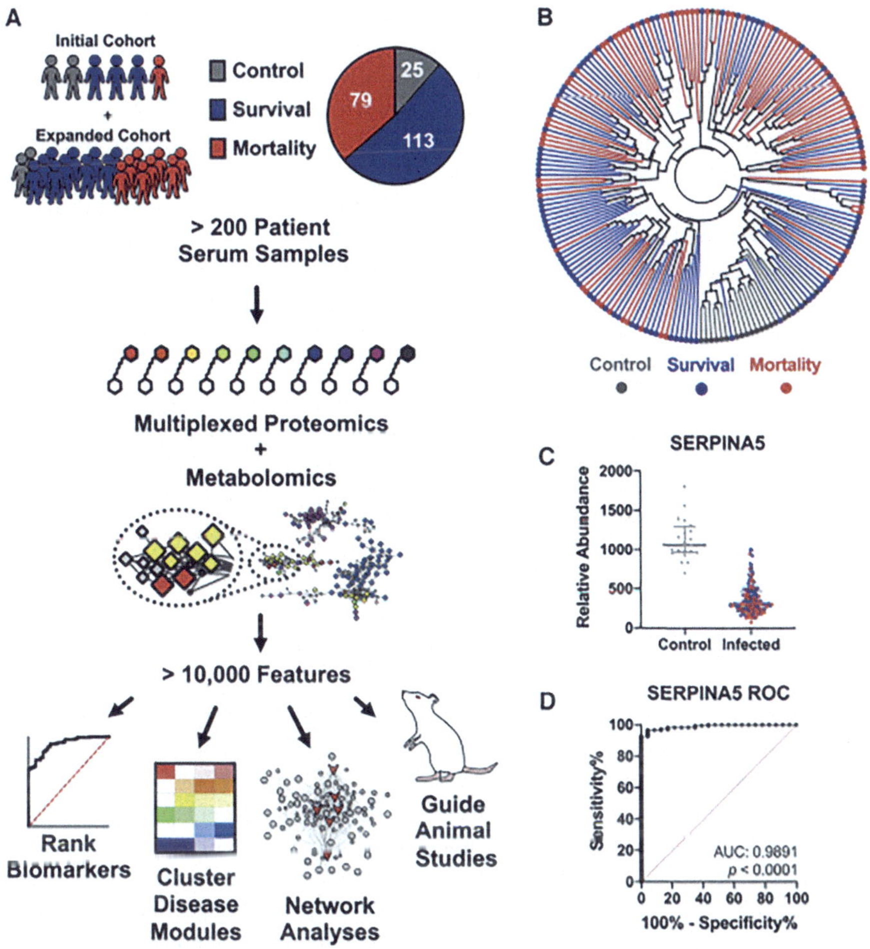

Fig. 1 Multi-omics analysis of SaB patient serum. (**a**) Workflow for SaB serum analysis. (**b**) Hierarchical clustering (Pearson) for proteins detected across all samples. (**c**) Abundance of SERPINA5 in control (gray, NN and HN) and infected samples (blue, HS, red, HM). (**d**) ROC curve of SERPINA5 (control versus infected). Error bars represent interquartile range (IQR). For all tests, significance values are denoted as follows: $^{****}p < 0.0001$; $^{***}p < 0.001$; $^{**}p < 0.01$; $^{*}p < 0.05$; ns, not significant [7]. (Reprinted with permission from Wozniak et al., *Cell*, 2020, 182, 1311–1327. copyright © *2020 Elsevier Inc.*)

detection and mapping of a wide-range biomolecular characterization of small metabolites, lipids, peptides, and proteins and exogenous (drugs and drug metabolites) species in histological tissue surfaces. Spatial information of analytes is traditionally lost through bulk omics analysis of tissue homogenates performed using liquid chromatography coupled to mass spectrometry (LC-MS).

IMS advantageously allows for the detection of the analytes of interest at the desired spatial resolution for its successful use in a multitude of clinical research applications ranging from oncology, pathology, neurological disorders, cardiology, and rheumatology, diagnostics, and surgery to the drug discovery and precision medicine. This chapter will focus on highlighting the significant technological and methodological improvements in the mass spectrometry that have contributed to pushing the limits of high-throughput, spatial resolution, and sensitivity of IMS, and the application of this innovative technology in cancer tissue imaging for classification of normal and tumor, biomarker discovery in clinical applications, and drug distribution. Three commonly used ionization techniques in IMS, secondary ion mass spectrometry (SIMS), matrix-assisted laser desorption ionization (MALDI), desorption electrospray ionization (DESI); liquid extraction methods; the matrix-free laser-based methods of ionization; new instrumental developments; and high-resolution mass analyzers are discussed that extend spatial resolution, mass resolving power, mass accuracy, and tandem-MS capabilities and offer the post ionization electrophoretic separation capabilities for imaging techniques. It will be shown how the success of MS imaging is continuously grown toward clinical research. Finally, supervised and unsupervised data analysis strategies for the large imaging data sets will be briefly discussed.

2 Metabolomics

Metabolomics is often referred to as the youngest of the omics. Unlike genes and proteins, whose function is subject to epigenetic regulation and posttranslational modifications in the biological cascade from genes to proteins to metabolites, the small molecule metabolites represent the closest chemical read out to the expressing phenotype. Metabolomics provides a direct chemical signature of biological activity and can be correlated to the phenotype. Metabolomic studies can lead to enhanced understanding of disease mechanisms and to develop new diagnostic markers as well as study the mechanisms for drug response to individuals for precision medicine [10–13]. Figure 2 shows the biosynthetic pathway of a cellular endogenous metabolite lactate and the multi-omics framework representation of the complexity and interconnectivity of omics data sources [6]. Metabolites are small biomolecules, such as amino acids, sugars, and lipids, which constitute precursors, intermediates, and products in cellular processes such as energy production and storage, signal transduction, and apoptosis [14]. In addition, there is increasing evidence that metabolites directly initiate cellular signaling cascades and modulate diverse biological processes such as epigenetic mechanisms and posttranslational modifications [15–18]. As a result, metabolomics profiling can link cellular pathways to metabolic reactions, and biological mechanisms in health and disease. Metabolomics has a major impact on understanding of the upstream biological processes and developing therapeutics [15, 19–21].

The omics measurement involves the identification and quantification of biomolecules, which is carried out with the overall aim of inferring the physiological state of an organism based on molecular type, location, and any change in abundance [2,

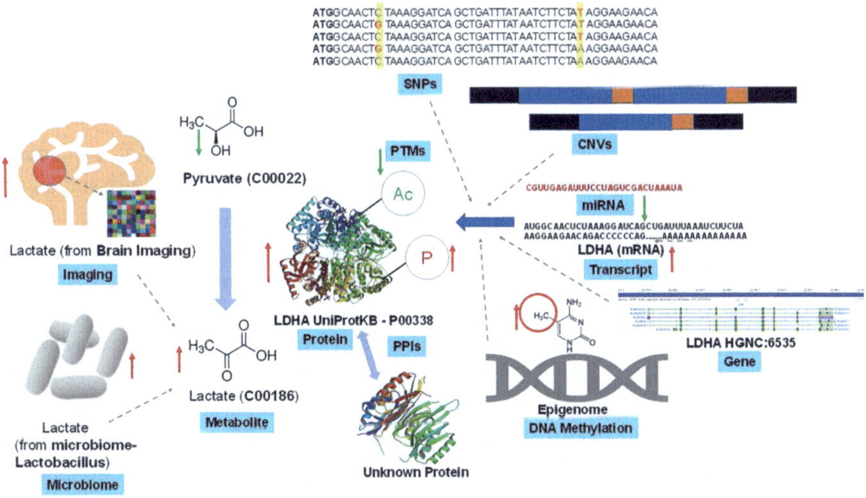

Fig. 2 Example of complexity and interconnectivity of omics data sources in a multi-omics framework. A simple cellular endogenous metabolite, lactate is biosynthesized enzymatically from pyruvate (another metabolite) with the help of lactate dehydrogenase (LDHA, a catalytic protein). In turn this LDHA can interact with several known and unknown proteins through protein-protein interactions to regulate its own function, and itself is subjected to diverse posttranslational modifications (PTMs) that regulate its catalytic function. Lactate measurement through techniques such as in vivo brain imaging in human or other model animals can generate lactate's spatial distribution. Gut microbiome via *Lactobacillus* and other microbes can synthesize lactate and release into human physiological systems to contribute to lactate levels. Lactate biosynthesis regulation can be due to various levels of genetic (e.g., SNPs, CNV), transcriptomic, post-transcriptomic (e.g., miRNA), and/or epigenetics (e.g., DNA methylation) changes on the LDHA gene. Though this is one of the well-studied sets of multi omics interactions, one can expect more complex and unknown interactions while integrating multi-omics data sets [6]. (Reprinted with permission from *Front. Genet.*, 2020, 11:610798. Copyright©2020 Krassowski, Das, Sahu, and Mishra)

15]. However, analysis of metabolites in biological samples is challenging due to the highly dynamic nature of metabolome, and the presence and abundance of metabolites are subjected to the response of cellular activities to the surrounding physical and chemical microenvironments [22–25]. In addition to the changes in metabolite concentrations, metabolite structures are highly diverse due to the presence of complex mixture of isomers and isobars that poses high demand for analytical technique with high resolution and high sensitivity. Metabolomics data sets contain enormous information of known and unknown metabolites, low and high abundant metabolites, while new metabolites are increasingly documented in the databases [26]. In the field of metabolomics, mass spectrometry (MS) has been invaluable as these methods are ubiquitously implemented for bulk characterization of metabolites extracted from homogenized tissue and cell lysates [27]. Recent advances in the mass spectrometry (MS) with increased high mass resolving power and sensitivity, in combination with the development of metabolite databases, provided an avenue for metabolomics studies [28].

2.1 Traditional Mass Spectrometry Analysis of Metabolome

The types of biological samples include tissues, cells, and biofluids and can be analyzed using metabolomics. Traditionally, tissue analysis involves homogenization and extraction of the sample into a liquid form, from which the metabolite changes are averaged across the different cell types and regions of the analyzed tissue using either a targeted or an untargeted approach [10]. Untargeted metabolomics provides a complex data set with a broad range of metabolites present in an extracted sample and analyzed without a prior knowledge of the metabolome. Untargeted metabolomics requires the development of computational methods to identify and correlate metabolites between samples and connect metabolic pathways in relation to the phenotype or abnormality. On the other hand, targeted metabolomics approach provides higher sensitivity and selectivity compared to untargeted metabolomics; data sets are analyzed based on a preselected metabolite and metabolic pathways of interest, whereby methods are developed and optimized for the analysis of specific metabolites. Thus, targeted metabolomics can be helpful to validate untargeted metabolomics workflow and the analysis [15, 29–31]. Although bulk metabolomics approach is widely accepted and successfully implemented to study the metabolic reprogramming in an organism or tissues, and bodily fluids [12, 13, 32], it cannot pinpoint the localization of overall metabolic changes to a particular tissue, organ, or spatially localized aberration such as a tumor [33]. To overcome this problem, bulk metabolomics can be performed on individual dissected parts which is time-consuming and hard to scale up the analysis [34–36]. In addition to this total tissue analysis, subregional, cellular, and even subcellular metabolite profiles can provide further insight into structure-to-function relationships which is particularly valuable in the case of heterogeneous tissues such as the brain and cancers [37].

3 Imaging Mass Spectrometry (IMS) for Spatial Metabolomics

Tissue analysis is perhaps the most powerful approach for studying localized and specific responses to stimuli and pathogenesis and drug treatment yielding explicit biochemical information about the mechanisms of disease. Addressing the need for in situ metabolomics and to discern the metabolic differences at a cellular level, researchers have developed an imaging mass spectrometry (IMS)/mass spectrometry imaging (MSI) technology enabling the spatial metabolomics analysis with simultaneous detection of multiple molecular distributions of small metabolites, fatty acids, lipids, sugars, oligonucleotides, peptides, proteins, and xenobiotics such as drugs and their metabolites across the surface of a sample with two-dimensional (2D) or two-dimensional (3D) spatial resolution on the scales of biological samples such as organisms, organs, tissues, and cells. Spatial metabolomics is a rapidly emerging field, fueled by the strong and ever-growing need in biology and medicine to characterize biological phenomena in situ and address a variety of biomedical questions, including the tumor molecular microenvironment, surgical margin evaluation, molecular histology and grading of tissue diagnosis, metabolic deregulation in cancer, neurodegeneration, cardiovascular diseases and other inflammation, metabolic regulation of epigenetics, functions of immune cells during homeostasis and immunotherapy, nanomedicine, host-microbiome interactions, and their contribution to the infections.

Despite the molecular images were obtained by mass spectrometers 40 years ago [38, 39], only recently new technologies have emerged that encompass a wide variety of ionization methods to produce ion maps of a biological sample. Imaging mass spectrometry (IMS) is a label-free molecular imaging technique, without the need for chemical labelling or antibodies which are the main strength of the method [40–42]. Over the past decade, this growing interest has stimulated rapid progress in the development of imaging mass spectrometry (IMS) and integration with the multimodal imaging techniques that have achieved unprecedented sensitivity, coverage, and robustness for its wide applicability in the scientific community. Figure 3 shows the growth of the imaging mass spectrometry field in terms of papers published over the past 34 years or so [42, 43], which shows that IMS has gained widespread applicability and use among biologists, chemists, and mass spectrometrists. Fortunately, now mass spectrometers with high-mass resolving power and increased sensitivity, in combination with the development of metabolite databases using computational approaches such as artificial intelligence (AI), machine learning, and deep learning used in IMS, have transformed the field with novel avenues for metabolomics [33, 44]. In this chapter, we will discuss three commonly used ionization techniques in IMS, secondary ion mass spectrometry

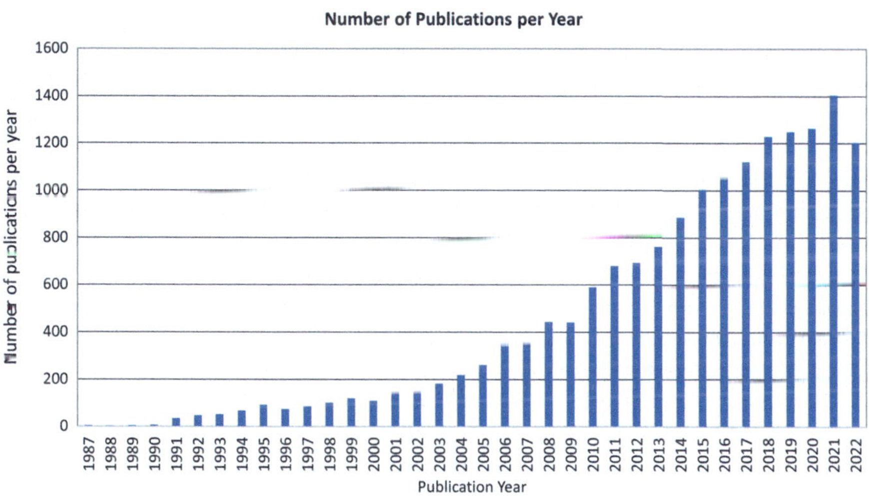

Fig. 3 Number of scientific papers per year obtained from a search containing topical keywords of imaging mass spectrometry using Web of Science data search on December 31, 2022

(SIMS), matrix-assisted laser desorption ionization (MALDI), desorption electrospray ionization (DESI), liquid extraction methods, and the matrix-free laser-based methods of ionization, sample preparation, new instrumental developments, and high-resolution mass analyzers are discussed that extend spatial resolution, mass resolving power, mass accuracy, and tandem-MS capabilities and offer the post ionization electrophoretic separation capabilities for imaging techniques. This chapter focuses on the applications of cancer tissue imaging for classification of normal and tumor, biomarker discovery in clinical applications, and drug distribution. Finally, data analysis using supervised and unsupervised methods for the large imaging data sets will be briefly discussed.

3.1 Imaging Mass Spectrometry (IMS): The State of the Art

Imaging mass spectrometry (IMS) has been demonstrated to be a powerful approach for label-free imaging of spatial distributions of molecular species in situ and does not need a priori knowledge of the potential target species. Although IMS is untargeted analysis by nature, the sample preparation can be tailored for targeted applications. MSI provides the distribution of both large and small molecules in sample systems ranging from cells to whole-body rat or mouse, and tissue sections of organs such as the brain, kidney, pancreas, and retina at high spatial resolution (as low as 1 μm) [45]. IMS enables the detection of thousands of molecules from both targeted or untargeted mode at the same time in a single experiment. There are two different modes of operation in imaging mass spectrometry, which differ significantly in a way how the spatial information is obtained from the tissue. The first one is called microprobe or scanning probe and the second is called microscope. The difference is schematically represented in Fig. 4 [44]. The basic workflow of imaging MS experiment involves the sequential acquisition of mass spectra of analytes simultaneously desorbed and ionized in situ by the ionizing probe from every pixel defined on the surface of the sample in a pre-defined raster pattern or array, termed microprobe mode [38, 46]. The molecular images are reconstructed from the individual mass spectra associated with each specific location that have been obtained from scanning the entire sample surface area. Alternatively, imaging MS in microscopy mode has been demonstrated, where a wide field of view is desorbed by an unfocused projectile following transfer and visualization of the desorbed ions using electromagnetic lenses [47]. Both methods generate the spatial intensity distribution maps of a distinct molecular species over the analyzed mass spectrometry analysis array of pixels [44].

The most prominent IMS techniques include matrix-assisted laser desorption ionization (MALDI) [38, 48], desorption electrospray ionization mass spectrometry (DESI) [49], nanodesorption electrospray ionization mass spectrometry (nano-DESI) [50], nanostructure-imaging mass spectrometry (NIMS) [51, 52], laser ablation electrospray ionization (LAESI) [53, 54], laser desorption

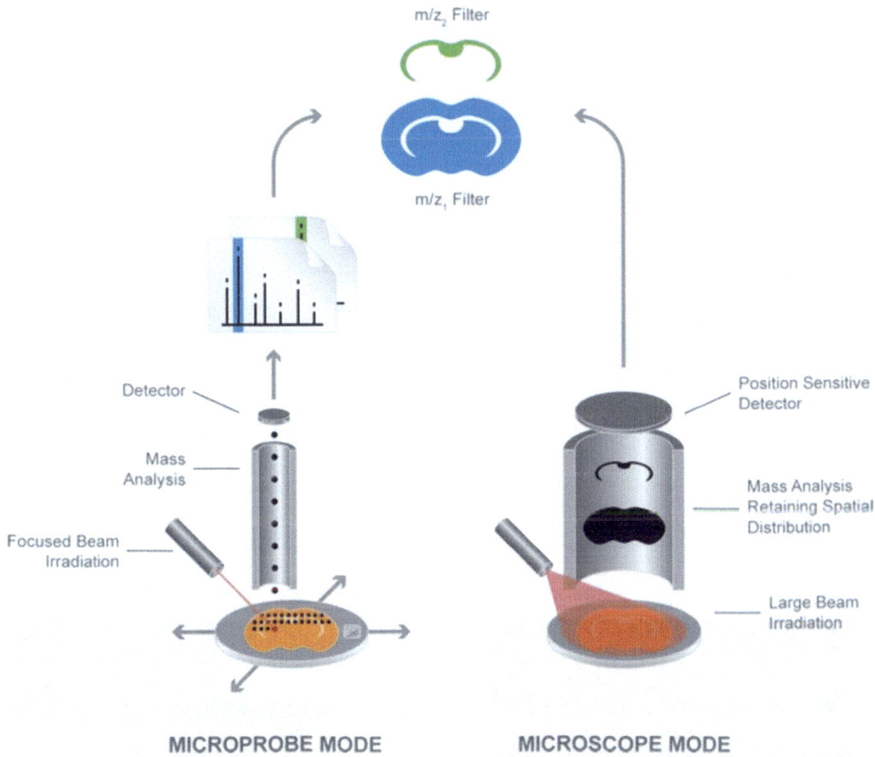

Fig. 4 Schematic representation of microprobe mode and microscope mode used in imaging mass spectrometry [44]. (Reprinted with permission from, *Histochem Cell Biol*, 2010, 134(5):423–43, *Copyright © 2010, Springer-Verlag*)

ionization (LDI) [55], laser ablation-inductively coupled plasma mass spectrometry (LA-ICP-MS) [56, 57], secondary ion mass spectrometry (SIMS) [58], and others that have been used to reveal the cellular localization of metabolites within the tissue sample. Each of these ionization techniques can be combined with different mass analyzers to obtain different spectral resolutions, increased dynamic ranges, or throughput; high-resolution mass spectrometers with mass analyzers such as time-of-flight (ToF) [59, 60], Fourier transform ion cyclotron resonance (FT-ICR) [61], Orbitrap [62], or triple quadrupole (QQQ) has revolutionized the development of imaging mass spectrometry and its potential for biomedical research and pharmaceutical applications. These combinations offer the ability to detect different molecular species with improved sensitivity and specificity and higher spatial resolution for imaging mass spectrometry (IMS) as shown in Fig. 5 [63].

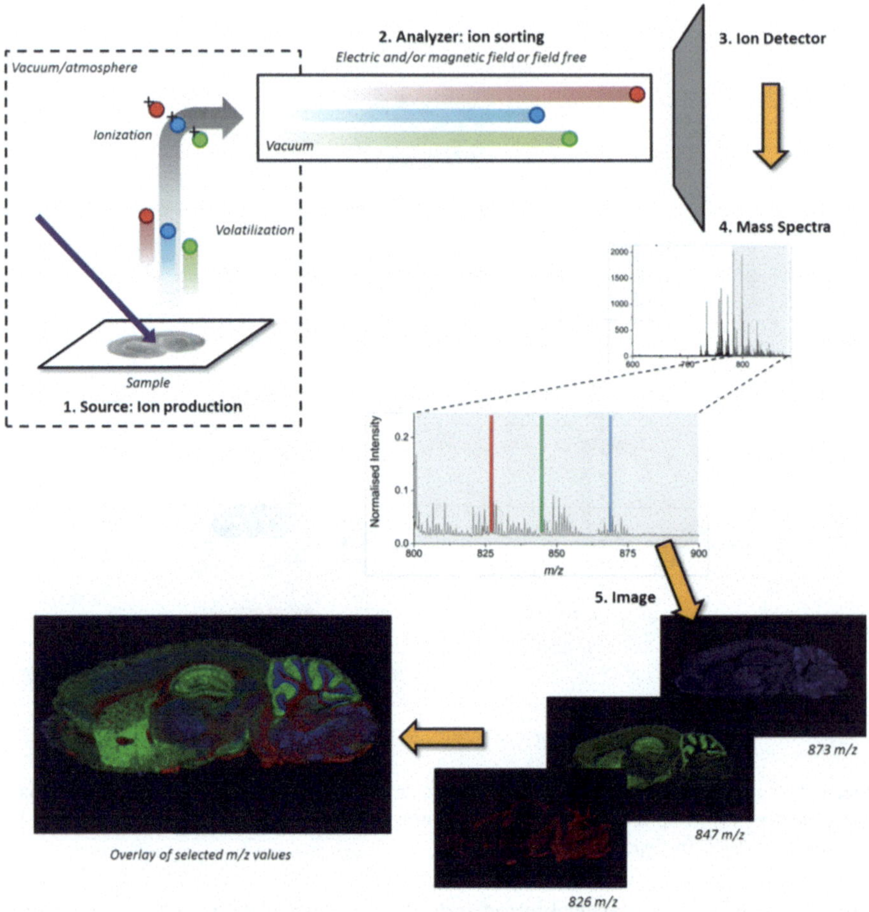

Fig. 5 Mass spectrometry imaging workflow. (1) Production of the charged ions by the ion source from the biological tissue section mounted on a slide. The ions are charged, positively or negatively, which allow them to penetrate the analyzer. (2) The ions penetrate the analyzer and are separated by their mass-to-charge ratio under vacuum condition by an electric/magnetic field. (3) The ions reach the detector; the lighter ions reach the detector first followed by the heavier ones. (4) Mass spectra are obtained with the different molecules corresponding to the peaks present in the spectra. (5) Ion images are generated from each peak corresponding to m/z value of interest in the mass spectrum, and displayed as a function of position in the tissue section and relative intensity. Hundreds of such images can be created from a single tissue section, and multiplexed for spatial mapping of molecules in the tissue. Each molecule can be selected to visualize its distribution in the tissue [63]. (Reproduced with permission from *Curr Pharm Des*, 2017, 23(13):1974-1984. Copyright© Bentham Science Publishers)

3.2 Sample Preparation for IMS: Tissue Embedding

Figure 6 shows the schematic representation of the complete workflow of imaging mass spectrometry (IMS). Sample preparation in imaging mass spectrometry (IMS) is effectively performed on fresh-frozen tissues that are prepared using powdered

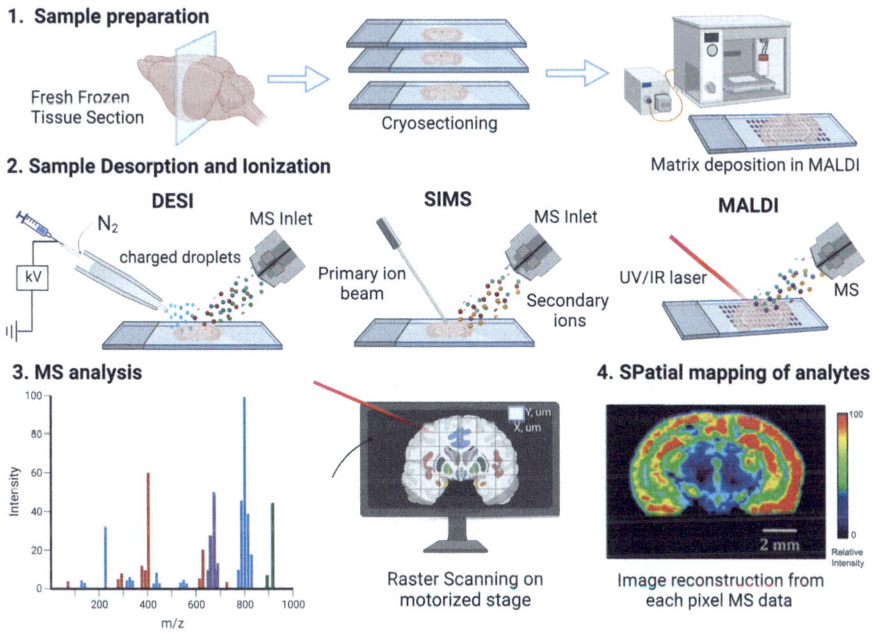

Fig. 6 Overview of IMS workflow using DESI, SIMS, and MALDI. (1) The organ is harvested from the animal or patient and snap-frozen or embedded in gelatin. Fresh frozen or embedded tissue is sliced (10 μm thickness) and mounted on a glass slide to perform SIMS or DESI. For MALDI, the tissue section is mounted on a conductive glass slide, and the matrix is deposited uniformly on top of the tissue using matrix sprayer. (2) In DESI, an impinging electrospray of primary charged droplets extracts the ions into secondary charged from the tissue for MS detection. In SIMS, a primary ion beam bombards the surface of the tissue, generating secondary ions for MS detection. In MALDI, a laser beam UV/IR incident on the sample with UV absorbing matrix generates ions for MS detection. (3) MS data acquisition from the pre-determined array of pixels in microprobe mode, and the image reconstruction from the differential distribution of analytes using MS spectra associated with spatial coordinates and ion abundance from every pixel on the entire surface. (4) Image reconstruction is carried by plotting the ion intensities as a function of the x and y coordinates on the tissue, ion images are generated using softwares such as Biomap, and MSiReader. (The workflow of IMS is Created with BioRender.com)

dry ice, liquid nitrogen, liquid nitrogen-chilled isopentane, etc. Among these fresh-freezing methods, the tissue section morphology appears to be well-maintained when samples are frozen by liquid nitrogen-chilled isopentane commonly used in clinical practice [64].

However, it is important to ensure that tissue section morphology is well maintained with the use of optimal freezing method. If the samples are small, fragile, or amorphous, or if the sample is a plant section with a solid cell wall [65], or if the intracellular space inside the sample is large and filled with water, the samples should be embedded with specific materials such as optimum temperature cutting (OCT) [66, 67], carboxymethyl cellulose (CMC) [68], gelatin [69, 70], or ice [71]. Embedding of the tissue samples in OCT allows maintenance of tissue morphology and precise sample sectioning. However, supporting materials are often smeared through tissue samples, and ionized during IMS data acquisition and poses ion suppressors of molecules of interest [67, 72]. For this reason, it is very important to choose the right embedding material prior to sectioning in IMS. CMC and gelatin compounds are more suitable for MSI than OCT. While CMC is generally selected as the embedding material in relatively large tissues, gelatin compounds are used mainly for small tissue samples [66, 73]. Formaldehyde fixation and paraffin embedding (FFPE) are the most used preservation techniques, which have limited use in IMS due to the formaldehyde-assisted protein cross-linking. The on-tissue proteolytic digestion method in which proteins are denatured, and digested by enzymes, has been developed to overcome the limitation. However, FFPE samples cannot be used for lipid imaging [74].

3.3 Sample Preparation for IMS: Cryosectioning, Mounting, and Freeze-Drying

A common method to section frozen tissue samples is via cryosectioning using a microtome. The frozen or frozen embedded tissue sample is placed in a chamber set at approximately −20 °C and then subsequently sliced to an optimal thickness using a precooled blade. Because the sectioning process is performed at a low temperature, the metabolic process in the tissue can be quenched, and the tissue sectioning can be rapidly performed. Recommended temperatures for sectioning frozen tissues is summarized in Table 1. In IMS, animal tissue sections are generally cut to a thickness of 10–20 μm [72]. Optimization of the temperature and thickness are required to produce high-quality sections for imaging mass spectrometry applications [75]. The most-used method for mounting biological samples is the thaw-mounted method. In this method, a thin frozen section is placed on a precooled (−20 °C) glass slide, or metal plate or indium-tin-oxide (ITO) glass. Then, the tissue is attached to the slide by transferring heat from finger pressing against the surface of the slide [66]. In the case of low-adhesive samples such as plant sections, tissue sample is mounted using an adhesive tape [76, 77]. The mounted samples are stored at −80 °C freezer until use. The mounted tissue sample should be dried prior to IMS analysis.

In general, samples are dried using freeze-drying method [78, 79], air-drying under nitrogen purging conditions or dehydration through solvent washing for further treatment of the tissue surface, followed by drying at room temperature. In freeze-drying method, commonly used for TOF-SIMS, the cold sample is placed immediately on a precooled glass plates inside a vacuum chamber, evacuating the system and slowly allowing the sample to warm up to room temperature at constant pumping ($<10^{-3}$ mbar) for 1–3 h, while water is removed from the tissue. Washing is often required for IMS analysis of peptides and proteins to remove the easily ionizable lipid species in the range of *m/z* 400–1000. The most common washing

Table 1 Recommended temperatures for sectioning frozen tissues

Tissue type	Working temperature
Brain	−12 °C
Liver	−14 °C
Lymph node	−14 °C
Kidney	−16 °C
Spleen	−16 °C
Muscle	−20 °C
Thyroid	−20 °C
Skin	−25 °C
Breast	−25 °C
Breast with fat	−30 °C or below
Adipose tissue	−30 °C or below
Fixed tissue	−12 °C to −17 °C

Adapted from the reference [75] and the resource http://www.ihcworld.com/_protocols/histology/frozen_section.htm

procedure used for MALDI-IMS of proteins is a fixation protocol; 70% ethanol for 30 s, 90% ethanol, and final wash in 95% ethanol for 30 s, followed by drying under ambient conditions [75, 80]. The washing method show some delocalization of the proteins and significantly enhanced signal to noise [81]. Dilapidation is also performed to further remove lipids from tissues using solvents like chloroform and xylene for enhanced detection of proteins directly from tissue. The washing method is optimized for the targeted imaging applications and for removing the staining matrix in MALDI-IMS. For example, 2,5-dihydroxybenzoic acid (DHB) can be removed by methanol after completion of MALDI-IMS analysis, and the tissue samples can be used from hematoxylin-eosin (H&E) staining [81, 82].

4 Ionization Techniques Commonly Used for IMS

The front runners of IMS include MALDI-based IMS, time-of-flight secondary ion mass spectrometry (ToF-SIMS) imaging, and desorption electrospray ionization IMS (DESI) that are currently being used in the widest range of applications [41, 63, 83–85]. In MALDI-IMS, tissue sections are precoated with a UV-light-absorbing matrix, and commonly a UV laser is used for ion desorption and ionization from the surface [38]. In ToF-SIMS, molecular species are desorbed and ionized using a focused beam of primary ions [86], while in DESI, mass spectrometry sampling is based on impinging spray of charged droplets focusing an electrospray onto the sample surface under atmospheric pressure shown in Fig. 5 [49]. The oldest of these three is the secondary ion mass spectrometry or SIMS, which has collected some of the first ion images of monolayers using ion microprobe mass analyzer [39, 58]. The different ionization methods employed in IMS are characterized by complementary

Table 2 Desorption and ionization methods in IMS

Source	Examples	Environment	Energy	Spot size (*d*)	Surface current	MW range (*m*/*z*)
Liquid metalion gun	Ga^+, In^+, Au^+ Au^{2+}, Au^{3+}	UHV	>25 eV	>1 µm	1–10 nA	0–3000
Solid-state gun	Cs^+	UHV	10 keV	2–3 µm	<10 nA	0–3000
C_{60}^+ cluster source	C_{60}^+	UHV	5 eV–40 keV	200 nm–200 µm		0–3000
MALDI	Nd:YAG, N_2 Nd:YLF	UHV, HVambient	100–200 µJ/pulse	5–300 µm	n/a	100–500,000
DESI	H_2O, MeOH, ACN, DMF	Ambient	n/a	>150 µm	0.5–50 nA	100–66,000

Adapted from the reference [88]

strengths and limitations, which concerns mainly the spatial resolution, mass accuracy and mass resolution, and chemical specificity and selectivity as well as molecular mass range of biological applications [87]. Table 2 summarizes the experimental conditions and important aspects of the desorption/ionization methods used in imaging mass spectrometry [88].

5 Secondary Ion Mass Spectrometry (SIMS)

Secondary ion mass spectrometry (SIMS) is the oldest mass spectrometry imaging (MSI) technique [58, 89, 90], which was traditionally applied in the domain of surface and solid-state physics [91]. SIMS primarily uses a focused ion beam of individual or clusters of high-energy particles, such as Bi^+[92], Au_3^+ [93, 94], and C_{60} [95, 96], where upon impact of a sample surface causes emission of secondary ions that are typically analyzed via a time-of-flight (TOF) mass analyzer as shown in Fig. 7a [97]. The particles are sputtered within 5–10 nm from the primary ion indent at high spatial resolution (~50 nm), but the mass range is limited to below 1000 Da in biological samples due to the small fraction of atoms and molecules sputtered from the sample surface being ionized (typically less than 1% of the total sputtered material) and to source-induced fragmentation of surface molecules [98–100]. The absence of matrix in typical TOF-SIMS analysis gains high spectral clarity in the spatial imaging of small ions such as Na^+, K^+, and even H^+ in the low mass range. However, the surface modifications in SIMS such as matrix-covered uppermost layer of the surface in matrix-enhanced SIMS (ME-SIMS) [101], the sputter-coated thin layer of metal in metal-assisted (MetA)-SIMS [102], and the incorporation of cluster ion sources [103, 104] have dramatically improved the

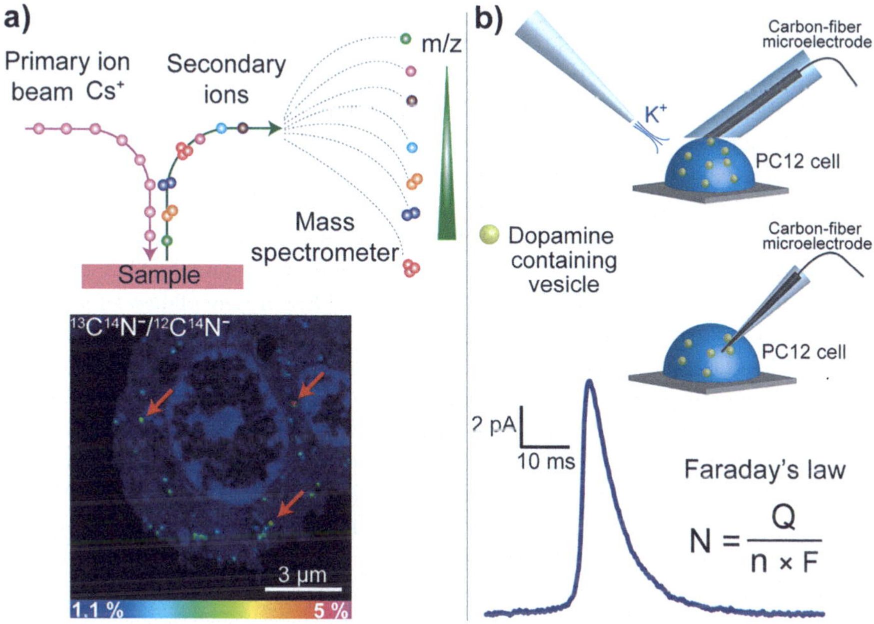

Fig. 7 NanoSIMS and its correlation to amperometric techniques. (**a**) Top: the principle of NanoSIMS analysis. Bottom: $^{13}C^{14}N^{-}/^{12}C^{14}N^{-}$ ratio image reveals the dopamine enrichment in single vesicles. Red arrows showing the three vesicles in the NanoSIMS image. (**b**) Amperometric techniques. Top: the principle of single-cell amperometry. Middle: the principle of intracellular vesicle electrochemical cytometry. Bottom: amperometric current transients allow, by applying Faraday's law, calculation of the mole amount of dopamine that is oxidized from each exocytotic release or individual vesicle (N). Q is the charge calculated from the time integral of current peak from the amperometric trace, n is the number of electrons exchanged in the oxidation reaction ($2e^{-}$ for dopamine), and F is the Faraday constant (96,485 C mol^{-1}) [106]. (Reprinted with permission from *ACS Nano*, 2017, 11, 4, 3446–3455. Copyright © 2016 American Chemical Society)

sputtering efficiency and the yields of secondary ion formation in biological samples to allow for sampling of peptides and oligonucleotides. However, the signal-to-noise ratio for molecules greater than 5 kDa drops off dramatically in ME-SIMS when compared with MALDI [105].

Advances in secondary ion mass spectrometry (SIMS) imaging, with improved detection limit and spatial resolution, have made the SIMS applicable for chemical imaging in biological and biomedical applications [97, 107–109], imaging cells [110], and tissue sections [111]. With different sample surface modifications, SIMS can analyze abundant lipids and small peptides [110, 112]. The new dynamic NanoSIMS ion probe with imaging capabilities at high spatial (50 nm) resolution has traditionally been used to analyze inorganic materials collecting either positively or negatively charged species [113]. The basic operation of NanoSIMS in negative mode involves a high-energy Cs + primary ion beam that scans the sample surface and sputters away secondary particles. The secondary ions are analyzed with a dual-focusing sector mass analyzer and separated based on their mass to charge ratio (m/z). The outstanding spatial resolution of NanoSIMS allows tracking the labeled molecules within single cells [106, 114]. Figure 7b shows the NanoSIMS imaging of the pheochromocytoma (PC12) cells loaded with dopamine biosynthesized from ^{13}C-L-DOPA inside of catecholamine vesicles, and its correlation to the electrochemical data to count the numbers of molecules in the vesicles, as well as those released in exocytosis. The transmitter release upon stimulation was measured with single-cell amperometry [115], whereas dopamine vesicle content was measured with an intracellular vesicle electrochemical cytometry [116]. Recently, the energy range of primary ions used in SIMS has been moved from keV into the MeV domain, which could desorb larger molecular fragments, and less fragmentation of molecular ions [117, 118]. One of the major limitations of SIMS is its low sensitivity to larger molecules, and ion suppression, thereby challenging to separate and identify the small isobaric species. This limitation has been overcome with the recent developments using the tandem MS imaging based on the precise monoisotopic selection of precursor ions from a TOF-SIMS secondary ion stream followed by the parallel and synchronous collection of the product ions. Thus, high-abundance sensitivity is achieved at low primary ion dose density with simultaneous surface screening of a complex matrix chemistry with TOF-SIMS (MS1) imaging and targeted identification of matrix components with MS/MS (MS2) [119].

6 Matrix-Assisted Laser Desorption/Ionization (MALDI)

The second and most widely used technique is MALDI, which first came onto the IMS of biological tissue sections in the late 1990s where it was for peptides and proteins imaging mass spectrometry [38]. MALDI has seen incredible technological advances in its applications to biological systems and proven as a powerful analytical tool for the analysis of biological and clinical tissue samples as shown in Fig. 8 [46, 71, 75, 120]. For example, MALDI imaging mass spectrometry (IMS) was used for characterization, localization, and relative quantification of striatal neuropeptides in a rat model of L-DOPA-induced dyskinesia (LID) in Parkinson's disease (Fig. 8) [120]. MALDI imaging mass spectrometry (IMS) is a soft ionization probe that analyzes proteins, peptides (both endogenous and enzymatically produced), lipids, and small molecules (such as drugs and endogenous metabolites) from the biological samples ranging from whole body [121], tissue sections [122–124], and cells [41, 85, 125–127]. In MALDI, an analyte is co-crystallized with a matrix in a molar ratio of 10^{-2}–10^{-6}. The sample is irradiated with UV or IR laser whose energy is absorbed by the matrix molecules and analytes are desorbed and ionized into the gas phase for MS analysis. Ions are produced by an excess of protons above the surface due to the fragmentation of the acidic matrix molecules [63, 128, 129]. Although the MALDI mechanism remains unclear, one of the accepted mechanisms of MALDI ionization processes involves gas-phase proton transfer between the analyte and the matrix molecules [128]. Therefore, protonated molecules are observed as the dominant ions in the positive ion mode, while sometimes sodiated ($[M + Na]^+$) and

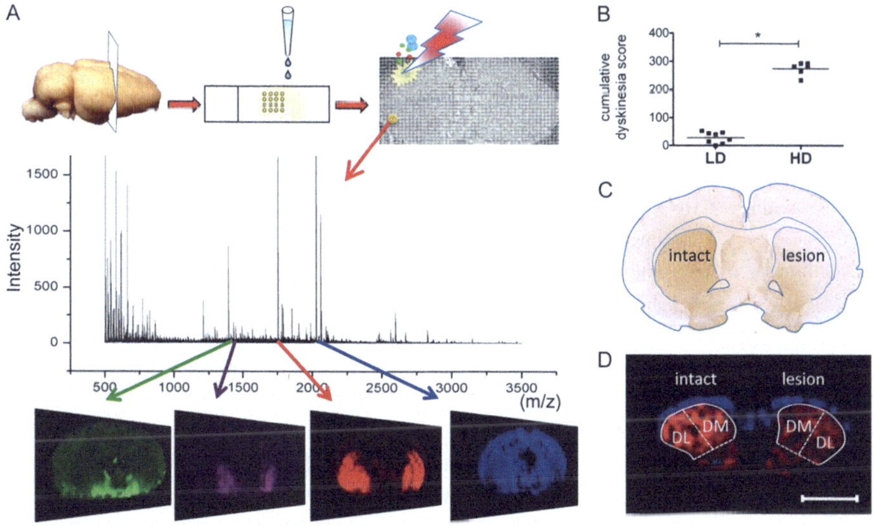

Fig. 8 MALDI imaging mass spectrometry. (**a**) Experimental workflow: sections of frozen rat brains are thaw-mounted onto MALDI compatible glass slides. The tissue section is covered with matrix as discrete spots in a quadratic pattern, irradiated by a pulsed laser beam, and the mass spectra from individual matrix deposits are acquired by means of MALDI TOF MS. MS images of different *m/z* peaks in tissue sections at different pixel size settings, and the intensity distribution of individual peaks is visualized using user-defined colors, here in *green* SP; in *violet* PEnk 220–229; in *red* PEP-19 fragment; and in *blue* MBP fragment. (**b**) Behavioral analysis: High dyskinetic (HD) animals accumulated a tenfold higher dyskinesia score than low dyskinetic (LD) animals. (**c**) Unilateral DA denervation: Unilateral 6-OHDA-lesion model of experimental PD results in selective dopamine-denervation of striatum, revealed by a loss of tyrosine hydroxylase immunoreactivity. (**d**) Regional analysis: Several markers were visualized to delineate anatomical features, including myelin basic protein (*m/z* 2028; *blue*) that is predominantly abundant in the corpus callosum and a PEP-19 fragment (*m/z* 1755; *red*) that is exclusively observed in the striatum. The striatum was divided into a dorsomedial (DM) and a dorsolateral (DL) region of interest (ROI) for MS analysis. * $p < 0.01$. Scale bar 2.5 mm [120]. (Reprinted with the permission from *Molecular & Cellular Proteomics*, 2011, 10, M111.009308. Copyright © The American Society for Biochemistry and Molecular Biology (2011))

potassiated ($[M + K]^+$) molecules are also seen in the MS experiments. In negative ion mode, deprotonated molecules ($[M{-}H]^-$) are observed in the MS experiments. The ionization process is influenced by the chemical properties of both the analyte and the matrix, the amount of matrix deposited with respect to the analyte, and the absorption of laser.

6.1 Application of Matrix for MALDI Imaging

By selecting different matrices and depending on the mass spectrometry instrumentation used (MALDI-TOF or high-resolution mass spectrometry MALDI-FT-ICR), the analyte classes can be chosen for imaging experiments [46, 75]. The

Table 3 Matrix selection for MALDI-MS

Matrix	Chemical name	Biomolecule specificity
DHB	2,5-dihydroxybenzoic acid	Lipids, peptides, <10 kDa proteins
DHB/aniline	DHB + aniline	Lipids, peptides, <10 kDa proteins
DHB/3-AP	DHB + 3-acetyl pyridine	Lipids, peptides, <10 kDa proteins
CHCA	α-Cyano-4-hydroxycinnamic acid	Peptides, small proteins (<10 kDa)
CHCA/aniline	CHCA + aniline	Peptides, <10 kDa proteins
SA	3,5-Dimethoxy-4-hydroxycinnamic acid	Proteins (>10 kDa)
SA/aniline	SA + aniline	Proteins (>10 kDa)
SA/3-AP	SA + 3-acetyl pyridine	Proteins (>10 kDa)
SA/HFIP	SA + 1,1,1,3,3,3-hexafluoro-2-propanol	Proteins (>30 kDa)
SA/TFE	SA + 2,2,2-trifluoroethanol	Proteins (>30 kDa)

Adapted from the references [88, 134]

three most commonly used matrices for MALDI-IMS are 2,5-dihydroxybenzoic acid (DHB) [130], α-cyano-4-hydroxycinnamic acid (CHCA) [131], and 3,5-dimethoxy-4-hydroxycinnamic acid (SA, Sinapinic acid) [132]. However, there are many other useful matrices that have been applied for tissue imaging experiments which are summarized below in the Table 3 [133, 134]. For a successful MALDI imaging experiment, homogeneous matrix deposition across the tissue surface is very important and can be achieved by spraying, spotting, or sublimation using either handheld sprayers, or the automated commercial systems, for example, Bruker ImagePrep, SunChrome SunCollect, HTX TM-Sprayer, tardo iMatrixSpray, and TransMIT SMALDIPrep [135–137].

In MALDI, two different laser types, N2 laser (337 nm) or neodymium-doped yttrium aluminum garnet (Nd: YAG; 355 nm) laser, with repetition rates of 200–1000 Hz and typical pulse lengths of $\leq$3 ns, are used for the analyte desorption/ionization. In the mass spectrometer, the tissue specimens are then raster-scanned under the focused laser with a reduced spot size from 100–150 μm to 20 μm, generating a mass spectrum for each measuring spot and spatial mapping at a resolution ranging from 200μm down to 20μm as shown in Fig. 9 [135, 138, 139]. Over time, MALDI imaging mass spectrometry allowing the direct imaging of single cells and tissues at subcellular spatial using advanced optics that focuses the laser beam to diameters less than 1 μm [140–142], and using a transmission geometry configuration [124]. In addition to achieving high spatial resolution, breakthroughs in acquisition speed have recently been made possible with a 10-kHz laser and two scanning mirrors (e.g., rapifleX MALDI TissuetyperTM, Bruker) that allow the laser beam to be rapidly moved across the sample with which acquisition rates can go up to 50 pixels/s [46]. Measurement speed plays an important role when comparing quantitative differences between two biological conditions, and higher speed enables data acquisition over biological replicates to obtain more reliable results.

Due to its versatility in terms of spatial resolution and molecular coverage, MALDI-MSI has become the most popular MSI technique over the past few years and has been successfully used in many clinical research studies. MALDI imaging

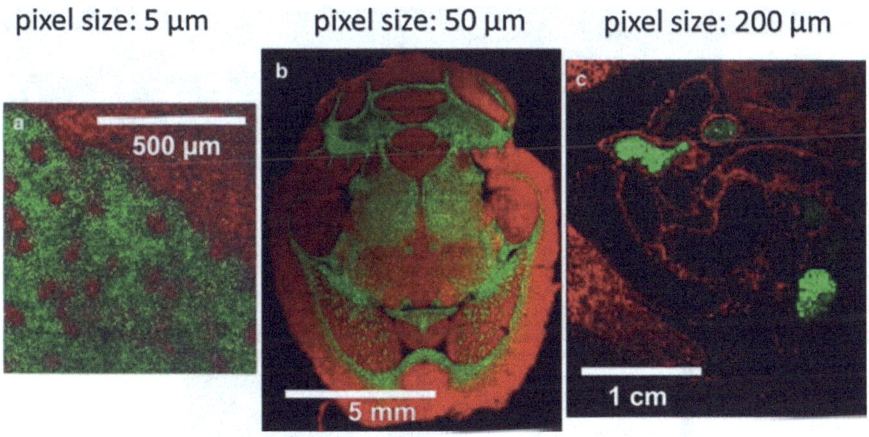

Fig. 9 (**a**) Mouse brain (coronal section), pixel size 5 µm, 170 × 170 pixels. (**b**) Mouse brain (*horizontal section*), pixel size 50 µm, 207 × 260 pixels. (**c**) Intestinal tract of rat (*part of whole-body section*), pixel size 200 µm, 128 × 150 pixels [138, 142]. (Reprinted with permission from *Analytical and Bioanalytical Chemistry*, 2011, 401, 65–73. Copyright © 2013, Römpp et al.)

combines the sensitivity and selectivity of mass spectrometry with the spatially descriptive characteristics of classic histology which has demonstrated potential for biological and clinical applications of cancer research [139, 143–145], neurodegenerative diseases [146], imaging of pharmaceuticals [147–149], metabolites [150], lipids [151, 152], peptides, and proteins [38, 153] including the distribution of neuropeptides [153, 154], tumor delineation in glands and N-linked glycans in tissues [155–157], molecular phenotyping of CNS glial cells in the mammalian brain tissues [158], microbial communities [159], and in nanomedicine [63]. For each of these applications and molecular classes of interest, specific sample preparation protocols have been optimized in MALDI-IMS which include solvent washes and recrystallization for low-abundance intact proteins [160, 161], enzymatic on-tissue digestion for peptide imaging [162–164], and in situ derivatization of analytes in order to increase their ionization efficiency [165–167].

6.2 *Enzymatic On-Tissue Digestion of FFPE Tissues for MALDI Imaging*

Enzymatic on-tissue digestion is particularly advantageous for analysis of formalin-fixed paraffin-embedded (FFPE) tissues because digestion liberates peptide fragments from the chemically cross-linked tissue. The size of FFPE archives and the importance of spatial proteome data have led to a growing demand for the application of MALDI-IMS to FFPE tissue, and the first being direct trypsin digestion followed by MS analysis [168, 169] and the second, antigen retrieval (AR) method,

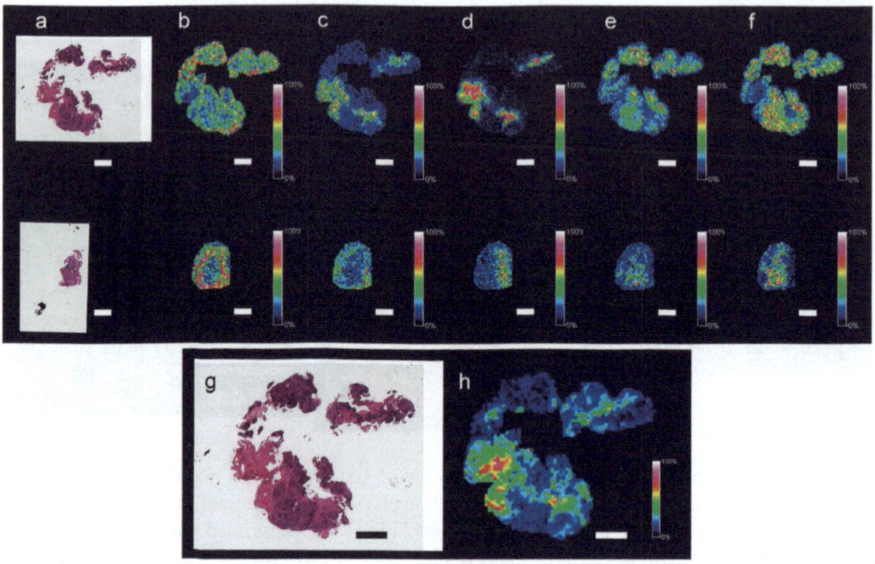

Fig. 10 Ion intensity maps generated for peptides observed on CAAR-treated FFPE tissue. An archived FFPE (2008 FFPE sample) section of human ovarian cancer (**a–f**, top row) was mounted onto a slide and treated with CAAR. A fresh-frozen section of ovarian cancer tissue (**a–f**, bottom row) from the same patient was mounted onto a separate slide and washed using a previously published method [171]. An ImagePrep station was used to nebulize trypsin followed by CHCA onto both slides. MS acquisition used an ultraflex III MALDI-TOF/TOF instrument. Hematoxylin and eosin (H&E) stains of the FFPE section and fresh-frozen section (**a**) are included as well as ion maps for *m/z* (**b**) 1143, (**c**) 1198, (**d**) 1210, (**e**) 1390, and (**f**) 1553. An (**g**) enlarged H&E stain and (**h**) ion intensity map for *m/z* 1267 are also included. Scale bars = 2 mm. Ion intensity color scales are included [163]. (Reprinted with permission from *J. Proteome Res.* 2010, 9, 9, 4315–4328. Copyright © 2010 American Chemical Society)

whereby high-temperature treatment in a buffer solution partially reverses the cross-linking in FFPE sections followed by in situ tryptic digestion and MALDI-TOF MS [161, 170]. On-tissue digestion coupled with MALDI-IMS enables the spatial proteome analysis and the investigation of molecular markers of disease. Figure 10 shows the imaging mass spectrometry analysis of an archived FFPE section of human ovarian cancer treated with citric acid antigen retrieval (CAAR) [163]. Figure 10 represents the spatial distribution of peptides observed at *m/z* 1143 (Fig. 10b), 1198 (Fig. 10c), 1210 (Fig. 10d), 1267, (Fig. 10h), 1390 (Fig. 10e), and 1553 (Fig. 10f) from both CAAR-FFPE (top row) and fresh-frozen ovarian tissue (bottom row). Hematoxylin and eosin (H&E)-stained images of the tissue section used for IMS are shown in Fig. 10a and g. As mentioned in the previous section, on-tissue digestion involves specific sample preparation protocol, which includes consecutive washes to avoid OCT embedding for IMS applications. Sample preparation appeared to have an effect of degree of delocalization as shown in Fig. 10b and Fig. 10c (bottom panels) [163].

7 Desorption Electrospray Ionization (DESI)

Unlike MALDI and SIMS, which are both performed in vacuum, several ambient ionization methods have been developed recently for different applications reviewed elsewhere [172, 173]. These ambient ionization methods involve sampling of unmodified and fresh-frozen sections in the open environment using a solvent spray, laser beam, or plasma probe to desorb and ionize the analytes from the surface. Desorption electrospray ionization mass spectrometry or DESI is the newest and soft ionization tissue imaging technique among the three MALDI, SIMS, and DESI [49, 174]. In DESI, pneumatically assisted stream of charged solvent droplets impinging on the tissue surface where, upon contact with the sample subsequent collisions of charged droplets with the wetted sample surface, it desorbs the analyte into secondary-charged droplets sprayed to the mass spectrometer and analyzed as shown in Fig. 11 [175–177]. The DESI imaging mass spectrometry setup is shown in Fig. 11b, and the optimal parameters such as the spray geometry, gas pressure, and the solvent flow rate used in the DESI imaging experiments to attain a spot size of 150–200 μm are summarized in the Table 4 [176]. Figure 11a shows an average mass spectrum of a mouse-brain tissue section obtained with DMF/EtOH as the solvent system with two distinctive *m/z* regions in the negative-ion-mode MS spectrum indicating the lipid distributions of the gray and white matter substructures of the mouse brain [178, 179]. Figure 11c–f represents high-quality two-dimensional (2D) DESI-MS ion images and comparison with the optical image of the same tissue section histologically stained with H&E after DESI-MS imaging (Fig. 11g).

While the sensitivity and spatial resolution for most applications of DESI are less than what is attainable by SIMS or MALDI, DESI offers a few unique characteristics that allow it to analyze samples that are otherwise impossible to analyze [41, 85]. For instance, DESI applications have been on cancer diagnostics [180–182] and grading [183], but it has also been used in other biological applications such as mapping changes in lipids associated with spinal cord trauma [184], analyzing pigments and alkaloids in plant tissue [185], metabolic exchange in bacterial imprints [186], antifungal chemical defense [187], and mapping cholesterol in the brain using reactive DESI imaging [185]. Recently, significant progress has been achieved in optimizing the solvent spray for achieving minimal or nondestructive nature of tissue imaging using DESI. In particular, the development of dimethylformamide (DMF)-based solvent combinations has enabled DESI-MS imaging, a nondestructive and histologically compatible technique to be performed with preserving morphological features, and the tissue can be subjected to histological and immunohistochemical staining post DESI-MS imaging experiment as shown in Fig. 11 [179]. Thus, DESI-MS imaging can be easily implemented into a workflow for tissue analysis.

New ionization techniques are continually emerging in the field of mass spectrometry, many of which are successfully applied to imaging mass spectrometry [188]. For instance, nanospray desorption electrospray ionization (Nano-DESI) improves DESI by substituting the charged aerosol with a continuous

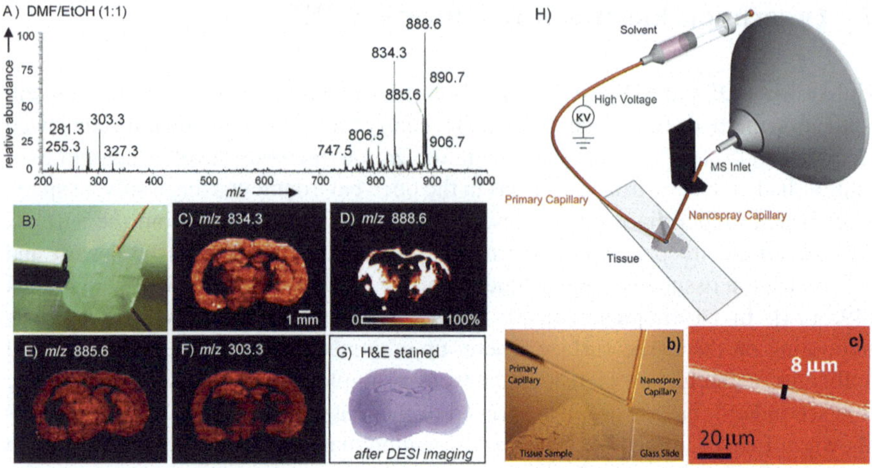

Fig. 11 (**a**) Average DESI-MS mass spectrum of a mouse-brain coronal section including regions of white and gray matter obtained by using DMF/EtOH as the solvent system. (**b**) Picture of the DESI-MS imaging experiment. DESI-MS ion images show the distribution of (**c**) *m/z* 834.3, PS(18:0/22:6); (**d**) *m/z* 888.6, ST(24:1); (**e**) *m/z* 885.6, PI(18:0/20:4); and (**f**) *m/z* 303.3, FA(20:4). (**g**) Optical image of the same tissue section after first being imaged by DESI-MS and then H&E stained [179]. (Reprinted with permission from *Chembiochem.* 2011, 12(14):2129 32.Copyright © 2011 WILEY-VCH Verlag GmbH & Co). (**h**) Schematic drawing of the nano-DESI ion source and (**i**) photograph of the nano-DESI probe taken during ambient imaging of a tissue sample on a glass slide (note that the liquid bridge is not visible to the eye); (**j**) optical image of a trace left by the nano-DESI probe on a rhodamine film on glass [50]. (Reprinted with permission from *Anal. Chem.* 2012, 84, 1, 141–148. Copyright © 2011 American Chemical Society)

micro-extraction liquid (microjunction or bridge) formed between two solvent capillaries and the surface leading to higher spatial resolutions of up to 7 μm, and the schematic of nano-DESI source is shown in Fig. 11h–j [50, 189–194]. Localized analyte extraction from a tissue section occurs within this microjunction, with a flow of liquid sampling the tissue, and the extracted analytes are analyzed by ESI. For native imaging mass spectrometry of proteins in tissue sections, liquid extraction surface analysis (LESA) based on liquid microjunction surface sampling (LMJ-SSP), is used to extract analytes from a liquid junction formed at the concentric capillary and the sample surface, and the extracted analytes into the solvent are aspirated and ionized using ESI [195, 196]. LESA was successfully applied for studying distributions of a wide range of analytes, including proteins at the millimeter scale [197, 198]. Both LESA and nano-DESI allow real-time atmospheric sampling and successfully implemented in proteomics, lipidomics, and the analysis of drugs and metabolites [199–204]. Proteoform imaging mass spectrometry by combining nano-DESI with individual ion detection enabled the proteoforms detection in the higher mass ranges up to 70 kDa at a spatial resolution of 80 μm [205–207].

Like DESI, laser ablation electrospray ionization mass spectrometry (LAESI-MS) is performed under atmospheric pressure and does not need the matrix application to the sample [53]. LAESI uses a combination of mid-infrared laser ablation for

Table 4 Optimum parameters in desorption electrospray mass spectrometry

Parameter	Optimal setting	
Analyte type	Peptides, proteins, carbohydrates, nucleic acids	Explosives, lipids, aromatic hydrocarbons
Electrospray voltage	1–4 kV	3–8 kV
Electrospray flow rate	0.1–3 μl/min	–
Nebulizing gas linear velocity	>350 m/s	–
Heated capillary temperature	200–350 °C	200 °C
Tube lens potential	200–250 V for proteins	30–150 V for small molecules
Capillary inlet sample distance	1–2 mm	2–8 mm
Tip sample distance	1–2 mm	5–8 mm
Incident angle (α)	60–90°	20–50°
Collection angle (β)	<10°	10–15°

Adapted from the reference [176]

desorption and an orthogonal electrospray for transferring multiple charges to the analytes, with subsequent MS/MS identification in a single run. LAESI has been applied to the molecular imaging of small molecules such as metabolites and lipids to larger biomolecules such as peptides and proteins. Finally, the matrix-free laser methods are laser-ablation inductively coupled plasma (LA-ICP), which enables analyzing elemental distribution in samples [208], and nanostructure initiator mass spectrometry (NIMS) [209] is also a worth-mentioning ambient ionization imaging mass spectrometry method. The emergence of novel techniques allowing near real-time analysis of patient material without extensive sample preparation and under ambient conditions should push MSI further in the fields of clinical research and diagnostics. Table 5 provides a list of the different MSI ionization methods.

8 IMS Instrumentation: Technical Developments in Mass Resolution, Accuracy, and Selectivity

Imaging mass spectrometry (IMS) has not only advanced in terms of speed, spatial resolution, and sample preparation but also improved mass analyzer technologies to enhance mass resolution and mass accuracy in IMS data [138, 141, 142, 210]. The key parameters that have the most profound impact on the IMS data are the sample preparation [211], source type, and polarity of the analyzer predetermine the molecular class that is desorbed and ionized, and thus highly inform the molecular content of the IMS data. However, these parameters do not directly affect the spectral properties of the data, but the mass analyzer type in the mass spectrometer has a strong impact on the spectral properties and the size of the data which makes most of the signal processing, data analysis, and molecular analysis [33, 212]. The mass

Table 5 Overview of the main techniques used for clinical MSI [83]

Ionization	Pretreatment	Molecular classes	Lateral resolution
MALDI Laser ablation and desorption/ionization within the ablated plume	Application of matrix solution	Dependent on the matrix; metabolites, lipids, neurotransmitters, peptides, or proteins	Commercial instruments <20 μm
SIMS Sputtering of sample with a focused primary ion beam	Not necessary, matrix/metal coatings used to increase yield of molecular ions	Static SIMS (<1% of surface analyzed by primary ion beam) – elements, fatty acids and lipids Dynamic SIMS – elements	Static SIMS >1 μm dynamic SIMS <1 μm
DESI Molecules collected from surface by impinging droplets then ionized by an electrospray like mechanism	None	Lipids, peptides, and proteins from standards. Mostly lipids and small metabolites from tissue	Generally >100 μm
LAESI Gas-phase particles generated by laser ablation ionized through capture by an electropsray	None	Metabolites, peptides, and proteins	<200 μm for imaging applications <50 μm in cell-by-cell LAESI
Nano-DESI Surface molecules sampled by liquid bridge then ionization by nanoESI	None	Dependent on solvent composition: metabolites, peptides, and proteins	<10 μm reported. Dependent on size of liquid bridge
LESA Surface molecules sampled by liquid bridge then ionization by micro/nanoESI	None	Dependent on solvent composition: metabolites, peptides, and proteins	>200 μm
MassTag/TAMSIM Antibodies functionalized with mass spectrometric reporter groups	Addition of functionalized antibodies (+ MALDI matrix for MassTag)	Proteins	Commercial instruments <20 μm

resolving power predetermines how well two molecules with similar *m/z* values can be resolved from each other, and the mass accuracy is, how accurately the measured centroids of the spectral peaks for a molecule represent its theoretical *m/z* value.

The mass resolving power and mass accuracy both are critical parameters in defining the size and the quality of the IMS data. The higher the spatial resolution, the smaller the histological features accessible in the study, whereas the higher the mass resolving power, the lower is the interferences between adjacent neighboring peaks which further helps improve the specificity of the spatial distribution information. For instance, Fig. 12a shows a mass spectrum from a single 50-mm pixel of the measurement of tryptic digestion of mouse brain tissue, acquired using atmospheric-pressure (AP) MALDI source attached to a linearion trap/Orbitrap mass spectrometer. Figure 12c and e represents the selected ion images corresponding to the peptide of the protein myelin (sequence HGFLPR, theoretical *m/z* 726.4036 for the $[M + H]^+$ion detected at a mass resolving power of R = 24,400 in this spectrum) (green) and the isotopomer of the ammonium adduct of the lipid phosphatidylserine 30:0 (theoretical *m/z* 726.5128, monoisotopic peak: *m/z* 725.5075 detected at a mass resolving power of R = 25,800) (red) generated with a bin width of $\Delta m/z = 0.01$, and Fig. 12d represents the selected ion image generated with a larger bin width of $\Delta m/z = 0.1$. In the case of the larger bin width, the fine structures of lipid and peptide were not resolved, leading to incorrect representation of the spatial distribution of the peptide [213].

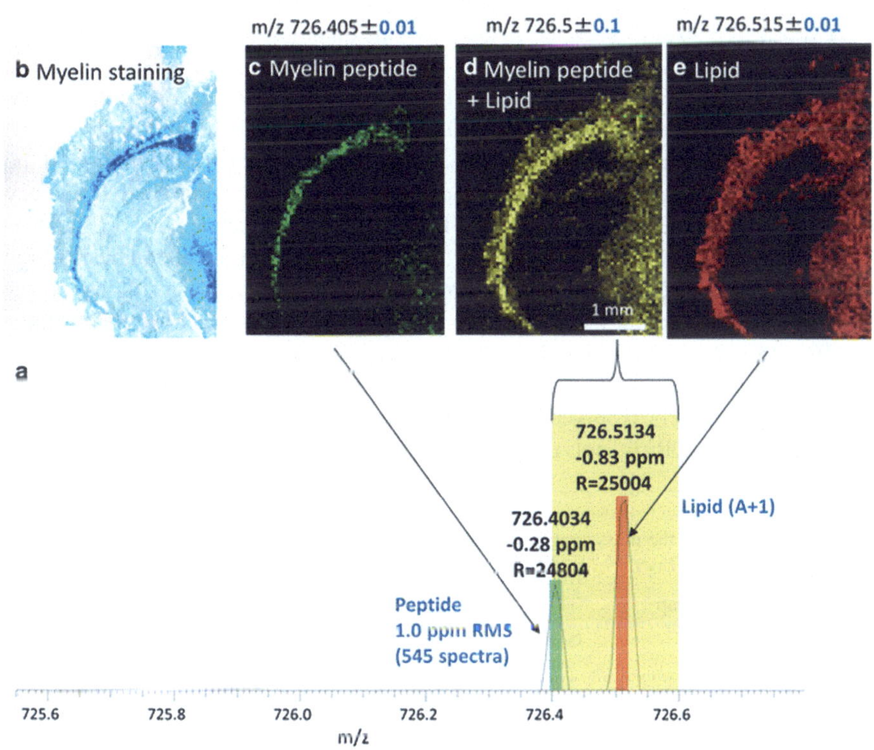

Fig. 12 Mouse brain (*coronal section*) after on-tissue tryptic digestion. (**a**) Mass spectrum from a single 50-µm pixel. (**b**) Optical image of adjacent section after staining for myelin (*Luxol fast blue*). (**c–e**) MS images, 50-µm pixel size, 92 × 128 pixels: (**c**) selected ion image of *m/z* = 726.40–726.41 corresponding to myelin peptide. (**d**) Selected ion image of *m/z* = 726.40–726.60. (**e**) Selected ion image of *m/z* = 726.51–726.52 corresponding to lipid isotopologue peak. (Details on method can be found in Schober et al. [138, 213]. Reprinted with permission from *Histochem Cell Biol*. 2013; 139(6): 759–783. Copyright © 2013, Römpp et al.)

9 Different Types of Mass Analyzers

Mass analyzer constitutes an important component of the mass spectrometer, which takes the ionized masses and separates them based on their mass-to-charge ratio (m/z). Types of mass analyzers include time-of-flight (ToF), quadrupole, magnetic sector, ion trap and Orbitrap or a combination system such as tandem mass spectrometry (MS). The overview of the characteristics, and performance of the most commonly used mass analyzers is listed in Table 6. [44, 214, 215] Interestingly, all six mass analyzers can be utilized for imaging mass spectrometry, but time-of-flight (TOF) is by far the most frequent analyzer currently used in imaging mass spectrometers [60, 216]. Particularly, the introduction of orthogonal acceleration geometry (oa-TOF) allows for hybridization between TOF and quadrupole mass analyzers, that is, QqTOF with mass resolving power up to 50,000 and with MS/MS ability have been demonstrated to the applications of small molecules [217–219]. TOF/TOF instrumentation that enables multiple fragmentation events per single-laser shot was also demonstrated to the on-tissue quantitation of drugs [220].

The other type of mass analyzers are Fourier transform ion cyclotron resonance (FT-ICR) [61] and Fourier transform (FT)-Orbitrap [62] in which the ions undergo a periodic motion with different frequencies that depend on their *m/z* values. External electrodes measure an image current induced by the periodic motion of the ions called *free induction decay* (FID) and converted from time domain to frequency domain using Fourier transformation (FT). The individual frequencies are recalculated to *m/z* values, while amplitudes of the signals represent the abundances of the ions in the mass spectrum. FT-ICR provides the highest mass resolving power (≫100,000), and accuracy (<1 ppm) which allows for multistage MS/MS capabilities [61, 142, 221, 222]. FT-ICR high-resolution MS imaging enables the acquisition of mass channels with a bin size as low as $\Delta m/z = 0.005$, thereby revealing new features that cannot be resolved with lower-resolution instrumentation [223]. On the other hand, FT-ICR is not routinely used for high-throughput/high-resolution

Table 6 Overview of the mass analyzers used in mass spectrometry

Quadrupole	Ion trap	Linear TOF (rTOF)	Magnetic sector	FT-ICR	FT-Orbitrap
Upper mass limit (m/z)	4000	6000	1,000,000 (10,000)	20,000	30,000
Mass resolution	2000	4000	5000 (20,000)	<100,000	500,000
Mass accuracy (ppm)	100	100	200 (10)	10	<5
Ion sampling	Continuous	Pulsed	Pulsed	Continuous	Pulsed
Tandem MS experiment	Yes, low-energy collision	Yes, multiple low-energy collision	Yes, low- or high-energy collision	Yes, high-energy collision	Yes, multiple low-energy collision

Adapted from the reference [44]

imaging due to the long in-cell accumulation. The Orbitrap, a compact but powerful mass analyzer [224], can produce a resolution of >100,000 at *m/z* 400 for a 1.5-s acquisition time, or 60,000 at a 1-Hz acquisition rate, which can resolve the localization of analytes with similar *m/z* values [210, 225–227].

Analyzer developments in SIMS such as the hybrid SIMS–C_{60}^+ Q-Star instrument, the buncher-TOF configuration of the J105 3D chemical imager, the SIMS-FT-ICR mass spectrometer, and the recently developed 3D OrbiSIMS and the improvements in the gas cluster ion sources (GCIB) have been successfully employed for the analysis on different tissues [212]. Figure 13 represents the different SIMS methods and TOF-SIMS modes along with the experimental conditions used for the analysis of human bone tissue samples. TOF-SIMS analysis is more surface-sensitive and faster and routinely provides a lateral resolution in the low- to below-micrometer range. However, relatively hard ionization suffers from the severe fragmentation of the analytes. Thus, analysis is mostly limited to lipid fragments at a lower mass resolution with routinely used time-of-flight mass analyzer. On the other hand, atmospheric pressure MALDI provides soft ionization and enables intact lipid characterization at high-mass resolution and high-mass accuracy with an Orbitrap mass analyzer. However, the analysis times for Orbitrap-SIMS measurements are very long when compared to that of ToF-SIMS as shown in Fig. 13.

10 Ion Mobility Separation

Since imaging mass spectrometry does not involve chromatographic separation step prior to imaging data acquisition, in situ desorption and ionization of analytes from the tissue generates a very complex data set which requires a multivariate data analysis algorithm. Recently, ion mobility separation-based field asymmetric waveform ion mobility (FAIMS) has been introduced as a post-sampling and post-ionization gas-phase separation technique in an MS imaging workflow. Ion mobility separation is an electrophoretic separation based on collisional cross-section differences between ions [228, 229]. FAIMS coupled to DESI has been shown to enhance the sensitivity and signal-to-noise of gangliosides and cardiolipins in DESI [230]. Ion mobility separation coupled to MALDI has demonstrated to discriminate between ions of interest and interfering biological matrix ions [231]. IMS enables the separation of nominally isobaric species that cannot be resolved by time-of-flight mass spectrometers (TOF-MS). The utility of ion mobility separation in imaging mass spectrometry has therefore increased considerably over the last few years. MALDI-MSI coupled with ion mobility separation has been applied to spatially localize and structurally identify acylcarnitines and other lipid species present in breast tumor xenograft models [232]. In DESI, ion mobility demonstrated its use in studying intact fragile lipid species such as multiply sialylated ganglioside and their acetylated forms that are detected directly from the murine brain tissue [233]. Ion

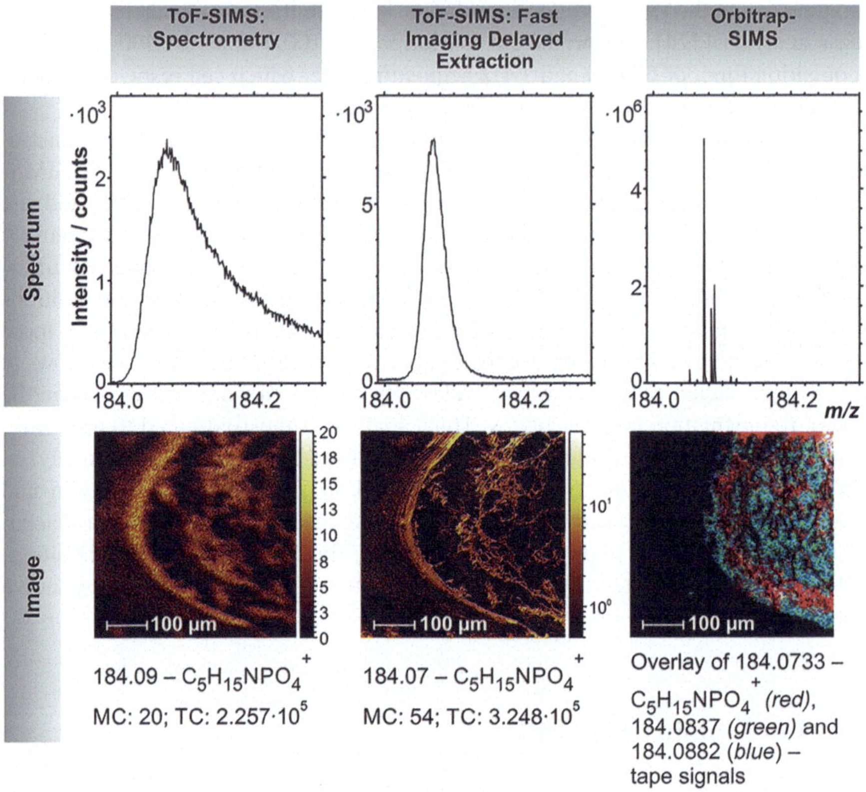

Measurement Conditions

	ToF-SIMS Spectrometry	ToF-SIMS Fast Imaging Delayed Extraction	Orbitrap-SIMS
Primary Ion Species	Bi_3^+	Bi_3^+	Ar_{2500}^+
Ion Dose	$1 \cdot 10^{11}$ ions·cm^{-2}	$2 \cdot 10^{12}$ ions·cm^{-2}	$1 \cdot 10^{15}$ ions·cm^{-2}
Pixels	256x256	1024x1024	250x250
Pixel Resolution	2 µm	488 nm	2 µm
Analysis Time	3 minutes	2.6 hours	8.7 hours
Mass Resolving Power (FWHM)	2,000	5,500	260,000
Mass Accuracy	106 ppm	1 ppm	0.3 ppm

Fig. 13 Comparison of different SIMS methods and TOF-SIMS modes. First and second rows: TOF-SIMS spectrometry mode averaged mass spectrum and ion image for $C_5H_{15}NPO_4^+$ at *m/z* 184 (left), in delayed extraction mode (middle) and Orbitrap-SIMS measurement (right, visualized in RGB mode). Two additional peaks were measured from the tape background with Orbitrap-SIMS but not with the surface-sensitive TOF-SIMS. Third row: measurement parameters. *m/z* values are measured data [212]. (Reprinted with permission from *Anal. Chem.* 2018, 90, 15, 8856–8864. Copyright © 2018 American Chemical Society)

mobility-enabled MS has been used for the analysis of small drugs, carbohydrates, and lipids to peptides, proteins, and protein complexes [135].

11 Tandem Mass Spectrometry

In MSI, tandem MS is particularly applied for targeted imaging and enables monitoring specific ion fragment transition(s) called selected reaction monitoring (SRM), thereby providing unprecedented selectivity and sensitivity for given compounds. Tandem MS has been performed directly from tissues to identify peptides or lipids using either TOF/TOF, Q-TOF, or Orbitrap analyzers [190, 195]. Despite their lower mass resolution, quadrupole and ion trap mass analyzers are particularly suited for targeted imaging of small molecules such as drug imaging studies with high selectivity and speed of SRM in MALDI and DESI [234–237]. In SIMS, parallel MS1 and MS2 have been introduced for tandem-MS imaging [119]. For instance, tandem-MS imaging is used to find out the distribution of the drug in the rat slice by analyzing the structure of the molecule based on their fragmentation and assigning the molecules observed in the mass spectrum [236]. Figure 14a represents a specific transition of the parent drug (*m/z* 468.5 → 252.0) which was detected in the lung, tracheae, and the stomach with the administration of the compound. By using targeted analysis, the file size remains very small and significantly accelerates data processing when compared with TOF data. Figure 14b shows the optical image of a rat slice before matrix coating. This study demonstrates the ability to acquire distribution images of a complete rat section in less than 15 min at a resolution of 1 mm.

12 Identification Strategies for Metabolites, Peptides, and Proteins

The availability of high-resolution mass spectrometers, tandem-MS analysis, and additional separation techniques such as ion mobility separations have greatly improved the identity (ID) of detected *m/z* species. Molecular identification and confirmation of the identified target is the key to predict novel biomarkers and therapeutic targets and also gain insights to the biological and chemical information of the pathological specimen. Tandem-MS analysis of compounds either performed by MS/MS fragmentation experiments directly from tissue, or by matching the measured *m/z* value to a database is the very step in the molecular identification. On-tissue MS/MS analysis often suffers from the small concentrations of analytes per pixel (e.g., in a 50-μm pixel, there are about 25 cells) [135]. Ionization procedures that could increase the ion yield per pixel would be valuable or separate proteomics/metabolomics experiments from tissue extracts into the appropriate solvents.

Ionization procedure in which a second laser to initiate a second ionization wave in the gas phase has been shown to increase ion yields by two orders of magnitude [238]. Overall, there are different identification procedures, which depend on the compounds of interest discussed elsewhere [135]. For small metabolites, and the lipids, the database approach is to first match the measured molecular weight with a small mass tolerance to a public database, such as human metabolome database (HMDB, http://www.hmdb.ca/), METLIN (https://metlin.scripps.edu), or lipid maps (http://www.lipidmaps.org/), and then check if the observed fragmentation spectra are matching with that of the proposed species. The final confirmation can be obtained by comparing the fragmentation pattern with a corresponding standard compound. For protein identification, protein fragmentation patterns were input into ProSight Lite software (http://prosightlite.northwestern.edu/) and compared with protein amino acid sequences that have been previously observed within the tissue types analyzed. The protein amino acid sequences are available from the UniProt database (https://www.uniprot.org/).

13 Image Generation and Data Analysis

Xcalibur RAW files were converted into mzML format using MSconvert (https://proteowizard.sourceforge.io/) and imzML format using (https://ms-imaging.org/imzml/) and then uploaded into the open-source imaging software packages such as MSiReader [239], Cardinal [240], Biomap [241], msiQuant [242], or SpectralAnalysis [243] for visualization of spatial distribution of analytes. All images are normalized to the maximum ion intensity within all the spectra (all pixels) used to create the image, and the raw data from each pixel were extracted for statistical analysis. The success of a MSI study is strongly dependent on the downstream data analysis of dozens to hundreds of samples to achieve the required statistical power. The data processing and handling of IMS data files in gigabyte range has become computationally demanding. Commercial software packages such as Multimaging (Imabiotech, France) or SCiLS (Bruker Daltonics, Germany) are developed to handle terabyte-sized, multi-sample data sets and include many statistical tools needed for biomarker discovery. Prior to data processing, several steps can be used to ensure accurate and efficient data analysis. These steps include normalization, baseline correction, spectra recalibration, smoothing, and data compression (unsupervised and supervised) [244, 245]. Depending on the goal of the experiment, researchers are continually developing data analysis techniques, which account for the very special nature of MSI data, for example, correlation of the multivariate patterns of a low-resolution MSI image with a higher-resolution optical image of the corresponding histological entities to create super-resolution IMS images [246]. Image fusion enhanced the resolution of an IMS image by a factor of ten. The first decision to be made in data analysis is to perform a supervised or an unsupervised analysis on the imaging MS data set.

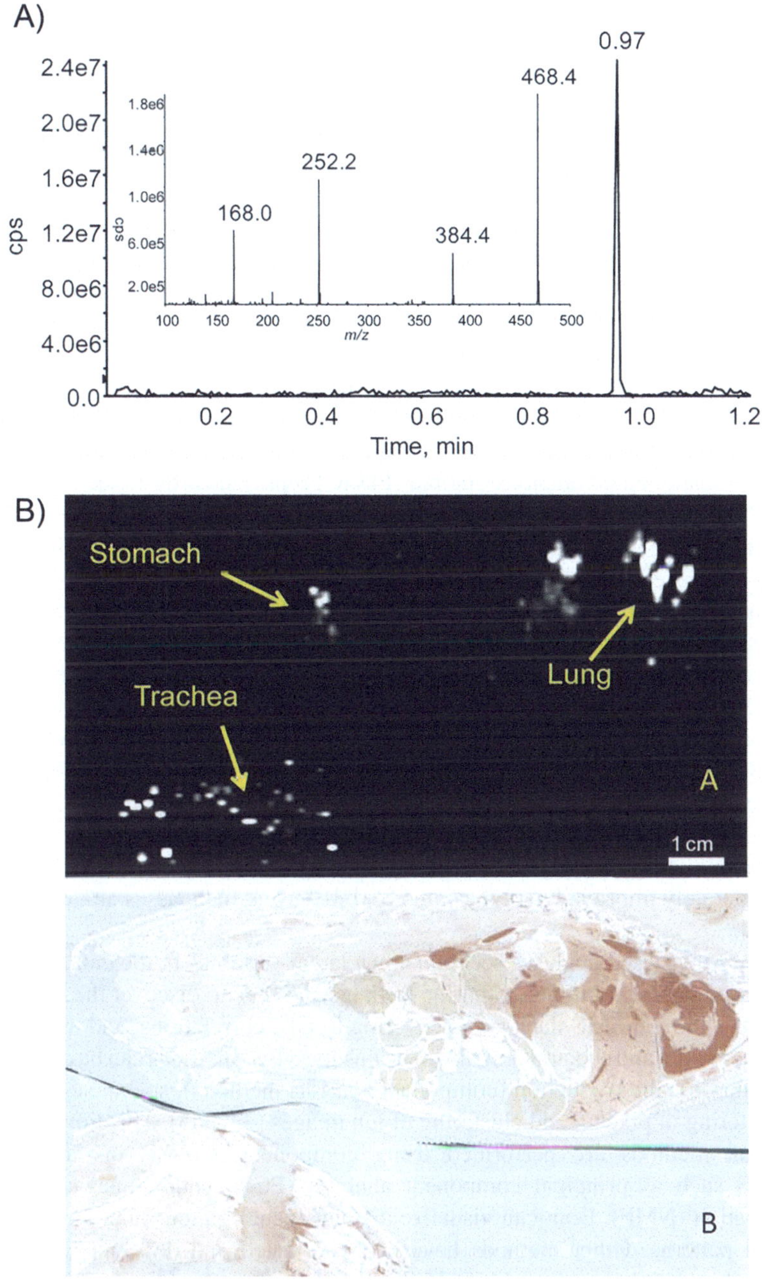

Fig. 14 The product ion spectrum of the [M + H] $^+$ precursor ion at *m/z* 468.4 obtained at a rastering speed of 1 mm/s. (**a**) SRM image (*m/z* 468.5 → 252.0) showing the compound distribution (intensity scale from black to white with increasing intensity). The acquisition of 106 lines corresponding to 10,000 data points (rastering mode: 1-mm distance between lines) took 15 min. (**b**) Optical image of the rat section. This figure is available in color online at www.interscience.wiley.com/journal/rcm [236]. (Reprinted with permission from *Rapid Commun. Mass Spectrom.* 2009;23: 733–736. *Copyright © 2009 John Wiley & Sons, Ltd.*)

In supervised analysis, prior knowledge about the samples is used to perform classification or tumor stratification by identifying profiles or specific biomolecular ions that discriminate the tumor from benign tissue. If information is known about patient outcome or response-to-treatment, a supervised analysis of the tumor-specific profiles is used to identify prognostic markers. In all instances it is essential that the model is validated using an independent validation set of imaging MS measurements. Statistical methods such as baseline least absolute shrinkage and selection operator (Lasso) [248], log-ratio Lasso [249–251], and a combined ridge regression linear model [252] are applied to select statistically significant molecular ion peaks in the imaging MS profiles to build a classifier that estimates the probability of an individual pixel in each tissue image to be normal or cancerous [247, 253, 254]. Figure 15 shows the application of imaging mass spectrometry (IMS) to assess the N-glycome of the human pancreas and pancreatic cancer in a cohort of patients with pancreatic ductal adenocarcinoma (PDAC) represented by tissue microarrays and whole-tissue sections [247]. In this study, analysis of the mass data was performed using hierarchical clustering which has resulted in clusters with distinct glycan properties and patterns for adjacent normal and cancer cores (Fig. 15a), and the supervised machine learning method of LASSO-regularized logistic regression resulted in a subset of predictors with minimal collinearity and had an abundance that was significantly different between the tumor and the adjacent-normal cores (Fig. 15b). LASSO finds classifier masses or the predictors that are complementary to the carbohydrate biomarkers such as carbohydrate antigen 19-9 (CA19-9) and sialylated tumor-related antigen (sTRA) and demonstrated that nine masses in total with seven masses being associated with tumor cores. The area under the curve (AUC) in receiver-operating characteristic (ROC) analysis was 0.939 using the combination, greatly improved over the value of 0.717 using the biomarkers alone and moderately improved over the value of 0.910 using the masses alone (Fig. 15c, d) [247].

The unsupervised analyses seek *unknown* latent variables in the data, which are not focused on a particular recognition task, instead seek to discover the underlying structure within an IMS data set, uncovering trends, correlations, and associations along the spatial and spectral domains. The unsupervised methods can be categorized into dimensionality reduction (component analysis methods), spatial segmentation (or clustering of pixels), and clustering of ion images [33, 255]. The dimensionality reduction methods are performed using component analysis or factorization methods such as principal component analysis (PCA) and nonnegative matrix factorization (NMF). Both can visualize distinct spatial regions and the associated spectral patterns. Other methods have been evaluated and developed which are discussed in detail elsewhere [255–257]. Spatial segmentation aims at partitioning a sample into distinct regions of molecularly similar pixels and is one of the most popular approaches for data representation in imaging MS. Clustering is a powerful tool in in imaging MS and has been demonstrated to be capable of differentiating tumor types, visualizing intratumor heterogeneity, and segmenting anatomical structures [258, 259]. To enable statistical testing for differences in ion intensities between the regions of interest (ROI) within the data set, conditional autoregressive

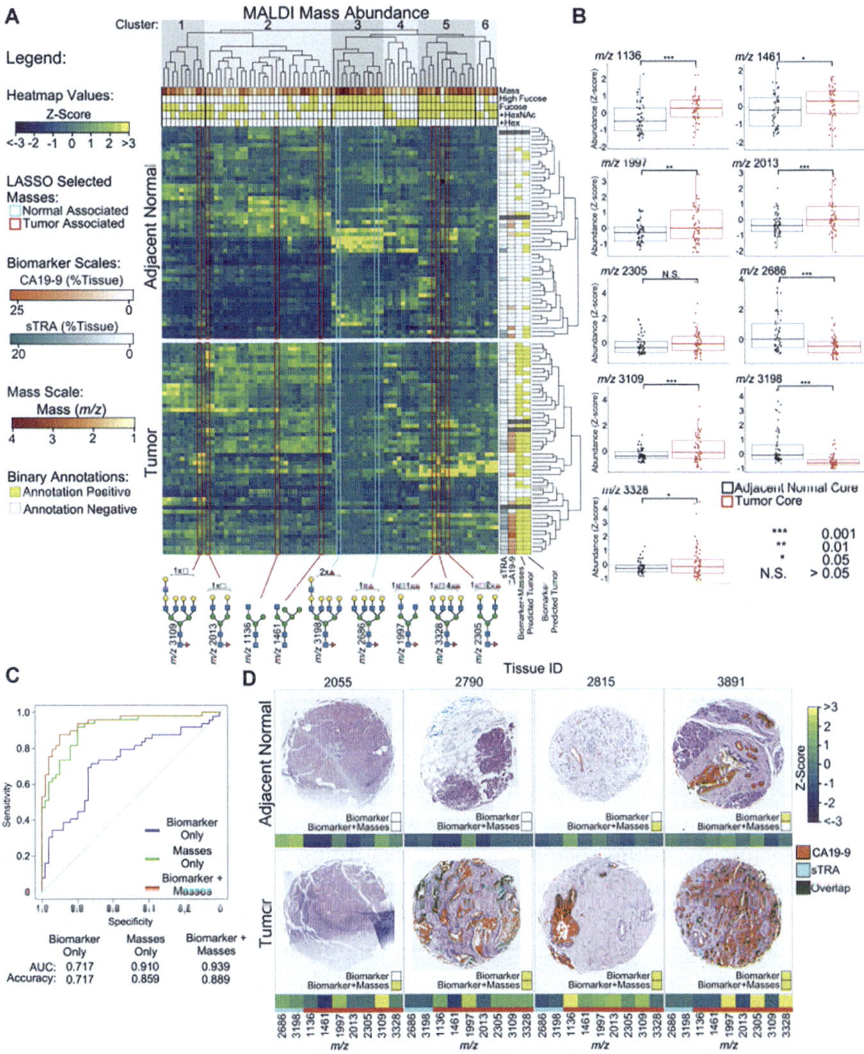

Fig. 15 Mass abundance indicates glycan structures with potential complementary value to current biomarkers. (**a**) Hierarchical clustering and LASSO-regularized logistic regression identified families of masses with potential complementary value to known PDAC biomarkers. Masses selected by LASSO regularization are highlighted and color-coded for tumor-core associated or normal-core associated. Cores predicted to be tumors are indicated by a *yellow box* on the right. (**b**) Individually, the masses selected by LASSO regularization used in the logistic regression model have significantly different abundances between adjacent-normal and tumor cores. (**c**) Receiver-operating characteristic curves of the use of the models to distinguish the tumor cores from the adjacent-normal cores indicate an improved accuracy using the combination of masses and biomarkers. (**d**) Example cores illustrate the complementary value of the masses and the biomarkers. The *yellow box* indicates prediction as a tumor core. The addition of the masses correctly identified a tumor core missed by the biomarkers (2055) and correctly identified a normal core falsely called tumor by the biomarkers (3891). In other instances, the combination produced no change (2790) and falsely predicted a normal core to be a tumor core (2815). LASSO, least absolute shrinkage and selection operator; PDAC, pancreatic ductal adenocarcinoma [247]. (Reprinted with permission from *Mol Cell Proteomics* 2021, 20, 10,001. Copyright © *2020* Colin T. McDowell et al. *Published by Elsevier Inc on behalf of American Society for Biochemistry and Molecular Biology*)

models, which account for the spatial autocorrelation of mass signals, within-sample statistical comparisons can be performed to determine significant differences in drug distributions within a tissue [135, 260]. Likewise, the project MetaSpace uses the spatial information of the detected isotopic patterns to increase the confidence of metabolite identification [33, 261]. The MetaSpace project is an open-data repository where every active user can see the data of other users and is a platform for artificial intelligence developments in the field of spatial metabolomics. MetaSpace provides a free cloud engine for metabolite annotation and encourages users to make their data public, thus creating an open knowledge base of spatial metabolomes.

14 Applications of Imaging Mass Spectrometry in Cancer Research

Imaging mass spectrometry has been gaining importance due to its ability to spatially localize the biomolecules such as metabolites, lipids, drugs, peptides, and proteins which can be used for biomarker discovery that can stratify patients according to their diagnosis, disease state (staging), and prognosis to enable a more personalized therapy [262]. In 2003, Yanagisawa et al. were the first to report the successful use of MALDI-MS tissue profiling histologically selected 1-mm-diameter regions of single frozen sections from lung cancer tissue. Class prediction models based on differentially expressed proteins were used to accurately determine lung cancer histology, distinguish primary tumors from metastases, classify nodal involvement, and predict the prognosis of patients [263]. This study was not an imaging study, but it is a landmark paper that gave evidence of the potential for MALDI-MS imaging on 112 biopsies from lung-tumor patients to discriminate adenocarcinoma from squamous cell carcinoma [161]. In the subsequent years, many oncological studies have exploited the potential of imaging mass spectrometry to find biomarkers for diagnosis, prognosis, or therapy response prediction [135, 264]. Some of the first cancer imaging MS studies using MALDI-IMS were finding protein fragment Reg alpha as the potential biomarker for ovarian cancer [265], and fragment of mitogen-activated protein kinase/extracellular signal-regulated kinase 2 (MEKK2, *m/z* 4355) to accurately discriminate cancer from normal tissue, and by confirming the discriminatory power of biomarkers orthogonally by IHC [266]. MALDI-IMS was used for the classification of Her2 receptor status in breast and gastric cancer tissues [267, 268]. In colon cancer tissues, MALDI-IMS was used to record the protein expression patterns to distinguish between patients with and without metastasis formation [269]. In breast cancer, MALDI-MS tissue profiling and imaging enabled the identification of proteins in pretreatment biopsies that predict response to neoadjuvant taxane-based therapy [270]. In another study on esophageal adenocarcinomas, a protein signature was first found to be correlated with the chemosensitivity to fluorouracil and cisplatin administration [271]. In a separate study on esophageal adenocarcinoma, Walch et al. reported low levels of the mitochondrial protein COX7A2 as being indicative of poor survival [271, 272].

In breast cancer study, MALDI-IMS was used to detect proteomic patterns associated with human epidermal growth factor receptor 2 (HER2) status in breast tumors, in a discovery set ($n = 30$) and a validation set ($n = 18$) [267]. The very same section measured by MALDI-IMS was stained for H&E afterward, and the co-registered imaging mass spectrometric data with the histological features allowed a histology-directed analysis of tissue samples to identify the regions of interest (ROIs) containing cancer cells or non-neoplastic tissue components (e.g., normal terminal ductal lobular units or stroma. Figure 16 shows an average MALDI-IMS mass spectra observed within the mass range of *m/z* 2400 to 25,000 that were shown to discriminate invasive ductal cancer from normal terminal ductal lobular unit. The peak at *m/z* 4309 clearly shows that this feature is specific for terminal ductal lobular units, while peak at *m/z* 22,490 is specific for invasive ductal cancer cells but not present in tumor stroma or other tissue components. In this study, Sandra Rauser et al. have shown that the overexpression of seven *m/z* species accurately define HER2-positive from HER2-negative tissues and identified the feature near *m/z* 8404 as cysteine-rich intestinal protein 1 (CRIP1) strongly associated with HER2 overexpression which demonstrates the potential of MALDI-IMS for tissue diagnostics and provides insights to the biologically significant molecular pathways which are not limited to traditional high-abundance proteins [267].

In addition to the use of IMS for proteomic biomarker discovery, it is also increasingly used to investigate lipids and other small metabolites as biomarkers of cancer [273]. Lipid metabolism undergoes a dramatic change in cancer where the metabolism converges to lipid synthesis [274]. More precisely there is a shift from lipid uptake to de novo lipogenesis, which has an impact on the fatty acid (FA), and lipid compositions of the entire cell. However, the mechanisms of cancer progression's and the metabolic reprogramming and the lipid biochemistry are still under investigation [275]. With the introduction of DESI in 2004, an increasing number of investigations focus on lipid and metabolic imaging mass spectrometry to characterize cancer and adjacent healthy tissues [176, 276, 277]. The biomarkers detected by DESI-MS include cancer-specific small metabolites, fatty acids (FAs), sphingomyelin (SM) and classes of phospholipids such as phosphatidylcholine (PC), phospha tidylserine (PS), phosphatidylinositol (PI), phosphatidylethanolamine (PE), phosphatidyl glycerides (PG), and cardiolipins (CL) [143]. For example, in a study on ovarian cancer, DESI-MSI was used to profile the lipidome of different epithelial ovarian carcinomas (EOC) and borderline ovarian tumors together with normal ovarian stroma and normal fallopian tube [278]. DESI-IMS revealed five distinct lipid classes which are responsible for the separation between the normal tissues and stroma/serous carcinoma tissues: phosphatidic acids (PA), PS, PE, PG, and ceramides (Cer). Figure 17 represents the distribution of the five different PAs identified, where the intensities are always greater in tumor tissue than in carcinoma-associated stromal tissue [278].

DESI imaging mass spectrometry for cancer diagnostics have been of pivotal focus in translating this technique from bench to bedside into the clinical setting as

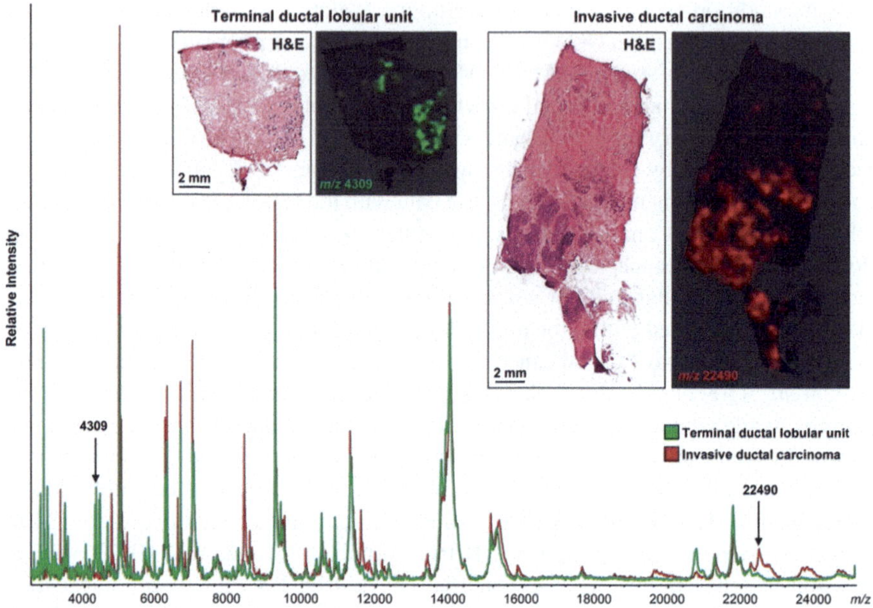

Fig. 16 Direct tissue mass spectrometric analysis reveals cell-type-specific profiles as shown in this comparison of neoplastic (invasive ductal cancer) and non-neoplastic (terminal ductal lobular unit) tissue from an individual patient tissue sample analyzed by MALDI-IMS. Average spectra in the mass range of 2400–25,000 Da are obtained from the non-neoplastic (green) and the cancerous (red) tissue areas. Two examples of differentially expressed masses (*m*/*z* 4309 and 22,490) are indicated by arrows. The visualization of *m*/*z* 4309 within this histological context (overlay with the optical microscopic image of the measured and subsequently H&E stained tissue section) clearly shows that this feature is specific for terminal ductal lobular units, while *m*/*z* 22,490 is specific for invasive ductal cancer cells but not present in tumor stroma or other tissue components. Statistical analysis revealed in this experiment of this individual patient sample 81 differentially expressed *m*/*z* species at significant levels between invasive ductal cancer and terminal ductal lobular units. Scale bars = 2 mm [265]. (Reprinted with permission from *J. Proteome Res.* 2010, 9, 4, 1854–1863. Copyright © 2010 American Chemical Society)

a validated cancer screening method [181, 182, 279]. Like histopathology examination, DESI requires surgical biopsy tissue cryosection into thin slices (5–20-μm thickness) and thaw mounted onto a glass slide and does not require pretreatment or matrix application. DESI-MS has been further exploited to diagnose various cancer types, based on the differences in lipid and metabolite profiles, for its application breast cancer [280, 281], prostate cancer [282, 283], gastric cancer [284], ovarian cancer [278, 285], brain tumors [286–288], thyroid oncolytic tumor [289], lymphoma [290], lymph node metastasis [291], pancreatic cancer [292], colorectal adenocarcinoma [293], and renal cell carcinoma [217, 250, 294, 295]. Air-flow-assisted desorption electrospray ionization-MSI (AFADESI-MSI) has demonstrated its application in spatial mapping of *EGFR*-mutation-associated biomarkers and visualized *EGFR* mutation spatial distribution in lung adenocarcinoma (LADC) tissues [296].

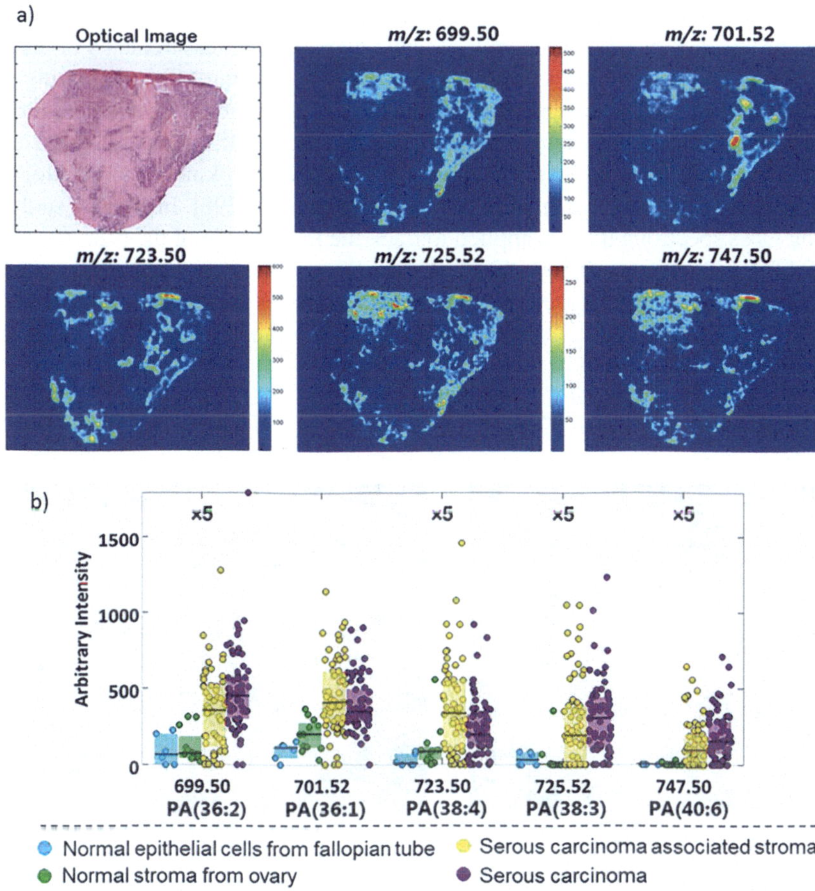

Fig. 17 Characterization of phosphatidic acid class (PA) in ovarian cancer. (**a**) Ion images of five different PA species in a serous ovarian carcinoma with (**b**) box plots of the same lipid species [278]. (Reprinted with permission from *Scientific Reports*, **6**, Article number: 39219 (2016). *Copyright © 2016,* Maria Luisa Dória et al.)

15 Applications of Imaging Mass Spectrometry in Drug Imaging

The combination of imaging mass spectrometry and histology is now extensively used for pharmacological research, which simultaneously images the distributions of the drug compounds and their metabolites [227, 297]. Figure 18 represents the high-resolution mass spectrometry imaging of drug distribution of the anticancer drug imatinib (a tyrosine kinase inhibitor) which was analyzed in mouse kidney at 35-μm-pixel size; Imatinib ($[M + H]^+$, *m/z* 494.2662) is displayed in green, while two phospholipids (red and blue) are included to indicate the histological features of the kidney [142]. The colocalization of imatinib and PC (40:6) indicates that imatinib is accumulated in the outer stripe of the outer medulla. With suitable calibration curves, MSI has been shown to provide quantitative measurements of drug compound concentrations [219, 235]. Pharmacological imaging MS is increasingly integrated with histology to localize the compounds within the often highly heterogeneous tumor tissue microenvironments [217]. Recently, IMS was combined with digital image analysis of the tissue sections to relate pharmacokinetics to histological features, in this case the degree of microvascularization [298]. In another study, imaging mass spectrometry was applied to track the localization of the drug-targeting BRAF in malignant melanoma [299]. The ability to determine drug uptake at the target sites, in the context of the tissue's histology, provides important opportunities for understanding the mode of action of drug activity within the disease microenvironment and the mechanism of drug resistance.

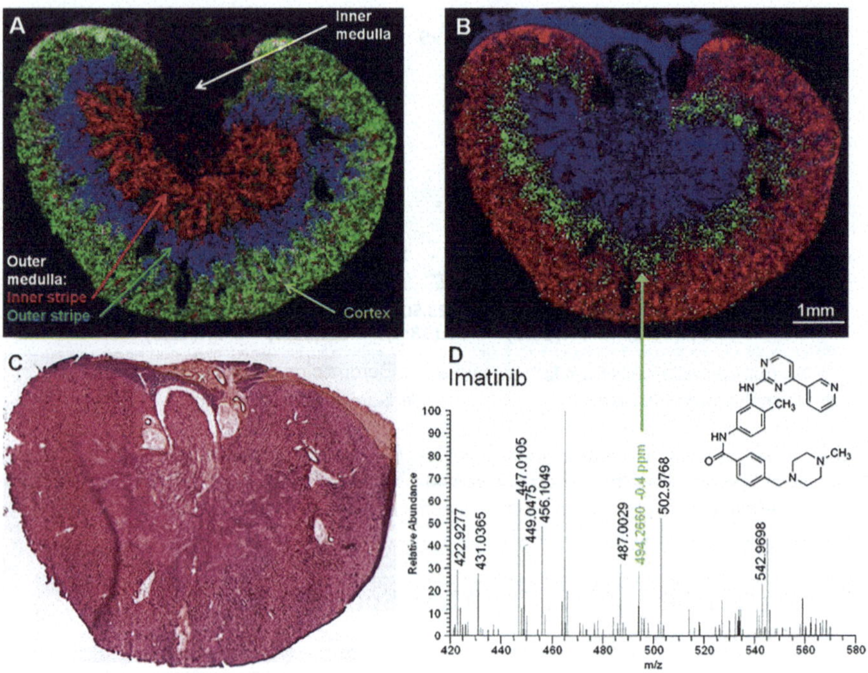

Fig. 18 (**a**) Overlay of selected ion images: *green*, $[PC(32:0) + K]^+ = 772.5253$ cortex; *blue*, $[PC(40:6) + K]^+ = 872.5566$ outer stripe outer medulla; and *red*, $[PC(38:5) + K]^+ = 846.5410$ inner stripe outer medulla; FTMS image, 225 × 150 pixels; 35-μm step size; bin width, $\Delta m/z = 0.01$. (**b**) Overlay of selected ion images: *red*, $[PC(32:0) + K]^+ = 772.5253$; *green*, imatinib $[M + H]^+ = 494.2662$; and *blue*, $[PC(34:1) + H]^+ = 760.5851$; FTMS image, 225 × 150 pixels; 35-μm step size; bin width, $\Delta m/z = 0.01$. (**c**) Optical image of the investigated mouse kidney section; H&E stained after MS imaging measurement. (**d**) Single-pixel FTMS spectrum of the outer stripe outer medulla of the mouse kidney section [142]. (Reprinted with permission from *Analytical and Bioanalytical Chemistry*, 401, 65–73 (2011). *Copyright © 2011,* Andreas Römpp et al., *Springer-Verlag*)

16 Concluding Remarks

For the successful biomedical application of imaging mass spectrometry as a molecular histological tool to highlight tissue regions with distinct molecular signatures, it will require further investigation using additional data mining and bioanalytical tools to explore the biological pathways and mechanisms of disease progression such as in-depth protein analysis for elucidating signaling pathways. More recently, imaging mass spectrometry has been used to develop novel clinical applications, such as imaging endogenous metabolites and neurotransmitters and the detection of intratumor heterogeneity. Rapid and intraoperative classification of tissues in the operation setting based on biomarker profiles for surgical margin evaluation, diagnosis, and prognosis and response to therapy shows the potential of IMS to be transformative from basic biomedical research to translational research. MSI developments with multimodality and high-resolution imaging at cellular length scales have been driven largely by academic and analytical chemists which will eventually be standardized at a large-scale multicenter validation study to demonstrate its clinical utility and significance.

References

1. Zhu C, Preissl S, Ren B. Single-cell multimodal omics: the power of many. Nature Methods. 2020;17(1):11–4.
2. Taylor MJ, Lukowski JK, Anderton CR. Spatially Resolved Mass Spectrometry at the Single Cell: Recent Innovations in Proteomics and Metabolomics. Journal of the American Society for Mass Spectrometry. 2021;32(4):872–94.
3. Cao Z-J, Gao G. Multi-omics single-cell data integration and regulatory inference with graph-linked embedding. Nature Biotechnology. 2022;40(10):1458–66.
4. Doran S, Arif M, Lam S, Bayraktar A, Turkez H, Uhlen M, et al. Multi-omics approaches for revealing the complexity of cardiovascular disease. Briefings in Bioinformatics. 2021;22(5).
5. Frey IM, Rubio-Aliaga I, Siewert A, Sailer D, Drobyshev A, Beckers J, et al. Profiling at mRNA, protein, and metabolite levels reveals alterations in renal amino acid handling and glutathione metabolism in kidney tissue of Pept2−/− mice. Physiological Genomics. 2007;28(3):301–10.
6. Krassowski M, Das V, Sahu SK, Misra BB. State of the Field in Multi-Omics Research: From Computational Needs to Data Mining and Sharing. Frontiers in Genetics. 2020;11.
7. Wozniak JM, Mills RH, Olson J, Caldera JR, Sepich-Poore GD, Carrillo-Terrazas M, et al. Mortality Risk Profiling of Staphylococcus aureus Bacteremia by Multi-omic Serum Analysis Reveals Early Predictive and Pathogenic Signatures. Cell. 2020;182(5):1311–27.e14.
8. Lyons J, Brubaker DK, Ghazi PC, Baldwin KR, Edwards A, Boukhali M, et al. Integrated in vivo multiomics analysis identifies p21-activated kinase signaling as a driver of colitis. Science Signaling. 2018;11(519):eaan3580.
9. Rubio T, Felipo V, Tarazona S, Pastorelli R, Escudero-García D, Tosca J, et al. Multi-omic analysis unveils biological pathways in peripheral immune system associated to minimal hepatic encephalopathy appearance in cirrhotic patients. Scientific Reports. 2021;11(1):1907.
10. Patti GJ, Yanes O, Siuzdak G. Innovation: Metabolomics: the apogee of the omics trilogy. Nat Rev Mol Cell Biol. 2012;13(4):263–9.

11. Kaddurah-Daouk R, Kristal BS, Weinshilboum RM. Metabolomics: A Global Biochemical Approach to Drug Response and Disease. Annual Review of Pharmacology and Toxicology. 2008;48(1):653–83.
12. Weiner J, Maertzdorf J, Sutherland JS, Duffy FJ, Thompson E, Suliman S, et al. Metabolite changes in blood predict the onset of tuberculosis. Nature Communications. 2018;9(1):5208.
13. Dinges SS, Hohm A, Vandergrift LA, Nowak J, Habbel P, Kaltashov IA, et al. Cancer metabolomic markers in urine: evidence, techniques and recommendations. Nature Reviews Urology. 2019;16(6):339–62.
14. Emara S, Amer S, Ali A, Abouleila Y, Oga A, Masujima T. Single-Cell Metabolomics. Adv Exp Med Biol. 2017;965:323–43.
15. Johnson CH, Ivanisevic J, Siuzdak G. Metabolomics: beyond biomarkers and towards mechanisms. Nature Reviews Molecular Cell Biology. 2016;17(7):451–9.
16. Sperber H, Mathieu J, Wang Y, Ferreccio A, Hesson J, Xu Z, et al. The metabolome regulates the epigenetic landscape during naive-to-primed human embryonic stem cell transition. Nature Cell Biology. 2015;17(12):1523–35.
17. Carey BW, Finley LWS, Cross JR, Allis CD, Thompson CB. Intracellular α-ketoglutarate maintains the pluripotency of embryonic stem cells. Nature. 2015;518(7539):413–6.
18. Yanes O, Clark J, Wong DM, Patti GJ, Sánchez-Ruiz A, Benton HP, et al. Metabolic oxidation regulates embryonic stem cell differentiation. Nature Chemical Biology. 2010;6(6):411–7.
19. Karlic H, Thaler R, Gerner C, Grunt T, Proestling K, Haider F, et al. Inhibition of the mevalonate pathway affects epigenetic regulation in cancer cells. Cancer Genetics. 2015;208(5):241–52.
20. Ulanovskaya OA, Zuhl AM, Cravatt BF. NNMT promotes epigenetic remodeling in cancer by creating a metabolic methylation sink. Nature Chemical Biology. 2013;9(5):300–6.
21. Schmidt DR, Patel R, Kirsch DG, Lewis CA, Vander Heiden MG, Locasale JW. Metabolomics in cancer research and emerging applications in clinical oncology. CA: A Cancer Journal for Clinicians. 2021;71(4):333–58.
22. Lyssiotis CA, Kimmelman AC. Metabolic Interactions in the Tumor Microenvironment. Trends in Cell Biology. 2017;27(11):863–75.
23. García-Cañaveras JC, Chen L, Rabinowitz JD. The Tumor Metabolic Microenvironment: Lessons from Lactate. Cancer Research. 2019;79(13):3155–62.
24. Lau AN, Heiden MGV. Metabolism in the Tumor Microenvironment. Annual Review of Cancer Biology. 2020;4(1):17–40.
25. Sullivan MR, Danai LV, Lewis CA, Chan SH, Gui DY, Kunchok T, et al. Quantification of microenvironmental metabolites in murine cancers reveals determinants of tumor nutrient availability. eLife. 2019;8:e44235.
26. Showalter MR, Cajka T, Fiehn O. Epimetabolites: discovering metabolism beyond building and burning. Curr Opin Chem Biol. 2017;36:70–6.
27. Gerber SA, Rush J, Stemman O, Kirschner MW, Gygi SP. Absolute quantification of proteins and phosphoproteins from cell lysates by tandem MS. Proc Natl Acad Sci U S A. 2003;100(12):6940–5.
28. Duncan KD, Fyrestam J, Lanekoff I. Advances in mass spectrometry based single-cell metabolomics. Analyst. 2019;144(3):782–93.
29. Cajka T, Fiehn O. Toward Merging Untargeted and Targeted Methods in Mass Spectrometry-Based Metabolomics and Lipidomics. Analytical Chemistry. 2016;88(1):524–45.
30. Manier SK, Keller A, Schäper J, Meyer MR. Untargeted metabolomics by high resolution mass spectrometry coupled to normal and reversed phase liquid chromatography as a tool to study the in vitro biotransformation of new psychoactive substances. Scientific Reports. 2019;9(1):2741.
31. Schrimpe-Rutledge AC, Codreanu SG, Sherrod SD, McLean JA. Untargeted Metabolomics Strategies-Challenges and Emerging Directions. J Am Soc Mass Spectrom. 2016;27(12):1897–905.
32. Wilmanski T, Rappaport N, Earls JC, Magis AT, Manor O, Lovejoy J, et al. Blood metabolome predicts gut microbiome α-diversity in humans. Nature Biotechnology. 2019;37(10):1217–28.

33. Alexandrov T. Spatial Metabolomics and Imaging Mass Spectrometry in the Age of Artificial Intelligence. Annual Review of Biomedical Data Science. 2020;3(1):61–87.
34. Garg N, Wang M, Hyde E, da Silva RR, Melnik AV, Protsyuk I, et al. Three-Dimensional Microbiome and Metabolome Cartography of a Diseased Human Lung. Cell Host & Microbe. 2017;22(5):705–16.e4.
35. Hudry B, de Goeij E, Mineo A, Gaspar P, Hadjieconomou D, Studd C, et al. Sex Differences in Intestinal Carbohydrate Metabolism Promote Food Intake and Sperm Maturation. Cell. 2019;178(4):901–18.e16.
36. Hui S, Ghergurovich JM, Morscher RJ, Jang C, Teng X, Lu W, et al. Glucose feeds the TCA cycle via circulating lactate. Nature. 2017;551(7678):115–8.
37. Ivanisevic J, Epstein Adrian A, Kurczy Michael E, Benton Paul H, Uritboonthai W, Fox Howard S, et al. Brain Region Mapping Using Global Metabolomics. Chemistry & Biology. 2014;21(11):1575–84.
38. Caprioli RM, Farmer TB, Gile J. Molecular Imaging of Biological Samples: Localization of Peptides and Proteins Using MALDI-TOF MS. Analytical Chemistry. 1997;69(23):4751–60.
39. Pacholski ML, Winograd N. Imaging with Mass Spectrometry. Chemical Reviews. 1999;99(10):2977–3006.
40. Petras D, Jarmusch AK, Dorrestein PC. From single cells to our planet—recent advances in using mass spectrometry for spatially resolved metabolomics. Current Opinion in Chemical Biology. 2017;36:24–31.
41. Watrous JD, Alexandrov T, Dorrestein PC. The evolving field of imaging mass spectrometry and its impact on future biological research. Journal of Mass Spectrometry. 2011;46(2):209–22.
42. Caprioli RM. Imaging Mass Spectrometry: A Perspective. J Biomol Tech. 2019;30(1):7–11.
43. Porta Siegel T, Ellis SR. Mass spectrometry imaging 2.0. Analytical and Bioanalytical Chemistry. 2021;413(10):2597–8.
44. Pól J, Strohalm M, Havlíček V, Volný M. Molecular mass spectrometry imaging in biomedical and life science research. Histochemistry and Cell Biology. 2010;134(5):423–43.
45. Ait-Belkacem R, Sellami L, Villard C, DePauw E, Calligaris D, Lafitte D. Mass spectrometry imaging is moving toward drug protein co-localization. Trends in Biotechnology. 2012;30(9):466–74.
46. Spraggins JM, Rizzo DG, Moore JL, Noto MJ, Skaar EP, Caprioli RM. Next-generation technologies for spatial proteomics: Integrating ultra-high speed MALDI-TOF and high mass resolution MALDI FTICR imaging mass spectrometry for protein analysis. PROTEOMICS. 2016;16(11–12):1678–89.
47. Luxembourg SL, Mize TH, McDonnell LA, Heeren RMA. High-Spatial Resolution Mass Spectrometric Imaging of Peptide and Protein Distributions on a Surface. Analytical Chemistry. 2004;76(18):5339–44.
48. Spraggins JM, Caprioli RM. High-Speed MALDI-TOF Imaging Mass Spectrometry: Rapid Ion Image Acquisition and Considerations for Next Generation Instrumentation. Journal of the American Society for Mass Spectrometry. 2011;22(6):1022–31.
49. Takáts Z, Wiseman JM, Gologan B, Cooks RG. Mass Spectrometry Sampling Under Ambient Conditions with Desorption Electrospray Ionization. Science. 2004;306(5695):471–3.
50. Laskin J, Heath BS, Roach PJ, Cazares L, Semmes OJ. Tissue Imaging Using Nanospray Desorption Electrospray Ionization Mass Spectrometry. Analytical Chemistry. 2012;84(1):141–8.
51. Wei J, Buriak JM, Siuzdak G. Desorption–ionization mass spectrometry on porous silicon. Nature. 1999;399(6733):243–6.
52. Northen TR, Yanes O, Northen MT, Marrinucci D, Uritboonthai W, Apon J, et al. Clathrate nanostructures for mass spectrometry. Nature. 2007;449(7165):1033–6.
53. Nemes P, Vertes A. Laser Ablation Electrospray Ionization for Atmospheric Pressure, in Vivo, and Imaging Mass Spectrometry. Analytical Chemistry. 2007;79(21):8098–106.

54. Giesen C, Wang HAO, Schapiro D, Zivanovic N, Jacobs A, Hattendorf B, et al. Highly multiplexed imaging of tumor tissues with subcellular resolution by mass cytometry. Nature Methods. 2014;11(4):417–22.
55. Dufresne M, Thomas A, Breault-Turcot J, Masson J-F, Chaurand P. Silver-Assisted Laser Desorption Ionization For High Spatial Resolution Imaging Mass Spectrometry of Olefins from Thin Tissue Sections. Analytical Chemistry. 2013;85(6):3318–24.
56. Becker JS, Dobrowolska J, Zoriy M, Matusch A. Imaging of uranium on rat brain sections using laser ablation inductively coupled plasma mass spectrometry: a new tool for the study of critical substructures affined to heavy metals in tissues. Rapid Communications in Mass Spectrometry. 2008;22(18):2768–72.
57. Becker JS, Zoriy M, Becker JS, Dobrowolska J, Matusch A. Laser ablation inductively coupled plasma mass spectrometry (LA-ICP-MS) in elemental imaging of biological tissues and in proteomics. Journal of Analytical Atomic Spectrometry. 2007;22(7):736–44.
58. Liebl H. Ion Microprobe Mass Analyzer. Journal of Applied Physics. 1967;38(13):5277–83.
59. Walch A, Rauser S, Deininger S-O, Höfler H. MALDI imaging mass spectrometry for direct tissue analysis: a new frontier for molecular histology. Histochemistry and Cell Biology. 2008;130(3):421–34.
60. Colliver TL, Brummel CL, Pacholski ML, Swanek FD, Ewing AG, Winograd N. Atomic and Molecular Imaging at the Single-Cell Level with TOF-SIMS. Analytical Chemistry. 1997;69(13):2225–31.
61. Comisarow MB, Marshall AG. The Early Development of Fourier Transform Ion Cyclotron Resonance (FT-ICR) Spectroscopy. Journal of Mass Spectrometry. 1996;31(6):581–5.
62. Hu Q, Noll RJ, Li H, Makarov A, Hardman M, Graham Cooks R. The Orbitrap: a new mass spectrometer. Journal of Mass Spectrometry. 2005;40(4):430–43.
63. Barre PYF, Heeren MAR, Potocnik ON. Mass Spectrometry Imaging in Nanomedicine: Unraveling the Potential of MSI for the Detection of Nanoparticles in Neuroscience. Current Pharmaceutical Design. 2017;23(13):1974–84.
64. Jehl B, Bauer R, Dörge A, Rick R. The use of propane/isopentane mixtures for rapid freezing of biological specimens. Journal of Microscopy. 1981;123(3):307–9.
65. Dong Y, Li B, Malitsky S, Rogachev I, Aharoni A, Kaftan F, et al. Sample Preparation for Mass Spectrometry Imaging of Plant Tissues: A Review. Frontiers in Plant Science. 2016;7.
66. Yoon S, Lee TG. Biological tissue sample preparation for time-of-flight secondary ion mass spectrometry (ToF–SIMS) imaging. Nano Convergence. 2018;5(1):24.
67. Zaima N, Hayasaka T, Goto-Inoue N, Setou M. Matrix-Assisted Laser Desorption/Ionization Imaging Mass Spectrometry. International Journal of Molecular Sciences. 2010;11(12):5040–55.
68. Hansen HT, Janfelt C. Aspects of Quantitation in Mass Spectrometry Imaging Investigated on Cryo-Sections of Spiked Tissue Homogenates. Analytical Chemistry. 2016;88(23):11513–20.
69. Gemperline E, Jayaraman D, Maeda J, Ané J-M, Li L. Multifaceted Investigation of Metabolites During Nitrogen Fixation in Medicago via High Resolution MALDI-MS Imaging and ESI-MS. Journal of the American Society for Mass Spectrometry. 2015;26(1):149–58.
70. Chen R, Hui L, Sturm RM, Li L. Three dimensional mapping of neuropeptides and lipids in crustacean brain by mass spectral imaging. Journal of the American Society for Mass Spectrometry. 2009;20(6):1068–77.
71. Khatib-Shahidi S, Andersson M, Herman JL, Gillespie TA, Caprioli RM. Direct Molecular Analysis of Whole-Body Animal Tissue Sections by Imaging MALDI Mass Spectrometry. Analytical Chemistry. 2006;78(18):6448–56.
72. Schwartz SA, Reyzer ML, Caprioli RM. Direct tissue analysis using matrix-assisted laser desorption/ionization mass spectrometry: practical aspects of sample preparation. Journal of Mass Spectrometry. 2003;38(7):699–708.
73. Goodwin RJA. Sample preparation for mass spectrometry imaging: Small mistakes can lead to big consequences. Journal of Proteomics. 2012;75(16):4893–911.

74. Djidja M-C, Francese S, Loadman PM, Sutton CW, Scriven P, Claude E, et al. Detergent addition to tryptic digests and ion mobility separation prior to MS/MS improves peptide yield and protein identification for in situ proteomic investigation of frozen and formalin-fixed paraffin-embedded adenocarcinoma tissue sections. PROTEOMICS. 2009;9(10):2750–63.
75. Norris JL, Caprioli RM. Analysis of Tissue Specimens by Matrix-Assisted Laser Desorption/Ionization Imaging Mass Spectrometry in Biological and Clinical Research. Chemical Reviews. 2013;113(4):2309–42.
76. Kawamoto T. Use of a new adhesive film for the preparation of multi-purpose fresh-frozen sections from hard tissues, whole-animals, insects and plants. Archives of Histology and Cytology. 2003;66(2):123–43.
77. Nissanov J, Bertrand L, Tretiak O. Cryosectioning distortion reduction using tape support. Microscopy Research and Technique. 2001;53(3):239–40.
78. Sjövall P, Lausmaa J, Johansson B. Mass Spectrometric Imaging of Lipids in Brain Tissue. Analytical Chemistry. 2004;76(15):4271–8.
79. Sjövall P, Johansson B, Lausmaa J. Localization of lipids in freeze-dried mouse brain sections by imaging TOF-SIMS. Applied Surface Science. 2006;252(19):6966–74.
80. Deutskens F, Yang J, Caprioli RM. High spatial resolution imaging mass spectrometry and classical histology on a single tissue section. Journal of Mass Spectrometry. 2011;46(6):568–71.
81. Lemaire R, Wisztorski M, Desmons A, Tabet JC, Day R, Salzet M, et al. MALDI-MS Direct Tissue Analysis of Proteins: Improving Signal Sensitivity Using Organic Treatments. Analytical Chemistry. 2006;78(20):7145–53.
82. Andersson M, Groseclose MR, Deutch AY, Caprioli RM. Imaging mass spectrometry of proteins and peptides: 3D volume reconstruction. Nature Methods. 2008;5(1):101–8.
83. Addie RD, Balluff B, Bovée JVMG, Morreau H, McDonnell LA. Current State and Future Challenges of Mass Spectrometry Imaging for Clinical Research. Analytical Chemistry. 2015;87(13):6426–33.
84. Porta Siegel T, Hamm G, Bunch J, Cappell J, Fletcher JS, Schwamborn K. Mass Spectrometry Imaging and Integration with Other Imaging Modalities for Greater Molecular Understanding of Biological Tissues. Molecular Imaging and Biology. 2018;20(6):888–901.
85. Watrous JD, Dorrestein PC. Imaging mass spectrometry in microbiology. Nature Reviews Microbiology. 2011;9(9):683–94
86. Fletcher JS, Vickerman JC, Winograd N. Label free biochemical 2D and 3D imaging using secondary ion mass spectrometry. Current Opinion in Chemical Biology. 2011;15(5):733–40.
87. McDonnell LA, Römpp A, Balluff B, Heeren RMA, Albar JP, Andrén PE, et al. Discussion point: reporting guidelines for mass spectrometry imaging. Analytical and Bioanalytical Chemistry. 2015;407(8):2035–45.
88. Amstalden van Hove ER, Smith DF, Heeren RMA. A concise review of mass spectrometry imaging. Journal of Chromatography A. 2010;1217(25):3946–54.
89. Macfarlane RD, Torgerson DF. Californium-252 Plasma Desorption Mass Spectroscopy. Science. 1976;191(4230):920–5.
90. Sundqvist B, Roepstorff P, Fohlman J, Hedin A, Håkansson P, Kamensky I, et al. Molecular Weight Determinations of Proteins by Californium Plasma Desorption Mass Spectrometry. Science. 1984;226(4675):696–8.
91. Bolbach G, Viari A, Galera R, Brunot A, Blais JC. Organic film thickness effect in secondary ion mass spectrometry and plasma desorption mass spectrometry. International Journal of Mass Spectrometry and Ion Processes. 1992;112(1):93–100.
92. Touboul D, Kollmer F, Niehuis E, Brunelle A, Laprévote O. Improvement of biological time-of-flight-secondary ion mass spectrometry imaging with a bismuth cluster ion source. Journal of the American Society for Mass Spectrometry. 2005;16(10):1608–18.
93. Novikov A, Caroff M, Della-Negra S, Depauw J, Fallavier M, Le Beyec Y, et al. The Au(n) cluster probe in secondary ion mass spectrometry: influence of the projectile size and energy

on the desorption/ionization rate from biomolecular solids. Rapid Commun Mass Spectrom. 2005;19(13):1851–7.
94. Benguerba M, Brunelle A, Della-Negra S, Depauw J, Joret H, Le Beyec Y, et al. Impact of slow gold clusters on various solids: nonlinear effects in secondary ion emission. Nuclear Instruments and Methods in Physics Research Section B: Beam Interactions with Materials and Atoms. 1991;62(1):8–22.
95. Piwowar AM, Keskin S, Delgado MO, Shen K, Hue JJ, Lanekoff I, et al. C60-ToF SIMS imaging of frozen hydrated HeLa cells. Surf Interface Anal. 2013;45(1):302–4.
96. Weibel D, Wong S, Lockyer N, Blenkinsopp P, Hill R, Vickerman JC. A C60 Primary Ion Beam System for Time of Flight Secondary Ion Mass Spectrometry: Its Development and Secondary Ion Yield Characteristics. Analytical Chemistry. 2003;75(7):1754–64.
97. Agüi-Gonzalez P, Jähne S, Phan NTN. SIMS imaging in neurobiology and cell biology. Journal of Analytical Atomic Spectrometry. 2019;34(7):1355–68.
98. Van Vaeck L, Adriaens A, Gijbels R. Static secondary ion mass spectrometry (S-SIMS) Part 1: methodology and structural interpretation. Mass Spectrometry Reviews. 1999;18(1):1–47.
99. Diehnelt CW, Van Stipdonk MJ, Schweikert EA. Effectiveness of atomic and polyatomic primary ions for organic secondary ion mass spectrometry. International Journal of Mass Spectrometry. 2001;207(1):111–22.
100. Benninghoven A, Hagenhoff B, Niehuis E. Surface MS: Probing Real-World Samples. Analytical Chemistry. 1993;65(14):630A–40A.
101. Wu KJ, Odom RW. Matrix-Enhanced Secondary Ion Mass Spectrometry: A Method for Molecular Analysis of Solid Surfaces. Analytical Chemistry. 1996;68(5):873–82.
102. Adriaensen L, Vangaever F, Gijbels R. Metal-Assisted Secondary Ion Mass Spectrometry: Influence of Ag and Au Deposition on Molecular Ion Yields. Analytical Chemistry. 2004;76(22):6777–85.
103. Winograd N, Garrison BJ. Biological Cluster Mass Spectrometry. Annual Review of Physical Chemistry. 2010;61(1):305–22.
104. Tian H, Sheraz née Rabbani S, Vickerman JC, Winograd N. Multiomics Imaging Using High-Energy Water Gas Cluster Ion Beam Secondary Ion Mass Spectrometry [(H2O)n-GCIB-SIMS] of Frozen-Hydrated Cells and Tissue. Analytical Chemistry. 2021;93(22):7808–14.
105. Heeren RMA, McDonnell LA, Amstalden E, Luxembourg SL, Altelaar AFM, Piersma SR. Why don't biologists use SIMS?: A critical evaluation of imaging MS. Applied Surface Science. 2006;252(19):6827–35.
106. Lovrić J, Dunevall J, Larsson A, Ren L, Andersson S, Meibom A, et al. Nano Secondary Ion Mass Spectrometry Imaging of Dopamine Distribution Across Nanometer Vesicles. ACS Nano. 2017;11(4):3446–55.
107. Fletcher JS, Rabbani S, Henderson A, Blenkinsopp P, Thompson SP, Lockyer NP, et al. A New Dynamic in Mass Spectral Imaging of Single Biological Cells. Analytical Chemistry. 2008;80(23):9058–64.
108. Boxer SG, Kraft ML, Weber PK. Advances in Imaging Secondary Ion Mass Spectrometry for Biological Samples. Annual Review of Biophysics. 2009;38(1):53–74.
109. Anderton CR, Gamble LJ. Secondary Ion Mass Spectrometry Imaging of Tissues, Cells, and Microbial Systems. Microscopy Today. 2016;24(2):24–31.
110. Passarelli MK, Ewing AG, Winograd N. Single-Cell Lipidomics: Characterizing and Imaging Lipids on the Surface of Individual Aplysia californica Neurons with Cluster Secondary Ion Mass Spectrometry. Analytical Chemistry. 2013;85(4):2231–8.
111. Angerer TB, Dowlatshahi Pour M, Malmberg P, Fletcher JS. Improved Molecular Imaging in Rodent Brain with Time-of-Flight-Secondary Ion Mass Spectrometry Using Gas Cluster Ion Beams and Reactive Vapor Exposure. Analytical Chemistry. 2015;87(8):4305–13.
112. Nygren H, Malmberg P. High-resolution imaging and proteomics of peptide fragments by TOF-SIMS. PROTEOMICS. 2010;10(8):1694–8.

113. Ahmed S, Titchmarsh JM, Kilburn MR, Grovenor CRM. Examination of the influence of boron on the microstructure and properties of low C ferritic steels using NanoSIMS and TEM. Applied Surface Science. 2006;252(19):7062–5.
114. Steinhauser ML, Bailey AP, Senyo SE, Guillermier C, Perlstein TS, Gould AP, et al. Multi-isotope imaging mass spectrometry quantifies stem cell division and metabolism. Nature. 2012;481(7382):516–9.
115. Wightman RM, Jankowski JA, Kennedy RT, Kawagoe KT, Schroeder TJ, Leszczyszyn DJ, et al. Temporally resolved catecholamine spikes correspond to single vesicle release from individual chromaffin cells. Proceedings of the National Academy of Sciences. 1991;88(23):10754–8.
116. Li X, Majdi S, Dunevall J, Fathali H, Ewing AG. Quantitative Measurement of Transmitters in Individual Vesicles in the Cytoplasm of Single Cells with Nanotip Electrodes. Angewandte Chemie International Edition. 2015;54(41):11978–82.
117. Nakata Y, Honda Y, Ninomiya S, Seki T, Aoki T, Matsuo J. Yield enhancement of molecular ions with MeV ion-induced electronic excitation. Applied Surface Science. 2008;255(4):1591–4.
118. Jenčič B, Jeromel L, Ogrinc Potočnik N, Vogel-Mikuš K, Kovačec E, Regvar M, et al. Molecular imaging of cannabis leaf tissue with MeV-SIMS method. Nuclear Instruments and Methods in Physics Research Section B: Beam Interactions with Materials and Atoms. 2016;371:205–10.
119. Fisher GL, Bruinen AL, Ogrinc Potočnik N, Hammond JS, Bryan SR, Larson PE, et al. A New Method and Mass Spectrometer Design for TOF-SIMS Parallel Imaging MS/MS. Analytical Chemistry. 2016;88(12):6433–40.
120. Hanrieder J, Ljungdahl A, Fälth M, Mammo SE, Bergquist J, Andersson M. l-DOPA-induced Dyskinesia is Associated with Regional Increase of Striatal Dynorphin Peptides as Elucidated by Imaging Mass Spectrometry*. Molecular & Cellular Proteomics. 2011a;10(10):M111.009308.
121. Chaurand P, Cornett DS, Angel PM, Caprioli RM. From whole-body sections down to cellular level, multiscale imaging of phospholipids by MALDI mass spectrometry. Mol Cell Proteomics. 2011;10(2):O110.004259.
122. Seeley EH, Caprioli RM. MALDI imaging mass spectrometry of human tissue: method challenges and clinical perspectives. Trends in Biotechnology. 2011;29(3):136–43.
123. Sun N, Trajkovic-Arsic M, Li F, Wu Y, Münch C, Kunzke T, et al. Native glycan fragments detected by MALDI mass spectrometry imaging are independent prognostic factors in pancreatic ductal adenocarcinoma. EJNMMI Research. 2021;11(1):120.
124. Zavalin A, Yang J, Hayden K, Vestal M, Caprioli RM. Tissue protein imaging at 1 μm laser spot diameter for high spatial resolution and high imaging speed using transmission geometry MALDI TOF MS. Analytical and Bioanalytical Chemistry. 2015;407(8):2337–42.
125. Balluff B, Schöne C, Höfler H, Walch A. MALDI imaging mass spectrometry for direct tissue analysis: technological advancements and recent applications. Histochemistry and Cell Biology. 2011;136(3):227–44.
126. Kaspar S, Peukert M, Svatos A, Matros A, Mock H-P. MALDI-imaging mass spectrometry – An emerging technique in plant biology. PROTEOMICS. 2011;11(9):1840–50.
127. Seeley EH, Schwamborn K, Caprioli RM. Imaging of Intact Tissue Sections: Moving beyond the Microscope*. Journal of Biological Chemistry. 2011;286(29):25459–66.
128. Karas M, Krüger R. Ion Formation in MALDI: The Cluster Ionization Mechanism. Chemical Reviews. 2003;103(2):427–40.
129. Chang WC, Huang LCL, Wang Y-S, Peng W-P, Chang HC, Hsu NY, et al. Matrix-assisted laser desorption/ionization (MALDI) mechanism revisited. Analytica Chimica Acta. 2007;582(1):1–9.
130. Strupat K, Karas M, Hillenkamp F. 2,5-Dihydroxybenzoic acid: a new matrix for laser desorption—ionization mass spectrometry. International Journal of Mass Spectrometry and Ion Processes. 1991;111:89–102.

131. Beavis RC, Chaudhary T, Chait BT. α-Cyano-4-hydroxycinnamic acid as a matrix for matrix-assisted laser desorption mass spectrometry. Organic Mass Spectrometry. 1992;27(2):156–8.
132. Beavis RC, Chait BT, Standing KG. Matrix-assisted laser-desorption mass spectrometry using 355 nm radiation. Rapid Communications in Mass Spectrometry. 1989;3(12):436–9.
133. Ma X, Fernández FM. Advances in mass spectrometry imaging for spatial cancer metabolomics. Mass Spectrometry Reviews.n/a(n/a):e21804.
134. Gustafsson JOR, Oehler MK, Ruszkiewicz A, McColl SR, Hoffmann P. MALDI Imaging Mass Spectrometry (MALDI-IMS)—Application of Spatial Proteomics for Ovarian Cancer Classification and Diagnosis. International Journal of Molecular Sciences. 2011a;12:773–94.
135. Vaysse P-M, Heeren RMA, Porta T, Balluff B. Mass spectrometry imaging for clinical research – latest developments, applications, and current limitations. Analyst. 2017;142(15):2690–712.
136. Gustafsson JO, Oehler MK, Ruszkiewicz A, McColl SR, Hoffmann P. MALDI Imaging Mass Spectrometry (MALDI-IMS)-application of spatial proteomics for ovarian cancer classification and diagnosis. Int J Mol Sci. 2011b;12(1):773–94.
137. Hankin JA, Barkley RM, Murphy RC. Sublimation as a Method of Matrix Application for Mass Spectrometric Imaging. Journal of the American Society for Mass Spectrometry. 2007;18(9):1646–52.
138. Römpp A, Spengler B. Mass spectrometry imaging with high resolution in mass and space. Histochemistry and Cell Biology. 2013;139(6):759–83.
139. Aichler M, Walch A. MALDI Imaging mass spectrometry: current frontiers and perspectives in pathology research and practice. Laboratory Investigation. 2015;95(4):422–31.
140. Spengler B, Hubert M. Scanning microprobe matrix-assisted laser desorption ionization (SMALDI) mass spectrometry: Instrumentation for sub-micrometer resolved LDI and MALDI surface analysis. Journal of the American Society for Mass Spectrometry. 2002;13(6):735–48.
141. Kompauer M, Heiles S, Spengler B. Atmospheric pressure MALDI mass spectrometry imaging of tissues and cells at 1.4-μm lateral resolution. Nature Methods. 2017;14(1):90–6.
142. Römpp A, Guenther S, Takats Z, Spengler B. Mass spectrometry imaging with high resolution in mass and space (HR2 MSI) for reliable investigation of drug compound distributions on the cellular level. Analytical and Bioanalytical Chemistry. 2011;401(1):65–73.
143. Holzlechner M, Eugenin E, Prideaux B. Mass spectrometry imaging to detect lipid biomarkers and disease signatures in cancer. Cancer Reports. 2019;2(6):e1229.
144. Chaurand P, Schwartz SA, Billheimer D, Xu BJ, Crecelius A, Caprioli RM. Integrating Histology and Imaging Mass Spectrometry. Analytical Chemistry. 2004;76(4):1145–55.
145. Andersen MK, Høiem TS, Claes BSR, Balluff B, Martin-Lorenzo M, Richardsen E, et al. Spatial differentiation of metabolism in prostate cancer tissue by MALDI-TOF MSI. Cancer & Metabolism. 2021;9(1):9.
146. Seeley EH, Caprioli RM. Molecular imaging of proteins in tissues by mass spectrometry. Proceedings of the National Academy of Sciences. 2008;105(47):18126–31.
147. Castellino S, Groseclose MR, Wagner D. MALDI imaging mass spectrometry: bridging biology and chemistry in drug development. Bioanalysis. 2011;3(21):2427–41.
148. Bartelink IH, Prideaux B, Krings G, Wilmes L, Lee PRE, Bo P, et al. Heterogeneous drug penetrance of veliparib and carboplatin measured in triple negative breast tumors. Breast Cancer Res. 2017;19(1):107.
149. Torok S, Rezeli M, Kelemen O, Vegvari A, Watanabe K, Sugihara Y, et al. Limited Tumor Tissue Drug Penetration Contributes to Primary Resistance against Angiogenesis Inhibitors. Theranostics. 2017;7(2):400–12.
150. Sun C, Li T, Song X, Huang L, Zang Q, Xu J, et al. Spatially resolved metabolomics to discover tumor-associated metabolic alterations. Proceedings of the National Academy of Sciences. 2019;116(1):52–7.

151. Paine MRL, Liu J, Huang D, Ellis SR, Trede D, Kobarg JH, et al. Three-Dimensional Mass Spectrometry Imaging Identifies Lipid Markers of Medulloblastoma Metastasis. Scientific Reports. 2019;9(1):2205.
152. Mao X, He J, Li T, Lu Z, Sun J, Meng Y, et al. Application of imaging mass spectrometry for the molecular diagnosis of human breast tumors. Scientific Reports. 2016;6(1):21043.
153. Hanrieder J, Zetterberg H, Blennow K. MALDI Imaging Mass Spectrometry: Neurochemical Imaging of Proteins and Peptides. In: Li KW, editor. Neuroproteomics. New York, NY: Springer New York; 2019. p. 179–97.
154. Shariatgorji M, Nilsson A, Fridjonsdottir E, Vallianatou T, Källback P, Katan L, et al. Comprehensive mapping of neurotransmitter networks by MALDI–MS imaging. Nature Methods. 2019;16(10):1021–8.
155. Altelaar AFM, Taban IM, McDonnell LA, Verhaert PDEM, de Lange RPJ, Adan RAH, et al. High-resolution MALDI imaging mass spectrometry allows localization of peptide distributions at cellular length scales in pituitary tissue sections. International Journal of Mass Spectrometry. 2007;260(2):203–11.
156. Briggs MT, Condina MR, Ho YY, Everest-Dass AV, Mittal P, Kaur G, et al. MALDI Mass Spectrometry Imaging of Early- and Late-Stage Serous Ovarian Cancer Tissue Reveals Stage-Specific N-Glycans. PROTEOMICS. 2019;19(21–22):1800482.
157. Drake RR, Powers TW, Jones EE, Bruner E, Mehta AS, Angel PM. Chapter Four - MALDI Mass Spectrometry Imaging of N-Linked Glycans in Cancer Tissues. In: Drake RR, McDonnell LA, editors. Advances in Cancer Research. 134: Academic Press; 2017. p. 85–116.
158. Hanrieder J, Wicher G, Bergquist J, Andersson M, Fex-Svenningsen Å. MALDI mass spectrometry based molecular phenotyping of CNS glial cells for prediction in mammalian brain tissue. Analytical and Bioanalytical Chemistry. 2011b;401(1):135–47.
159. Dunham SJB, Ellis JF, Li B, Sweedler JV. Mass Spectrometry Imaging of Complex Microbial Communities. Accounts of Chemical Research. 2017;50(1):96–104.
160. Yang J, Caprioli RM. Matrix Sublimation/Recrystallization for Imaging Proteins by Mass Spectrometry at High Spatial Resolution. Analytical Chemistry. 2011;83(14):5728–34.
161. Groseclose MR, Massion PP, Chaurand P, Caprioli RM. High-throughput proteomic analysis of formalin-fixed paraffin-embedded tissue microarrays using MALDI imaging mass spectrometry. PROTEOMICS. 2008;8(18):3715–24.
162. Groseclose MR, Andersson M, Hardesty WM, Caprioli RM. Identification of proteins directly from tissue: in situ tryptic digestions coupled with imaging mass spectrometry. J Mass Spectrom. 2007;42(2):254–62.
163. Gustafsson JO, Oehler MK, McColl SR, Hoffmann P. Citric acid antigen retrieval (CAAR) for tryptic peptide imaging directly on archived formalin-fixed paraffin-embedded tissue. J Proteome Res. 2010a;9(9):4315–28.
164. Cillero-Pastor B, Heeren RMA. Matrix-Assisted Laser Desorption Ionization Mass Spectrometry Imaging for Peptide and Protein Analyses: A Critical Review of On-Tissue Digestion. Journal of Proteome Research. 2014;13(2):325–35.
165. O'Neill KC, Dueñas ME, Larson E, Forsman TT, Lee YJ. Enhancing Metabolite Coverage for Matrix-Assisted Laser Desorption/Ionization Mass Spectrometry Imaging Through Multiple On-Tissue Chemical Derivatizations. Methods Mol Biol. 2022;2437.197–213.
166. Sun C, Liu W, Geng Y, Wang X. On-Tissue Derivatization Strategy for Mass Spectrometry Imaging of Carboxyl-Containing Metabolites in Biological Tissues. Analytical Chemistry. 2020;92(18):12126–31.
167. Esteve C, Tolner EA, Shyti R, van den Maagdenberg AMJM, McDonnell LA. Mass spectrometry imaging of amino neurotransmitters: a comparison of derivatization methods and application in mouse brain tissue. Metabolomics. 2016;12(2):30.
168. Lemaire R, Desmons A, Tabet JC, Day R, Salzet M, Fournier I. Direct Analysis and MALDI Imaging of Formalin-Fixed, Paraffin-Embedded Tissue Sections. Journal of Proteome Research. 2007a;6(4):1295–305.

169. Stauber J, Lemaire R, Franck J, Bonnel D, Croix D, Day R, et al. MALDI Imaging of Formalin-Fixed Paraffin-Embedded Tissues: Application to Model Animals of Parkinson Disease for Biomarker Hunting. Journal of Proteome Research. 2008;7(3):969–78.
170. Gustafsson JOR, Oehler MK, McColl SR, Hoffmann P. Citric Acid Antigen Retrieval (CAAR) for Tryptic Peptide Imaging Directly on Archived Formalin-Fixed Paraffin-Embedded Tissue. Journal of Proteome Research. 2010b;9(9):4315–28.
171. Chaurand P, Norris JL, Cornett DS, Mobley JA, Caprioli RM. New Developments in Profiling and Imaging of Proteins from Tissue Sections by MALDI Mass Spectrometry. Journal of Proteome Research. 2006;5(11):2889–900.
172. Harris GA, Galhena AS, Fernández FM. Ambient Sampling/Ionization Mass Spectrometry: Applications and Current Trends. Analytical Chemistry. 2011;83(12):4508–38.
173. Eberlin LS, Liu X, Ferreira CR, Santagata S, Agar NYR, Cooks RG. Desorption Electrospray Ionization then MALDI Mass Spectrometry Imaging of Lipid and Protein Distributions in Single Tissue Sections. Analytical Chemistry. 2011a;83(22):8366–71.
174. Wiseman JM, Ifa DR, Song Q, Cooks RG. Tissue Imaging at Atmospheric Pressure Using Desorption Electrospray Ionization (DESI) Mass Spectrometry. Angewandte Chemie International Edition. 2006;45(43):7188–92.
175. Venter A, Sojka PE, Cooks RG. Droplet Dynamics and Ionization Mechanisms in Desorption Electrospray Ionization Mass Spectrometry. Analytical Chemistry. 2006;78(24):8549–55.
176. Takáts Z, Wiseman JM, Cooks RG. Ambient mass spectrometry using desorption electrospray ionization (DESI): instrumentation, mechanisms and applications in forensics, chemistry, and biology. Journal of Mass Spectrometry. 2005;40(10):1261–75.
177. Costa AB, Graham Cooks R. Simulated splashes: Elucidating the mechanism of desorption electrospray ionization mass spectrometry. Chemical Physics Letters. 2008;464(1):1–8.
178. Eberlin LS, Ifa DR, Wu C, Cooks RG. Three-Dimensional Visualization of Mouse Brain by Lipid Analysis Using Ambient Ionization Mass Spectrometry. Angewandte Chemie International Edition. 2010;49(5):873–6.
179. Eberlin LS, Ferreira CR, Dill AL, Ifa DR, Cheng L, Cooks RG. Nondestructive, Histologically Compatible Tissue Imaging by Desorption Electrospray Ionization Mass Spectrometry. ChemBioChem. 2011b;12(14):2129–32.
180. Vijayalakshmi K, Shankar V, Bain RM, Nolley R, Sonn GA, Kao C-S, et al. Identification of diagnostic metabolic signatures in clear cell renal cell carcinoma using mass spectrometry imaging. International Journal of Cancer. 2020a;147(1):256–65.
181. King ME, Zhang J, Lin JQ, Garza KY, DeHoog RJ, Feider CL, et al. Rapid diagnosis and tumor margin assessment during pancreatic cancer surgery with the MasSpec Pen technology. Proceedings of the National Academy of Sciences. 2021;118(28):e2104411118.
182. Tzafetas M, Mitra A, Paraskevaidi M, Bodai Z, Kalliala I, Bowden S, et al. The intelligent knife (iKnife) and its intraoperative diagnostic advantage for the treatment of cervical disease. Proceedings of the National Academy of Sciences. 2020;117(13):7338–46.
183. Calligaris D, Norton I, Feldman DR, Ide JL, Dunn IF, Eberlin LS, et al. Mass spectrometry imaging as a tool for surgical decision-making. Journal of Mass Spectrometry. 2013;48(11):1178–87.
184. Girod M, Shi Y, Cheng J-X, Cooks RG. Mapping Lipid Alterations in Traumatically Injured Rat Spinal Cord by Desorption Electrospray Ionization Imaging Mass Spectrometry. Analytical Chemistry. 2011;83(1):207–15.
185. Talaty N, Takáts Z, Cooks RG. Rapid in situ detection of alkaloids in plant tissue under ambient conditions using desorption electrospray ionization. Analyst. 2005;130(12):1624–33.
186. Watrous J, Hendricks N, Meehan M, Dorrestein PC. Capturing Bacterial Metabolic Exchange Using Thin Film Desorption Electrospray Ionization-Imaging Mass Spectrometry. Analytical Chemistry. 2010;82(5):1598–600.
187. Lane AL, Nyadong L, Galhena AS, Shearer TL, Stout EP, Parry RM, et al. Desorption electrospray ionization mass spectrometry reveals surface-mediated antifungal chemical defense

of a tropical seaweed. Proceedings of the National Academy of Sciences of the United States of America. 2009;106(18):7314–9.
188. Wu C, Dill AL, Eberlin LS, Cooks RG, Ifa DR. Mass spectrometry imaging under ambient conditions. Mass Spectrometry Reviews. 2013;32(3):218–43.
189. Hsu C-C, Chou P-T, Zare RN. Imaging of Proteins in Tissue Samples Using Nanospray Desorption Electrospray Ionization Mass Spectrometry. Analytical Chemistry. 2015;87(22):11171–5.
190. Yang M, Unsihuay D, Hu H, Nguele Meke F, Qu Z, Zhang Z-Y, et al. Nano-DESI Mass Spectrometry Imaging of Proteoforms in Biological Tissues with High Spatial Resolution. Analytical Chemistry. 2023.
191. Lanekoff I, Thomas M, Laskin J. Shotgun Approach for Quantitative Imaging of Phospholipids Using Nanospray Desorption Electrospray Ionization Mass Spectrometry. Analytical Chemistry. 2014;86(3):1872–80.
192. Yin R, Burnum-Johnson KE, Sun X, Dey SK, Laskin J. High spatial resolution imaging of biological tissues using nanospray desorption electrospray ionization mass spectrometry. Nature Protocols. 2019;14(12):3445–70.
193. Lanekoff I, Thomas M, Carson JP, Smith JN, Timchalk C, Laskin J. Imaging Nicotine in Rat Brain Tissue by Use of Nanospray Desorption Electrospray Ionization Mass Spectrometry. Analytical Chemistry. 2013;85(2):882–9.
194. Abdelmoula WM, Carreira RJ, Shyti R, Balluff B, van Zeijl RJM, Tolner EA, et al. Automatic Registration of Mass Spectrometry Imaging Data Sets to the Allen Brain Atlas. Analytical Chemistry. 2014;86(8):3947–54.
195. Sarsby J, Martin NJ, Lalor PF, Bunch J, Cooper HJ. Top-Down and Bottom-Up Identification of Proteins by Liquid Extraction Surface Analysis Mass Spectrometry of Healthy and Diseased Human Liver Tissue. Journal of the American Society for Mass Spectrometry. 2014;25(11):1953–61.
196. Van Berkel GJ, Sanchez AD, Quirke JME. Thin-Layer Chromatography and Electrospray Mass Spectrometry Coupled Using a Surface Sampling Probe. Analytical Chemistry. 2002;74(24):6216–23.
197. Wisztorski M, Desmons A, Quanico J, Fatou B, Gimeno J-P, Franck J, et al. Spatially-resolved protein surface microsampling from tissue sections using liquid extraction surface analysis. PROTEOMICS. 2016;16(11–12):1622–32.
198. Sarsby J, Griffiths RL, Race AM, Bunch J, Randall EC, Creese AJ, et al. Liquid Extraction Surface Analysis Mass Spectrometry Coupled with Field Asymmetric Waveform Ion Mobility Spectrometry for Analysis of Intact Proteins from Biological Substrates. Analytical Chemistry. 2015;87(13):6794–800.
199. Hale OJ, Cooper HJ. Native Mass Spectrometry Imaging of Proteins and Protein Complexes by Nano-DESI. Analytical Chemistry. 2021;93(10):4619–27.
200. Unsihuay D, Qiu J, Swaroop S, Nagornov KO, Kozhinov AN, Tsybin YO, et al. Imaging of triglycerides in tissues using nanospray desorption electrospray ionization (Nano-DESI) mass spectrometry. International Journal of Mass Spectrometry. 2020;448:116269.
201. Ryan DJ, Patterson NH, Putnam NE, Wilde AD, Weiss A, Perry WJ, et al. MicroLESA: Integrating Autofluorescence Microscopy, In Situ Micro-Digestions, and Liquid Extraction Surface Analysis for High Spatial Resolution Targeted Proteomic Studies. Analytical Chemistry. 2019;91(12):7578–85.
202. Li X, Hu H, Yin R, Li Y, Sun X, Dey SK, et al. High-Throughput Nano-DESI Mass Spectrometry Imaging of Biological Tissues Using an Integrated Microfluidic Probe. Analytical Chemistry. 2022;94(27):9690–6.
203. Chen C-L, Kuo T-H, Chung H-H, Huang P, Lin L-E, Hsu C-C. Remodeling nanoDESI Platform with Ion Mobility Spectrometry to Expand Protein Coverage in Cancerous Tissue. Journal of the American Society for Mass Spectrometry. 2021;32(3):653–60.

204. Bergman H-M, Lundin E, Andersson M, Lanekoff I. Quantitative mass spectrometry imaging of small-molecule neurotransmitters in rat brain tissue sections using nanospray desorption electrospray ionization. Analyst. 2016;141(12):3686–95.
205. Su P, McGee JP, Durbin KR, Hollas MAR, Yang M, Neumann EK, et al. Highly multiplexed, label-free proteoform imaging of tissues by individual ion mass spectrometry. Science Advances. 2022;8(32):eabp9929.
206. Yang M, Hu H, Su P, Thomas PM, Camarillo JM, Greer JB, et al. Proteoform-Selective Imaging of Tissues Using Mass Spectrometry**. Angewandte Chemie International Edition. 2022;61(29):e202200721.
207. Kafader JO, Melani RD, Durbin KR, Ikwuagwu B, Early BP, Fellers RT, et al. Multiplexed mass spectrometry of individual ions improves measurement of proteoforms and their complexes. Nature Methods. 2020;17(4):391–4.
208. Sussulini A, Becker JS, Becker JS. Laser ablation ICP-MS: Application in biomedical research. Mass Spectrometry Reviews. 2017;36(1):47–57.
209. Yanes O, Woo H-K, Northen TR, Oppenheimer SR, Shriver L, Apon J, et al. Nanostructure Initiator Mass Spectrometry: Tissue Imaging and Direct Biofluid Analysis. Analytical Chemistry. 2009;81(8):2969–75.
210. Piehowski PD, Zhu Y, Bramer LM, Stratton KG, Zhao R, Orton DJ, et al. Automated mass spectrometry imaging of over 2000 proteins from tissue sections at 100-μm spatial resolution. Nature Communications. 2020;11(1):8.
211. Shimma S, Sugiura Y. Effective Sample Preparations in Imaging Mass Spectrometry. Mass Spectrom (Tokyo). 2014;3(Spec Issue):S0029.
212. Schaepe K, Bhandari DR, Werner J, Henss A, Pirkl A, Kleine-Boymann M, et al. Imaging of Lipids in Native Human Bone Sections Using TOF–Secondary Ion Mass Spectrometry, Atmospheric Pressure Scanning Microprobe Matrix-Assisted Laser Desorption/Ionization Orbitrap Mass Spectrometry, and Orbitrap–Secondary Ion Mass Spectrometry. Analytical Chemistry. 2018;90(15):8856–64.
213. Schober Y, Guenther S, Spengler B, Römpp A. High-resolution matrix-assisted laser desorption/ionization imaging of tryptic peptides from tissue. Rapid Communications in Mass Spectrometry. 2012;26(9):1141–6.
214. De Hoffmann E, Stroobant V. Mass spectrometry: principles and applications: Wiley; 2007.
215. Gross JH. Mass spectrometry: a textbook: Springer Science & Business Media; 2006.
216. Stoeckli M, Farmer TB, Caprioli RM. Automated mass spectrometry imaging with a matrix-assisted laser desorption ionization time-of-flight instrument. Journal of the American Society for Mass Spectrometry. 1999;10(1):67–71.
217. Prideaux B, Stoeckli M. Mass spectrometry imaging for drug distribution studies. Journal of Proteomics. 2012;75(16):4999–5013.
218. Nilsson A, Goodwin RJA, Shariatgorji M, Vallianatou T, Webborn PJH, Andrén PE. Mass Spectrometry Imaging in Drug Development. Analytical Chemistry. 2015;87(3):1437–55.
219. Hatsis P, Brombacher S, Corr J, Kovarik P, Volmer DA. Quantitative analysis of small pharmaceutical drugs using a high repetition rate laser matrix-assisted laser/desorption ionization source. Rapid Communications in Mass Spectrometry. 2003;17(20):2303–9.
220. Prentice BM, Chumbley CW, Caprioli RM. Absolute Quantification of Rifampicin by MALDI Imaging Mass Spectrometry Using Multiple TOF/TOF Events in a Single Laser Shot. Journal of the American Society for Mass Spectrometry. 2017;28(1):136–44.
221. Taban IM, Altelaar AFM, van der Burgt YEM, McDonnell LA, Heeren RMA, Fuchser J, et al. Imaging of peptides in the rat brain using MALDI-FTICR mass spectrometry. Journal of the American Society for Mass Spectrometry. 2007;18(1):145–51.
222. Robichaud G, Barry JA, Garrard KP, Muddiman DC. Infrared Matrix-Assisted Laser Desorption Electrospray Ionization (IR-MALDESI) Imaging Source Coupled to a FT-ICR Mass Spectrometer. Journal of the American Society for Mass Spectrometry. 2013a;24(1):92–100.

223. Bowman AP, Blakney GT, Hendrickson CL, Ellis SR, Heeren RMA, Smith DF. Ultra-High Mass Resolving Power, Mass Accuracy, and Dynamic Range MALDI Mass Spectrometry Imaging by 21-T FT-ICR MS. Analytical Chemistry. 2020;92(4):3133–42.
224. Makarov A, Denisov E, Kholomeev A, Balschun W, Lange O, Strupat K, et al. Performance Evaluation of a Hybrid Linear Ion Trap/Orbitrap Mass Spectrometer. Analytical Chemistry. 2006;78(7):2113–20.
225. Rubakhin SS, Sweedler JV. A mass spectrometry primer for mass spectrometry imaging. Methods Mol Biol. 2010;656:21–49.
226. Manicke NE, Dill AL, Ifa DR, Cooks RG. High-resolution tissue imaging on an orbitrap mass spectrometer by desorption electrospray ionization mass spectrometry. J Mass Spectrom. 2010;45(2):223–6.
227. Moreno-Gordaliza E, Esteban-Fernández D, Lázaro A, Humanes B, Aboulmagd S, Tejedor A, et al. MALDI-LTQ-Orbitrap mass spectrometry imaging for lipidomic analysis in kidney under cisplatin chemotherapy. Talanta. 2017;164:16–26.
228. Kiss A, Heeren RMA. Size, weight and position: ion mobility spectrometry and imaging MS combined. Analytical and Bioanalytical Chemistry. 2011;399(8):2623–34.
229. Paglia G, Angel P, Williams JP, Richardson K, Olivos HJ, Thompson JW, et al. Ion Mobility-Derived Collision Cross Section As an Additional Measure for Lipid Fingerprinting and Identification. Analytical Chemistry. 2015;87(2):1137–44.
230. Feider CL, Elizondo N, Eberlin LS. Ambient Ionization and FAIMS Mass Spectrometry for Enhanced Imaging of Multiply Charged Molecular Ions in Biological Tissues. Analytical Chemistry. 2016;88(23):11533–41.
231. Jackson SN, Ugarov M, Post JD, Egan T, Langlais D, Schultz JA, et al. A study of phospholipids by ion mobility TOFMS. Journal of the American Society for Mass Spectrometry. 2008;19(11):1655–62.
232. Chughtai K, Jiang L, Greenwood TR, Glunde K, Heeren RMA. Mass spectrometry images acylcarnitines, phosphatidylcholines, and sphingomyelin in MDA-MB-231 breast tumor models[S]. Journal of Lipid Research. 2013;54(2):333–44.
233. Škrášková K, Claude E, Jones EA, Towers M, Ellis SR, Heeren RMA. Enhanced capabilities for imaging gangliosides in murine brain with matrix-assisted laser desorption/ionization and desorption electrospray ionization mass spectrometry coupled to ion mobility separation. Methods. 2016;104:69–78.
234. Thiery-Lavenant G, Zavalin AI, Caprioli RM. Targeted Multiplex Imaging Mass Spectrometry in Transmission Geometry for Subcellular Spatial Resolution. Journal of the American Society for Mass Spectrometry. 2013;24(4):609–14.
235. Pirman DA, Yost RA. Quantitative Tandem Mass Spectrometric Imaging of Endogenous Acetyl-l-carnitine from Piglet Brain Tissue Using an Internal Standard. Analytical Chemistry. 2011;83(22):8575–81.
236. Hopfgartner G, Varesio E, Stoeckli M. Matrix-assisted laser desorption/ionization mass spectrometric imaging of complete rat sections using a triple quadrupole linear ion trap. Rapid Communications in Mass Spectrometry. 2009;23(6):733–6.
237. Porta T, Grivet C, Kraemer T, Varesio E, Hopfgartner G. Single Hair Cocaine Consumption Monitoring by Mass Spectrometric Imaging. Analytical Chemistry. 2011;83(11):4266–72.
238. Soltwisch J, Kettling H, Vens-Cappell S, Wiegelmann M, Müthing J, Dreisewerd K. Mass spectrometry imaging with laser-induced postionization. Science. 2015;348(6231):211–5.
239. Robichaud G, Garrard KP, Barry JA, Muddiman DC. MSiReader: An Open-Source Interface to View and Analyze High Resolving Power MS Imaging Files on Matlab Platform. Journal of the American Society for Mass Spectrometry. 2013b;24(5):718–21.
240. Bemis KD, Harry A, Eberlin LS, Ferreira C, van de Ven SM, Mallick P, et al. Cardinal: an R package for statistical analysis of mass spectrometry-based imaging experiments. Bioinformatics. 2015;31(14):2418–20.

241. Robbe M-F, Both J-P, Prideaux B, Klinkert I, Picaud V, Schramm T, et al. Software Tools of the Computis European Project to Process Mass Spectrometry Images. European Journal of Mass Spectrometry. 2014;20(5):351–60.
242. Källback P, Nilsson A, Shariatgorji M, Andrén PE. msIQuant – Quantitation Software for Mass Spectrometry Imaging Enabling Fast Access, Visualization, and Analysis of Large Data Sets. Analytical Chemistry. 2016;88(8):4346–53.
243. Race AM, Palmer AD, Dexter A, Steven RT, Styles IB, Bunch J. SpectralAnalysis: Software for the Masses. Analytical Chemistry. 2016;88(19):9451–8.
244. Norris JL, Cornett DS, Mobley JA, Andersson M, Seeley EH, Chaurand P, et al. Processing MALDI mass spectra to improve mass spectral direct tissue analysis. International Journal of Mass Spectrometry. 2007;260(2):212–21.
245. Abdelmoula WM, Lopez BG-C, Randall EC, Kapur T, Sarkaria JN, White FM, et al. Peak learning of mass spectrometry imaging data using artificial neural networks. Nature Communications. 2021;12(1):5544.
246. Van de Plas R, Yang J, Spraggins J, Caprioli RM. Image fusion of mass spectrometry and microscopy: a multimodality paradigm for molecular tissue mapping. Nature Methods. 2015;12(4):366–72.
247. McDowell CT, Klamer Z, Hall J, West CA, Wisniewski L, Powers TW, et al. Imaging Mass Spectrometry and Lectin Analysis of N-Linked Glycans in Carbohydrate Antigen–Defined Pancreatic Cancer Tissues. Molecular & Cellular Proteomics. 2021;20:100012.
248. Tibshirani R. Regression shrinkage and selection via the lasso: a retrospective. Journal of the Royal Statistical Society Series B: Statistical Methodology. 2011;73(3):273–82.
249. Bates S, Tibshirani R. Log-ratio lasso: Scalable, sparse estimation for log-ratio models. Biometrics. 2019;75(2):613–24.
250. Vijayalakshmi K, Shankar V, Bain RM, Nolley R, Sonn GA, Kao CS, et al. Identification of diagnostic metabolic signatures in clear cell renal cell carcinoma using mass spectrometry imaging. Int J Cancer. 2020b;147(1):256–65.
251. Banerjee S, Zare RN, Tibshirani RJ, Kunder CA, Nolley R, Fan R, et al. Diagnosis of prostate cancer by desorption electrospray ionization mass spectrometric imaging of small metabolites and lipids. Proc Natl Acad Sci U S A. 2017a;114(13):3334–9.
252. Tibshirani R. Regression shrinkage and selection via the lasso. Journal of the Royal Statistical Society: Series B (Methodological). 1996;58(1):267–88.
253. Margulis K, Chiou AS, Aasi SZ, Tibshirani RJ, Tang JY, Zare RN. Distinguishing malignant from benign microscopic skin lesions using desorption electrospray ionization mass spectrometry imaging. Proceedings of the National Academy of Sciences. 2018;115(25):6347–52.
254. Feider CL, Woody S, Ledet S, Zhang J, Sebastian K, Breen MT, et al. Molecular Imaging of Endometriosis Tissues using Desorption Electrospray Ionization Mass Spectrometry. Scientific Reports. 2019;9(1):15690.
255. Verbeeck N, Caprioli RM, Van de Plas R. Unsupervised machine learning for exploratory data analysis in imaging mass spectrometry. Mass Spectrometry Reviews. 2020;39(3):245–91.
256. Buchberger AR, DeLaney K, Johnson J, Li L. Mass Spectrometry Imaging: A Review of Emerging Advancements and Future Insights. Analytical Chemistry. 2018;90(1):240–65.
257. Jones EA, Deininger S-O, Hogendoorn PCW, Deelder AM, McDonnell LA. Imaging mass spectrometry statistical analysis. Journal of Proteomics. 2012;75(16):4962–89.
258. Widlak P, Mrukwa G, Kalinowska M, Pietrowska M, Chekan M, Wierzgon J, et al. Detection of molecular signatures of oral squamous cell carcinoma and normal epithelium – application of a novel methodology for unsupervised segmentation of imaging mass spectrometry data. PROTEOMICS. 2016;16(11–12):1613–21.
259. Alexandrov T, Kobarg JH. Efficient spatial segmentation of large imaging mass spectrometry datasets with spatially aware clustering. Bioinformatics. 2011;27(13):i230–i8.
260. Cassese A, Ellis SR, Ogrinc Potočnik N, Burgermeister E, Ebert M, Walch A, et al. Spatial Autocorrelation in Mass Spectrometry Imaging. Analytical Chemistry. 2016;88(11):5871–8.

261. Palmer A, Phapale P, Chernyavsky I, Lavigne R, Fay D, Tarasov A, et al. FDR-controlled metabolite annotation for high-resolution imaging mass spectrometry. Nature Methods. 2017;14(1):57–60.
262. Cole LM, Clench MR. Mass spectrometry imaging tools in oncology. Biomarkers in Medicine. 2015;9(9):863–8.
263. Yanagisawa K, Shyr Y, Xu BJ, Massion PP, Larsen PH, White BC, et al. Proteomic patterns of tumour subsets in non-small-cell lung cancer. The Lancet. 2003;362(9382):433–9.
264. Lou S, Balluff B, Cleven AHG, Bovée JVMG, McDonnell LA. An experimental guideline for the analysis of histologically heterogeneous tumors by MALDI-TOF mass spectrometry imaging. Biochimica et Biophysica Acta (BBA) – Proteins and Proteomics. 2017;1865(7):957–66.
265. Lemaire R, Ait Menguellet S, Stauber J, Marchaudon V, Lucot J-P, Collinet P, et al. Specific MALDI Imaging and Profiling for Biomarker Hunting and Validation: Fragment of the 11S Proteasome Activator Complex, Reg Alpha Fragment, Is a New Potential Ovary Cancer Biomarker. Journal of Proteome Research. 2007b;6(11):4127–34.
266. Cazares LH, Troyer D, Mendrinos S, Lance RA, Nyalwidhe JO, Beydoun HA, et al. Imaging Mass Spectrometry of a Specific Fragment of Mitogen-Activated Protein Kinase/Extracellular Signal-Regulated Kinase Kinase Kinase 2 Discriminates Cancer from Uninvolved Prostate Tissue. Clinical Cancer Research. 2009;15(17):5541–51.
267. Rauser S, Marquardt C, Balluff B, Deininger S-O, Albers C, Belau E, et al. Classification of HER2 Receptor Status in Breast Cancer Tissues by MALDI Imaging Mass Spectrometry. Journal of Proteome Research. 2010;9(4):1854–63.
268. Balluff B, Elsner M, Kowarsch A, Rauser S, Meding S, Schuhmacher C, et al. Classification of HER2/neu Status in Gastric Cancer Using a Breast-Cancer Derived Proteome Classifier. Journal of Proteome Research. 2010;9(12):6317–22.
269. Meding S, Balluff B, Elsner M, Schöne C, Rauser S, Nitsche U, et al. Tissue-based proteomics reveals FXYD3, S100A11 and GSTM3 as novel markers for regional lymph node metastasis in colon cancer. The Journal of Pathology. 2012;228(4):459–70.
270. Bauer JA, Chakravarthy AB, Rosenbluth JM, Mi D, Seeley EH, De Matos Granja-Ingram N, et al. Identification of Markers of Taxane Sensitivity Using Proteomic and Genomic Analyses of Breast Tumors from Patients Receiving Neoadjuvant Paclitaxel and Radiation. Clinical Cancer Research. 2010;16(2):681–90.
271. Aichler M, Elsner M, Ludyga N, Feuchtinger A, Zangen V, Maier SK, et al. Clinical response to chemotherapy in oesophageal adenocarcinoma patients is linked to defects in mitochondria. The Journal of Pathology. 2013;230(4):410–9.
272. Elsner M, Rauser S, Maier S, Schöne C, Balluff B, Meding S, et al. MALDI imaging mass spectrometry reveals COX7A2, TAGLN2 and S100-A10 as novel prognostic markers in Barrett's adenocarcinoma. Journal of Proteomics. 2012;75(15):4693–704.
273. Leung F, Musrap N, Diamandis EP, Kulasingam V. Advances in mass spectrometry-based technologies to direct personalized medicine in ovarian cancer. Translational Proteomics. 2013;1(1):74–86.
274. Menendez JA, Lupu R. Fatty acid synthase and the lipogenic phenotype in cancer pathogenesis. Nature Reviews Cancer. 2007;7(10):763–77.
275. Zaidi N, Lupien L, Kuemmerle NB, Kinlaw WB, Swinnen JV, Smans K. Lipogenesis and lipolysis: The pathways exploited by the cancer cells to acquire fatty acids. Progress in Lipid Research. 2013;52(4):585–9.
276. Perez CJ, Bagga AK, Prova SS, Yousefi Taemeh M, Ifa DR. Review and perspectives on the applications of mass spectrometry imaging under ambient conditions. Rapid Communications in Mass Spectrometry. 2019;33(S3):27–53.
277. He MJ, Pu W, Wang X, Zhang W, Tang D, Dai Y. Comparing DESI-MSI and MALDI-MSI Mediated Spatial Metabolomics and Their Applications in Cancer Studies. Frontiers in Oncology. 2022;12.

278. Dória ML, McKenzie JS, Mroz A, Phelps DL, Speller A, Rosini F, et al. Epithelial ovarian carcinoma diagnosis by desorption electrospray ionization mass spectrometry imaging. Scientific Reports. 2016;6(1):39219.
279. Zhang J, Rector J, Lin JQ, Young JH, Sans M, Katta N, et al. Nondestructive tissue analysis for ex vivo and in vivo cancer diagnosis using a handheld mass spectrometry system. Science Translational Medicine. 2017;9(406):eaan3968.
280. Calligaris D, Caragacianu D, Liu X, Norton I, Thompson CJ, Richardson AL, et al. Application of desorption electrospray ionization mass spectrometry imaging in breast cancer margin analysis. Proceedings of the National Academy of Sciences. 2014;111(42):15184–9.
281. Guenther S, Muirhead LJ, Speller AVM, Golf O, Strittmatter N, Ramakrishnan R, et al. Spatially Resolved Metabolic Phenotyping of Breast Cancer by Desorption Electrospray Ionization Mass Spectrometry. Cancer Research. 2015;75(9):1828–37.
282. Banerjee S, Zare RN, Tibshirani RJ, Kunder CA, Nolley R, Fan R, et al. Diagnosis of prostate cancer by desorption electrospray ionization mass spectrometric imaging of small metabolites and lipids. Proceedings of the National Academy of Sciences. 2017b;114(13):3334–9.
283. Rice MA, Kumar V, Tailor D, Garcia-Marques FJ, Hsu E-C, Liu S, et al. SU086, an inhibitor of HSP90, impairs glycolysis and represents a treatment strategy for advanced prostate cancer. Cell Reports Medicine. 2022;3(2):100502.
284. Eberlin LS, Tibshirani RJ, Zhang J, Longacre TA, Berry GJ, Bingham DB, et al. Molecular assessment of surgical-resection margins of gastric cancer by mass-spectrometric imaging. Proceedings of the National Academy of Sciences. 2014;111(7):2436–41.
285. Sans M, Gharpure K, Tibshirani R, Zhang J, Liang L, Liu J, et al. Metabolic Markers and Statistical Prediction of Serous Ovarian Cancer Aggressiveness by Ambient Ionization Mass Spectrometry Imaging. Cancer Research. 2017;77(11):2903–13.
286. Jarmusch AK, Alfaro CM, Pirro V, Hattab EM, Cohen-Gadol AA, Cooks RG. Differential Lipid Profiles of Normal Human Brain Matter and Gliomas by Positive and Negative Mode Desorption Electrospray Ionization – Mass Spectrometry Imaging. PLoS One. 2016;11(9):e0163180.
287. Pirro V, Alfaro CM, Jarmusch AK, Hattab EM, Cohen-Gadol AA, Cooks RG. Intraoperative assessment of tumor margins during glioma resection by desorption electrospray ionization-mass spectrometry. Proceedings of the National Academy of Sciences. 2017;114(26):6700–5.
288. Calligaris D, Feldman DR, Norton I, Brastianos PK, Dunn IF, Santagata S, et al. Molecular typing of meningiomas by desorption electrospray ionization mass spectrometry imaging for surgical decision-making. International Journal of Mass Spectrometry. 2015;377:690–8.
289. Zhang J, Yu W, Ryu SW, Lin J, Buentello G, Tibshirani R, et al. Cardiolipins Are Biomarkers of Mitochondria-Rich Thyroid Oncocytic Tumors. Cancer Research. 2016;76(22):6588–97.
290. Jarmusch AK, Kerian KS, Pirro V, Peat T, Thompson CA, Ramos-Vara JA, et al. Characteristic lipid profiles of canine non-Hodgkin's lymphoma from surgical biopsy tissue sections and fine needle aspirate smears by desorption electrospray ionization – mass spectrometry. Analyst. 2015;140(18):6321–9.
291. Abbassi-Ghadi N, Veselkov K, Kumar S, Huang J, Jones E, Strittmatter N, et al. Discrimination of lymph node metastases using desorption electrospray ionisation-mass spectrometry imaging. Chemical communications. 2014;50(28):3661–4.
292. Eberlin LS, Margulis K, Planell-Mendez I, Zare RN, Tibshirani R, Longacre TA, et al. Pancreatic Cancer Surgical Resection Margins: Molecular Assessment by Mass Spectrometry Imaging. PLOS Medicine. 2016;13(8):e1002108.
293. Gerbig S, Golf O, Balog J, Denes J, Baranyai Z, Zarand A, et al. Analysis of colorectal adenocarcinoma tissue by desorption electrospray ionization mass spectrometric imaging. Analytical and Bioanalytical Chemistry. 2012;403(8):2315–25.
294. Alfaro CM, Jarmusch AK, Pirro V, Kerian KS, Masterson TA, Cheng L, et al. Ambient ionization mass spectrometric analysis of human surgical specimens to distinguish renal cell carcinoma from healthy renal tissue. Analytical and Bioanalytical Chemistry. 2016;408(20):5407–14.

295. Shroff EH, Eberlin LS, Dang VM, Gouw AM, Gabay M, Adam SJ, et al. MYC oncogene overexpression drives renal cell carcinoma in a mouse model through glutamine metabolism. Proceedings of the National Academy of Sciences. 2015;112(21):6539–44.
296. Zhang M, He J, Li T, Hu H, Li X, Xing H, et al. Accurate Classification of Non-small Cell Lung Cancer (NSCLC) Pathology and Mapping of EGFR Mutation Spatial Distribution by Ambient Mass Spectrometry Imaging. Front Oncol. 2019;9:804.
297. Prideaux B, Dartois V, Staab D, Weiner DM, Goh A, Via LE, et al. High-Sensitivity MALDI-MRM-MS Imaging of Moxifloxacin Distribution in Tuberculosis-Infected Rabbit Lungs and Granulomatous Lesions. Analytical Chemistry. 2011;83(6):2112–8.
298. Huber K, Feuchtinger A, Borgmann DM, Li Z, Aichler M, Hauck SM, et al. Novel Approach of MALDI Drug Imaging, Immunohistochemistry, and Digital Image Analysis for Drug Distribution Studies in Tissues. Analytical Chemistry. 2014;86(21):10568–75.
299. Sugihara Y, Végvári Á, Welinder C, Jönsson G, Ingvar C, Lundgren L, et al. A new look at drugs targeting malignant melanoma—An application for mass spectrometry imaging. PROTEOMICS. 2014;14(17–18):1963–70.

Future Perspectives of Metabolomics: Gaps, Planning, and Recommendations

Vijay Soni, Nicholas Bartelo, Annalise Schweickart, Yogesh Chawla, Ankita Dutta, and Sonia Jain

Abbreviations

AD	Alzheimer's disease
ANN	Artificial neural network
AUC	Area under the curve
CE-MS	Capillary electrophoresis - Mass Spectrometry
CGM	Continuous glucose monitor
CNN	Convolutional neural network
DI-MS	Direct infusion-mass spectrometry
DL	Deep learning
EI-MS	Electron ionization mass spectrometry

Authors: "Vijay Soni" and "Nicholas Bartelo" have equally contributed to this chapter.

V. Soni (✉)
Division of Infectious Diseases, Weill Department of Medicine, Weill Cornell Medicine, New York, NY, USA
e-mail: vis2032@med.cornell.edu

N. Bartelo
Department of Physiology and Biophysics, Weill Cornell Medicine, Institute for Computational Biomedicine, Englander Institute for Precision Medicine, New York, NY, USA

A. Schweickart
The HRH Prince Alwaleed Bin Talal Bin Abdulaziz Alsaud Institute for Computational Biomedicine, Weill Cornell Medicine, New York, NY, USA

Tri-Institutional Computational Biology and Medicine Program, Weill Cornell Medicine, New York, NY, USA

Y. Chawla
Department of Hematology, Mayo Clinic, Rochester, MN, USA

A. Dutta
International AIDS Vaccine Initiative, Gurugram, Haryana, India

S. Jain
Department of Radiation Oncology, Mayo Clinic, Rochester, MN, USA

V. Soni, T. E. Hartman (eds.), *Metabolomics*,
https://doi.org/10.1007/978-3-031-39094-4_14

ELISA	Enzyme-linked immunosorbent assay
FDR	False discovery rate
GC-FID	Gas chromatography-flame ionization detection
GC-MS	Gas chromatography-mass spectrometry
hCG	Human chorionic gonadotropin
HDL	High-density lipoprotein
IEM	Inborn errors of metabolism
LAESI	Laser ablation electrospray ionization
LC-HRMS	Liquid chromatography coupled to high-resolution mass spectrometry
LC-MS	Liquid chromatography-mass spectrometry
LDL	Low-density lipoprotein
LDTs	Laboratory-developed tests
MALDI	Matrix-assisted laser desorption/ionization
mGWAS	Metabolic genome-wide association studies
ML	Machine learning
mQTL	Metabolites and quantitative trait loci
MSI	Metabolomics Standards Initiative
NMR	Nuclear magnetic resonance
PMRN	Pharmacometabolomics Research Network
QSP	Quantitative and systems pharmacology
SIMS	Secondary ion mass spectrometry
SNPs	Single nucleotide polymorphisms
VOC	Volatile organic compound

1 Introduction

Metabolites are the connecting link between the genome and the environment. With the development of novel technologies and bioinformatics approaches, it is now possible to study the global metabolic changes in any organisms and cells. Metabolomes represent an organism's physiological state and can be used to help diagnose and treat a variety of diseases. Metabolomics as a field emerged in the late 1990s with the advent of proteomics [1] and is now rapidly evolving. It is the study of the metabolome that comprises the entire repertoire of small molecules with molecular weights of <1000 Da or <1500 Da excluding biopolymers like proteins or nucleic acids [2]. The small molecules are also referred to as metabolites and are present in human cells, tissues, and body fluids. They can be studied using large-scale detection, quantification, and analysis methodologies. Metabolites are organic and inorganic chemicals and are either reactants, intermediates, or end products generated during biological enzymatic reactions or may be of xenobiotic origin (i.e., the chemicals that are found in living organisms however are not produced by them) and are known to bridge gene functions and nongenetic or phenotypic end points [3–5]. They exhibit variable chemical properties that range from polar hydrophobic compounds and hydrophilic compounds including carbohydrate moieties to nonpolar hydrophobic molecules such as lipids.

Metabolomics has made significant progress in the past two decades. Nevertheless, several aspects of this field are still in the development phase and restrict its application in various domains. These limitations include a restricted detection range, bulk analysis with precise molecular features, a lack of chromatographic methods for better resolution, and the high cost of analytical devices such as mass spectrometers and NMR. However, despite these constraints, metabolomics continues to be a valuable research tool in translational biology, pharmacological medicine, biomarker discovery, and diagnosis. By analyzing the metabolic profiles of patients with different diseases, researchers can gain insights into the underlying mechanisms and identify potential targets for therapy. Furthermore, it has tremendous potential to monitor disease progression and treatment response and detect the side effects of therapies. Besides, the broad range of metabolomics is enabling research in other areas such as agriculture, environmental surveillance, and nutritional biology.

In the coming years, advancements in basic research and healthcare technologies are expected to surpass our current understanding of living organisms. Metabolomics is one such field that holds great promise. Therefore, in this chapter, we aim to provide an overview of the future perspectives of metabolomics in various areas, including precision medicine, personalized nutrition, disease diagnosis, biomarker discovery, single-cell metabolism, the development of novel AI-/ML-based tools for data integration, applications in translational biology, and therapeutic development. Additionally, we will touch upon the topic of metabolic sensors and wearables for disease surveillance. Finally, we have provided recommendations to consider while developing new technologies using metabolomics results. Overall, the future of metabolomics looks bright, with the potential to revolutionize our understanding of biological processes and identify new strategies for enhancing human welfare.

2 Metabolomics for the Masses

While the field of metabolomics has many applications in a clinical setting, there is also a clear use for bringing metabolomics into everyday life. Technologies such as pregnancy tests, breathalyzers, and blood glucose monitors are familiar examples of tools implemented for simple readouts of our physiological state from biomolecular readouts. However, these technologies are only designed for the detection of single molecules: pregnancy tests detect human chorionic gonadotropin (hCG) in urine, breathalyzer tests measure alcohol levels from a breath,[1] and blood glucose monitors measure instantaneous plasma glucose levels. While these narrow searches for

[1] Breathalyzer tests have also been designed for diagnosis of viral and bacterial infections through volatile organic compound detection. Secondary electrospray ionization-mass spectrometry (SESI-MS) on mouse breath could detect infection as well as distinguish between different pathogens and strains [6], and a diagnostic breath test using gas chromatography-mass spectrometry (GC-MS) was approved for emergency use in the Covid-19 pandemic [7, 8].

specific molecules achieve the designed task, the conclusions that can be drawn using these technologies are equally as narrow.

Recently, there has been an emergence of consumer products that expand biomolecule detection through large metabolomic analyses to provide wider insights into our physiological state. The promise of these products includes early disease detection, lifestyle recommendations to improve metabolic health, and future health outcome predictions, all without the need to visit a doctor's office. While consumer products may bring metabolomics closer to everyday life, many challenges still need to be addressed to scale up these analyses.

2.1 Physiological Assessment Through Endogenous Molecules

The accessibility of high-throughput metabolomics has grown, and therefore associations between individual, endogenously derived metabolites and physiological states have become easier to identify [9]. These findings can be leveraged for inference on a new individual's physiology. However, high-throughput metabolomics on the same biological source must be used for the inference to be accurate.

Biofluids commonly used for such analyses are blood (either plasma or serum [10]), saliva, and urine [11]. While the latter two fluids are the easiest to obtain noninvasively, saliva is highly affected by external factors such as hygiene and food intake, and urine is a waste product, meaning its metabolic contents will heavily represent molecules being excreted rather than those being actively used [12]. Blood is therefore the fluid most reflective of bodily processes. Blood circulates in all organs and tissues, making it a "reasonably good metabolic proxy for the entire organism" [13]. As such, businesses building analytical pipelines to bring metabolomics to the masses have opted for blood draw devices to collect samples for further analysis.

Reflecting the methods of published metabolomics-disease associations again, blood is then profiled through liquid chromatography-mass spectrometry (LC-MS) methods, and the resulting spectra are mined for known physiological and disease biomarkers. From plasma draws, there are ways to discern the general function of our organs as each has a unique metabolomic footprint. Organs consume and excrete different metabolites, such as the liver's production of bile acids, the thyroid's production of thyroxine, or the adrenal glands' production of epinephrine and cortisol [13]. Such unique metabolomic footprints allow health-based conclusions to be drawn per organ from metabolomics analysis on a plasma sample.

There have been numerous studies identifying metabolites as biomarkers for non-organ-derived disease as well [14–16]. While such studies have faced criticism due to low sample size and statistical power [17], a 2021 study from Pietzner et al. established robust metabolomic effects in a population of just under 12,000 individuals [9]. Associations were found between both identified and unidentified metabolites and metabolic, heart, and lung diseases, as well as an array of cancer types. The detection of abnormally high or low levels of these metabolites in an individual's blood could potentially lead to the early detection of disease. This prospect is what drives the establishment of metabolomics tools for the masses – with a single blood draw, a person could discover potentially life-saving information and seek intervention much earlier than they would with classical screening techniques. However, this promise has yet to be realized, and only time will tell if the masses are willing to participate in and act on metabolomics-based health insights.

2.2 *Limitations of Current Technologies*

2.2.1 Exogenous Molecule Identification

Beyond endogenous compounds, exogenous compounds that enter the body through the environment can also provide insight into the disease state, as discussed in chapter "Exploring Ecometabolomics Landscapes: Progress, Applications, Challenges, and Future Recommendations". In fact, most complex human diseases, from cancer to cardiovascular disease, can be attributed to the environment and the interplay between an individual's genes and their environment [18, 19]. Unfortunately, these environmentally derived compounds, collectively known as our exposome, come from a vastly larger pool of chemicals than endogenous metabolites, and only a small fraction of these are known and have been measured in human tissues [20].

To make matters worse, the exposome is spatiotemporally dependent, meaning exogenous molecules in the blood will depend on the location and time of biofluid collection. Additionally, the physiological response to these molecules varies widely between individuals [21]. This, along with the only very recent development of high-throughput exposure data collection and experimental pipelines, means the number of known associations between exogenous molecules and disease lags behind those of endogenous molecules [22, 23]. The exposome contains contaminants, toxins, pollutants, and carcinogens, all molecules with potentially serious consequences to human health. Due to the aforementioned technological difficulties, potential exogenous biomarkers for disease will be missed with current tools. Therefore, technologies for exposome data collection, identification, and quantification are needed to bring the further benefit of metabolomics to the masses.

2.2.2 Commercial Affordability and Interest

Full metabolomics screens are not currently available as a part of everyday healthcare, and therefore access by individuals to their full metabolomics profiles must come from a direct-to-consumer business. From a commercial perspective, to provide helpful metabolomics-derived information to the masses, the masses must want, and therefore pay for, their metabolomics readouts. However, a viable business model for a direct-to-consumer metabolomics product has yet to emerge. The multi-omics profiling company *Arivale* shut its doors in 2019, four years after its founding, due to an inability to fund its consumer-based scientific wellness program which included genome, metabolome, and microbiome profiling [24]. Beyond the high technological costs, CEO Clayton Lewis also cited the high costs of customer acquisition and a lack of interest in this data as a regular part of healthcare [25]. While competitors have either lowered consumer costs or made licensing agreements with large healthcare companies that could make molecular profiling more widespread, the future availability of metabolomics for the masses is going to depend on translational utility and the willingness of the masses to independently invest in health insights derived from metabolomics data [26].

3 Future Opportunities and Challenges in Translational Metabolomics

The emerging focus of personalized medicine is greatly due to the explosion of omics data: genomics, transcriptomics, proteomics, metabolomics, etc. While bountiful research has shown the potential for biomarker detection, disease subtyping, drug repurposing and discovery, and other useful applications for patient care, the development of widespread healthcare tools has not followed suit equally for each type of omics data. Genomics has dominated clinical implementation, with more than 75,000 genetic tests available by 2017 [27]. On the contrary, 2 years later in 2019, transcriptomics and proteomics were the basis for only five assays and one assay, respectively [28], in a clinical setting. Even more astounding, metabolomics is still without an FDA-approved test in 2023 [29]. While there are no current clinical tools for metabolomics, there are many areas in healthcare research conducting extensive experiments that can benefit from their development.

Numerous branches of medicine have recognized metabolomics as a potential strategy to identify predictive, diagnostic, or prognostic markers of disease. Oncology has already made strides in this respect, as the ability to find metabolite biomarkers in serum and image-based applications has been explored. In breast cancer patients, the metabolome is representative of over 30 endogenous metabolites, characteristic of low glucose, low glycerophosphocholine, and increased tCho levels [30]. The mapping of metabolic signatures has been conducted for additional cancers including ovarian [31], lung [32], and endometrial [33]. In addition, the detection of breast cancer tissue from noninvolved adjacent tissue using metabolomics with simultaneous measurement of tumor size, lymph node status, hormone status, and histology was determined with accuracy, sensitivity, and specificity all around 90% [34]. As an increasing number of studies are performed to validate existing and discover new biomarkers of disease, the chemical fingerprint of the phenotype may become increasingly specific. In addition to diagnostic usage, metabolomics has been used to predict treatment outcomes, such as sensitivity and resistance of either chemotherapy- or hormonal therapy-treated samples of human glioma cell cultures [35]. This study demonstrates the ability to create diagnostic tools using metabolomics that are detectable before changes in phenotype are evident using conventional imaging techniques [36].

Type 2 diabetes and other diseases under the umbrella of endocrinology have bountiful potential for metabolomic techniques to aide in our understanding of their etiology. Due to the current global type 2 diabetes epidemic and its worrying projections, a large focus has been placed on developing diagnostic biomarkers. An example of type 2 diabetes-associated metabolite, 2-aminoadipic acid, was reported by Wang et al. [37]. Using the Framingham Heart Study, 2-aminoadipic acid was discovered to be increased in diseased individuals up to 12 years before onset, and was not well correlated with other metabolites, suggesting a distinct metabolic pathway for risk assessment of phenotype outcome. As the nature vs nurture debate continues in the diabetes field, it has been demonstrated that metabolic markers are more predictive of type 2 diabetes development than genome-wide association studies or other genetic data [28]. This proposes a greater environmental role for disease onset, suggesting lifestyle changes are a viable solution. In rheumatology, 20 metabolites have been identified to decently discriminate rheumatoid arthritis from ankylosing spondylitis, Behcet's disease, and gout with an area under the receiver operating characteristic curve (AUC) of 0.812 [38]. As additional metabolomics studies targeting the classification between two groups are conducted with larger sample sizes, the validation and refinement of metabolic profiles associated with specific diseases will be attainable.

In neurological disorders, such as Alzheimer's disease (AD) and Parkinson's disease, the quest for metabolic biomarkers for early diagnosis and subtyping is of interest to many. As early AD diagnosis and AD treatment have had very limited success, metabolomics may provide novel insights into the underlying mechanisms driving AD development and progression. Metabolomics of plasma samples from AD cases compared with controls identified a higher abundance of free cholesterol in small HDL associated with a lower risk of AD and higher levels of glutamine associated with increased AD risk [39]. Additional metabolites were discovered to be correlated with general cognition. These results demonstrate potential biomarkers for further study which could be indicative of AD development and cognitive decline. If these metabolites are valid, it may be possible to create tools to aid in the diagnosis of AD and other cognition-centered conditions. In addition to biomarker detection in neurological disease, the stratification between patients with Parkinson's disease versus controls, and Parkinson's with dementia versus Parkinson's without dementia, has been demonstrated by Han et al. with AUC of 0.955 and 0.862, respectively [40]. The ability to distinguish between neurodegenerative stages provides a clinical application for diagnosing disease severity and necessary treatment.

A pediatric study focusing on the volatile organic compound (VOC) abundance for children with and without asthma has concluded an 80–100% accuracy of diagnosis with a combination of VOCs [41]. While still in its infancy, further focus on VOCs may lead to better risk assessment for identifying children with the greatest risk of adverse events. The metabolic profile of cardiovascular disease has shown promise for the discovery of new biomarkers for the diagnosis of heart-related adverse events. A significant increase in the discrimination between 150 individuals developing atherosclerosis and 1445 not developing atherosclerosis between a 6-year time interval resulted when low-density lipoprotein (LDL) cholesterol, high-density lipoprotein (HDL), docosahexaenoic acid, and tyrosine were added to traditional biomarkers [42]. This conclusion demonstrates the capability of metabolites to assist in the predictive power of traditional diagnostic methods.

As the metabolome is composed of thousands of compounds, there is a high potential for the discovery of metabolites indicative of early detection for many diseases. Vallejo et al. revealed perfect separation between plasma samples from patients with atherosclerosis and controls, and patients with acute coronary syndrome and controls, using gas chromatography-mass spectrometry [43]. This suggests a metabolite-specific assay could be created to confidently identify patients with these heart problems. Heart failure patients and controls have been demonstrated to be separable based on the measurement of pseudouridine, 2-oxoglutarate, 2-hydroxy 2-methyl propanoic acid, erythritol, and 2,4,6-trihydroxy pyrimidine [44]. This research concluded that there are novel metabolic biomarkers of heart failure which can be further investigated to discover their potential to be used in prognosis.

Taken together, metabolomics has been utilized for many applications in a wide variety of medical fields. A list of metabolite biomarkers for use as diagnostics is located on the Mayo Clinic website (https://www.mayocliniclabs.com/) [28]. As the deployment of metabolomics continues to offer promising results with respect to the identification of novel predictive, diagnostic, and prognostic biomarkers that aid in the overall understanding of the biological mechanisms underlying a phenotype, new studies will result in the precise identification and refinement of a metabolic fingerprint of many diseases, which can be measured to make clinical assessments.

As interest in metabolomics increases in research, industrial efforts are aiding in the future translational capability by focusing on the creation of simpler and better LC-MS/MS systems [28]. Both Sciex and Waters have created instruments for use in clinical laboratories. In addition, enzyme-linked immunosorbent assay (ELISA) is a method by which targeted metabolites can be quantified for clinical use, although there are still limitations with this approach that need to be addressed [45]. Also, as the use of mobile device data and wearable data rapidly grows, metabolomic measurements can accompany these sources to create a foundation for the metabolome of both diseased and healthy individuals on a massive scale. The influx of this data may be used to create personalized recommendations for numerous applications including exercise and nutrition.

As mentioned earlier, diseases such as type 2 diabetes have been suggested to be driven, on average, more by environmental than by genetic components. Therefore, the culmination of multiple modalities of data for a large population could lead to the identification of necessary lifestyle requirements for the prevention of disease development. Nutritional metabolomics focuses on how chronic or acute food intake causes a response in an organism's metabolism [46]. Research involving medical foods and dietary supplements has shown promise as a solution to treat many inborn errors of metabolism, dietary deficiency diseases (such as rickets, scurvy, and goiter), and other medical conditions such as coeliac disease (through gluten-free diets) and epilepsy (through ketogenic diets) [47].

Although cancer prevention guidelines have stated the association between the consumption of red meat, processed meat, and sugary drinks with the development of cancer, these products are still overeaten by some of the US population. Research focused on the resulting metabolic changes may aid in the discovery of direct mechanisms responsible for the correlation between these foods and drinks and cancer. As with smoking, as the evidence amounts to and becomes popularized, an incentive for the government to intervene and create a policy limiting the ability of the population to consume these harmful products may be implemented. As more studies are conducted with a focus on medicinal food, an additional application may be the incorporation of a supplemented diet in addition to standard treatment in healthcare. As nutrition influences overall well-being, the potential of precision nutrition to create a healthier population has an enormous beneficial consequence and may promise a large market as technological advances demonstrate a positive impact.

Personalized medicine, drug discovery, and minimization of risk for blood contamination are potential candidates for translational use of metabolomics. Laboratory-developed tests (LDTs) are defined by the FDA as "in vitro diagnostic tests that are manufactured by and used within a single laboratory," which can measure either individual or multiple analytes [29]. Abnormalities in metabolic pathways and biomarkers unable to be detected by other means are measured by Metabolon's Meta UDx™ test. For hereditary metabolic disorders, the diagnostic tests Meta IMD™ and Meta IMD™ + (Plus) were developed. Although not approved by the FDA, these LDTs may provide information for new tests that can be used in the clinic to gain more information about a patient than the current standard of care.

With the development of the chemical fingerprints of metabolic changes resulting from disease development, tailored recommendations can guide treatment of an individual given the metabotype of the patient [28]. There are many opportunities to expand metabolomics in the future to new sources, such as cerebrospinal fluid, human saliva, bronchoalveolar lavage, sweat, feces, semen, and amniotic fluid [48]. These studies will provide answers to current questions in multiple healthcare fields and may lead to the ability to investigate new topics. Also, by comparing metabolite abundance before and after drug treatment and studying the resulting phenotype, insight to the mechanistic impacts of drugs can be elucidated. This could also guide drug developers to create more effective therapeutics, as direct experimental evidence would provide a more comprehensive understanding of the drug mechanism [49].

Additionally, a treatment's level of toxicity may be confidently measured by metabolomics, aiding in the development of optimal medications for a patient and also providing an avenue for the creation of predictive modeling of drug toxicity for the creation of new therapeutics [50]. Bacterial problems arising from contamination and antibiotic resistance may be solved through the study of metabolism. As pathogens can be transmitted by human blood and blood-derived products, metabolomics may be a tool to minimize or eradicate this risk due to its sensitivity of measurement through clinical screening [48]. Also, the rise in antimicrobial resistance due to antibiotic use could be alleviated by the determination of metabolomic biomarkers of resistance and the creation of methods for metabolic changes that can kill the bacteria. The implementation of these tools in healthcare could prevent the spread of resistant bacteria and save many lives in the future.

Although there are an impressive number of applications for translational metabolomics, there are numerous challenges that must be overcome to create valid, precise techniques for analyzing patient samples. Many logistical challenges provide impediments to an optimal workflow for the utilization of metabolomics in the clinic. As a relatively recent approach to omics technologies, its publicity is much less than others and often clouded by the successes more advanced omics have achieved [51]. One bottleneck is the high cost of the instruments needed to measure the samples and the laborious sample preparation methodologies [36]. The resources and armamentarium needed to store and measure metabolites before they undergo transformation and/or degradation may not be available at many healthcare institutions and must be purchased to allow for metabolomic analyses to be made possible [51]. As trends in other omics have shown decreased time and cost for the generation of data [52], it is possible that as new measurement techniques are developed, economic and temporal barriers will be less of a factor. An additional issue is a current need for the culmination of experts in different research areas including biologists, analytical chemists, statisticians, data scientists, and bioinformaticians to successfully conduct and interpret a metabolomics-based experiment in its entirety [28]. The reason for this is that the data output from traditional metabolomics platforms is rich and complex.

To be used in clinical settings, the number of metabolites must be reduced greatly for a clear interpretation of the results, thus making a risk assessment, diagnosis, and prognosis easier for the clinician. To select biomarkers for use in healthcare from large metabolomic datasets, there is a need for a user-friendly platform that can process, statistically interpret, and determine straightforward conclusions about data, demonstrating the direct effect of a change in phenotype on the metabolome. The market will also drive the availability of clinical tests. To make a product commercially viable, it must be profitable, which will depend on an estimate of how many people will use it [28]. To increase the probability of incorporating new technology in healthcare, researchers can work with industrial organizations to develop easy-to-use, clinician-approved tools.

Numerous technical aspects of metabolomics need to be overcome to create reliable metabolite biomarkers for unique metabotyping of disease. One major obstacle for untargeted metabolomics is overcoming its semiquantitative nature. As data generation relies on the normalization of a signal, the definition of the normal concentration of metabolites is needed for reliable conclusions regarding the ability of a compound to be used as a biomarker of a phenotype [48]. A showcase example is the comparison of two studies analyzing roughly 45 total Crohn's disease and ulcerative colitis patients. One concluded that there is no discrimination using metabolomics between ulcerative colitis and Crohn's disease [53], while the other concluded that choline, lipoprotein, and N-acetylated glycoprotein levels were able to separate the conditions significantly with an AUC greater than 0.9 [54].

Due to the wide range of biological variation in the metabolome, different cohorts may exhibit wide-ranging metabolic profiles. Therefore, absolute quantification of metabolite concentration is necessary for accurate benchmarking [28]. To ensure bias is not impacting the results of benchmarking experiments, validation needs to be repeated in multiple populations with large sample sizes. Currently, many biomarkers determined through metabolomics are results from studies limited in validity, statistical robustness, and experimental design [48]. Quantification coupled with repeated validation must be performed to develop an understanding of the true metabolic pathway alterations characteristic of a disease. Another major challenge to translational metabolomics is the inability to identify metabolites and the difficulty of pathway mapping [28]. There are many metabolites unavailable for measurement in commercial products and/or cannot be identified using current spectral libraries such as METLIN or mzCloud. Thus, current methods lack the complete metabolome as a whole. An extension of this is the inability to identify metabolic pathways perturbed by a disease, hindering the potential use of metabolomics for the proper determination of changing biological mechanisms, biomarker identification, and therapeutic development.

The future of translational metabolomics is contingent on the creation of standardized protocols for experimental design and measurement, simplification of data analysis and results, and the development of robust quantization methods leading to the reliable identification of metabolites. As groups such as the Metabolomics Standards Initiative (MSI) continue to meet and perfect the current procedures in metabolomic analysis, metabolomics continually progresses toward translational applications. Still, in its infancy, numerous unanswered questions in biology will be elucidated by metabolomics as it develops and its utilization increases, making it one of the most exciting technologies of the present. There is an enormous opportunity for the study of the metabolome to influence global healthcare.

4 Metabolomics as a tool to Accelerate Therapeutics and Novel Drug Discoveries

To enable capturing of the diverse array of metabolites and their dynamic cellular concentrations, the detection of these molecules is primarily based on two technologies, namely, nuclear magnetic resonance (NMR) spectroscopy (1H or 13C) and mass spectrometry (MS). Mass spectrometry is often coupled with capillary electrophoresis (CE-MS), gas chromatography (GC-MS), gas chromatography-flame ionization detection (GC-FID), direct infusion-mass spectrometry (DI-MS), or liquid chromatography (LC-MS). Due to the large chemical diversity and limited knowledge on metabolism despite the implementation of these technologies coupled with a range of analytical methods, less than 5% of the metabolome is annotated [55].

As discussed in previous chapters, there are primarily two approaches for investigating the metabolome, i.e., targeted and global approach (also referred to as non-targeted). The former refers to the identification and measurement of the well-defined groups of chemically characterized metabolites that have been biochemically annotated as well using appropriate internal standards. Measurement is quantitative, and metabolite concentrations are expressed in molar units. This approach enables studying novel associations between the metabolites under variable physiological conditions [56]. Owing to the high sensitivity of this approach, it is often used for studying the flux metabolic pathways and for the detection of well-defined chemical compounds or validation of known biomarkers [57, 58]. Alternatively, the untargeted approach provides opportunities for comprehensive data analysis as it is possible to detect all measurable compounds or analytes within a given sample including putative annotated metabolites or chemically unknown samples. Due to this variability in detection, the data is examined in a semiquantitative or relative manner using multivariate analysis, wherein the extensive dataset is divided into smaller datasets of manageable signals. Owing to the relative quantification of the readouts as chromatographic peak areas, the data generally is expressed in terms of the intensity of ions or arbitrary units [56–59].

Moreover, these peak areas are determined by the experimental conditions under which the detection is performed using NMR, GC-MS, or liquid chromatography coupled to high-resolution mass spectrometry (LC-HRMS). Therefore, these variations make it difficult to directly compare the data from experiments done at different time points within the same laboratory or by different laboratories. The current challenges associated with the non-targeted approach include the nonavailability of standardized workflows primarily for data generation, complexity of signatures detected, lower sensitivity of detection owing to lower abundance, identification of metabolites (only a small proportion of metabolites and their annotated features are known), automated processing of data through feature detection and integration with other omics data, and finally the availability of only limited well-defined interoperability frameworks. Moreover, the current platforms for detection and analysis are highly expensive. Therefore, these together lead to nonoptimal reuse or interoperability of the data. Additionally, the identification of medically important molecular signatures and the demand for participatory medicine will also impact the establishment of methodologies for simplifying the complex data and accelerating research in the field, and will subsequently catalyze the development of affordable and accessible alternative analytical methods for the nonspecialized end users [11, 60]. Currently, the non-targeted approach is being used for the discovery of biomarkers.

There are multifaceted applications of metabolomics in various fields of biology [61]. For example, in environmental research, it is being used for addressing ecotoxicological issues [62], and in plant biology and agricultural science, it is being used for understanding cellular functions and discovering biomarkers, for diagnostics and phenotyping (*specific metabolites and quantitative trait loci (mQTL) & metabolic genome-wide association studies (mGWAS)*), and for predicting the metabolite-genome correlations [63, 64]. The current applications of metabolomics have also expanded to microbiome research [65–68]; animal health [69–71]; human healthcare including toxicology [72], epidemiology [73], cancer biology [36, 74,

75], cardiovascular diseases [36, 76], gastrointestinal diseases, [36, 77–79], aging research [80], and infectious diseases [81–83]; and nutrimetabolomics [83–85]. Toward biomedical research, metabolomics has furthered systems biology and systems medicine which has paved a pathway for personalized medicine (also referred to as precision medicine).

Personalized medicine or precision medicine is aimed at developing both disease prevention and clinical care strategies that account for variability in individuals that is affected by their environment, genetics, lifestyle, and molecular phenotype (determined by both genotype & metabolome) [60, 86]. It follows the concept of "P4 medicine," i.e., preventive, predictive, personalized, and participatory [87] in nature. The approach relies on the characterization of genetic, epigenetic, and clinical information of individuals and provides adept tailored medical treatment, which will consequently be safe and effective. As a trickle-down effect, it may enable reducing time and financial expenditure on healthcare, improve quality of life, and reduce side effects of the given treatment. Personalized medicine may have several implications including early detection of disease using medically relevant biomarkers and identification of key genetic and epigenetic parameters during the initiation and progression of the disease [60, 86, 88]. Overall, it is promising in providing deeper insights into the mechanism of disease emergence and progression and facilitates using noninvasive methods and easy-to-obtain clinical samples like body fluids (blood, sweat, urine, etc.) or volatile breath components for diagnostic purposes and stratifying disease propensity. Consequently, it is promising in laying a foundation for pharmacogenomics and targeted drug discovery, thereby enabling the measurement of well-being.

Medical decision-making is based on the examination of biochemical parameters, clinical assays, imaging scans, and rarely genetic markers. The drugs used either for treatment or for the drug discovery process are based on cellular proteins, for example, enzymes, receptors, transporters, etc. Thus, comprehensive measurement of metabolites generated through this process may provide deeper insights [89]. Moreover, these metabolic signatures have started to emerge as new biomarkers for diseases and for responding to treatment [90–109]. Furthermore, the discovery of new biomarkers can also be based on the co-metabolism of the gut microbiome along with that in humans, which have shown to modulate the levels of drugs in the blood and their effects, i.e., altering their pharmacokinetic (PK) profiles. Thus, these may be useful signatures for PK studies and PK modeling studies [110].

Toward the implementation of personalized medicine, research in pharmacometabolomics has accelerated in the last decade. It is aimed at the identification of the detailed biochemical roadmap to facilitate understanding intraindividual heterogeneity for a given disease (especially for depression and cardiovascular disorders) and their variation in response to drug treatment [98, 100, 102, 106, 111–115]. It has been also shown that urinary drug metabolite profile before treatment or at baseline can inform about the metabolism of the drugs and their toxicity [116]. One of the key initiatives has been funded by the National Institutes of Health (NIH) through the Pharmacometabolomics Research Network (PMRN) (http://pharmacometabolomics.duhs.duke.edu/) in partnership with the Pharmacogenomics Research

Network (PGRN; https://www.pgrn.org/) [89]. Studies from the same have shown that a patient's genetic and metabolic data alone or their combinations are crucial in informing the treatment outcomes as well as the underlying cause for their variation in response to treatment including the contribution of ethnicity, sex, etc. [117].

Other studies have demonstrated that metabolomics can effectively complement genomics data for assessing the risk of a given disease and for its monitoring and management to enable precision care [118]. This has led to the emergence of pharmacometabolomics-informed pharmacogenomics for addressing various diseases [100], wherein metabolic profiles are analyzed and further linked to the clinical phenotypic manifestations and with relevant genetic variants (single nucleotide polymorphisms (SNPs)) to enable the identification of novel genetic variants or SNPs that are associated with these varied drug response phenotypes [115]. Moreover, this approach seems to be useful, especially in the case of complex diseases, where similar phenotypes may arise owing to pathophysiologic processes and information from genomics data is not sufficient [115].

The upscaling of data generated in clinical pharmacology and the integration of knowledge from systems biology have led to the emergence of quantitative and systems pharmacology (QSP) [119]. This was led by the National Institute of General Medical Sciences (NIGMS) with the engagement of domain experts in pharmacology, systems biology, pharmacokinetics/pharmacodynamics, and computer modeling. QSP is enriched by data from both pharmacometabolomics and pharmacogenomics datasets [89, 115]. Owing to the low success rate of the drugs that progress from preclinical to first in human studies, the data from QSP is based on an understanding of biological pathways, disease progression, and drug mechanisms. This feeds into informing this translation that is critical for pharmaceutical R&D [119]. Thus, the combination of information from the metabolic and genetic markers can be used as unique identifiers for novel biomarker discovery.

Thus, metabolomics and the recent tools being developed are crucial in facilitating the identification of diseases through unique metabolic fingerprints or signatures. The culmination of this information with genomic data will contribute to novel biomarker discovery. Besides, the examination of clinical characteristics and their variability will enable patient stratification for informing personalized drug treatment and inform clinical trial designs including their inclusion criteria. Additionally, the clinical characterization may enable the identification of new pathways for therapeutic discovery as well as provide novel insights into mechanisms of drug actions. This will provide scope for enhancing treatment outcomes by integration of the metabolomics data with fluxomics. Advanced methodologies and the integration of omics data along with computational methods and systems biology may enable higher success rates critical in drug discovery, development, and translation. Moreover, a comparative analysis of the metabolome under baseline versus the treatment and environmental variations (including knowledge of host gut microbiome) would further provide confirmatory results for the success of personalized medicine or treatment.

5 AI-/ML-Based Approaches for Metabolomics Data Mining and Analysis

Machine learning (ML) and deep learning (DL) applications encompass everyday life, including product recommendations, spam filtering, language translation, and even customer service chatbots. The artificial intelligence revolution has gained considerable interest in healthcare, and its implementation of omics data is evident from a multitude of studies [120–122]. However, metabolomics analysis pipelines are much less developed than other omics, such as genomics and transcriptomics, which have a plethora of validated databases and tools at their disposal [123, 124]. While many challenges need to be overcome in the field of computational metabolomics, numerous applications of ML and DL are working toward solutions. In addition, research has already begun to reveal the utility of ML and DL using metabolomics, and future work has the potential to transform our understanding of health and nature.

Two similar issues facing metabolite measurements are the inability to annotate metabolites and the misidentification of similar metabolites from the raw spectral data output by the mass spectrometer. The study of the entire metabolome is ongoing, and there are many databases with metabolomic information (https://metabolomicsna.org/index.php/resources/databases). However, a majority of small compounds have not yet been added to the databases. This leads researchers to a dilemma, and the following questions are inevitable: Does one throw away metabolites unable to be annotated given current databases to increase statistical power for identifiable metabolites? Alternatively, does one include all metabolites and keep them as m/z ratios when reporting results, discussing the need for unidentifiable compounds to be found in future analyses? Regardless of the researcher's decision, valid identification of metabolites is paramount to the ability to interpret the findings and create diagnostic tools.

By employing ML and DL, researchers are devising different strategies to increase the robustness of metabolomic annotations. D.D. Matyushin, A.Y. Sholokhova, and A.K. Buryak created a deep convolutional neural network (CNN) to rank small molecules for identification using low-resolution electron ionization mass spectrometry (EI-MS) [125]. This model used the NIST 17 database to train the CNN, and the validation sets were the Golm Metabolome Database, Human Metabolome Database, and FiehnLib. The CNN outperformed other methods in ranking the metabolites. This work demonstrates the superiority of DL approaches over other methods for this specific case of metabolite identification. Multiple reviews mention convolutional neural networks developed to automate the peak-picking process [126, 127]. Kantz et al. created a CNN which removed about 90% of false positives from a conventional peak-picking pipeline [128]. By precisely identifying true metabolites in noisy mass spec data, robust biological findings can be more readily discovered. Lauren M. Petrick and Noam Shomron discuss multiple ML and DL models for peak picking, including ML models WiPP and MetaClean, and DL models Peakonly, NeatMS, NPFimg, and Eva [129]. However, some of these models were developed to work best for specific types of peaks and therefore may not be generalizable to all analyses. Nevertheless, the DL model Eva obtained a classification accuracy for good and bad peak shapes greater than 90% when applied to 22 publicly available LC-MS metabolomics datasets. This result is very promising and shows the power of DL when large datasets have culminated. Additional models have been developed to predict the absence of a mass spectrum in a database [130, 131]. This type of analysis allows for enhanced sensitivity of metabolite annotation, as only matched metabolites are kept for further analysis. If researchers are only interested in the metabolites that are already annotated, there will be widespread use of these tools and their successors. It is possible that attributes from all these strategies using ML and DL will be combined to create a robust metabolomic analysis preprocessing pipeline.

One area of enormous potential that metabolomic analyses using ML and DL have already impacted is the safety and optimization of food. Wang et al. created a deep artificial neural network (ANN) to classify pathogenic and nonpathogenic microbes commonly found in food [132]. Although the plot of the linear dimensionality reduction PCA showed overlap between different microbes, the ANN was able to discriminate all microbe types in a model using only metabolite signals that increase during the incubation time with an accuracy of 99.2% [132]. Through the screening of microbes in food, it is possible to greatly reduce the risk of illness due to the large-scale distribution of infected food. Asakura et al. performed metabolomic profiling on eight fish species and, using an ensemble deep neural network, revealed that there were metabolites that correlated with fish size [133]. This study demonstrates the potential to engineer animals with desired traits using the metabolomic composition as biomarkers. Therefore, future applications using other animals could reshape current farming techniques and increase food supply.

Healthcare diagnoses and biomarker discovery using ML and DL with metabolomic data have the potential to revolutionize healthcare standards. A widely cited study of DL using metabolomics was conducted by Alakwaa et al. [134]. One DL and six ML models were fit to ER+ and ER- breast cancer tissue metabolites. The DL model significantly outperformed the ML models, achieving an area under the receiver operating characteristic curve of 0.93. It also identified important metabolites for separation of ER+ and ER- samples which were not identified by the ML algorithms, signifying DL's superiority to find more complex relationships within the data. The tumor microenvironment is a topic of increasing interest. Metabolomic sampling can provide a snapshot of the small compounds in contact with a patient's tumor. As databases become larger, it may be possible to identify better diagnostic markers of cancer and develop personalized therapies specific to cancer progression.

Breast cancer is one of the many phenotypes in which researchers have used ML or DL approaches with metabolomics data to classify samples by group. This type of research has already been conducted in tuberculosis [135], preterm delivery [136], colorectal cancer [137], influenza [138], renal cancer [139], acute myocardial ischemia diagnosis [140], systemic lupus erythematosus [141], NAFLD [142], Covid-19 [143], and depression [144]. Many of these analyses follow a similar format such that metabolites from the optimal ML or DL approach which are most influential to the discrimination of the groups studied are validated by conducting literature searches for previous work correlating the metabolites with the disease. Proper use of ML and DL with sufficient sample sizes can produce powerful results because they quantify the extent to which potential sets of metabolites can be used as diagnostic markers of disease. After repeated validation using multiple datasets, diagnostic tools should be developed and deployed to better stratify risk for patients, and tailored treatments should be designed to provide patients with optimal recovery.

While the potential benefit of DL and ML using metabolomic data in the future is compelling, there are challenges that the field must overcome to advance to its greatest potential. As in all omics data analyses, there must be a comprehensive understanding of the precise question being asked and how the data will be measured to answer this question. In addition, the possibility of batch effects and confounding due to different species, sex, etc. within samples must be considered. ML and DL papers should provide and discuss numerous evaluation metrics. For example, classification problems should report area under the receiver operating characteristic curve, area under the precision-recall curve, F1 score, accuracy, sensitivity, specificity, positive predictive value, negative predictive value, etc. to grant the audience a better ability to understand how the model correctly makes predictions and where it lacks predictive power.

One reason metabolomics has fallen behind other omics is that there are not large, standard benchmarking datasets for many of the analyses. Therefore, it has been difficult to reliably compare different DL and ML techniques for data analysis, making a standardized pipeline, such as DESeq2 for transcriptomics, nonexistent. A universal feature of DL models is the need for large amounts of data. Large studies with metabolomic data are not widely available and are not easy to use or lack a high coverage of the metabolome. While there are openly accessible data on platforms such as MetaboLights and Metabolomics Workbench, the abundance matrices are often not shared, resulting in time-consuming replication using the raw data. Without an understanding of the files necessary to process, and which tools are available for processing the raw data, it is extremely difficult to correctly reproduce the data used in the study's analysis. A recommendation would be to require the abundance matrix to be added to all studies, along with any necessary metadata to replicate the analysis that has been performed. This would likely bring a lot of attention to the metabolomics field from bioinformaticians and computer scientists who are interested in applying their knowledge to a new discipline, greatly accelerating the advancement of new techniques, technology, and the overall use of metabolomics. As an increasing attraction to the field of metabolomics continues, an already AI influenced world will inevitably incorporate new metabolomic technologies that will contribute to a healthier, happier population.

6 Single-Cell Metabolomics

Cells are widely recognized as the most minimal, basic units of life. In biological systems, cells differentiate based on genetic expression to create heterogeneous populations which can then intercommunicate and organize from complex structures like tissues [145]. Biological function can vary greatly across cells, and therefore bulk analyses of pooled cells lose the ability to discern the differential function of distinct populations. To avoid losing this information, single-cell omics analyses have become increasingly prevalent as they have the potential to detect cellular heterogeneity within tissues [146–148]. Single-cell omics allow for the extraction and measurement of biomolecules specific to individual cells, which can then be compared to other cells to identify discrete populations that are invisible in bulk analyses [149]. While single-cell transcriptomics has seen the most rapid development of all single-cell omics technologies, there has been a push for single-cell proteomics and metabolomics data to make the functional connection between single-cell genotype and molecular phenotype [150].

6.1 Current Single-Cell Metabolomics Technologies

Probe-based mass spectrometry, also known as mass spectrometry imaging, has emerged as the most useful technique for biomolecular profiling in single-cell metabolomics. MSI can detect the levels and localization of biomolecules using a probe, such as an ion beam or laser, to perform in situ chemical desorption and/or ionization [145, 151]. By overlaying MSI probe ablation coordinates with cell images from the same sample slide, mass spectra can be assigned to cells in tissue [152]. This circumvents the need for single-cell isolation, a costly, time-consuming process on which single-cell transcriptomic methods heavily rely [153].

In using MSI for biological applications, the spatial resolution of the probe is incredibly important as cells vary greatly in scale: most eukaryotic cells are 10–20 μm, while bacterial cells are only 1–2 μm [145]. In addition, single cells can contain a large variety of metabolites at very low abundances as compared to bulk analyses, such that ion competition among molecules could lead to the detection of only the most abundant metabolites in the cell. Table 1 gives an overview of current MSI-based techniques to perform single-cell metabolomics measurements, along with their resolutions, sensitivities, and current areas for development.

Table 1 An overview of MSI-based techniques for single-cell metabolomics

Technique	Resolution (μm)	Sensitivity(fmol)	Technique description	Areas for development and application
Secondary ion mass spectrometry (SIMS)	0.05–200	$>10^{-4}$	A primary beam of positive or negative ions is focused on a sample, providing energy to ionize molecules in its focus. These "secondary ions" are then accelerated into a mass spectrometer [154]	Primary ion impact energies are high compared to bond energies within the analytes. This leads to molecular fragmentation, which complicates downstream data analysis [154] The development of high lateral resolution SIMS (NanoSIMS) makes this technology the best for smaller organisms such as microbes, with spatial resolution as low as 30 nm [155]
Matrix-assisted laser desorption/ionization (MALDI)	1–25	>1	A sample is covered with a chemical matrix. A laser, generally ultraviolet (infrared in IR-MALDI), is then focused on a point in the sample. The matrix absorbs energy from the laser, causing analytes to be desorbed and ionized into the gas phase, which are then measured by a mass spectrometer [156]	MALDI is the most popular technique for biological application due to its "soft" ionization technique that reduces fragmentation, leading to measurements of biomolecules with a wide range of molecular weights [157] Methods to reduce probe size for a higher resolution in MALDI include transmission geometry MALDI (TG-MALDI) and scanning microprobe MALDI (SMALDI, [145] Methods to increase sensitivity of MALDI include MALDI-2 which integrates laser post-ionization to ionize molecules in the gas phase [158]

(continued)

Table 1 (continued)

Technique	Resolution (μm)	Sensitivity(fmol)	Technique description	Areas for development and application
Laser ablation electrospray ionization (LAESI)	>30	>0.6	A mid-infrared laser is focused on a sample. The resulting ablation plume is intercepted by a highly charged aqueous spray (electrospray) to post-ionize the ablated molecules, which are then funneled into a mass spectrometer [159]	LAESI allows for sampling in ambient conditions as compared to SIMS and MALDI which occur in a vacuum The IR wavelengths used by LAESI lead to large probe diameters. To increase resolution, LAESI uses optical fiber (f-LAESI) for IR laser transmission to the sample surface [160]
Desorption electrospray ionization (DESI)	>50	>0.5	Under ambient conditions, an electrospray is aimed at a sample where it desorbs and ionizes analyte molecules on the sample surface. These now ionized analytes then travel through the air into a mass spectrometer [161]	DESI is a combination of electrospray (ESI) and desorption ionization Nanospray-DESI (NanoDESI) uses capillary action to desorb analytes which improves sensitivity and lateral resolution [162]

6.2 The Future of Single-Cell Metabolomics

The past decade has shown rapid technological advancement in the realm of MSI, addressing concerns about spatial resolution and biomolecular sensitivity, and recent developments are bringing this advancement into the coming decade. Preliminary research has already shown alternative ionization approaches for laser desorption through nanostructured surfaces and stable isotope tracking to detect metabolite incorporation into metabolic pathways at the single-cell level [145].

Beyond the development of single-cell metabolomics technologies, their application is inevitable. Already, the most popular of these MSI techniques, MALDI, has produced mass spectra on tens of thousands of cells to define subpopulations with distinct metabolic states in human hepatocytes [152]. The distinction of metabolic differences between cell populations could impact the fields of cancer research, as tumors are composed of different cell types, each with cell-type-specific metabolism [163, 164], and viral infection, as elucidating the cell-specific metabolic pathways required in viral replication could propose new therapeutic targets for antiviral mechanism [153]. Single-cell metabolomics is the next step in understanding cellular diversity in complex biological organisms. With the development of these technologies, we will finally be able to see the genetic and phenotypic profiles of individual cells in tandem.[2]

7 Metabolic Sensors and the Future of Healthcare

The development of metabolomic-based technology is leading to a new revolution in healthcare. Recent research has identified metabolites whose abundances are indicative of changes in phenotype. For example, one group of compounds receiving focus as diagnostic markers is volatile organic compounds (VOCs). This trend has been accompanied by sensors engineered to quickly measure unique metabolites. The combination of these advances has the potential to produce personalized recommendations for nutrition, early detection of disease, food desirability, and many other applications.

VOCs are produced by a change in normal physiology and metabolic pathways in disease-affected tissues of the GI tract [166]. VOCs are measured using noninvasive techniques and could be key elements in the early detection of many diseases. Electronic nose (e-nose) instruments are tools developed that can measure many VOCs, utilizing many different sensor arrays. GI tract diseases detected using e-noses include colorectal cancer, Crohn's disease, ulcerative colitis, irritable bowel syndrome, and cholera [166]. The development of disease-specific e-nose devices has increased specificity and sensitivity. A preprint describes work by Gladding et al. in which they demonstrate VOC patterns of heart failure using a unique breath sensor that was optimized to detect acetone [167]. As acetone is an early signal of future heart failure, this technology provides a noninvasive, inexpensive diagnostic tool that can be used to assess a user's risk. Panebianco et al. conducted a study to compare the results of an untargeted GC-MS approach to GC-olfactometry (GC-O), a faster biomarker identification, on healthy and gastrointestinal cancer patients [168]. Their analyses showed that GC-O identified differentially abundant odor-active compounds that were not discovered using the GC-MS method. The targeted approach of GC-O exemplifies an increased sensitivity to compounds of interest, resulting in the identification of more biomarker candidates. These studies illustrate the potential for VOCs to be used in finding metabolites that correlate with a disease, leading to the ability to create screening and early diagnosis of several diseases.

We have already incorporated wearables, such as Fitbit, into our daily lives. These devices have sparked an interest, and sometimes an obsession, with personalized health, as consumers can continuously track some health markers. The perception of increased longevity, the ability to live a longer, healthier life, through personal tracking of biomarkers, has become the forefront with no age

[2] Metabolomics analysis has been performed on an isolated mouse-embryonic fibroblast cell by sucking a cell's contents into a nano-electrospray ionization tip and sent through a mass spectrometer to measure compounds of low molecular weight [165].

discrimination. Whereas previously the sick and elderly had access to continuous monitoring of biomarkers, all ages now have the ability, although still in a limited capacity compared to the standard of care in a medical center, to understand their current health and how it changes over time. With a focus on metabolomics, new technology is under development that will transform our understanding of personalized health through the identification of biomarkers for various purposes, leading to recommendations for fitness, nutrition, early detection of disease, and optimal treatments for a consumer.

As a pioneer of healthcare wearables, one of the major successes is the continuous glucose monitor (CGM) [169]. These wearables are typically inserted into the interstitial fluid (ISF) in the skin and repeatedly measure the consumer's glucose level at regular intervals. A CGM is traditionally worn by patients with type 1 or late-stage type 2 diabetes. Patients with diabetes must monitor their blood glucose levels because their pancreas does not produce insulin efficiently, resulting in the need for insulin intake through injection. A CGM monitors the healthy range for a patient, and some designs can send alerts when the blood glucose level is predicted to move outside the desired range. This intervention has reshaped the treatment landscape for diabetes, as patients can receive real-time readings of their glucose and, depending on the CGM provider, receive personalized information and recommendations about diet and exercise. As increasing interest has been given to CGM devices, there has now been a noninvasive CGM designed, called GlucoWatch, which uses reverse iontophoresis to obtain glucose samples on the skin [170]. The development of a noninvasive CGM lays the foundation for future wearables which measure biomarkers traditionally through blood to design new ways to record these markers.

As CGMs have provided clear evidence of the success of metabolomic sensor deployment worldwide, new wearable technologies are being developed, targeting salivary and tear fluid metabolites. Both vectors are of great interest because they also provide noninvasive means of biomarker measurements. Mannoor et al. measured bacteria in saliva using a dental tattoo [171], demonstrating the ability to noninvasively detect harmful pathogens. Kim et al. have created multiple biosensors using noninvasive mouthguards, which have successfully measured lactate [172] and uric acid [173]. Google entered the CGM space in 2014 using tear fluid metabolite measurements [174]. The demand for new wearables is evident, and there is a large potential benefit for both societal health and profits in emerging biotechnology. This research is pioneering the metabolomics-centered wearable field with success. As studies are validated, and new questions are asked and solved, an explosion of biomarker technology is likely imminent.

The wearables movement has largely been driven by private companies, as opposed to government agencies. As a benefit, the data that is collected can be used by these companies for internal research, new algorithms can be quickly developed, and more personalized recommendations for lifestyle changes can be given as a result. Conversely, consumers are, sometimes unknowingly, agreeing to share their health information with a source that could use this information to negatively affect them, for example, through increased healthcare costs if the company shares information with insurance companies. We should be conversing about the potential benefits and drawbacks of continuous health monitoring and discussing possible regulations that should be enforced to keep consumer data secure and to help the consumer optimize health.

In addition to wearables, new devices which measure metabolites will impact a multitude of fields. To make metabolite measurement faster than current methods, Heinemann et al. created a microfluidic system that consists of a metabolite extraction chip (MEC) integrated with an automatic sampler, micropumps, and LC-MS detection [175]. Whole blood and urine samples can be analyzed in 7 min and 5 min, respectively. This innovation could severely enhance the possibility of clinical metabolomics becoming a reality, as samples could be taken, results could be received, and diagnoses could be made in the same patient visit. Measuring stool samples allows for a noninvasive method to determine a patient's current nutritional status and future needs. Auggi, a startup acquired by Seed Health, aims to create an AI algorithm that uses the collection of stool samples over time to create connections between a consumer's triggers and symptoms to suggest dietary needs [176]. This platform has the potential to increase biological knowledge about the influence of diet on overall gut health and demonstrate the use of stool as a method of biomarker development. It is possible that as more companies like Auggi are created, personalized healthcare usage of stool could become normalized as health benefits are discovered. Another area metabolomic devices have infiltrated is the criminal justice system. Abdelshafi et al. created a miniaturized device that can detect cocaine in bodily fluids using saliva and urine samples [177]. This technology may influence similar diagnostic tests to be developed for other drugs, increasing law enforcement's ability to correctly determine users under the influence. Finally, one amazing application of metabolomic devices in food is called ripeSense® [178]. This tool is the world's first sensor that changes color to indicate how ripe a fruit is, allowing consumers to choose the fruit which is most suitable for their eating schedules. Through the engineering of new technologies designed to solve numerous problems using metabolomics, an applied metabolomics revolution is likely to begin as these exciting developments become part of our daily lives.

8 Recommendations

In the foreseeable future, major scientific endeavors will be focused on personalized care. Personalized care assumes that each of us has a biomolecular variation pattern determining the disease outcome entailing personalized medical interventions. This is best highlighted in twin studies; e.g., a multi-year study comparing pairs of monozygotic and dizygotic twins found quantifiable differences in selected features of plasma proteome, which could not be explained, alone, by genetic similarity [179]. Similar studies have been conducted on metabolites [180], suggesting longitudinal and inter-individual phenotypic variability to differing degrees. It is now believed that certain biomolecular features with complex variation patterns may be discovered which make a different person respond to treatments differently. The impact of the metabolomic approach versus the traditional biochemical approach can be appreciated by the work of Liu N. et al. wherein the use of untargeted metabolomics is employed to identify inborn errors of metabolism (IEM) [181]. Our next frontier to conquer in personalized medicine will be to identify physiological features which can reasonably predict the upcoming pathological changes to optimize individual health. This would be achieved by both preventative biomarker utilization to therapy response and monitoring biomarker discovery. Wearable devices which monitor heart rate, physical activity, sleep pattern, and other parameters are a start to this journey.

Metabolomics arguably holds more potential than other omics-based technologies, in detecting features (metabolites) mirroring the physiological state, with metabolites being a natural culmination of DNA-RNA-enzyme-metabolite dogma. However, we must be aware that metabolites change not only in disease vs healthy conditions but by age (citrate levels increase with age even in healthy controls), sex (hormonal differences along with level change with age), food habits, and population niche as well. Something seemingly simple such as increased water uptake by the subject may alter the relative concentration of crucial metabolites and can have bearing on the interpretation of the data. This requires adopting a more careful approach, which can provide us with more information mirroring the person's health.

To advance the field further, improvements need to be made at every step of the process. This involves sample collection, data acquisition, data processing, analysis, data storage, and sharing. Automation and standardization in sample collection practices must be followed. Sample type, collection method, storage conditions, and processing reagents all play a crucial role in the final output of experiments. Currently, absolute quantitation of metabolites (to achieve molar differences in key metabolites like cAMP or cGMP in healthy versus disease state) and untargeted metabolomics (to achieve complete metabolome endeavor) are two challenging aspects of metabolomics research. Absolute quantitation will open the field of biomarker discovery wherein key metabolites get perturbed in healthy versus disease state. Finding an array of key signature metabolites, which are altered, will be determining aspects of the success of metabolomics translation into clinics.

Further, to move the field forward, untargeted quantitative metabolomics is going to be of primary focus. For total metabolite detection and absolute metabolite quantification, the approach of generating synthetic metabolite standards, including isomeric metabolites, will be a key aspect. In this endeavor, MSI is going to play a central role. Untargeted metabolomics is pursued to expand the breadth and totality of metabolome profiling. This can be realized with a rigorous and exhaustive pool of reference metabolites. Future work should lean toward collaborative approaches for metabolite synthesis, thorough characterization by atomic spectroscopy and/or NMR, and then inclusion into the reference metabolites list. A consortium with worldwide access will be essential to make these reference metabolites available for analytical, and reference material used. This would accelerate the field of biomarker discovery by identifying unknown metabolites, secondary metabolites, and personalized metabolic profiles for disease prognosis and treatment. To achieve this, the research community, sample collection centers, analytical facilities/companies, statutory organizations, private device companies, and overseeing committees (e.g., bodies like MSI) need to come together and work for a common outcome. False discovery rate (FDR) is another factor that must be minimized by increasing the sample size to appropriate numbers (minimum 5, as recommended by MSI), with least perturbation by sample collection and preparation. This study discusses at length the promises and challenges in untargeted metabolomics [182].

Second, during the scientific discovery period, data acquisition of samples should be performed by utilizing different mass spectrometry modalities, which can increase the coverage of diverse types of metabolites from hydrophobic to hydrophilic, uncharged to charged, cationic to anionic, and to different isomeric metabolites, so that an exhaustive pool is generated, making the repository more exhaustive. This will feed into the system, and more and more metabolites will be discovered and characterized. We must appreciate that no single acquisition platform can achieve a global/total breadth of the metabolome. Hence, new discoveries ranging from diverse chromatography techniques, ultrasensitive mass spectrometers, and technological advancement will take us closer to our aim of profiling near-complete metabolome.

Third, the analysis of obtained data is one of the most crucial aspects of success in the metabolomics endeavor. For this, filtering out the most common metabolites and performing longitudinal studies wherein key metabolite is absent or more abundant in condition A versus B. For this, user-friendly software, training human resource, and automated data processing are the way forward. Open-sourcing the platforms and powerful analysis software would make it robust and conclusive. Further, combining the results from three different omics approaches (transcriptomics, proteomics, and metabolomics) and integrating them to understand biologically relevant questions are crucial, and platforms like MetaCore, MetaboAnalyst, InCroMAP, and 3Omics are useful tools to analyze the metabolomic data in a standalone or integrated manner. More robust open and connected platforms with robust statistical methodologies will greatly enhance the reach of the metabolomics approach.

Further, accessibility to the masses will be a key theme to bring metabolomics closer to life. For this, small yet sensitive instrumentation and pocket-size devices with the availability of reliable and easy-to-use detection kits will be a key advance. Metabolomics has a huge role to play in public safety like airports, sports administration, and control (from detecting controlled/banned substances to measuring athlete performance markers). This can be achieved by miniaturizing the instruments and making them available at the site of use. This will greatly translate the metabolomic prowess to real-world use. Many of these current challenges, metabolomic technology updates, and opportunities for the future are discussed in good detail in different review articles for further reading [28, 61, 183].

9 Conclusion

According to Gary Patti (one of the leading scientists in the field), "Metabolomics is like a molecular-level snapshot of what's happening inside a cell or organism. It provides a unique perspective on the metabolic pathways that are active at a particular moment in time" [57]. It provides a molecular-level understanding of biology and connects it with the environment. A single human cell contains more than 42,000 metabolites [184], and most of them are still uncharacterized. The use of advanced analytical technologies, coupled with the increasing availability of large-scale datasets, has enabled researchers to identify novel pathways that are associated with a wide range of biological phenomena and diseases. With these facts in mind, we believe that metabolomics would have a spectacular impact on biology and healthcare developments.

One of the key trends in metabolomics research is precision medicine, which aims to provide personalized treatment options based on an individual's unique genetic makeup and metabolic profile. In this chapter, we have summarized the potential of metabolomics in precision medicine by identifying biomarkers that can be used to predict an individual's response to different treatments. In the same direction, personalized nutrition is another area where metabolomics can have a significant impact by identifying dietary biomarkers that can be used to develop personalized dietary recommendations. We have also reviewed the development of novel computational tools and their applications to integrate the metabolomics data with other omics (such as genomics, transcriptomics, and proteomics) and provide a more comprehensive understanding of biological systems. A comprehensive segment delineated the connection between translational biology and therapeutic progress, stemming from the comprehension of metabolic levels in diverse disease models and pathogens.

We hold the belief that interdisciplinary partnerships are vital for the advancement of metabolomics research. Collaborating with experts from various fields, such as biology, chemistry, computer science, mathematics, and engineering, can uncover fresh insights into biological systems and facilitate the identification of novel biomarkers and pathways linked to disease. Moreover, the significance of metabolomics research in public health cannot be overstated. It has the potential to revolutionize healthcare by allowing for early disease detection, more precise diagnosis, customized treatment, and nutrition options. Nonetheless, some obstacles must be overcome to fully realize the potential of metabolomics in enhancing human health. These challenges involve standardizing sample collection, data acquisition, data processing, data analysis, data storage, and sharing. To sum up, the prospects for metabolomics research are bright and full of promise. With the continuous progress of advanced analytical technologies, along with the growing availability of large-scale datasets and the use of artificial intelligence and machine learning, researchers and clinicians will be able to make faster decisions.

References

1. Oliver, S.G., et al., *Systematic functional analysis of the yeast genome.* Trends Biotechnol, 1998. **16**(9): p. 373–8.
2. Junot, C., et al., *High resolution mass spectrometry based techniques at the crossroads of metabolic pathways.* Mass Spectrom Rev, 2014. **33**(6): p. 471–500.
3. Shin, S.Y., et al., *An atlas of genetic influences on human blood metabolites.* Nat Genet, 2014. **46**(6): p. 543–550.
4. Suhre, K., et al., *Human metabolic individuality in biomedical and pharmaceutical research.* Nature, 2011. **477**(7362): p. 54–60.
5. Foster, M., et al., *Uncovering PFAS and Other Xenobiotics in the Dark Metabolome Using Ion Mobility Spectrometry, Mass Defect Analysis, and Machine Learning.* Environ Sci Technol, 2022. **56**(12): p. 9133–9143.
6. Zhu, J., et al., *Detecting bacterial lung infections: in vivo evaluation of in vitro volatile fingerprints.* J Breath Res, 2013. **7**(1): p. 016003.
7. Exline, M.C., et al., *Exhaled nitric oxide detection for diagnosis of COVID-19 in critically ill patients.* PLoS One, 2021. **16**(10): p. e0257644.
8. McKinney, J., *Coronavirus (COVID-19) Update: FDA Authorizes First COVID-19 Diagnostic Test Using Breath Samples*, in *Test provides results in less than three minutes*. 2022, FDA.gov.
9. Pietzner, M., et al., *Plasma metabolites to profile pathways in noncommunicable disease multimorbidity.* Nat Med, 2021. **27**(3): p. 471–479.
10. Yu, Z., et al., *Differences between human plasma and serum metabolite profiles.* PLoS One, 2011. **6**(7): p. e21230.
11. Trivedi, D.K., K.A. Hollywood, and R. Goodacre, *Metabolomics for the masses: The future of metabolomics in a personalized world.* New Horiz Transl Med, 2017. **3**(6): p. 294–305.
12. Do, K.T., et al., *Network-based approach for analyzing intra- and interfluid metabolite associations in human blood, urine, and saliva.* J Proteome Res, 2015. **14**(2): p. 1183–94.
13. Wishart, D.S., *Metabolomics for Investigating Physiological and Pathophysiological Processes.* Physiol Rev, 2019. **99**(4): p. 1819–1875.
14. Hollywood, K., D.R. Brison, and R. Goodacre, *Metabolomics: current technologies and future trends.* Proteomics, 2006. **6**(17): p. 4716–23.
15. Armitage, E.G. and C. Barbas, *Metabolomics in cancer biomarker discovery: current trends and future perspectives.* J Pharm Biomed Anal, 2014. **87**: p. 1–11.
16. Kell, D.B., *Metabolomic biomarkers: search, discovery and validation.* Expert Review of Molecular Diagnostics, 2014. **7**(4): p. 329–333.
17. Xia, J., et al., *Translational biomarker discovery in clinical metabolomics: an introductory tutorial.* Metabolomics, 2013. **9**(2): p. 280–299.
18. Cui, Y., et al., *The Exposome: Embracing the Complexity for Discovery in Environmental Health.* Environ Health Perspect, 2016. **124**(8): p. A137–40.
19. Rappaport, S.M., *Genetic Factors Are Not the Major Causes of Chronic Diseases.* PLoS One, 2016. **11**(4): p. e0154387.
20. Andra, S.S., et al., *Trends in the application of high-resolution mass spectrometry for human biomonitoring: An analytical primer to studying the environmental chemical space of the human exposome.* Environ Int, 2017. **100**: p. 32–61.
21. Gao, P., et al., *Precision environmental health monitoring by longitudinal exposome and multi omics profiling.* Genome Res, 2022. **32**(6): p. 1199–1214.
22. Jiang, C., et al., *Dynamic Human Environmental Exposome Revealed by Longitudinal Personal Monitoring.* Cell, 2018. **175**(1): p. 277–291 e31.
23. Jiang, C., et al., *Decoding personal biotic and abiotic airborne exposome.* Nat Protoc, 2021. **16**(2): p. 1129–1151.
24. Arivale. 2019; Available from: http://www.arivale.com/.

25. Bishop, T. and J. Thorne. *Why Arivale failed: Inside the surprise closure of an ambitious 'scientific wellness' startup*. 2019 [cited 2022; Available from: https://www.geekwire.com/2019/arivale-shut-doors-inside-surprise-closure-ambitious-scientific-wellness-startup/.
26. Paquette, D. *Molecular You Signs Agreement for Roche to Use its Molecular Profiling Technology*. 2021 [cited 2022; Available from: https://www.biospace.com/article/molecular-you-signs-agreement-for-roche-to-use-its-molecular-profiling-technology/.
27. Phillips, K.A., et al., *Genetic Test Availability And Spending: Where Are We Now? Where Are We Going?* Health Aff (Millwood), 2018. **37**(5): p. 710–716.
28. Pinu, F.R., S.A. Goldansaz, and J. Jaine, *Translational Metabolomics: Current Challenges and Future Opportunities*. Metabolites, 2019. **9**(6).
29. Trifonova, O.P., et al., *Current State and Future Perspectives on Personalized Metabolomics*. Metabolites, 2023. **13**(1).
30. Spratlin, J.L., N.J. Serkova, and S.G. Eckhardt, *Clinical applications of metabolomics in oncology: a review*. Clin Cancer Res, 2009. **15**(2): p. 431–40.
31. Turkoglu, O., et al., *Metabolomics of biomarker discovery in ovarian cancer: a systematic review of the current literature*. Metabolomics, 2016. **12**(4).
32. Yu, L., K. Li, and X. Zhang, *Next-generation metabolomics in lung cancer diagnosis, treatment and precision medicine: mini review*. Oncotarget, 2017. **8**(70): p. 115774–115786.
33. Troisi, J., et al., *Metabolomic Signature of Endometrial Cancer*. J Proteome Res, 2018. **17**(2): p. 804–812.
34. Bathen, T.F., et al., *Feasibility of MR metabolomics for immediate analysis of resection margins during breast cancer surgery*. PLoS One, 2013. **8**(4): p. e61578.
35. Glunde, K., et al., *Choline metabolism-based molecular diagnosis of cancer: an update*. Expert Rev Mol Diagn, 2015. **15**(6): p. 735–47.
36. Ashrafian, H., et al., *Metabolomics: The Stethoscope for the Twenty-First Century*. Med Princ Pract, 2021. **30**(4): p. 301–310.
37. Wang, T.J., et al., *2-Aminoadipic acid is a biomarker for diabetes risk*. J Clin Invest, 2013. **123**(10): p. 4309–17.
38. Kim, S., et al., *Global metabolite profiling of synovial fluid for the specific diagnosis of rheumatoid arthritis from other inflammatory arthritis*. PLoS One, 2014. **9**(6): p. e97501.
39. van der Lee, S.J., et al., *Circulating metabolites and general cognitive ability and dementia: Evidence from 11 cohort studies*. Alzheimers Dement, 2018. **14**(6): p. 707–722.
40. Han, W., et al., *Profiling novel metabolic biomarkers for Parkinson's disease using in-depth metabolomic analysis*. Mov Disord, 2017. **32**(12): p. 1720–1728.
41. Licari, A., et al., *Asthma Endotyping and Biomarkers in Childhood Asthma*. Pediatr Allergy Immunol Pulmonol, 2018. **31**(2): p. 44–55.
42. Wurtz, P., et al., *High-throughput quantification of circulating metabolites improves prediction of subclinical atherosclerosis*. Eur Heart J, 2012. **33**(18): p. 2307–16.
43. Vallejo, M., et al., *Plasma fingerprinting with GC-MS in acute coronary syndrome*. Anal Bioanal Chem, 2009. **394**(6): p. 1517–24.
44. Dunn, W.B., Broadhurst, D.I., Deepak, S.M. et al., *Serum metabolomics reveals many novel metabolic markers of heart failure, including pseudouridine and 2-oxoglutarate*. Metabolomics 2007. **3**: p. 413–426.
45. Zhang, S., et al., *Predicting detection limits of enzyme-linked immunosorbent assay (ELISA) and bioanalytical techniques in general*. Analyst, 2014. **139**(2): p. 439–45.
46. Gibney, M.J., et al., *Metabolomics in human nutrition: opportunities and challenges*. Am J Clin Nutr, 2005. **82**(3): p. 497–503.
47. Wishart, D.S., *Emerging applications of metabolomics in drug discovery and precision medicine*. Nat Rev Drug Discov, 2016. **15**(7): p. 473–84.
48. Di Minno, A., et al., *Challenges in Metabolomics-Based Tests, Biomarkers Revealed by Metabolomic Analysis, and the Promise of the Application of Metabolomics in Precision Medicine*. Int J Mol Sci, 2022. **23**(9).

49. Mastrangelo, A., et al., *Metabolomics as a tool for drug discovery and personalised medicine. A review.* Curr Top Med Chem, 2014. **14**(23): p. 2627–36.
50. Nicholson, J.K., et al., *Metabonomics: a platform for studying drug toxicity and gene function.* Nat Rev Drug Discov, 2002. **1**(2): p. 153–61.
51. Sengupta, A., A. Uppoor, and M.B. Joshi, *Metabolomics: Paving the path for personalized periodontics – A literature review.* J Indian Soc Periodontol, 2022. **26**(2): p. 98–103.
52. D'Adamo, G.L., J.T. Widdop, and E.M. Giles, *The future is now? Clinical and translational aspects of "Omics" technologies.* Immunol Cell Biol, 2021. **99**(2): p. 168–176.
53. Dawiskiba, T., et al., *Serum and urine metabolomic fingerprinting in diagnostics of inflammatory bowel diseases.* World J Gastroenterol, 2014. **20**(1): p. 163–74.
54. Williams, H.R., et al., *Serum metabolic profiling in inflammatory bowel disease.* Dig Dis Sci, 2012. **57**(8): p. 2157–65.
55. da Silva, R.R., P.C. Dorrestein, and R.A. Quinn, *Illuminating the dark matter in metabolomics.* Proc Natl Acad Sci U S A, 2015. **112**(41): p. 12549–50.
56. Roberts, L.D., et al., *Targeted metabolomics.* Curr Protoc Mol Biol, 2012. **Chapter 30**: p. Unit 30 2 1–24.
57. Patti, G.J., O. Yanes, and G. Siuzdak, *Innovation: Metabolomics: the apogee of the omics trilogy.* Nat Rev Mol Cell Biol, 2012. **13**(4): p. 263–9.
58. Dunn, W.B., et al., *Systems level studies of mammalian metabolomes: the roles of mass spectrometry and nuclear magnetic resonance spectroscopy.* Chem Soc Rev, 2011. **40**(1): p. 387–426.
59. Zhang, X., et al., *Non-targeted and targeted metabolomics approaches to diagnosing lung cancer and predicting patient prognosis.* Oncotarget, 2016. **7**(39): p. 63437–63448.
60. Castelli, F.A., et al., *Metabolomics for personalized medicine: the input of analytical chemistry from biomarker discovery to point-of-care tests.* Anal Bioanal Chem, 2022. **414**(2): p. 759–789.
61. Yang, Q., et al., *Metabolomics biotechnology, applications, and future trends: a systematic review.* RSC Adv, 2019. **9**(64): p. 37245–37257.
62. Dumas, T., et al., *Environmental Metabolomics Promises and Achievements in the Field of Aquatic Ecotoxicology: Viewed through the Pharmaceutical Lens.* Metabolites, 2022. **12**(2).
63. Wu, W., et al., *Emerging applications of metabolomics in food science and future trends.* Food Chem X, 2022. **16**. p. 100500.
64. Sakurai, N., *Recent applications of metabolomics in plant breeding.* Breed Sci, 2022. **72**(1): p. 56–65.
65. Tang, Z.Z., et al., *Multi-Omic Analysis of the Microbiome and Metabolome in Healthy Subjects Reveals Microbiome-Dependent Relationships Between Diet and Metabolites.* Front Genet, 2019. **10**: p. 454.
66. Nguyen, Q.P., et al., *Associations between the gut microbiome and metabolome in early life.* BMC Microbiol, 2021. **21**(1): p. 238.
67. Bauermeister, A., et al., *Mass spectrometry-based metabolomics in microbiome investigations.* Nat Rev Microbiol, 2022. **20**(3): p. 143–160.
68. Lee-Sarwar, K.A., et al., *Metabolome-Microbiome Crosstalk and Human Disease.* Metabolites, 2020. **10**(5).
69. Tran, H., M. McConville, and P. Loukopoulos, *Metabolomics in the study of spontaneous animal diseases.* J Vet Diagn Invest, 2020. **32**(5): p. 635–647.
70. Goldansaz, S.A., et al., *Livestock metabolomics and the livestock metabolome: A systematic review.* PLoS One, 2017. **12**(5): p. e0177675.
71. Zhang, G., et al., *A Multi-Platform Metabolomics Approach Identifies Urinary Metabolite Signatures That Differentiate Ketotic From Healthy Dairy Cows.* Front Vet Sci, 2021. **8**: p. 595983.
72. Ramirez, T., et al., *Metabolomics in toxicology and preclinical research.* ALTEX, 2013. **30**(2): p. 209–25.

73. Playdon, M.C., et al., *Metabolomics Analytics Workflow for Epidemiological Research: Perspectives from the Consortium of Metabolomics Studies (COMETS).* Metabolites, 2019. **9**(7).
74. Dougan, M.M., et al., *Metabolomic profiles in breast cancer: a pilot case-control study in the breast cancer family registry.* BMC Cancer, 2018. **18**(1): p. 532.
75. De Preter, V. and K. Verbeke, *Metabolomics as a diagnostic tool in gastroenterology.* World J Gastrointest Pharmacol Ther, 2013. **4**(4): p. 97–107.
76. McGarrah, R.W., et al., *Cardiovascular Metabolomics.* Circ Res, 2018. **122**(9): p. 1238–1258.
77. Han, L., et al., *Altered metabolome and microbiome features provide clues in understanding irritable bowel syndrome and depression comorbidity.* ISME J, 2022. **16**(4): p. 983–996.
78. Vernocchi, P., F. Del Chierico, and L. Putignani, *Gut Microbiota Profiling: Metabolomics Based Approach to Unravel Compounds Affecting Human Health.* Front Microbiol, 2016. **7**: p. 1144.
79. Zhao, Z., et al., *Application of metabolomics in osteoporosis research.* Front Endocrinol (Lausanne), 2022. **13**: p. 993253.
80. Adav, S.S. and Y. Wang, *Metabolomics Signatures of Aging: Recent Advances.* Aging Dis, 2021. **12**(2): p. 646–661.
81. Wendt, C.H., et al., *Metabolite profiles associated with disease progression in influenza infection.* PLoS One, 2021. **16**(4): p. e0247493.
82. McCreath, G., et al., *A Metabolomics approach for the diagnosis Of SecondAry InfeCtions in COVID-19 (MOSAIC): a study protocol.* BMC Infect Dis, 2021. **21**(1): p. 1204.
83. Ulaszewska, M.M., et al., *Nutrimetabolomics: An Integrative Action for Metabolomic Analyses in Human Nutritional Studies.* Mol Nutr Food Res, 2019. **63**(1): p. e1800384.
84. Reisdorph, N.A., et al., *Nutrimetabolomics reveals food-specific compounds in urine of adults consuming a DASH-style diet.* Sci Rep, 2020. **10**(1): p. 1157.
85. Rangel-Huerta, O.D. and A. Gil, *Nutrimetabolomics: An Update on Analytical Approaches to Investigate the Role of Plant-Based Foods and Their Bioactive Compounds in Non-Communicable Chronic Diseases.* Int J Mol Sci, 2016. **17**(12).
86. Mathur, S. and J. Sutton, *Personalized medicine could transform healthcare.* Biomed Rep, 2017. **7**(1): p. 3–5.
87. Hood, L. and M. Flores, *A personal view on systems medicine and the emergence of proactive P4 medicine: predictive, preventive, personalized and participatory.* N Biotechnol, 2012. **29**(6): p. 613–24.
88. Ginsburg, G.S. and K.A. Phillips, *Precision Medicine: From Science To Value.* Health Aff (Millwood), 2018. **37**(5): p. 694–701.
89. Kaddurah-Daouk, R., R. Weinshilboum, and N. Pharmacometabolomics Research, *Metabolomic Signatures for Drug Response Phenotypes: Pharmacometabolomics Enables Precision Medicine.* Clin Pharmacol Ther, 2015. **98**(1): p. 71–5.
90. Hesse, J., et al., *Profound inhibition of CD73-dependent formation of anti-inflammatory adenosine in B cells of SLE patients.* EBioMedicine, 2021. **73**: p. 103616.
91. Lin, X., et al., *Metabolic biomarker signature for predicting the effect of neoadjuvant chemotherapy of breast cancer.* Ann Transl Med, 2019. **7**(22): p. 670.
92. Ye, Z., et al., *A 13-Gene Metabolic Prognostic Signature Is Associated With Clinical and Immune Features in Stomach Adenocarcinoma.* Front Oncol, 2021. **11**: p. 612952.
93. Yousf, S., et al., *Metabolic signatures suggest o-phosphocholine to UDP-N-acetylglucosamine ratio as a potential biomarker for high-glucose and/or palmitate exposure in pancreatic beta-cells.* Metabolomics, 2019. **15**(4): p. 55.
94. Kaddurah-Daouk, R., et al., *Pretreatment metabotype as a predictor of response to sertraline or placebo in depressed outpatients: a proof of concept.* Transl Psychiatry, 2011. **1**(7): p. e26–.
95. Zhu, H., et al., *Pharmacometabolomics of response to sertraline and to placebo in major depressive disorder – possible role for methoxyindole pathway.* PLoS One, 2013. **8**(7): p. e68283.

96. Kaddurah-Daouk, R., et al., *Pharmacometabolomic mapping of early biochemical changes induced by sertraline and placebo.* Transl Psychiatry, 2013. **3**(1): p. e223.
97. Yao, J.K., et al., *Associations between purine metabolites and clinical symptoms in schizophrenia.* PLoS One, 2012. **7**(8): p. e42165.
98. Yerges-Armstrong, L.M., et al., *Purine pathway implicated in mechanism of resistance to aspirin therapy: pharmacometabolomics-informed pharmacogenomics.* Clin Pharmacol Ther, 2013. **94**(4): p. 525–32.
99. Abo, R., et al., *Merging pharmacometabolomics with pharmacogenomics using '1000 Genomes' single-nucleotide polymorphism imputation: selective serotonin reuptake inhibitor response pharmacogenomics.* Pharmacogenet Genomics, 2012. **22**(4): p. 247–53.
100. Ji, Y., et al., *Glycine and a glycine dehydrogenase (GLDC) SNP as citalopram/escitalopram response biomarkers in depression: pharmacometabolomics-informed pharmacogenomics.* Clin Pharmacol Ther, 2011. **89**(1): p. 97–104.
101. Perroud, B., et al., *Pharmacometabolomic signature of ataxia SCA1 mouse model and lithium effects.* PLoS One, 2013. **8**(8): p. e70610.
102. Lewis, J.P., et al., *Integration of pharmacometabolomic and pharmacogenomic approaches reveals novel insights into antiplatelet therapy.* Clin Pharmacol Ther, 2013. **94**(5): p. 570–3.
103. Trupp, M., et al., *Metabolomics reveals amino acids contribute to variation in response to simvastatin treatment.* PLoS One, 2012. **7**(7): p. e38386.
104. Kaddurah-Daouk, R., et al., *Metabolomic mapping of atypical antipsychotic effects in schizophrenia.* Mol Psychiatry, 2007. **12**(10): p. 934–45.
105. McEvoy, J., et al., *Lipidomics reveals early metabolic changes in subjects with schizophrenia: effects of atypical antipsychotics.* PLoS One, 2013. **8**(7): p. e68717.
106. Ellero-Simatos, S., et al., *Pharmacometabolomics reveals that serotonin is implicated in aspirin response variability.* CPT Pharmacometrics Syst Pharmacol, 2014. **3**(7): p. e125.
107. Cooper-Dehoff, R.M., et al., *Is diabetes mellitus-linked amino acid signature associated with beta-blocker-induced impaired fasting glucose?* Circ Cardiovasc Genet, 2014. **7**(2): p. 199–205.
108. Wikoff, W.R., et al., *Pharmacometabolomics reveals racial differences in response to atenolol treatment.* PLoS One, 2013. **8**(3): p. e57639.
109. Kaddurah-Daouk, R., et al., *Lipidomic analysis of variation in response to simvastatin in the Cholesterol and Pharmacogenetics Study.* Metabolomics, 2010. **6**(2): p. 191–201.
110. Li, H. and W. Jia, *Cometabolism of microbes and host: implications for drug metabolism and drug-induced toxicity.* Clin Pharmacol Ther, 2013. **94**(5): p. 574–81.
111. Ji, Y., et al., *Pharmacogenomics of selective serotonin reuptake inhibitor treatment for major depressive disorder: genome-wide associations and functional genomics.* Pharmacogenomics J, 2013. **13**(5): p. 456–63.
112. Ji, Y., et al., *Citalopram and escitalopram plasma drug and metabolite concentrations: genome-wide associations.* Br J Clin Pharmacol, 2014. **78**(2): p. 373–83.
113. Gupta, M., et al., *TSPAN5, ERICH3 and selective serotonin reuptake inhibitors in major depressive disorder: pharmacometabolomics-informed pharmacogenomics.* Mol Psychiatry, 2016. **21**(12): p. 1717–1725.
114. Mitchell, B.D., et al., *The genetic response to short-term interventions affecting cardiovascular function: rationale and design of the Heredity and Phenotype Intervention (HAPI) Heart Study.* Am Heart J, 2008. **155**(5): p. 823–8.
115. Neavin, D., R. Kaddurah-Daouk, and R. Weinshilboum, *Pharmacometabolomics informs Pharmacogenomics.* Metabolomics, 2016. **12**(7).
116. Clayton, T.A., et al., *Pharmaco-metabonomic phenotyping and personalized drug treatment.* Nature, 2006. **440**(7087): p. 1073–7.
117. Beger, R.D., et al., *Metabolomics enables precision medicine: "A White Paper, Community Perspective".* Metabolomics, 2016. **12**(10): p. 149.
118. Guo, L., et al., *Plasma metabolomic profiles enhance precision medicine for volunteers of normal health.* Proc Natl Acad Sci U S A, 2015. **112**(35): p. E4901–10.

119. Leil, T.A. and R. Bertz, *Quantitative Systems Pharmacology can reduce attrition and improve productivity in pharmaceutical research and development.* Front Pharmacol, 2014. **5**: p. 247.
120. Martorell-Marugan, J., et al., *Deep Learning in Omics Data Analysis and Precision Medicine*, in *Computational Biology*, H. Husi, Editor. 2019: Brisbane (AU).
121. Sammut, S.J., et al., *Multi-omic machine learning predictor of breast cancer therapy response.* Nature, 2022. **601**(7894): p. 623–629.
122. Reel, P.S., et al., *Using machine learning approaches for multi-omics data analysis: A review.* Biotechnol Adv, 2021. **49**: p. 107739.
123. Tercan, B., & Leblebici, A, *Data Resources and Machine Learning for Transcriptomics Data Analysis.* Current Studies in Basic Sciences, Engineering and Technology, 2021: p. 70–85.
124. Félix Raimundo, L.M.-P., Céline Vallot, Jean-Philippe Vert, *Machine learning for single-cell genomics data analysis.* 2021. **26**: p. 64–71.
125. Pomyen, Y., et al., *Deep metabolome: Applications of deep learning in metabolomics.* Comput Struct Biotechnol J, 2020. **18**: p. 2818–2825.
126. Matyushin, D.D., A.Y. Sholokhova, and A.K. Buryak, *Deep Learning Driven GC-MS Library Search and Its Application for Metabolomics.* Anal Chem, 2020. **92**(17): p. 11818–11825.
127. Liebal, U.W., et al., *Machine Learning Applications for Mass Spectrometry-Based Metabolomics.* Metabolites, 2020. **10**(6).
128. Kantz, E.D., et al., *Deep Neural Networks for Classification of LC-MS Spectral Peaks.* Anal Chem, 2019. **91**(19): p. 12407–12413.
129. Petrick, L.M. and N. Shomron, *AI/ML-driven advances in untargeted metabolomics and exposomics for biomedical applications.* Cell Rep Phys Sci, 2022. **3**(7).
130. Sotnezova, K.M., Samokhin, A.S. & Revelsky, I.A., *Use of PLS Discriminant Analysis for Revealing the Absence of a Compound in an Electron Ionization Mass Spectral Database.* Anal Chem, 2017.
131. Samokhin, A., K. Sotnezova, and I. Revelsky, *Predicting the absence of an unknown compound in a mass spectral database.* Eur J Mass Spectrom (Chichester), 2019. **25**(6): p. 439–444.
132. Wang, D., P. Greenwood, and M.S. Klein, *Deep Learning for Rapid Identification of Microbes Using Metabolomics Profiles.* Metabolites, 2021. **11**(12).
133. Asakura, T., Y. Date, and J. Kikuchi, *Application of ensemble deep neural network to metabolomics studies.* Anal Chim Acta, 2018. **1037**: p. 230–236.
134. Alakwaa, F.M., K. Chaudhary, and L.X. Garmire, *Deep Learning Accurately Predicts Estrogen Receptor Status in Breast Cancer Metabolomics Data.* J Proteome Res, 2018. **17**(1): p. 337–347.
135. Hu, X., et al., *Combining metabolome and clinical indicators with machine learning provides some promising diagnostic markers to precisely detect smear-positive/negative pulmonary tuberculosis.* BMC Infect Dis, 2022. **22**(1): p. 707.
136. Bahado-Singh, R.O., et al., *Artificial intelligence and amniotic fluid multiomics: prediction of perinatal outcome in asymptomatic women with short cervix.* Ultrasound Obstet Gynecol, 2019. **54**(1): p. 110–118.
137. Troisi, J., et al., *A Metabolomics-Based Screening Proposal for Colorectal Cancer.* Metabolites, 2022. **12**(2).
138. Hogan, C.A., et al., *Nasopharyngeal metabolomics and machine learning approach for the diagnosis of influenza.* EBioMedicine, 2021. **71**: p. 103546.
139. Prade, V.M., et al., *The synergism of spatial metabolomics and morphometry improves machine learning-based renal tumour subtype classification.* Clin Transl Med, 2022. **12**(2): p. e666.
140. Cao, J., et al., *Combined metabolomics and machine learning algorithms to explore metabolic biomarkers for diagnosis of acute myocardial ischemia.* Int J Legal Med, 2022.
141. Niu, J., et al., *Joint deep learning for batch effect removal and classification toward MALDI MS based metabolomics.* BMC Bioinformatics, 2022. **23**(1): p. 270.

142. Oh, T.G., et al., *A Universal Gut-Microbiome-Derived Signature Predicts Cirrhosis.* Cell Metab, 2020. **32**(5): p. 878–888 e6.
143. Shen, B.e.a., *Proteomic and Metabolomic Characterization of COVID-19 Patient Sera.* Cell, 2020. **182**: p. 59–72.
144. Takahashi, Y., et al., *Improved metabolomic data-based prediction of depressive symptoms using nonlinear machine learning with feature selection.* Transl Psychiatry, 2020. **10**(1): p. 157.
145. Taylor, M.J., J.K. Lukowski, and C.R. Anderton, *Spatially Resolved Mass Spectrometry at the Single Cell: Recent Innovations in Proteomics and Metabolomics.* J Am Soc Mass Spectrom, 2021. **32**(4): p. 872–894.
146. Pelkmans, L., *Cell Biology. Using cell-to-cell variability--a new era in molecular biology.* Science, 2012. **336**(6080): p. 425–6.
147. Lee, M.C., et al., *Single-cell analyses of transcriptional heterogeneity during drug tolerance transition in cancer cells by RNA sequencing.* Proc Natl Acad Sci U S A, 2014. **111**(44): p. E4726–35.
148. Russell, A.B., C. Trapnell, and J.D. Bloom, *Extreme heterogeneity of influenza virus infection in single cells.* Elife, 2018. **7**.
149. Altschuler, S.J. and L.F. Wu, *Cellular heterogeneity: do differences make a difference?* Cell, 2010. **141**(4): p. 559–63.
150. Lanekoff, I., V.V. Sharma, and C. Marques, *Single-cell metabolomics: where are we and where are we going?* Curr Opin Biotechnol, 2022. **75**: p. 102693.
151. Hansen, R.L. and Y.J. Lee, *High-Spatial Resolution Mass Spectrometry Imaging: Toward Single Cell Metabolomics in Plant Tissues.* The Chemical Record, 2018. **18**(1): p. 65–77.
152. Rappez, L., et al., *SpaceM reveals metabolic states of single cells.* Nat Methods, 2021. **18**(7): p. 799–805.
153. Kumar, R., et al., *Single Cell Metabolomics: A Future Tool to Unmask Cellular Heterogeneity and Virus-Host Interaction in Context of Emerging Viral Diseases.* Front Microbiol, 2020. **11**: p. 1152.
154. Mas, S., et al., *Cluster TOF-SIMS imaging: a new light for in situ metabolomics?* Proteomics, 2008. **8**(18): p. 3735–45.
155. Behrens, S., A. Kappler, and M. Obst, *Linking environmental processes to the in situ functioning of microorganisms by high resolution secondary ion mass spectrometry (NanoSIMS) and scanning transmission X-ray microscopy (STXM).* Environ Microbiol, 2012. **14**(11): p. 2851–69.
156. Aichler, M. and A. Walch, *MALDI Imaging mass spectrometry: current frontiers and perspectives in pathology research and practice.* Lab Invest, 2015. **95**(4): p. 422–31.
157. Norris, J.L. and R.M. Caprioli, *Analysis of tissue specimens by matrix-assisted laser desorption/ionization imaging mass spectrometry in biological and clinical research.* Chem Rev, 2013. **113**(4): p. 2309–42.
158. Soltwisch, J., et al., *Mass spectrometry imaging with laser induced postionization.* Science, 2015. **348**(6231): p. 211–5.
159. Shrestha, B. and A. Vertes, *In situ metabolic profiling of single cells by laser ablation electrospray ionization mass spectrometry.* Anal Chem, 2009. **81**(20): p. 8265-71.
160. Taylor, M.J., et al., *Ambient Single-Cell Analysis and Native Tissue Imaging Using Laser-Ablation Electrospray Ionization Mass Spectrometry with Increased Spatial Resolution.* J Am Soc Mass Spectrom, 2021. **32**(9): p. 2490–2494.
161. Arentz, G., et al., *Applications of Mass Spectrometry Imaging to Cancer.* Adv Cancer Res, 2017. **134**: p. 27–66.
162. Roach, P.J., J. Laskin, and A. Laskin, *Nanospray desorption electrospray ionization: an ambient method for liquid-extraction surface sampling in mass spectrometry.* Analyst, 2010. **135**(9): p. 2233–6.
163. Kim, J. and R.J. DeBerardinis, *Mechanisms and Implications of Metabolic Heterogeneity in Cancer.* Cell Metab, 2019. **30**(3): p. 434–446.

164. Lau, A.N., et al., *Dissecting cell-type-specific metabolism in pancreatic ductal adenocarcinoma.* Elife, 2020. **9**.
165. Tsuyama, N., et al., *Live single-cell molecular analysis by video-mass spectrometry.* Anal Sci, 2008. **24**(5): p. 559–61.
166. Wilson, A.D., *Application of Electronic-Nose Technologies and VOC-Biomarkers for the Noninvasive Early Diagnosis of Gastrointestinal Diseases (dagger).* Sensors (Basel), 2018. **18**(8).
167. Patrick A. Gladding, R.Y., Maxine Cooper, Suzanne Loader, Kevin Smith, Erica Zarate, Saras Green, Silas G. Villas-Boas, Phillip Shepherd, Purvi Kakadiya, Eric Thorstensen, Christine Keven, Margaret Coe, Mia Jüllig, Edmond Zhang, Todd T. Schlegel, *Metabolomics and a Breath Sensor Identify Acetone as a Biomarker for Heart Failure.* 2021.
168. Panebianco, C., et al., *Cancer sniffer dogs: how can we translate this peculiarity in laboratory medicine? Results of a pilot study on gastrointestinal cancers.* Clin Chem Lab Med, 2017. **56**(1): p. 138–146.
169. Gao, W., G.A. Brooks, and D.C. Klonoff, *Wearable physiological systems and technologies for metabolic monitoring.* J Appl Physiol (1985), 2018. **124**(3): p. 548–556.
170. Potts, R.O., J.A. Tamada, and M.J. Tierney, *Glucose monitoring by reverse iontophoresis.* Diabetes Metab Res Rev, 2002. **18 Suppl 1**: p. S49–53.
171. Mannoor, M.S., et al., *Graphene-based wireless bacteria detection on tooth enamel.* Nat Commun, 2012. **3**: p. 763.
172. Kim, J., et al., *Non-invasive mouthguard biosensor for continuous salivary monitoring of metabolites.* Analyst, 2014. **139**(7): p. 1632–6.
173. Kim, J., et al., *Wearable salivary uric acid mouthguard biosensor with integrated wireless electronics.* Biosens Bioelectron, 2015. **74**: p. 1061–8.
174. DiaTribe. *Google Secures Patent for Glucose-Sensing Contact Lens.* 2015; Available from: https://diatribe.org/google-secures-patent-glucose-sensing-contact-lens.
175. Heinemann, J., et al., *Analysis of Raw Biofluids by Mass Spectrometry Using Microfluidic Diffusion-Based Separation.* Anal Methods, 2017. **9**(3): p. 385–392.
176. 2021; Available from: https://www.auggi.ai/.
177. Lapizco-Encinas, B.H. and Y.V. Zhang, *Microfluidic systems in clinical diagnosis.* Electrophoresis, 2022.
178. Jacobs, D.M., M.A. van den Berg, and R.D. Hall, *Towards superior plant-based foods using metabolomics.* Curr Opin Biotechnol, 2021. **70**: p. 23–28.
179. Liu, Y., et al., *Quantitative variability of 342 plasma proteins in a human twin population.* Mol Syst Biol, 2015. **11**(1): p. 786.
180. Bermingham, K.M., et al., *Genetic and Environmental Contributions to Variation in the Stable Urinary NMR Metabolome over Time: A Classic Twin Study.* J Proteome Res, 2021. **20**(8): p. 3992–4000.
181. Liu, N., et al., *Comparison of Untargeted Metabolomic Profiling vs Traditional Metabolic Screening to Identify Inborn Errors of Metabolism.* JAMA Netw Open, 2021. **4**(7): p. e2114155.
182. Gertsman, I. and B.A. Barshop, *Promises and pitfalls of untargeted metabolomics.* J Inherit Metab Dis, 2018. **41**(3): p. 355–366.
183. Barron, R., et al., *Twin metabolomics: the key to unlocking complex phenotypes in nutrition research.* Nutr Res, 2016. **36**(4): p. 291–304.
184. Fessenden, M., *Metabolomics: Small molecules, single cells.* Nature, 2016. **540**(7631): p. 153–155.

Index

V. Soni, T. E. Hartman (eds.), *Metabolomics*,
https://doi.org/10.1007/978-3-031-39094-4

The manufacturer's authorised representative in the EU is Springer Nature Customer Service Centre GmbH, Europaplatz 3, 69115 Heidelberg, Germany. If you have any concerns regarding our products, please contact ProductSafety@springernature.com

Printed and bound by CPI Group (UK) Ltd, Croydon, CR0 4YY
23/09/2026
02221538-0002